Molecular Markers, Natural History, and Evolution

Second Edition

Molecular Markers, Natural History, and Evolution

Second Edition

JOHN C. AVISE

University of Georgia

Sinauer Associates, Inc. Publishers
Sunderland, Massachusetts

About the Cover

Clockwise from the top: Autoradiograph of an electrophoretic gel showing microsatellite DNA bands from a wild population of *Peromyscus* mice. (From the Avise lab) • Quaking aspens, *Populus tremuloides*, in the Rocky Mountains of Colorado. (Copyright © Bob Thompson/In Image Photography) • An evolutionary tree. • Shoal of juvenile striped catfish, *Plotosus lineatus*, photographed in the tropical ocean off Mabul, Malaysia. (Copyright © Matthew Oldfield/SPL/Photo Researchers, Inc.).

Cover design by John C. Avise.

Molecular Markers, Natural History, and Evolution *Second Edition*

Sinauer Associates, Inc.
23 Plumtree Road
Sunderland, MA 01375 USA

www.sinauer.com

email: publish@sinauer.com, orders@sinauer.com
FAX 413-549-1118

Library of Congress Cataloging-in-Publication Data
Avise, John C.
Molecular markers, natural history, and evolution / John C. Avise.— 2nd ed.
        p. ; cm.
  Includes bibliographical references and index.
  ISBN 0-87893-041-8 (pbk.)
        1. Biochemical markers. 2. Molecular evolution.
  [DNLM: 1. Genetic Markers. 2. Evolution, Molecular.
QH 438.4.B55 A958m 2004] I. Title.
QH438.4.B55A95 2004
572.8′6—dc22   2004003130

Printed in U.S.A.
5 4 3 2 1

Contents

Preface to the
Second Edition

This treatment is an updated and expanded version of a book that was first published in 1994. Much has transpired in the intervening decade: new laboratory methods for uncovering molecular markers have been introduced and refined, statistical and conceptual approaches for estimating intraspecific genealogy and interspecific phylogeny have been improved, and a vast armada of empirical examples has been added to a burgeoning scientific literature. In some topical areas (e.g., fossil DNA and horizontal genetic transmission), earlier scientific thought has been completely overturned by molecular findings over the past 10 years; and knowledge on numerous other topics (e.g., vertebrate mating systems, ecological speciation, and life's deep phylogeny) has expanded greatly. On the other hand, the major types of questions tackled by molecular ecologists, behaviorists, and evolutionists remain much the same. Researchers still employ molecular markers to estimate and interpret evolutionary relationships of organisms along a temporal continuum ranging from clonality, genetic parentage and genealogy in the most recent generations, to phylogenetic affinities in ancient branches of the Tree of Life. This revised edition will further document how molecular markers reveal otherwise hidden aspects of behavior, natural history, ecology, and the evolutionary histories of plants, animals, and microbes in the wild.

Why is a treatment of this topic necessary when several excellent texts in molecular ecology or evolution already exist? Most of these books have focused on: proteins and DNA as primary objects of interest in their own right (e.g., Graur and Li 2000; Li 1997; Li and Graur 1991); broad conceptual issues regarding patterns, processes, or mechanisms of molecular evolution (Ayala 1976a; Nei and Koehn 1983; Selander et al. 1991b; Takahata and Clark 1993); statistical or mathematical aspects of population-genetic or phylogenetic theory (Nei and Kumar 2000; Page and Holmes 1998); or detailed methodological procedures of data acquisition and analysis (Baker

2000; Ferraris and Palumbi 1996; Hillis et al. 1996; Karp et al. 1998). Some textbooks and edited volumes have approached more closely what is attempted here (Baker 2000; Caetano-Anollés and Gresshoff 1997; Carvalho 1998; Hoelzel 1992; Hoelzel and Dover 1991a), but most of them are either popularized accounts (Avise 2001a, 2002) or else are restricted to research topic, laboratory method, or taxonomic group (Avise 2000a; Hollingsworth et al. 1999; Kocher and Stepien 1997; Mindell 1997; Phillips and Vasil 2001; Sibley and Ahlquist 1990; Soltis et al. 1992). No other classroom textbook or reference work quite fills the niche toward which this book is aimed: the wide world of biological applications for molecular genetic markers in the contexts of ecology, behavior, natural history, evolution, and organismal phylogeny.

The first edition of *Molecular Markers* included references to about 2,200 studies from the then-neophyte fields of molecular ecology and evolution, and this second edition approximately doubles that total count of citations from the primary literature. Thus, this compendium is again intended to provide a thorough introduction to relevant research that can serve both as an educational tool and stimulus for students, and an extensive reference guide for practicing researchers. Despite this coverage, an encyclopedic treatment of all relevant studies is no longer feasible because of the explosive growth of molecular ecology and evolution since the early 1990s. Thus, by necessity I have been selective in the choice of additional examples to illustrate various topics. I also retained many of the citations and examples (albeit updated) from the first edition, in part to provide historical perspective (research approaches in molecular ecology and evolution have themselves evolved), and in part to give due credit to pioneering works that should not be forgotten. Indeed, an important goal of this book is to describe not only the current state of biological knowledge derived from molecular markers, but also to trace how that current state of affairs has come to be.

Like its predecessor, this second edition is not intended to be a detailed "how to" book on laboratory details and analytical methods of molecular ecology and evolution (although sufficient background is provided for beginners). Rather, this book is more of a "what-has-been-and-can-be-done" treatment intended to stimulate ideas and pique the research curiosity of young biology students and seasoned professionals alike. I hope this renovated edition will be read and enjoyed in this imaginative spirit of scientific adventure.

Dedication

This book is dedicated in part to my current and former graduate students, postdocs, and research technicians: Charles Aquadro, Marty Ball, Eldredge Bermingham, Brian Bowen, Robert Chapman, Beth Dakin, Andrew DeWoody, Michael Douglas, Anthony Fiumera, Matt Hare, Glenn Johns, Adam Jones, Steve Karl, Lou Kessler, Trip Lamb, Mark Mackiewicz, Judith Mank, Joe Neigel, Bill Nelson, Guillermo Ortí, John Patton, Devon Pearse,

Brady Porter, Paulo Prodöhl, Joe Quattro, Carol Reeb, Nancy Saunders, Kim Scribner, DeEtte Walker, and Kurt Wollenberg. Without them, my own involvement in molecular ecology and evolution would hardly have been possible, and not nearly so much fun. DeEtte Walker in particular has been of invaluable assistance in all phases of this book's preparation. I also want to thank Drs. Jeff Mitton and Loren Rieseberg for helpful suggestions on this second edition. Over the years, my laboratory has been supported by grants primarily from the National Science Foundation, the University of Georgia, the Sloan Foundation, and most recently the Pew Foundation.

I want to dedicate this book also to my family—Joan, Jennifer, Edith, and Dean—all of whom have given unwavering support.

Preface to the First Edition*

I never cease to marvel that the DNA and protein markers magically emerging from molecular-genetic analyses in the laboratory can reveal so many otherwise hidden facets about the world of nature. Can individual plants sometimes exist as genetic mosaics derived from multiple zygotes? Is reproduction by unicellular organisms predominantly sexual or clonal? What is the typical evolutionary lifespan of parthenogenetic all-female lineages, given that they lack recombinational genetic variation that otherwise might enable them to respond to changing environments? What is the genetic makeup of social groups within various species of insects, fishes, mammals, and other organisms whose behaviors might have evolved under the influence of kin selection? In birds and other taxa, how often does intraspecific brood parasitism occur, wherein females surreptitiously "dump" eggs into the nests of soon-to-be foster parents? Do migratory marine turtles return to their natal sites for nesting? How often has carnivory evolved among plants? What are the evolutionary origins of cytoplasmic genomes within eukaryotic cells? How old are the fossils from which DNA can be extracted? How and how often have horizontal gene transfers taken place between distant forms of life? Have demographic bottlenecks diminished genetic variability to the extent that some populations can no longer adapt to environmental challenges? How useful is the criterion of phylogenetic distinctiveness as a guide to prioritizing taxa and regional biotas for conservation efforts? These are but a small sample of the diverse problems addressed and answered (at least provisionally) through the use of molecular genetic markers.

This treatment of molecular natural history and evolution is written at a level appropriate for advanced undergraduates and graduate students, or

---

*Reprinted with slight modifications from the First Edition (1994).

for professional ecologists, geneticists, ethologists, molecular biologists, population biologists, conservationists, and others who may wish a readable introduction or refresher to the burgeoning application of molecular markers in their disciplines. I hope to have captured and conveyed the genuine sense of excitement that can be brought to such fields when molecular genetic markers with known patterns of inheritance are applied to questions about nature and evolution. I also hope to have provided a wellspring of research ideas for people entering the field. My goal is to present material in a manner that is technically straightforward, without sacrificing the richness of underlying concepts and biological applications. For the reader, the only necessary prerequisites are an introductory knowledge of genetics and an acute interest in the natural biological world.

The fields of molecular ecology and evolution are at a stage where reflection on the past half-century may provide useful historical perspective, as well as a springboard to the future. The mid-1960s witnessed the first explosion of interest in molecular techniques with the seminal introduction of protein-electrophoretic approaches to population genetics and evolutionary biology. In the late 1970s, attention shifted to methods of DNA analysis primarily through restriction enzymes, and in the 1980s, mitochondrial DNA assays as well as various nuclear-DNA fingerprinting approaches gained great popularity. Beginning in the late 1980s, the introduction of PCR-mediated DNA sequencing helped to provide the first ready access to the "ultimate" genetic data: nucleotide sequences themselves. Nonetheless, it would be naive to suppose that direct DNA sequence information invariably provides the preferred or most accessible pool of genetic markers for all biological applications. Because of ease, cost, the amount or nature of genetic information accessed, or simplicity of data interpretation, several alternative assay methods continue to be the techniques of choice for many ecological and evolutionary questions. Biologists sometimes are unaware of the arguments for and against various molecular-genetic methodologies, and one goal of this book is to clarify these issues.

In scientific advance, timing and context are all-important. Imagine for the sake of argument that DNA sequencing methods had been widely employed for the past half-century and then protein electrophoresis was introduced. No doubt a headlong rush into allozyme techniques would ensue, on justifiable rationales that: the methods are inexpensive and technically simple; observable variants reflect independent Mendelian polymorphisms at several loci scattered around the genome (rather than as linked polymorphisms in particular stretches of DNA); and amino acid replacements uncovered by protein electrophoresis (as opposed to the silent nucleotide changes often revealed in DNA assays) might bring molecular evolutionists closer to the real "stuff" of adaptive evolution. To carry the argument farther, suppose that laboratory molecular genetics had been conducted throughout the last century but that some brave entrepreneurial scientist then ventured outdoors and discovered organisms, complete with phenotypes and behaviors! At last, the interface of gene products with the environment would have been revealed. Imagine the sense of excitement and the research prospects.

These fanciful scenarios emphasize a point— molecular approaches carry immense popularity now, but nonetheless they provide just one of many avenues toward understanding the natural histories and evolutionary biologies of organisms. Studies of morphology, ecology, and behavior undeniably have shaped the great majority of scientific perceptions about the natural world. Molecular approaches are especially exciting at this time in the history of science because they have opened new empirical windows and novel insights on more traditional biological subjects.

In this book, I have attempted to identify and highlight select case histories where molecular methods have made significant contributions to natural history, ecology, and evolutionary biology. The treatment cannot be exhaustive because many thousands of studies have utilized genetic markers. Rather, I have tried to choose classic, innovative, or otherwise interesting examples illustrative of the best that molecular methods, both old and new, have to offer. Overall, I have attempted to retain a balanced taxonomic perspective that includes examples from plants, animals, and microbes, and indeed I hope that common threads will be evident that tie together the similar classes of biological questions that frequently apply to such otherwise disparate organisms.

This book is organized into two parts. Part I provides introductory material and background: the rationale for molecular approaches in natural history and evolution; the history of molecular phylogenetics; introductory outlines of various laboratory methods and the nature of genetic data that each molecular method provides; and brief descriptions of some interpretive tools of the trade, including molecular clock concepts and phylogenetic methods as applied to molecular data.

Part II departs significantly from most other books in molecular ecology and evolution by emphasizing significant biological applications via a plethora of empirical examples. Topics are arranged along a genealogical continuum from micro- to macro-evolutionary: assessment of genetic identity/non-identity and parentage; kinship and intraspecific genealogy; speciation, hybridization, and introgression; and assessment of mid-depth and deep phylogeny in the evolutionary Tree of Life. A concluding chapter deals with the relevance of molecular studies to conservation biology and the preservation of genetic diversity.

PART I

Background

1

Introduction

The stream of heredity makes phylogeny; in a sense, it is phylogeny.
Complete genetic analysis would provide the most priceless data for the
mapping of this stream.

G. G. Simpson (1945)

This book is about molecular markers and their role in genetic studies of population biology, natural history, and evolution. Researchers routinely utilize the hereditary information in biological macromolecules (proteins and nucleic acids) to address questions concerning organismal behavior, kinship, and phylogeny. When used to best effect, molecular data are integrated with information from ecology, observational natural history, ethology, comparative morphology, physiology, historical geology, paleontology, systematics, and other time-honored disciplines. Each of these traditional areas of science has been enriched, if not rejuvenated, by contact with the field of molecular genetics.

Interest in molecular ecology and evolution can center either on particular genes or proteins themselves (i.e., in genetic variation per se and its functional roles in development, physiology, and metabolism) or on the utility of molecules as genealogical markers for analyses of natural history and phylogeny. This book primarily addresses the second of these arenas. However, functional and genealogical understandings are mutually informative. For example, knowledge of the precise molecular basis and mode of hereditary transmission of a genetic polymorphism can be crucial to proper interpretation of molecular markers in a population context; conversely, patterns of population variation and divergence in molecular markers can be highly enlightening about the causal forces impinging on molecular as well as organismal evolution.

With exceptional research effort, it is sometimes possible to identify and characterize the actual genes or chromosomal regions contributing to population variation in particular organismal features and adaptations. Such molecular-level dissections can give fresh insight into the mechanistic basis, as well as the evolutionary origins and maintenance, of phenotypic variety (Jackson et al. 2002; Lynch and Walsh 1998; Purugganan and Gibson 2003). Another research objective, however, is to examine patterns of genetic variation in appropriate "randomly chosen" proteins or segments of DNA. When these naturally occurring molecular tags are interpreted as genealogical markers, they offer extraordinary power to illuminate such topics as wildlife forensics, genetic parentage, reproductive modes, mating systems, kinship, population structure, dispersal and gene flow, intraspecific phylogeography, speciation, hybridization, introgression, phylogeny, taxonomy, systematics, and conservation biology.

Phylogeny is evolutionary history—that is, topology in the proverbial "Tree of Life." All organisms share certain features (most notably, nucleic acids as hereditary material) that suggest a single or monophyletic origin on Earth between 3 and 4 billion years ago. The proliferation of life has involved successive branching and occasional anastomosis (secondary joining) of hereditary lineages, with organisms alive today representing twig-tips in the now-outermost canopy of the phylogenetic tree. A complete understanding of phylogeny requires knowledge of both branching order (cladogenetic splitting of lineages) and branch lengths (anagenetic changes within lineages through time). Occasional instances of lateral DNA transfer between branches (reticulate evolution), mediated by such evolutionary processes as interspecific hybridization, establishment of endosymbiotic associations among genomes, or other means of horizontal gene flow must also be considered (see Chapters 7 and 8).

Nearly all studies that utilize molecular markers can be viewed as attempts to estimate genetic relationships somewhere along a hierarchical continuum of evolutionary divergences ranging from recent to distant (Figure 1.1): genetic identity versus non-identity (as in distinguishing clonemates from non-clonemates in species that can reproduce both asexually and sexually), genetic parentage (biological maternity and paternity), extended kinship within the pedigree of a local deme, genealogical affinities among geographic populations of a species, genetic divergence among species that separated recently, to phylogenetic connections at intermediate and ancient branches in the Tree of Life. Different types of molecular assays provide genetic information ideally suited to different temporal horizons in this hierarchy, and a continuing challenge is to develop and utilize methods appropriate for each particular biological question (Parker et al. 1998).

It is also befitting to orient this book around genealogy because of the central importance of historical factors and nonequilibrium outcomes in ecology and evolution. As noted by Hillis and Bull (1991), "Virtually all comparative studies of biological variation among species depend on a phylogenetic framework for interpretation." If Dobzhansky's (1973) famous dic-

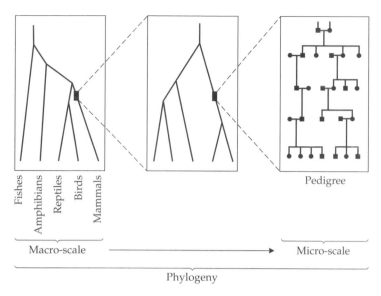

Fishes

Amphibians

Reptiles

Birds

Mammals

Pedigree

Macro-scale

Micro-scale

Phylogeny

Figure 1.1   The hierarchical nature of phylogenetic assessment. (After Avise et al. 1987a.)

tum that "nothing in biology makes sense except in the light of evolution" is correct, then it might be appended that "much in evolution makes even more sense in the light of historical genealogy." Brooks and McLennan (1991, 2002) have repeatedly emphasized the need for phylogenetic analyses in ecology and ethology, as have many others for more than two decades (e.g., Eldredge and Cracraft 1980; Harvey and Pagel 1991). Such calls have increasingly been heeded, and molecular phylogenetic analyses are now an integral part of modern appraisals of population genealogy (Avise 2000a), speciation (Barraclough and Vogler 2000), and interspecific evolution (Farrell 1998; Lutzoni and Pagel 1997). With genealogical relationships of individuals and species properly sorted out via molecular markers, the phylogenetic origins and histories of all other organismal traits, as well as the ecological and evolutionary processes that have forged them, usually become much clearer.

Why Employ Molecular Genetic Markers?

In the pre-molecular era, standard approaches to estimating kinship and phylogeny necessarily entailed comparisons of phenotypic data from morphology, physiology, behavior, or other organismal characteristics amenable to observation. Molecular ecologists and evolutionists also employ the comparative method, but the comparisons now include direct or indirect genotypic information from nucleic acid and protein sequences. Why are such molecular features of special significance?

Molecular data are genetic

Molecular data provide genetic information. This simple truism is of over-riding significance. Because phylogeny is "the stream of heredity," only genetic traits are genealogically informative. Molecular assays reveal not only detailed features of DNA (or, sometimes, their protein products), but also variable character states whose particular genetic bases and modes of transmission can be precisely specified. Thus, from explicit knowledge of the amount and nature of genetic information assayed, statements of rela-tive confidence can be placed on molecular-based genealogical conclusions.

This situation contrasts with the insecurity of knowledge concerning the precise genetic bases of conventional characters used to address organ-ismal relationships (Barlow 1961; Boag 1987). Seldom can scientists specify particular genes or alleles that govern the morphological, physiological, or behavioral traits traditionally surveyed in phylogenetic assessments. Indeed, some such taxonomic traits have been shown to be affected by environmental as well as hereditary factors. In plants, phenotypic or devel-opmental plasticity (wherein individuals can assume different forms dur-ing ontogeny in response to varying environmental milieus, ranging from intracellular to ecological) has long been recognized as a potent source of phenotypic variation (Clausen et al. 1940). The phenomenon is pervasive in the animal world as well (see review in West-Eberhard 2003), involving fea-tures ranging from leg forms in barnacles (Marchinko 2003) to gill-raker numbers in fish (Loy et al. 1999) to phenotypic components of mate choice in moths and birds (Ohlsson et al. 2002; Rodriguez and Greenfield 2003). For example, a significant fraction of the variance in morphometric features among taxonomic subspecies of the red-winged blackbird (*Agelaius phoeniceus*) proved to be due to nestling rearing conditions, as became evi-dent when progeny hatched from experimentally transplanted eggs con-verged on some of the morphological traits of their foster parents (James 1983). Similarly, cross-fostered great tits (*Parus major*) were shown to have partially converged on the carotenoid-based plumage colorations of their foster fathers (Fitze et al. 2003). Although certainly important in ecology and evolution, such phenotypic variation per se can be misleading if inter-preted as providing genetic characters of immediate utility in kinship assessment or phylogeny estimation.

Molecular methods open the entire biological world for genetic scrutiny

Prior to the introduction of molecular approaches, most genetic studies were confined to a small handful of species that could be reared and crossed in the laboratory or garden: bacteria such as *Escherichia coli* and their phages, Mendel's pea plants (*Pisum sativum*), corn (*Zea mays*), fruit flies (genus *Drosophila*), and house mice (*Mus musculus*). From hereditary patterns across generations, the genetic bases of particular morphological or physiological traits in these species were deduced. However, such analyses could hardly be expected to capture the full richness of diversity among the multitudinous genes within these study organisms, much less to embrace the broader flavor of genetic diversity across the Earth's other biota. In contrast, molecular assays can provide direct physical evidence on essentially any DNA sequence or protein, and they can be applied to the genetics of any and all creatures, from microbes to whales.

Molecular methods access a nearly unlimited pool of genetic variability

The information content of a genome is enormous. For example, a typical mammalian genome consists of some 3 billion nucleotide pairs in a composite sequence roughly 100 times longer than the total string of letters in a 20-volume encyclopedia. Each genome truly is an encyclopedic repository of information, not only encoding the ribonucleic acids and proteins that are the working machinery of cellular life, but also retaining within its nucleotide sequence a detailed historical record of phylogenetic links to other forms of life. The genomes of various bacterial species range in size from about half a million to more than 10 million base pairs (bp); those of unicellular protists range from 20 million to more than 500 billion bp, and those of multicellular fungi, plants, and animals range from about 10 million to more than 200 billion bp (Cavalier-Smith 1985; Graur and Li 2000; Sparrow et al. 1972). Molecular assays employed in ecology and evolution involve sampling, often more or less at random, dozens to thousands of genetic markers from such vast hereditary archives.

The levels of genetic variation within most species also are incredible. Consider, for example, the 3 billion bp human genome, the first full exemplar of which was draft-sequenced in 2001 (Lander et al. 2001; Venter et al. 2001). From this and other less exhaustive molecular appraisals, it appears that randomly drawn pairs of homologous DNA sequences from the human gene pool typically differ at about 0.1% of nucleotide positions (Chakravarti 1998; Stephens et al. 2001b). Thus, if a second human genome were to be sequenced fully, it would probably differ from the first at roughly 3 million nucleotide sites. Furthermore, the magnitude of nucleotide diversity in humans is near the lower end of the scale, compared with that reported in many animal and plant species (Li and Sadler 1991). Indeed, because of the recent "peopling of the planet" and the lack of long-term population structure in extant humans (see Chapter 6), total genetic diversity within *Homo sapiens* is even lower than that within our closest primate relatives, chimpanzees and gorillas (Ruvolo et al. 1993, 1994).

Complete DNA sequences are now available for numerous model species, including more than 100 prokaryotic microbes and a growing list of multicellular organisms of special interest in medicine, epidemiology, and comparative genomics (Hedges and Kumar 2002). However, for most applications in population and evolutionary biology, full genomic sequencing is unnecessary because, with far less intensive laboratory effort, molecular markers can be obtained that display ample variability for even the most refined forensic diagnoses and phylogenetic appraisals.

Nearly 100 protein and blood group polymorphisms already had been surveyed among the major human races more than a decade ago (Nei and Livshits 1990), and in excess of 2,000 DNA polymorphisms in the human gene pool had been uncovered by restriction enzyme analyses by that time (Stephens et al. 1990a; Weissenbach et al. 1992). The numbers of available molecular markers have increased dramatically since then (Boyce and Mascie-Taylor 1996; Cavalli-Sforza 2000; Donnelly and Tavaré 1997). For example, a recent analysis examined the statistical distributions of 1.4 million SNPs (single nucleotide polymorphisms) in sample databases from the human genome (Kendal 2003). For the sake of extremely conservative argument, suppose that just 30 marker polymorphisms in humans were available for examination, each with the minimum possible two alleles (many loci in fact have multiple alleles). Rules of Mendelian heredity show that the theoretical number of different human genotypes that could arise from genetic recombination would then be 3^{30}, or about 200 trillion. Approximately 6 billion people are alive today, and roughly 13 billion people have inhabited the planet since the origin of *Homo sapiens*. Thus, even with this unrealistically small number of genetic polymorphisms, the potential number of distinct human genotypes would vastly exceed the number of individuals who have ever lived, and no two people (barring monozygotic twins) in the past, present, or foreseeable future would likely be identical at all loci. In human forensic practice, standardized assays of even modest numbers of highly allelic Mendelian polymorphisms (typically from microsatellite loci) provide "DNA fingerprints" so powerful that courts of law now routinely accept the results as definitive genetic evidence of individual identification and biological parentage (see Chapter 5). Molecular polymorphisms in other species likewise permit endless opportunities in wildlife forensics.

Molecular data can distinguish homology from analogy

A central challenge of phylogenetics is to distinguish the component of biological similarity that is due to descent from a common ancestor (homology) from that due to evolutionary convergence from different ancestors (analogy). Evolutionary classifications should be reflective of homologous traits that genuinely register genealogical descent. However, particular morphological, behavioral, or other phenotypic features (the conventional data of systematics) often evolve independently as selection-mediated responses to common environmental challenges.

For example, Old World and New World vultures share several adaptations for carrion feeding (soaring food-searching behavior, featherless head, and powerful hooked beak) that formerly were thought to indicate that these birds had close evolutionary ties to each other and to other diurnal birds of prey (Falconiformes). However, extensive molecular data later prompted a competing hypothesis that New World and Old World vultures are only distant evolutionary relatives, and that carrion feeding probably evolved more than once, independently (Seibold and Helbig 1995; Sibley and Ahlquist 1990; Wink 1995). Many other such cases have been unveiled by molecular phylogenetic appraisals. For example, several species pairs of cichlid fishes from Africa's Lake Malawi and Lake Tanganyika are closely similar in external appearance, but molecular markers proved that the resemblance in each case evolved in convergent fashion from separate cichlid ancestors (Kocher et al. 1993). Likewise, molecular data showed that multiple adaptive radiations of *Anolis* lizards on various Caribbean islands entailed repeated convergent evolution of particular morphological attributes (Jackman et al. 1997; Losos et al. 1998).

Referring to molecular phylogenetic approaches, the late paleontologist Stephen J. Gould wrote, "I do not fully understand why we are not proclaiming the message from the housetops … We finally have a method that can sort homology from analogy." Gould (1985) was reveling in the fact that when species are assayed for perhaps hundreds or thousands of molecular characters, any widespread and intricate similarities that are present in these biological macromolecules are unlikely to have arisen by convergent evolution and, thus, must reflect true phylogenetic descent. With species' phylogenies properly sorted out via molecular markers, the evolutionary origins and histories of other organismal phenotypes usually become far more apparent. In other words, molecular phylogenies provide an archival road map of biodiversity.

This is not to say that particular molecular characters invariably are free from homoplasy (convergences, parallelisms, or evolutionary reversals that muddy the historical record). Indeed, some molecular features, considered individually, may be quite prone to homoplasy due to a small number of interconvertible character states and a sometimes rapid rate of change among them. For example, one of only four different character states (adenine, guanine, thymine, or cytosine) can occupy each nucleotide position in DNA, and one of only 20 different character states can occupy each amino acid position in a protein sequence. Thus, the phylogenetic power of macromolecular sequences resides not so much in specific sites or residues, but rather in the extraordinary amount of *cumulative* information provided by vast numbers of ordered positions in lengthy chains of molecular sequence. Furthermore, some types of molecular character states, such as specific duplications, deletions, or rearrangements of DNA, are rare or unique events likely to be of single (monophyletic) evolutionary origin. These too can offer tremendous power, even individually, as informative road signs along the trail of phylogenetic history.

Molecular data provide common yardsticks for measuring divergence

A singularly important aspect of molecular data is that they allow direct comparisons of relative levels of genetic differentiation among essentially any taxa (Avise and Johns 1999; Wheelis et al. 1992). Suppose, for example, that one wished to quantify evolutionary differentiation within a taxonomic family or genus of fishes as compared with that within a taxonomic counterpart in birds. The kinds of morphological traits traditionally employed in fish systematics (e.g., numbers of lateral line scales, fin rays, or gill rakers or the position of the swim bladder) clearly have no direct utility for comparisons with avian taxonomic characters (plumage features, structure of the syrinx, or arrangement of toes on the feet). Thus, in traditional systematics, no universal criteria were available to standardize comparisons between the fields of ichthyology and ornithology, much less across more disparate disciplines such as entomology and bacteriology. However, birds, fishes, insects, and microbes (as well as nearly all other forms of life) do share numerous types of molecular traits. Ribosomal RNAs (rRNAs) and transfer RNAs (tRNAs) are examples of molecules with widespread, if not ubiquitous, taxonomic distributions, as are various genes encoding enzymes involved in central metabolic and biochemical pathways for the respiration and synthesis of carbohydrates, fats, amino acids, and the replication and expression of nucleic acids.

The general notion of universal yardsticks in comparative molecular evolution is introduced in Figures 1.2 and 1.3. These graphs summarize reported levels of genetic divergence as estimated, respectively, by electrophoresis of several proteins encoded by nuclear genes and by sequencing of one mitochondrial gene among recognized taxa representing five vertebrate classes. By these empirical molecular criteria, the assayed congeners and confamilial genera of birds showed less genetic divergence than did many other vertebrate species of identical taxonomic rank (an unanticipated result, given birds' often high anatomical differentiation; Wyles et al. 1983). Perhaps these avian taxa separated more recently, on average, than did many of their non-avian taxonomic counterparts, or perhaps they evolved more slowly at the molecular level. Whatever the explanation, comparative molecular genetic studies of this sort can be expanded to include almost any number of taxa and genes, and such exercises frequently raise exciting conceptual issues about evolutionary processes that would not have been evident from traditional phenotypic or taxonomic inspections alone.

In an early example of this "common yardstick" perspective, King and Wilson (1975) reviewed evidence that the assayed protein and nucleic acid sequences of humans and chimpanzees are only about as divergent as are those of morphologically similar species of fruit flies or rodents, but much

Figure 1.2 One early exploration of a "universal genetic yardstick" for the verte- ▶
brates. Plotted on a common scale are means and ranges of genetic distance (codon substitutions per locus, as estimated from multi-locus protein electrophoretic data) among congeneric species (in parentheses are numbers of pairwise species comparisons) within each of five vertebrate classes. Note the relatively small genetic distances in assayed bird genera compared with those of many other taxa. (After Avise and Aquadro 1982.)

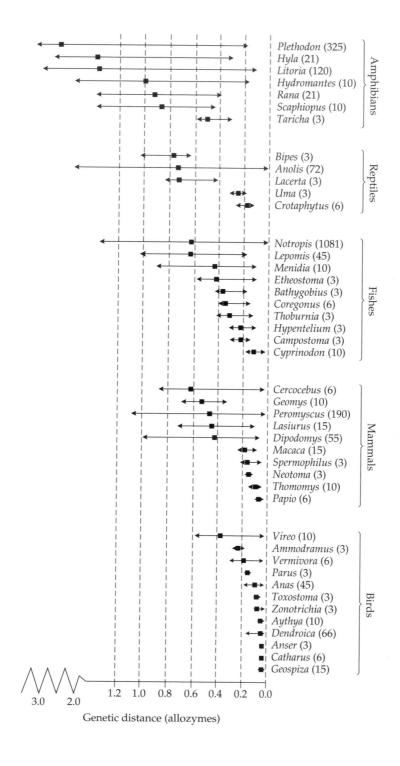

Genetic distance (allozymes)

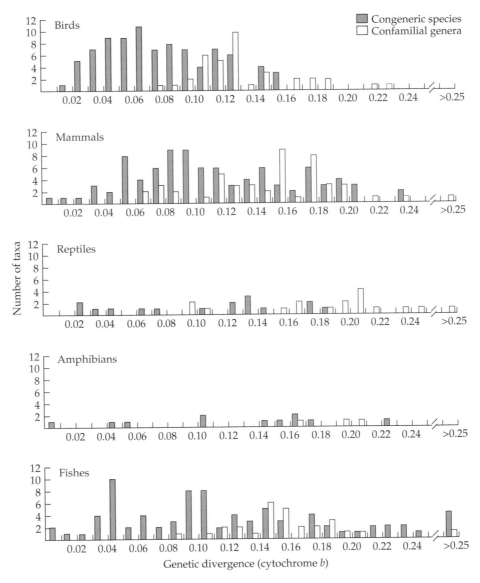

Figure 1.3 Another potential genetic yardstick for diverse taxa. Shown on a common scale are genetic distances (in this case, sequence divergence in the mitochondrial cytochrome *b* gene) observed among congeneric and confamilial species within each of five vertebrate classes. Each data point in a histogram represents the average genetic distance among species within a genus or family. Assayed bird taxa showed significantly less genetic divergence, on average, than their non-avian taxonomic counterparts. (After Johns and Avise 1998a.)

less divergent than those of many amphibian congeners. They speculated that the morphological differences between humans and chimpanzees, which led in part to the traditional placement of these species in different taxonomic families (Hominidae and Pongidae), might be due to evolutionary changes at a few key sites of gene regulation with major phenotypic effects. Years later, researchers used microarray techniques and related molecular assays to address one prediction of this gene regulation hypothesis: that gene expression patterns might be better predictors than most structural genes of important differences in organismal morphology, behavior, and cognition (Oleksiak et al. 2002). To test one aspect of this hypothesis, Enard et al. (2002a) compared transcriptional levels in various tissues of humans, chimpanzees, and other primates, and found that species-specific changes in gene expression had been greatest in the human brain. Further suggestive evidence for the importance of gene regulation and positive natural selection in human evolution came from recent analyses that focused on detailed expression profiles of particular gene regions (Enard et al. 2002b; Hellmann et al. 2003) and from molecular findings of extensive local repatterning of hominoid chromosomes (Locke et al. 2003).

Another possibility, however, is that the perceived morphological distinctness of humans from chimpanzees and other primates has been exaggerated by anthropocentric bias. In a fascinating classic paper titled "Frog perspective on the morphological difference between humans and chimpanzees," Cherry et al. (1978) employed the anatomical traits normally used to discriminate among frogs (eye–nostril distance, forearm length, toe length, etc.) to quantify the morphological separation between humans and chimpanzees. By these criteria, morphological divergence between the two primates was large even by frog standards (whereas molecular divergence was not), a result interpreted as consistent with the postulate that morphological and molecular evolution can proceed at very different rates. It is ironic, yet understandable, that this pioneering attempt to evaluate a comparative yardstick for morphological evolution came from a research laboratory otherwise devoted to molecular biology, where genetic comparisons across diverse biota tend to come more naturally.

Quite apart from helping to evaluate the probable importance of gene regulation (as well as nonregulatory changes) in organismal evolution, the comparative information content of molecular markers raises other important issues for taxonomy and systematics. Using extensive DNA and protein sequences (interpreted in conjunction with paleontological evidence), might it soon become possible to adopt a universally standardized and quantifiable scheme for classifying all forms of life (see Chapter 8)? If so, this would represent a dramatic departure from traditional systematic practices, in which both the empirical data and their interpretation have often been quite idiosyncratic to each taxonomic group. This is not to imply that the overall magnitude of genetic divergence between species is necessarily the only, or even the best, guide to phylogenetic (i.e., cladistic) relationships within particular taxonomic groups, but it is a potentially important means of standardizing and quantifying inter-group comparisons in ways that simply were not possible prior to the molecular revolution in systematics.

Molecular approaches facilitate mechanistic appraisals of evolution

Ever since Darwin and Mendel, assessments of gross phenotype have been crucial in elucidating the general nature of spontaneous mutations, natural selection, and other evolutionary genetic forces. Today, comparative genomics provides previously inaccessible information about the fundamental mechanistic basis of evolutionary transitions among phenotypes. For example, through DNA sequencing and other laboratory approaches, various morphological and physiological mutations in *Drosophila* fruit flies and many other species have been characterized at the molecular level and shown to be attributable to specifiable molecular events, such as point mutations in coding regions, mutations in flanking and non-flanking regulatory domains, and insertions of transposable elements (Carroll et al. 2001; Gerhart and Kirschner 1997; B. Lewin 1999). Homeotic genes are another class of loci in which genetic alterations well characterized at the molecular level can be of special phenotypic importance, in this case with respect to the evolution of fundamental body plans in metazoan animals (Box 1.1). As cogently stated by Lenski (1995), "Molecules are more than markers."

Such mechanistic appraisals fall somewhat outside the subject matter of this book, but a few examples nonetheless can be mentioned in which data of relevance to functional biology arise as a direct or indirect by-product of molecular genealogical analyses. For example, in quantitative genetics (the study of complex phenotypes), a now-popular enterprise introduced by Paterson et al. (1988; see also Lander and Botstein 1989; Lander and Schork 1994) involves the use of DNA markers in conjunction with experimental crosses to map genomic positions of loci underlying polygenic traits (i.e., those influenced by multiple "quantitative trait loci" or QTLs) (Box 1.2). Also, increasing numbers of phenotypic attributes, especially in model species, are yielding to detailed molecular-level characterizations informed by phylogeny (e.g., Long et al. 1998; Mackay 2001; Peichel et al. 2001). For example, genealogical reconstructions based on molecular data from the alcohol dehydrogenase locus in *Drosophila melanogaster* revealed that a mutation conferring a higher capacity to utilize or detoxify environmental alcohols probably arose within the last million years (Aquadro et al. 1986; Stephens and Nei 1985). In house mice, sequence analyses of introns at *t*-loci on chromosome 17 indicated that particular chromosomal inversions affecting embryonic development originated about 3 million years ago and accumulated recessive lethality factors that spread globally within the last 800,000 years (Morita et al. 1992). Detailed molecular analysis of an esterase locus in mosquitoes indicated that the worldwide distribution of an insecticide resistance allele had resulted from the migrational spread of a single mutation, rather than independent

BOX 1.1 Homeotic Genes in Metazoan Animals

Homeotic loci, first identified in *Drosophila*, are developmental genes that play a key ontogenetic role by regulating the identity of body regions, such as particular thoracic or abdominal segments. Their salient effect on morphology is perhaps best registered when things go wrong: Mutations in homeotic genes sometimes cause the developmental transformation of one body region into the likeness of another, such as converting an antenna into a leg that protrudes from a fruit fly's head, or converting a two-winged into a four-winged fly. Although most such mutations are quickly eliminated by natural selection, they nonetheless evidence the magnitude of the morphotypic influence routinely exercised by homeotic loci during normal development.

Families of homeotic genes have proved to be widespread in metazoan animals, and their evolutionary histories have been elucidated by comparative molecular analyses. Best characterized is the Hox gene family, which apparently arose early in metazoan evolution, then expanded greatly in number of loci during the radiation of bilateral animals, and again with a probable tetraploidization event early in vertebrate evolution. The net result of these repeated gene duplications is the presence in various modern taxa of as many as a dozen oft-linked genes specialized to orchestrate the development of specific body regions.

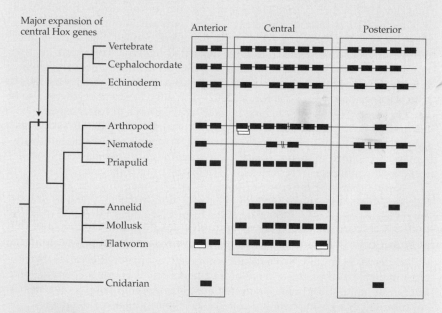

Number and arrangement of Hox loci in representative metazoan animals. Each rectangle represents a Hox gene influencing anterior, central, or posterior body segments. Horizontal lines indicate gene arrangements (when known) from mapping data. On the left is a phylogenetic tree for these metazoans based on ribosomal DNA sequences. (After Carroll et al. 2001 and de Rosa et al. 1999.)

BOX 1.2 QTL Mapping

Quantitative traits are phenotypic features influenced by multiple loci, or poly-genes. A popular exercise in recent years is the employment of large banks of molecular markers to identify the numbers and chromosomal locations of quantitative trait loci (QTLs) that contribute to genetic variation in particular phenotypic attributes. Such polygenic traits might be levels of acidity in tomatoes, rates of senescence in fruit flies, or components of reproductive isolation between closely related species (see Chapter 7).

The QTL mapping approach requires the availability of legions of molecular markers, scattered throughout the genome, that have been ordered (mapped) along the chromosomes of the species of interest. Data banks consisting of dozens to thousands of such molecular markers are now available for increasing numbers of model species, such as *Mimulus* monkeyflowers (Bradshaw et al. 1998), *Helianthus* sunflowers (Rieseberg et al. 1995a,c), *Drosophila* fruit flies (Macdonald and Goldstein 1999), and *Homo sapiens* (Weiss and Clark 2002). A typical experimental approach is as follows: Two pure-breeding strains or species that differ in many such molecular markers are crossed to produce F_1 progeny, and these hybrids are then backcrossed to one or the other parental form. The idea is then to monitor whether specific molecular markers tend to co-associate ("co-segregate") in this backcross generation with specific phenotypes of interest that also distinguish the parental forms (Paterson 1998; Tanksley 1993). When particular polymorphic markers of known chromosomal location explain significant proportions of the variance among phenotypes in these backcross progeny, the deduction is that genes contributing to those phenotypes must be closely linked to those molecular markers. The approximate numbers and locations of genes underlying polygenic phenotypes can then be estimated, often with the assistance of computer programs for analyzing the statistical associations (e.g., Basten et al. 2002). The same basic rationale can also be used to identify QTLs by searching for statistical patterns of co-segregation between phenotypes and molecular markers through known family pedigrees extending across multiple generations. Some QTL analyses in the literature are remarkably refined in their capacity to pinpoint the chromosomal locations of loci exerting influence over particular phenotypic traits (Luo et al. 2002).

Notwithstanding the current popularity of QTL mapping, the approach has some limitations. First, any molecular polymorphisms that tend to co-segregate with phenotypes of interest are not necessarily mechanistically responsible for those phenotypes. Rather, they are merely physically linked to the responsible chromosomal regions, within which there may be hundreds of candidate genes. Second, only polygenes with relatively major effects (i.e., that explain perhaps 10%–20% or more of the phenotypic variance, depending on the power of the study) can be detected by QTL mapping. Third, polygenes contributing to a given phenotype often exert their influence differentially depending on the particular genetic backgrounds examined (e.g., Devlin et al. 2001; Hardy et al. 2003; Muir and Howard 2001; Spencer et al. 2003). Such epistasis (inter-locus interaction) between QTL loci and their genetic backgrounds emphasizes the desirability of conducting QTL mapping using a variety of different strains.

mutations of different resistance alleles (Raymond et al. 1991). A similar conclusion was reached for the global spread of a methicillin resistance gene in a
pathogenic bacterium, *Staphylococcus aureus* (Kreiswirth et al. 1993).

Medicine has also benefited from genealogical insights from molecular
analysis (see Avise 1998a; McKusick 1998; Rannala and Bertorelle 2001;
Scriver et al. 2000). In addition to their routine use in the clinical diagnosis of
numerous genetic disorders, molecular markers have been employed to
assess whether specific genetic diseases (e.g., phenylketonuria in Yemenite
Jews, Huntington's chorea in Afrikaners, or fragile X syndrome) are of monophyletic or polyphyletic evolutionary origin (Avigad et al. 1990; Diamond
and Rotter 1987; Hayden et al. 1980; Richards et al. 1992). For example, DNA-
level markers interpreted in conjunction with historical accounts revealed
that about 90% of cases of variegate porphyria in South Africa trace to a single distinctive gene mutation that arose in Cape Town in the late 1600s (Hift
et al. 1997). Molecular analyses of the distributions of specific sets of repetitive DNA elements likewise permitted researchers to identify the phylogenetic roots and approximate evolutionary ages of three human genetic diseases (involving the *LPL*, *ApoB*, and *HPRT* genes; Martinez et al. 2001).
Additional examples of how DNA markers can inform epidemiology and
medicine appear in later chapters.

In general, by mapping variable phenotypic traits of species and taxonomic groups onto phylogenies estimated from molecular markers, scientists are transforming modes of inquiry into the evolutionary origins and
histories of numerous organismal features (see Chapter 8).

Molecular approaches are challenging and exciting

A tremendous appeal of molecular phylogenetics is the sheer intellectual
challenge this discipline provides. Many discoveries in molecular biology
clearly affect the practice of genealogical assessment, and some molecular-
level phenomena now taken for granted were undreamed of even a few
years ago. For example, nucleotide sequences in many multi-gene families
tend to evolve in concert within a species and thereby remain relatively
homogeneous. This process of "concerted evolution" (Ohta 2000; Zimmer et
al. 1980), first noted by Brown et al. (1972), is due to the homogenizing
effects of unequal crossing over among tandem repeats and to gene conversion events even among unlinked loci (Arnheim 1983; Dover 1982; Ohta
1980, 1984; Smithies and Powers 1986). Concerted evolution means that
multiple copies of a gene within such families do not provide the independent bits of phylogenetic information formerly assumed (Ohno 1970).
Particular rRNA gene families, for example, are employed routinely as
informative markers of phylogeny, a task that would be far more difficult or
impossible if each of the hundreds of tandem gene sequences within a family evolved independently of all others. Thus, concerted evolution makes
genes within multi-locus families far more tractable for phylogenetic analysis than would otherwise be the case.

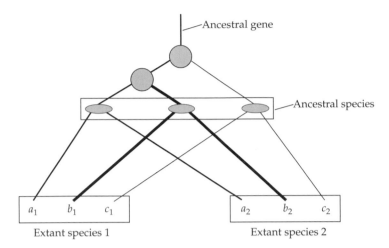

Figure 1.4   Possible allelic relationships within a multi-gene family. The two circles indicate gene duplication events from an ancestral locus, producing three extant genes a, b, and c. The three ellipses represent allelic separations, leading to the extant alleles a_1, b_1, and c_1 in species 1 and to a_2, b_2, and c_2 in species 2. Genetic comparisons between a_1 and a_2, b_1 and b_2, or c_1 and c_2 are orthologous, whereas all other comparisons (e.g., between a_1 and b_2, a_1 and c_2, or a_1 and b_1) are paralogous. Orthologous similarities generally date to times near the speciation event (but see later chapters and also the following for additional distinctions between a gene tree and a species tree: Hey 1994; W. P. Maddison 1995; Page and Charleston 1997, 1998; Slowinski and Page 1999). By contrast, paralogous similarities date to relevant gene duplication events, which could vastly pre-date speciation times of the organisms compared. However, under strong concerted evolution (see text), all or portions of a_1, b_1, and c_1 would appear more closely related to one another than to their respective allelic counterparts in species 2.

However, the sequences of multi-copy genes within a species do not always evolve in concert subsequent to the duplications from which they arose. This fact makes the fundamental distinction between the concepts of orthology (sequence similarity tracing to a speciation event) and paralogy (sequence similarity tracing to a gene duplication) important (Figure 1.4). Indeed, when estimating phylogenetic relationships from sequence data on multi-gene families, trying to disentangle the orthologous from the paralogous similarities, and then draw proper genealogical conclusions accordingly, is a challenging intellectual and empirical exercise (Cotton and Page 2002; Page 1998, 2000).

Another example of how molecular data can offer exciting new perspectives on phylogeny relates to the introduction of mitochondrial (mt) DNA approaches to population genetics in the late 1970s. Prior to that time, most biologists viewed intraspecific evolution primarily as a process of shifting allele frequencies, a perspective that fit well with the traditional language and framework of population genetics but failed to focus adequately on the genealogical component of population history (Avise 1989a; Wilson et al. 1985). By providing the first accessible data on "gene trees" at the intraspecific level, mtDNA methods forged an empirical and conceptual bridge that now connects the formerly separate realms of microevolutionary analysis (population genetics and ecology) and interspecific macroevolution (the traditional arena of phylogenetic biology). The notion of gene trees has also raised intriguing conceptual challenges regarding the meaning of "organismal phylogeny," which in a very real sense can be thought of as an emergent or composite property of multitudinous gene genealogies that have trickled through an extended sexual pedigree under the vagaries of Mendelian (and sometimes non-Mendelian) inheritance (see Chapters 3 through 7).

Overall, phylogenetic studies on mtDNA have stimulated a wide variety of formerly unorthodox (but now mainstream) notions about evolutionary processes (Table 1.1). Similar claims can be made for molecular characterizations of homeotic genes (see Box 1.1; Erwin et al. 1997; Knoll and Carroll 1999), transposable elements (Box 1.3; see also Chapter 8), introns (Gilbert 1978; Li 1997), and several other molecular genetic systems, all of which are now appreciated to play huge but formerly unimagined roles in organismal evolution.

BOX 1.3 Transposable Elements

Perhaps the most unexpected and revolutionary finding in all of molecular evolution is that the genomes of most species are riddled with roving pieces of DNA (Sherratt 1995), commonly known as transposable elements (TEs), mobile elements, or "jumping genes." These elements come in two broad categories: class I elements (retrotransposable elements, or RTEs), which transpose *proliferatively* by making RNA copies of themselves and reverse-transcribing those copies into DNA, which then inserts into other genomic locations; and class II elements, which move by excising themselves from one genomic site and reinserting themselves into another. Class I elements are especially abundant in eukaryotes (organisms whose cells have distinct nuclei separated from cytoplasm), whereas class II elements tend to be relatively more abundant in bacteria and lower eukaryotes. Class I RTEs come in various structural families and subfamilies. One common distinction, for example, is whether an element is flanked by two long terminal repeat sequences (LTRs; see figure) or not (as in LINEs, an acronym for long interspersed nuclear elements).

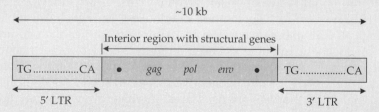

General structure of one type of LTR retrotransposon. Shown is a gypsy-like element from *Drosophila*, in which long terminal repeats (LTRs) flank genes, in this case, for capsid protein (*gag*), reverse transcriptase (*pol*), and envelope protein (*env*). (After McCarthy and McDonald 2003.)

Retrotransposable elements are interesting for evolutionary as well as functional reasons. They are similar in structure and mode of replication to infectious retroviruses (Coffin et al. 1997), and their proliferate nature makes them quintessential "selfish" or "parasitic" elements within cells. Although quite variable in relative abundance, they and other classes of mobile elements often constitute huge fractions of plant and animal genomes (Brosius 1999), making up 50%–80% of the corn genome and 90% of the wheat genome (Flavell 1986; SanMiguel et al. 1996), for example, and about 40% or more of the genomes of many mammals (Smit 1999), including humans (Yoder et al. 1997). Mobile elements tend to induce mutations in host genomes when they jump from spot to spot, and this factor, together with the suspected metabolic burden of their maintenance, produces conflicts of interest with their host cells. In this coevolutionary war, host genomes occasionally win battles too, as evidenced by the fact that some former jumping genes appear to have been recruited to various cellular tasks that benefit their host (McDonald 1990, 1998).

Two kinds of algorithms can be employed in computer-based searches of available genomic sequence for the presence of particular families of TEs (or any other specified gene sequences). In the traditional method (often implemented in a BLAST program; Altschul et al. 1997), a researcher compares a specific nucleotide sequence of interest (the "query") with one or more sequences in the database, looking for significant matches, often arbitrarily defined as 90% or more sequence similarity. The second method involves scanning the database for defined structural features of particular sequences of interest. In the case of RTEs, these structural signatures can be two long stretches of nucleotide sequence (putative LTRs) that are (1) highly similar to each other, (2) in fairly close proximity in the genome, and (3) themselves flanked by short target repeat sites (McCarthy and McDonald 2003).

Why Not Employ Molecular Genetic Markers?

Against these advantages of molecular genetic methods appear to stand only two major disadvantages: Considerable training is required of practitioners, and monetary costs are rather high (but also variable across methods) by traditional systematics standards (Weatherhead and Montgomerie 1991). However, a fact sometimes overlooked is that most molecular genealogical assessments have proved to support (rather than contradict) earlier phylogenetic hypotheses based on morphology or other phenotypic characteristics. Thus, a complete molecular reanalysis of the biological world is unnecessary for phylogenetic purposes. In such genealogical applications, molecular markers are used most intelligently when they address controversial areas or when they are employed to analyze problems in natural history and evolution that fall beyond the purview or capabilities of traditional nonmolecular observation.

TABLE 1.1 **Twelve unorthodox perspectives on evolution prompted by molecular genetic findings on animal mitochondrial DNA**

1. **Asexual Transmission (Chapter 3)**
   Cytoplasmic genes within sexually reproducing species normally exhibit clonal (uniparental, non-recombinational) transmission.

2. **A New Level in the Population Hierarchy**
   Entire populations of mtDNA molecules inhabit somatic and germ cell lineages within each individual (Birky et al. 1989).

3. **Non-Universal Code (Chapter 8)**
   Genetic codes in mtDNA sometimes differ among taxa, and also differ from the nuclear code formerly thought to be universal.

4. **Conserved Function, Rapid Evolution (Chapter 3)**
   Considerations in addition to functional constraint are required to explain the rapid pace of animal mtDNA evolution.

5. **Lack of Mobile Elements, Introns, Repetitive DNA**
   Genes with selfish motives gain no fitness advantage by becoming repetitive within an asexually transmitted genome (Hickey 1982).

6. **Endosymbiotic Origins (Chapter 8)**
   Eukaryotic organisms are genetic mosaics containing interacting nuclear and organelle genomes that are descended from what had been independent forms of life early in Earth's history.

7. **Intergenomic Conflicts of Interest**
   Because of their contrasting modes of biparental versus uniparental inheritance, nuclear and cytoplasmic genomes have inherent evolutionary conflicts of interest (in addition to their evident requirements for functional collaboration) (Avise 2003a; Eberhard 1980).

8. **Intergenomic Cooperation**
   Multitudinous interactions between products of cytoplasmic and nuclear genes lead to expectations of functional coevolution between the different genomes within a cell (Kroon and Saccone 1980).

9. **Matrilineal Genealogy (Chapter 6)**
   Mutational differences among mtDNA haplotypes record the phylogenetic histories of female lineages within and among species.

10. **Gene Trees versus Organismal Phylogenies (Chapters 4, 6, and 7)**
    In sexually reproducing organisms, pedigrees contain multitudinous gene genealogies (gene trees) that differ in topological details from locus to unlinked locus, and may also differ from a composite population-level or species-level phylogeny. Thus, a species tree or cladogram is in actuality a statistical "cloudogram" (Maddison 1997), with a variance, of semi-independent gene trees.

11. **The Historical, Nonequilibrium Nature of Microevolution (Chapter 6)**
    Genealogical signals from various molecular markers indicate that historical idiosyncrasies and nonequilibrium population genetic outcomes are a sine qua non of intraspecific (as well as interspecific) evolution.

12. **Degenerative Diseases**
    Genetic defects in mitochondrial oxidative phosphorylation provide a new paradigm for the study of aging and degenerative diseases (Wallace 1992; Wallace et al. 1999).

Source: After Avise 1991a.

2

The History of Interest in Genetic Variation

In 1951, the problematic of population genetics was the description and explanation of genetic variation within and between populations. That remains its problematic 40 years later ...

R. C. Lewontin, 1991

Since their inception in the latter half of the twentieth century, the fields of molecular ecology and molecular evolution have been preoccupied with the functional role and the possible adaptive significance of genetic variation. This focus led to compelling conceptual and empirical debates that captured nearly everyone's interest, but also served to divert attention from what many researchers perceived as mundane applications of molecules as "mere" genetic markers. Thus, with relatively few notable exceptions before the mid-1980s (e.g., Selander 1982), most of the early research programs that employed molecular assays were preoccupied with uncovering functional variation and illuminating how natural selection operates at the levels of proteins and DNA. Only gradually did molecular markers come to be appreciated on their own merit (even if many of them might be selectively neutral) for the empirical and conceptual richness they can bring to studies of organismal behavior, natural history, and phylogenetic relationships. This chapter traces the history of scientific interest in natural selection's role in maintaining molecular variation. It also describes why an understanding of that role, while extremely important, is seldom a precondition for employing molecules as genealogical markers.

The Classical–Balance Debate

Classical versus balance views of genome structure

Prior to the molecular era that began in the mid-1960s, the magnitude of genetic variability in animal and plant genomes was the subject of a long-standing controversy. Evolution has been defined as temporal changes in the genetic composition of populations (Dobzhansky 1937). Genetic variation is prerequisite for this process. Thus, a central empirical challenge for population genetics has always been to measure genetic variability under the rationale that such quantification would help to reveal the operation of natural selection as well as mutation, genetic drift, and other evolutionary forces (Gillespie 1987; Kimura 1991; Kreitman 1987; Li 1978; Ohta and Tachida 1990). Unfortunately, the exact genes or alleles responsible for the phenotypic variation routinely observable within and among natural populations rarely could be specified explicitly. This problem of empirical insufficiency plagued population genetics throughout the first half of the twentieth century, as evidenced by the establishment of two diametrically opposed scientific opinions about magnitudes of genetic variation in nature. Advocates of the "classical" school maintained that genetic variability in most species was low, such that conspecific individuals typically were homozygous for the same "wild-type" allele at nearly all genes. Proponents of the "balance" view maintained that genetic variation was high—that most loci were polymorphic, and most individuals were heterozygous at a large fraction of genetic loci.

Several corollaries and ramifications stem from these opposing schools of thought (Lewontin 1974). Under the classical view, natural selection was seen as a purifying agent, cleansing the genome of inevitable mutational variation. Deleterious recessive alleles in heterozygotes might escape elimination temporarily, but were prevented from reaching high frequencies in populations because of their negative fitness consequences when homozygous. The classicists did not deny adaptive evolution, but they felt that the process was due to occasional selectively advantageous mutations that would quickly sweep through a species to become the new wild-type alleles. Because little genetic variation was available to be shuffled into novel multi-locus allelic associations, recombination was viewed as a rather insignificant process compared with mutation. Furthermore, any genetic differences that might be uncovered between populations or species must be of profound importance (because of the low within-population component of variability). Central to the classical school was the concept of genetic load (Wallace 1970, 1991): the notion that genetic variation produces a heavy burden of diminished fitness, which in the extreme might even cause population extinction. This perception of genetic variation as a curse was forcefully summarized by Muller (1950), who predicted from genetic load calculations that only one locus in 1,000 (0.1%) would prove to be heterozygous in a typical individual.

The balance school, by contrast, viewed natural selection as favoring genetic polymorphisms through balancing mechanisms such as the fitness superiority of heterozygotes (Dobzhansky 1955), variation in genotypic fitness among habitats, or frequency-dependent fitness advantages (Ayala and Campbell 1974). Genetic variability was thought to be both ubiquitous and adaptively relevant. Deleterious alleles were not ruled out, but presumably were held in check by natural selection and contributed little to heterozygosity. Because high variability was predicted for sexually reproducing species, no allele could properly be termed wild-type. Genetic recombination, therefore, assumed far greater significance than de novo mutation in producing inter-individual fitness variation from one generation to the next. Furthermore, genetic differences among populations were perhaps of less importance (because of the high within-population component of overall variability). How much genetic variation was predicted under the balance view? Wallace (1958) raised a proposal that seemed extreme at the time, but not at all unreasonable today: "The proportion of heterozygosis among gene loci of representative individuals of a population tends towards 100 percent."

The balance hypothesis gained support from several indirect lines of evidence: extensive phenotypic variation, which in wild populations of several well-studied species often proved to be genetically influenced and adaptively relevant (e.g., Ford 1964); a genetic underpinning for many naturally occurring morphological variants and fitness characters in populations that could be experimentally manipulated (e.g., by inbreeding, or through "common garden" experiments in which the fraction of phenotypic variation attributable to genetic influence could be estimated by controlling for environmental effects); and fast genetic responses to artificial selection exhibited by numerous traits of many domestic animals and plants (reviewed in Ayala 1982a). However, none of these or related observations permitted direct answers to the fundamental question: What fraction of genes is heterozygous in an individual and polymorphic in a population?

An answer to this question requires that variation be assessed at many independent loci, chosen without bias with respect to magnitude of genomic variability. But this requirement introduces a catch-22 for any appraisal based on conventional Mendelian genetic approaches: Genes underlying a particular phenotype can be identified only when they carry segregating polymorphisms. In other words, genetic assignments for phenotypic features traditionally were inferred from segregation patterns of allelic variants through organismal pedigrees, but this also meant that invariant loci escaped detection, and no accumulation of such data could provide an uncolored estimate of overall genetic variation. Other means were needed to screen genetic variability more directly, and in a manner that allowed assay of an unbiased sample of polymorphic *and* monomorphic loci.

Molecular input to the debate

A fundamental breakthrough occurred in 1966, when independent research laboratories published the first estimates of genetic variability based on multi-locus protein electrophoresis (Harris 1966; Johnson et al. 1966; Lewontin and Hubby 1966). This method involves separation of non-denatured proteins by their net charge under the influence of an electric current, followed by application of histochemical stains to reveal enzymatic or other protein products of particular, specifiable genes (see Chapter 3). Because invariant as well as variant proteins are revealed, this approach represented the first serious attempt to obtain unbiased estimates of genomic variability at a reasonable number (usually 20–50) of genetic loci. The empirical results were clear: Genomes of fruit flies and humans harbored a wealth of variation, with 30% or more of assayed genes polymorphic in a population, and roughly 10% of loci heterozygous in a typical individual (Box 2.1). Especially over the next two decades, multi-locus electrophoretic surveys were conducted on hundreds of plant and animal species, and they likewise revealed levels of genetic variation that were often high, but also quite variable among

BOX 2.1 Measures of Genetic Variability within a Population

For multi-locus protein electrophoretic data (or other comparable classes of information, such as data from microsatellite loci), one useful measure of genetic diversity is population heterozygosity (H), defined as the mean percentage of loci heterozygous per individual (or equivalently, the mean percentage of individuals heterozygous per locus). Estimates of H can be obtained by direct count from a raw data matrix, the body of which consists of observed diploid genotypes, as in the following hypothetical example involving eight loci (A–H) scored in each of five individuals (i):

i	A	B	C	D	E	F	G	H	h_i
					Locus (j)				
1	*aa*	*aa*	*aa*	*aa*	**ab**	*aa*	*aa*	*aa*	0.125
2	*bb*	**ab**	**ab**	**ab**	*bb*	*aa*	**ab**	*aa*	0.500
3	*cc*	**ac**	**bc**	**bd**	*dd*	**ad**	**cd**	**bc**	0.750
4	*aa*	*aa*	*aa*	*aa*	*aa*	*aa*	*aa*	*aa*	0.000
5	*cc*	*cc*	*cc*	*cc*	*cc*	*cc*	*cc*	*cc*	0.000
h_j	0.0	0.4	0.4	0.4	0.2	0.2	0.4	0.2	$H = 0.275$

Here, diploid genotypes are indicated by italic lowercase letters (each letter representing an allele), and heterozygotes are boldfaced. In this example, 11 of 40 assayed genotypes are heterozygous ($H = 0.275$). Equivalently, H may be

species (Figure 2.1; see seminal reviews by Hamrick and Godt 1989; Nevo 1978; Powell 1975; Ward et al. 1992).

Protein electrophoretic techniques were not entirely new in 1966—indeed, crude methods had been available for nearly 30 years (see Brewer 1970). Rather, the scientific impact of the landmark allozyme surveys lay primarily in the manner in which methods and data from the seemingly alien field of molecular biology were applied for the first time to long-standing issues in population genetics. After the mid-1960s, contacts between molecular genetics and population genetics would only expand, and at an ever-faster pace. Today, these disciplines are thoroughly wedded.

The data and conceptual orientations of the protein electrophoretic era exerted an overriding influence on research goals in population genetics for the next quarter century. Although the allozyme revolution provided original information on genetic variation in natural populations, it also inhibited a stronger molecular genealogical focus in at least two ways. First, at each locus, protein electrophoresis reveals qualitative allelic products (electromorphs) whose phylogenetic order cannot be inferred safely from the observable property—band mobility on a gel. Much of traditional population

interpreted as the mean of the row or column totals, which represent direct-count heterozygosities for single individuals (h_i) or single loci (h_j), respectively.

Heterozygosities also may be estimated from observed frequencies of alleles (rather than genotypes), assuming that the population is in Hardy–Weinberg equilibrium. Thus, $h_j = 1 - \Sigma q_k^2$, where q_k is the frequency of the kth allele. Other common measures of population variability for such data are the mean number of alleles per locus and the proportion of polymorphic loci (P), which is 0.6 in the above example. To avoid an expected positive correlation between P and sample size, a locus usually is considered polymorphic only if the frequency of the most common allele falls below an arbitrary threshold, typically 0.99 or 0.95.

For molecular data that involve restriction sites or nucleotide sequences along a particular stretch of DNA that can be thought of as one locus, a useful statistic summarizing heterozygosity at the base-pair level is nucleotide diversity (Nei and Li 1979; Nei and Tajima 1981), or the mean sequence divergence between haplotypes (alleles): $p = \Sigma f_i f_j p_{ij}$, where f_i and f_j are the frequencies of the ith and jth haplotypes in a population, and p_{ij} is the sequence divergence between them. Another informative measure is haplotype diversity: $h = 1 - \Sigma f_i^2$. This measure is a DNA-level analogue of h (defined above for protein electrophoretic data) because its calculation entails no assessment of the magnitude of genetic divergence between the alleles involved. Depending on the loci and species surveyed, nucleotide diversities within a population often fall in the range of 0.001–0.020 (Stephan and Langley 1992), and haplotype diversities may be above 0.5 for rapidly evolving markers such as microsatellites or animal mtDNA.

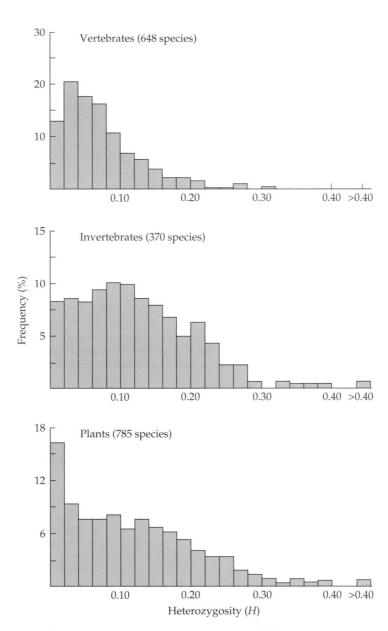

Figure 2.1   Allozyme-based estimates of mean multi-locus heterozygosity.
Shown are estimates of mean heterozygosity (*H*), per species, based on empirical
surveys of nearly 2,000 species of vertebrate and invertebrate animals and plants.
An average of more than 20 loci were scored per study. (Data for animals from
Ward et al. 1992; data for plants from Hamrick and Godt 1989.)

genetic theory, built around the formative works of Fisher (1930), Wright (1931), and Haldane (1932), can be couched in the language of the expected frequency dynamics of such phylogenetically unordered character states under the separate or joint evolutionary forces of natural selection, genetic drift, gene flow, mutation, and recombination. Thus, protein electrophoresis produced data that could be interpreted using the language and perspectives of traditional population genetic theory, but at the price of diverting attention from the phylogenetic orientation that increasingly characterized other areas of evolutionary biology. Thus, 20 years after the allozyme revolution began, Lewontin (1985) concluded, "Population genetics is conceptually the study of gene lineages, [but] until now, the data to study such lineages have not really been available."

Second, by necessarily focusing on issues of genetic variation per se (rather than on the genealogical content of the molecular information), the protein electrophoretic era stimulated new quests to refine and interpret data-based estimates of molecular variability. The field of empirical population genetics soon became further preoccupied with characterizing and quantifying genetic variation, first via more refined comparisons of proteins and later directly at the level of DNA.

Questions of empirical refinement

How much cryptic variation at the protein level remained hidden beyond the resolving power of conventional gel electrophoresis? To address this question, two general experimental protocols were followed. In "backward experiments" (Selander and Whittam 1983), protein variants of known amino acid sequence were subjected to electrophoresis to determine the proportion of known alleles that were detectable (Ramshaw et al. 1979). Unfortunately, this approach could be applied only in the few instances in which proteins already had been well characterized by other methods. In "forward experiments," assay conditions (buffer concentrations, thermal regimes, gel-sieving media, etc.) were varied so as to further discriminate alleles within the electromorph classes identified in the original tests. This approach often uncovered additional protein variants (particularly at loci that were polymorphic in the initial assays), and it left a general impression that the original electrophoretic methods had revealed only the tip of the genetic variation iceberg (Aquadro and Avise 1982a, 1982b; Ayala 1982b; Bernstein et al. 1973; Bonhomme and Selander 1978; Coyne 1982; Johnson 1976a, 1977; McDowell and Prakash 1976; Milkman 1976; Prakash 1977). Unfortunately, data from these forward experiments were often difficult to interpret because the particular genetic bases of the polymorphisms seldom were clear from the assays. Thus, none of these refined methods gained more than transient popularity, nor did they offer good sources of utilitarian genetic markers.

How representative of the genome were the variability estimates derived from protein electrophoretic loci? Because the availability of histochemical stains was a deciding criterion for the inclusion of particular proteins in most electrophoretic surveys, dehydrogenases and other enzymes of the glycolytic pathway and citric acid cycle were disproportionately represented in the assays. An initial concern was whether these proteins might provide misrepresentative estimates of genetic variation at other protein-coding genes. Thus, for a brief time in the late 1970s and early 1980s, attention shifted to abundant proteins (membrane-associated, ribosomal, and others) as revealed by nonspecific stains and newly introduced two-dimensional electrophoresis (which separates proteins by isoelectric focusing in one dimension, and then by molecular weight in the second; O'Farrell 1975). For several species, somewhat lower heterozygosities were revealed than had been estimated in the original allozyme surveys (Aquadro and Avise 1981; Leigh Brown and Langley 1979; Racine and Langley 1980; Smith et al. 1980).

However, any lingering thoughts that genomes might harbor little genetic variation were conclusively dispelled by the subsequent explosion of direct DNA-level information. Indeed, researchers soon discovered that at most protein-coding genes, synonymous or "silent-site" nucleotide polymorphisms (those not translated into amino acid variations) greatly outnumber the non-synonymous or replacement substitutions that in effect had been the subject of earlier protein electrophoretic surveys. Furthermore, in most species, nucleotide polymorphism was found to be a nearly ubiquitous feature of a wide variety of DNA sequences, both protein-coding and otherwise (Li 1997; Nei and Kumar 2000).

The Neutralist–Selectionist Debate

Strangely, the discoveries of extensive molecular variation did not clinch the case for the philosophical perspective on genetic variation embodied in the balance school of thought. Instead, they prompted development of an alternative explanation for molecular variability that was to assume a prominent role in population genetics to the present time. Under this strict "neutral mutation theory," alternative alleles confer no differential fitness effects on their bearers. The theory, as summarized by Kimura (1991), holds that "the great majority of evolutionary mutant substitutions at the molecular level are caused by random fixation, through sampling drift, of selectively neutral (i.e., selectively equivalent) mutants under continued mutation pressure." As applied to intraspecific variability, neutrality theory predicted that most molecular polymorphisms are maintained by a balance between mutational input and random allelic extinction by genetic drift. Neutralists did not deny the existence of high molecular genetic variation, but rather questioned its relevance to organismal fitness. Because neutralists and classicists shared the views that balancing selection plays little role in maintaining molecular polymorphism, and that most selection is directional or "purifying" selection against deleterious alleles, neutrality theory also was referred to as neoclassical theory (Lewontin 1974).

Several points should be made clear at the outset. First, neutrality theory did not suggest that most genes or allelic products are dispensable (of course they are not). Rather, it proposed that different alleles at a locus are functionally equivalent, such that organismal fitness does not vary as a function of the particular genotypes possessed. Second, neutrality theory did not deny that many de novo mutations are deleterious and are eliminated by purifying selection. Rather, the focus was on the supposed neutrality of segregating polymorphisms that escape selective elimination. Indeed, one cornerstone of neutrality theory was that functionally unconstrained nucleotide positions or genic regions are those most likely to harbor neutral variation and to exhibit the most rapid pace of allelic substitution. Third, neutrality theory did not fundamentally challenge the Darwinian mode of adaptive evolution for organismal morphologies and behaviors (although some extensions of neutrality theory did propose a significant role for genetic drift at these phenotypic levels also; Kimura 1990). Rather, neutrality theory developed in response to the intellectual challenge posed by the unexpectedly high levels of observed molecular variability.

Neutrality concepts were introduced in the late 1960s (Kimura 1968a,b) and gained immediate widespread attention, due in part to a paper by King and Jukes (1969) provocatively entitled "Non-Darwinian evolution: Random fixation of selectively neutral mutations." Indeed, the theory directly challenged the prevailing approach of naively extending to the molecular level neo-Darwinian views on the adaptive significance of nearly all differences in organismal phenotype (see Gould and Lewontin 1979). It is quite remarkable that within a decade, Kimura's neutrality theory (and its intellectual offshoot, the "nearly-neutral theory"; Ohta 1992a) gained almost universal acceptance as molecular evolution's gigantic "null hypothesis"— the simplest possible conceptual framework for interpreting molecular variability, and the basic theoretical construct whose predictions must be falsified before alternative proposals invoking balancing or other forms of selection were to be seriously entertained. This is not to say, however, that the neutralist–selectionist debate is fully resolved.

The neutrality school has strong roots in the quantitative tradition of theoretical population genetics developed earlier in the twentieth century (Fisher 1930; Haldane 1932; Wright 1931). An elegant and elaborate theory predicts the amount of genetic variability within a given population as a function of mutation rate, gene flow (where applicable), and population size (Kimura and Ohta 1971). Conspicuously absent from the calculations are selection coefficients, because alleles are assumed to be neutral. Under strict neutrality theory, molecular variability is a function of neutral mutation rate and evolutionary effective population size, N_e (Box 2.2). For example, the magnitude of heterozygosity expected for electrophoretically detectable alleles at equilibrium between mutation and genetic drift is given by

$$H = 1 - \frac{1}{(1 + 8N_e\mu)^{\frac{1}{2}}} \tag{2.1}$$

where μ is the per locus per generation mutation rate to neutral alleles (Ohta and Kimura 1973). Figure 2.2 plots this expected relationship between H and N_e for reasonable neutral mutation rates, and also shows the range of allozyme heterozygosities empirically observed for numerous animal species with indicated population census sizes. (Such comparisons should deal with species-level effective population sizes because the theory involves equilibrium expectations over longer-term evolution.)

BOX 2.2 Effective Population Size

Not all individuals in a population contribute gametes to the next generation with equal probability. This realization has led to the concept of *effective population size* (N_e), originally proposed by Wright (1931). The effective population size refers to the number of individuals in an idealized population that would have the same genetic properties (such as inter-generational variance in allele frequencies due to chance sampling error) observed in the real population. Usually, N_e is much smaller than N (the census size) for one or more of the following reasons:

1. *Separate sexes.* In organisms with separate sexes, one sex may be more common than the other. Let N_M and N_F be census numbers of males and females in such a population. Then, effective population size due to this disparity alone is

$$N_e = \frac{4N_M N_F}{N_M + N_F}$$

   This equation shows that, unless $N_M = N_F$, N_e is less than the total census count ($N_M + N_F$).

2. *Fluctuations in population size.* Most populations in nature probably fluctuate greatly in size due to disease, changes in habitat quality, predation, and so forth. The effective population size due to such fluctuations is equal to the harmonic mean of breeding population sizes across generations. A harmonic mean is a function of the mean of reciprocals, or in this case,

$$N_e = \frac{n}{\Sigma(1/N_i)}$$

   where N_i is the population size in the ith generation and n is the number of generations. A harmonic mean is closer to the smaller rather than to the larger of a series of numbers being averaged, so N_e can be much lower than most population censuses. A severe reduction in population size, called a "population bottleneck," can greatly depress evolutionary N_e.

3. *Combination of separate sexes and fluctuating population size.* If census population sizes of males and females are known across multiple generations, joint

effects of the above factors on N_e can be determined. For each generation, census sizes of males and females are converted to an effective size for that generation, as in (1). Then, the equation in (2) is employed to take the harmonic mean of the single-generation estimates.

4. *Variation in progeny numbers.* Even in a non-fluctuating population with equal numbers of males and females, some individuals may leave many more progeny than others, creating a large fitness variance across families. Only when offspring numbers follow a Poisson distribution, with a mean (and hence variance) of 2.0 per family, does $N_e = N$. In more realistic situations, in which the variance often exceeds the mean, N_e is smaller than the census breeding population size (Crow 1954). Organisms with extremely high fecundity are particularly prone to gross disparities between N_e and N due to high variability in fertility across individuals (Hedgecock et al. 1992; Hedgecock and Sly 1990).

5. *Other factors.* Various other factors can also reduce N_e relative to N. For example, in a species composed of many subpopulations, each of which is subject to periodic extinction and recolonization, the species as a whole can have a far lower N_e than might have been predicted had only composite census sizes for specific generations been available (Maruyama and Kimura 1980).

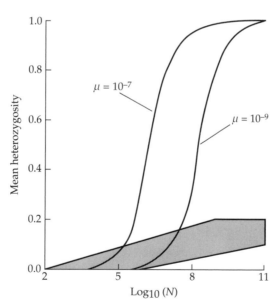

Figure 2.2   Predicted relationship between species effective population size and mean protein electrophoretic heterozygosity under neutrality theory. Expectations for two plausible neutral mutation rates (μ) are presented. Also shown (shaded area) is the general range of observed allozyme heterozygosities for numerous animal species as a function of present-day census population size (N, logarithmic scale). (After Soulé 1976.)

BOX 2.3 Mean Times to Shared Allelic Ancestry

Another way to formulate neutrality theory regarding the association between genetic variability and population size is through consideration of the expected frequency distribution of times to common ancestry among alleles. Imagine an idealized population with non-overlapping generations and large constant size N. Suppose further that in each generation, individuals contribute to a gamete pool from which $2N$ nuclear gametes are drawn at random (effectively with replacement) to produce individuals of the next generation. The probability that two gametes carry copies of the same allele from the prior generation is $1/2N$. This is also the probability that the time to common ancestry of two alleles is one generation ago ($G = 1$). The probability that a pair of alleles is not identical from the prior generation is $1 - 1/2N$. Thus, the probability that these latter alleles trace to an identical copy two generations ago is $(1 - 1/2N)(1/2N)$. From an extension of such reasoning, the probability that two randomly chosen alleles derive from a common ancestral allele that existed G generations ago is

$$f(G) = (1/2N)(1 - 1/2N)^{G-1}$$

or approximately

$$(1/2N)\, e^{-(G-1)/2N}$$

This equation gives the probability distribution of times to common ancestry in terms of the number of generations (Tajima 1983). The distribution is geometric, with mean approximately $2N$. The mean time to shared haplotype ancestry for mtDNA genes can be derived similarly (Avise et al. 1988), but in theory is only one-fourth as large as for autosomal nuclear genes, the difference being attributable to a twofold effect due to the haploid transmission of mtDNA and another twofold effect due to mtDNA's normal pattern of uniparental transmission.

The above theory assumes that times to common ancestry for allelic pairs are independent. Therefore, in interpreting empirical data for any particular species against these expectations (see Chapter 6), caution must be exercised because the history of lineage coalescence within a real population imposes a severe correlation on pairwise comparisons (Ball et al. 1990; Felsenstein 1992; Hudson 1990; Slatkin and Hudson 1991).

Except perhaps for some of the least abundant species, observed values of H have proved to be much lower than neutrality theory might predict. This conclusion generally holds even when mutations are assumed to be mildly deleterious (Nei 1983). One likely explanation for the relative paucity of genetic variation is that long-term effective population sizes for most species are vastly smaller than otherwise might be supposed, given present-day census sizes. Another possibility involves repeated selective sweeps (see below). If large numbers of genes or sites are under directional selection even occasionally during evolution (as argued, for example, by Smith and Eyre-Walker 2002), then other loci physically linked to them will also tend to show reduced variation via hitchhiking effects, a phenomenon sometimes referred to as "genetic draft" (Gillespie 2001).

The dearth of variation from the neutralist perspective extends to the level of some DNA sequences as well (Box 2.3). For example, with regard to mitochondrial DNA alleles (which are maternally inherited), the expected mean time to common ancestry under neutrality theory is approximately

$$G = N_{F(e)} \tag{2.2}$$

where G is the number of generations and $N_{F(e)}$ is the female evolutionary effective population size (Avise et al. 1988). Figure 2.3 plots values of $N_{F(e)}$ for various species as estimated from observed mtDNA haplotype differences, using conventional calibrations of evolutionary rate for the mtDNA molecule (Brown et al. 1979). Most observed values fall orders of magnitude below neutrality expectations extrapolated from present-day census population sizes. In other words, despite extensive molecular genetic heterogeneity, less mtDNA polymorphism is observed than neutrality theory might have predicted, at least for species that are numerically abundant today. Either the neutral mutation rate in mtDNA is much lower than normally believed, or (more likely) evolutionary effective population sizes are vastly lower than contemporary census population sizes in most species (for any or all of the reasons outlined in Box 2.2).

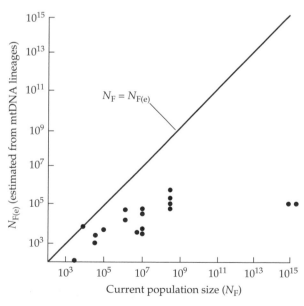

Figure 2.3   **Current census numbers of females (N_F) versus evolutionary effective female population sizes ($N_{F(e)}$)** for 18 marine species with high gene flow. Effective population sizes were estimated from empirical data on mtDNA nucleotide diversities. Both axes are on a logarithmic scale. Most observed values fall well below the hypothetical line if N_F and $N_{F(e)}$ were equal. (After Avise 2000a.)

Such discoveries raise a remarkable irony about the neutrality–selection debate that stems from the historical precedents of the classical–balance controversy. When extensive allozyme variation was uncovered in the seminal protein electrophoretic surveys, selectionists interpreted the observations as consistent with the balance view, and they sought (as described in the next section) to discover the selective forces responsible. At the same time, neutralists were facing the conceptual question of how to account for the *paucity* of molecular polymorphism relative to neutrality expectations, given the mutation rates and population sizes thought to characterize most species. In summarizing this state of affairs, Nei and Graur (1984) concluded that "polymorphism is actually much lower than the neutral expectation and that if the bottleneck effect is not sufficient for explaining the observed level, the type of selection to be considered is not diversity-enhancing selection but diversity-reducing selection." The irony of this neutralist perspective in the history of the classical–balance debate is still not universally appreciated by proponents of balancing selection.

Another important aspect of the neutral mutation theory concerns predictions about molecular evolutionary rates. Two aspects of these rates must be distinguished. *With regard to shifts in frequencies of preexisting alleles*, rates of neutral evolution should be greater in small populations. Genetic drift refers to random changes in allele frequency due to sampling variation of gametes from generation to generation, and it is a special case of the general phenomenon of statistical sampling error, which is inversely related to sample size. However, *with regard to the origin and substitution of new alleles*, the rate of neutral evolution is, in principle, independent of population size and depends only on the mutation rate to neutral alleles.

This latter conclusion can be demonstrated as follows. In a diploid population of size N, there are $2N$ allelic copies of each nuclear gene. In time, descendants of only one of these copies will survive (i.e., only one is destined for fixation). The chance that any newly arisen neutral mutation will undergo random fixation is simply $1/2N$. On the other hand, the probability that a new neutral mutation will arise in a population is $2N\mu$, where μ is again the mutation rate to neutral alleles. It follows that the rate of fixation of new neutral mutations is the product of the origination rate of mutations and their probability of fixation once present, or $2N\mu \times 1/2N = \mu$. In other words, the rate of nucleotide substitution in evolution under strict neutrality equals the rate of mutation to neutral alleles. This simple conclusion was the theoretical basis for the important neutrality prediction that biological macromolecules could provide reliable "molecular clocks" (see Chapter 3), irrespective of population size.

Multi-locus allozyme heterozygosity and organismal fitness

Empirically, does molecular variability matter to organismal fitness? Considering the protein electrophoretic results in the historical context of the classical–balance and neutralist–selectionist debates, it is hardly surprising that many empirical population geneticists soon began to address how natural selection might serve to maintain the newly discovered stores of molecular polymorphism. This problem was attacked in several ways.

In the "multi-locus approach," searches were launched for correlations between mean allozyme heterozygosity and life history attributes or fitness components. One early issue was whether protein variation might be correlated with environmental heterogeneity (Hedrick 1986; Levene 1953; Soulé and Stewart 1970), and some intriguing associations were reported. For example, Nevo and Shaw (1972) observed low heterozygosity (H) in burrowing mole-rats (*Spalax ehrenbergi*) and attributed this outcome to selection for homozygosity in their supposedly constant subterranean niche. Similarly, Avise and Selander (1972) reported much lower values of H in cave-dwelling than in surface-dwelling forms of the fish *Astyanax mexicanus*, although they provisionally attributed this outcome to effects of genetic drift in small cave populations (similar conclusions were later drawn from microsatellite data for these fishes; Strecker et al. 2003). Two influential studies using experimental cages of *Drosophila* fruit flies reported significantly higher allozyme heterozygosities in populations maintained under variable than under uniform environmental regimes (McDonald and Ayala 1974; Powell 1971).

More generally, Selander and Kaufman (1973a) suggested that genic heterozygosity was high in small sedentary animals that experience environments as coarse-grained patches of alternative habitat types (Levins 1968), and significantly lower in large mobile animals that perceive environments as fine-grained. Smith and Fujio (1982) concluded that allozyme heterozygosities in marine fishes were correlated with the degree of habitat specialization. Powell and Taylor (1979) summarized evidence that environmental heterogeneity in conjunction with habitat choice contributed to genotypic diversity, whereas Valentine and Ayala (1974) favored an environmental selection model consistent with an observed correlation in marine invertebrates between low allozyme variation and trophic resource stability over time (Ayala et al. 1975a; Valentine 1976).

On the other hand, Sage and Wolff (1986) suggested that patterns of genic heterozygosity might be attributable not to varying environmental selection pressures per se, but rather to environment-dependent population histories and genetic drift effects. For example, several species of large mammals in glaciated regions were noted to have lower allozyme variation than their counterparts in temperate and tropical regions, purportedly due to serial population bottlenecks accompanying recolonizations of northern latitudes following retreats of the Pleistocene glaciers. Such observations, although debatable in interpretation, raised an important general point: The extant standing crop of genetic variation must be a function of both the genetic diversity originally available to a species (its phylogenetic legacy) and contemporary processes, such as selection, gene flow, and the mating system, that further govern how that available variation is partitioned within and among populations.

Positive correlations also were noted between magnitudes of allozyme heterozygosity and particular life history attributes, such as short generation times and small body and egg sizes in bony fishes (Mitton and Lewis 1989; but see Waples 1991a) and high fecundity, outcrossing modes of reproduction, pollination by wind, and long generation times in plants (Hamrick et al. 1979). Within species, positive correlations were reported between individual heterozygosity estimates and a wide variety of phenotypic characters arguably associated with fitness (reviews by Mitton 1993, 1994, 1997), such as exploratory behavior in mice (Garten 1977), antler characteristics in deer (Scribner and Smith 1990), shell shape in bivalve mollusks (Mitton and Koehn 1985), herbivory resistance in pines (Mopper et al. 1991), disease resistance and other phenotypes in salmonid fishes (Ferguson and Drahushchak 1990; Wang et al. 2002), and growth rates in several animal and plant species (Ferguson 1992; Garton et al. 1984; Koehn et al. 1988; Ledig et al. 1983; Mitton and Grant 1984; Pierce and Mitton 1982; Singh and Zouros 1978). Of course, caution is indicated in interpreting such observations because various of these physiological and developmental characteristics might be functionally interrelated or statistically non-independent. Correlations were also noted between allozyme heterozygosity and developmental stability, the latter supposedly evidenced by lower phenotypic variance between individuals (Lerner 1954; Zink et al. 1985) or by lower "fluctuating asymmetry" (the difference between bilateral features) within individuals (Allendorf and Leary 1986; Leary et al. 1985; Palmer and Strobeck 1986; Van Valen 1962).

Another aspect of the multi-locus approach involved searches for molecular or metabolic features correlated with allozyme heterozygosity. Among the examined factors arguably associated with genic variation were: molecular size of the enzyme (Eanes and Koehn 1978); quaternary structure of the protein (Solé-Cava and Thorpe 1989; Ward 1977; Zouros 1976); frequency of intragenic recombination (Koehn and Eanes 1976); physiological role in regulating flux through metabolic pathways (Johnson 1976b); and enzymatic action on intracellular versus extracellular substrates (Ayala and Powell 1972a; Gillespie and Langley 1974; Kojima et al. 1970; see also reviews in Koehn and Eanes 1978; Selander 1976).

Several difficulties attended attempts to interpret such multi-locus associations, beyond the obvious point that correlation by itself cannot prove causality. First, there is likely to be a reporting bias in favor of positive correlations, and the number of variables that can be examined is essentially limitless. Second, mean heterozygosity as estimated from small numbers of protein loci may not accurately rank-order specimens with respect to genome-wide variability (Chakraborty 1981; Mitton and Pierce 1980), unless, perhaps, individuals vary dramatically along an outbred-inbred continuum due to demographic cycles, fine demic structure, or mating behaviors (Mitton 1993; Scribner 1991; Smith et al. 1975; Smouse 1986). This point led some authors to conclude that associations of individual heterozygosity with personal fitness were attributable not to differing levels of genome-wide variation, but rather to physiological advantages stemming from heterozygosity at the particular glucose-metabolizing or other enzymes surveyed (or at tightly linked genes in the chromosomal blocks that they mark; Koehn et al. 1983; Mitton 1997). Third, most of the heterozygosity correlates listed above involved weak trends for which exceptions could be cited or alternative explanations advanced. For example, with regard to the possible relationship between mean H and environmental heterogeneity, high genetic variability nonetheless was found to characterize some species inhabiting proverbially stable environments such as underground caves and the deep sea (Dickson et al. 1979), and low genetic variability certainly can result from demographic population contractions in any environment.

Such correlational approaches to studies of allozyme heterozygosity were summarized by Ward et al. (1992). From a survey of the literature on more than a thousand animal species, they concluded that approximately 21%–34% of the variance in mean protein heterozygosity could be attributed to taxonomic effects (e.g., fishes tend to have low H values, amphibians the highest such values), and that 41%–52% of the variance was related to protein effects (including size of the protein molecule, subunit number, enzyme function, etc.). In another such meta-analysis of protein electrophoretic studies, Britten (1996) concluded that "selection, including overdominance, has at most a weak effect at allozyme loci, and [this] casts some doubt on the widely held notion that heterozygosity and individual fitness are strongly correlated." This conclusion generally echoed a sentiment expressed 20 years earlier by Selander (1976): "Notwithstanding the immense amount of effort expended in surveying (allozyme) variation … , the sample sizes of loci are generally inadequate for satisfactory analyses … : molecular heterogeneity is too great." The various correlations involving heterozygosity listed above are intriguing, but their interpretations remain controversial.

In recent years, a few studies have revisited multi-locus heterozygosity–fitness issues using polymorphic DNA-level markers (e.g., Hansson et al. 2001; Thelen and Allendorf 2001). For example, heterozygosity at multiple microsatellite loci proved to be correlated with birth weight in red deer (*Cervus elatus*; Slate and Pemberton 2002) and with birth weight and neonatal survival in harbor seal pups (*Phoca vitulina*; Coltman et al. 1998a). An accumulation of such analyses might someday help to disentangle three primary hypothesis for such oft-observed heterozygosity–fitness correlations: "overdominance" precisely at the loci scored (i.e., the scored genes themselves affect fitness), "associative overdominance" (in which other loci linked to those scored are actually under balancing selection), and genome-wide heterozygosity fitness effects. However, most available data remain inconclusive on such issues (Coltman and Slate 2003; Hansson and Westerberg 2002). At least as important will be the use of molecular markers in conjunction with experimental studies that move beyond empirical correlations and attempt to test alternative causal mechanisms. For example, by surveying microsatellite loci in experimentally inbred lines of *Crassostrea* oysters, Launey and Hedgecock (2001) uncovered evidence for an abundance of deleterious recessive mutations as well as segregation distortions in F_2 hybrids, two results predicted for high genetic load under models of associative overdominance. Because it documented this basis for inbreeding depression experimentally, this study was quite relevant to the 25-year-long debate concerning correlations between growth rate and heterozygosity in bivalve mollusks.

Single-locus allozyme variation and the vertical approach

Frustration with such multi-locus approaches to assessing natural selection's role led other researchers to the "vertical" approach, wherein protein polymorphisms at specific loci were studied at multiple levels ranging from biochemistry, physiology, and developmental expression to transmission patterns, population dynamics, and ecological associations (Clarke 1975; Koehn and Hilbish 1987; McDonald 1983). Intensive studies of several such model systems all uncovered convincing evidence for differences between allozyme genotypes upon which selection probably operates (Table 2.1).

TABLE 2.1 Examples of major research programs that employed the "vertical" approach in studies of the possible adaptive significance of allozyme polymorphisms[a]

Protein	Organism
Alcohol dehydrogenase	*Drosophila* (fruit flies)
α-Glycerophosphate dehydrogenase	*Drosophila*
Carboxylesterase	*Drosophila*
Glucose-6 phosphate dehydrogenase; 6-phosphogluconate dehydrogenase	*Drosophila*
Glucose-phosphate isomerase	*Colias* (butterflies)
Glutamate pyruvate transaminase	*Tigriopus* (copepods)
Lactate dehydrogenase	*Fundulus* (killifish)
Leucine aminopeptidase	*Mytilus* (mollusks)

[a]In each case, documented kinetic differences between allelic products suggested that the polymorphisms were maintained by natural selection.

Unfortunately, rather few polymorphisms were analyzed so intensively. Furthermore, most of the allozyme loci studied had been identified a priori as likely candidates for natural selection; thus, polymorphisms analyzed by the vertical approach undoubtedly constituted a biased sample with regard to issues of selective maintenance.

Selection at the level of DNA

As data at the level of DNA sequences increasingly became available during the 1980s and later, they too fed into ongoing neutrality–selection debates in molecular evolution (Orr 2002). These data came not only from nuclear genes, but also from loci housed in cytoplasmic organelles (McCauley 1995; Rand 2001). Because DNA-level assays typically were conducted one locus at a time (but see above), attention usually shifted from estimates of composite or "genome-wide" heterozygosity (as in the multi-locus allozyme era) to possible molecular signatures of natural selection or other evolutionary forces on particular sequences of linked nucleotide sites. Many of these efforts involved detailed statistical analyses of DNA sequences. For example, from

Evidence for natural selection	Key references
Kinetic differences between isozymes associated with differences in survivorship, developmental time, and environment	Aquadro et al. 1986; Clarke 1975; van Delden 1982
Kinetic differences correlated with flight metabolism, power output, and environmental temperature	Miller et al. 1975; O'Brien and MacIntyre 1972; Oakeshott et al. 1982
Differences in enzyme activity associated with reproduction	Gilbert and Richmond 1982; Richmond et al. 1980
Differences in metabolic flux associated with differences in function	Barnes and Laurie-Ahlberg 1986; Cavener and Clegg 1981; Hughes and Lucchesi 1977
Kinetic differences correlated with mating success and survivorship	Watt 1977; Watt et al. 1983, 1985
Enzyme activity differences associated with differential responses to hyperosmotic stress	Burton and Feldman 1983
Differences in kinetic and other biochemical properties associated with differences in metabolism and fitness	DiMichele et al. 1986, 1991; Place and Powers 1979, 1984
Enzyme activity differences associated with osmoregulation and fitness	Hilbish and Koehn 1985; Hilbish et al. 1982; Koehn and Immerman 1981

a sliding-window analysis of nucleotide sequence divergence along the length of the ADH gene in *Drosophila melanogaster*, Kreitman and Hudson (1991) identified a sharp spike of variation consistent with the operation of balancing selection on one specific portion of the ADH locus.

Dramatic evidence for long-term balancing selection also emerged from fine-scale molecular characterizations of genes of the MHC (major histocompatibility complex) in mammals (Klein et al. 1993, 1998; Takahata et al. 1992) and a self-incompatibility mating locus in plants (Clark 1993). In each case, different alleles in extant populations proved to have much deeper evolutionary separations than neutrality theory predicted. Indeed, the MHC complex in particular has become a paradigm molecular system for developing and evaluating statistical analyses of genetic data for evidence of natural selection (Garrigan and Hedrick 2003). Another such example is represented in Figure 2.4. Two other notable early quests to identify the molecular footprints of natural selection involved detailed analyses of genes encoding an esterase (Cooke and Oakeshott 1989; Odgers et al. 1995) and a superoxide dismutase (Hudson et al. 1997; Lee et al. 1985) in *Drosophila*.

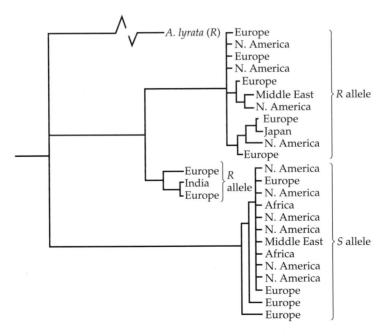

Figure 2.4   Phylogeny of alleles at the *Rpm1* gene. The *Rpm1* gene is involved in pathogen recognition in *Arabidopsis thaliana* plants. Note that in several parts of the world, this species retains highly divergent alleles conferring pathogen resistance (R) and sensitivity (S), suggesting that long-term balancing selection has played a role in allelic maintenance. Note also the two distinct clusters of R alleles, divergence between which is apparently due to a historical recombination event between the R and S alleles at one end of the sequenced region. (After Stahl et al. 1999.)

Today, several statistical approaches are fairly standard practice for deducing the probable action of various forms of natural selection on DNA sequences (Bustamante et al. 2002; Ford 2002). One of the first such tests, proposed by Hughes and Nei (1988) in their analyses of MHC variation, asks whether replacement substitutions in a protein-coding gene outnumber silent substitutions. If so, positive selection on advantageous alleles is implicated, because neutrality theory predicts that synonymous substitutions in a gene should be more common than those that result in amino acid replacements. The Hughes-Nei test is highly conservative, however, because it normally detects only strong instances of positive selection (Sharp 1997). Another popular statistical method, introduced by Tajima (1989a), compares numbers of segregating sites and the mean number of nucleotide differences estimated from pairwise DNA sequence comparisons *within* a population. The resulting statistic (Tajima's D) is often used to test selective neutrality of DNA sequences under an "infinite-site" model (Watterson 1975), but particular outcomes can also be affected by (i.e., indicative of) historical demographic events such as dramatic expansions in population size (Aris-Brosou and Excoffier 1996; Tajima 1989b). For addressing historical population growth explicitly, Fu (1997) introduced a statistical test that distinguishes excesses of low-frequency alleles in an expanding population as compared with the number expected in a static population.

Two other popular statistical approaches involve comparing patterns of DNA sequence variation within and between populations or species. The HKA (Hudson, Kreitman, and Aguadé) test (Hudson et al. 1987) addresses whether levels of DNA polymorphism and divergence are significantly correlated, as might be predicted under neutrality theory because both the extent of polymorphism of a gene and its evolutionary rate are primarily functions of mutation rates to neutral alleles. The test requires an extensive series of DNA sequences both within and among populations, and it assumes that effective population sizes have remained roughly constant throughout the relevant evolutionary time frame. Examples of empirical outcomes can be found in Moriyama and Powell (1996) and Wells (1996).

The MK test (McDonald and Kreitman 1991) is similar in its basic rationale, but it examines, for any protein-coding gene, whether the ratio of non-synonymous to synonymous nucleotide differences between related species is the same as that within species. If the ratios differ, departures from neutrality may be indicated. For example, if replacement substitutions are relatively more frequent as fixed differences between species than as polymorphisms within them, an implication is that the nucleotide fixations were selectively promoted; but if the ratio of replacement to silent substitutions is significantly lower in the fixed differences between species than in intraspecific polymorphisms, then purifying selection against replacement substitutions may have taken place (for further explanation, see Nei and Kumar 2000; Sawyer and Hartl 1992). The MK test has been applied widely, with various forms of sequence-level selection sometimes deduced (and sometimes not; Eanes 1994; Hey and Kliman 1993) in the genes of protists (Escalante et al.

1998), plants (Purugganan and Suddith 1998), and animals (Eanes et al. 1993). Caution again is indicated, however, because the assumptions underlying the MK test (e.g., lack of multiple substitutions at the same site, absence of codon usage bias, and lack of temporal variation in the mode of selection) could, if violated, compromise the interpretations (Akashi 1995; Eyre-Walker 1997; Ohta 1993; Whittam and Nei 1991). In general, all such statistical models entail assumptions (sometimes subtle) that often make their results less than definitive.

Reservations notwithstanding, these and other statistical tests for selection on DNA sequences have proved extremely helpful in identifying putative selective footprints in the genome (Golding 1994; Olson 2002). Especially noteworthy is empirical evidence for strong positive selection on replacement substitutions in certain protein-coding loci that apparently must deal with highly variable selective challenges. Examples of such genes are those that encode species-specific egg–sperm interactions in marine invertebrates (Galindo et al. 2003; Metz and Palumbi 1996; Swanson and Vacquier 1998), self-fertilization avoidance mechanisms in plants (Clark and Kao 1991), disease resistance operations in mammals (Tanaka and Nei 1989; Zhang et al. 1998), and digestive functions in primates (Messier and Smith 1997).

Recent selection may also be signaled by abnormally low population variation in specific islands of linked nucleotide sites, an outcome that could evidence either "hitchhiking via selective sweeps" or "background selection" (Figure 2.5). Especially when recombination rates are low, any positively selected mutation that sweeps through a population to fixation inevitably carries along other tightly linked markers with which it happened to be associated at its time of origin. This genetic hitchhiking temporarily purges a chromosomal region of preexisting sequence variety, an effect that dissipates only gradually as genetic variation is restored by mutation (Dorit et al. 1995; Nurminsky et al. 1998). Similarly, background selection against deleterious alleles can reduce the level of polymorphism in a DNA region with a low recombination rate (Charlesworth et al. 1993). Conversely, balancing selection on a target locus or nucleotide position can buffer against allelic extinction and thereby elevate levels of genetic polymorphism at linked neutral sites.

Of course, any genomic signature of natural selection deduced strictly from statistical inference about nucleotide patterns is merely a starting point for understanding how natural selection mechanistically operates on gene sequences. Further exploration can then focus more precisely on particular environmental or cellular selective forces that may be involved.

The unresolved status of the controversy

Neutrality–selection debates have dominated both the theoretical and empirical sides of population genetics for several decades (see Ayala 1976a; Gillespie 1991; Lewontin 1974, 1991; Mitton 1994; Nei and Koehn 1983 for

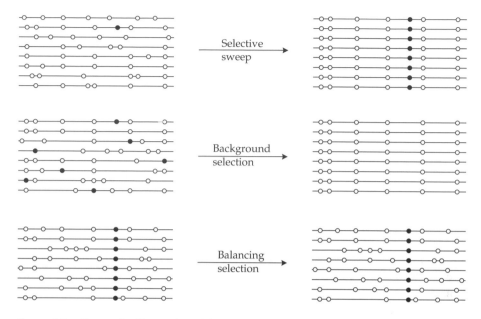

Figure 2.5   General effects of positive, negative, and balancing selection on levels of genetic variation at neutral sites adjacent to true targets of selection. Each diagram shows linked markers in a population of homologous chromosomes before and after the indicated mode of selection. Open circles are neutral positions; solid circles are mutations actually under functional selection.

seminal reviews). Much has been learned, not least about the surprising intractability of the problem and the difficulty of obtaining the hard evidence required for its resolution. It has become far clearer, for example, that correlational studies (such as comparisons of multi-locus heterozygosity levels across environments or taxa), intriguing though they may be, are inadequate; that "vertical" gene-by-gene analyses of specific polymorphisms are necessary, but extraordinarily difficult, and are likely to yield a biased sample of outcomes; and that statistical analyses of DNA sequences can help to detect particular departures from neutrality, but alone, fall far short of yielding what is ultimately desired: mechanistic descriptions of evolutionary forces that sustain genetic variation and govern molecular change.

There are further reasons why these debates are not yet fully resolved. First, selection theory and neutrality are both immensely powerful conceptual constructs in the sense that they can accommodate a wide variety of molecular outcomes simply by varying key assumptions and parameters. Because of the multitudinous ways in which natural selection can operate, falsification of all selectionist scenarios for a given empirical data set is nearly impossible (indeed, this was a prime motivation for development of a quantitative neutrality theory that specifies expectations explicitly). But neutrality theory also can yield a nearly limitless array of predictions by altering parameters (N_e, μ, and selection intensities for or against alleles that may be nearly neutral) that are notoriously difficult to measure in natural populations. Furthermore, many apparent departures from neutrality expectations probably are due to idiosyncratic historical events that have removed populations and species from the equilibrium conditions often assumed in various neutral models.

Another challenge underlying the neutrality–selection controversy is in defining exactly what is meant by natural selection. For example, is the phenomenon of meiotic drive (whereby certain alleles appear to "cheat" during meiosis by distorting Mendelian segregation ratios in their favor) to be viewed as a form of natural selection at the gametic level? In general, are "selfish genes" that compete for transmission within an organismal lineage to be interpreted as evolving under the influence of selection? Holmquist (1989) and Avise (2001b) have argued that in sexual species, multitudinous quasi-independent genes and other DNA sequences (such as transposable elements) that co-inhabit an extended cell lineage form an interactive community of oft-competing as well as collaborating entities, rather like individuals in a miniature ecological community. This vast molecular interplay is quite different from what traditionally has been meant by natural selection at the organismal level (Ohta 1992b), but it emphasizes the emerging view that natural selection can entail differential reproduction (i.e., differential replicative fitness) at hierarchical levels below (Dawkins 1989) as well as above (Gould 1980; Wilson 1980) that of individual organisms.

The issue of exactly what is meant by natural selection also arises in the context of interpreting genomic patterns. For example, even a single mutation under positive selection can promote a selective sweep that leaves a long-lasting footprint or genomic signature across a multitude of adjacent nucleotide sites (see Figure 2.5). Thus, the evolutionary dynamics and levels of genetic variation in a whole chromosomal region with a low recombination rate can be governed or influenced by natural selection even if the vast majority of pre- and post-sweep variants within that region are absolutely neutral in functional or operational terms. These two distinct senses or meanings of natural selection (mechanistic versus evolutionary dynamic) must be distinguished carefully in all discussions of molecular evolution, including those involving molecular markers.

Notwithstanding these many difficulties of interpretation, it has become clear that the final truth about selection versus neutrality lies somewhere between the two polarized views. One emerging consensus seems to be that natural selection often works most effectively on exon sequences, but typically less so on introns and many spacer regions between genes. Another emerging view is that some DNA sequences unquestionably have been subject at times to natural selection of various forms (balancing, positive, and purifying), whereas segregating alleles at many other loci have been mechanistically neutral, or nearly so, throughout most or all of their evolutionary durations. What remains contentious are the overall fractions and compositions of DNA sequences falling into these categories, the frequency distributions of selection intensities, and (in most cases) the exact operational connections between molecular variation and organismal fitness.

Must Molecular Markers Be Neutral To Be Informative?

Do the continuing uncertainties about the relative roles of selected and neutral mutations in evolution seriously compromise the utility of molecular polymorphisms as genetic markers? A common sentiment is that molecular markers gain their informativeness by being neutral, or nearly so, but this is a considerable oversimplification. In many microevolutionary applications, such as forensic identification and genetic parentage analysis, any non-neutrality of chosen markers is normally inconsequential to the outcome. In other applications, such as in assessing population structure and gene flow, genetic markers under intense selection could be misleading if interpreted under the assumption of neutrality (see Chapter 6). For example, strong balancing selection via heterosis (fitness superiority of heterozygotes) can inhibit population differentiation in allele frequencies by random drift and result in uniform spatial patterns that could be misinterpreted as evidence for high gene flow under neutral models. Conversely, strong habitat-specific or diversifying selection on particular marker loci could promote a false illusion of pronounced population isolation when actually there had been considerable genetic interchange.

With regard to estimating species' phylogenies, many considerations apply, depending on the nature of selection, its effect on character state distributions, and the methods of phylogenetic reconstruction. Balancing selection can again complicate the effort by acting, for example, to retain particular molecular polymorphisms across successive speciation nodes. Furthermore, different intensities of directional or diversifying selection can compromise some types of data analysis by generating significant rate heterogeneity in DNA sequence evolution. However, different mutation rates at neutral loci also generate rate heterogeneity, yet these are dealt with routinely in appropriate phylogenetic analyses. Indeed, even dramatic rate variation across genes or nucleotide sites can be of genuine benefit, because it offers researchers the opportunity to choose molecular markers ideally suited to the particular phylogenetic questions being addressed (see Chapter 3 and elsewhere throughout this book).

When Lewontin (1991) asked whether the introduction of molecular methods had been a milestone or a millstone for evolutionary studies, it was primarily in the context of illuminating the nature of fitness and adaptation. From the perspective of providing utilitarian markers that open a world of novel opportunities for studying organismal relationships and behaviors in nature, the molecular revolution has been an unqualified success, as I hope this book will testify.

The Molecule–Morphology Debate

Especially in the 1960s and 1970s, many systematists trained in traditional organismal disciplines viewed the nascent molecular revolution with considerable skepticism, if not consternation. Molecular biology and genetics seemed like alien fields, and their rapid growth threatened the long-standing dominance of comparative morphology and behavior in systematics and phylogenetic biology. In some circles, this created an underlying tone of antagonism toward molecular approaches that persisted for quite some time. And it certainly did not help that many molecular biologists were ill-informed about ecology and evolution.

In recent years, a more appreciative attitude has emerged with the recognition that molecular and organismal data can be reciprocally informative, and indeed, often require each other's services (Hillis 1987). For example, a new enterprise in comparative evolution may be termed "phylogenetic character mapping" (see Chapter 8). This approach typically involves plotting the taxonomic distributions of morphological (or other) characters along molecule-inferred phylogenies, the intent being to uncover the evolutionary origins and histories of organismal attributes (Harvey et al. 1996). Phylogenetic character mapping also can be conducted in the reverse direction—that is, by plotting the distributions of molecular characters along a morphology-inferred phylogeny. Several instances of horizontal gene transfer between otherwise unrelated organisms have been revealed under the compelling logic of this approach (see Chapter 8). Such enlightened analyses that attempt to capitalize on the cross-comparative information content of multiple classes of data are now leading to a more contented marriage between molecule-based and morphology-based approaches to ecology and evolution. Under this developing perspective, the interplay between alternative lines of evidence becomes of greater interest and value than either data source considered alone.

One entrenched notion in biology is that mean rates of morphological and molecular evolution are largely uncoupled: Different taxa sometimes evolve at grossly different rates with respect to phenotype (Simpson 1944), whereas molecular evolution often proceeds at a fairly steady pace (Wilson et al. 1977). Notable examples of this disconnect include "living fossils," such as species of horseshoe crabs that have remained nearly static in phenotype for tens of millions of years, yet are highly divergent at the molecular level (Avise et al. 1994); and conversely, phylogenetic groups such as the cichlid fishes in Africa's Lake Victoria that in recent times have radiated into arguably hundreds of morphologically recognizable species (Turner et al. 2001) that nonetheless often remain nearly indistinguishable at the molecular level (Meyer et al. 1990). Such apparent discrepancies in evolutionary rates for molecules and morphology helped to motivate the idea that changes in gene regulation may play a hugely disproportionate role in phenotypic evolution (Carroll et al. 2001; King and Wilson 1975). More recently, however, a broad review of empirical evidence led Omland (1997) to conclude that rates of molecular and morphological evolution usually *are* correlated, and that earlier researchers understandably had emphasized the interesting exceptions. In another such meta-analysis, but with a different outcome, Bromham et al. (2002) found no evidence for an association between morphological and molecular evolutionary rates.

Such rate controversies notwithstanding, the fact remains that most molecule-based phylogenies mirror quite closely their predecessors based on morphology, particularly within well-studied taxonomic groups. Why should this be, given that much (certainly not all) of phenotypic evolution is Darwinian (adaptive), whereas much (certainly not all) of molecular evolution seems to be nearly neutral? Two factors probably contribute.

First, when integrated over long sweeps of time and across large numbers of characters (morphological or molecular), overall magnitudes of divergence between species should tend to be correlated, at least crudely, with times elapsed since common ancestry. If molecules and morphology both march appreciably in step with the passage of time (either by accumulation of neutral mutations or via changing selection regimes), their histories will appear correlated, even if these two suites of characters themselves are functionally mostly independent. To the extent that functional links also exist (as they certainly must) between molecules and morphology, such correlations between the two would only be elevated.

Second, overall phenetic or genetic divergence is not the only, or even necessarily the best, guide to phylogeny (see Chapter 3). Most phylogenetic algorithms in use today give added weight to specific character states that (for logically defensible reasons) should be most indicative of historical genealogy. Thus, when applied to either morphological or molecular traits, such algorithms are likely to converge on the one-and-true organismal phylogeny.

Molecular Phylogenetics

While the grand controversies described above were being played out, other researchers adopted the more pragmatic approach of simply applying molecular markers to resolvable problems in natural history and genealogical assessment. Beginning as a subsidiary endeavor in molecular ecology and evolution, this previously neglected approach has grown steadily and now occupies a position of central prominence in biology, as this book will attest.

Advances in the molecular marker field have proceeded as a series of major waves, each initiated by the development of a new laboratory method. A typical pattern is as follows: A novel assay technique for proteins or DNA is introduced, and a flood of evaluative activity follows. Methods that fail to live up to advance billing (e.g., because of technical difficulty, poor repeatability in outcomes, or ambiguities of genetic interpretation) are abandoned. Approaches that survive the initial evaluations are then employed to address broad conceptual topics in the research paradigms of molecular evolution described above. For example, questions inevitably arise about what role, if any, natural selection plays in maintaining molecular variation revealed by the new laboratory techniques. Observational or experimental data are gathered and evaluated against the predictions of neutrality theory. Discussions also ensue about how best to analyze and interpret the new classes of molecular data.

In the meantime, genetic markers provided by each new method are applied to interesting problems in natural history or evolution where their use appears appropriate. Success in such endeavors stimulates further interest, and the more utilitarian molecular approaches eventually gain wide popularity. Usually, after a period of several years, the enthusiasm crests, and a new wave of interest may focus on another newly introduced assay procedure. Typically, the earlier methods are not abandoned, but merely become incorporated into the growing pool of molecular techniques that find continued application in studies of organismal biology, natural history, and evolution.

Protein electrophoresis was the first molecular approach employed widely in the field. The utilitarian allozyme markers it reveals are protein variants that behave in straightforward Mendelian fashion and, hence, are interpretable as simple allelic products of a gene. (The term "isozyme" refers to the broader class of all protein variants observed on electrophoretic gels, including heteromeric products of multiple loci, post-translational variants, and other protein alterations.) Allozyme methods were introduced in the mid-1960s and dominated molecular ecology and systematics for the next 10 years (see early reviews in Avise 1974, 1983; Buth 1984; Gottlieb 1977; Whitt 1983, 1987). Today, protein electrophoresis remains a simple and useful method for generating molecular markers.

The next technique to become popular in molecular ecology and evolution involved analyses of "restriction fragment-length polymorphisms" (RFLPs) in DNA. For both technical and conceptual reasons, mitochondrial (mt) DNA received most of the early attention. Mitochondrial approaches dominated the field during the late 1970s and 1980s (see early reviews in Avise 1986, 1991a; Avise and Lansman 1983; Birley and Croft 1986; Harrison 1989; Moritz et al. 1987; Palmer 1990; Wilson et al. 1985), much as had allozyme studies a decade earlier. Today, strong interest in mtDNA continues, although the markers usually now come from direct sequencing of particular mitochondrial genes. In the middle and late 1980s, another wave of excitement attended RFLP analyses as applied to hypervariable nuclear DNA regions known as minisatellites. Because of the power of these multilocus assays to distinguish individuals, this class of procedures became known as "DNA fingerprinting" (Burke 1989; Hill 1987; Jeffreys et al. 1985a,b, 1988a; Kirby 1990).

Another revolution began at about that same time with the introduction of the PCR (polymerase chain reaction) technique for in vitro amplification of specific DNA fragments (Erlich and Arnheim 1992; Erlich et al. 1991; Mullis 1990; Mullis et al. 1986; Saiki et al. 1988; White et al. 1989). This discovery spurred at least three major breakthroughs in marker acquisition. First, when coupled with further development of amplification primers (Kocher et al. 1989) and improved laboratory methods for rapidly sequencing DNA fragments (Innis et al. 1988; Scharf et al. 1986; Wrischnik et al. 1987), PCR-based approaches afforded direct access to the vast phylogenetic information content of nucleotide sequences. The many thousands of PCR "primers" that are now available greatly facilitate direct sequencing of a wide variety of nuclear and cytoplasmic loci. Second, when used to amplify an abundant class of newly discovered microsatellite loci (Litt and Luty 1989; Tautz 1989; Weber and May 1989), PCR methods could be used to tap another vast wellspring of genetic polymorphism. Microsatellite loci contain variable numbers of tandem repeat units, each about 2–5 nucleotide pairs long, and they typically far surpass allozyme loci in heterozygosity levels and numbers of alleles per locus (Figure 2.6). Accordingly, these molecular

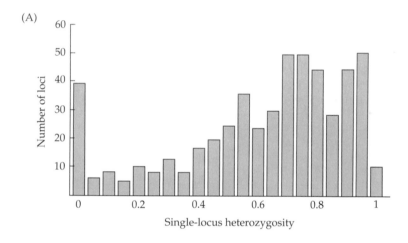

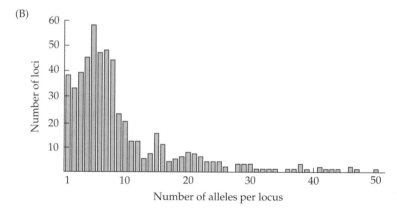

Figure 2.6   Genetic variation at microsatellite loci. Distributions are shown for (A) heterozygosity values and (B) numbers of alleles per locus at each of 524 microsatellite loci surveyed within local natural populations of 78 animal species. Note in particular the high heterozygosity levels compared with those typifying most allozyme loci (see Figure 2.1). (After DeWoody and Avise 2000.)

BOX 2.4 A Brief Chronology of Some Important Events Relevant to the (Remarkably Recent) History of Molecular Markers

1944   Avery, MacLeod, and McCarty provide experimental evidence that DNA is the genetic material.

1953   Watson and Crick propose a molecular model for DNA structure.

1955   Smithies uses starch-gel electrophoresis to identify protein polymorphisms.

1963   Margoliash determines amino acid sequences for cytochrome *c* in several taxa and generates the first phylogenetic tree for a specific gene product.

1966   Several independent researchers use electrophoretic methods and histochemical enzyme stains to assess levels of genetic variability in animal populations and humans.

1967   Sarich and Wilson provide an early application of protein immunological methods and discover a far more recent shared ancestry for humans and great apes than previously suspected.

1968   Kimura proposes the neutral theory of molecular evolution. Meselson and Yuan isolate and characterize the first specific restriction enzyme. Britten and Kohne use DNA hybridization methods to characterize animal genomes.

1971   Publication of the first periodical devoted to molecular evolution (*Journal of Molecular Evolution*).

1975   Southern describes a method for the transfer of DNA fragments to nitrocellulose filters, hybridization to radioactive probes, and detection of fragments by autoradiography.

1977   Maxam and Gilbert and also Sanger and colleagues describe laboratory methods for DNA sequencing.

1978   Maniatis and colleagues develop a procedure for gene isolation that involves construction and screening of cloned libraries of eukaryotic DNA.

markers have found wide application in wildlife forensics, parentage analyses, and other microevolutionary studies (Goldstein and Schlötterer 1999). Finally, because the PCR can amplify DNA sequences from minuscule amounts of tissue, or even from some well-preserved fossils, it has extended molecular applications to a much wider biological arena (see Chapters 6, 8, and 9).

As elaborated in Chapter 3, several other laboratory methods have also made significant contributions to molecular ecology and evolution. Prominent among these have been protein immunological comparisons, which provided some of the first evidence for molecular clocks (Benjamin et al. 1984; Goodman 1963; Maxson and Maxson 1986, 1990; Sarich and Wilson 1966, 1967; Wilson et al. 1977), and DNA–DNA hybridization methods

1979	Avise and colleagues and also Brown and colleagues introduce mtDNA approaches to analyses of natural populations.
1981	Palmer and colleagues help initiate an important series of papers utilizing cpDNA for phylogenetic reconstruction in plants.
1983	The journal *Molecular Biology and Evolution* is launched.
1985	Jeffreys and colleagues develop multi-locus DNA fingerprinting and point out its potential for forensic science. Saiki, Mullis, and colleagues report the enzymatic in vitro amplification of DNA via the polymerase chain reaction.
1987	Avise and colleagues characterize phylogeography as a new approach to population genetics.
1989	Kocher and colleagues report the discovery of conserved PCR primers that can be employed to amplify mtDNA segments from many species. Several seminal papers identify the utility of microsatellite loci as a source of highly polymorphic molecular markers.
1992	Periodicals devoted to evolutionary applications for molecular markers, such as *Molecular Ecology, Molecular Phylogenetics and Evolution,* and *Molecular Marine Biology and Biotechnology,* proliferate.
1996	Edited volumes by Avise and Hamrick and by Smith and Wayne emphasize roles for molecular markers in conservation biology.
2000	The journal *Conservation Genetics* is launched. The first textbook summarizing the field of phylogeography (by Avise) is published.
2001	Draft sequences of the human genome are published by Lander and colleagues and by Venter and colleagues.

Note: See King and Stansfield (1990) for an expanded history of genetic discoveries.

(Britten and Kohne 1968; Doty et al. 1960), which had special influences on the systematics of particular taxonomic groups, such as insects (Caccone and Powell 1987; Caccone et al. 1988a,b), birds (Sibley and Ahlquist 1990), marsupial mammals (Springer and Kirsch 1991), and hominoid primates (Caccone and Powell 1989; Sibley and Ahlquist 1987).

Given the burgeoning interest today in molecular ecology and phylogenetics, it is useful to remain cognizant of the remarkably shallow history of these scientific disciplines. An abbreviated chronology of salient events in the development and application of molecular markers is provided in Box 2.4.

SUMMARY

1. The study of ecology and evolution from a molecular perspective is a fairly recent enterprise, beginning in substantive form only in the latter half of the twentieth century.

2. Several fundamental controversies have dominated molecular evolution and related fields. These controversies include the classical–balance debate on the magnitude of genetic variation, the neutrality–selection debate on the adaptive significance of molecular variation, and arguments over the relative utility of molecular versus morphological characters in phylogenetics and systematics.

3. These grand controversies raised issues that certainly are germane, but are seldom crucial, to the interpretation of molecular polymorphisms as genetic markers in ecology and evolution. Indeed, to a considerable extent, the debates diverted attention from the many utilitarian applications for protein and DNA polymorphisms in investigating natural history and organismal phylogeny. Only in relatively recent years have large numbers of researchers begun to shift their primary focus to these latter, highly informative arenas.

4. Several waves of excitement in molecular ecology and evolution have followed the introduction of new laboratory techniques. Among the most influential methods have been protein electrophoresis in the late 1960s and 1970s, RFLP analyses of mitochondrial DNA in the late 1970s and 1980s, multi-locus DNA fingerprinting in the late 1980s, and PCR-mediated DNA sequencing as well as microsatellite assessments, beginning mostly in the 1990s.

3

Molecular Techniques

Perhaps nowhere has the power of the scientific method been more brilliantly demonstrated than in the development of procedures for the study of the chemistry of life.

M. O. Dayhoff and R. V. Eck (1968)

Many different classes of laboratory assays can be employed to reveal molecular genetic markers. Detailed protocols are published elsewhere (key literature is cited herein), but in practice there is no substitute for hands-on training "at the bench" under the guidance of an experienced practitioner. Thus, this chapter will merely outline laboratory methods, focusing on the most popular and the most historically important. These methods will be treated roughly in the chronological order of their introduction to the field. Another intent of this chapter is to emphasize the conceptual basis of, the nature of the genetic information provided by, and the rationale underlying each of the various molecular approaches that has had a major impact on molecular ecological and evolutionary studies.

Protein Immunology

Crude immunological assays (using unpurified proteins in simple precipitin tests) were introduced a century ago (Nuttall 1904), but the development of micro-complement fixation (MCF) methods in the 1960s allowed outcomes to be quantified in terms of immunological distances (ID values), which proved to be phylogenetically informative. The MCF procedure is outlined in Figure 3.1 (details are in Champion et al. 1974; Maxson and Maxson 1990). First, a protein such as albumin is purified from a reference species using standard biochemistry.

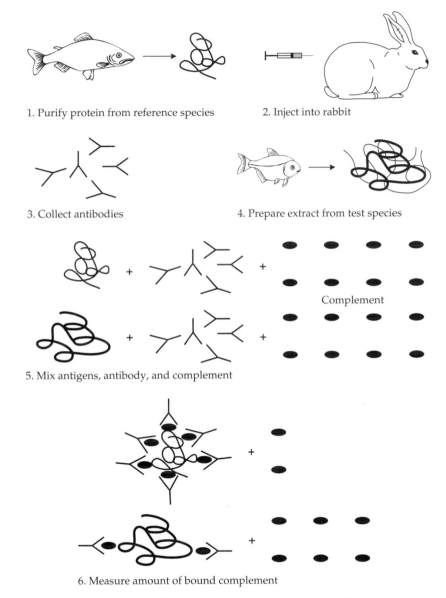

1. Purify protein from reference species

2. Inject into rabbit

3. Collect antibodies

4. Prepare extract from test species

Complement

5. Mix antigens, antibody, and complement

6. Measure amount of bound complement

Figure 3.1   General protocol for immunological comparison of proteins by micro-complement fixation. (After Wilson 1985.)

This highly purified protein is injected several times into a host species (often a rabbit) over several months. One week after the last injection, host antiserum is collected and standardized to a given level of reactivity under specified MCF conditions. When an MCF assay is to be performed, the antiserum is mixed with varying concentrations of antigen from diluted plasma (purified protein is not required) of a test species. Included in each reaction is a group of proteins ("complement"), normally found in vertebrate serum, that become trapped in developing lattices of the antigen–antibody complex. A spectrophotometer is then used to quantify the amount of complement "fixed": The greater the cross-reactivity between antibody and antigen, the more fixation. The MCF technique has been shown to be capable of detecting even single amino acid replacements in the antigenic site of a challenging protein. For the albumin molecule in particular, ID as measured by MCF has been shown to be a linear function of the number of amino acid substitutions between the reference and test species (Maxson and Maxson 1986; Prager and Wilson 1993).

Albumin was the usual protein of choice in vertebrate MCF studies for several reasons: it is abundant, ubiquitous, and easily purified; it consists of one subunit encoded by a single gene; the molecule has many (25–50) major antigenic sites at which amino acid substitutions were detectable by MCF (Benjamin et al. 1984); and it typically proved useful for phylogenetic studies at the taxonomic levels of families or genera. Other proteins sometimes used in MCF assays included lysozyme, ovalbumin, and transferrin in vertebrates (Leone 1964; Prager and Wilson 1976; Wright 1974), acid phosphatase, glycerophosphate dehydrogenase, and larval proteins in invertebrates (Beverley and Wilson 1985; Collier and MacIntyre 1977; MacIntyre et al. 1978), and alkaline phosphatase in bacteria (Cocks and Wilson 1972; Maxson and Maxson 1990).

For a complete assessment of the taxa under consideration, MCF requires that antibodies be produced from each species. Otherwise, missing elements in the pairwise ID matrix compromise phylogenetic reconstruction. Generating and testing antisera for multiple species is time-consuming, but offers the added advantage of permitting evaluations of reciprocity (anti-A vs. B; anti-B vs. A). Differences between reciprocal outcomes provide a measure of experimental error in the MCF procedure (Maxson and Wilson 1975).

Protein immunology (especially via the MCF technique) was one of the earliest methods of molecular phylogenetics and provided some of the first evidence for molecular clocks. Although the technique has been superseded by other approaches since the late 1980s, it remains of interest both for its historical significance and for its conceptual and operational distinctness from other molecular methods.

Protein Electrophoresis

Protein electrophoretic techniques were introduced in the mid-1960s, and they still remain a simple, popular, and powerful workhorse for generating Mendelian nuclear markers in many ecological and evolutionary applications. Detailed descriptions of laboratory techniques can be found in Baker (2000), Harris and Hopkinson (1976), Müller-Starck (1998), Murphy et al. (1996), Selander et al. (1971), and Shaw and Prasad (1970).

Protein electrophoresis takes advantage of the fact that non-denatured proteins with different net charges migrate at different rates through starch or acrylamide gels (or other supporting media such as cellulose acetate strips) to which an electric current is applied (Figure 3.2). These charge features stem primarily from the three amino acids with positive side chains (lysine, arginine, and histidine) and the two with negative side chains (aspartic acid and glutamic acid). A protein's net charge, which varies with the pH of the running condition, determines the protein's movement toward the positive pole (anode) or negative pole (cathode) in a gel. Protein size and shape can also interact with a gel's pore size to influence migration properties.

Because of their low cost, safety, and ease of use, gels made from hydrolyzed potato starch are popular. The starch-gel electrophoresis (SGE) procedure begins with extraction of unpurified water-soluble proteins from a particular source (leaves, roots, liver, heart, blood, skeletal muscle, etc.). The extract from each specimen is absorbed onto a paper wick, and 20 or more such wicks are placed side by side along a slit (the origin) in the gel.

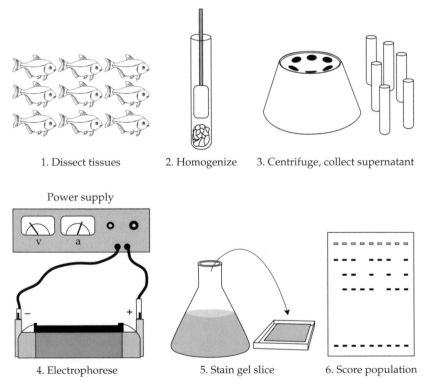

1. Dissect tissues   2. Homogenize   3. Centrifuge, collect supernatant

Power supply

4. Electrophorese   5. Stain gel slice   6. Score population

Figure 3.2   General protocol for protein electrophoresis.

The gel is placed in a buffer tray connected to an electrical power supply, and electrophoresis proceeds over several hours. The gel is then removed and sliced horizontally, and each wafer-thin slice is incubated with a histo-chemical stain specific for a particular enzyme (Hunter and Markert 1957). Each stain contains a commercially available substrate for the enzyme, nec-essary cofactors, and an oxidized salt (usually nitro-blue tetrazolium, or NBT). For example, the staining solution for lactate dehydrogenase (LDH) includes lactic acid (the substrate), nicotinamide adenine dinucleotide (the cofactor), phenazine methosulfate (an intermediary catalyst), and NBT. At each position in the gel to which LDH from a specimen migrated, a reaction is catalyzed whereby lactic acid is oxidized to pyruvic acid and the salt is reduced to a blue precipitate visible to the naked eye as a discrete band. Such bands collectively are the enzyme's "zymogram" pattern, and they usually can be interpreted in simple genetic terms.

Histochemical stains single out the products of particular genes from among the thousands of other undetected proteins also migrating through a gel. When coupled with improvements in electrophoretic procedures and media, such stains eliminated the need for laborious protein purification procedures that had always precluded direct sequencing of amino acids for most population applications. Recipes for more than a hundred allozyme stains are available, but not all enzymes resolve well for a given taxon, so a typical multi-locus SGE survey involves successful assay of about 10–30 enzymes, perhaps encoded by 15–50 loci (some enzymes are encoded by multiple genes).

Mendelian markers

Often, it is quite feasible to assay hundreds or even thousands of individu-als in a given SGE study. One starch gel carrying extracts from about 25 indi-viduals can be sectioned into perhaps five replicate slices, and each slice can be incubated with a different stain. Twenty such gels per day can be run in an active laboratory. Thus, a total of 2,500 genotypes (25 individuals × 20 gels × 5 enzyme stains) could be scored with just several hours of effort. This is a conservative estimate because many enzyme stains reveal genotypes for two, three, or four genes whose products catalyze the same reaction. Such masses of genetic data are incredible by the standards of the pre-molecular era, when elucidation of even a handful of Mendelian genotypes in a few specimens required monitoring cross-generation inheritance patterns. Indeed, shortly after the onset of the allozyme revolution in the mid-1960s, vastly more genotypic data from natural populations had been gathered than in all the preceding 100 years since Gregor Mendel.

Zymogram patterns are normally interpretable as Mendelian genotypes at specific loci. The Mendelian basis of an observed polymorphism may be verified in several ways. First, most enzymes have a known quaternary structure that predicts characteristic gel-band signatures for Mendelian variants (Figure 3.3). For example, phosphoglucomutase (PGM) is a monomer, meaning that it is composed of a single polypeptide with enzymatic activity. Thus, a PGM homozygote shows one band on gels, and a heterozygote shows two bands (one produced by each of the two PGM alleles). Glucose-6-phosphate isomerase (GPI) is an example of a dimer (a molecule whose catalytic activity requires the union of two polypeptide subunits). Thus, a GPI heterozygote displays a three-band gel profile (with the middle band about twice the intensity of the flanking bands), reflecting random associations between polypeptides produced by the two alleles. Purine-nucleoside phosphorylase (PNP) is an example of a trimer, and LDH is a tetramer, such that heterozygotes for such loci can exhibit four-band and five-band zymograms, respectively, with characteristic band intensities. Sometimes, polypeptide subunits encoded by two loci join together to form active enzymes. Zymogram patterns in such cases are more complex, but nonetheless readily interpretable using an extension of the logic described above.

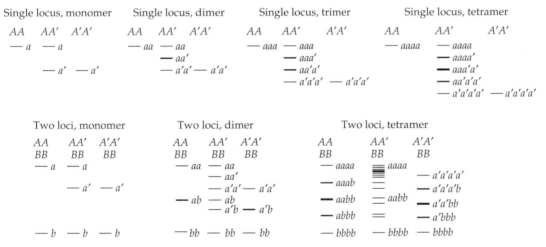

Figure 3.3   Single-locus and multi-locus zymogram patterns. Lowercase letters indicate polypeptide subunits produced by alleles *A* or *A'* at locus 1, and by allele *B* at locus 2; uppercase letters indicate diploid genotypes. Shown in each central lane (column) is the expected zymogram pattern for a heterozygous individual when the enzyme in question is monomeric, dimeric, trimeric, or tetrameric, and in cases in which either one or two loci are involved. For example, for a tetrameric protein encoded by a single gene, a heterozygote should exhibit five gel bands with the following subunit compositions: *aaaa*, *aaaa'*, *aaa'a'*, *aa'a'a'*, and *a'a'a'a'*. If the two alleles produce similar polypeptide concentrations and subunit assembly is random, then intensities of the five respective bands should appear in the approximate ratio 1:4:6:4:1.

A second approach to verifying a Mendelian basis for zymogram variation entails experimental crosses. For example, heterozygous progeny with predictable zymogram patterns should emerge from a cross between two alternative homozygotes, and backcrosses of such progeny to either parent should produce heterozygotes and appropriate homozygotes in nearly equal frequency. In the early years of protein electrophoretic surveys, such experimental validation of genotypes was important, but as experience with the simple genetic bases of banding patterns for commonly used enzymes grew, routine direct corroboration became less critical. Finally, population genetic considerations can help verify a Mendelian basis for zymogram variation. For outcrossed populations that are free from microspatial subdivision, observed frequencies of allozyme genotypes normally agree with frequencies predicted under Hardy–Weinberg equilibrium.

Most allozyme polymorphisms revealed by electrophoresis are probably attributable to nucleotide substitutions causing replacements of charged amino acids, but direct molecular confirmation of this is seldom available. If distinct allozyme alleles at a locus are not further characterized at the molecular level, they must be viewed as qualitative multi-state traits whose phylogenetic order cannot be safely inferred from the observable property: electrophoretic mobility. Allozyme data consist, then, of specified genotypes at each of perhaps dozens of typically unlinked nuclear loci (Pasdar et al. 1984; Shows 1983; Wheat et al. 1973). Apart from their many applications as Mendelian markers in such areas as parentage assessment and gene flow estimation, allozyme frequencies at multiple loci can also be employed to compute quantitative estimates of genetic distance between populations or species.

Idiosyncratic protein features

Occasionally, protein electrophoresis reveals additional kinds of phylogenetic markers. For example, among 26 avian orders and more than 40 assayed taxonomic families, only woodpeckers (Picidae), honeyguides (Indicatoridae), barbets (Capitonidae), and toucans (Rhamphastidae) consistently exhibited a three-band zymogram pattern for malate dehydrogenase (MDH). This unique gel profile, perhaps attributable to a gene duplication event (Avise and Aquadro 1987), helped settle a long-standing debate about whether these superficially different birds are indeed phylogenetically allied, as their traditional placement within Piciformes would suggest. However, apart from identifying this single clade (monophyletic group), the MDH marker was of no further use for avian phylogenetic assessment (unlike allozyme allele frequency data, which provided a more comprehensive picture of piciform relationships; Lanyon and Zink 1987). Thus, the rarity of idiosyncratic allozyme characters is both a phylogenetic blessing and a curse—rarity suggests monophyly and implies that evolutionary clades earmarked by eccentric molecular features may be quite secure, but it also means that few such genetic markers will normally be present in a given data set.

Many enzymes are encoded by two or more loci that arose through gene duplications via polyploidy, aneuploidy, or regional intra-chromosome duplication (Buth 1983; MacIntyre 1976; Ohno 1970). Their zymogram patterns usually are interpretable from rules governing polypeptide assembly into functional enzymes with known quaternary structures (see Figure 3.3). Using such evidence as well as more detailed DNA-level characterizations of various diploid species of *Clarkia* plants, several gene duplications have been discovered (Ford and Gottlieb 1999; Ford et al. 1995; Gottlieb 1988; Soltis et al. 1987) and used to identify putative clades (Gottlieb and Ford 1996). On the other hand, such phylogenetic exercises (Sytsma and Smith 1992) entail several possible complications: the possibility of convergent origin of a duplicate gene in independent lineages; post-duplication gene silencing (perhaps on multiple occasions in different lineages) of either member of a duplicate pair (Ford and Gottlieb 2002; Gottlieb and Ford 1997); and the possibility that a duplicated locus is a retained ancestral condition at the taxonomic level examined. The latter two possibilities have indeed been documented in *Clarkia* (Gottlieb 1988). All three possibilities emphasize the importance of distinguishing orthology from paralogy (see Chapter 1) when drawing phylogenetic conclusions from multi-gene families.

The protein products of duplicated genes usually diverge in structure and regulatory control after the duplication event and may show striking ontogenetic changes or tissue specificities of potential relevance to phylogenetic assessment. For example, whereas most vertebrates express a single GPI gene, bony fishes express two unlinked GPI loci, one predominantly in skeletal muscle and the other in liver, where they perform somewhat different functions (Avise and Kitto 1973; Whitt et al. 1976). All vertebrates examined (with the exception of lampreys) have muscle- and heart-specific LDH expression involving two genes (Markert et al. 1975). As gauged by zymograms, the assembly of heterotetramers between these LDH loci is sometimes nonrandom, presumably due to taxon-specific genetic regulatory influences (Murphy 1988; Sites et al. 1986). Some birds (doves) and mammals have a third LDH gene expressed only in primary spermatocytes (Blanco and Zinkham 1963; Matson 1989; Zinkham et al. 1969). Bony fishes also carry a third LDH gene, expressed in a variety of tissues in primitive fish species but predominantly in the eyes or liver of advanced teleosts (Horowitz and Whitt 1972; Markert and Faulhaber 1965; Shaklee et al. 1973; Whitt et al. 1975). It is doubtful that this fish gene is orthologous to the third locus in birds or mammals (Fisher et al. 1980; Quattro et al. 1993). Other multi-locus allozyme systems that have been studied extensively with regard to gene expression patterns and phylogeny include malate dehydrogenase, glycerol-3-phosphate dehydrogenase, creatine kinase (Buth et al. 1985; Fisher and Whitt 1978; Fisher et al. 1980; Philipp et al. 1983a), and the globin superfamily of oxygen-carrying molecules (Dayhoff 1972; Doolittle 1987).

Duplicate genes are also subject to evolutionary silencing or loss, and these outcomes can be phylogenetically informative. For example, Ferris and Whitt (1978, 1979) used patterns of enzyme loss and change in gene expression to reconstruct phylogeny in catostomid suckers, a group of freshwater fishes that underwent a polyploidization event approximately 50 million years ago and subsequently became "diploidized" at approximately 50% of assayed structural genes. The rationale for this phylogenetic reconstruction was as follows. Immediately following the polyploidization event, all loci in the ancestral sucker genome must have been duplicated, and the "primitive" condition from that point forward became presence of each duplicate gene. As mutations accumulated, some genes lost expression (becoming pseudogenes), whereas others diverged in structure and function (Ferris and Whitt 1977). These processes presumably were nearly irreversible (but see Buth 1979, 1982), such that taxa sharing the derived states (loss or alteration of gene expression) probably are monophyletic (assuming that identical changes did not occur independently in separate evolutionary lineages).

Not all idiosyncratic gene expression patterns are phylogenetically informative, however. For example, Mindell and Sites (1987) assayed tissue expression profiles for about 30 allozyme loci in species representing two avian orders (Charadriiformes and Passeriformes), and they observed numerous overt discrepancies with accepted taxonomy. These authors concluded that "widespread homoplasy, opposite polarities and limited predictive capability for the isozyme tissue expression patterns suggest that most may be more useful in studies of gene regulation than in higher level taxonomy."

DNA–DNA Hybridization

DNA–DNA hybridization relies on the double-stranded nature of duplex DNA and the fact that paired nucleotides on the two complementary strands are held together by hydrogen bonds (two coupling each adenine–thymine base pair and three coupling each guanine–cytosine). These hydrogen bonds are the weakest links in DNA, so when native DNA is heated to 100°C in solution, the duplexes dissociate or "melt" into single strands. When the sample is cooled, these strands collide by chance, and those of complementary sequence reassociate into duplex molecules as hydrogen bonds re-form between matched bases. A rapidly reassociating component represents repetitive DNA, because these homologous strands are most numerous and collide most frequently. This fraction is removed. The remaining fraction is composed of single- or low-copy sequences in the genome. These DNA strands are then added to a mixture under conditions permitting duplex formation. The mixture may contain DNA strands from a single sample or species (yielding homoduplexes) or it may contain strands from two species (yielding heteroduplex molecules). The final step in the hybridization protocol involves characterizing the thermal stabilities of these homo- and heteroduplexes by gradually raising the temperature and monitoring the course of molecular dissociation to single strands. The thermal stability exhibited by any duplex depends largely on the similarity of nucleotide sequences in its two strands, because only properly paired bases form hydrogen bonds. The measured difference in thermal stability between homoduplexes and heteroduplexes provides a quantitative estimate of the genetic divergence between the species examined.

More details about DNA hybridization are as follows (Figure 3.4; exact methods are described by Werman et al. 1996):

- First, DNA is extracted from the nuclei of cells, separated from RNA and proteins, and physically sheared into fragments averaging 500 nucleotides in length (to reduce viscosity and to permit subsequent fractionation of repetitive from low-copy DNA).

- The sheared fragments are boiled, cooled, and their reassociation kinetics employed to remove the highly repetitive fraction (Britten and Kohne 1968; Britten et al. 1974). This is accomplished by incubating the DNA in solution at about 50°C for a short time, such that repetitive sequences preferentially anneal and most low-copy sequences remain unpaired.

- This solution is passed through a hydroxyapatite column that binds double-stranded DNA only. Single-stranded DNA that passes through the column is labeled with radioactive iodine and becomes known as the tracer.

- Tracer DNA is then mixed with a much larger amount of unlabeled DNA (called the driver) from the same or a different species. This mixture is incubated at 60°C for several days to form hybrid molecules that have one labeled and one unlabeled strand.

- The sample is then placed on a hydroxyapatite column and gradually heated at 2.5°C increments over a 60–95°C range. At each temperature increment, additional duplexes that have melted (a function of degree of base-pair mismatch) are washed from the column into a beaker. Counts of radioactivity in the beakers record the amount of duplex DNA that melted at various temperatures.

Raw data from DNA–DNA hybridization consist of "thermal elution profiles" (Figure 3.5) that summarize observed percentages of dissociated, single-stranded DNA as a function of melting temperature. From these melting curves, quantitative estimates of the degree of base-pair mismatch between DNAs under comparison can be derived (Britten 1986; Kirsch et al. 1990; Sarich et al. 1989; Sheldon and Bledsoe 1989). One such measure of genetic divergence is based on the temperature (T_m) at which 50% of the hybrid molecules remain in duplex condition. The difference in T_m values between homoduplex and heteroduplex melting profiles (ΔT_m) provides a

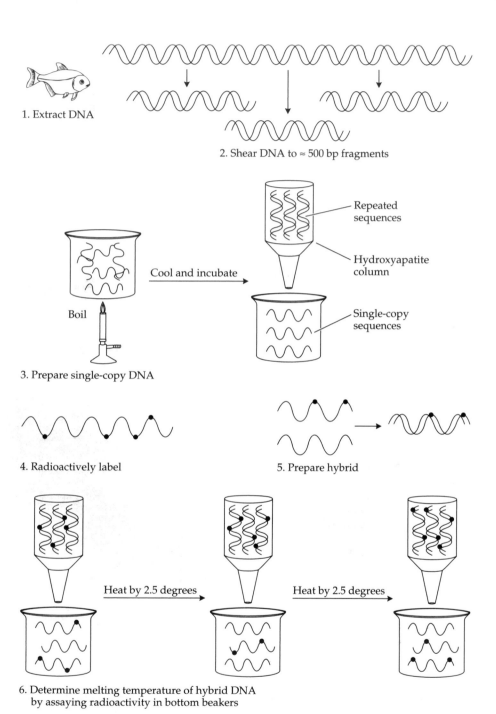

1. Extract DNA

2. Shear DNA to ≈ 500 bp fragments

Boil

Cool and incubate

Repeated sequences

Hydroxyapatite column

Single-copy sequences

3. Prepare single-copy DNA

4. Radioactively label

5. Prepare hybrid

Heat by 2.5 degrees

Heat by 2.5 degrees

6. Determine melting temperature of hybrid DNA by assaying radioactivity in bottom beakers

Figure 3.4   General protocol for DNA–DNA hybridization.
(After Sibley and Ahlquist 1986.)

(A)

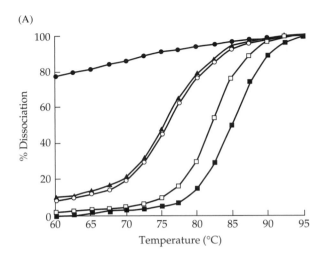

(B)

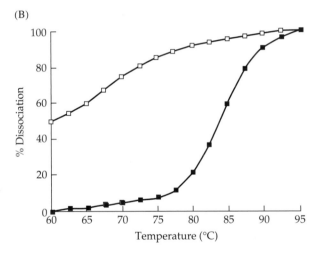

Figure 3.5 Thermal elution profiles from DNA–DNA hybridization. Shown are cumulative melting curves for the single-copy fraction of nuclear DNA in several flightless ratite birds. (A) Homoduplex DNA of emu (*Dromaius novaehollandiae*) (solid squares), and heteroduplexes between that species and southern cassowary (*Casuarius casuarius*) (open squares), greater rhea (*Rhea americana*) (open circles), ostrich (*Struthio camelus*) (solid triangles), and chicken (*Gallus gallus*—a non-ratite outgroup) (solid circles). In these assays, cassowary appears genetically closest to emu, followed in order by rhea, ostrich, and chicken. (B) Homoduplex DNA of ostrich (solid squares) and its heteroduplex with rhea (open squares). Note that although the melting curves involving rhea and ostrich are nearly identical when compared with emu (panel A), this does not necessarily imply that rhea and ostrich are genetically close to one another. Indeed, differences between the melting curves in panel B indicate a large genetic distance between rhea and ostrich. (After Sibley and Ahlquist 1990.)

genetic distance estimate. When such estimates are available for pairwise comparisons among three or more species, they can be used as a basis for phylogenetic reconstruction.

The relationship between ΔT_m and percent base-pair mismatch is approximately linear (Britten et al. 1974; Caccone et al. 1988b; Kohne 1970), as shown by studies of thermal stability of synthetic oligonucleotides or other sequences of known base composition (Bautz and Bautz 1964; Hutton and Wetmur 1973; Laird et al. 1969; Springer et al. 1992). As a working rule of thumb, each increase in ΔT_m of 1°C translates to an additional 1% or 2% base-pair mismatch in DNA (Britten 1986; Caccone et al. 1988b; Koop et al. 1986; Powell et al. 1986).

Because the DNA hybridization method in effect yields a mean genetic difference across a large fraction (the low-copy portion) of any two genomes compared, it was sometimes promoted as one of the strongest possible sources of phylogenetic information (Sibley and Ahlquist 1990). Indeed, inter-taxon genetic distances from DNA hybridization have been shown to corre-late reasonably well with those from direct sequencing of some mitochondri-al and nuclear genes (van Tuinen et al. 2000). However, reservations about the approach were voiced as well: The raw data consist solely of distance values, rather than discrete character states amenable to cladistic analysis (see below), and the factors that can affect the kinetics of hybridization (such as differences in base composition, DNA fragment size, and genome size) are incompletely understood. Some of these factors nonetheless are partially controlled in most DNA hybridization studies. For example, effects of base composition differ-ences (numbers of A-T versus C-G pairs) among sequences can be ameliorat-ed by use of chaotropic solvents (Werman et al. 1996), and genetic compar-isons can be confined to particular taxonomic groups (such as birds), within which confounding variables such as pronounced differences in genome structure or base composition should be minimized.

The development of automated thermal elution devices (e.g., the "DNAnalyzer" of Sibley and Ahlquist 1981) greatly expedited the process of gathering DNA hybridization data. Indeed, the honor for the largest set of animals included in any molecular systematic survey might still belong to Sibley and Ahlquist (1990), who conducted nearly 30,000 DNA–DNA hybridizations on 1,700 avian species. Although a few research laboratories were devoted to DNA hybridization assays, the technique was not other-wise widely employed by molecular systematists. The method nevertheless had a large impact on ornithology and some other fields, and it remains of interest for the contrasts it provides with most other genetic approaches (which typically focus on molecular details at far fewer loci).

Restriction Analyses

The discovery of restriction endonucleases (Linn and Arber 1968; Meselson and Yuan 1968) revolutionized molecular biology. Type II restriction enzymes (Kessler 1987) cleave duplex DNA at particular oligonucleotide sequences, usually either 4, 5, or 6 base pairs in length (Roberts 1984). For example, *Eco*RI (named after the bacterium *Escherichia coli* from which it was isolated) acts like a precise scalpel that cuts double-stranded DNA wherever the non-methylated sequence 5'–GAATTC–3' occurs. Several hundred such enzymes, most with different recognition sequences, have been isolated and characterized from various bacterial strains and are commercially available. In bacteria, these enzymes protect against invasion by foreign DNA (host DNA is protected by bacteria-specific methylation systems). In DNA laboratories, restriction enzymes find wide application in assays of restriction fragment-length polymorphisms (RFLPs).

RFLP analyses involve cutting (restricting) DNA with one or more endonucleases, separating the resulting fragments according to molecular weight by gel electrophoresis, and visualizing the size-sorted fragments. Differences among individuals in these "digestion profiles" may result from base substitutions within cleavage sites, additions or deletions of DNA, or sequence rearrangements, with each source of variation producing characteristic banding changes. Three important and partially interrelated variables in these assays are the electrophoretic media employed, the means of fragment visualization, and the choice of DNA to be analyzed. These general considerations will be discussed first, and methodological details for particular applications will be added afterward.

The usual electrophoretic media are agarose or acrylamide gels. These gels form dense matrices through which, under the influence of an electric current, small DNA fragments migrate faster than large fragments. At neutral pH, DNA is negatively charged, and thus moves toward the anode at rates determined by molecular size. Agarose gels are used to separate DNA fragments in the size range of about 300–20,000 base pairs (bp), whereas acrylamide gels optimally separate restriction fragments about 10–1,000 bp long. To facilitate estimates of restriction fragment lengths, researchers normally include molecular size standards (commercially available) in each gel.

DNA fragments can be visualized in several ways. Some electrophoretic assays begin with highly purified DNA isolated from particular sources (such as mitochondria), in which case DNA fragments in the gel are revealed by chemical stains or radioactivity. When DNA amounts are high (>50 ng per gel band), ethidium bromide provides a convenient agent for fragment detection. This chemical binds to DNA in such a way that staining intensities are proportional to fragment sizes, so digestion profiles are stoichiometric. Silver staining is similar and provides greater sensitivity in detecting small DNA quantities (< 100 pg; Guillemette and Lewis 1983). In highly sensitive "end-labeling" procedures, DNA digestion fragments are labeled radioactively with ^{32}P- or ^{35}S-tagged nucleotides prior to electrophoretic separation. After a gel is run, it is vacuum-dried and overlaid by X-ray film, whose development as an autoradiograph reveals the positions to which the DNA fragments migrated. With end-labeling, band intensities are independent of fragment size (because all fragments have two labeled ends), and the method is therefore especially useful in revealing small fragments when DNA amounts are limited.

Other RFLP assays begin with DNA of heterogeneous classes (e.g., total nuclear DNA), with elucidation of DNA fragments from particular genes accomplished after electrophoresis by the technique of "Southern hybridization" (Southern 1975). In this method, DNA fragments in a gel are denatured in a basic solution and then transferred as single strands (by capillary action or electrophoresis) to a nylon or nitrocellulose membrane (Figure 3.6). This membrane is incubated with a single-stranded "probe"—DNA previously isolated, purified, and radioactively labeled—under conditions in which strands that are complementary to those of the probe hybridize with the probe to form radioactive duplexes in the membrane. Under high-stringency conditions, hybridization with distantly related or non-homologous DNA is avoided. Thus, the probe in effect picks sequences that are complementary and (ideally) homologous to itself from the multitude of undetected fragments that also migrated through the gel. Those fragments with sequence similarity to the probe then are visualized by autoradiography of the "Southern blot."

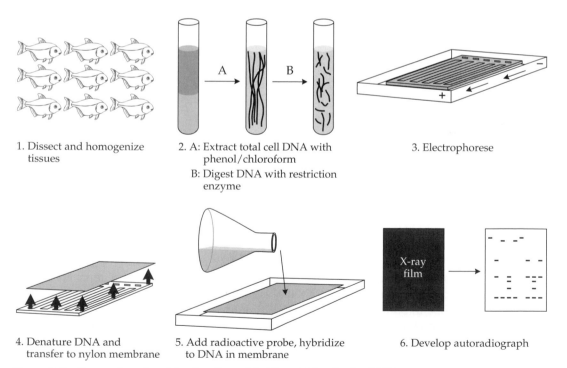

1. Dissect and homogenize tissues

2. A: Extract total cell DNA with phenol/chloroform
   B: Digest DNA with restriction enzyme

3. Electrophorese

4. Denature DNA and transfer to nylon membrane

5. Add radioactive probe, hybridize to DNA in membrane

6. Develop autoradiograph

Figure 3.6   General protocol for Southern blotting. (After Burke 1989.)

The probe in Southern hybridizations thus identifies particular pieces of DNA. This probe may constitute, for example, a single gene from a nuclear or cytoplasmic genome, a non-coding stretch of DNA sequence, or an entire animal mtDNA. If the probe contains DNA that is present in multiple copies in the genome, the Southern blot reveals fragments from all members of the family to which the probe hybridized. In some such cases, Southern blots reveal highly complex digestion profiles wherein nearly all individuals are distinguished by their "DNA fingerprints." The probes used in Southern hybridizations may come from DNA highly purified by physical means (e.g., mtDNA isolated via CsCl gradient centrifugation) or from cloning of particular genes through biological vectors (Sambrook et al. 1989). For rapidly evolving sequences, a probe's utility may be confined to assays of closely related species, whereas probes for slowly evolving sequences may cross-hybridize across broad taxonomic assemblages.

Restriction analyses therefore encompass a wide diversity of technical approaches, details of which are described by Dowling et al. (1996c), Hames and Higgins (1985), Hoelzel (1992), Karl and Avise (1993), Lansman et al. (1981), Quinn and White (1987), Sambrook et al. (1989), and Watson et al. (1992). Furthermore, different classes of DNA differ dramatically with respect to the nature of genetic information provided by restriction analysis, so they will be discussed separately in the following sections.

Animal mitochondrial DNA

Restriction analyses of mtDNA dominated the field of phylogeography from the late 1970s until about 1990, when direct mtDNA sequencing made them somewhat passé. Nonetheless, most of the key evolutionary properties of mtDNA were discovered in this earlier era, so RFLP assays are of historical interest for the lessons they provide as well as for their similarities and contrasts with modern direct mtDNA sequencing.

In most RFLP assays, a crucial initial step in isolation of animal mtDNA is the efficient separation of cytoplasm (where mitochondria are housed) from cell nuclei (Figure 3.7). Soft tissue such as heart, liver, or ovary is minced, gently homogenized, and centrifuged at low speed ($700 \times g$) to remove nuclei and cellular debris. Subsequent centrifugation at higher speed ($20,000 \times g$) pellets mitochondria, which are then washed and lysed. The next step involves CsCl-EtBr gradient centrifugation at speeds in excess of $160,000 \times g$. Mitochondrial DNA, which appears as a discrete band in the resulting gradient, is removed by hypodermic needle and separated from remaining contaminants by dialysis. Purified mtDNA can then be used as a probe in Southern blots (to reveal mtDNA bands in samples containing heterogeneous DNA), or the purification process described above can be repeated for each specimen and the RFLPs elucidated directly by chemical stains or radioactive labeling. A rate-limiting step in these mtDNA analyses is the lengthy centrifugation process, but various shortcuts have been developed for special circumstances, such as when mitochondria-rich cells are

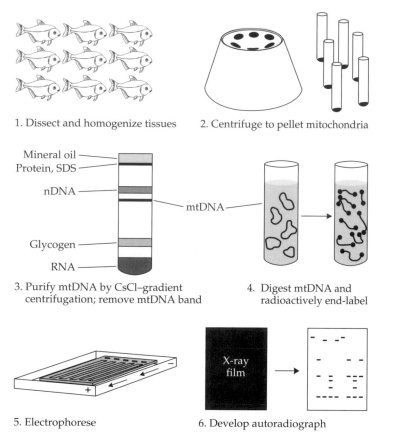

1. Dissect and homogenize tissues

2. Centrifuge to pellet mitochondria

Mineral oil
Protein, SDS

nDNA

mtDNA

Glycogen

RNA

3. Purify mtDNA by CsCl–gradient centrifugation; remove mtDNA band

4. Digest mtDNA and radioactively end-label

X-ray film

5. Electrophorese

6. Develop autoradiograph

Figure 3.7   General protocol for mtDNA restriction site analysis by radioactive end-labeling.

available or only small amounts of mtDNA are needed (Carr and Griffith 1987; Chapman and Powers 1984; Jones et al. 1988; Palva and Palva 1985; Powell and Zuninga 1983).

In the restriction enzyme era, most researchers working with animal mtDNA employed end-labeling procedures. For plant mitochondrial and chloroplast genomes, which are much larger, procedures usually involved Southern blotting using cloned genes or subsets of the genome as probes. More recently, PCR-mediated sequencing of cytoplasmic genomes has largely supplanted these earlier approaches. Interestingly, however, a serious complication has been discovered that can affect Southern blotting and PCR-based approaches, but not assays of physically isolated DNA. It turns out that both recent and ancient transfers of organelle DNA to the nucleus have been relatively common events in evolution, such that many sequences related to cytoplasmic genes now also exist as functionless derivatives in the nuclear genome (see Chapter 8). Unless these paralogous pseudogenes are properly recognized for what they are (sometimes a tricky proposition; e.g., Schneider-Broussard and Neigel 1997), they can potentially confound phylogenetic analyses. It is rather fortuitous that in the RFLP era, mtDNA was typically physically purified before assay, such that the unwanted complication of mtDNA pseudogenes in the nucleus simply was not encountered.

Long before whole-genome sequencing of animal mtDNA became almost routine (e.g., Miya et al. 2003), the general structure and genetic basis of variation in this molecule already were well elucidated (see seminal reviews in Attardi 1985; Brown 1985; Cantatore and Saccone 1987; Gray 1989; Wallace 1982, 1986; Wolstenholme 1992). With few exceptions, animal mtDNA is a closed circular molecule, typically 15–20 kilobases (kb) in length, and composed of 37 genes (Figure 3.8A) coding for 22 tRNAs, 2 rRNAs, and 13 mRNAs specifying proteins involved in electron transport and oxidative phosphorylation. Nearly the entire mtDNA genome is involved in coding function; introns, large families of repetitive DNA, pseudogenes, and even sizable spacer sequences between genes are rare or lacking in most cases. A "control region" of about 1 kb initiates replication and transcription. Gene arrangement in animal mtDNA is evolutionarily conserved, although differences in gene order and content often do distinguish higher taxa, and these features have proved useful in macrophylogeny estimation (see Chapter 8). Similarly, the detailed structures of mitochondrial tRNA and rRNA genes and their products (Cantatore et al. 1987; Wolstenholme et al. 1987), as well as differences among taxa in the mtDNA genetic code, have proved to be phylogenetically informative (see Chapter 8).

With regard also to the general mode of animal mtDNA evolution, much was learned from early population surveys (see seminal reviews by Avise and Lansman 1983; Birley and Croft 1986; Harrison 1989; Moritz et al. 1987; Wilson et al. 1985). For example, notwithstanding extensive intraspecific sequence variation, most individual animals proved to be homoplasmic, or nearly so, meaning that a single mtDNA sequence predominates in all cells and tissues of a given specimen (although many exceptions are known; e.g., Bermingham et al. 1986; Hale and Singh 1986; Moritz and Brown 1987, and references therein). Possible reasons for this characteristic sequence homogeneity within individuals remain poorly understood, but bottlenecks in mtDNA numbers in germ cell lineages probably are involved (Birky et al. 1989; Chapman et al. 1982; Clark 1988; Laipis et al. 1988; Rand and Harrison 1986; Solignac et al. 1984, 1987; Takahata 1985). Regardless of how such sequence homogeneity mechanistically arises, this phenomenon is pragmatically crucial in virtually all genealogical applications for mtDNA.

It was also discovered early on that animal mtDNA evolves rapidly at the sequence level, due in part to inefficient mutation repair mechanisms (Brown et al. 1979; Wilson et al. 1985). Some sequences within the control region evolve even faster and are therefore of special utility in high-resolution analyses of shallow (recent) population structure (e.g., Stoneking et al. 1991). Addition/deletion changes in mtDNA are not rare, but most differences

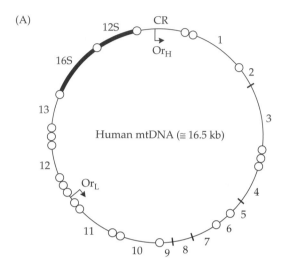

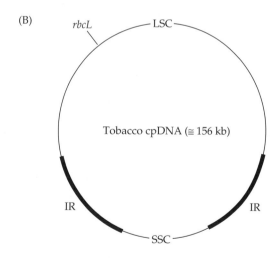

Figure 3.8   Major structural features of animal mitochondrial DNA and plant chloroplast DNA. (Molecules are not drawn to the same scale.) (A) Human mtDNA, composed of a control region (CR) and genes encoding 2 rRNAs (12S and 16S), 22 tRNAs (open circles), and 13 functional polypeptides. Also shown are sites (Or_H and Or_L) at which replication is initiated along complementary DNA strands. (B) Tobacco (*Nicotiana tabacum*) cpDNA, composed of large and small single-copy regions (LSC, SSC) and a large inverted repeat (IR). Also shown is the position of the *rbcL* gene, DNA sequences of which have figured prominently in phylogenetic analyses of plants (see Chapter 8).

between sequences reflect point mutations, with a strong initial bias for transitions over transversions (Aquadro and Greenberg 1983; Brown and Simpson 1982; Brown et al. 1982; Greenberg et al. 1983).

Perhaps most important for genetic marker purposes was the discovery that mtDNA is transmitted predominantly through maternal lines in most species (Avise and Vrijenhoek 1987; Dawid and Blackler 1972; Giles et al. 1980; Gyllensten et al. 1985a; Hutchison et al. 1974; see review in Birky 1995). Only a few departures from predominant maternal inheritance have been uncovered (Avise 1991b; Gyllensten et al. 1991; Kondo et al. 1990), the most notable being a "doubly uniparental" hereditary mode in *Mytilus* and related bivalves wherein females transmit their mtDNA to both sons and daughters, whereas males transmit their mtDNA to sons only (Hoeh et al. 1991, 1997, 2002; H.-P. Liu et al. 1996; Zouros et al. 1992, 1994). In this mollusk system, genetic recombination among mtDNA molecules has also been documented (Ladoukakis and Zouros 2001). The *Mytilus* case has been of special interest precisely because it is so exceptional (Zouros 2000).

In most other taxa, uniparental inheritance clearly limits the opportunity for evolutionarily significant genetic recombination between mtDNA molecules. Nonetheless, reports of occasional "paternal leakage" of animal mtDNA into zygotes have raised the specter of possible recombination between distinctive mtDNA genotypes, as have some interpretations of disequilibrium patterns in population genetic data for this molecule (Awadalla et al. 1999; Eyre-Walker et al. 1999; Lunt and Hyman 1997). However, this latter class of evidence has been challenged (Arctander 1999; Elson et al. 2001; Kivisild et al. 2000; Merriweather and Kaestle 1999), and the recombination issue has not been definitively resolved. What remains uncontested is that animal mtDNA genotypes are mostly maternally transmitted, and that if physical recombination does occur, it is unusual or rare in most taxa. Thus, mtDNA molecules provide matrilineal markers that are transmitted asexually through the pedigrees of what may otherwise be sexually reproducing species. For simplicity, mtDNA genotypes are therefore often referred to as molecular clones or haplotypes, and their inferred evolutionary interrelationships are interpreted as estimates of "matriarchal phylogeny" (Avise et al. 1979b). From a functional perspective, mtDNA consists of about 37 genes, but from a phylogenetic perspective, the entire molecule is one linked genealogical unit (i.e., supergene) with numerous alleles.

The raw data in most mtDNA restriction surveys consisted of fragment-length profiles produced by each of a dozen or more restriction enzymes (Figure 3.9). Because mtDNA is a closed circle, the number of linear fragments equals the number of restriction sites recognized by each endonuclease. A useful check on gel scoring is provided by mtDNA genome size within a given species, to which observed fragment lengths should sum. This feature also facilitates direct comparisons of RFLP profiles across studies or among different laboratories (a useful characteristic not fully shared by allozyme methods, in which meaningful cross-study comparisons of allelic products require that known electromorph standards be run in all gels). A typical mtDNA popula-

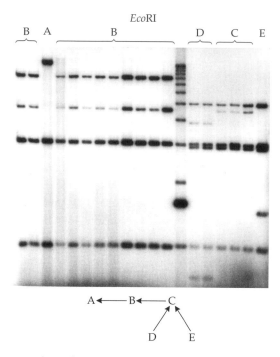

Figure 3.9 Interpretation of mtDNA digestion profiles. Shown is an autoradiograph of *Eco*RI digests of mtDNA from 18 eels (genus *Anguilla*). The seventh lane from the right is a molecular size standard in which the darkest band is 1.6 kb in size; successive bands above it are approximately 2, 3, 4, 5, 6, 7, 8, ... kb, and the band below it is 1.0 kb. Five *Eco*RI patterns (A–E) are evident, their interrelationships summarized in a parsimony network shown under the radiograph (arrows indicate direction of restriction site loss, not necessarily the direction of evolution). For example, pattern A differs from B by loss of an *Eco*RI restriction site, which converts the 4.6-kb and 8.0-kb fragments in the B profile to the 12.6-kb fragment in A. In turn, C differs from B by gain of an *Eco*RI site, which converts B's 8.0-kb fragment to C's fragments of sizes 5.1 kb and 2.9 kb. Pattern E apparently has a "doublet" (two fragments of indistinguishable molecular weight) at 3.1 kb. (From Avise 1987.)

tion survey revealed about 50–100 restriction fragments per individual. Because the restriction enzymes employed were commonly five- and six-base cutters, these results were equivalent in information content to assaying 250–600 bp of recognition sequence per specimen. The larger mtDNA population surveys sometimes included many hundreds of individuals.

Most differences among mtDNA digestion profiles proved to arise from point mutations that had created or destroyed enzyme recognition sequences. Often, the number of mutations distinguishing particular digestion profiles could be deduced directly from single-enzyme gel patterns, using information about mtDNA fragment sizes (see Figure 3.9). Such data could be accumulated across restriction enzymes and used to generate composite mtDNA haplotype descriptions. Data were also recorded as binary characters and summarized in presence-absence matrices of restriction sites across individuals or mtDNA clones (Box 3.1). Some studies went further by mapping the positions of sites relative to one another or to landmarks on the mtDNA genome, using double-digestion or partial-digestion procedures (Figure 3.10). Normally, however, sites could be mapped only to within a few tens or hundreds of base pairs of their true locations.

Some mtDNA RFLPs were sufficiently complex that mutational pathways distinguishing haplotypes were difficult or impossible to deduce from the gel patterns alone. Researchers could nonetheless count percentages of shared (and presumably homologous) fragments, but caution was indicated because, unlike site changes, not all fragment changes are independent. For example, a point mutation that creates a restriction site results in the correlated appearance of two smaller fragments with same total molecular weight

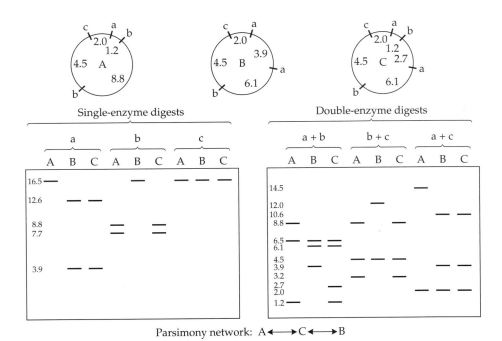

Figure 3.10   Scoring and mapping of mtDNA restriction sites. Shown are mtDNA haplotypes A, B, and C as revealed by digestion by restriction endonucleases a, b, and c. The restriction maps at the top were unknown at the outset, but were deduced from observed gel profiles produced in single- and double-enzyme digests. Fragment sizes (in kb) are indicated. A parsimony network at the bottom summarizes the likely pathway of evolutionary interconversion between haplotypes with respect to these assayed sites. Note in this case that restriction site changes and the parsimony network (but not the full restriction site map) could also have been deduced directly from single-enzyme digestion profiles.

as the one larger fragment lost (see Figure 3.9). Statistical methods were developed to take into account such correlated fragment changes in converting percentages of shared fragments to estimates of sequence divergence (e.g., Nei and Li 1979; Upholt 1977). Additional sources of RFLP variation stemmed from occasional differences in mtDNA size, most often due to variation in copy number of localized tandem repeats in or near the control region of the molecule (Bermingham et al. 1986; Harrison et al. 1985; Moritz and Brown 1986, 1987). These localized repeat regions ranged in size from a few base pairs to more than 1 kb, and the larger ones were readily distinguished from restriction site changes because they concordantly altered the total lengths of restriction fragments in all digestion profiles (smaller fragment size differences could be overlooked, however, particularly when they resided in high-molecular-weight gel bands).

With the advent of PCR-mediated DNA sequencing (described below), the laboratory methods for analyzing mtDNA have changed, but the general nature of the information provided (on matrilines) and the classes of biological problems that can be addressed (especially at the intraspecific level) remain much the same. Direct sequencing has also expanded opportunities for delivering mtDNA data in a form suitable for phylogenetic reconstruc-

---

BOX 3.1 Restriction Site Matrix

The example shown here involves 96 restriction sites (0, absent; +, present) in 10 different mtDNA haplotypes (*a–e* and *p–t*) in sharp-tailed sparrows, *Ammodramus caudacutus*.

a   +++++++++++00+0+++++0+++++++++++++++++0000++++++++++++++++++0++++0+++++0+0+0++++++++++0+++++

b   +++++++++++00+0+++++0+++++++++++++++++0000++++++++++++++++++00++++0+++++0+0+0++++++++++0+++++

c   +++++++++++00+0+++++0++++++++++++++++++000++++++++++++++++++0++++0+++++0+0+0++++++++++0+++++

d   +++++++++++00+0+++++0+++++++++++++++++0+00++++++++++++++++++0++++0+++++0+0+0++++++++++0+++++

e   +++++++++++00+0++++++++++++++++++++++0000++++++++++++++++++00++++0+++++0+0+0+++++++++++++++++

p   +++++++++++00+0+++++0++++++++++++++++++0+00+++++0++++++++++0++++++++++00+0+++++++00+++0+++++

q   +++++++++++00+0+++++0++++++++++++++++++0+00+++++0++++++++++0++++++++++00++++++++++00+0+0+++++

r   +++++++++++00++++++++0++++++++++++++++++0+0+++++0++++++++++0++++++++++00+0+0++++++00+++0+++++

s   +++++++++++00+0+++++0++++0+++++++++++++0+0+++++0++++++++++0++++++++++00+0+0++++++00+++0+++++

t   +++++++++++00+0+++++0++++0++++++++++++0+0+++++0++++++++++0++++++++++00+++0++++++00+++0+++++

Source: From a larger data set in Rising and Avise 1993.

tion at supraspecific levels. Assays of animal mtDNA remain among the most powerful and popular of all molecular approaches in ecology and evolution (Randi 2000).

Plant organelle DNA

Several evolutionary features of animal mtDNA were completely unanticipated, among them the predominance of a single mtDNA sequence (homoplasmy) within single specimens despite extensive between-individual sequence differences, and the rapid pace of nucleotide substitution despite what would seem to be severe functional constraints on mtDNA, as judged by the molecule's "genetic economy" (Attardi 1985). However, after the major evolutionary features of animal mtDNA were revealed, it might have been supposed that these attributes would apply to other cytoplasmic genomes as well. Surprisingly, this did not prove to be the case.

PLANT MITOCHONDRIAL DNA. Plant mtDNA is highly variable in size, ranging from about 200 kb to 2,500 kb across species (Palmer 1985; Pring and Lonsdale 1985; Ward et al. 1981). Within an individual, mtDNA sequences typically exist as a heterogeneous collection of circles arising from extensive recombination that interconverts between sub-genomes and higher-order multimers (Backert et al. 1996, 1997; Hanson and Folkerts 1992; Palmer and Herbon 1986; Palmer and Shields 1984). Inheritance is often, but not invariably, maternal (Birky 1978; Forsthoefel et al. 1992; Havey et al. 1998; Kondo et al. 1998). Although plant mtDNA is generally similar to animal mtDNA with regard to gene content and general function, its evolutionary pattern differs diametrically (Birky 1988; Palmer 1992): Plant mtDNA evolves rapidly with respect to gene order, but about a hundredfold more slowly than animal mtDNA with respect to nucleotide sequence (Palmer and Herbon 1988). These properties, as well as the technical difficulties of laboratory assay, have conspired to limit the utility of plant mtDNA in molecular systematics and population biology (but see Desplanque et al. 2000; Huang et al. 2001; Olson and McCauley 2002).

CHLOROPLAST DNA. Plant chloroplast DNA offers yet another story, as emphasized in an influential early review by Palmer (1985). It is transmitted maternally in most species (Birky 1978; Gillham 1978; Hachtel 1980; Havey et al. 1998), biparentally in some (e.g., Harris and Ingram 1991; Metzlaff et al. 1981; Shore and Triassi 1998), and paternally in various others (Chat et al. 1999; Yang et al. 2000), including most gymnosperms (e.g., Dong et al. 1992; Kondo et al. 1998; Sperisen et al. 2001; Szmidt et al. 1987; Wagner et al. 1987). This circular molecule (Figure 3.8B) varies greatly in size among species (from approximately 120 to 217 kb in photosynthetic land plants, for example), due largely to the extent of reiteration of a large inverted repeat that includes genes for rRNA subunits (Zurawski and Clegg 1987). With some possible exceptions (Milligan et al. 1989; Wagner et al. 1987), the rate of cpDNA evolution is slow, in terms of both primary nucleotide sequence (mean silent substitution rates have been estimated at only three to four times greater than those of plant mtDNA; Wolfe et al. 1987) and gene rearrangement (Curtis and Clegg 1984; dePamphilis and Palmer 1989; Palmer 1990; Palmer and Thompson 1981; Ritland and Clegg 1987).

Seminal studies on cpDNA involved RFLP analyses (Palmer and Zamir 1982), but in later years most molecular analyses entailed direct sequencing of particular cpDNA genes. Because of cpDNA's leisurely pace of evolution, such data have proved especially valuable for estimating plant phylogeny at higher taxonomic levels (e.g., Clegg et al. 1986; Palmer 1987; Palmer et al. 1988a; Zurawski and Clegg 1987). Several unique structural features of cpDNA (see Chapter 8) have further contributed to the identification of various plant clades (e.g., Downie and Palmer 1992; Jansen and Palmer 1987). Particular cpDNA sequences have also been tapped as a valuable source of information on intraspecific phylogeography (see Chapter 6). Interestingly, the first restriction site appraisals of cpDNA (Atchison et al. 1976; Vedel et al. 1976) were contemporaneous with early studies on animal mtDNA RFLPs, and they sometimes uncovered at least modest variation within as well as among closely related plant species (Banks and Birky 1985). Nonetheless, phylogeographic studies of plants generally lagged far behind those of animals, a situation that only recently has become partially rectified (Petit and Vendramin 2003; Schaal et al. 2003).

Single-copy nuclear DNA

As applied to single-copy loci (scnDNA) or low-copy genes in the nucleus, RFLP analyses traditionally relied on Southern blotting (see Figure 3.6), with the probes representing DNA sequences cloned into a biological vector such as lambda phage or a bacterial plasmid (Kochert 1989; Figure 3.11). The probes were particular genes of known function or anonymous sequences drawn at random from a genomic library (i.e., from a collection of cloned DNA fragments).

Probes for known-function genes are often derived from "complementary" DNA (cDNA); that is, sequences produced by reverse transcription of a particular messenger RNA. Anonymous single-copy probes can be generated as follows: Total cell DNA is extracted and digested with a restriction enzyme. Fragments of size 500–5,000 bp are isolated by electrophoresis and cloned into a vector, thereby generating a DNA library. This library is then screened for single-copy sequences by dot-blot hybridization (Figure 3.12), whereby each clone's DNA is hybridized under controlled conditions with radioactively labeled total cell DNA. The radioactive signal intensity of each dot in the blot is then assessed: A strong signal indicates clones carrying repetitive DNA, and a weak signal may identify clones with low-copy sequence. One important distinction between cDNA and genomic clones as probes is that the former represent processed coding sequences for a transcribed gene product, whereas the latter can also include gene-flanking

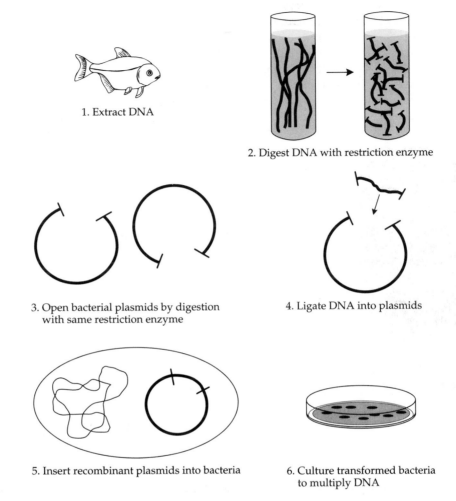

1. Extract DNA

2. Digest DNA with restriction enzyme

3. Open bacterial plasmids by digestion with same restriction enzyme

4. Ligate DNA into plasmids

5. Insert recombinant plasmids into bacteria

6. Culture transformed bacteria to multiply DNA

Figure 3.11   General protocol for DNA cloning and genomic library construction.

regions as well as introns (non-coding sequences interspersed with the exons that specify a gene's amino acid sequence). Some of these non-coding sequences evolve rapidly and provide additional classes of genetic markers (Friesen 2000).

Later-developed methods for generating scnRFLPs took advantage of the polymerase chain reaction (PCR). For scnDNA identified in a nuclear genomic library, PCR primers were generated and used to amplify homologous DNA from each individual. These amplified DNAs were then digested by restriction enzymes, electrophoresed, and chemically stained

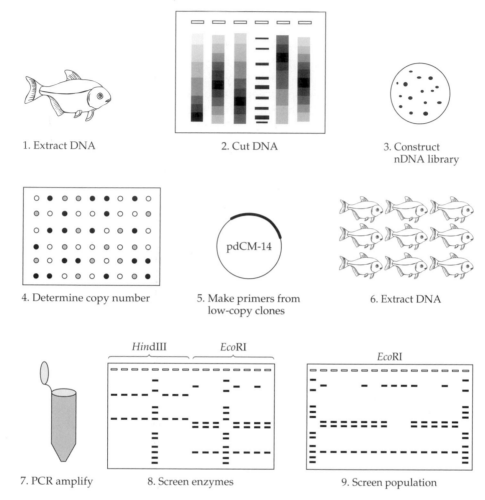

1. Extract DNA
2. Cut DNA
3. Construct nDNA library

4. Determine copy number
5. Make primers from low-copy clones
6. Extract DNA

pdCM-14

HindIII   EcoRI

EcoRI

7. PCR amplify
8. Screen enzymes
9. Screen population

Figure 3.12   General protocol for surveys of scnDNA RFLPs using the polymerase chain reaction.

(Saperstein and Nickerson 1991). The process is summarized in Figure 3.12. This PCR-based RFLP method offered some advantages over Southern blotting (Karl and Avise 1993): it requires only a small amount of template DNA; that DNA is of defined length (bounded by primers), so size differences underlying RFLPs can readily be distinguished from restriction site differences; PCR amplifies DNA in unmethylated condition, so natural DNA methylation is not a potential confounding source of variation in restriction digests; and the method bypasses any need for radioactive isotopes and autoradiography. On the negative side, great effort is entailed in constructing genomic libraries and screening them for low-copy sequences.

Restriction analyses of scnDNA are intended to reveal scorable RFLPs at individual loci. Raw data are in many respects analogous to those provided by allozymes: diploid specimens can be described as homozygous or heterozygous (Figure 3.13); the Mendelian nature of polymorphisms can be verified by pedigree studies or by agreement of genotypic frequencies with Hardy-Weinberg expectations; a population may exhibit multiple alleles at a locus; and genotypic descriptions can be accumulated across loci. In principle, two major advantages over protein electrophoresis are the nearly unlimited pool of genetic loci that might be tapped (thousands of low-copy regions exist in most genomes), and the fact that polymorphisms include silent as well as replacement substitutions. In practice, however, these methods proved much more demanding than protein electrophoresis and were used infrequently in molecular ecology and evolution, although they had considerable impact in related research areas such as mapping of disease genes and quantitative trait loci (Botstein et al. 1980; B. Martin et al. 1989; Paterson et al. 1988; Weller et al. 1988), and in breeding studies and strain verification in domestic species (Apuya et al. 1988; Beckmann et al. 1986).

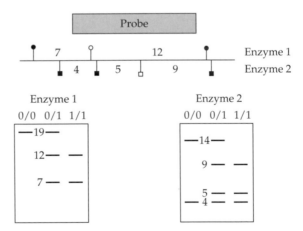

Figure 3.13   Interpretation of gel profiles for nuclear RFLP polymorphisms.
Illustrated is the method for diploid organisms, as assayed by Southern blotting using a scnDNA probe. Restriction site positions along a stretch of DNA are shown for each of two enzymes, with solid symbols indicating invariant restriction sites and open symbols indicating variable sites either present (1) or absent (0) in various chromosomes in the population. Heterozygous individuals in lanes of the gels are "0/1"; homozygotes are "0/0" and "1/1." Other numbers indicate sizes (in kb) of various restriction fragments.

Moderately repetitive gene families

When a DNA probe with homology to repetitive genomic sequences is used in Southern blotting, the probe hybridizes to all such sequences and simultaneously reveals restriction profiles at multiple members of the gene family. For example, ribosomal RNA genes in the nuclei of eukaryotic cells usually exist as tandemly repeated elements, with each repeat unit composed of a highly conserved coding sequence with a total length of about 6 kb, plus shorter and more variable non-coding spacer regions (Figure 3.14). These rDNA modules may occur at one or several chromosomal sites, with the total number of rDNA copies per genome varying from several hundred in some mammals and insects to many thousands in plants (Long and Dawid 1980).

The ready availability of probes for rRNA genes prompted many Southern blotting studies of population variation and differentiation in these genetic regions (Appels and Dvorak 1982; Arnold et al. 1991; Davis et al. 1990; Rieseberg et al. 1990a,b; Rogers et al. 1986; Saghai-Maroof et al. 1984; Schaal et al. 1987; Williams et al. 1985). These studies revealed RFLP markers that often distinguished related species and sometimes conspecific populations. Most of the genetic differences involved varying lengths of the repeat unit due to heterogeneity in the size of the spacer regions (see Figure 3.14), with additional variation occasionally reflecting restriction site changes in both the coding and spacer regions (Schaal 1985).

A central difficulty in interpreting genetic markers provided by any multi-gene family lies in understanding the degree to which concerted evolution may have homogenized the repeated DNA sequences (Arnheim 1983; Ohta 1980; Ohta and Dover 1983). From one perspective, an ideal situation would be concerted evolution that was so pronounced that all copies of a

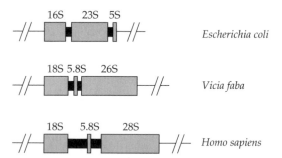

Figure 3.14   Structural features of rDNA repeat modules. Drawn to approximate scale are representative structures in a bacterium, plant, and animal. Shaded regions indicate loci encoding small (16S and 18S) and large (23S, 26S, 28S) subunits of ribosomal RNA, as well as 5S rRNA elements. Black regions indicate internal transcribed spacers, which often differ in length. (After Appels and Honeycutt 1986.)

repeat within each individual or local population were quickly homogenized, such that each specimen carried a single, unambiguous genotype. On the other hand, extreme concerted evolution of this sort would also, in effect, confine the information content in a family of sequences to that of merely one gene. Conversely, a paucity or absence of concerted evolution would mean that a gene family carries multi-locus phylogenetic information, but it would also make that information more difficult to retrieve and interpret, due ultimately to the complications of distinguishing orthology from paralogy (see Chapter 1). Empirically, Williams et al. (1987) showed that rDNA variants in *Drosophila* have a nonrandom distribution between X and Y chromosomes, suggesting in that case that concerted evolution within chromosomes is more pronounced than concerted evolution between them. In another such analysis, Arnold et al. (1988) showed that biased gene conversion had influenced distributions of rDNA sequences in a grasshopper hybrid zone. Thus, the use of nuclear rRNA genes (and other repetitive DNA families) as genetic markers in microevolutionary studies can be problematic (Schaal et al. 1991). On the other hand, Hamby and Zimmer (1992) concluded that "the most remarkable feature of rDNA is the overall sequence homogeneity among members of the gene family." Concerted evolution, plus a slow pace of sequence divergence, have in fact been crucial to the widespread use of various rDNA sequences in higher-level systematics (see Chapter 8).

Despite these potential complications, RFLP markers from rDNAs have contributed to studies of geographic population structure and patterns of introgression in hybrid zones (Arnold et al. 1987; Baker et al. 1989; Cutler et al. 1991; Learn and Schaal 1987). Southern blotting procedures have also been employed to assess levels of genetic variability in other multi-gene families (e.g., Gibbs et al. 1991). For example, the major histocompatibility complex (MHC) is a family of tightly linked loci that encodes cell surface antigens involved in immunological responses (Edwards et al. 2000). In many mammals, particular MHC genes are known to be highly polymorphic, some with scores of alleles (Hedrick et al. 1991; Hughes and Nei 1988, 1989; Klein 1986). In one early example of MHC's utility in microevolutionary studies, feline probes homologous to one class of MHC loci were employed to assess molecular variation in two cat populations (African cheetahs and Asiatic lions; Winkler et al. 1989; Yuhki and O'Brien 1990) suspected by other criteria to possess low genome-wide variability due to historical bottlenecks in population size (see Chapter 9).

In recent years, RFLP approaches applied to moderately repetitive nuclear gene families (like those applied to mtDNA, cpDNA, and scnDNA) have mostly been supplanted by more efficient methods of direct DNA sequencing.

Minisatellites and DNA fingerprinting

The genetic complexity inherent in repetitive DNA families sometimes can be turned to advantage in terms of providing individual-specific genetic markers. The term "DNA fingerprinting" usually is associated with a molecular approach introduced by Jeffreys et al. (1985a), in which Southern blot assays of hypervariable DNA regions reveal complex gel-band profiles that distinguish most or all individuals (barring monozygotic twins) within a sexually reproducing species (Bruford et al. 1992). The DNA probes originally employed by Jeffreys (1987) hybridize to conserved core sequences (10–15 bp long) scattered in numerous arrays about the human genome as part of a system of "dispersed tandem repeats," also referred to as minisatellite loci or VNTRs (variable number of tandem repeats; Figure 3.15). Each repetitive unit within an array is about 10–70 bp long. Increases and decreases in the lengths of particular arrays often result from changes in tandem repeat copy number arising from high rates of unequal crossing over during meiosis (indeed, minisatellite sequences may be genomic hotspots for recombination; Jarman and Wells 1989).

Jeffreys's original probes were isolated from a human myoglobin intron and were applied to problems in human forensics (Dodd 1985; Gill et al. 1985; Jeffreys et al. 1985b,c), but these probes soon were shown to cross-hybridize to reveal DNA profiles in other mammals (Hill 1987; Jeffreys and Morton 1987; Jeffreys et al. 1987), as well as birds (Brock and White 1991; Burke and Bruford 1987; Hanotte et al. 1992a; Meng et al. 1990), fishes (Baker et al. 1992), and even some invertebrates, such as corals and snails (Coffroth et al. 1992; Jarne et al. 1990, 1992). Probes for additional hypervariable minisatellites were then identified (such as one from M13 phage) that behaved similarly in providing complex DNA fingerprints in various vertebrate taxa (Georges et al. 1987; Longmire et al. 1990, 1992; Vassart et al. 1987), invertebrate animals (Zeh et al. 1992), plants (Rogstad et al. 1988), and microbes (Ryskov et al. 1988).

The complex gel profiles characteristic of multi-locus DNA fingerprints appear when a restriction enzyme is employed that cleaves DNA outside the tandem repeat arrays (see Figure 3.15). From each homologous chromosome position in an individual, either one or two DNA gel bands is revealed, depending on whether the specimen is homozygous or heterozygous with respect to the number of tandem repeats in that array. The presence of several such arrays scattered about the genome results in composite digestion profiles typically consisting of 20 or more scorable bands per individual in the 4–25-kb size range. In many animal populations, including humans, dozens of alleles of different lengths may segregate at each chromosomal position. The multi-allelic plus multi-locus nature of the data result in elaborate, individual-specific gel profiles that are powerful in revealing genetic identity versus non-identity. They are also useful in assessing genetic parentage because, barring spontaneous de novo mutation (Jeffreys et al. 1988b), each band in an individual's DNA fingerprint derives from either its biological mother or father (see Chapter 5).

The genetic complexity that makes multi-locus DNA fingerprints advantageous for distinguishing individuals becomes a liability in other contexts. Generally, it remains unknown which bands in a fingerprint belong to which locus (array), so whether individuals are homozygous or heterozygous at

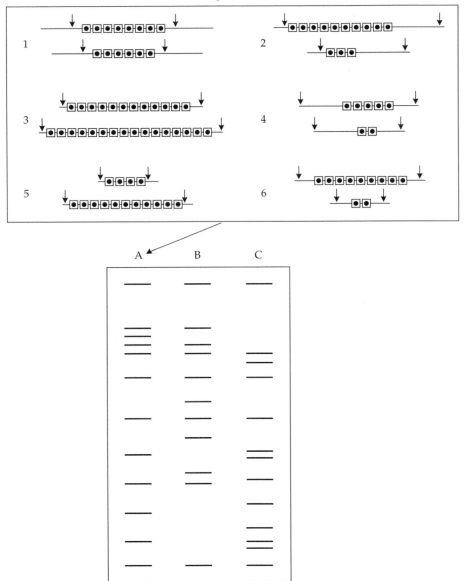

Figure 3.15 **DNA fingerprinting using VNTR loci.** Shown at the top of the figure are six genomically dispersed DNA regions (on the same or different chromosomes), each of which in a population harbors variable numbers of tandem repeat elements. Solid circles within repeat units indicate a conserved core sequence to which a probe hybridizes in a Southern blot. A restriction enzyme that cuts DNA (at positions shown by arrows, outside each repeat array) reveals a complex digestion profile for individual A on a gel autoradiograph (bottom part of the figure). Other individuals (e.g., B and C) are likely to have different DNA fingerprints when similarly assayed.

particular loci seldom is ascertained, nor can allelic or genotypic frequencies in a population be specified. These problems seriously compromise attempts to estimate genetic relatedness or other population-level parameters (such as gene flow) from DNA fingerprints' measurable attribute: the proportion of shared marker bands (Kuhnlein et al. 1990; Lynch 1988; Packer et al. 1991).

These shortcomings of multi-locus DNA fingerprinting for applications other than genetic identity and parentage assessment thus prompted development of methods to analyze minisatellite loci one at a time (Carter 2000). These methods included use of refined DNA probes and more stringent hybridization conditions in Southern blotting, as well as PCR-based methods (Horn et al. 1989), to reveal genetic variation tied to specific minisatellite arrays (Higgs et al. 1986; Jarman et al. 1986; Jeffreys et al. 1988a; Nakamura et al. 1987; Wong et al. 1986, 1987). Each VNTR locus displays just one or two bands in a DNA digestion profile, depending on whether the specimen assayed is homozygous or heterozygous for number of repeats (or other features) situated between the restriction sites. Another variant of this approach was to characterize patterns of base substitution (in addition to repeat copy number) within particular minisatellite loci (Jeffreys et al. 1990, 1991). For a brief time, minisatellite assays were the method of choice for forensic practices in humans (Balazs et al. 1989; Budowle et al. 1991; see Chapter 5) and wildlife (Burke et al. 1991; Hanotte et al. 1991, 1992b), but they too soon were largely supplanted, in this case by microsatellite assays (described below).

Polymerase Chain Reaction

Invention of the polymerase chain reaction (PCR; Mullis and Faloona 1987; Saiki et al. 1985, 1988) revolutionized not only molecular biology, but also the fields of organismal and population biology (Arnheim et al. 1990), by stimulating many powerful new approaches to genetic marker acquisition. Basically, PCR enables researchers to quickly amplify or clone, in a test tube, assayable quantities of almost any desired piece of DNA from almost any biological source. Technical descriptions of PCR are given by Birt and Baker (2000), Innis et al. (1990), Palumbi (1996a), and Palumbi et al. (1991).

The PCR technique involves three main steps (Figure 3.16): denaturation of double-stranded DNA by heating; annealing of primers to sites flanking the region to be amplified; and primer extension, in which strands complementary to the region between flanking primers are synthesized under the influence of a thermostable DNA polymerase (*Taq*). These three steps are repeated 20 or more consecutive times, all in automated thermocycler devices that are now standard apparatus in genetics laboratories. During each cycle of denature-anneal-extend, the target sequence roughly doubles in quantity, so it soon assumes overwhelming preponderance. This purified product can then be assayed by any of several molecular procedures to be described shortly (as well as by the various kinds of RFLP analyses already mentioned; Morales et al. 1993).

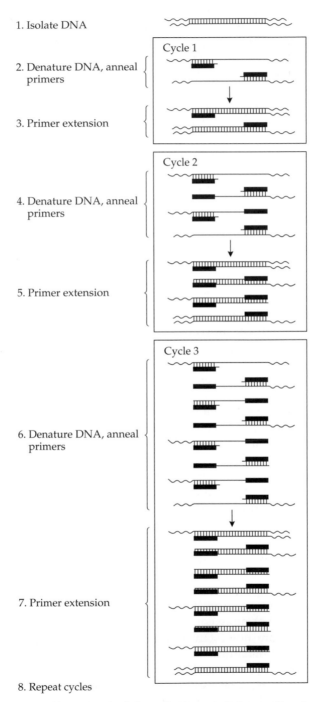

Figure 3.16   General protocol of the polymerase chain reaction for amplifying DNA. (After Oste 1988.)

The primers employed to initiate any PCR reaction typically are short sequences (ca. 20–30 nucleotides long) that exhibit high sequence similarity (especially at the 3′ end) to regions flanking the target sequence. They normally are generated in special machines that are programmed (based on a specific known DNA sequence) to synthesize particular oligonucleotides. Depending upon the primers employed, sequences amplified by PCR can be coding or non-coding DNA, and they can come from any location in nuclear or organelle genomes. Amplified sequences are normally a few hundred bp to about 1 kb in length, but "long-PCR" procedures have also been developed (Cheng et al. 1994; Cohen 1994; Li et al. 1995) to amplify much larger fragments, such as the whole 16-kb animal mtDNA genome (Nelson et al. 1996).

Some PCR primers have wide taxonomic latitude, meaning that they successfully amplify not only target DNA from the species from which they were developed, but also homologous DNA across broader taxonomic groups. Other PCR primers work well only within a target species and perhaps among its close relatives. Following the introduction of PCR, vast research effort has gone into primer identification and development. Published compilations of highly conserved primers flanking particular loci, such as various mtDNA genes in insects (Simon et al. 1994; Zhang and Hewitt 1997), fishes (Meyer 1994a; Normark et al. 1991), birds (Sorenson et al. 1999a), and even across broader taxonomic groups (Kocher et al. 1989) are available. At the other end of the spectrum are species-specific or genus-specific PCR primers, such as those typically employed to amplify microsatellite loci. In 2001, a new journal (*Molecular Ecology Notes*) was launched largely to accommodate papers reporting PCR primers and microsatellite assay conditions for hundreds of individual plant and animal species.

Another great advantage of PCR over earlier approaches is that it enables recovery of assayable quantities of DNA from even tiny bits of biological material, such as can be obtained from single feathers (Taberlet and Bouvet 1991), hairs (Morin et al. 1992; Taberlet and Bouvet 1992; Vigilant et al. 1989), dung (Fernando et al. 2003a; Kohn and Wayne 1997; Palomeres et al. 2002), dried or ethanol-preserved museum material (Higuchi et al. 1984), and even some fossils up to tens of thousands of years old (see Chapter 8). Table 3.1 describes some remarkable and even bizarre sources from which various DNA sequences have been successfully amplified and assayed.

However, PCR-based approaches are not entirely free of technical difficulties. One issue is the degree of fidelity of PCR amplification (Dunning et al. 1988; Ennis et al. 1990; Pääbo and Wilson 1988; Saiki et al. 1988). Any misincorporation of nucleotides, especially in early rounds of amplification, can result in an amplified sequence that differs at least slightly from the original template. Such low-frequency misincorporation has been observed, but its effects in most biological applications probably are negligible (Kwiatowski et al. 1991). A second potential difficulty, especially when amplifying loci from polyploids or members of multi-gene families, is PCR-mediated

TABLE 3.1 Examples of some remarkable biological sources from which DNA has been successfully amplified using PCR

Source	Biological context
Sloughed skin	Sex and individual identification in whales
Plucked or shed hair	Individual identification and parentage analysis
Saliva	Forensic identification of crime suspects
Egg shells and membranes	Forensic identification in birds
Soil or clay (bound DNA)	Detection of microbes or other remains
Subterranean groundwater	Detection of bacteria
Otoliths ("ear stones")	Population genetic structure in fishes
Old scales	Temporal population genetics in fishes
Old baleen	Temporal population genetics in whales
Dried blood, semen	Forensic identification of crime suspects
Arthropod blood meals	Identification of mosquito hosts
Extracted blood	Detection of malaria parasites in birds and lizards
Retail caviar or meat	Species identification of illegal wildlife products
Stomach contents	Identification of species consumed
Stomach contents	Paternity assessment of cannibalized offspring
Regurgitated owl pellets	Dietary analysis for studies of mammal abundance
Feces	Individual identification of excrement producer
Feces	Species identification of excrement producer
Feces	Sex (etc.) identification of excrement producer
Feces	Foods consumed by excrement producer
Carcasses	Individual identification in an endangered species
Single diploid cells	Genome analysis, genetic aberrations, cancers
Single eggs, embryos	Identification of marine organisms
Single larvae	Identification of marine organisms
Single sperm cells	Genetic mapping and other medical applications
Fossils	Phylogeography and phylogeny estimation
Fine plant roots	Genetic assignment to tree species
Pollen on arthropods	Identification of orchids pollinated by insects
Dry wood	Tree identification
Ancient wood	Tree identification

recombination among the amplification products (Cronn et al. 2002). Third, PCR reactions sometimes fail for a variety of reasons, thereby producing "null" alleles that can add complications in particular applications, such as genetic parentage assessments via microsatellite assays. However, the most serious challenges posed by PCR probably stem directly from what is also one of the technique's greatest blessings: its extreme sensitivity. Thus, sample contamination (by microbes, physical handling, or any source of contact with even minuscule amounts of foreign DNA) can sometimes result in amplification of molecules other than those intended, and this can occasionally result in serious interpretive errors (see Chapter 8).

Representative references

Baker et al. 1991; Valsecchi et al. 1998
Goossens et al. 1998; Morin et al. 2001
Allen et al. 1998
Pearce et al. 1997; Arnold et al. 2003
Alvarez et al. 1998; Porteous et al. 1997
Pedersen et al. 1996
Hutchinson et al. 1999; Ruzzante et al. 2001
E. E. Nielsen et al. 1997
Rosenbaum et al. 1997
See Chapter 5
Gokool et al. 1993
Feldman et al. 1995; Perkins et al. 1998
See Chapter 9
Jarman et al. 2002; Rosel & Kocher 2002; Scribner & Bowman 1998; Symondson 2002
DeWoody et al. 2001
Taberlet and Fumagalli 1996
Constable et al. 1995; Höss et al. 1992; Vege and McCracken 2001
Paxinos et al. 1997; Tikel et al. 1996
Reed et al. 1997; Wasser et al. 1997
Farrell et al. 2000; Hofreiter et al. 2000; Poinar et al. 1998
Banks et al. 2003; Dallas et al. 2003
Klein et al. 1999; Sun et al. 1995
Cary 1996
Evans et al. 1998
Navidi and Arnheim 1999
See Chapters 6 and 8
I. Brunner et al. 2001; Linder et al. 2000
Widmer et al. 2000
Deguilloux et al. 2002, 2003
Dumolin-Lapegue et al. 1999b

RAPDs

Any PCR reaction per se is merely prelude to some form of DNA assay. One well-known example is a methodology to reveal RAPDs (pronounced "rapids," short for "randomly amplified polymorphic DNAs"). This technique involves the use of short (ca. 10 bp) PCR primers of arbitrary sequence to amplify anonymous genomic sequences. When the resulting products are separated in suitable electrophoretic gels (Welsh et al. 1991; Williams et al. 1990), DNA-level polymorphisms (usually in primer recognition sites) can be uncovered. (Detailed laboratory methods are provided in Edwards 1998.) The RAPD approach was widely employed in population biology, especially during the 1990s (e.g., Hadrys et al. 1992; Hedrick 1992; Rieseberg 1996), but it developed a reputation for poor reproducibility in some cases (e.g., Pérez et al. 1998), and it also has the disadvantage of normally revealing only dominant markers (Figure 3.17). For these reasons, its use has generally waned, especially with the rise in popularity of microsatellite markers, which are also highly polymorphic, but provide more informative codominant markers. Nonetheless, RAPD techniques are still employed by many laboratories, as they do offer a quick and easy way of screening potential molecular markers from many loci (Ritland and Ritland 2000).

STRs (microsatellites)

PCR-based assays of "short tandem repeat" loci (STRs, or "microsatellites") have become probably the most popular and powerful of the current methods for identifying highly polymorphic Mendelian markers (Li et al. 2002; Scribner and Pearce 2000). Each microsatellite locus consists of reiterated short sequences (usually di-, tri-, or tetranucleotides) that are tandemly arrayed at a particular chromosomal location (Hamada et al. 1984), with variation in repeat copy number often underlying a profusion of distinguishable alleles (sometimes 20 or more) within a population (see Figure 2.6). Thus, a microsatellite array is reminiscent of a minisatellite array, except that each of its repeat units is much shorter. This means that alleles

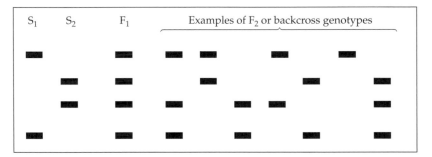

Figure 3.17 Nature of data generated by RAPD procedures. In this approach, short random primers are employed to amplify anonymous DNA segments (in this case, from four unlinked gene regions) by PCR. The presence of a band indicates successful amplification; absence indicates unsuccessful amplification, perhaps resulting from naturally occurring mutations in the primer recognition site. Thus, RAPDs in diploid organisms usually behave in dominant/recessive fashion (although cases of codominant inheritance are known; e.g., Fritsch and Rieseberg 1992). As exemplified in this figure, the RAPD approach can be especially useful in characterizing first-generation or later-generation hybrids between genetically distinct species (here, S_1 and S_2).

at STR loci can be distinguished in acrylamide (rather than agarose) gels, and without need for arbitrary binning (because even similar-sized alleles normally differ in readily detected increments of 2, 3, or 4 base pairs). Microsatellite loci were discovered in the late 1980s (Litt and Luty 1989; Tautz 1989; Weber and May 1989) and soon were shown to be characteristic features scattered abundantly throughout the nuclear genomes of most plants (Morgante et al. 1998; Nybom et al. 1992; Squirrell et al. 2003) and animals (Ellegren 1991; Fries et al. 1990; Stallings et al. 1991), including humans (Valdés et al. 1993). They also occur in cytoplasmic genomes, including animal mtDNA (Lunt et al. 1998), although the conventional term "STRs" is normally taken to imply nuclear markers.

Microsatellite or STR variants are sometimes also referred to as simple-sequence length polymorphisms (SSLPs; Schlötterer et al. 1991). Techniques for their assay are outlined in Figure 3.18. The first and most laborious step is primer development, which necessitates constructing a genomic library for the target species, screening that library for clones that contain microsatellite repeats, sequencing the inserts from those positive clones, and using the information from unique sequences flanking each repeat region to synthesize PCR primers. Once primers are available, however, large numbers of individuals can be readily screened for Mendelian genotypes at specific STR loci displaying codominant alleles (Figure 3.19). Further details of laboratory protocols and analysis methods for microsatellites are given in Ciofi et al. (1998), Goldstein and Schlötterer (1999), and Zane et al. (2002).

The discrete genotypic data provided by microsatellites are in several ways analogous to those provided by allozymes. However, population variation is typically much higher at STR loci (compare Figures 2.1 and 2.6) due to the high mutation rate of microsatellites: often about 10^{-3} or 10^{-4} per locus per gamete per generation (Primmer et al. 1996; Schug et al. 1997; Weber and Wong 1993). Indeed, de novo microsatellite mutations are occasionally uncovered in genetic parentage analyses (e.g., Jones et al. 1999a). Many, but not all, of these mutations result in alleles of the adjacent size classes, so that the mutational process tends to be imperfectly stepwise or ladderlike. Much discussion in the literature has centered on whether genetic distances based on microsatellite data should incorporate not only the frequencies of alleles, but also their size relationships to one another (e.g., Goldstein et al. 1995a; Nauta and Weising 1996; Ruzzante 1998). Another important point is that because of the rapid evolutionary pace of microsatellites and the underlying nature of their mutational processes, alleles that are identical in state (size) are not necessarily identical by descent (Angers and Bernatchez 1997; Estoup et al. 1995; Garza and Freimer 1996; Ortí et al. 1997b; van Oppen et al. 2000; see review in Estoup et al. 2002). This characteristic of microsatellites can often cause serious interpretive difficulties, especially in studies of geographic population structure (Balloux and Lugon-Moulin 2002).

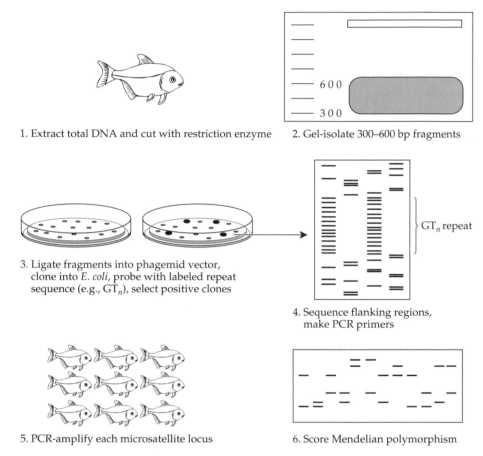

1. Extract total DNA and cut with restriction enzyme   2. Gel-isolate 300–600 bp fragments

3. Ligate fragments into phagemid vector, clone into *E. coli*, probe with labeled repeat sequence (e.g., GT_n), select positive clones

GT_n repeat

4. Sequence flanking regions, make PCR primers

5. PCR-amplify each microsatellite locus   6. Score Mendelian polymorphism

Figure 3.18   General protocol for microsatellite assays.

AFLPs

The technology for revealing "amplified fragment-length polymorphisms" (AFLPs; Vos et al. 1995) has elements of both PCR-based and RFLP-based methods. The laboratory protocols, described in Matthes et al. (1998), are rather complicated, but the basic outline is as follows. The goal is to selectively amplify a subset of restriction fragments from a complex mixture of fragments produced by digestion of genomic DNA by two restriction endonucleases, one with a 4-bp and one with a 6-bp recognition site. These fragments are linked to adapter sequences and biotinylated in such a way that subsequent PCR amplifies only a small subset of them, thus reducing what would otherwise be a hopelessly heterogeneous mix of different-sized genomic fragments to a manageable level. Even so, the resulting band

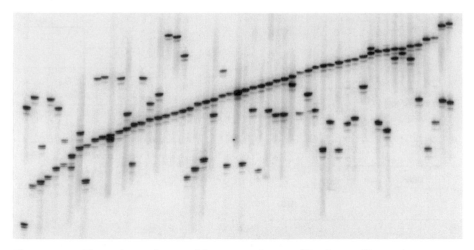

Figure 3.19   Extensive polymorphism at a microsatellite locus. Shown is an autoradiograph of a gel displaying more than 50 alleles in a local population of deer mice (unpublished data, Avise laboratory). All individuals shown are heterozygous (each displays two primary bands). Specimens were purposely arranged from left to right such that one of the two alleles in each contributes to a steplike series of consecutive-sized alleles. Actually, most microsatellite assays today involve the use of fluorescent dyes and computer gel scans to score alleles and genotypes. Several microsatellite loci can sometimes be "multiplexed" and run on the same gel, with their alleles distinguished by use of a different fluorescent dye for each.

profiles on polyacrylamide gels are complex, reminiscent of those in multilocus minisatellite DNA fingerprints. Polymorphisms are then scored as differences in the lengths of amplified fragments, which may be due to base substitutions in or near the restriction sites or to sequence insertions or deletions. Unfortunately, most of these molecular polymorphisms, like RAPDs, display genetic dominance. Nonetheless, the APLF approach has found application in various genomic and forensic analyses that require large numbers of qualitative, mostly unlinked polymorphisms (Mueller and Wolfenbarger 1999). When the dominant markers can be suitably analyzed, this technique (like RAPDs) can also find application in estimating pairwise relatedness between individuals from large numbers of loci (Hardy 2003).

SINEs

"Short interspersed elements" (SINEs) have proved to be superb phylogenetic markers for identifying monophyletic groups (clades). Typically, each SINE is a tRNA-derived retropseudogene (Nei and Kumar 2000) residing at a specific chromosomal location. Large numbers of SINEs are dispersed

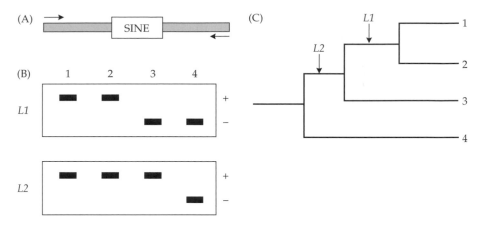

Figure 3.20   Methodology and logic underlying the use of SINEs for phylogenetic purposes. (A) A SINE flanked by unique genomic sequences from which PCR primers (represented by arrows) are generated. (B) Diagram of electrophoretic gels showing PCR products in which two different SINE loci (*L1* and *L2*) are assayed for presence (+) versus absence (–) of the expected fragment size in four different host taxa (gel lanes 1–4). (C) Phylogenetic summary corresponding to the results in B. Taxa 1, 2, and 3 apparently share a common ancestor that had the SINE *L2* inserted into its genome, and taxa 1 and 2 share a common ancestor that must have later acquired the SINE *L1* insertion. (After Shedlock and Okada 2000.)

throughout the genomes of eukaryotic organisms, but each occupied site presumably represents a single evolutionary insertion event. The trick is to design PCR primer pairs specific to the unique flanking regions of a given SINE, then repeat the process for each of many independent SINEs. This battery of primers is then employed in PCR amplification applied to genomic DNA isolated from species of interest. The PCR products are electrophoresed, and each taxon is then scored for presence versus absence of each SINE element (Figure 3.20).

Taxa that prove to share even one or a few independent SINEs almost certainly belong to a clade because the presence of a given SINE probably signals a single evolutionary acquisition. SINEs are stable once inserted, so SINE absence at a given chromosomal site is presumably the original ancestral condition (see Figure 3.20). Accordingly, a strong case has been made that SINEs are among the most powerful of all molecular markers for phylogenetics (Shedlock and Okada 2000). Although the SINE approach has been employed in relatively few laboratories, it has proved to be extremely useful for phylogenetic reconstruction at diverse evolutionary time frames in several taxonomic groups, such as fishes (Hamada et al. 1998; Murata et al. 1996; Takahashi et al. 1998) and mammals (Nikaido et al. 1999, 2003; Shimamura et al. 1997).

SSCPs

Especially in population genetic applications, the expense and effort of fully sequencing PCR products (see next section) for large numbers of individuals is sometimes prohibitive, but shortcuts are available. One such approach, known as "single-strand conformational polymorphism" (SSCP; Orita et al. 1989), takes advantage of the fact that single-stranded (denatured) DNA molecules a few hundred bp in length often assume different conformations even when differing by as little as one base pair. These distinctive conformations can be detected by electrophoresing PCR-amplified molecules through neutral polyacrylamide gels (Sunnucks et al. 2000). The SSCP approach is usually employed in conjunction with spot sequencing of representative PCR products so that conformational bands on SSCP gels can be related to particular full-length DNA sequences. The technique can also be used to separate haplotypes as a prelude to their further characterization by other techniques, such as by direct sequencing (Ortí et al. 1997a). Another approach with similar potential applications is DGGE, or denaturing gradient gel electrophoresis (Lessa 1993; Myers et al. 1986). Lessa and Applebaum (1993) review and compare these and other related shortcut screening methods for detecting allelic variation in DNA sequences.

SNPs

Another commonly used acronym refers not so much to any particular assay approach, but rather to "single nucleotide polymorphisms," regardless of how they are revealed (Brookes 1999). Several of the techniques described above (such as various RFLP and AFLP analyses) in effect often reveal SNPs, but not at the detailed molecular level, and not necessarily in distinction from small sequence insertions or deletions (indels). True SNPs are specific base-pair variants, and they are abundant in most genomes (approximately 1.5 million are known in humans, for example; Kendal 2003).

In principle, and sometimes in practice, SNPs provide a wellspring of molecular markers well suited to genomic analyses such as studies of linkage and linkage disequilibrium (Stephens et al. 2001a). In recent years, several laboratory approaches have been developed explicitly to screen large numbers of SNPs in well-characterized model genomes, such as those of yeast (Winzeler et al. 1998) and humans (See et al. 2000; D.G. Wang et al. 1998). Some of these methods (Gilles et al. 1999; Syvänen 1999) incorporate microarray or microchip technologies (microchips are miniaturized holding platforms for nucleic acids), which have found broad application especially in meta-analyses of gene expression patterns. Because methods for massive screening of SNPs have not yet been widely developed and employed in ecological and evolutionary studies, they will not be elaborated upon here, but these technologies may be adapted for some such purposes in the future.

HAPSTRs and SNPSTRs

A new approach, still mostly in the developmental phase, is to utilize a combination of molecular techniques that weds the strengths of different kinds of molecular markers. For example, STR loci are typically highly polymorphic, but also suffer from extensive parallel and convergent evolution due to recurrent mutations to finite numbers of length-defined allelic states. By contrast, unique sequences that flank these microsatellites might provide a clearer genealogical signal, but they suffer the disadvantage (especially for microevolutionary reconstructions) of slow evolutionary rates. In an attempt to combine the advantages of clarity in phylogenetic signal and high polymorphism, Hey et al. (2004) introduced a laboratory protocol that involves separating STR alleles and their long flanking sequences by size, cutting these alleles from electrophoretic gels, and then sequencing flanking DNA to reveal individual haplotypes for genealogical appraisal. The approach was nicknamed HAPSTR (for haplotypes at STR regions). A similar protocol that combines SNP analysis at short autosomal sequences with STR assays was given the acronym SNPSTR (Mountain et al. 2002).

DNA sequencing

Two DNA sequencing methods have been available since the mid-1970s (Figure 3.21). One approach, introduced by Maxam and Gilbert (1977, 1980), relies on chemical cleavage reactions specific to individual nucleotides (A, T, C, or G). Ends of a targeted stretch of DNA are radioactively labeled, and the DNA is divided into four subsamples, which then are treated with different chemical reagents that cleave at base-specific positions. For example, one subsample is treated with dimethyl sulfate and piperidine, which results in DNA cleavage only at G positions. Reactions are carried out under conditions such that only a small, random fraction of sites is cleaved in any molecule, so that the composite digestion contains a collection of molecular fragments of varying lengths, each terminated at a G position. The fragments are then separated electrophoretically in a polyacrylamide gel and visualized by autoradiography. Parallel reactions specific for the other three bases are likewise carried out, and fragments are separated on adjacent lanes of a gel. Then, DNA sequence is read directly from ladderlike bands in the autoradiograph.

A different sequencing procedure, introduced at the same time by Sanger et al. (1977), is the forerunner of techniques employed today. This method relies on controlled interruption of in vitro DNA replication (Figure 3.21). Double-stranded DNA is denatured to single strands, and a short DNA segment (the primer) known to be complementary to a sequence in targeted DNA is annealed to that target sample. This primer/template mixture is divided into four subsamples, each of which is subjected to a primer extension reaction catalyzed by DNA polymerase. Each reaction mixture

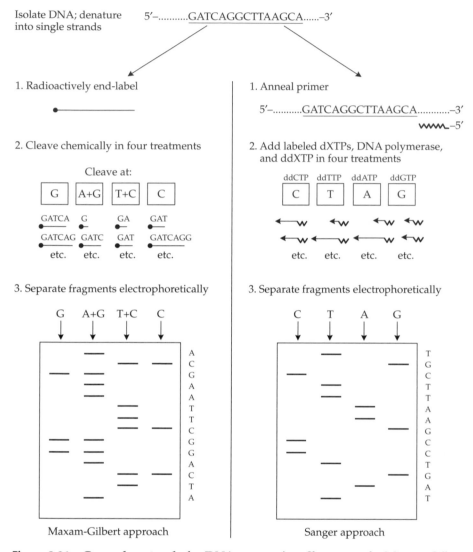

Isolate DNA; denature into single strands 5′–...........GATCAGGCTTAAGCA.......–3′

Maxam-Gilbert approach

1. Radioactively end-label

2. Cleave chemically in four treatments

Cleave at:

| G | A+G | T+C | C |

GATCA G GA GAT

GATCAG GATC GAT GATCAGG

etc. etc. etc. etc.

3. Separate fragments electrophoretically

G A+G T+C C

A C G A A T T C G G A C T A

Sanger approach

1. Anneal primer

5′–...........GATCAGGCTTAAGCA............–3′
                                      ⌁⌁⌁–5′

2. Add labeled dXTPs, DNA polymerase, and ddXTP in four treatments

ddCTP ddTTP ddATP ddGTP

| C | T | A | G |

etc. etc. etc. etc.

3. Separate fragments electrophoretically

C T A G

T G C T T A A G C C T G A T

Figure 3.21 General protocols for DNA sequencing. Shown are the Maxam-Gilbert (left panel) and Sanger (right panel) approaches. (After Hillis et al. 1990.)

contains the four deoxynucleotides (dA, dC, dG, and dT), plus a single dideoxynucleotide (ddN), which is a nucleotide that lacks the 3′ OH group present in deoxynucleotides. The newly synthesized DNA strand is made radioactive, either by labeling the end of the primer or by incorporating a labeled deoxynucleotide during synthesis. DNA sequence extension occurs by attachment of nucleotides to a free 3′ OH, so that wherever ddNTP has been incorporated into the growing strand, further extension is arrested. The polymerase reaction is carried out under conditions such that incorporation of ddNTPs is rare and random. Thus, different DNA molecules in a subsample achieve varying lengths before termination at a particular base. As under the Maxam-Gilbert approach, fragments from the four subsamples are separated electrophoretically in a polyacrylamide gel and visualized by autoradiography, and the DNA sequence is read directly.

Before PCR was invented, genes normally had to be laboriously cloned through biological vectors (see Figure 3.11) to make them suitable for direct sequencing. Otherwise the gene would merely be part of a complex mixture of many sorts of DNA sequences isolated from a given tissue, and its sequence would have been at insufficient concentration to prime the sequencing reactions. With PCR, both of these interrelated difficulties are overcome in one simple, straightforward, and fast set of DNA cycling conditions. Furthermore, amplification primers for PCR can be employed as primers in the sequencing reactions, allowing PCR and DNA sequencing to be directly coupled (methodological details are described in Hillis et al. 1996; Volckaert et al. 1998). PCR and sequencing procedures have become increasingly automated and are now carried out routinely with PCR thermocyclers coupled to sequencing apparatus. The output typically consists of computer printouts of DNA sequences such as that shown in Figure 3.22.

PCR-mediated DNA sequencing has made some (but certainly not all) earlier methods for generating molecular markers, especially for phylogenetic purposes, nearly obsolete (e.g., Hillis et al. 1996; Miyamoto and Cracraft 1991). In recent years, DNA sequence information has increased explosively, such that by the beginning of 2003 nearly 25 million sequences, representing 115,000 taxa, already had been deposited in GenBank (a standard computer repository for such information).

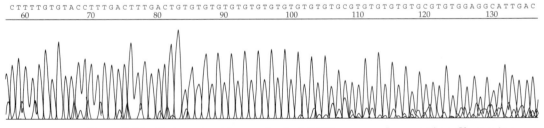

Figure 3.22 Data output from a typical DNA sequencing reaction. Shown is a computer printout in which successive peaks in the graph (normally color-coded) describe bases (labeled along the top of the graph) at successive nucleotide positions in one of the DNA strands. In this example, the sequence from positions 83 to 118 is a microsatellite region (composed mostly of tandem copies of a GT repeat) flanked by unique sequences that could serve as a basis for developing microsatellite PCR primers.

Categorical Breakdowns of Molecular Methods

In a sense, any assay method that stops short of obtaining DNA sequence can be thought of as providing an indirect and incomplete picture of genetic information at the loci screened. Indeed, nucleotide sequence data permit recovery of genetic information at less detailed levels (amino acid sequences, RFLP maps, etc.), whereas the converse is clearly not true (Box 3.2). This is not to imply that DNA sequencing is always the preferred method for obtaining molecular markers suitable for a given research problem in molecular ecology and evolution. For logistic reasons, DNA sequences usually are gathered from only one or a few genes in a given study and from relatively small numbers of individuals. Thus, sequence data provide high-resolution pictures of molecular differences (transitions versus transversions, synonymous versus non-synonymous substitutions, nucleotide substitutions versus insertions or deletions, changes in coding versus non-coding regions, etc.), but typically at the expense of sacrificing genetic information from a broad base of loci and large numbers of individuals. Furthermore, because single nucleotide positions can assume only four alternative states (A, T, C, or G), they are individually far more subject to phylogenetic noise (homoplasy; see below) than are some other kinds of molecular markers (such as SINE insertions). Researchers must weigh these and additional considerations when contemplating biological applications for molecular markers (Zhang and Hewitt 2003).

BOX 3.2 The Nature of Nucleotide Sequence Data

The tables shown on the next two pages compare DNA sequences from the mitochondrial cytochrome *b* gene in 14 samples (A–N) from several marine turtle species. The full data set (Bowen et al. 1993a) involved more than 500 nucleotide positions per sample, but only 63 positions are shown here. Dashes indicate a state identical to that of the reference sample "a."

Any of these coded data sets could be analyzed phylogenetically by appropriate computer programs. Note that when data are coded as purines versus pyrimidines (2), only transversions are counted in the resulting phylogenetic estimates; and when data are coded as amino acid sequences (3), only replacement substitutions are counted. These latter two treatments exemplify a trade-off common to most sequencing studies: Although they weight heavily for mutational events that are rare, and thus are less likely to show homoplasy over short evolutionary time, much information of potential phylogenetic significance (especially at lower sequence divergence levels) is lost by neglecting silent transitions.

Partial DNA and protein sequence data for mitochondrial cytochrome *b* in 14 marine turtle samples [a]

(1) Nucleotide sequences

a	ACC	GGA	ATC	TTC	TTG	GCA	ATA	CAC	TAT	TCA
b	- - -	- - -	- - -	- - -	C - A	- - -	- - -	- - T	- - C	- - -
c	- - -	- - -	- - -	- - -	C - A	- - -	- - -	- - T	- - C	- - -
d	- - T	- - -	- - -	- - -	C - A	- - -	- - -	- - T	- - C	- - -
e	- - T	- - -	- - -	- - -	C - A	- - -	- - -	- - T	- - C	- - -
f	- - T	- - -	G - -	- - -	C - A	- - -	- - -	- - T	- - C	- - -
g	- - T	- - -	- - -	- - -	C - A	- - -	- - -	- - T	- - C	- - -
h	- - -	- - -	- - -	- - -	C - A	- - -	- - -	- - -	- - C	- - -
i	- - -	- - -	- - -	- - T	C - A	- - -	- - -	- - -	- - C	- - -
j	- - -	- - -	- - -	- - T	C - A	- - -	- - -	- - -	- - C	- - -
k	- - -	- - -	- - -	- - T	C - A	- - -	- - -	- - -	- - C	- - -
l	- - -	- - -	- - -	- - -	C - A	- - -	- - -	- - -	- - C	- - -
m	- - -	- - -	- - -	- - -	- - A	- - -	- - -	- - -	- - -	- - -
n	- - -	- - -	- - -	- - -	- - A	- - -	- - -	- - -	- - -	- - -

(2) Same data coded as purines ("0") versus pyrimidines ("1")

a	011	000	011	111	110	010	010	101	101	110
b	- - -	- - -	- - -	- - -	- - -	- - -	- - -	- - -	- - -	- - -
c	- - -	- - -	- - -	- - -	- - -	- - -	- - -	- - -	- - -	- - -
d	- - -	- - -	- - -	- - -	- - -	- - -	- - -	- - -	- - -	- - -
e	- - -	- - -	- - -	- - -	- - -	- - -	- - -	- - -	- - -	- - -
f	- - -	- - -	- - -	- - -	- - -	- - -	- - -	- - -	- - -	- - -
g	- - -	- - -	- - -	- - -	- - -	- - -	- - -	- - -	- - -	- - -
h	- - -	- - -	- - -	- - -	- - -	- - -	- - -	- - -	- - -	- - -
i	- - -	- - -	- - -	- - -	- - -	- - -	- - -	- - -	- - -	- - -
j	- - -	- - -	- - -	- - -	- - -	- - -	- - -	- - -	- - -	- - -
k	- - -	- - -	- - -	- - -	- - -	- - -	- - -	- - -	- - -	- - -
l	- - -	- - -	- - -	- - -	- - -	- - -	- - -	- - -	- - -	- - -
m	- - -	- - -	- - -	- - -	- - -	- - -	- - -	- - -	- - -	- - -
n	- - -	- - -	- - -	- - -	- - -	- - -	- - -	- - -	- - -	- - -

(3) Same data translated into amino acid sequences (by reference to the mitochondrial genetic code)

a	thr	gly	ile	phe	leu	ala	met	his	tyr	ser
b	—	—	—	—	—	—	—	—	—	—
c	—	—	—	—	—	—	—	—	—	—
d	—	—	—	—	—	—	—	—	—	—
e	—	—	—	—	—	—	—	—	—	—
f	—	—	val	—	—	—	—	—	—	—
g	—	—	—	—	—	—	—	—	—	—
h	—	—	—	—	—	—	—	—	—	—
i	—	—	—	—	—	—	—	—	—	—
j	—	—	—	—	—	—	—	—	—	—
k	—	—	—	—	—	—	—	—	—	—
l	—	—	—	—	—	—	—	—	—	—
m	—	—	—	—	—	—	—	—	—	—
n	—	—	—	—	—	—	—	—	—	—

Source: After Bowen et al. 1993a.
[a] Dashes represent a character state identical to that of the reference sample "a."

CCA	GAT	ACT	TCC	CTG	GCA	TTC	TCA	TCA	ATC	ATC
- - -	- - C	- T C	- - -	A - A	- - C	- - T	- - -	- - -	- - T	- C -
- - -	- - C	- T C	- - -	A - A	- - C	- - T	- - -	- - -	- - T	- C -
- - -	- - -	- T C	- - -	A - A	- - C	- - T	- - -	- - -	- - -	T C -
- - -	- - -	- T C	- - -	A - A	- - C	- - T	- - -	- - -	- - -	T C -
- - -	- - -	- T C	- - -	A - A	- - C	- - T	- - -	- - -	- - -	T C -
- - -	- - -	- T C	- - T	A - A	- - C	- - -	- - -	- - -	- - -	- C -
- - -	- - C	- T C	- - -	A - A	- - T	- - T	- - -	- - -	- - T	G C T
- - -	- - C	- T C	- - -	A - A	- - T	- - T	- - -	- - -	- - T	G C T
- - -	- - C	- T C	- - -	A - A	- - T	- - T	- - -	- - -	- - T	G C T
- - -	- - C	- T C	- - -	A - A	- - T	- - T	- - -	- - -	- - T	G C T
- - -	- - -	- T C	- - -	A - A	- - T	- - -	- - -	- - -	- - -	G C -
- - -	- - C	- T -	- - -	- - C	- - C	- - -	- - -	- - -	G - T	G C T
- - -	- - C	- T -	- - -	- - C	- - C	- - -	- - -	- - -	G - T	G C T

110	001	011	111	110	010	111	110	110	011	011
- - -	- - -	- - -	- - -	0 - -	- - 1	- - -	- - -	- - -	- - -	- - -
- - -	- - -	- - -	- - -	0 - -	- - 1	- - -	- - -	- - -	- - -	- - -
- - -	- - -	- - -	- - -	0 - -	- - 1	- - -	- - -	- - -	- - -	1 - -
- - -	- - -	- - -	- - -	0 - -	- - 1	- - -	- - -	- - -	- - -	1 - -
- - -	- - -	- - -	- - -	0 - -	- - 1	- - -	- - -	- - -	- - -	1 - -
- - -	- - -	- - -	- - -	0 - -	- - 1	- - -	- - -	- - -	- - -	- - -
- - -	- - -	- - -	- - -	0 - -	- - 1	- - -	- - -	- - -	- - -	- - -
- - -	- - -	- - -	- - -	0 - -	- - 1	- - -	- - -	- - -	- - -	- - -
- - -	- - -	- - -	- - -	0 - -	- - 1	- - -	- - -	- - -	- - -	- - -
- - -	- - -	- - -	- - -	0 - -	- - 1	- - -	- - -	- - -	- - -	- - -
- - -	- - -	- - -	- - -	0 - -	- - 1	- - -	- - -	- - -	- - -	- - -
- - -	- - -	- - -	- - -	- - 1	- - 1	- - -	- - -	- - -	- - -	- - -
- - -	- - -	- - -	- - -	- - 1	- - 1	- - -	- - -	- - -	- - -	- - -

pro	asp	thr	ser	leu	ala	phe	ser	ser	ile	ile
—	—	ile	—	met	—	—	—	—	—	thr
—	—	ile	—	met	—	—	—	—	—	thr
—	—	ile	—	met	—	—	—	—	—	ser
—	—	ile	—	met	—	—	—	—	—	ser
—	—	ile	—	met	—	—	—	—	—	ser
—	—	ile	—	met	—	—	—	—	—	thr
—	—	ile	—	met	—	—	—	—	—	ala
—	—	ile	—	met	—	—	—	—	—	ala
—	—	ile	—	met	—	—	—	—	—	ala
—	—	ile	—	met	—	—	—	—	—	ala
—	—	ile	—	met	—	—	—	—	—	ala
—	—	ile	—	—	—	—	—	—	val	ala
—	—	ile	—	—	—	—	—	—	val	ala

One initial way to organize our thinking about alternative approaches for acquiring genetic markers is to subdivide the diverse types of molecular data into broad categories for which shared philosophical approaches to analysis and interpretation may pertain. The following are some of these partitions (Figure 3.23).

Protein versus DNA information

An obvious consideration is whether a molecular technique reveals variation at the level of proteins or at the level of DNA. Methods in the former category primarily unmask genetic changes in coding regions that alter amino acid sequences. By contrast, the various nucleic acid techniques provide access to a far greater panoply of architectural changes in both coding and non-coding regions of the genome. Thus, informative markers can come from synonymous as well as non-synonymous (amino acid-altering) nucleotide substitutions in protein-coding sequences, genetic changes in introns and gene-flanking regions, additions and deletions of genetic material, sequence rearrangements, and other such DNA-level features.

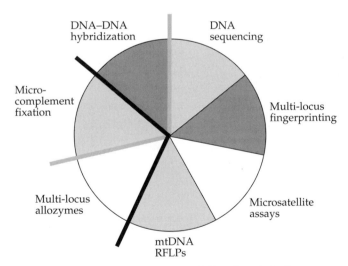

Figure 3.23 Alternative conceptual ways to "slice the pie" of representative classes of molecular genetic data. The heavy black line separates protein assays from those dealing directly with DNA. The heavy gray line divides the methods according to whether raw data consist of qualitative character states or distance values only. Lightly shaded slices of the pie indicate techniques that normally supply information from only one gene (linkage group) at a time, as opposed to the remaining methods, which usually access genetic data from multiple loci simultaneously (darkly shaded slices) or cumulatively (open slices).

Discrete versus distance data

Some molecular techniques, notably protein immunology (via MCF) and DNA–DNA hybridization, provide raw information solely in the form of numerical or quantitative distance estimates between taxa (Springer and Krajewski 1989). Other molecular approaches provide raw data in various forms of qualitative character states (Table 3.2), such as electromorph alleles, restriction sites or fragments, PCR-amplified gel bands, or nucleotide sequences. Data from such discrete characters can be converted to quantitative estimates of genetic distance, if so desired, but the converse is not true; that is, discrete character states cannot be recovered from distance data alone. This distinction between qualitative and distance data is important for two reasons. First, many biological applications, such as forensic identification, parentage assessment, kinship appraisal, gene flow estimation, and characterization of hybrids, require qualitative genetic markers, whereas other applications, such as phylogeny estimation, can employ either discrete or distance data. Second, several phylogenetic tree-building algorithms require discrete data, whereas others utilize matrices of genetic distances among taxa.

The concept of genetic distance is fundamental to molecular systematics. A genetic distance between two sequences, individuals, or taxa is a quantitative estimate of how divergent they are genetically. Units of distance depend on the nature of the molecular information used. For example,

TABLE 3.2   Comparisons among several molecular techniques for generating qualitative markers at particular loci in a population genetic context

Criterion	AFLP	RAPD	STR	nRFLP	mtRFLP	Allozymes
Number of loci typically assayed	Many	Many	Several	Few	"One"	Many
Many alleles often identifiable per locus?	No	No	Yes	Yes	Yes	Yes
Replicability of assays	High	Variable	High	High	High	High
Resolution of genetic differences	High	Moderate	High	High	High	Moderate
Nature of markers (most often)	Dominant	Dominant	Codominant	Codominant	Maternal	Codominant
Ease of use and development	Moderate	Easy	Difficult	Difficult	Moderate	Easy
Laboratory development time	Short	Short	Long	Long	Short	Short

Note: For similar but not always identical appraisals, see also Hillis et al. 1996; Karp and Edwards 1997; Mueller and Wolfenbarger 1999.

Nei's (1972) D for protein electrophoretic data is interpreted as the net number of codon substitutions per locus that have accumulated between populations since their separation, and its values can be either corrected or uncorrected for presumed multiple substitutions ("hits") at particular amino acid sites. An analogous measure for DNA restriction site or sequence data is p, the estimated number of base substitutions per nucleotide site (or percent sequence divergence if uncorrected for multiple hits). Genetic distance values (ΔT_m) from DNA–DNA hybridization have no immediate molecular interpretation beyond measured difference in thermal stability per se, although independent information typically permits calibration of ΔT_m to magnitude of sequence divergence. Similarly, immunological distance (ID) values from micro-complement fixation merely describe antigen–antibody cross-reactivity (although with additional empirical information they can be calibrated to numbers of amino acid substitutions). Much debate has centered on which particular statistical estimators of genetic distance are most appropriate for various classes of molecular information (e.g., Kalinowski 2002; Tomiuk and Loeschcke 2003). Definitions of some standard distance measures for protein and DNA data are summarized in Box 3.3.

The converse of genetic distance is genetic similarity (or the degree of genetic "identity"; Tomiuk et al. 1998). Thus, when genetic distance is low, genetic similarity is high, and vice versa. In the early literature of molecular evolution, it was customary to refer to "percent homology" between DNA sequences or other molecular characters under comparison. This practice is now discouraged. The word "homology" properly refers to organismal features (such as particular genes) that trace to a shared ancestral condition, so that sequences are either homologous or they are not. (Complications can arise, however, when a sequence includes regions of both homologous and non-homologous origin.) In any event, truly homologous sequences among an array of organisms can exhibit a wide range of genetic similarities, the values depending in large part on how long ago the extant taxa separated from common ancestors.

Discrete character data are usually presented as a matrix that assigns a character state to each taxon for each character (e.g., Boxes 3.1 and 3.2). In an allozyme survey, for example, the gene for LDH could be considered a character, with its different states being the observed electromorphs. In a DNA sequence, a character might be a particular nucleotide position, with possible states A, T, C, or G. Thus, allozyme loci and nucleotide sites are examples of multi-state characters that can display three or more variable conditions. Such characters may also be defined at a more inclusive level, examples being a gene sequence with many alleles, or even an entire mtDNA genome that may exhibit a whole collection of different haplotypes in a given population. Binary characters, by contrast, are those described in such a way that they can assume only two states, such as presence versus absence of a RFLP restriction site at a particular map location, or purine versus pyrimidine at a given nucleotide position.

Box 3.3 Examples of Genetic Distance Statistics

Allozymes

For protein-electrophoretic data, the two distance measures that have been employed most commonly are Rogers's distance (Rogers 1972) and Nei's standard genetic distance (Nei 1972, 1978). Several other such distance estimates for protein-electrophoretic data also have been proposed (see Nei 1987), but all tend to be highly correlated (although they may differ in absolute magnitude, especially at larger values).

Rogers's distance. For a given locus, let x_i and y_i be frequencies of the ith allele in populations X and Y, respectively. Rogers's D is then defined as

$$D = [0.5 \, \Sigma \, (x_i - y_i)^2]^{0.5} \tag{3.1}$$

where the summation is over all alleles. When data from more than one gene are considered, the arithmetic mean of such values across loci becomes the overall genetic distance. Rogers's D can take values between zero and one. Rogers's genetic similarity (the converse of distance) is $S = 1 - D$.

Nei's standard genetic distance. At any locus, Nei's genetic "identity" (similarity) is defined as

$$I = \frac{\Sigma x_i y_i}{(\ \Sigma x_i^2 \ \Sigma y_i^2)^{0.5}} \tag{3.2}$$

and for multiple loci, the composite identity or similarity is

$$I = \frac{J_{xy}}{(J_x J_y)^{0.5}} \tag{3.3}$$

where J_x, J_y, and J_{xy} are arithmetic means across loci of Σx_i^2, Σy_i^2, and $\Sigma x_i y_i$, respectively. Nei's I can assume values between zero and one. Standard genetic distance is calculated as

$$D = -\ln I \tag{3.4}$$

Nei's D can range from zero to infinity, and its values are interpreted as mean numbers of codon substitutions per locus, corrected for multiple hits.

DNA restriction fragments

Upholt (1977) was the first to derive a relationship between the proportion of fragments shared in mtDNA digestion profiles and an estimate of nucleotide sequence divergence. Let N_x and N_y be the number of restriction fragments observed in sequences X and Y, and N_{xy} be the number of fragments shared by X and Y. The overall proportion of shared fragments is calculated as

$$F = \frac{2N_{xy}}{(N_x + N_y)} \tag{3.5}$$

Then, an estimate of the number of base substitutions per nucleotide (or, approximately, the percentage of nucleotides substituted) is given by

$$p = 1 - \{0.5[-F + (F^2 + 8F)^{0.5}]\}^{1/r} \tag{3.6}$$

where r is the number of base pairs in the enzymes' recognition sites. Values of p are computed separately for enzymes recognizing four-, five-, and six-base sequences, and the final distance value is a weighted average of these estimates (weighted by total numbers of fragments produced by these respective enzyme classes). Using a different approach, Nei and Li (1979) derived a relationship between F and p essentially identical to that of Upholt.

DNA restriction sites

Let N_x and N_y be the number of restriction sites observed in sequences X and Y, and N_{xy} be the number of sites shared by X and Y. The proportion of sites shared is

$$S = \frac{2N_{xy}}{(N_x + N_y)} \tag{3.7}$$

and the number of base substitutions per nucleotide is then estimated by either

$$p = -\ln S / r \tag{3.8}$$

or

$$p = -(3/2) \ln [(4S^{1/(2r)} - 1)/3] \tag{3.9}$$

Equation 3.8 treats original restriction sites restored by back mutations as new sites, whereas equation 3.9 considers reverted sites as identical to the originals. Values of p again must be calculated separately for enzymes cleaving at four-, five-, and six-base recognition sites and the overall distance computed as a weighted mean.

Nucleotide sequences

Only the simplest case will be considered, in which sequences of same length can be aligned without ambiguity. Let z_d be the number of nucleotides that differ between two sequences, and z_t be the total number of nucleotides compared. Percent sequence difference is then

$$p = z_d / z_t \tag{3.10}$$

For sequences exhibiting little divergence, p is a close approximation to the accumulated number of nucleotide substitutions per site because no correction is needed for multiple substitutions at a site. When sequence divergences are larger, corrections for multiple hits are more important. One simple correction was provided by Jukes and Cantor (1969):

$$D = -0.75 \ln (1 - 4/3p) \qquad (3.11)$$

which assumes that substitutions at any nucleotide position occur with equal probability to any other nucleotide. Note that the maximum expected value of D under this equation is 0.75. Other corrections relax these assumptions. For example, the "two-parameter" model of Kimura (1980) does so by accommodating different probabilities for transitions versus transversions:

$$D = 0.5 \ln [1/(1 - 2T - V)] + 0.25 \ln [1/(1 - 2V)] \qquad (3.12)$$

where T and V are the observed proportions of transitions and transversions, respectively. Many additional considerations in deriving and interpreting distance estimates for DNA sequences (and other molecular data) are detailed in Swofford et al. (1996).

Note that unlike allozyme genetic distances that are based on population allele frequencies and provide distances between populations or species, the sequence divergence estimates just described apply to separations between particular genes or alleles. If the sequences come from haploid individuals (as is effectively true for uniparentally inherited cytoplasmic genomes), calculated values can also be interpreted as between-individual distances (in that case, typically with respect to matrilines). When many such sequences within a population are assayed, mean genetic distance (or nucleotide diversity; see Box 2.1) is then estimated by

$$\text{mean } p = \Sigma f_i f_j \, p_{ij} \qquad (3.13)$$

where f_i and f_j are frequencies of the ith and jth sequences in the population sample, and p_{ij} is the estimated sequence divergence between the ith and jth sequences (Nei 1987). Net sequence divergence (p_n) between populations can then be estimated by correcting for within-population polymorphism:

$$p_n = p_{xy} - 0.5(p_x + p_y) \qquad (3.14)$$

where p_{xy} is the mean pairwise genetic distance between individuals in population X versus population Y, and p_x and p_y are mean distances among individuals within these populations. This correction assumes that sequence diversity within extant populations is similar to the magnitude of sequence variation present in the ancestral population from which they diverged (which may not always be true).

Many computer programs are available to estimate genetic distances from sequence data, a fine example being MEGA (molecular evolutionary genetic analysis) software, recent versions of which include an expanded repertoire of distance estimation options that account for heterogeneity of substitution patterns when correcting for multiple hits (Kumar et al. 2000).

Multi-state characters may be unordered or ordered. Electromorphs at an allozyme locus are examples of unordered character states because their evolutionary interrelationships cannot be deduced directly from their observed electrophoretic mobilities. Similarly, alternative states at a given nucleotide position normally are considered unordered because there is no a priori reason to assume a particular evolutionary pathway for interconversion among A, T, C, and G (although transitions in many cases may be more frequent than transversions, for example). On the other hand, mtDNA haplotypes usually occur as ordered multi-state characters because their probable evolutionary transformations can be deduced by reasonable criteria such as parsimony (see Figures 3.9 and 3.10). However, phylogenies estimated from such composite characters are evolutionary inferences (hypotheses) ultimately derived from information accumulated across lower-level character-state descriptions (restriction sites or nucleotide positions, in these cases). Thus, character-state matrices for most computer-based phylogenetic algorithms consist of binary or multi-state data coded at these more fundamental levels (see Boxes 3.1 and 3.2).

Polarized characters are those for which ancestral and descendant states have been determined. Thus, polarity refers to the *direction* of character-state evolution and goes beyond the concept of character order, which can be described even for non-polarized states. Properly rooting a phylogenetic tree is important in helping to establish character-state polarities.

Detached versus connectable information

Some types of molecular data can be readily connected across studies, others less so, or not at all. DNA sequences are good examples of connectable data. Once nucleotide sequences are available for any gene or species, newly obtained sequences can be compared against the originals without need to repeat earlier assays. Other kinds of information are impossible to link directly across studies. For example, a ΔT_m value (from DNA–DNA hybridization) between genomes A and B is of no immediate service in assessing their relationships to genomes C and D, for which another ΔT_m value might be available.

The distinction between connectable and detached data is not the same as that between discrete versus distance information. For example, protein electrophoresis provides qualitative genotypic data, but the electromorphs themselves are distinguished by gel mobilities relative to one another. Thus, it is difficult to compare particular electromorph genotypes reported in one study with those of another (unless shared standards have been employed in both). Another point is that data comparability does not necessarily imply that the phylogenetic analyses it permits will be easy. Serious computational challenges arise in describing the vast combinational properties of connectable data. DNA sequencing assays, for example, have become so prolific that methods of data acquisition sometimes outstrip current systems of data management (Clegg and Zurawski 1992). This situation has given an urgency to the development of faster computer algorithms in comparative genomics and phylogenetics.

Single-locus versus multi-locus data

As normally applied, some molecular techniques (such as immunological methods and DNA sequencing) entail data acquisition from individual loci, whereas others (such as DNA–DNA hybridization and multi-locus finger-printing) inherently access genetic information from multiple independent gene regions. This distinction is important because the amount and nature of genetic information influences the interpretations that can be drawn from data. An inherent advantage of DNA–DNA hybridization is that the method encapsulates information from multitudinous sequences. Similarly, the power of multi-locus DNA fingerprinting in forensic applications stems in large part from independent assortment among the dispersed polymorphic arrays from which the method captures information.

For most applications of genetic markers, the number of functional genes assayed is less important than the number of linkage groups represented, which influences how many independent bits of phylogenetic information are revealed. For example, animal mtDNA is usually composed of 37 functional genes, but all of these are transmitted as a non-recombining unit primarily through female lines. Thus, from a phylogenetic perspective, the entire 16-kb mtDNA molecule is a single locus.

Multi-locus assays can be categorized further into those that assess information from multiple loci simultaneously (e.g., DNA–DNA hybridization or multi-locus DNA fingerprinting) versus sequentially (e.g., multi-locus protein electrophoresis or assays of microsatellites). Only the latter type of assay normally provides information that is interpretable in simple genetic terms; that is, as Mendelian genotypes at particular loci. Although the number of genes included in an allozyme or microsatellite survey is typically small or moderate, even a handful of interpretable genetic polymorphisms, considered in aggregate, provide remarkable power in applications such as forensics, parentage assessment, gene flow estimation, and characterization of hybrids.

Utility of data along the phylogenetic hierarchy

Another way to slice the molecular techniques pie is with regard to the level of evolutionary separation at which the various methods best apply (Box 3.4). Most assays provide an empirical window of opportunity that is fairly narrow relative to the broad field of potential phylogenetic applications. For example, protein immunological and DNA–DNA hybridization methods

BOX 3.4 Representative Molecular Approaches and Phylogenetic Resolution

The chart below indicates various levels of the evolutionary hierarchy at which a given molecular approach normally provides optimal phylogenetic resolution. More asterisks indicate higher suitability; dashes indicate that the technique is not particularly useful at that hierarchical level. However, few of these characteriza-

Hierarchical level	DNA fingerprinting		Protein electrophoresis
	Minisatellites	Microsatellites	
Clonality	***	***	*
Parentage	*	***	**
Populations	—	**	***
Related species	—	*	**
Deeper evolutionary history	—	—	—

typically are suitable for phylogenetic studies at intermediate taxonomic levels, where species' separations may date to approximately 2–100 million years ago (mya). At the microevolutionary end of the continuum, DNA fingerprinting methods using mini- or microsatellites are powerful for individual forensics and parentage analysis. Studies of mtDNA (RFLPs or sequencing) have been highly fruitful at the levels of conspecific populations and closely related species, as have allozyme surveys. Among the available molecular methods, only DNA sequencing can find application at virtually any taxonomic level. This flexibility stems from the fact that different DNA sequences evolve at highly different rates, such that the choice of sequence to be examined can be tailored to each research question. Nonetheless, because of the labor and expense involved, obtaining DNA sequences from large numbers of individuals and large numbers of genes in a population context is not particularly cost-effective, and sequencing studies usually are conducted at intermediate or higher levels of the phylogenetic hierarchy.

Of course, the volume of genetic data obtained also influences the resolution obtainable in a given application. For example, restriction site or sequencing studies of animal mtDNA at the population level commonly involved assays of about 500 bp per individual. At the conventional mammalian sequence divergence rate of 2% per million years, roughly one in 500 bp is expected to have changed after 100,000 years of matrilineal separation, thus establishing 100 millennia as an approximate lower limit of resolving power for de novo mutations with this level of effort. If the full 16,000-bp mtDNA genome were assayed in each individual, the ability to detect de

tions are absolute. For example, some protein-electrophoretic characters, such as presence versus absence of duplicate gene products, can be phylogenetically informative about deeper nodes in a phylogenetic tree, as noted in the text.

DNA sequencing		cpDNA	DNA–DNA	Protein	SINEs
mtDNA	Nuclear genes	assays	hybridization	immunology	
—	—	—	—	—	—
—	—	—	—	—	—
***	—	**	—	—	—
***	*	***	*	*	***
**	***	***	***	**	***

novo sequence divergence would increase by more than 30-fold, pushing a lower limit on resolving power to just a few thousand years. These statements refer strictly to the accumulation of novel mutations, and not to significant shifts in allele frequencies from preexisting ancestral polymorphisms (such shifts could take place in as little as one generation by genetic drift, natural selection, or migration).

SUMMARY

1. Many laboratory methods exist for revealing genetic markers. In rough order of historical appearance, the most important of these have been immunological methods (especially micro-complement fixation), multi-locus starch-gel electrophoresis (SGE) of proteins, and a succession of DNA-level approaches ranging from DNA–DNA hybridization to restriction assays to minisatellite DNA fingerprinting. More recently, numerous PCR-based approaches have been developed to reveal AFLPs, RAPDs, SSCPs, SINES, SNPs, STRs (microsatellites), and other polymorphic features of the genome. Direct DNA sequencing techniques have been available since the mid-1970s, but as they have become increasingly automated and coupled with PCR, they have vastly increased the ease with which great volumes of information can be obtained directly at the nucleotide level.

2. The wide variety of laboratory methods available can be categorized for heuristic purposes into several different groupings according to the general nature of genealogical information each technique provides: protein-level versus DNA-level data; discrete character states versus genetic distances only; information that is detached as opposed to readily connectable between separate studies; and single-locus versus multi-locus data.

3. No single molecular technique is ideally suited to all research endeavors. Rather, each assay method is fruitfully applied to specific tasks somewhere along a genealogical hierarchy. At the microevolutionary end of this hierarchy are analyses of clonality, genetic parentage, and close kinship. Then come assessments of extended intraspecific genealogy and phylogeography, speciation and hybridization, and finally, at the macroevolutionary end of the scale, phylogeny estimation at intermediate and deeper branches in the Tree of Life.

4

Philosophies and Methods of Molecular Data Analysis

Thus the hereditary properties of any given organism could be characterized by a long number written in a four-digital system.

G. Gamow (1954)

Molecular markers lend themselves to a wide variety of data analysis methods, depending on the particular research problem being addressed. Many specific approaches and their applications (e.g., in gene flow estimation, genetic parentage analyses, cytonuclear dissections of hybrid zones) are described where relevant in later chapters, but one rather generic class of applications and its historical backdrop merits introduction here: phylogenetic analysis. This chapter presents thumbnail sketches of the history and some underlying principles of phylogenetic reconstruction, brief descriptions of the primary techniques employed, and some key references to a vast literature on this topic. For a comprehensive and advanced overview of phylogenetic methods, I recommend Swofford et al. (1996), and for a "how-to" treatment with empirical examples and computer exercises, see Hall (2004).

Cladistics versus Phenetics

Until the mid-1900s, following the tradition of Linnaeus (1759), organisms were classified mostly by qualitative gestalt appraisals of morphology. Specialists typically devoted years of study to a particular group, such as birds or beetles, and based on accumulated experience, classified their creatures into a hierarchical taxonomy. This approach enabled scientists to organize and catalogue the otherwise bewildering diversity of the natural world, and the endeavor resulted in most of the biological classifications still followed today. Shortcomings of the approach stemmed from the lack of unifying or standardized classification methods (either conceptual or operational), often with the following consequences: a centering of systematic authority within a small number of research experts for each taxonomic group; the lack of formalized procedures for corroboration or refutation of a proposed classification; the absence of a uniform measure by which classifications for different taxonomic groups might meaningfully be compared; and the lack of a clear philosophy on precisely what aspects of evolution a particular classification reflected.

In the 1960s, explicit concern with these shortcomings of traditional practice prompted the rise of numerical taxonomy, or the phenetic approach to systematics (Sokal and Sneath 1963). Pheneticists proposed that organisms should be grouped and classified according to overall similarity (or its converse, distance), as measured by defined rules and using as many organismal traits as possible. Among the principles guiding numerical taxonomy were the following (Sneath and Sokal 1973): the best classifications result from analyses based on many organismal features or characters; at least at the outset, every

BOX 4.1 Terminology and Concepts Relevant to Cladistic–Phenetic Discussions

I. Classes of organismal resemblance

Phenetic similarity: The overall resemblance between organisms.

Patristic similarity: The component of overall similarity that is due to shared ancestry.

Homoplastic similarity (homoplasy): The component of overall similarity that is due to convergence from unrelated ancestors. (The term *homoplasy* is also frequently used to describe "extra steps" implied in a phylogenetic network beyond those that distinguish taxa in a raw data matrix. In this latter usage, homoplasy may arise from convergence, parallelism, or evolutionary reversals in character states.)

II. Classes of character states used to characterize organismal resemblance

Plesiomorphy: An ancestral character state (i.e., one present in the common ancestor of the taxa under study).

Symplesiomorphy: An ancestral character state shared by two or more descendant taxa.

Apomorphy: A derived or newly evolved character state (i.e., one not present in the common ancestor of the taxa under study).

Synapomorphy: A derived character state shared by two or more descendant taxa.

Autapomorphy: A derived character state unique to a single taxon.

character is afforded equal weight; classifications should be based on quantitative measures of overall (phenetic) similarity or distance between taxa; and patterns of character correlation can be used to recognize distinct taxa and draw systematic inferences. Philosophically, numerical taxonomists aimed to develop methods that were "objective, explicit, and repeatable" (Sneath and Sokal 1973). Operationally, its practitioners quantitatively estimated phenetic similarity, examined character correlations, and grouped taxa accordingly.

Numerical taxonomy provided a valuable service to science by critically scrutinizing traditional systematic practices that had been needlessly opaque. Nonetheless, pheneticists were criticized on several fronts (Hull 1988), most notably by cladists, who proposed an alternative philosophy and protocol for phylogenetic reconstruction and classification (Eldredge and Cracraft 1980). Under the tenets of the cladistic school, phylogeny should be appraised not by overall similarity between organisms, but rather by a subset of similarity attributable to synapomorphic or shared-derived traits (Box 4.1; Figure 4.1). Cladists typically focused almost exclusively on the branch-

III. Other relevant definitions

Monophyletic group or *clade*: An evolutionary assemblage that includes a common ancestor and all of its descendants.

Paraphyletic group: An artificial assemblage that includes a common ancestor and some, but not all, of its descendants.

Polyphyletic group: An artificial assemblage derived from two or more distinct ancestors.

Outgroup: A taxon phylogenetically outside the clade of interest.

Sister taxa: Taxa stemming from the same node in a phylogeny.

Phenetic resemblance may be due to patristic and/or homoplastic similarity. Patristic similarity may arise from symplesiomorphic and/or synapomorphic character states. Cladists attempt to distinguish between symplesiomorphic and synapomorphic similarity and to identify clades on the basis of synapomorphies only (see Figure 4.1). Pheneticists usually make no such attempts to distinguish sources of resemblance. Phylogenetic reconstructions based on either cladistic or phenetic principles can be compromised by extensive homoplasy.

Because cladists try to distinguish symplesiomorphies from synapomorphies, much effort is devoted to the elucidation of evolutionary "polarities" (derived versus ancestral conditions) of character states. The following are some criteria that have been suggested to indicate primitiveness for a character:

1. Presence in fossils

2. Commonness among an array of taxa

3. Early appearance in ontogeny

4. Presence in an outgroup

The fourth criterion is most widely employed today, as the others have proved misleading or incorrect in many instances (Stevens 1980).

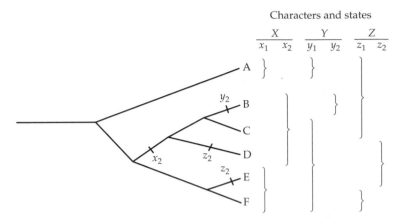

Figure 4.1 Philosophical rationale underlying Hennigian cladistic attempts to distinguish sources of similarity. Shown is the true (but unknown to the researcher) phylogeny for taxa A–F and the distribution of observed binary states for characters X, Y, and Z. Suppose that taxon A is known to be an outgroup for B–F, whose phylogeny is to be reconstructed. Character states x_1, y_1, and z_1, possessed by various taxa and the outgroup, are symplesiomorphies (shared ancestral states), and hence identify no clades. In particular, y_1 and z_1 could be positively misleading in amalgamating ingroup members (C–F and B, C, F, respectively) had their ancestral status gone unrecognized. Character state y_2 defines no multi-taxon clade because it is an autapomorphy, and z_2 could be misleading as a putative clade marker because it evolved in parallel (convergent) fashion in taxa D and E. Only x_2 is a valid synapomorphy in this example, correctly identifying the true clade composed of taxa B, C, and D.

ing component (cladogenetic aspect) of evolutionary trees, rather than on branch lengths (accumulated changes within lineages through time, or anagenesis). Their ultimate goal was to develop organismal classifications based on correctly inferred cladogenesis.

As sometimes practiced, cladistic approaches themselves were not entirely immune from criticism. For example, one widely held belief was that "one *true* synapomorphy is enough to define a unique genealogical relationship" (Wiley 1981). This sentiment is incorrect, however, if "genealogical relationship" is taken to imply organismal relationship (as it often was), because it fails to recognize the fundamental distinction between a gene tree (i.e., a character phylogeny) and an organismal tree (see below). An unfortunate consequence was that some cladists occasionally remained dogged in advocating putative organismal clades that had received support from only one or a few presumptive morphological synapomorphies, even if these conflicted with volumes of other information (e.g., from molecular sequences). Thus, unless many independent characters were assayed (as advocated by pheneticists), there was a potential danger in cladistics of the kinds of authoritarianism that had plagued systematics earlier in the century and had prompted the original rise of numerical taxonomy.

Indeed, there was little justification for the rancor of cladistic attacks on phenetics, for at least two reasons. First, cladistics owed a deep debt to numerical taxonomy for opening new discussions and viewpoints on traditional systematic practices. Second, cladistic methods as applied to large data sets often came rather close to numerical taxonomy's procedures (because for such data, character conflicts in clade delineation almost inevitably arise, thereby requiring some form of numerical tallying of putative synapomorphies). Thus, numerical cladistics and numerical phenetics were not as distinct operationally as they at first appeared. The important element that cladistics added to systematics was its explicit attempt to distinguish between alternative sources of evolutionary similarity and thereby account for specified character-state distributions in terms of phylogenetic history.

The original "bible" of the cladistic school, published in 1950 by the German entomologist Willi Hennig, was published in an English version entitled *Phylogenetic Systematics* in 1966. In the ensuing decades, cladistic methods based on Hennig's insights revolutionized systematic practice as applied to traditional taxonomic characters. Cladograms were generated for many taxonomic groups, and these typically included explicit descriptions of character polarities (ancestral versus derived conditions), character transformations, and temporal orders of appearance of various morphological, physiological, or behavioral character states. Thus, a major advantage of cladistic approaches was that they often resulted in testable hypotheses regarding evolutionary origins of and conversions among particular character states, including molecular ones (Buth 1984; Patton and Avise 1983).

The philosophical pillar of the cladistic school—that shared-derived traits are the appropriate basis for clade delineation—is now almost universally accepted. Most questions center instead on operational issues: How reliably can synapomorphies be identified? What kinds of characters are best used? How are character conflicts resolved when putative clades identified by different presumptive synapomorphies disagree? How is a phylogeny to be translated into a classification? All such questions apply to molecular markers as well as to phenotypic traits.

In a historical sense, the cladistic–phenetic debate that began in the 1960s had at least two important ramifications for molecular evolution, which also was beginning its rise then. First, the debate gave renewed energy to morphology-based systematics at a time when some traditionally trained systematists felt threatened by the emergence of molecular biology. One unfortunate consequence of this timing is that molecular and morphological approaches to systematics often were viewed as being in opposition, a perception with no valid basis. Second, the cladistic–phenetic war, although waged primarily in the context of morphology-based systematics, occasionally spilled over such that molecular phylogenetics was caught in the crossfire. For example, strict Hennigian approaches cannot be applied to raw data consisting solely of numerical distance values between taxa, and as a result some cladists automatically discredited all such information derived from the important immunological and nucleic acid hybridization methods of molecular biology. Attacks also were mounted against the widespread practice of summarizing molecular data using some of the algorithms of numerical taxonomy, despite the fact that an important assumption of phenetic procedures (a constant evolutionary rate across phenogram branches) appeared to mesh well with growing independent evidence of clocklike behavior for many biological macromolecules. Furthermore, many molecular data, even those in the form of qualitative character states (such as allozyme alleles, restriction sites, or SNPs) are not particularly well suited for strict Hennigian cladistic analysis because of a high risk of homoplasy at individual electromorphs or nucleotide character states (Straney 1981).

Apparent conflicts among characters in clade delineation arise primarily from homoplasy (see Box 4.1). To minimize the number of ad hoc hypotheses required to resolve character conflicts along a phylogeny, principles of "maximum parsimony" were soon developed as an extension of cladistic principles (Felsenstein 1983; Sober 1983). As applied to phylogenetic inference, parsimony algorithms operate by estimating evolutionary trees of minimum total length (i.e., trees that minimize the number of evolutionary transformations among character states required to explain a given data set). Although general notions of parsimony (as in Occam's razor) have long been a key part of much biological reasoning, the emergence of cladistic philosophy provided an important historical step in the further elaboration of parsimony and other approaches in molecular phylogenetic reconstruction.

Molecular Clocks

Zuckerkandl and Pauling (1965) were the first to propose that particular proteins and DNA sequences might evolve at roughly constant rates over time, and might thereby provide internal biological timepieces for dating past evolutionary events. The concept of molecular clocks fits well with neutrality theory because, as discussed in Chapter 2, the rate of neutral evolution in genetic sequences is, in principle, equal to the mutation rate to neutral alleles. However, clock concepts need not be incompatible with selection scenarios: If large numbers of assayed genes are acted upon by multifarious selection processes over long periods of time, short-term fluctuations in selection intensities could tend to average out, such that overall magnitudes of genetic distance between taxa might well correlate strongly with times elapsed since common ancestry.

Few concepts in molecular evolution have been more contentious (or abused) than molecular clocks. At the outset, several general points must be understood. First, the debate is not whether molecular clocks behave metronomically, like a working timepiece—they do not. If molecular clocks exist, both neutralists and selectionists predict at best a "stochastically constant" behavior, somewhat like radioactive decay (Ayala 1982c; Fitch 1976). Second, not all genealogical applications for genetic markers hinge critically on the reliability of molecular clocks. For example, genetic characters likely to be of monophyletic origin (such as specific gene duplications, changes in gene arrangement, or SINE insertions) provide powerful phylogenetic markers regardless of rate of evolution at the DNA sequence level; and appraisals of genetic identity, parentage, and kinship hardly depend at all on a steady pace of DNA sequence evolution.

Third, most tree-building algorithms (see next section) relax assumptions of rate homogeneity among genes and lineages. In principle and practice, branching orders in phylogenies can be inferred directly from distributions of qualitative character states using techniques such as cladistic, parsimony, or maximum likelihood analyses, which remain valid irrespective of whether molecules evolve in strictly time-dependent fashion. Even when genetic distance matrices are the starting point for phylogenetic reconstructions, most tree-building methods can accommodate heterogeneity in molecular evolutionary rates at least to some extent.

A fourth basic point about molecular clocks is that different DNA sequences empirically do evolve at markedly different rates (see review in Graur and Li 2000). Rate heterogeneity is apparent at many levels: across nucleotide positions within a codon (where, for example, mean rates for synonymous substitutions normally are several times higher than for non-synonymous substitutions involving changes in protein-coding regions; Figure 4.2A,B); among non-homologous genes within a lineage (in which non-synonymous rates can vary by orders of magnitude, as between the slowly evolving histones and rapidly evolving relaxins; Figure 4.2A); among classes of DNA within a genome (e.g., introns and pseudogenes evolve more rapidly than do non-degenerate sites in protein-coding genes; Figure 4.2C); and among different genomes within an organismal lineage (for example, synonymous substitution rates in cpDNA are severalfold lower than mean rates in plant nuclear genomes, and mtDNA in many vertebrate animals evolves about 5–10 times faster than typical single-copy nuclear DNA).

Under neutrality theory (Kimura 1983), such rate heterogeneity across nucleotide sites, genes, and genomes within a phylogenetic lineage is interpreted to reflect varying intensities of purifying selection associated with differing levels of functional constraint on DNA sequences, perhaps in conjunction with variation in the underlying rate of mutation to strictly neutral alleles (Britten 1986; but see Kumar and Subramanian 2002). Ironically, extreme rate heterogeneity can be highly beneficial for phylogenetic studies by permitting choice of appropriate DNA regions geared to the time frame of a particular phylogenetic problem (see Box 3.4). For example, the slowly evolving sequences of ribosomal RNA genes have been extremely informative in

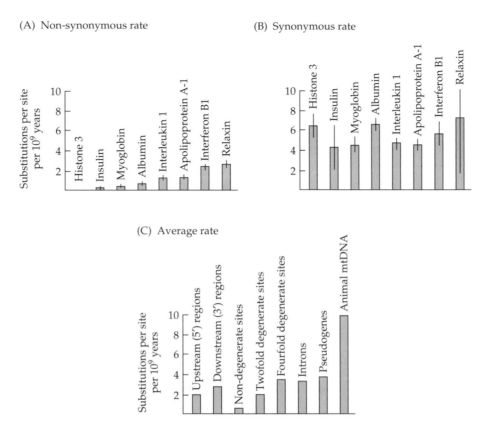

Figure 4.2   Estimated rates of nucleotide substitution in various genes and gene regions. (A) Non-synonymous and (B) synonymous substitution rates in eight protein-coding genes sequenced from humans and rodents. Calculations are based on the assumption that observed sequence differences accumulated over 80 million years. (C) Average substitution rates in different parts of these and other genes. Non-degenerate nucleotide sites are those at which all possible substitutions are non-synonymous. At fourfold degenerate sites, all possible substitutions are synonymous; at twofold degenerate sites, one of three possible nucleotide changes is synonymous and the other two are non-synonymous. (Data from Li and Graur 1991; see also Graur and Li 2000; Li 1997.)

reconstructing deep branches in the Tree of Life (see Chapter 8), whereas rapidly evolving mtDNA sequences have revolutionized phylogeographic studies of animals at the intraspecific level (see Chapter 6). Within a given gene or molecule, different classes of sites are also informative over different time frames. In animal mtDNA, for example, slowly accumulating replacement substitutions and transversions are often most useful for addressing phylogeny at the level of species, genera, or taxonomic families, whereas rapidly accumulating silent substitutions and transitions provide many more markers for analyzing local populations within a species.

History of clock calibrations and controversies

For purposes of phylogenetic reconstruction, an even more controversial and problematic form of rate heterogeneity involves differences in the evolutionary tempo of homologous DNA sequences across organismal lineages. For securely dating past separation events, approximate uniformity in evolutionary rates would clearly be desirable (A. C. Wilson et al. 1987). Indeed, standard rate calibrations have been suggested for a variety of genes and assay methods (Figure 4.3), but in general, these have also been criticized as being far less than universally applicable.

ANIMAL MITOCHONDRIAL DNA AS A PROTOTYPE.   A conventional calibration for the evolutionary rate of animal mtDNA (derived from the initial slope of the linear portion of the divergence curve; Figure 4.3A) is about 2% sequence divergence per million years (or 2×10^{-8} substitutions per site per year) between pairs of mammalian lineages that have been separated for less than 10 million years (Brown et al. 1979). In referring to mtDNA in higher animal taxa, Wilson et al. (1985; see also Shields and Wilson 1987a) concluded that "no major departures from this rate are known for the molecule as a whole."

However, this conclusion became controversial for the following empirical reasons. First, *ratios* of mean mtDNA/scnDNA divergence rates soon were shown to differ significantly between animal taxa (Caccone et al. 1988a; DeSalle et al. 1987; Powell et al. 1986; Vawter and Brown 1986), although whether this was due to rate variation in mtDNA, scnDNA, or both was unclear. Second, different nucleotide positions and genes within mtDNA were shown to evolve at varying rates within a lineage (Brown et al. 1982; Gillespie 1986; Moritz et al. 1987), and particular mtDNA genes (such as cytochrome oxidase) reportedly showed rate differences as high as fivefold across taxa (Brown and Simpson 1982; Crozier et al. 1989). Third, many researchers began to report significant variation in mean mtDNA evolutionary rates among a variety of organismal lineages, including different mammalian orders (Hasegawa and Kishino 1989), vertebrate classes (Avise et al. 1992a; Bowen et al. 1993a), homeothermic versus heterothermic vertebrates (Kocher et al. 1989), major fish groups (Krieger and Fuerst 2002), different clades of Hawaiian *Drosophila* (DeSalle and Templeton 1988), and various invertebrate taxa such as scleractinian corals vis-à-vis the vertebrate norm (Romano and Palumbi 1996; van Oppen et al. 1999).

In attempts to understand such apparent differences among animal groups, some authors noted that mtDNA evolutionary rates seemed to be positively correlated with organisms' basal metabolic rates and negatively correlated with generation lengths or body sizes (A. P. Martin et al. 1992; Rand 1994; Thomas and Beckenbach 1989). A related hypothesis involved the concept of nucleotide generation time: the mean absolute time elapsed between successive episodes of DNA replication or repair (Martin and Palumbi 1993). The idea was that nucleotide positions with higher turnover (shorter replication intervals) might be subject to more mutational opportunities per unit of

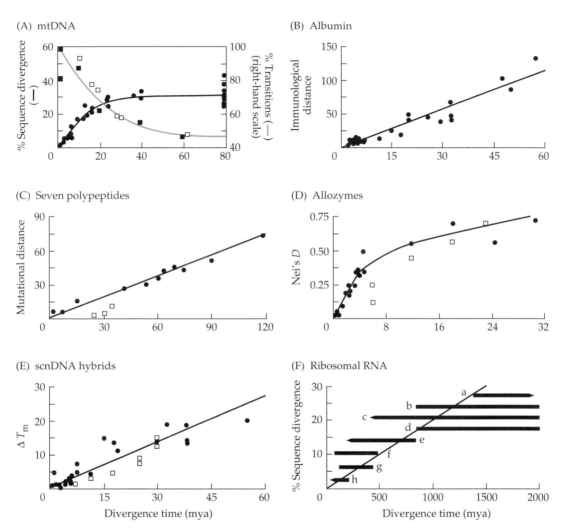

Figure 4.3   Early examples of clock calibrations reported for various types of molecular genetic data. All dates along the abscissa came from fossil or biogeographic evidence. (A) Dark line: mtDNA sequence divergence for various mammals (after Brown 1983). The slope of the linear portion of this curve gives the conventional mtDNA clock calibration of 2% sequence divergence per million years between recently separated lineages. Beyond about 15–20 million years, mtDNA sequence divergence begins to plateau, presumably as the genome becomes saturated with substitutions at the variable sites. Light gray line: percentage (right-hand axis) of observed mtDNA transitions for various mammals (open squares; after Moritz et al. 1987) and *Drosophila* (solid squares; after DeSalle et al. 1987). (B) Albumin immunological distances (as estimated by micro-complement fixation) for various carnivorous mammals and ungulates. (After Wilson et al. 1977.) (C) Accumulated codon substitutions in seven proteins (cytochrome *c*, myoglobin,

sidereal time, such that organisms with higher metabolic rates and briefer life spans might tend to have shorter nucleotide generation lengths and higher absolute rates of molecular evolution. Additional possibilities (not mutually exclusive) are that species might differ in their inherent fidelity of mtDNA replication, that DNA repair mechanisms (which are known to be deficient in mitochondria) might vary across taxa, that differences in mtDNA base composition and codon usage across taxa contribute to variation in evolutionary rates, and that mtDNA in different organisms might be exposed to different concentrations of mutagens (such as oxygen radicals generated within mitochondria). These and other possibilities and their effects on phylogenetic estimation are reviewed by Mindell and Thacker (1996) and Arbogast et al. (2002).

One difficulty in wholeheartedly accepting any such hypothesis by itself is that at least a few "troubling" exceptions seem to crop up in all suspected trends. For example, Stanhope et al. (1998a) reported that mitochondrial genes in some long-generation mammals, such as elephants and humans, evolve faster than their homologues in short-generation rodents. Likewise, any correlation between metabolic and evolutionary rates seems to be less than universal when mtDNA sequences from widely diverse taxa are considered (e.g., Seddon et al. 1998). The bottom line is that inter-taxon differences in the pace of mtDNA nucleotide evolution remain poorly understood mechanistically, and multiple interacting factors probably contribute to the outcomes.

OTHER GENES AND GENOMES. Examples of reported calibrations for several other putative molecular clocks are summarized in Figure 4.3. Again, controversies soon arose over the validity and universality of such results, particularly as applied across taxonomic groups. The inherent appeal of molecular clocks occasionally led to some egregious claims. For example, the early protein electrophoretic literature conveyed a strong impression that allozyme distances reliably date speciation events. Many papers concluded

α- and β-hemoglobin, fibrinopeptides A and B, and insulin) for various mammalian species. The three open squares involve primate comparisons. (After Langley and Fitch 1974; Nei 1975.) (D) Codon substitutions per locus (Nei's D) based on allozyme comparisons of carnivores (solid circles) and primates (open squares). (After Wayne et al. 1991a.) (E) ΔT_m values from DNA–DNA hybridization involving carnivores (closed circles) and primates (open squares). (After Wayne et al. 1991a.) (F) Percent sequence divergence in 16S ribosomal DNA for various eubacterial forms: (a) cyanobacteria; (b) chloroplasts; (c) microaerophiles; (d) mitochondria; (e) obligate aerobes; (f) *Photobacterium*; (g) *Rhizobium* and *Bradyrhizobium*; and (h) *Escherichia*. Wide horizontal bars indicate high uncertainty about divergence times from nonmolecular evidence (the slope of the line is arbitrarily drawn to represent an evolutionary rate of 1% sequence divergence per 50 million years). (After Ochman and Wilson 1987.)

that observed genetic distances were consistent with suspected separation times for particular species as gauged by nonmolecular evidence (such as fossil ages or the presence of geographic barriers). It turns out, however, that different authors had employed (perhaps unwittingly) allozyme rate calibrations that differed from one another by more than 20-fold (see review in Avise and Aquadro 1982). Given such a huge range of potential clock calibrations in the literature, it is difficult to imagine any observed allozyme distance that could not have been accommodated with a given fossil-based or biogeography-based scenario. Ironically, if these researchers individually were correct that a molecular clock had been ticking *within* each taxonomic group, then collectively no single allozyme clock could apply *across* these same taxa. Ayala (1986) provided another early summary of evidence for the erratic behavior of particular protein clocks across lineages, as did Britten (1986) and Brunk and Olson (1990) for overall rates of scnDNA sequence divergence as gauged by DNA–DNA hybridization.

Some investigators concluded that although mean molecular rates in nuclear genomes vary among taxa, they do so in a predictable or at least consistent fashion. For example, molecular evolution in nDNA was argued to be slower in primates than in rodents (in contrast to mtDNA) and was thought to be especially slow in hominoids (Goodman et al. 1971; Koop et al. 1989; Li and Tanimura 1987; Li et al. 1987; Maeda et al. 1988; but see also Caccone and Powell 1989; Easteal 1991; Kawamura et al. 1991; Sibley and Ahlquist 1987). Several published interpretations were similar to those presented above for mtDNA. For example, Catzeflis et al. (1987) and Sibley et al. (1988) presented evidence that short generation times or increased numbers of germ line cell divisions were associated with higher mean rates of scnDNA sequence evolution in birds and mammals [thereby contradicting Sibley and Ahlquist's (1984) prior advocacy of a "uniform rate of DNA evolution"].

Other researchers provisionally interpreted apparent variation in nuclear substitution rates among taxa to differences in generation length (Gaut et al. 1992; Kohne 1970; Laird et al. 1969; Li et al. 1987), numbers of DNA replications in germ line cells (Hurst and Ellegren 1998; Wu and Li 1985), repair efficiencies during DNA replication (Britten 1986), or magnitudes of exposure to mutagenic agents including free radicals, whose level of DNA damage appears to be correlated with metabolic rate differences among species (Adelman et al. 1988). Most such proposed factors presumably would tend to be more similar in closely related than in distantly related taxa, producing a "phylogenetic legacy" that is both good and bad for molecular analyses. On the positive side, such a historical legacy opens interesting possibilities for estimating "the rate of evolution of the rate of molecular evolution" within a phylogenetic tree (Thorne et al. 1998). On the negative side, it means that rate estimates from different pairs of species are statistically non-independent, and also that molecular clock calibrations are likely to be "local" rather than universal (Sanderson 2002).

On the other hand, some leading researchers consistently maintained that sidereal time was the single best predictor of genetic divergence, and that molecular clocks could be calibrated universally across taxa (Wilson 1985). As stated by A. C. Wilson et al. (1987), "Molecular evolutionary clocks have ticked at much the same rate per year in many eubacterial genes as in the nuclear genes of animals and plants." These authors also proposed an intriguing explanation for this conclusion based on the following assumptions: most nucleotide substitutions involve neutral mutations; the mutation rate per year is higher in short-generation organisms; the fraction of effectively neutral mutations is lower in larger populations (because of the greater effect there of deterministic forces, including natural selection); and species with shorter generations tend to have larger populations. To the extent that these assumptions hold, a greater mutation rate in short-generation species might be counterbalanced by a lower fraction of effectively neutral mutations, such that molecular evolutionary rates overall would remain fairly constant among diverse taxa.

A somewhat related proposal is that if a great many genes are surveyed in the taxa under consideration, any rate variation across loci may tend to average out by the law of large numbers. Thus, even if large statistical errors are associated with single-gene sequences, reliable estimates of divergence times should emerge from multi-locus appraisals. In one empirical exploration of this thesis, Kumar and Hedges (1998) found excellent agreement between fossil-based estimates of divergence times for major vertebrate lineages and combined molecular divergence estimates from sequence appraisals of more than 650 nuclear genes (Figure 4.4).

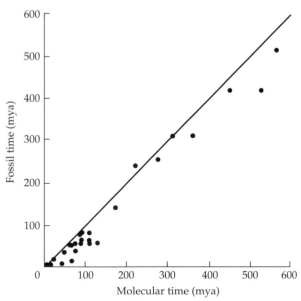

Figure 4.4 Agreement between composite molecular estimates and fossil-based estimates of divergence times for major vertebrate lineages. Composite molecular estimates were derived from 658 nuclear genes. (After Kumar and Hedges 1998.)

Absolute and relative rate comparisons

Interpretive controversies aside, how are molecular evolutionary rates assessed, and how is it that debates about rate heterogeneity continued for so long without final resolution? One difficulty is that molecular distance measures often become nonlinear with time (due to "saturation effects") at increased evolutionary depths (e.g., Figure 4.3A,D). Some arguments against molecular clocks have stemmed from appraisals at inappropriate regions of the divergence curves. Another problem is the difficulty of determining confidence limits for genetic distance estimates (Nei 1987). Thus, any distance in Figure 4.3 is a point value with (often large) estimation errors. Another aspect of statistical concern is whether, for a given molecule, the mean substitution rate among lineages is equal to the variance in rate, as predicted under neutrality theory for a Poisson-like process. Several early studies concluded that the empirical variance in rates somewhat exceeds the mean (Langley and Fitch 1974; Ohta and Kimura 1971), and this was interpreted as evidence against a uniform clock (Gillespie 1986, 1988; Takahata 1988).

Perhaps the most serious difficulty in calibrating molecular distance against sidereal time is that firm independent knowledge from fossil or biogeographic evidence is also required, at least initially. Unfortunately, such information is uncertain or lacking for many taxa (Kidwell and Holland 2002). Indeed, if this were not true, little motivation would exist for phylogenetic reappraisals based on molecular data. All separation dates in Figure 4.3, for example, came from fossil evidence, and the range in estimates of divergence time was in some cases extremely wide (e.g., Figure 4.3F). Fossils seldom provide the solid ground-truthing of separation dates that molecular biologists sometimes suppose because preserved remains are often scanty and confined to a few phenotypic attributes whose phylogenetic relevance is suspect. These problems are especially acute for morphologically simple creatures such as bacteria. Furthermore, even under the best of preservation circumstances, the earliest known appearance of a fossil provides only a minimum date for the true evolutionary origin of the lineage that it represents.

Biogeographic evidence can also be difficult to interpret, even in the cleanest of instances. To cite one example, it is well documented that the Isthmus of Panama rose above the sea about 3 million years ago, and it must therefore have curtailed any former gene flow between tropical marine faunas in the eastern Pacific and western Atlantic oceans after that time. Today, green turtles (*Chelonia mydas*) are circumtropically distributed, but their populations show a clear genealogical distinction in mtDNA between the Atlantic and Pacific (see Chapter 6). The estimated magnitude of *net* mtDNA sequence divergence between these two clades (0.6%), however, is tenfold lower than expected, assuming that the "conventional" mammalian mtDNA evolutionary rate (2% sequence divergence per million years; Figure 4.3A) has applied across the supposed 3 million years of evolutionary separation. One possibility is that mtDNA evolution in these turtles is slower than in

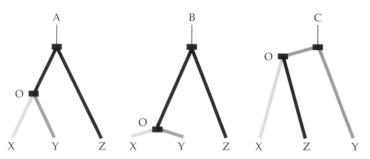

Figure 4.5 Relative rate test. (A) The rationale behind the relative rate test. (B, C) Potential difficulties of this test. (See text for details.)

mammals by an order of magnitude (Avise et al. 1992a). Or perhaps turtle mtDNA evolves at the standard mammalian pace, but Atlantic and Pacific populations were in recent genetic contact via dispersal of animals around South America or Africa during interglacial episodes of the Pleistocene. Similar molecular surveys have been conducted on more than 30 marine "geminate" species pairs inhabiting the Atlantic versus Pacific sides of Panama (see review in Lessios 1998). These pairs include numerous fishes (Bermingham et al. 1997; Grant 1987; Vawter et al. 1980), sea urchins (Bermingham and Lessios 1993; Lessios 1979, 1981), and shrimps (Knowlton et al. 1993). However, even these extensive molecular analyses, conducted in a superbly favorable biogeographic setting, have yielded uncertain conclusions about the magnitudes of possible rate variation in homologous classes of genetic markers (Bermingham et al. 1997; Marko 2002).

   To circumvent the requirement of firm separation dates from fossil or biogeographic evidence, molecular evolutionists also developed relative rate tests that do not depend on knowledge of absolute divergence times (Margoliash 1963; Sarich and Wilson 1973). Each test requires at least two related species (X and Y) and an outside reference species (Z) that branched off prior to the separation of X and Y. The rationale is illustrated in Figure 4.5A. By definition, the true evolutionary distance between X and Y (d_{XY}) is equal to the sum of their branch lengths from a common ancestor at point O (i.e., $d_{XY} = d_{OX} + d_{OY}$). Similarly,

$$d_{XZ} = d_{OX} + d_{OZ}$$

or rearranged,

$$d_{OX} = d_{XZ} - d_{OZ} \tag{4.1}$$

and

$$d_{YZ} = d_{OY} + d_{OZ}$$

or rearranged,

$$d_{OY} = d_{YZ} - d_{OZ} \tag{4.2}$$

Subtracting Equation 4.2 from Equation 4.1 yields

$$d_{OX} - d_{OY} = d_{XZ} - d_{YZ} \tag{4.3}$$

According to a molecular clock, d_{OX} and d_{OY} should be equal ($d_{OX} - d_{OY} = 0$) and, hence, $d_{XZ} = d_{YZ}$. Genetic distances d_{OX} and d_{OY} cannot be measured empirically, but d_{XZ} and d_{YZ} can. The relative rate test asks whether d_{XZ} and d_{YZ} as estimated by genetic data are sufficiently different as to be incompatible with a strict molecular clock.

By similar logic, branches leading to extant sister taxa X and Y can also be deduced from empirical data to have different (or perhaps identical) lengths, even though neither of their genetic distances to the common ancestor can be measured directly. Suppose the following hypothetical distances were observed among three extant taxa: $d_{XY} = 0.08$, $d_{XZ} = 0.19$, and $d_{YZ} = 0.17$. Solution of three equations with three unknowns (using the logic of Fitch and Margoliash 1967) yields the desired branch lengths as follows:

$$d_{OX} + d_{OY} = 0.08$$

plus

$$d_{OX} + d_{OZ} = 0.19$$

yields

$$\left(2d_{OX} + d_{OY} + d_{OZ} = 0.27\right)$$

minus

$$\left(d_{OY} + d_{OZ} = 0.17\right)$$

yields

$$2d_{OX} = 0.10 \text{ or } d_{OX} = 0.05$$

The unique solutions are $d_{OX} = 0.05$, $d_{OY} = 0.03$, and $d_{OZ} = 0.14$. Note that d_{OX} and d_{OY} differ despite the fact that the same amount of time has elapsed since X and Y shared a common ancestor. Note also that the sums of all branch lengths in this little tree agree perfectly with the empirical distances between X, Y, and Z, meaning that the data have not been distorted in the reconstructed phylogeny. However, this seldom proves true when more than three taxa are considered.

Many relative rate tests for molecular data have been conducted (Graur and Li 2000). For example, based on their DNA–DNA hybridization studies, Sibley and Ahlquist (1986) reported that "genetic distances between the outlier and each of the other species … are always equal, within the limits of experimental error" and that "thousands of such trios of species … yield the same result and attest to the uniform average rate of the DNA clock in birds." Mice (*Mus*), rats (*Rattus*), and hamsters (*Cricetulis* and *Mesocricetus*) were likewise among the many organisms that provisionally passed relative rate tests early on (Li et al. 1987; O'hUigin and Li 1992). On the other hand, there have also been many failures to pass relative rate tests, and these failures have contributed to such conclusions as that nDNA sequences evolve faster in rodents than in primates (Li et al. 1987; Wu and Li 1985) and that particular cpDNA sequences evolve more rapidly in annual plants than in perennials (Bousquet et al. 1992; Gaut et al. 1992).

Relative rate tests are not without difficulties, however, as illustrated in Figure 4.5. If X and Y separated very recently compared with their separation from Z (Figure 4.5B), d_{XZ} and d_{YZ} may appear equal in an empirical test even under a highly erratic clock (because the vast majority of evolution has taken place in the long and shared OZ branch). Conversely, if the separation of X and Y was temporally close to that of their common ancestor with Z (Figure 4.5C), an incorrect assumption about the branching order of the three species might exist, such that d_{XZ} and d_{YZ} could appear different even under a nearly constant clock. Thus, errors both of false acceptance and false rejection of evolutionary clocks can be envisioned in particular tests of relative rates. Additional nuances of the relative rate test are discussed in Tajima (1993) and Graur and Li (2000).

Closing thoughts on clocks

What can be concluded from the vast effort expended on assessment of molecular clocks? It is now undeniable that some, and probably most, molecular systems evolve at heterogeneous rates, not only among classes of sites within a given molecule but also across taxonomic groups. Thus, if precise clocks exist, they are local rather than universal, both with respect to different classes of molecular features and different phylogenetic lineages. Nonetheless, time elapsed since common ancestry remains an important, and arguably the single best, predictor of molecular divergence, especially when genetic distance is measured across large numbers of loci. What justifies this latter statement? The evidence is mostly indirect, inconclusive in individual instances, but cumulatively compelling. First, nodes in numerous molecular phylogenetic trees generally seem to be at least roughly consistent with independent time estimates (see Figures 4.3 and 4.4), uncertain as these dates of separation may sometimes be. Second, molecular evolution often proceeds mostly independently of morphological and phenotypic evolution, in which rates can vary wildly under the influence of different selection regimes. Finally, given current understanding of the mechanisms underlying DNA sequence evolution, it would be most surprising if mean genome-wide sequence divergence did not generally tend to increase with time.

In the near future, it will become commonplace to base phylogenetic conclusions on DNA sequences from many genes, rather than just one or a few genes. The study by Kumar and Hedges (1998), which analyzed more than 650 nuclear genes to estimate phylogeny for mammalian orders and major vertebrate lineages (see Figure 4.4), was among the pioneering efforts in this regard, but as data from DNA sequences exponentially increase, such massive composite analyses may soon be standard practice in the field of molecular phylogenetics. Such analyses should tend to smooth the high rate variation often seen in the individual genes that traditionally were considered one at a time in phylogenetic appraisals.

Furthermore, in many instances, even molecular timepieces with less than full precision can provide significant improvements over phylogenetic understanding gained from nonmolecular data. Consider, for example, single-celled microbes, whose phylogeny was totally unknown prior to the application of molecular information. Phylogenetic patterns in rRNA genes and other loci have revealed stunning genetic relationships and subdivisions among microbial taxa, including those that early in evolution entered into endosymbiotic relationships with proto-eukaryotic cells (see Chapter 8). Molecular clocks keep far from perfect time, but to dismiss the inherent time-dependent nature of molecular evolution out of hand would be to deny empirical access to an invaluable, and sometimes the sole, source of information on temporal history.

Phylogenetic Reconstruction

Phylogenetic trees (rooted) or networks (unrooted) are graphical representations consisting of nodes and branches (pathways connecting nodes) that summarize evolutionary relationships among particular taxa (Figure 4.6). In most molecular studies, the units of phylogenetic analysis (operational taxonomic units, or OTUs) are species or higher taxa (actually, the analyzed genetic material that they house), but they can also be conspecific populations, indi-

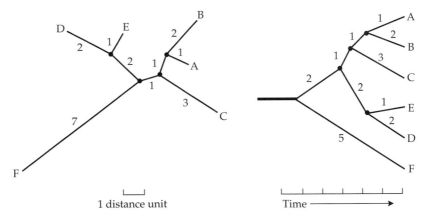

Figure 4.6   Alternative representations of phylogenetic relationships for six extant taxa (A–F). *Left*: Unrooted network with scaled branches. *Right*: Rooted tree (the heavy line is the root) with branches that are only roughly scaled. Internal nodes in both drawings are indicated by black dots. Note that branch angles have no meaning because branches may be rotated freely about any internal node without materially affecting network or tree topology.

vidual organisms, or non-recombined alleles of a specific gene, such as mtDNA. The key requirement or assumption underlying a phylogenetic representation is that tree branches be mostly non-reticulate or non-anastomose (Chapter 8 describes exceptions), such that the tree faithfully captures ancestral–descendant evolutionary history. External nodes in a phylogenetic tree or network are normally extant OTUs, and internal nodes are deduced ancestral units. Peripheral branches lead to external nodes; interior branches connect internal nodes. Branch lengths reflect the number of evolutionary changes along each ancestral–descendant pathway. If the genetic distance between each pair of OTUs exactly equals the sum of the branch lengths connecting that pair, the tree is said to be strictly additive (Waterman et al. 1977). Departures from additivity provide one measure of the degree to which a depicted phylogeny may have been distorted by homoplasy (convergence, parallelisms, or reversals) in the molecular data, or perhaps by improper behavior of the distance measure or the phylogenetic algorithm employed.

Phylogenetic representations may be graphed in several ways (see Figure 4.6). A tree or network is scaled when its branches are proportional in length to the numbers of genetic changes along them; otherwise, it is unscaled or partially scaled (although branch lengths may be indicated numerically along the diagram). A tree is rooted when an internal node is specified that represents the common ancestor of all OTUs under study; otherwise, the diagram is unrooted and is commonly referred to as a network. A tree is bifurcating when two immediate descendant lineages come from each node, and multifurcating when three or more lineages do so.

The process of estimating a phylogenetic tree can be remarkably challenging, in large part because even small numbers of OTUs can, in principle, be connected by astronomical numbers of different trees, only one of which is presumably the correct representation of actual evolutionary history. For n OTUs, the theoretical number of different bifurcating unrooted networks (N_{Tu}) is

$$N_{Tu} = \frac{(2n-5)!}{2^{n-3}(n-3)!} \tag{4.4}$$

and the theoretical total number of different bifurcating rooted trees (N_{Tr}) is

$$N_{Tr} = \frac{(2n-3)!}{2^{n-2}(n-2)!} \tag{4.5}$$

(Felsenstein 1978a). Thus, the number of possible tree structures increases dramatically as the number of taxa increases, and even the small value of n = 10 yields N_{Tr} = 34,459,425.

Many phylogenetic algorithms work by searching among possible trees for those that exhibit desirable properties according to some specified optimality criterion (e.g., shortest total branch length under parsimony). However, it is currently impossible for even the faster computers to exhaustively search all possible trees when n is moderate or large. Thus, truncated search procedures must be implemented with the hope that they will adequately explore the vast parameter space to identify correct tree(s) according to some specified optimality criterion. A second complication is that the optimality criterion itself may or may not be a valid representation of evolution. A third fundamental difficulty is that a true phylogeny is seldom known with certainty from independent evidence (Atchley and Fitch 1991 and Hillis et al. 1992 describe rare exceptions), so empirical appraisals of alternative phylogenetic methods normally rest on indirect evidence. The net result of these difficulties in evaluating alternative tree-building approaches has been a continuing debate over which method of phylogenetic reconstruction is "best."

A vast scientific literature, far beyond the scope of this book, addresses the merits and demerits of a wide variety of tree-building procedures for molecular (and other) data. Some key books on the topic include those by Hall (2004), Hillis et al. (1996), Li (1997), and Nei and Kumar (2000), to which interested readers are referred. What follows are merely brief descriptions and the rationales of the most commonly used algorithms.

Distance-based approaches

All distance-based approaches begin with an OTU × OTU matrix, the body of which consists of estimated pairwise genetic distances between taxa (Farris 1972). For n OTUs, there are $n (n - 1)/2$ such distances (excluding "self" comparisons along the matrix diagonal). Clearly, because OTUs have phylogenetic connections, such estimates cannot be treated as independent values in a statistical sense. Indeed, the historical connections that genetic distances register *are* the primary focus. Table 4.1 presents a hypothetical distance matrix for five OTUs (ten pairwise comparisons) that will serve to illustrate two of the most widely used distance algorithms for constructing phenograms (also called dendrograms or, loosely, "trees").

UPGMA CLUSTER ANALYSIS. Cluster analyses (of which there are several variants), which group OTUs according to overall similarity or distance, are the simplest methods computationally (Sneath and Sokal 1973). Under the "unweighted pair group method with arithmetic averages" (UPGMA), a

TABLE 4.1	Hypothetical genetic distance matrix for five OTUs			
	A	**B**	**C**	**D**	**E**
A	—	0.08	0.19	0.70	0.65
B		—	0.17	0.75	0.70
C			—	0.80	0.60
D				—	0.12
E					—

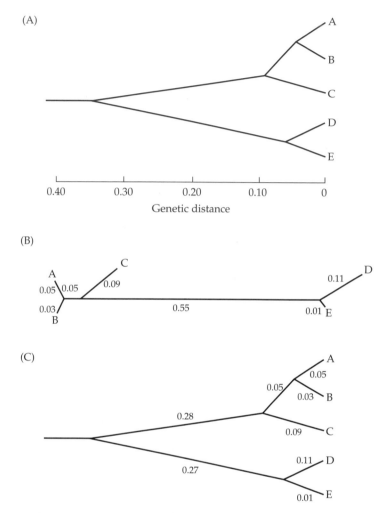

Figure 4.7 Phenograms produced from the genetic distance matrix in Table 4.1 by various distance-based approaches: (A) UPGMA dendrogram; (B) neighbor-joining network, unrooted; (C) neighbor-joining tree, rooted and right-justified.

distance matrix (such as that in Table 4.1) is scanned for the smallest distance element, and the OTUs involved are joined at an internal node drawn in an appropriate position along a distance axis (Figure 4.7A). In this example, OTUs A and B are joined first, at distance level $d = 0.04$ (because the sum of branch lengths connecting A and B is the observed $d = 0.08$). This distance element in the matrix is then discarded. The matrix is scanned again for the smallest remaining distance, which in this case is $d = 0.12$ between D and E. These OTUs are clustered at level $d = 0.06$. The next smallest distance in the matrix is $d = 0.17$ between C and B. However, B is already part of a previously formed cluster with A, so C cannot be joined directly to B, but rather must be connected through the A–B internal node. This clustering level is determined by the arithmetic mean of the distances between C and OTUs in the previous cluster [$d = (0.19 + 0.17) / 2 = 0.18$]. Thus, C joins the A–B group at $d = 0.09$. All that remains is to join the A–B–C cluster with the D–E cluster, at a depth determined from the mean of all pairwise distances between the OTUs they contain [$d = (0.70 + 0.65 + 0.75 + 0.70 + 0.80 + 0.60)/6/2 = 0.35$]. Note that this exhausts all distances in the matrix. The final UPGMA phenogram is shown in Figure 4.7A.

Some points about UPGMA should be clarified. In each cycle of the clustering procedure, OTUs or previously formed clusters are grouped according to the smallest *mean* distance between the taxa involved (rather than smallest single distance element remaining in the matrix). Each OTU contributes equally to these mean distances (hence the term "unweighted"). All extant OTUs are depicted as "right-justified" along a genetic distance axis (Figure 4.7A). Finally, the dendrogram is implicitly rooted at the point where the deepest clusters join. The major assumption (and practical limitation) of UPGMA clustering is that evolutionary rates are equal along all dendrogram branches. Even so, UPGMA often performs surprisingly well in recovering proper tree structures in computer simulation tests (Nei et al. 1983; Sourdis and Krimbas 1987; Tateno et al. 1982). This seems to be because estimates of genetic distance are subject to large stochastic errors, and the distance-averaging feature of UPGMA tends to reduce these effects (Nei 1987).

NEIGHBOR-JOINING METHOD.   The neighbor-joining ("N-J") method (Saitou and Nei 1987) is conceptually related to cluster analysis, but allows for unequal rates of molecular change among branches. It does so by constructing, at each step of the analysis, a transformed distance matrix that has the net effect of adjusting branch lengths between pairs of nodes on the basis of mean divergence from all other nodes. The procedure is detailed in Box 4.2, again using the distance matrix in Table 4.1. The resulting unrooted network, showing deduced branch lengths as well as topology, is shown in Figure 4.7B. This network can also be rooted, for example by placing an ancestral node at the midpoint of the longest total set of branch lengths between extant OTUs (Figure 4.7C). Note the close similarity in this example between the structures of the N-J and UPGMA phenograms (Figure 4.7A,C). This kind of agreement between alternative clustering algorithms is commonly observed for real molecular data sets. Readers interested in further details and the formal steps of the N-J operation, which is probably the most popular of the distance approaches in use today, should consult Studier and Keppler (1988) or Swofford et al. (1996).

COMPARISON OF DISTANCE MATRIX METHODS.   Much debate has concerned which of these (or other) distance algorithms produces the "best" phenogram. One basis for this choice is goodness of fit, a measure of how well the inferred distances in the "tree" match the empirical distance values in the original matrix. In a molecular phylogenetic context, Prager and Wilson (1978) were among the first to apply one such measure:

$$F = \frac{100 \sum_{i=1}^{s} |I_i - O_i|}{\sum_{i=1}^{s} I_i} \tag{4.6}$$

Box 4.2 Cycling Operation of the Neighbor-Joining Algorithm

Each value of r is the sum of observed distances between the OTU of that row and other extant OTUs or nodes. All values were rounded to two decimal points.

	A	B	C	D	E	r	$r/3$
A	—	0.08	0.19	0.70	0.65	1.62	0.54
B	−1.03	—	0.17	0.75	0.70	1.70	0.57
C	−0.94	−0.99	—	0.80	0.60	1.76	0.59
D	−0.63	−0.61	−0.58	—	0.12	2.37	0.79
E	−0.58	−0.56	−0.68	−1.36	—	2.07	0.69

Distance D to node 1 = 0.12/2 + (0.79 − 0.69)/2 = 0.11
Distance E to node 1 = 0.12 − 0.11 = 0.01

	A	B	C	Node 1	r	$r/2$
A	—	0.08	0.19	0.62	0.89	0.44
B	−0.82	—	0.17	0.66	0.91	0.46
C	−0.75	−0.79	—	0.64	1.00	0.50
Node 1	−0.78	−0.76	−0.82	—	1.92	0.96

Distance C to node 2 = 0.64/2 + (0.50 − 0.96)/2 = 0.09
Distance node 1 to node 2 = 0.64 − 0.09 = 0.55

	A	B	Node 2	r	$r/1$
A	—	0.08	0.08	0.16	0.16
B	−0.26	—	0.10	0.18	0.18
Node 2	−0.26	−0.26	—	0.18	0.18

Distance A to node 3 = 0.08/2 + (0.16 − 0.18)/2 = 0.03
Distance node 2 to node 3 = 0.08 − 0.03 = 0.05

	B	Node 3
B	—	0.05
Node 3		—

where for the s pairwise comparisons among OTUs, I and O are input distances from the matrix and output distances from the tree, respectively. Smaller values of F indicate better fit (although not necessarily the correct tree). Other such measures include a co-phenetic correlation (Sneath and Sokal 1973) and percent standard deviation (Fitch and Margoliash 1967). As expected, procedures such as neighbor-joining that explicitly adjust tree branches to improve fit to an additive tree generally outperform methods such as UPGMA that do not (Avise et al. 1980; Berlocher 1981; Prager and Wilson 1978).

A second basis for choice among distance-based tree-building algorithms involves the degree of congruence among trees derived from different data sets. Because a given array of species presumably has a single general phylogenetic typology along which all characters have evolved (although not necessarily through the exact same transmission routes), methods of data analysis producing more highly congruent trees might be judged superior (Farris 1971). Several measures for evaluating levels of congruence among trees have been suggested (e.g., Farris 1973; Mickevich 1978; Swofford et al. 1996).

Another approach for comparing the performance of distance-based or other phylogenetic algorithms involves computer simulations of molecular change along trees generated under some specified model of evolution. Genetic distances among extant computer OTUs are estimated, and each algorithm's performance is evaluated by how well it recovers the known tree (Fiala and Sokal 1985; Jin and Nei 1991; Saitou and Nei 1987). Potential difficulties of this approach lie in assessing the biological plausibility of the model's assumptions and in the risk of circular reasoning when the best phylogenetic algorithm proves to be the one whose assumptions most closely match those underlying the simulation. All phylogenetic algorithms involve assumptions (transparent or opaque).

A powerful method for evaluating algorithm performance was introduced by Hillis et al. (1992). They serially propagated bacteriophage T7 in the presence of a mutagen, experimentally dividing the culture at various time intervals, such that a known phylogeny was produced. Terminal lineages were then assayed for restriction site maps, and the data were used to infer evolutionary history by various phylogenetic methods. All five algorithms employed, which included N-J and UPGMA as well as a qualitative parsimony approach, produced the correct branching order of the known topology, but they differed slightly in ability to recover correct branch lengths. Of course, such direct appraisals of phylogenetic methods using living organisms are logistically possible in only a few systems, such as T7, in which mutation rates are high and thousands of generations transpire each year.

UPGMA automatically produces a rooted "tree." To root an N-J tree or another such distance network, two procedures may be followed. First, one or more outgroup taxa (see Box 4.1) can be included in the analysis, in which case the root is placed between an outgroup and the node leading to ingroup members. Alternatively, if an approximate uniform rate of evolution is assumed for long time periods, the network may be rooted at the midpoint of the longest pathway between any extant OTUs (as was done in Figure 4.7C).

Character-state approaches

For most kinds of molecular markers, including DNA sequences, discrete character states permit phylogenetic analyses to be performed directly on the qualitative raw data, if so desired, without the requirement of first constructing a distance matrix. Several such approaches are available.

MAXIMUM PARSIMONY.   A maximum parsimony (MP) tree is one that requires the smallest number of evolutionary changes to explain observed differences among OTUs. Consider Figure 4.8, which presents a parsimony network for ten extant OTUs based on the depicted states of nine variable

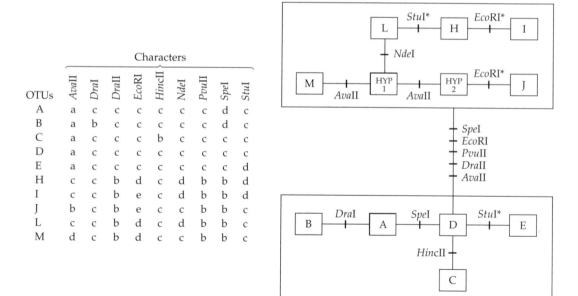

OTUs	*Ava*II	*Dra*I	*Dra*II	*Eco*RI	*Hinc*II	*Nde*I	*Pvu*II	*Spe*I	*Stu*I
A	a	c	c	c	c	c	c	d	c
B	a	b	c	c	c	c	c	d	c
C	a	c	c	c	b	c	c	c	c
D	a	c	c	c	c	c	c	c	c
E	a	c	c	c	c	c	c	c	d
H	c	c	b	d	c	d	b	b	d
I	c	c	b	e	c	d	b	b	d
J	b	c	b	e	c	c	b	b	c
L	c	c	b	d	c	d	b	b	c
M	d	c	b	d	c	c	b	b	c

Figure 4.8   Estimate of an unrooted parsimony network (*right*) based on an OTU × character matrix (*left*) for a subset of mtDNA clones observed in green turtles. Lowercase letters represent character states (mtDNA digestion profiles produced by nine restriction enzymes; adjacent letters of the alphabet denote profiles that differed by a single restriction site). Inferred restriction site changes along branches of the network are indicated. Asterisks indicate probable instances of homoplasy. (After Bowen et al. 1992.)

characters (mtDNA restriction site patterns in this case). Construction of the network from this simple data matrix can be done by hand. The process is initiated by connecting any OTU to its nearest genetic neighbors via reference to the character-state matrix. For example, haplotype D is one mutation step removed from each of three other haplotypes (A, C, and E), which are two steps removed from one another. Haplotype B, in turn, is one step from A, two steps from D, and three steps each from C and E. Thus, the distribution of character states for OTUs A–E yields the singular most parsimonious network shown at the bottom right of Figure 4.8. Note that this portion of the network is strictly additive.

Generation of the complete network for all ten extant OTUs illustrates some complications that can arise. First, not all nearest-neighbor OTUs are a single mutation step apart, so in this case two hypothetical OTUs (HYP1 and HYP2) were arbitrarily added to make all branches in the upper portion of the tree of unit length. Second, there is a large genetic gap distinguishing OTUs A–E from the assemblage H, I, J, L, M, so where the branch connecting these groups should be placed is not initially obvious. Here, D and HYP2 are joined because they differ by five steps, whereas any other intergroup branches would involve six steps or more. Third, some genetic character states appear in different (presumably distantly related) portions of the network. For example, the character state *Stu*I-d appears in some representatives of both the upper and lower OTU groups, probably due to polyphyletic origins from *Stu*I-c. Similarly, the state *Eco*RI-e appears in OTUs I and J, which are not adjacent genetically as judged by the other assayed characters. Such character states contribute to homoplasy (indicated by asterisks) by introducing additional steps along network branches beyond those that differentiate OTUs in the original character-state matrix. Nonetheless, in this example, the sum of all pairwise output distances in the network (256) is only slightly greater that the sum of all input distances (250), indicating strong goodness of fit between network and data.

In usual practice, data are analyzed by computer programs that search vast numbers of alternative trees for minimum total length. PAUP* (phylogenetic analysis using parsimony) by Swofford (2000), and PHYLIP (phylogeny inference package) by Felsenstein (available at evolution.genetics.washington.edu/phylip.html) have been among the industry standards for such analyses. Sometimes, many MP trees of different topology prove equally parsimonious or require similar numbers of steps. Nonetheless, such networks constitute only a small fraction of the vast universe of potential trees, most of which require many more steps and can therefore be eliminated from further serious consideration.

Actually, parsimony approaches constitute a large family of related methods incorporating varying assumptions about how character-state transformations occur (Swofford et al. 1996). Under Wagner parsimony (Farris 1970; Fitch 1971; Kluge and Farris 1969), for example, free evolutionary reversibility of character states is allowed, with changes in either direction equally likely. By contrast, Dollo parsimony (Farris 1977) assumes that each non-ancestral character is uniquely derived (although multiple reversions to the ancestral condition are allowed), and Camin-Sokal (1965) parsimony assumes that all evolutionary change is irreversible. Ideally, the assumptions employed in a particular analysis should match the true nature of evolutionary change in the molecular markers utilized. For example, Wagner parsimony might be appropriate for analyzing data that differ by transitional substitutions that interconvert readily; Camin-Sokal parsimony might work well for SINEs, which uniquely insert into genomes but once present are usually retained indefinitely; and Dollo parsimony might be suitable for some restriction data in which each particular site loss is mechanistically more likely by severalfold than its gain (DeBry and Slade 1985; Templeton 1983, 1987). Caution is always indicated, however, so it is probably best to attempt and compare multiple approaches, including distance-based analyses, before drawing firm biological conclusions from any reconstructed phylogenies.

MAXIMUM LIKELIHOOD.   Maximum likelihood (ML) covers a large class of procedures united by the principle that reconstructed trees should maximize the probability of observing an available data set under a particular model of evolutionary change. Of course, there are countless different models of molecular evolutionary change, but the idea is to choose one that is plausible based on data-informed or theory-informed knowledge about the system. A simple model might assume that the rate of nucleotide substitution is the same between all nucleotide pairs, that the substitution rate is the same throughout the tree, and that the expected number of substitutions along each tree branch is a function of this substitution rate and the length of the branch. A computer-based search then seeks to identify a tree that, under those assumptions, is most likely to explain the data that initiated the search.

Advantages of ML are that it allows the user to specify an evolutionary model and that it usually identifies a single ML tree (although not necessarily one that is statistically or biologically better than alternatives that are close, or sometimes even grossly different, in structure). A practical disadvantage is that ML is considerably slower than parsimony and distance-based approaches and can easily exceed the capacity of even high-speed desktop computers. The computer programs DNAML (introduced by Felsenstein 1981a) and TREE-PUZZLE (Strimmer and von Haeseler 1996) implement maximum likelihood procedures for molecular sequence data, as does PAUP* (Swofford 2000).

BAYESIAN ANALYSIS.   The hottest new approach in phylogenetics was made practicable by the computer program MrBayes (Huelsenbeck 2000), which again requires that the user postulate a model of evolution. Bayesian analysis is actually a variant of likelihood methods, but instead of seeking a single tree with the greatest likelihood of observing the data, it produces best sets of trees and entire probability distributions of likelihoods given the data and the evolutionary model specified (Rannala and Yang 1996). Like ML, it does so by searching a landscape of multitudinous possible trees, moving from point to point on the likelihood surface in pursuit of higher vantages (i.e., more likely trees). But unlike ML, which can get trapped on a local hill instead of the globally highest mountains, the algorithm in MrBayes (a Markov chain Monte Carlo process; Mau et al. 1999) allows the search to leap valleys and thereby gain not only a better perspective on the full likelihood landscape, but also a better opportunity to ascend its highest peaks.

Conclusions about phylogenetic procedures

The plethora of data analysis methods in molecular phylogenetics can be confusing, but it is important to remember a few major points. First, it is always desirable to attempt a match between the assumptions of a phylogenetic procedure and the nature of evolution in the molecular characters assayed. Second, where possible, it is often advisable to include both a distance-based and a character-based approach in an analysis for comparison. Distance-based methods are relatively straightforward and often provide a simple overview. Character-based approaches can be more information-rich and testable (*sensu* Popper 1968) because they provide hypotheses about character-state distributions along tree branches (Avise 1983; Baverstock et al. 1979; Patton and Avise 1983). Consider, for example, Figure 4.8, in which hypothesized character-state changes are explicitly summarized as an inherent aspect of the reconstruction process (such detail is impossible in distance analyses alone). Furthermore, if other evidence suggests that portions of the tree topology are suspect , the characters responsible for the difficulty can be identified. Perhaps they were scored incorrectly, or perhaps their states were polyphyletic, in which case they might be subjected to further analysis to assess the molecular basis of the apparent homoplasy.

A third general point is that outgroup taxa should be included in phylogenetic analyses because they facilitate tree rooting and help to establish character-state polarities. Fourth, it is important to include confidence statements about putative clades. The most common approach is bootstrapping (Felsenstein 1985a; Hedges 1992), which involves three steps: sample (with replacement) from the existing data; generate a new tree from the re-sampled data; and then repeat the process perhaps hundreds of times to assess how frequently particular groups or clades appear in these pseudo-replicate trees. Thus, bootstrapping indicates how well putative clades in a tree are supported by the existing data (although not necessarily how well the available data represent genes not assayed). Interestingly, bootstrapping is unnecessary under Bayesian analyses because the probability of a given clade is already evident from its frequency in the set of Bayesian trees with high likelihood.

Fifth, the brouhaha about finding the "best" tree is often somewhat moot, because even less than perfect phylogenetic reconstructions still usually capture the major features of a tree that may be of primary biological interest. For example, a common observation in phylogeographic surveys (see Chapter 6) is that regional sets of populations are tightly allied genealogically, yet are separated from one another by pronounced phylogenetic breaks. These gaps, which are often the historical footprints of Pliocene or Pleistocene separation events, are evident regardless of the method used to reconstruct the phylogenetic trees.

One final point about phylogenetic reconstruction will provide a segue into the next section, on gene trees. All phylogenetic methods discussed above are based on the assumption that the characters analyzed are "independent," but this concept warrants elaboration. Characters are independent in a mechanistic sense if changes in one character occur independently of those in another, such that their states do not covary because of pleiotropic effects in the underlying mutational process. For example, appearances and disappearances of restriction fragments tend to covary across digestion profiles, whereas the responsible restriction site changes do not; for this reason, it is preferable to code RFLP data as presence versus absence of particular restriction sites rather than fragments (or at least to accommodate the covariance of fragments in the phylogenetic analysis). The assumption of independence, critical to most computational algorithms, is probably valid for most molecular characters (unlike the situation for many morphological traits) and, indeed, is a major strength of multi-character molecular approaches.

However, there is another sense in which molecular characters may be partially non-independent in evolution. When molecular characters are tightly linked, molecular states tend to covary *in transmission* across organismal generations. Such is the case for nearby nucleotides in a nuclear gene, or for any and all nucleotides in a non-recombining organelle genome. Thus, although such character states are independent in the mechanistic sense of mutational origins, they are not independent with regard to genealogy. Recognition of this fact led to the important distinction between a "gene tree" and a "species tree" (Avise 1989a; Doyle 1992; Neigel and Avise 1986; Nichols 2001; Pamilo and Nei 1988; Tajima 1983; Tateno et al. 1982; Wilson et al. 1985).

Gene Trees versus Species Trees

When OTUs are multiple alleles of a complex locus (e.g., haplotype sequences in mtDNA genomes), a reconstructed phylogeny represents a gene tree. Any group of organisms has a single true pedigree that extends back through time as an unbroken chain of parent–offspring genetic transmission, but due to the non-deterministic nature of biparental Mendelian heredity in sexually reproducing species, not all genes will have trickled through this organismal pedigree in identical fashion. Thus, gene trees inevitably differ somewhat in topology from one unlinked locus to the next (Ball et al. 1990), both within and between related species. Thus, even in the absence of introgression (see Chapter 7) or horizontal gene transfer (see Chapter 8), a given gene tree within any set of species will differ in structure from others, as well as from the consensus population tree or species tree of which it is a part. These differences result from "lineage sorting" processes that may be exemplified as follows.

Consider a single population pedigree through which haplotypes have descended. A simple conceptual case involves mtDNA inherited through female lines, but the principles apply to any non-recombined haplotypes of particular nuclear genes as well. As shown in Figure 4.9, some females, by chance, leave no daughters (their mtDNA lineages terminate), whereas others produce one or more daughters that may, in turn, contribute mtDNA to

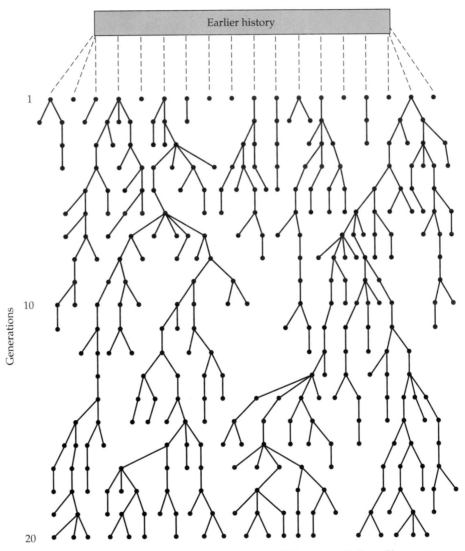

Figure 4.9   The allelic lineage sorting process within a population. Shown is an mtDNA gene tree through 20 generations. Each node represents an individual female, and branches lead to daughters. The tree was generated by assuming a Poisson distribution of progeny numbers with a mean of one daughter per female. (After Avise 1987.)

successive generations. Thus, as an inevitable consequence of differential organismal reproduction, any gene tree (cytoplasmic or nuclear) continually self-prunes: some branches are lost as others proliferate. At equilibrium, the expected frequency distribution of times to common ancestry can be approximated if the population demography is specified (see Box 2.3). In general, for relatively stable populations, it is unlikely that two or more founding lineages will survive beyond $4N$ generations, where N is the population size (Figure 4.10). In Chapter 6, more will be said about lineage sorting and related concepts of coalescent theory as they relate to population size.

Now consider the lineage sorting process extended to two daughter (or sister) taxa, A and B, that stem from the same ancestral population. With regard to a gene tree within these sister populations or species, three categories of phylogenetic outcomes are possible (Figure 4.11): (I) *reciprocal monophyly*, in which all alleles within each sister taxon are genealogically closer to one another than to any heterospecific alleles; (II) *polyphyly*, wherein some alleles in each taxon are genealogically closer to heterospecific alleles than to homospecific alleles; and (III) *paraphyly*, in which all alleles within one daughter taxon are one another's closest relatives, but some alleles in the second taxon are genealogically closer to heterospecific alleles. Category I in Figure 4.11 also illustrates how the depths of gene trees can vary even when their branching topologies agree with the species tree (e.g., the extant alleles in B trace to an ancient node "b," whereas the alleles in A trace to a recent node "a"). Categories II and III in Figure 4.11 illustrate how a gene tree can differ in fundamental branching order from a species tree. These discor-

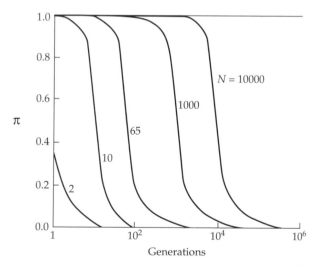

Figure 4.10   Probabilities (π) of survival of two or more founding lineages through time. Shown are probability curves for populations of various size (N) in which females produce daughters according to a Poisson distribution with mean 1.0. (After Avise et al. 1984a.)

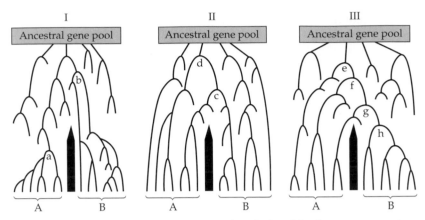

Figure 4.11   Three categories of phylogenetic relationships between two sister taxa (A and B) are possible with respect to an allelic genealogy. Lowercase letters point out important ancestral nodes to which extant alleles or haplotypes trace. Solid dark bars indicate barriers to reproduction (extrinsic or intrinsic). The phylogenetic categories in the gene tree are as follows: I, reciprocal monophyly; II, polyphyly; III, paraphyly of A with respect to B. (After Avise et al. 1983.)

dances arise because many allelic separations predate the species split (unless the ancestral form went through an extreme population bottleneck just prior to speciation). Figure 4.12 shows diagrammatically how these three categories of phylogenetic relationships can characterize the same pair of sister taxa at different times following their separation. This occurs because lineage sorting typically converts any initial condition of genealogical polyphyly to one of paraphyly and eventually to one of reciprocal monophyly.

If evolutionary genetic distances among haplotypes are measured in units of time since common ancestry, these three categories of phylogenetic relationships between sister taxa (with respect to a gene tree) may be defined by the formal inequalities in Table 4.2. Neigel and Avise (1986; see also Hudson and Coyne 2002; Rosenberg 2003) employed replicated computer simulations to monitor the probabilities of each phylogenetic status of sister taxa with respect to mtDNA lineages (Figure 4.13). Shortly after their separation, it is highly likely that sister taxa will exhibit a polyphyletic gene tree status. At intermediate times since speciation (typically $N - 3N$ generations, where N is the population size of each sister taxon), probabilities of polyphyly, paraphyly, and monophyly are intermediate as well. Only after about $4N$ generations do sister taxa finally have a high probability of becoming reciprocally monophyletic. Similar results apply to nuclear genes (Nei 1987), although times to monophyly are extended accordingly because of the expected fourfold larger effective population sizes for nuclear loci (see Box 2.3).

Such models assume selective neutrality at the locus per se (plus some specified set of population demographic conditions), but dramatically different outcomes can arise if strong selection occurs at or near the locus under examination. For example, when balancing selection maintains haplotype

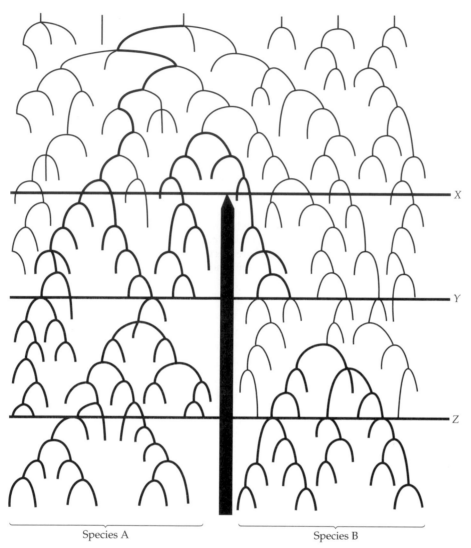

Figure 4.12   The lineage sorting process extended to two sister taxa. Shown are distributions of allelic lineages at a single gene through an ancestral population subdivided at time X into two daughter populations or species. With respect to this gene tree, these sister species are polyphyletic between times X and Y, are paraphyletic between times Y and Z, and are reciprocally monophyletic beyond time Z. (After Avise and Ball 1990.)

polymorphisms within a species, expected times to reciprocal monophyly in a gene tree can be much longer than those expected under a neutral model because lineage sorting at that locus is effectively inhibited. Beginning with several studies published in the late 1980s on two such balanced polymorphisms—involving major histocompatibility loci in mammals (Figueroa et al. 1988; Lawlor et al. 1988; McConnell et al. 1988; Takahata and Nei 1990) and a

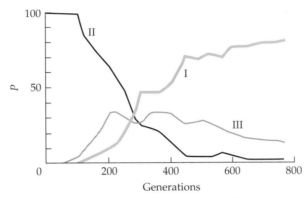

Figure 4.13 **Probabilities of reciprocal monophyly (I), polyphyly (II), and para-phyly (III)** for two sister taxa at indicated numbers of generations following a simulated speciation. In each of 100 replicate computer runs, daughter species were founded by 20 and 30 individuals respectively and allowed to grow rapidly to carrying capacity $N = 200$. (After Neigel and Avise 1986.)

self-incompatibility locus in plants (Ioerger et al. 1990)—it soon became apparent that some such balanced polymorphisms can persist for millions of years and be maintained across sequential speciation events.

Discordances between species splitting patterns and the topologies of gene trees can also characterize taxa that separated anciently, but whose speciations occurred close together in time (Figure 4.14). The same kinds of lineage sorting processes are responsible: The lineages from the polymorphic ancestral gene pool that happen to have reached fixation in the descendant taxa may, by chance, be those that produce a gene tree/species tree discordance (Takahata 1989; Wu 1991). Such discordances can create problems for

TABLE 4.2 **Definitions of the phylogenetic status of two sister taxa with respect to a gene tree they contain**

Phylogenetic category	Phylogenetic status	Distance relationship[a]
I	A and B monophyletic	max d_{AA} < min d_{AB} and max d_{BB} < min d_{AB}
II	A and B polyphyletic	max d_{AA} > min d_{AB} and max d_{BB} > min d_{AB}
IIIa	A paraphyletic with respect to B	max d_{AA} > min d_{AB} and max d_{BB} < min d_{AB}
IIIb	B paraphyletic with respect to A	max d_{AA} < min d_{AB} and max d_{BB} > min d_{AB}

Source: Neigel and Avise 1986.
[a] Maximum evolutionary distances within either taxon (max d_{AA} or max d_{BB}) versus minimum distance between sister taxa (min d_{AB}) are the deciding criteria (assuming that these genealogical distances are linearly related to time; see Figure 4.11 and text).

phylogeny estimation at the species or population level. For example, intense debate about the relationships of humans, chimpanzees, and gorillas has centered on which phylogenetic tree is "true": humans and chimps as sister taxa, or perhaps chimps and gorillas, or perhaps humans and gorillas. The branching pattern in this related triad of species clearly consists of two closely spaced nodes like those in Figure 4.14. Many molecular assays have been applied to this question, but not all genetic data have yielded exactly the same phylogenetic outcome (see Chapter 8). From the theory outlined above concerning quasi-independent gene trees and the idiosyncrasies of lineage sorting across closely spaced phylogenetic nodes, perhaps no single outcome should be expected.

These perspectives stemming from molecular research reveal several points of qualitative importance to phylogenetics, beyond the immediate fact that gene trees and species trees can differ in branching topology. First, with regard to haplotype relationships, the phylogenetic status of a given pair of species is itself an evolutionarily dynamic characteristic, with a usual time course subsequent to speciation being polyphyly or paraphyly preceding reciprocal monophyly (see Figures 4.12, 4.13). Second, the phylogenetic

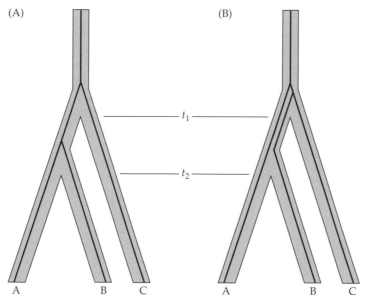

Figure 4.14 Gene genealogies within a species phylogeny. Shown are two topologically distinct gene trees (thin dark lines) possible within a species phylogeny (broader shaded branches) that consists of two sister taxa and an outgroup. In (A), the gene tree and the species tree have the same branching pattern, whereas in (B) the branching topologies differ. For neutral alleles, the probability of the discordance exemplified in (B) is given by $2/3e^{-T/2N_e}$ (Nei 1987), where t_1 is the time of the first speciation, t_2 is the time of the second speciation, $T = t_1 - t_2$, and N_e is the effective population size.

status of species is a function both of the pattern of population splitting and of historical demography within the populations involved (Avise et al. 1984a). For example, sister species with large effective population sizes (see Box 2.2) will tend to retain a polyphyletic or paraphyletic status for longer times than will species with small N_e, all else being equal. Third, in accounting for the appearance of "heterospecific" alleles within a given species, it now is clear that possibilities involving lineage sorting from an ancestral gene pool must be considered (in addition to the usual scenarios of interspecies transfer mediated by introgressive hybridization; see Chapter 7).

BOX 4.3 Isolation of DNA Haplotypes

The table enumerates some special genetic systems and approaches for the isolation of DNA haplotypes. When such systems also exhibit high genetic variation

Approach	Rationale
Organelle genomes	Most individuals effectively homoplasmic and haploid; non-recombinational heredity
Sex chromosomes (e.g., X or Y in mammals, Z or W in birds)	Heterogametic sex is haploid; limited recombination, particularly in the sex-specific Y or W chromosomes
Species with haplo-diploid sex determination	Males are haploid in many hymenopteran insects
Species with prominent haploid phase of life cycle, or of a particular tissue	Gametophyte stage of mosses, for example, is haploid; endosperm in seeds of gymnosperms is a haploid product (gametophyte) from the female parent
Haploid species	Haploid microorganisms should be suitable, provided that sexual reproduction and recombination are limited
Selfing diploid species	Highly inbred strains (natural or artificial) usually carry identical-by-descent alleles

Because of its rapid pace of evolution, its "haploid" packaging within most organisms, and its non-recombining mode of transmission, mtDNA has provided the vast majority of empirical data suitable for estimating gene trees over microevolutionary time scales (see Chapter 6). In principle, data from nuclear loci could be exploited similarly, and some successful empirical examples do exist (Antunes et al. 2002). In general, however, three major practical difficulties arise. First is the technical problem of isolating individual haplotypes of a nuclear locus from diploid organisms. Box 4.3 describes several genetic systems and experimental approaches that might circumvent this difficulty. A second potential complication, especially at the intraspecific

but little or no inter-allelic recombination, they provide suitable potential opportunities for construction of gene trees.

Comments	Early (or otherwise key) reference
By far the most widespread source for gene-tree data	Avise 1989a
Relatively few genes identified or surveyed so far	Bishop et al. 1985; Hurles and Jobling 2001; Vulliamy et al. 1991; see Chapters 6 and 7
Not yet widely capitalized upon for full gene trees	Hall 1990
Not yet widely capitalized upon for full gene trees; does not apply to triploid endoperms of angiosperms	McDermott et al. 1989
Relatively few attempts	Nelson et al. 1991
Often used to advantage in *Arabidopsis*, for example	Stahl et al. 1999

(continued)

Approach	Rationale
PCR amplification in vitro from single sperm or egg, or single DNA molecule	Gametes are haploid; each molecule represents one haplotype
DNA amplification in vivo (in biological vectors)	Cloning passes DNA through a bottleneck of one molecule
Allele-specific PCR amplification and related approaches	If primers can be identified as allele-specific, they can be used to amplify haplotypes even from heterozygous loci
SSCP and DGGE gels	Physical separation of haplotypes at a locus following PCR amplification
HAPSTR and SNPSTR assays	Another method to physically separate nuclear haplotypes, in this case in regions surrounding STR loci or specific SNPs
Extraction of individual chromosomes	Use of inbred strains, or of controlled crosses producing individuals with chromosomes identical by descent; especially powerful when applied to genes within chromosome inversion systems where recombination is limited or absent

level, is intragenic recombination, which over time in a population can mix and match various pieces of otherwise separate haplotypes at a nuclear locus. A third complication is gene conversion, wherein particular DNA sequences in effect are converted to those of another allele, or even to those of another locus in the same gene family (e.g., Popadic and Anderson 1995; Popadic et al. 1995).

Any shuffling of genetic material among alleles by intragenic recombination or gene conversion, if frequent over time frames relevant to a genealogical reconstruction, will obscure the otherwise linear evolutionary histories of particular haplotypes within a species (Hudson 1990). Although

Comments	Early (or otherwise key) reference
Technical hurdles high, but accomplished successfully in humans and a few other species	Boehnke et al. 1989; Grewal et al. 1999; H. Li et al. 1988; Lien et al. 1999; Navidi and Arnheim 1999; Ruano et al. 1990; Stephens et al. 1990; Sun et al. 1995
Highly laborious; concerns about PCR misincorporation	Scharf et al. 1986
Potentially of wide nuclear applicability, but primer design is labor-intensive	Fullerton et al. 1994; Harding et al. 1997; Newton et al. 1989
Potentially useful, but seldom employed to date in a gene-tree context	See section on SSCPs in Chapter 3
Still mostly in development	See relevant section in Chapter 3
Methods most readily available in *Drosophila*	Aquadro et al. 1986, 1991

statistical analyses of patterns of nonrandom association (disequilibrium) among tightly linked polymorphic markers can help to reveal whether recombination has been frequent in the history of a gene region (Clark 1990; Crandall and Templeton 1999; Kuhner et al. 2000; Stephens and Nei 1985; Stephens et al. 2001a,b), these and related methods (e.g., McGuire et al. 1997) usually encounter serious limitations in reconstructing haplotypes when recombination has been other than rare (Posada 2002; but see also Maynard Smith and Smith 1998). Thus, if the intent is to recover a non-anastomose gene tree, attention must normally be confined to DNA segments with little or no recombination (Box 4.4).

BOX 4.4 Intraspecific Gene Trees for Nuclear Loci

Most early attempts to study nuclear gene trees at the intraspecific level involved *Drosophila* species because experimental crosses could be conducted to "extract" individual chromosomes from wild diploid fruit flies. Each use of this breeding procedure resulted in a pure strain whose members were "identical by descent" for a particular chromosomal haplotype (see Box 4.3). Such haplotypes could then be assayed for DNA sequences at particular loci, and the resulting data could be used in attempts to reconstruct allelic genealogies (haplotype trees or gene trees).

One classic application of this approach resulted in a successful estimate of a gene tree for *Adh* (alcohol dehydrogenase) in *D. melanogaster*, as summarized in Figure A. Any clean gene tree reconstruction of this sort requires that the haplotypes involved have had a history of little or no recombination over the time scales of the phylogeny. When such stretches of DNA with limited internal recombination can be identified, the gene genealogies that they imply can also be used to map phenotypes associated with that chromosomal region (Templeton et al. 1992). An example involving phylogenetic placements of the "fast" (F) and "slow" (S) protein electromorphs at *Adh* is shown in Figure A. The rationale for this type of endeavor is that any phenotypes dictated by alleles within the non-recombining marker region would be embedded in the same evolutionary history that is reflected in the gene tree.

However, it has proved difficult to predict which gene regions are likely to be sufficiently free of inter-allelic recombination to permit gene tree reconstructions. For example, similar attempts to construct an intraspecific allelic phylogeny for *Adh* in *D. pseudoobscura* were mostly thwarted by an absence of strong nonrandom associations among linked restriction sites, apparently due to a history of more frequent inter-allelic exchange (Schaeffer and Miller 1992; Schaeffer et al. 1987; but see also Schaeffer et al. 2001). The xanthine dehydrogenase region of *D. pseudoobscura* also showed few nonrandom associations among restriction sites (Riley et al. 1989), and the same proved true for several other loci (e.g., *notch*, *white*, *zeste-tko*, and perhaps *amylase*) in *D. melanogaster* (Aguadé et al. 1989a; Langley and Aquadro 1987; Langley et al. 1988; Schaeffer et al. 1988).

Aquadro et al. (1991) capitalized on the recombination suppression properties of chromosomal inversions in *D. pseudoobscura* to generate a phylogeny for an amylase gene (*Amy*) that is contained within the inverted region of the third chromosome. The reduction in effective recombination in inversion heterozygotes is dramatic and occurs because crossing over inside the inverted region normally produces dysfunctional duplication and deficiency products that are shunted to polar bodies, where they fail to participate in zygote formation (also, *Drosophila* males lack recombination). The biological significance of recombination suppression in this system is that it facilitates the maintenance of linked and apparently adaptive epistatic complexes of genes within the inverted regions (Schaeffer et al. 2003). A gene tree based on restriction site maps for 28 *Amy* haplotypes is presented in Figure B; the karyotypic parsimony network is shown in Figure C.

The following conclusions emerged from comparisons of this tree and network (Aquadro et al. 1991): restriction site differences are greater among gene

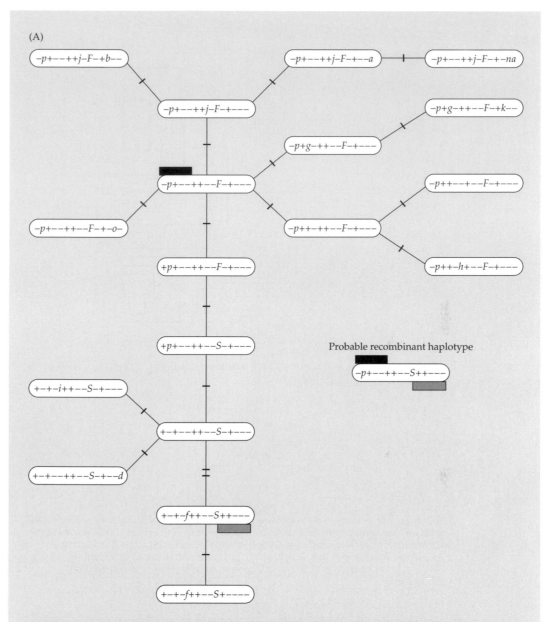

(A) Classic example of an intraspecific nuclear gene tree involving the alcohol dehydrogenase (*Adh*) gene in *Drosophila melanogaster*. Shown is a parsimony network for 18 haplotypes (in ovals) as identified by 15 restriction sites and other DNA sequence characters (coded from left to right in the 5′ → 3′ direction). This case also illustrates how inter-allelic recombination at a nuclear locus can sometimes be deduced from haplotype data. Note from the black and gray bars that the 5′ end of a probable recombinant haplotype appears to stem from one portion of the parsimony network and the 3′ end from the other. From this phylogeny and additional genetic evidence, it was also surmised that the "fast" allozyme allele (*F*) probably evolved recently from a "slow" (*S*) ancestral allele (Aquadro et al. 1986; Ashburner et al. 1979; Stephens and Nei 1985). (From Avise 2000a, using data from a broader genealogy presented in Aquadro et al. 1986.)

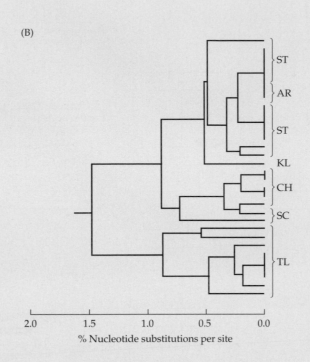

(B)

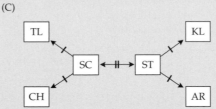

(C)

(B, C) Molecular and karyotypic phylogenies in *Drosophila pseudoobscura*. (B) Gene tree for 28 molecular haplotypes observed at the amylase (*Amy*) locus. Also indicated are the chromosomal inversion types from which these *Amy* haplotypes were extracted: ST (Standard); AR (Arrowhead); KL (Klamath, from *D. persimilis*); CH (Chiricahua); SC (Santa Cruz); and TL (Treeline). (After Aquadro et al. 1991.) (C) Cytogenetic phylogeny of these same gene arrangements.

arrangements than among haplotypes within the same gene arrangement; the gene phylogeny based on *Amy* is generally concordant with the inversion phylogeny as gauged by gross karyotype; and from application of a molecular clock, the inversion polymorphism is old (perhaps about 2 million years). A follow-up study utilized direct DNA sequence data at *Amy* to deduce that ST could be excluded as the ancestral chromosomal condition, TL could not, and SC appeared to be the most likely ancestral arrangement (Popadic and Anderson 1994).

Outside such inverted regions (as well as within them; Navarro et al. 1997, 2000), many demographic as well as molecular factors can influence the history of effective recombination within a gene. For example, haplotype contents of species that are spatially subdivided should tend to exhibit greater linkage disequilibrium (nonrandom associations) than those in high-gene-flow species in which alleles are routinely brought together such that recombination among them is at least possible (Baum and Shaw 1995). Different chromosomal regions also are known to differ inherently in recombination rates, one consequence again being different patterns of linkage disequilibrium between markers (Aguadé and Langley 1994; Aguadé et al. 1989b; Aquadro et al. 1994; Begun and Aquadro 1992). For these and additional reasons, different genes within the same organismal phylogeny can and often do show very different patterns of genetic variation and genealogical relationships (e.g., Hasson and Eanes 1996; Machado et al. 2002).

Unfortunately, the technical and biological challenges of estimating autosomal gene trees at the intraspecific level (such as those in the accompanying figures) have seldom been overcome in sexual diploid species, so relatively few good examples exist in the current literature (see Chapters 6 and 7).

Gene trees and species trees are equally "real" phenomena, merely reflecting different aspects of the same phylogenetic process. Thus, occasional discrepancies between the two need not be viewed with consternation as sources of "error" in phylogeny estimation. When a species tree is of primary interest, gene trees can assist in understanding the population demographies underlying the speciation process, as well as the species splitting patterns themselves (see Chapters 6, 7, and 8). Of course, for such purposes it would be desirable to include information from multiple gene genealogies. Each gene tree also is of inherent interest because it describes the evolutionary history of genetic changes within a localized bit of the genome. In studies of the ages and origins of specific genetic adaptations (or disorders), for example, such single-locus reconstructions can become the primary foci of attention (see Box 4.4).

SUMMARY

1. Population genetic theory and phylogenetic theory are highly germane to analyses and interpretations of molecular data. Many methods of data analysis (described throughout this book) are idiosyncratic to particular research questions, but this chapter outlines some major principles and methods for estimating phylogenetic trees within and among species. In rough order of historical appearance, these include phenetic methods of numerical taxonomy, Hennigian cladistics as applied to small numbers of discrete characters with alternative states, various forms of parsimony analysis (a logical outgrowth of cladistics) as applied to more complicated data sets, maximum likelihood methods, and Bayesian approaches (the latest and most powerful approach in phylogenetic analyses of DNA sequences).

2. Empirical studies have documented substantial variation of several kinds in molecular evolutionary rates: across nucleotide positions within a codon; among non-homologous genes within a lineage; among classes of DNA within a genome; among different genomes (nuclear versus organelle) within an organismal lineage; and among organismal lineages with respect to particular classes of homologous genes and characters. Nonetheless, a general time-dependent nature of molecular evolution is also evident. The concept of molecular clocks has played a major role in molecular phylogenetics. Both absolute and relative rate tests have been widely employed in the assessment of molecular evolutionary tempos in different taxa.

3. Important distinctions exist between a gene tree and a species tree. These two aspects of genealogy provide different but mutually informative phylogenetic perspectives on evolutionary processes.

PART II

Applications

5

Individuality and Parentage

With the recognition ... that the ... genome is replete with DNA sequence polymorphisms such as RFLP's, it was only a small leap to imagine that DNA could, in principle, provide the ultimate identifier.

E. S. Lander (1991)

Most species of sexually reproducing organisms harbor sufficient genetic variation that appropriate molecular assays can distinguish each individual from all others with near certainty. Furthermore, polymorphic molecular markers with known pathways of hereditary transmission afford powerful opportunities to identify parent–offspring links. Issues of genetic identity versus non-identity and of biological parentage (genetic maternity and paternity) fall at the extreme microevolutionary end of the genealogical continuum. Especially suitable for such analyses are highly variable nuclear genome markers with specifiable modes of Mendelian inheritance. These markers include the allozyme products of numerous protein-coding genes as well as DNA-level "fingerprints" such as those provided by minisatellite loci, RAPDs, and (especially in recent years) microsatellites.

Human Forensics

Several of the techniques underlying molecular forensics were developed initially for human applications and only later modified and adapted for application to wildlife issues. Thus, this treatment will begin with a discussion of DNA-based forensics as applied to people.

History of laboratory approaches

The earliest efforts in human molecular forensics involved typing various blood groups and serum proteins, but these markers registered only modest genetic variation and hence offered limited evidence on individual identity and uniqueness. At the DNA level, the first widely employed approach entailed RFLP analyses of minisatellite sequences, typically probed one locus at a time. Especially in the late 1980s and early 1990s, several private companies (e.g., Cellmark and Lifecodes) and governmental agencies (e.g., the U.S. Federal Bureau of Investigation) began routinely typing genomes using minisatellite loci. Several such loci in human populations proved to display numerous alleles distinguished from one another (in gel mobility) by virtue of variable numbers of tandem repeat sequences (VNTRs) (Figure 5.1).

However, DNA fragments at these minisatellite loci were typically several kilobases in size and were measured with some error, so determining the number of distinct allelic classes from a quasi-continuous distribution of fragment lengths was problematic (Devlin et al. 1991, 1992). In practice, grouping procedures were employed to pool fragments of similar length into allelic "bins," whose widths reflected magnitudes of experimental error across replicates (Budowle et al. 1991). Even so, each VNTR locus often exhibited 10–30 or more differentiable bins of alleles in a typical population sample. Most of these alleles were uncommon (Table 5.1), and nearly all individuals were heterozygous. This extensive variation carried a key consequence: At any locus, the probability of a genetic match between randomly

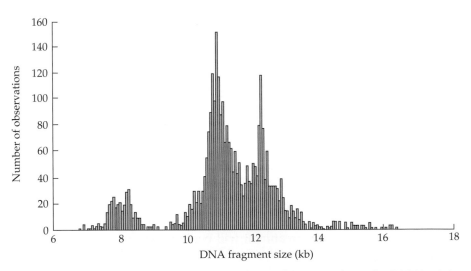

Figure 5.1 Frequency distribution of restriction fragments from the *D2S44* minisatellite locus in Caucasian samples. Data are from Lifecodes, Inc. (After Devlin et al. 1992.)

chosen individuals was low. As illustrated in Box 5.1, genotypic frequencies from several unlinked VNTR loci could then be statistically combined to calculate probabilities of observing any particular multi-locus DNA profile in a random person drawn from a baseline population for which allele frequencies were known.

In the ensuing years, most human forensic laboratories switched to assays of the short tandem repeat (STR) sequences of microsatellite markers, and this remains the primary method in use today. Microsatellites are also highly variable, but they offer several advantages over minisatellite sequences: an effectively unlimited supply of loci for examination; shorter fragments and

TABLE 5.1   Frequencies of alleles (bins) at four hypervariable VNTR loci in Caucasians

| | *VNTR locus D1S7* | | *Binned allele frequencies in Caucasian samples at VNTR loci* | | |
| | *Frequency in* | | | | |
Binned allele	*Caucasians (n = 605)*	*Afr. Amer.[a] (n = 372)*	*D2S44 (n = 802)*	*D17S79 (n = 563)*	*D4S139 (n = 460)*
1	0.004	0.007	0.005	0.010	0.004
2	0.006	0.009	0.003	0.003	0.010
3	0.009	0.011	0.016	0.007	0.006
4	0.012	0.007	0.024	0.004	0.014
5	0.011	0.016	0.046	0.015	0.033
6	0.014	0.020	0.034	0.223	0.024
7	0.010	0.011	0.123	0.199	0.040
8	0.029	0.035	0.106	0.263	0.047
9	0.021	0.023	0.084	0.200	0.054
10	0.014	0.030	0.049	0.029	0.071
11	0.028	0.030	0.083	0.032	0.108
12	0.031	0.026	0.039	0.010	0.190
13	0.046	0.044	0.041	0.006	0.129
14	0.067	0.069	0.039		0.095
15	0.057	0.065	0.087		0.036
16	0.061	0.073	0.089		0.036
17	0.069	0.054	0.075		0.103
18	0.055	0.051	0.022		
19	0.060	0.047	0.018		
20	0.063	0.063	0.008		
21	0.079	0.062	0.008		
22	0.077	0.060			
23	0.077	0.074			
24	0.032	0.017			
25	0.019	0.027			
26	0.050	0.071			

Source: These data were introduced by the Federal Bureau of Investigation into a criminal case in Athens, Georgia in May, 1991. They are part of a larger database that includes frequencies from additional VNTR loci in Caucasians, African Americans, and Hispanics.
[a] Shown for comparison are allele frequencies at the *D1S7* locus in a sample of African Americans.

BOX 5.1 Probabilities of Single-Locus and Multi-locus DNA Profiles

Based on the Caucasian VNTR data in Table 5.1, these calculations assume random associations of alleles both within loci (Hardy–Weinberg equilibrium) and among loci (gametic phase equilibrium).

(a) Probability that an individual is heterozygous at the *D1S7* locus:

$$h = 1 - \sum [(0.004)^2 + (0.006)^2 + \ldots + (0.050)^2] = 0.945$$

(b) Examples of probabilities of particular allelic combinations at individual loci:

5/22	Heterozygote at *D1S7*	$0.011 \times 0.077 \times 2.0$	= 0.001694
6/8	Heterozygote at *D2S44*	$0.034 \times 0.106 \times 2.0$	= 0.007208
5/9	Heterozygote at *D17S79*	$0.015 \times 0.200 \times 2.0$	= 0.006000
12/12	Homozygote at *D4S139*	0.190×0.190	= 0.036100

(c) Probability of the multi-locus DNA profile in (b):

$$0.001694 \times 0.007208 \times 0.006000 \times 0.036100 = 3 \times 10^{-9}$$

Suppose a crime suspect exhibited the multi-locus genotype shown in (b). Then, if the assumptions of the model are met, the probability of a match with a randomly drawn genotype from the Caucasian population is about one in 333 million.

---

smaller tandem repeat units (2–4 bp each), such that appropriate acrylamide gels can cleanly distinguish all alleles differing in size (i.e., there is no need for artificial binning); and a PCR basis for the procedure, meaning that even minuscule amounts of source tissue (a smidgen of dried blood, a hair, or a drop of saliva) suffices. Once a database on microsatellite allele frequencies is available for a reference population, the logic and procedures of data interpretation in a forensic context are basically identical to those described in Table 5.1 and Box 5.1 for minisatellite data.

By the year 1990, more than 2,000 court cases in 49 states had used DNA evidence in civil litigation or criminal proceedings (Chakraborty and Kidd 1991), and more than one legal expert predicted that DNA typing would be for the late twentieth century and beyond what traditional fingerprinting was to the nineteenth (Melson 1990). That prediction proved to be correct, and DNA fingerprinting has today become an integral component of modern crime laboratories around the world (Butler 2001). Molecular analyses have revolutionized human forensic practice. However, as with conventional fingerprinting, DNA typing merely provides physical evidence that assigns particular individuals to, or excludes them from, particular tissue samples and, thus, must be used in conjunction with additional lines of evidence to resolve legal issues such as criminal guilt or innocence.

History of controversies

If relevant tissue samples left at a crime scene have yielded assayable DNA, two evidential outcomes are then possible: the samples do not match the suspect, in which case the forensic evidence may be declared exculpatory; or the tissues are declared a match. In the Western judicial tradition, where a suspect is considered innocent until proven guilty, the latter situation clearly focuses attention on the following question: What is the probability of a spurious DNA match? At face value, such probabilities calculated from DNA forensic data are infinitesimally small (3×10^{-9} in the example in Box 5.1). Thus, a perfect multi-locus match is usually interpreted as establishing genetic identity "beyond reasonable doubt."

However, this type of conclusion is based on some key assumptions whose validity was questioned, thus placing DNA fingerprinting itself on trial, not long after the technique's debut (Lander 1989; Lewontin and Hartl 1991). Regarding the probability of a spurious genotypic match, one contentious issue was the premise that genotypic frequencies are independent across loci, an assumption that could be violated if (for example) there was pronounced nonrandom mating and population subdivision. Human populations are not entirely homogeneous, but rather exhibit genetic structure that can produce allelic correlations due to historical or cultural separations. (For example, alleles for blond hair and blue eyes, although independent in genetic transmission and mode of action, nonetheless are highly correlated in human populations due to historical associations and nonrandom mating.) Thus, researchers soon analyzed the molecular data sets employed in human forensics for possible genetic correlations within or among loci (Risch and Devlin 1992; Weir 1992). Overt correlations were not found, at least for most of the genetic markers analyzed.

However, the effect of population substructure on forensic conclusions is a matter of degree (Nichols and Balding 1991). Consider an extreme example in which a suspect belongs to a small inbred community that differs dramatically in allele frequency from North American Caucasians overall. Use of a Caucasian database (as in Table 5.1) as a reference for calculating genotypic probabilities clearly would be inappropriate, and the direction of error could work against an innocent defendant (e.g., the likelihood of a genotypic match between an innocent suspect and another member of the local community who may actually have committed the crime is much higher than the probability of a match within a broader Caucasian population). To circumvent this problem, each relevant human "subgroup" might be specified separately and appropriate probabilities of a genetic match calculated accordingly for each specific case. Unfortunately, such extensive genetic characterization is infeasible logistically, even if appropriate subgroups could somehow be identified. Such concerns led Lewontin and Hartl (1991) to conclude that then-current applications of DNA fingerprinting had serious flaws as forensic evidence.

The degree of human population substructure should not be overstated, however. In terms of allozyme and blood group polymorphisms, Lewontin (1972) had earlier argued that more than 90% of total genetic diversity in humans occurred within (rather that between) races and concluded that "our perception of relatively large differences between human races and subgroups… is indeed a biased perception, and that, based on randomly chosen genetic differences, human races and populations are remarkably similar to each other." Much the same conclusion applies to at least some DNA fingerprint loci as well (Balazs et al. 1989, 1992; Krane et al. 1992), as illustrated by the similar spectra of allele frequencies at D1S7 in Caucasian and African American populations (see Table 5.1). This and other observations led Morton (1992) to conclude that Lewontin and Hartl's (1991) objections to genotypic probability calculations were themselves "absurdly" conservative in favor of the defense.

The scientific brouhaha over DNA fingerprinting in legal forensics led to two major reports by the National Research Council (1992, 1996). In these reports, it was pointed out that a variety of conservative calculation procedures could be followed to diminish any bias against the defense. These procedures included the use of wider bins for grouping minisatellite fragments and the employment of observed rather than expected frequencies of single-locus genotypes in the reference population (to circumvent the assumption of Hardy–Weinberg equilibrium). The 1992 NRC report proposed another solution that would be highly conservative in favor of the defense: the "ceiling principle." Under this suggestion, allele frequencies would be estimated at all marker loci in 15–20 human populations representing a diversity of ethnic groups. For each allele, its highest frequency in any population or 0.05, whichever is higher, would then be employed in an estimate of expected genotypic frequencies against which to evaluate a suspect's genotypic profile. This method overestimates the expected frequency of genotypes in the reference database, but the effect is such that any error introduced is in the direction of decreasing the chance that an innocent suspect is falsely convicted.

Apart from these and other statistical issues, the NRC reports also addressed a variety of technical matters, such as proper handling and processing of samples at all steps in the investigation, standardization and validation of molecular procedures and data, and accreditation and monitoring of forensic laboratories. Fortunately, all of these concerns have become less worrisome over the years as laboratory techniques and background databases have improved and expanded and tighter standards for quality control at all levels have been widely adopted. Furthermore, the genetic variation now available for typing at microsatellite loci is sufficiently high to have muted former criticisms about probability estimates of genotypic matches. In the late 1990s in the United States, a national database known as CODIS (Combined DNA Index System) was implemented. It involves 13 core STR loci that are now employed widely in forensic analysis and which collectively yield an average random match probability of less than one in a trillion among unrelated individuals (Chakraborty et al. 1999). Mitochondrial and Y-linked markers are also employed routinely when information is required about particular human matrilines or patrilines. Today, DNA typing via standardized batteries of molecular markers is routine practice in human forensics, and scientific controversies about the evidentiary power of DNA have mostly faded to a distant memory (Lander and Budowle 1994).

Empirical examples

In the United States, one of the first legal cases (*Pennsylvania v. Pestinikas*) to admit DNA as evidence came in 1986, and it involved use of PCR-based assays to analyze tissue samples from an exhumed corpse (Moody 1989). The first criminal conviction in the United States based in part on DNA evidence came in a 1987 rape trial: *State v. Andrews, Orange County, Florida*. This case established a legal precedent for the use of DNA typing to link a suspect to biological material (blood, semen, or hair follicles) left at a crime scene (Kirby 1990; Roberts 1991). One of the earliest examples of DNA typing in a homicide case was also unusually bizarre: It involved a mortuary worker accused of killing and incinerating his estranged wife at a crematorium in Kansas (see Kirby 1990). Circumstantial evidence had implicated the worker in his wife's death, but he staunchly maintained that she had not been at the mortuary near the time of her disappearance. However, bloodstains discovered on the side of the crematorium proved by DNA typing to match other remaining tissue from the deceased woman. The mortuary worker was convicted of aggravated kidnapping and first-degree homicide. In another early example of the power of DNA typing methods, Hagelberg et al. (1991) used PCR to amplify DNA sequences from the 8-year-old skeletal remains of a murder victim. By comparing microsatellite DNA markers in the remains with those of the presumptive parents, the victim's identity was established.

Not all forensic applications of DNA typing involve crimes this macabre, but molecular genetic methods have provided powerful physical evidence in thousands of homicides, rapes, burglaries, assaults, hit-and-run accidents, missing persons, identifications of war-atrocity victims, and other cases. Forensic DNA methods, validation studies, and empirical examples are the focus of no less than three major scientific journals: *Forensic Science International* (Elsevier Science), *International Journal of Legal Medicine* (Springer-Verlag), and *Journal of Forensic Sciences* (American Society for Testing and Materials). Several interesting high-profile cases in human DNA forensics are summarized in Table 5.2.

TABLE 5.2   Uses of DNA typing in high-profile human forensic investigations

General description and results of investigation	Reference
O. J. Simpson "trial of the century." A famous football player was acquitted of the 1994 murder of his wife, despite DNA fingerprint evidence that by itself yielded an unequivocal match of the defendant's blood to the crime scene.	Levy 1996
Clinton–Lewinsky affair. A semen stain on a blue dress matched a U.S. president's genotype, thereby contradicting claims of no concrete evidence for a sexual liaison; this evidence played a role in impeachment proceedings.	Grunwald and Adler 1999
The Russian Czar. Skeletal remains presumed to be those of Nicholas II, the Russian Czar killed in the Bolshevik revolution of 1918, were exhumed, genotyped, and compared with surviving members of the family; results confirmed the identity of the bones.	Gill et al. 1994
Arlington Cemetery, Tomb of the Unknown Soldier. By comparing the genotype of skeletal remains in this famous monument to members of candidate surviving families, the identity of a "Vietnam Unknown" in the Tomb was confirmed.	Holland and Parsons 1999
Armed Forces forensics. In 1992, personnel entering the U.S. Army through six basic training sites donated samples for DNA typing. This program soon expanded dramatically. Several million such specimens have been collected; they have been used to identify remains, for example, in the Gulf Wars and in terrorist bombings of military personnel.	www.afip.org/Departments/ oafme/dna/
Branch Davidian fire. In the first mass-disaster investigation involving DNA evidence, charred remains were used to genotype and thereby identify victims of a federal assault on the Branch Davidian compound in Waco, Texas.	Clayton et al. 1995
SwissAir Flight 111. DNA typing helped to identify all 229 people who died in the 1998 crash of this trans-Atlantic flight.	Butler 2001
Jefferson–Hemings affair. Y-chromosome markers established that Thomas Jefferson sired children by one of his slaves, Sally Hemings. The approach involved tracing paternal lineages back from living descendants of the Jefferson and Hemings families.	Foster et al. 1998
Human rights abuses. Nuclear and mtDNA markers have solved missing-persons and other cases of human rights abuses in politically repressive regimes and in war-torn areas throughout the world, from Argentina to the Balkans.	Owens et al. 2002

Note: For more details, see the references cited and Butler 2001.

Ramets and Genets

Background

Many species of invertebrate animals and plants exhibit facultative asexual (clonal) as well as sexual reproduction (Jackson et al. 1985). For example, each colony of staghorn coral (*Acropora cervicornis*) consists of numerous asexually derived polyps that are genetically identical to one another and to the sexually produced planula larvae from which they arose. These polyps are housed jointly in a secreted calcareous skeleton that breaks occasionally, producing physically disjunct "daughter" colonies whose members are also genetically identical. Some coral species also produce dispersive asexual larvae (Stoddart 1983a). In various other invertebrates, mechanisms of asexual proliferation may include clonal production of larva-like propagules, somatic fragmentation, polyembryony (production of multiple individuals by division of an early embryo or zygote), or parthenogenesis (Blackwelder and Shepherd 1981; Eaves and Palmer 2003; Jackson 1986). Similarly, in plants, clonal proliferation can involve runners, stolons, rhizomes, bulbs, root or stem suckers, plant fragments, or even highly dispersive asexual (apomictic) seeds, which can arise when a non-meiotic cell in the ovarian wall initiates seed formation or when failure of a reduction division in the germ line produces eggs with a full complement of chromosomes from the maternal parent (Cook 1980; Vielle-Calzada et al. 1996). One example of asexual reproduction involves quaking aspen trees (*Populus tremuloides*), which produce sexual seeds, but also can proliferate vegetatively via buds that sprout from roots of mature specimens. Death of the mother stem may then result in the physical disconnection of clonemates. As phrased by Harper (1985), "It is the nature of many plant and animal growth forms that the organism dies in bits and continues growth as separated parts."

In species with such mechanisms of clonal reproduction, challenging questions arise, such as, What constitutes an individual? What are the units of selection? (Buss 1983, 1985). Harper (1977) defined the genetic individual, or "genet," to include all entities (however physically organized) that have descended from a single sexually produced zygote and, hence, are genetically identical to one another (barring mutation). By contrast, a "ramet" is an individual in a physical or functional sense—a physiologically or morphologically coherent module having arisen through clonal replication. Thus, a genet may consist of many modular ramets, asexually derived. Many evolutionary interpretations of field data hinge critically on the correct distinction of clonemates from non-clonemates. For example, secure genetic knowledge of which ramets ultimately derive from the same zygote is necessary for drawing proper inferences about sex ratios within sexual–asexual populations, magnitudes and patterns of effective gene flow, degrees of outcrossing and the mating system, extents of interclonal competition, and the evolutionary ages of clones (Cook 1983, 1985).

In many cases in which a species' mode of reproduction is unknown, but asexuality is suspected, molecular genetic markers can help to settle the issue. The evidence often involves family data. For example, by showing that assayed siblings were genetically identical to their parent plant at several polymorphic allozyme loci, Roy and Rieseberg (1989) confirmed the occurrence of apomixis (reproduction without fertilization) in the mustard plant *Arabis holboellii*, a species that had been suspected of clonality from other evidence (e.g., occurrence of pollen with unreduced chromosome number). Conversely, some species suspected of clonal reproduction prove upon molecular examination to be capable of outcrossing, as was demonstrated by the observation of recombinant RAPD genotypes in a benthic freshwater byrozoan, *Cristatella mucedo* (Jones et al. 1994). In other cases, molecular evidence for clonal reproduction may involve population genetic data. For example, Nybom and Schaal (1990) used DNA fingerprinting assays to show that a population of blackberry (*Rubus pensylvanicus*)—a species suspected of frequent asexual reproduction—had many fewer recombinant genotypes than a congener with predominant sexual reproduction (the black raspberry, *R. occidentalis*). Similarly, Graham et al. (1997) used RAPD markers to show that the primary mode of reproduction in the red raspberry (*R. idaeus*) is sexual.

A wide variety of species have been subject to such molecular appraisals of clonality. Many marine benthic algae release spores into the water column, but whether these are sexual or asexual propagules remains uncertain. For one such species (*Enteromorpha linza*), Innes and Yarish (1984) documented from allozyme markers that the spores are clonal. Conversely, sexual reproduction was documented by allozyme markers in both a free-living amoeba, *Naegleria lovaniensis* (Pernin et al. 1992), and a fungal pathogen, *Crumenulopsis sororia* (Ennos and Swales 1987). Many marine invertebrates brood their young, and it is of interest to know whether these larvae are the products of sexual or asexual reproduction. Using allozyme assays, Black and Johnson (1979) showed that brooded young of the intertidal anemone *Actinia tenebrosa* were genetically identical to their parents, indicating asexual reproduction. Similarly, Ayre and Resing (1986) documented asexual reproduction for two coral species (*Tubastraea diaphana* and *T. coccinea*). On the other hand, in two other coral species assayed for allozymes (*Acropora palifera* and *Seriatopora hystrix*), nonparental genotypes were detected in a majority of larval broods, thus indicating sexual recombination.

For many animal populations, the first suggestion of parthenogenesis, wherein progeny develop directly from an unfertilized and unreduced female gamete (Soumalainen et al. 1976), comes from the indirect evidence of a strongly female-biased sex ratio. Clonal reproduction then may be confirmed with genetic markers: True ameiotic parthenogens derived from a single female are genetically uniform, barring post-formational mutations (Hebert and Ward 1972), and clonal parthenogenetic populations that arose through recent hybridization exhibit "fixed heterozygosity" at loci distinguishing the parental species (Dessauer and Cole 1986). For example, Echelle and Mosier (1981) used allozyme evidence to confirm that a population of silverside fishes in the *Menidia clarkhubbsi* complex reproduces by clonal means, as did Dawley (1992) for two killifish populations that proved to have arisen through crosses between the sexual species *Fundulus heteroclitus* and *F. diaphanus*. In some populations of the parthenogenetic aphids *Myzus persicae* and *Sitobion avenae*, DNA fingerprinting assays revealed characteristic genetic signatures that confirmed suspected modes of clonal reproduction (Carvalho et al. 1991; Simon et al. 1999). As discussed below, molecular markers have likewise substantiated clonal reproduction in many other vertebrate and invertebrate animals, as well as in microbes.

The genetic hallmark of clonal reproduction is the stable transmission of genotypes across generations, without the shuffling effects of genetic recombination (the only source of variation therefore being mutation). Thus, the phrase "clonal reproduction" is sometimes also used to describe genetic transmission in self-fertilizing hermaphrodites, in which the intense inbreeding that characterizes this reproductive mode can result over the generations in near homozygosity at most loci. Although selfing organisms may retain meiosis and syngamy (union of gametes, in this case from a single parent), genetic segregation and recombination in effect are suppressed once homozygosity through inbreeding is achieved.

This latter phenomenon is well illustrated by a cyprinodontid fish, *Rivulus marmoratus*—the only known vertebrate hermaphrodite with regular self-fertilization (Harrington 1961). This species exists in nature as highly homozygous "clones," as was initially revealed by intraclonal fin graft acceptances (indicating near-identity at histocompatibility loci; Harrington and Kallman 1968; Kallman and Harrington 1964) and complete homozygosity at more than 30 allozyme loci (Vrijenhoek 1985). Subsequent DNA-level studies based on minisatellite loci (Turner et al. 1990, 1992) revealed the presence of additional clonal genotypes, most of which probably arose via recombination, suggesting that outcrossing had also taken place, albeit infrequently at most sites (Laughlin et al. 1995; Lubinski et al. 1995). More variation was then uncovered in molecular assays of MHC loci, but the heterozygosity in this case was probably due to long-term retention of alleles in different strains that had arisen from ancestral outcrossing forms (Sato et al. 2002).

These kinds of genetic observations highlight a cautionary note that applies to studies of "clonal diversity" in strongly inbreeding species (as well as in truly asexual taxa): The absolute number of recognized clones can depend on the discriminatory power of the molecular (or other) assay used as well as the reproductive biology and evolutionary history of a species. Thus, it is not enough to be concerned with mere tallies of identifiable genotypes. Of much greater importance are establishing historical relationships among genotypes and understanding the biological processes (including selection as well as reproductive mode) that may have forged and maintained whatever genetic variation is observed.

Spatial Distributions of Clones

The geographic distribution of genotypes in a facultatively asexual species is influenced by the relative frequencies of sexual versus clonal proliferation and by the dispersive characteristics of propagules produced under each reproductive mode. Early attempts to illuminate these population genetic structures utilized indirect phenotypic criteria for clonal identification, such as distributions of morphological attributes or phenologies in plants (Barnes 1966), agonistic behaviors in sea anemones (Sebens 1984), and histocompatibility responses (acceptance versus rejection of tissue grafts) in marine invertebrates and unisexual vertebrates (Cuellar 1984; Neigel and Avise 1983a; Schultz 1969). Subsequent attempts to assign ramets to genets and map the spatial distributions of clones often involved direct molecular genetic assays (Ellstrand and Roose 1987; Silander 1985).

PLANTS AND ALGAE.   Maddox et al. (1989) used allozyme genotypes at four polymorphic loci to map the microspatial distributions of goldenrod (*Solidago altissima*) clones in fields of various ages. In this case, different allozyme genotypes almost certainly reflected sexual recruitment (via seeds) into the population, whereas multiple ramets of the same genotype registered asexual proliferation via rhizomes. Patterns of dispersion of goldenrod genotypes differed among fields: Clones were localized in the youngest plots (Figure 5.2), whereas older fields exhibited greater spatial intermixture of clones and fewer remaining rhizome connections among ramets. Apparently, colonization of a field by sexually produced seeds is followed by ramet proliferation and eventual spatial mixing of clones over microgeographic scales. In other such allozyme assessments of local clonal reproduction, Burke et al. (2000) discovered that approximately 50% of genets in two populations of *Iris* hybrids consisted of multiple ramets originating from vegetative proliferation. Murawski and Hamrick (1990) found that clonal growth in the bromeliad *Aechmea magdalenae* had resulted in the spread of ramets over distances of several meters. In the columnar cactus *Lophocereus schottii*, Parker and Hamrick (1992) found that most clonemates were tightly aggregated, but also that a few individuals were separated from their clonemates by more than 70 m (probably as a result of detached stem pieces washing downstream during floods). Using AFLP markers, Suyama et al. (2000) identified a large clone of bamboo (*Sasa senanensis*) in Japan that occupied an area about 300 meters in diameter.

Molecular markers have also been used to study clonal distributions in various bush and tree species. Torimaru et al. (2003) found from RAPD analyses that patches of holly (*Ilex leucoclada*) were composed partly of different genets and partly of multiple ramets (stems in this case) within a given clone. Applying allozyme approaches to arctic dwarf birch (*Betula glandulosa*), Hermanutz et al. (1989) identified single clones encompassing areas of at least 50 m². Although normally a sexual species, dwarf birch at the study site had apparently reproduced by "vegetative layering," wherein prostrate

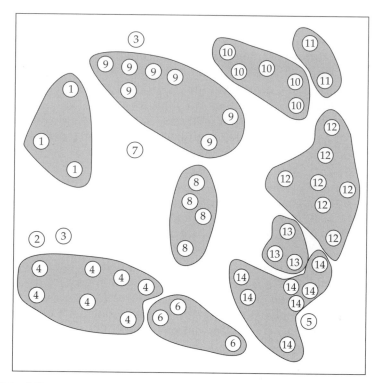

Figure 5.2   Microspatial map of allozyme-identified clones in *Solidago altissima*. Each circle denotes a current living ramet; total map area 0.75 m². Different numbers indicate distinct electrophoretic genotypes, and shaded areas encompass multiple nearby ramets apparently belonging to each genet. (After Maddox et al. 1989.)

branches beneath the moss layer produced new ramets vegetatively. In a local population of scrub oaks (*Quercus geminata*) on Merritt Island, Florida, Ainsworth et al. (2003) used microsatellite markers to identify and map clones that arose from vegetative proliferation via suckers.

A form of apomictic reproduction in plants and algae that could result in unusually widespread dispersion of clones is agamospermy, the formation of unreduced spores, seeds, or embryos by asexual processes (Bayer 1989; Hughes and Richards 1989). For example, in an agamospermous marine alga (*Enteromorpha linza*) that can produce water-dispersed asexual spores, particular allozyme-identified clones were distributed over the entire survey transect of more than 150 shoreline kilometers (Innes 1987). Within each of two obligate agamospermous populations of dandelions (*Taraxacum* sp.), all individuals proved to be genetically identical at 15 allozyme loci, whereas related sexual populations were highly diverse genetically (Hughes and Richards 1988). Based on additional molecular evidence, other apomictic dandelion populations showed considerable genetic variation, perhaps evidencing the coexistence of multiple clones (Ford 1985;

Ford and Richards 1985). In general, however, a serious complication in interpreting clonal diversity in ancient and widespread "agamospecies" is in distinguishing sexually derived genetic variation from that which may have arisen via post-formational mutations (Brookfield 1992; see below). In a molecular study of *Taraxacum* dandelions in Norway, Mes et al. (2002) attributed much of the genetic variation observed in STRs and AFLPs to be the result of mutation accumulation within an ancient clone. On the other hand, using rDNA and cpDNA analyses of triploid dandelions in North America, King (1993) uncovered high genetic variation, much of which she attributed to multiple hybridization events that produced these apomicts.

According to Cook (1980), the record holder for size and age of a plant clone may be the quaking aspen. Based on a distinctive morphological appearance and spatial arrangement, one suspected genet was represented by more than 47,000 ramets (covering 107 acres) that might have traced to a single seed perhaps deposited several thousand years ago at the close of the Wisconsin glaciation (Kemperman and Barnes 1976). Looks might be deceiving, however. In allozyme surveys of other quaking aspen populations, Cheliak and Patel (1984) found that several "clones" provisionally identified by morphology actually were composed of several distinct electrophoretic genotypes that probably had arisen through recombination (and hence sexual reproduction). They concluded that environmental influences on phenotype invalidate morphological appraisals of aspens as a reliable guide to clone identification. In contrast, DNA-level assays *can* provide definitive information on clonal identities and distributions in this species (Rogstad et al. 1991).

FUNGI.   Molecular documentation of clonal identity is also available for the honey mushroom, *Armillaria bulbosa*, in which one gigantic clone identified by mtDNA and nuclear RAPD markers was claimed at the time to be one of the Earth's largest and oldest individuals of any species (M. L. Smith et al. 1990, 1992). This pathogenic fungus of tree roots lives in mixed hardwood stands, where it can spread vegetatively by cordlike aggregations of hyphae that weave across the forest floor. The molecular markers revealed that one presumably interconnected clone of *A. bulbosa* in northern Michigan had spread across 37 acres, weighed in aggregate more than 90,000 kg (about the size of an adult blue whale), and was perhaps 1,500 years old. A smaller clone nearby covered a mere 5 acres. No doubt even larger and older clones in this or other fungal species remain to be discovered.

INVERTEBRATE ANIMALS.   Most sea stars (Asteroidea, Echinodermata) can reproduce asexually by fission, whereby detached arms regenerate new bodies. Johnson and Threlfall (1987) used allozyme markers to estimate the occurrence of fission versus sexual reproduction in the sea star *Coscinasterias calamaria* in Western Australia. On local scales, clonal reproduction proved to predominate, such that many individuals within 50 m of one another were clonemates, but sexual recruitment was important as well, as evidenced by the fact that distinct genotypes were present in different parts of the study site.

In the soft coral *Alcyonium rudyi*, members of allozyme-documented clones produced by binary fission were typically grouped within 50 cm of one another on the same rock surface (McFadden 1997). Hard coral colonies can also proliferate clonally, in this case by fragmentation. An allozyme survey of *Pavona cactus* revealed the significance of this asexual process in distributing clonemates over distances of up to nearly 100 meters along reefs in eastern Australia (Ayre and Willis 1988). Some corals, such as *Pocillopora damicornis*, also produce dispersive planula larvae by asexual means, which presumably accounts for an observation by Stoddart (1984a,b) that particular clones identified by protein electrophoresis were dispersed over distances up to several kilometers.

In a sea anemone (*Actinia tenebrosa*) that can produce brooded young asexually, particular clonal genotypes identified by allozyme assays were distributed over hundreds of meters of shoreline in Australia (Ayre 1984). In lagoons surrounding the United Kingdom, RAPD assays revealed that about 60% of surveyed individuals of the anemone *Nematostella vectensis* were genetically identical, probably as a result of clonal proliferation in conjunction with historical population bottlenecks (Pearson et al. 2002). On the other hand, allozyme surveys of the intertidal sea anemone *Oulactis muscosa* (a species that can reproduce by fission) revealed a population genetic structure consistent with recruitment almost exclusively by sexual reproduction (Hunt and Ayre 1989). In yet another sea anemone species similarly surveyed (*Metridium senile*), various populations in northeastern North America evidenced large differences among sites in the frequencies with which sexual versus clonal recruitment had taken place (Hoffman 1986).

In several marine invertebrate groups, clones were traditionally distinguished using various nonmolecular sources of information, such as aggressive behaviors among sea anemones or morphotypic appearances of corals and sponges (Ayre 1982; Ayre and Willis 1988; Solé-Cava and Thorpe 1986). Another source of data on putative clonal identities was histocompatibility responses. Within many coral and sponge species, for example, artificial grafts between colony branches exhibit either an acceptance or rejection reaction, and indirect evidence suggested that these two responses signal clonal identity and non-identity, respectively (Hildemann et al. 1977; Neigel and Avise 1983b; review in Avise and Neigel 1984). However, this possibility was not always fully corroborated in more direct genetic analyses (Curtis et al. 1982; Neigel and Avise 1985; Resing and Ayre 1985; see also Hunter 1985). Instead, reports appeared of occasional graft rejections between colonies that were identical in multi-locus allozyme genotype and of graft acceptances between some colonies that differed in allozyme genotype (Table 5.3). The former observation may simply reflect the likelihood that small numbers of surveyed allozyme loci failed to distinguish all clones. The latter observation may reflect imperfect discriminatory power at histocompatibility genes themselves. Like the histocompatibility systems of vertebrates (Parham and Ohta 1996) and the self-incompatibility systems of many

TABLE 5.3 Levels of agreement between histocompatibility response and allozyme genotype as possible indicators of clones in local populations of marine invertebrates

Allozyme genotype	Tissue graft response	
	Accept	Reject
Identical		
Niphates	23	5
Montipora	26	12
Non-identical		
Niphates	5	32
Montipora	21	35

Note: Results shown are for a sponge, *Niphates erecta*, assayed in 65 pairs of colonies at three polymorphic loci (data from Neigel and Avise 1985); and the corals *Montipora dilatata* and *M. verrucosa*, assayed in a total of 94 pairs of colonies at one polymorphic allozyme locus (data from Heyward and Stoddart 1985). For both *Niphates* and *Montipora*, associations between histocompatibility response and allozyme genotype were statistically highly significant, but nonetheless imperfect.

outcrossing plants (Charlesworth 1995; Richman and Kohn 1996), tissue recognition loci in several marine invertebrates are known to be highly polymorphic (Grosberg 1988; Grosberg et al. 1996, 1997), yet they too can fall somewhat short of perfection in distinguishing genetic self from nonself (see the section on genetic chimeras below).

Various DNA fingerprinting methods are especially powerful in studies of clonal population structure. In an early example, Coffroth et al. (1992) utilized minisatellite probes to study DNA fingerprints and clonal structure in a gorgonian coral (*Plexaura* sp.) that reproduces by fragmentation as well as by sexual production of dispersive larvae. Among 73 scrutinized colonies on seven reefs in Panama, 29 different genotypes were identified in these molecular assays. Identical DNA fingerprints were observed only in adjacent colonies on a reef. In some cases, both DNA fingerprinting and histocompatibility assays were conducted, and these approaches revealed similar numbers of putative clones: 17 and 13, respectively. Also worthy of mention were experimental controls demonstrating that multiple samples from a single colony produced identical DNA fragment profiles, and that symbiotic zooxanthellae (algae) living within the corals' tissues were not the source of the DNA gel bands scored. A follow-up DNA fingerprinting study by Coffroth and Lasker (1998) provided spatial maps of particular *Plexaura* clones on a Panama reef (Figure 5.3).

Many invertebrates can proliferate clonally by parthenogenesis. For example, 17 of 33 North American species in the earthworm family Lumbricidae exhibit a parthenogenetic reproductive mode that apparently evolved from an ancestral hermaphroditic condition (Jaenike and Selander 1979). In one such species (*Octolasion tyrtaeum*), Jaenike et al. (1980) used allozyme markers to identify eight distinct clones, two of which were wide-

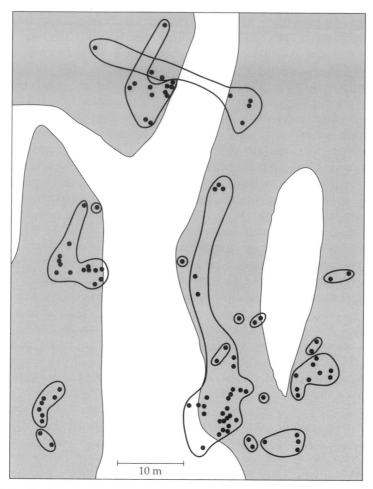

Figure 5.3   Underwater mesospatial map of *Plexaura kuna* **clones** on a Panama reef, as identified by DNA fingerprints. Shaded areas denote reef structure; light areas are sand channels. Each black dot is a living ramet, and continuous lines encircle ramets that belong to a given genet. (After Coffroth and Lasker 1998.)

spread and common in diverse soil types over several thousand square kilometers surveyed in the eastern United States. Thus, some asexual lineages apparently can occupy broad niches and achieve great success, at least over short ecological time. Many freshwater gastropods (snails) likewise reproduce parthenogenetically (Jarne and Delay 1991). Allozyme studies of the polyploid parthenogen *Thiara balonnensis* in Australia revealed that each local population generally consisted of only one clone, with genetic distance among clones correlated with geographic distance (Stoddart 1983b). This pattern of variation was postulated to result from the gradual evolution of new clones by mutational processes (as opposed to occasional sexual reproduction) in conjunction with geographic separations.

A converse example of high clonal diversity over microgeographic scales was encountered in early studies of another species—the cladoceran *Daphnia pulex*—that exhibits obligate parthenogenesis throughout much of its range (Hebert and Crease 1980). Initial surveys revealed a total of 22 allozyme-identified clones in eleven populations, with up to seven genotypes coexisting in a single lake. Subsequent studies of *D. pulex* using mtDNA and allozyme markers revealed many additional clones and also demonstrated that obligate parthenogenesis had a polyphyletic origin from facultative parthenogenesis within this species (Crease et al. 1989; Hebert et al. 1989). In a related species, *D. magna*, tens to hundreds of clones sometimes coexist within a pond (Hebert 1974a,b; Hebert and Ward 1976). Where *D. magna* occurs in temporary habitats, it is a cyclic parthenogen, producing drought-resistant sexual eggs (requiring fertilization) each year. These zygotes reestablish populations that then are maintained by two or three generations of clonal parthenogenesis until the pond dries up again. Interestingly, in permanent habitats, *D. magna* reproduces continually by parthenogenesis and tends to exhibit fewer clonal types (Hebert 1974c). Several more molecular surveys of *Daphnia* species have appeared in recent years (see Colbourne et al. 1998 and citations therein).

VERTEBRATE ANIMALS. Many vertebrate species, including humans, also produce clonemates occasionally—that is, whenever a pregnancy involves identical (monozygotic) twins. Although such instances are usually sporadic, in one mammalian taxonomic group—armadillos in the genus *Dasypus*—this reproductive mode is constitutive (Loughry et al. 1998a). In *D. novemcinctus*, for example, each litter typically consists of four pups that are genetically identical to one another, but distinct from both parents. This clonal mode of reproduction, known as polyembryony, differs from classic asexual reproduction in that clonemates are intra- rather than intergenerational. This reproductive mode is also evolutionarily puzzling because it entails the production of "carbon copies" of a new and previously untested genotype (analogous to parents buying multiple raffle tickets with the same number).

One hypothesis to account for polyembryony in armadillos involves the notion of nepotism (favoritism toward kin). Perhaps armadillo clonemates within a litter help one another build dens, find food, or detect predators. Any genes responsible for polyembryony might have been favored across the generations if strong littermate cooperation was the norm and led to higher mean survival and reproduction. Such close cooperation would entail tight spatial associations among clonemates, an idea that was critically tested using the clone-discriminating power of polymorphic microsatellites. In a large study population in northern Florida, Prodöhl et al. (1996) used these molecular markers to map the mesospatial distribution of armadillo clonal sibships (as well as to assess genetic parentage; Prodöhl et al. 1998). Armadillo clonemates (especially adults) proved not to be spatially clustered. This finding, together with direct behavioral evidence (Loughry et al. 1998a,b), argues against the nepotism hypothesis for this remarkable instance of vertebrate polyembryony.

Ages of clones

A common belief is that populations of clonally reproducing organisms must have short evolutionary life spans, due either to "Muller's ratchet" (the accumulation of deleterious mutations and gene combinations that in the absence of recombination cannot readily be purged from populations; Muller 1964) or to a presumed lack of sufficient recombinant variety to allow adaptive responses to environmental challenges (Darlington 1939; Felsenstein 1974; Maynard Smith 1978; Williams 1975). However, the evidence cited above suggests that clonal lineages in some plants and invertebrate animals can achieve fairly wide distributions and enjoy at least moderate-term ecological success. Nevertheless, virtually all extant asexual taxa represent only the outermost twig-tips (rather than deeper branches or major limbs) in the Tree of Life, indicating that they are geologically young and, in general, that asexual lineages are evolutionarily ephemeral. Among metazoans, primary exceptions to this rule *might* involve bdelloid rotifers and darwinulid ostracods, as described next.

BDELLOID ROTIFERS.   In the class Bdelloidea of the phylum Rotifera, all 360 described species (belonging to 18 genera and four families) appear to lack males, hermaphrodites, meiosis, or any of the other trappings typically associated with sexual reproduction and genetic exchange (Butlin 2000; Mayr 1963). Instead, these tiny freshwater creatures reproduce parthenogenetically via direct-developing eggs produced by mitotic cell divisions, without reduction in chromosome number and without fertilization. Based on fossil evidence including amber-preserved specimens, bdelloid rotifers arose at least 35 million years ago. Has their long evolutionary survival and success truly been without benefit of sexual reproduction? If so, this would be somewhat of an "evolutionary scandal" (Maynard Smith 1986) because conventional wisdom holds that genetic recombination has long-term as well as short-term adaptive importance (Butlin 2002).

By examining nucleotide sequences at each of several nuclear loci in various species of bdelloid rotifers and comparing the results with those for sexual species of rotifers and some other invertebrates, Mark Welch and Meselson (2000, 2001) tested the ancient asexuality hypothesis. Long-term asexual reproduction in the bdelloids was presumably evidenced by several genetic signatures, including: high sequence divergence between allelic pairs (the "Meselson effect") even within single specimens, as expected if non-recombining alleles within an asexual diploid lineage have been maintained independently for long periods of time; and a relative paucity of transposable elements (Arkhipova and Meselson 2000), as might also be expected because, in theory, selfish transposable elements that proliferate within an asexual genome gain none of the personal fitness advantages they typically enjoy when housed in sexually reproducing hosts (Hickey 1982). On the other hand, in statistical analyses of these same sequence data, Gandolfi et al. (2003) advanced what they provisionally interpreted as lasting footprints of otherwise cryptic genetic recombination in the past, so perhaps even the bdelloid rotifers have not been entirely free of sex.

OSTRACODS. Tiny crustaceans in the family Darwinulidae offer another likely example of unisexual complexes with great evolutionary antiquity, perhaps as much as 200 million years (Martens et al. 2003). DNA sequence analyses of several nuclear and mitochondrial regions in the asexual ostracod *Darwinula stevensoni* documented that genetic diversity in this taxon was quite low (Schön and Martens 2002, 2003), in contrast to the situation in bdelloid rotifers. At face value, this finding seems to contradict an expectation for ancient asexuals—that is, that genetic variation should be high between alleles within lineages as well as between separate lineages. Low allelic diversity does not necessarily negate the possibility of ancient asexuality for the darwinulids, however, because several other factors might account for this result (Butlin 2000). For example, Schön and Martens (1998) provisionally favor a hypothesis of ancient asexuality in *D. stevensoni* accompanied by the postulated evolution of highly efficient DNA repair mechanisms that might have enabled these organisms to circumvent Muller's ratchet.

In another ostracod family (Cyprididae), the freshwater species *Cyprinotus incongruens* displays a diversity of diploid, triploid, and tetraploid parthenogenetic clones; it also has sexual relatives (Turgeon and Hebert 1994). Molecular genetic studies have suggested that transitions to polyploidy have been common in this taxon and that asexuality in the complex has persisted for at least several million years (Chaplin and Hebert 1997). However, these latter authors also emphasize how difficult it is to eliminate the possibility of independent and recent transitions to asexuality from closely related sexual ancestors, a caveat that applies to many studies that have proclaimed discoveries of ancient asexual lineages (Judson and Normark 1996; Little and Hebert 1996; see below).

VERTEBRATE ANIMALS. Apart from polyembryony (discussed above), other modes of clonal or quasi-clonal reproduction are known in about 70 vertebrate species (Dawley and Bogart 1989), for which the term "biotype" is often preferred because traditional species concepts hardly apply. These biotypes typically consist solely of females that propagate by parthenogenesis or related reproductive modes (Figure 5.4). Essentially all unisexual vertebrates arose through hybridization between related sexual species; this aspect of their phylogenetic histories will be deferred to Chapter 7. Here we consider the evolutionary ages of vertebrate unisexual lineages as inferred from molecular data (primarily from mtDNA).

Two conceptual approaches to assessing vertebrate clonal ages have been attempted. The first involves estimating the genetic distance between a unisexual biotype and its closest sexual relative. In a review of 24 unisexual vertebrate lineages (Avise et al. 1992c), 13 (54%) proved indistinguish-

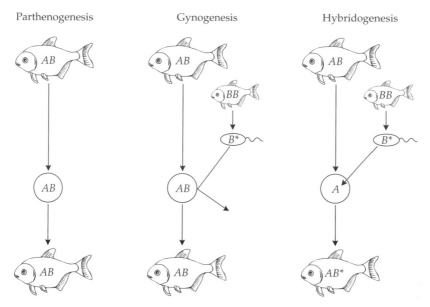

Figure 5.4   Three modes of unisexual reproduction in vertebrates. In partheno-genesis, the female's nuclear genome is transmitted intact to the egg, which then develops into an offspring genetically identical to the mother. In gynogenesis, the process is similar, except that heterospecific sperm (indicated by an asterisk) from a related sexual species is required to stimulate egg development. In hybridogenesis, an ancestral genome from the maternal line is transmitted to the egg without recombination, whereas paternally derived chromosomes are discarded pre-meioti-cally, only to be replaced each generation via fertilization by heterospecific sperm (indicated by asterisk) from a related sexual species. (After Avise et al. 1992c.)

able in mtDNA assays from an extant genotype in the related sexual taxon, indicating extremely recent evolutionary origins for these unisexuals. Five additional unisexual lineages differed from their closest sexual relatives by less than 1% in mtDNA sequence, suggesting origination times within the last 500,000 years under a standard mtDNA clock (Figure 5.5A). A few unisexual haplotypes showed greater sequence differences from related sex-ual forms, and these differences translated into literal estimates of evolu-tionary durations of perhaps a few million years. However, there is a seri-ous reservation about the relevance of such estimates: Closer relatives with-in the sexual progenitor lineage may have gone extinct after the separation of the unisexual biotype, or otherwise remained unsampled, such that uni-sexual ages could be grossly overestimated by this approach. Indeed, because of the low mtDNA lineage diversity observed within most unisex-ual taxa relative to their sexual cognates (Figure 5.5B), most authors have concluded that nearly all unisexuals arose very recently, even when geneti-cally close mtDNA lineages were not observed among the sexual relatives sampled (e.g., Vyas et al. 1990).

(A)

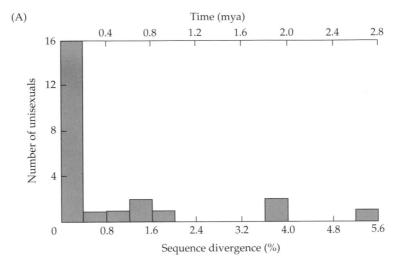

(B)

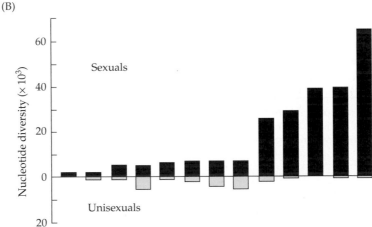

Figure 5.5 Genetic patterns in unisexual vertebrates. (A) Frequency distribution of the smallest genetic distances observed between mtDNA genotypes in 24 unisexual biotypes and their closest assayed sexual relatives. Also shown are associated evolutionary ages of the unisexuals based on the conventional mtDNA clock calibration of 2% sequence divergence per million years between lineages. (B) Nucleotide diversities in mtDNA within 13 sexual species (above horizontal axis) and their respective unisexual derivatives (below horizontal axis), arranged in rank order from left to right by the magnitude of variation within the various sexual taxa. (After Avise et al. 1992c.)

One example in which an ancient clonal age *was* promulgated involved gynogenetic mole salamanders in the genus *Ambystoma*. From comparisons of mtDNA sequences in the unisexuals and in their extant sexual relatives, Hedges et al. (1992a) and Spolsky et al. (1992) estimated that the gynogens had had evolutionary durations of about 4 to 5 million years. However, this evidence for clonal persistence comes with a serious caveat: namely, that these salamanders are unusual among vertebrate unisexuals in that their evolution may not be strictly clonal. Other molecular data suggest that they continually acquire nuclear DNA from sexual species, presumably via occasional incorporation of sperm DNA into the egg. If so, the antiquity of these "clonal" salamander biotypes applies strictly only to the mtDNA lineages that they contain.

A second approach to estimating clonal ages involves assessing the scope of genetic variation within particular unisexual clades that, according to independent evidence, are monophyletic (i.e., originated from a single hybridization event, in this case). In principle, this method avoids confounding post-formational mutational variation that might be indicative of ancient lineage age with genetic diversity that collectively arose from separate hybridization events. Quattro et al. (1992a) examined mtDNA and allozyme variability within a hybridogenetic clade of fishes in northwestern Mexico (*Poeciliopsis monacha-occidentalis*) that, according to independent zoogeographic evidence and tissue graft analyses, was indeed monophyletic. The molecular data confirmed the clade's monophyly and also documented considerable genetic diversity within it, including the accumulation of several mitochondrial and allozyme mutations in the *monacha* portion of its genome (which comes from the female parent). From the magnitude of this post-formational diversity, the authors estimated that the unisexual clade was more than 100,000 generations old. However, concerning the relevance of this conclusion to broader arguments about clonal persistence, there is again at least one reservation: These fishes reproduce by hybridogenesis, which means that only the maternal component of their genetic heritage is strictly clonal. The nuclear genome receives fresh but transient sexual input each generation from a sire, *P. occidentalis*, such that the overall genetic system is "hemiclonal."

In any event, Maynard Smith (1992) argued that 100,000 generations in evolutionary terms "is but an evening gone," and that the molecular findings for *Poeciliopsis* therefore do not contradict the conventional wisdom that organismal clones are short-lived. Regardless of one's perspective on whether such time scales are "long" or "short" in the context of clonal persistence debates, molecular data have provided the first critical information regarding the evolutionary duration of vertebrate as well as invertebrate lineages (or portions thereof) that lack recombination.

Clonal reproduction in microorganisms

PROTOZOANS.   Eukaryotic protozoans such as the agents of malaria, sleeping sickness, Chagas' disease, and leishmaniasis infect more than 10% of the world's human population and account for tens of millions of deaths every year. The classic assumption was that these parasites (most of which are diploid) routinely engage in sexual reproduction because recombination among strains had been observed in the laboratory and because a sexual phase in the life cycle was thought to occur in particular host species. However, in the early 1990s, studies using molecular markers began to demonstrate that several parasitic protozoans can and often do propagate clonally in nature (Tibayrenc et al. 1990, 1991a). For example, analyses of scnRFLPs revealed that virulent strains of the protozoan *Toxoplasma gondii* from around the world are surprisingly homogeneous genetically, probably because of frequent clonal reproduction and evolutionary origins from a small number of nonvirulent strains that exhibit moderate polymorphism and are capable of sexual reproduction (Sibley and Boothroyd 1992). Recent molecular evidence has elaborated on this scenario by indicating the global predominance of three clonal lineages in *T. gondii* that all may have emerged within the last 10,000 years after a single genetic cross (Su et al. 2003). Such findings are of tremendous medical importance because of their relevance in diagnostic tests for disease agents, as well as in strategies for developing vaccines and curative drugs (Tibayrenc et al. 1991b).

BOX 5.2 Population Genetic Criteria Suggestive of Clonal Reproduction

This table presents two classes of inferences, based on four types of genetic obser-vations (criteria a–d) that would tend to indicate clonal reproduction in natural

---

Inferences (from observations)

I. *Absence of meiotic segregation at particular marker loci*

   (a) Fixed heterozygosity (most or all individuals appear
         heterozygous, at least in some populations or subpopulations)

   (b) Significant deficit in frequencies of some expected
         diploid genotypes; other deviations from Hardy–Weinberg
         equilibrium

II. *Absence of recombination among multiple marker loci*

   (c) Overrepresented, widespread identical genotypes; significant
         deficit of expected recombinant genotypes; non-random
         associations of alleles (gametic phase disequilibrium; see Box 5.3)

   (d) Correlation between independent sets of genetic markers

---

Source: After Tibayrenc et al. 1991b.

The case for clonal reproduction in nature is particularly strong for *Trypanosoma cruzi*, the agent of Chagas' disease. Extensive molecular studies of *T. cruzi* have involved strains isolated from humans, insect vectors, and mammals sampled throughout the range of the disease in Central and South America (Barnabé et al. 2000; Oliveira et al. 1998; Tibayrenc and Ayala 1987, 1988; Tibayrenc et al. 1986; Zhang et al. 1988). These surveys have revealed several population genetic signatures characteristic of prevalent clonal reproduction (Box 5.2): fixed heterozygosity and other evidence for an absence of segregation genotypes at individual loci; an overrepresentation of identical multi-locus genotypes, often geographically widespread; and significant correlations (disequilibrium) between independent sets of genetic markers (Box 5.3), even after accounting for the effects of population subdivision (Tibayrenc 1995). Furthermore, molecular markers indicate that two widespread groups of clonal genotypes in *T. cruzi* were the result of ancient hybridization events (Brisse et al. 2000; Machado and Ayala 2001). These findings overturned the

populations. "Comments and caveats" are given in regard to eliminating the possibility of frequent sexual reproduction.

Comments and caveats

Observation also incompatible with self-fertilization; must consider possibility of gel mis-scoring due, for example, to gene duplication or polyploidy

Missing heterozygotes also consistent with self-fertilization; must exclude effects of population subdivision, assortative mating, selection, etc.

Should consider possible effects of selection or population subdivision; must take into account low expected frequencies of multi-locus genotypes when allelic variation is high

Should consider possible effects of population subdivision or correlated selection pressures

BOX 5.3 Gametic Phase Disequilibrium

Gametic phase disequilibrium is the nonrandom association between alleles at different loci. "Nonrandom" means that the multi-locus combinations of alleles depart significantly from expectations based on products of the single-locus allele frequencies. Consider the simplest possible case, involving two autosomal nuclear loci (A and B), each with two alleles (A_1, A_2 and B_1, B_2), whose frequencies are p_1, p_2, q_1, and q_2, respectively. Four di-locus gametic genotypes (or haplotypes) are possible:

	Alleles at locus A	
	A_1 (p_1)	A_2 (p_2)
B_1 (q_1)	A_1B_1 (p_1q_1)	A_2B_1 (p_2q_1)
B_2 (q_2)	A_1B_2 (p_1q_2)	A_2B_2 (p_2q_2)

Alleles at locus B

If alleles are associated at random in haplotypes (gametic phase equilibrium), the expected frequencies of these di-locus genotypes are p_1q_1, p_1q_2, p_2q_1, and p_2q_2. One quantitative measure of a departure from this expectation is given by the gametic phase disequilibrium parameter, defined as

$$D = P_{11}P_{22} - P_{12}P_{21}$$

where P_{11} and P_{22} are the observed frequencies of haplotypes in the "coupling" phase, and P_{12} and P_{21} are observed frequencies of haplotypes in the "repulsion" phase. It can be shown that in a large randomly mating population, any initial disequilibrium [$D(0)$] among neutral alleles tends to decay toward zero (provided that $c \neq 0$) according to the equation

$$D(G) = (1 - c)^G D(0)$$

where $D(G)$ is the disequilibrium remaining at generation G, and c is the probability of a recombination event between the two loci each generation (or the "recombination fraction"). Thus, for unlinked loci ($c = 0.5$), disequilibrium decays by one-half each generation. Disequilibrium decays more slowly as the recombination fraction decreases. The loci examined may also involve one nuclear gene and one cytoplasmic gene, in which case the analogous nonrandom gametic associations are referred to as "cytonuclear disequilibria" (as described later in Box 7.4). Because nuclear and cytoplasmic genes are unlinked, any initial disequilibria between such loci are likewise expected to decay monotonically to zero by one-half per generation in a randomly mating population.

Gametic phase disequilibrium can arise from any historical or contemporary process that has restricted recombination among loci. These processes may include physical linkage of genes on a chromosome or other factors that also can generate nonrandom associations among the alleles of unlinked loci, such as population subdivision, founder effects, mating systems with close inbreeding (see Chapter 6), or selection favoring particular multi-locus allelic combinations (Lewontin 1988).

Genomes or portions thereof characterized by a rarity or absence of recombination may be viewed as linked "supergenes" with regard to evolutionary dynamics. One likely consequence involves "genetic hitchhiking," wherein alleles that are mechanistically neutral nonetheless may spread through a population because of chance association with an allele favored by natural selection at another locus. Such hitchhiking on rare favorable mutations could lead to "periodic selection" that has the net effect of purging population genetic variability in nonrecombining systems (Levin 1981). The "selective sweeps" involved in periodic selection may account in part for the observation that genetic variation in mtDNA (see Chapter 2) and in some bacterial species (Milkman 1973) is vastly lower than might otherwise have been expected (given suspected mutation rates to neutral alleles and apparent population sizes in these systems). Such selective sweeps may also contribute to the observation for eukaryotic nuclear genomes that chromosomal regions characterized by low recombination rates sometimes have reduced nucleotide diversity (Aquadro and Begun 1993; Aquadro et al. 1994; Begun and Aquadro 1992; but see also Hamblin and Aquadro 1999 for exceptions and interpretive complications).

conventional wisdom that this species was an undifferentiated quasi-panmictic entity, and they carried major medical ramifications as well (Revollo et al. 1998). On the other hand, further molecular evidence has emerged supporting the contention that some natural populations of *T. cruzi* also experience regular occurrences of genetic exchange (Gaunt et al. 2003).

The "clonal theory" for parasitic protozoans was rather heretical in the early 1990s, but it has become widely accepted as an important element (but seldom the whole story) of reproduction in several of these microbes (Table 5.4). Depending on the particular life cycle and other biological factors, many parasitic protozoan species also engage occasionally, if not regularly, in a variety of modes of sexual or recombinational genetic exchange, including horizontal gene transmission (see Chapter 8). This also means that when "clonal genotypes" occur in such quasi-sexual species, they can have varied evolutionary origins and diverse ages. One way to acknowledge such heterogeneity for partially clonal taxa, especially in medical applications, is to employ additional clarifying terms. One such moniker is "clonet," defined by Tibayrenc and Ayala (1991) as a set of lineages that appears identical strictly according to a specified set of genetic markers. Another is "discrete typing units," defined by Tibayrenc (1998) as recognizably stable genetic subdivisions of a species as assessed from total evolutionary evidence.

TABLE 5.4   Population genetic evidence for significant clonal reproduction in several parasitic protozoans and fungi that are important agents of human disease

Organism	Criterion[a]				Evidence for clonality[b]
	(a)	(b)	(c)	(d)	
Protozoans					
Entamoeba histolytica	0	0	+	0	Moderate
Giardia spp.	0	0	+	+	Moderate
Leishmania guyanensis	0	0	+	+	Strong
Leishmania infantum	0	0	+	+	Strong
Leishmania tropica	+	0	+	+	Strong
Leishmania major	0	0	+	0	Strong
Leishmania spp.	0	0	+	+	Strong
Naegleria australiensis	+	0	+	0	Weak
Naegleria fowleri	+	0	0	0	Weak
Naegleria gruberi	+	0	0	0	Weak
Plasmodium falciparum	0	0	+	0	Weak
Toxoplasma gondii	0	0	+	0	Weak
Trichomonas foetus	0	0	+	0	Weak
Trichomonas vaginalis	0	0	+	0	Weak
Trypanosoma brucei	+	+	+	+	Strong
Trypanosoma congolense	+	0	+	0	Moderate
Trypanosoma cruzi	+	+	+	+	Strong
Trypanosoma vivax	0	0	+	0	Strong
Fungi					
Candida tropicalis complex	0	0	+	0	Weak
Candida albicans	+	+	+	+	Strong
Cryptococcus neoformans	0	0	+	0	Weak
Saccharomyces cerevisiae	0	0	+	0	Weak

Source: Expanded from Tibayrenc et al. 1991b; Tibayrenc and Ayala 2002.

[a] Criteria for clonality are described in Box 5.2. +, criterion is satisfied; 0, data not available. Even where clonal reproduction has been firmly documented, this does not necessarily exclude occasional or even frequent sexual reproduction, which often has been evidenced as well.

[b] Overall weight of available molecular evidence. Data in various taxa have come from allozymes, RAPDs, microsatellites, gene sequences, or other molecular markers.

FUNGI.   Molecular markers have also been used to address questions about modes of asexual and sexual reproduction in several microbial fungi, including pathogenic forms (see Table 5.4). In one early study, Newton et al. (1985) employed allozymes and cytoplasmically transmitted RNAs (from mycoviruses inside the fungal cells) to assay numerous accessions of the cereal rust *Puccinia striiformis* from around the world. Representatives of one widespread group of wheat-attacking forms (*P. s. tritici*) proved completely uniform, whereas some other related species showed much higher levels of genetic variation. The cereal rust has no known sexual stage, so the genetic results are probably attributable to prevalent clonal transmission.

Candida albicans is a commensal diploid yeast normally inhabiting human mucosal epithelia, but in immunocompromised patients it can become a lethal pathogen. Traditionally this species was thought to be asexual, but its genome was recently found to possess a fungal mating-type-like (*MTL*) gene (Hull and Johnson 1999), and cell fusions have been observed (Hull et al. 2000; Magee and Magee 2000), raising the possibility that sexual reproduction might be a normal feature of its poorly known life cycle. To address these issues, various molecular markers, including allozymes, RAPDs, RFLPs, microsatellites and others, have been employed. Results indicate that the population genetics of this species is characterized by a mixture of occasional recombination *and* extensive clonality (Fundyga et al. 2002; Gräser et al. 1996; Pujol et al. 1993; Xu et al. 1999). Apart from contributing to knowledge of basic biology in this species, these molecular markers have also found clinical application in distinguishing various *Candida* strains and species (Boerlin et al. 1996; Mannarelli and Kurtzman 1998; Pinjon et al. 1998).

BACTERIA.   Following the discoveries in the mid-twentieth century of conjugation, transformation, and transduction in laboratory strains of *Escherichia coli*, a notion developed that recombinational genetic exchange might be prevalent in bacterial taxa in the wild (Hedges 1972). However, traditional research into bacterial systematics, epidemiology, and pathogenicity involved analyses of gross phenotypes (physiological, serological, etc.) that seldom could be tied to specific alleles or loci. Thus, reliable assessments of population genetic structure and reproductive mode remained elusive (Selander et al. 1987a). This situation started to change with protein electrophoretic surveys beginning in the early 1970s (Milkman 1973, 1975). Initial results from population analyses of allozyme markers suggested that, in addition to various mechanisms for genetic exchange, clonal reproduction is an important element underlying the genetic structure in natural populations of many bacterial taxa. Most bacteria are haploid, so the evidence for clonality usually consists primarily of criteria (c) and (d) in Box 5.2.

For example, early work on *E. coli* revealed that despite high allozyme variation ($H \cong 0.50$—an order of magnitude greater than values typifying higher eukaryotes), the number of distinctive protein electrophoretic profiles was unexpectedly constrained. Furthermore, presumptive clonal lineages (those identical or closely similar in multi-locus allozyme genotype) were sometimes observed even in samples from geographically widespread, unassociated hosts (Ochman and Selander 1984; Selander and Levin 1980; Whittam et al. 1983a,b). Such findings of strong disequilibrium across loci (see Box 5.3) soon led to a view that chromosomal transmission in *E. coli* is basically clonal, albeit with occasional exchanges of sequence causing partial reticulation among lineages (Milkman and Bridges 1990; Milkman and Stoltzfus 1988). These conclusions applied strictly to chromosomal DNA; extrachromosomal sequences (notably plasmids, which are often of adaptive relevance to the bacterium) did appear to be commonly exchanged among *E. coli* strains (Hartl and Dykhuizen 1984; Valdés and Piñero 1992).

This new paradigm favoring frequent clonal inheritance in *E. coli* carried several biological ramifications. Within an individual host, the presence of multiple clones might reflect successful invasions of independent founder lineages (Caugant et al. 1981), rather than novel genotypes arising solely via recombination among a few pioneering strains. Conversely, some specifiable clonal lineages could be distributed widely, perhaps even globally. It also became somewhat easier to deduce the phylogeny of various strains of *E. coli* and its suspected relatives, such as *Shigella* (Whittam et al. 1983b). Furthermore, to the extent that recombination in *E. coli* was less frequent than formerly imagined, its genome would constitute a low-recombination genetic system subject to some special evolutionary dynamic forces (see Box 5.3). Notable among these forces is "periodic selection" (sequential replacement of clonal lineages by genotypes displaying higher fitness), which can have the net effect of truncating overall population genetic variability both in neutral markers and in the selected genes to which they are linked.

Similar evidence for clonal chromosomal inheritance in other bacterial species, including pathogenic forms, soon led in the 1980s to a molecular genetic revolution in bacterial taxonomy and epidemiology (Selander and Musser 1990). The following discoveries are just a few examples (Selander et al. 1987b). Based on the distinctiveness of sets of clones earmarked by more than 50 different multi-locus allozyme genotypes, *Legionella* was shown to consist of two distinct species that had masqueraded as *L. pneumophila* (Selander et al. 1985). Some of the electrophoretic types (ETs) occurred worldwide, and one particular lineage caused Legionnaires' disease and Pontiac fever. In *Bordetella*, a pathogen responsible for a variety of respiratory diseases in animals and whooping cough in humans, numerous clones and several genetically distinct species were shown to exhibit strong host specificities (Musser et al. 1987): clone ET-1 of *B. bronchiseptica* is a pig specialist, ET-6 is a dog specialist, and the named taxa *B. parapertussis* and *B. pertussis* are other clonal forms of *B. bronchiseptica* that have become specialized as human pathogens. In *Haemophilus influenzae*, certain clones were found to be distributed worldwide, and one distinctive clonal group (ET-91-94) was pinpointed as the cause of meningitis and septicemia in human neonates. Another clone (ET-1) was found to have increased greatly in frequency in the United States between 1939 and 1954, and by 1990 caused about 30%–40% of all disease cases, whereas other *Haemophilus* clonal groups exhibited no clear association with particular disease conditions (Musser et al. 1985, 1986). In *Neisseria meningitidis*, high clonal variation was found using molecular markers, but only a few among the hundreds of multi-locus genotypes may have been responsible for most major pandemics in the twentieth century. For example, an epidemic that started in Norway in the mid-1970s and spread through Europe was caused by one group of clones in the ET-5 complex (Caugant et al. 1986). So too was another severe epidemic that appeared in the late 1970s in Cuba, from which it was spread by Cuban refugees to Miami, where another outbreak was initiated in 1980–1981.

These and other pioneering studies of clonal structure in bacteria exemplified the power of molecular markers in addressing evolutionary problems of diagnostic and epidemiologic relevance. On the other hand, not all bacterial taxa proved to be predominantly clonal. *Neisseria gonorrhoeae* appeared to have rampant genetic exchange, as judged by random associations of alleles among loci (Maynard Smith et al. 1993). One collection of wild *Bacillus subtilis* displayed a great diversity of allozyme and RFLP genotypes, leading Istock et al. (1992) to conclude that recombination must be frequent in this species also, perhaps due to its proclivity for spontaneous transformation, whereby DNA is exchanged by cell-to-cell contact. Such examples notwithstanding, population genetic data generally indicated that recombination events and gene flow within many bacterial species were far too rare to produce random allelic associations throughout a taxonomic species. However, even rare recombination can be of huge evolutionary significance if it occasionally generates high-fitness genotypes that then for a time are mostly clonally propagated. Indeed, for many bacterial taxa, direct evidence for chromosomal recombination accrued from allozymes and other molecular markers (DuBose et al. 1988; Selander et al. 1991a). Thus, any particular bacterial strain probably has its chromosomes derived in bits and pieces from multiple ancestors, the degree of historical heterogeneity depending on such factors as ancestral remoteness, the magnitude of geographic population structure due to restricted gene flow, the closeness of physical linkage of the genes involved, and, of course, the primary reproductive mode, which also significantly affects the number of past recombinational events (Hartl and Dykhuizen 1984).

A second revolution in the study of pathogenic bacteria, beginning mostly in the 1990s, accompanied the enormous influx of information from DNA sequencing. Genomes of hundreds of bacterial taxa have now been fully sequenced. A powerful new approach in evolutionary medicine is to analyze multiple strains within a disease-causing species to establish the historical origins of virulence and the molecular basis of pathogenesis (Fitzgerald et al. 2001; see reviews in Fitzgerald and Musser 2001; Whittam and Bumbaugh 2002). For example, by phylogenetically analyzing DNA sequences from multiple pathogenic strains of *E. coli*, Reid et al. (2000) demonstrated the following: recombination had not completely obscured the chromosomes' ancestral conditions; an evolutionary diversification of virulent clones probably began about 9 million years ago; a virulent strain (O157:H7) responsible for epidemics of food poisoning originally separated from a common ancestor of *E. coli* type K-12 as long as 4.5 million years ago; and some old lineages of *E. coli* apparently acquired the same virulence factors in parallel.

The virulence factors that can convert harmless *E. coli* to pathogenic forms include gene sequences encoded on mobile genetic elements such as plasmids and bacteriophages, as well as on distinct "pathogenicity islands" integrated into the bacterial chromosome (McGraw et al. 1999). Some of these pathogenicity islands have been well characterized and may involve, for example, genes encoding outer membrane proteins that mediate intimate attachment of bacteria to eukaryotic cells. Various islands may themselves be legacies of past integrations of nucleic acids following horizontal gene transfer events (see Chapter 8) between otherwise distinct bacterial strains. The net result is that a given bacterial clonal lineage is really a historical mosaic of genes with different evolutionary sources.

Genetic chimeras

A genetic chimera is an "individual" composed of a mixture of genetically different cells—that is, cells typically stemming from separate zygotes. The phenomenon is quite rare in the biological world, having been documented only in miscellaneous protists, plants, and animals representing about ten phyla (Buss 1982). Far more normally in nature, each multicellular individual is composed of genetically identical cells all tracing back asexually through mitotic cell divisions to one fertilized egg (Grosberg and Strathmann 1998; Michod 1999). This makes evolutionary sense, because multicellularity is the ultimate expression of inter-cell collaboration attributable to kin selection stemming from high genetic relatedness (see Chapter 6). Normally, close kinship is prerequisite for the extraordinary levels of cooperation and self-sacrifice displayed by somatic cells, which toil on behalf of their potentially immortal germ line kin without prospect of self-perpetuation per se (Maynard Smith and Szathmáry 1995; Queller 2000). From similar logic, true genetic chimeras are of special evolutionary interest: Why would unrelated cells ever join collaborative forces to constitute a multicellular individual?

Proper identification of genetic chimeras is important for several other reasons. First, cross-pollinations in plants (or cross-fertilizations in animals) might occur between genotypes within a chimeric individual, thus influencing the genotypes of progeny and the perceived mating system of a species. Second, if chimeras are common in particular species, the number of genets in a population could be seriously underestimated by a mere census of ramet numbers, with consequences extending to any parameters that are influenced by effective population size (such as expected magnitudes of genetic drift). The occurrence of genetic chimeras also raises important issues regarding the degree of physiological and functional integration of composite individuals.

MICROBES AND PLANTS.   One of the best-studied chimeras—the slime mold or social amoeba, *Dictyostelium discoideum*—occurs at one stage in the creature's life cycle. Individual free-living cells inhabit the forest floor, where they consume bacteria, but when the food supply becomes scarce, these haploid cells gather together to form a sluglike amoeboid structure that migrates to a more favorable spot and then forms a fruiting body that releases spores. Studies based on microsatellite markers have shown unequivocally that the amoeboid form often consists of genetically distinct cells and, hence, is a genuine chimera (Fortunato et al. 2003; Strassmann et al. 2000b).

Additional analyses of *Dictyostelium* have shown one potential advantage of unlike cells joining forces: Amoeboid aggregations with many cells (as opposed to fewer, as would be the case for any amoeba that excluded genetically different cells) can travel farther, thereby enhancing the aggregation's overall prospects for survival and reproduction in times of food scarcity (Foster et al. 2002; Queller et al. 2003).

Chimeras occur rarely in diploid metazoans as well. In plants, what appears to be a single ramet occasionally proves to be two or more genets that have fused into one morphologically (and perhaps physiologically) integrated module. A case in point involves strangler figs (*Ficus* spp.). These trees typically begin growth when bird-deposited or mammal-deposited seeds germinate in humus-filled crotches of a host tree. Shoots grow upward and roots downward around the host. The roots eventually cross and fuse to form a unified woody sheath around the host, which then may die, so that only the fig tree remains. Thompson et al. (1991) showed by allozyme analyses that "individual" fig trees often were genetic chimeras consisting of multiple genotypes. Thirteen of 14 sampled trees showed detectable genetic differences among branches, such that at least 45 genetic individuals were represented altogether. Presumably, the chimerism was attributable to post-germination fusions among roots that traced to multiple seeds deposited in the host tree.

INVERTEBRATE AND VERTEBRATE ANIMALS.   Fusions among ramets are common in many marine invertebrates, including sponges, cnidarians, bryozoans, and colonial ascidians (Grosberg 1988; Jackson 1985). These somatic fusions may involve recently settled larvae (Hidaka et al. 1997) or mature colonies (Neigel and Avise 1983a), with the participants normally being asexual products of a single genet. However, in some situations, the fusing entities are known or suspected to be sexually produced siblings or more distant kin, thus generating genetic chimeras (Barki et al. 2002; Maldonado 1998). For example, although fusions among clonemates are normal in the colonial hydroid *Hydractinia symbiolongicarpus*, they also occur occasionally between sexually produced full-sib and even half-sib pairs (Grosberg et al. 1996; Hart and Grosberg 1999).

The ascidian *Botryllus schlosseri* likewise can form chimeric colonies typically involving close relatives (Stoner and Weissman 1996). Microsatellite assays have been used to identify genetically different cells and monitor their fates in laboratory settings (Pancer et al. 1995; Stoner et al. 1999). Another ascidian (*Diplosoma listerianum*) has been documented by molecular markers to have rampant chimerism, sometimes even involving several distinct genotypes amalgamated from unrelated individuals (Bishop and Sommerfeldt 1999; Sommerfeldt and Bishop 1999). The authors speculate that chimerism is favored in this species because the phenomenon produces large colonies that survive and reproduce better than small ones (an explanation that closely parallels the documented advantages discussed above for chimerism in *Dictyostelium* slime molds).

In some marine invertebrate species, aggregation and fusion among kin may often be facilitated by ecological factors such as co-settlement of non-dispersive larvae. Even then, however, chimerism remains the exception rather than the rule because cell rejection responses mediated by polymorphic allorecognition genes typically are highly effective (Grosberg and Hart 2000; Grosberg and Quinn 1986).

Chimerism is known even in vertebrate animals, most notably in marmosets and tamarins (Benirschke et al. 1962; Haig 1999). These primates normally give birth to two fraternal (non-identical, or dizygotic) twins per pregnancy. In the first month of pregnancy, however, the tiny embryos partially fuse for a time inside the uterus, exchanging blood and some other body cells. Although the fetuses physically separate again before birth, molecular fingerprinting assays (of the common marmoset, *Callithrix jacchus*) have confirmed that each individual continues to be a chimera of its own blood cells plus those from its genetically distinct sibling (Signer et al. 2000).

In a quite different sense, the cells of all eukaryotic organisms can be viewed as genetic chimeras consisting of amalgamations of nuclear and organelle genomes that had independent evolutionary histories prior to their ancient endosymbiotic mergers (Margulis 1970). This topic will be deferred to Chapter 8.

Gender Ascertainment

Ascertainment of an individual's sex can prove difficult in many situations, such as early life history stages, species with little dimorphism in secondary sexual characters, or species with internal gonads (such as birds). Yet knowledge of sexual identity is crucial in many ethological studies, in estimation of population sex ratios, in management of matings among endangered captive animals (e.g., Millar et al. 1997), and in several other areas of population biology. In some taxonomic groups, such as reptiles, sex is often influenced by the temperature at which eggs are incubated (Bull 1980; Shine 1999), but gender in most other taxa is genetically "hard-wired." For these latter species, molecular assays of gender-associated ("sex-linked") DNA markers provide a powerful approach to sex identification at any stage of life (Griffiths 2000).

A flurry of such pioneering studies, especially on avian species, began to appear by the early 1990s. In birds, females are the heterogametic gender, possessing ZW sex chromosomes in contrast to the male ZZ condition. Thus, W-specific molecular markers have been a prime target for sex-typing. For example, Quinn et al. (1990) isolated a segment of DNA homologous to the W (female-specific) chromosome of the snow goose (*Chen caerulescens*) and employed this molecular probe to determine the sex of more than 150 birds, using blood samples. Similarly, Griffiths and Holland (1990) isolated a W-linked repetitive DNA marker for the herring gull (*Larus argentatus*), as did Rabenold et al. (1991) for stripe-backed wrens (*Campylorhynchus nuchalis*). In DNA fingerprinting assays, Millar et al. (1992) uncovered W-specific bands in the brown skua (*Catharacta lonnbergi*) and used them to document significantly different sex ratios in adults versus chicks.

Most of these early genetic screens involved Southern blotting techniques (so that fairly large tissue samples were needed) and were based on rapidly evolving repetitive sequences (so that the W-linked markers were of limited taxonomic range). A possibility also existed that occasional cross-homology between the Z and W chromosomes (as is known for portions of the mammalian X and Y; Burgoyne 1986; Ohno 1967; Page et al. 1982, 1984) might compromise some of these assays. Thus, later approaches moved toward PCR-based assays (as in RAPDs; Lessells and Mateman 1998) and toward assays of slower-evolving and better-characterized W-linked genes that might have broader taxonomic scope. The W chromosome of birds (like the Y chromosome of mammals) is small and carries few functional loci, but Ellegren (1996) did identify one useful W-linked gene (*CHD-W*) that provided a near-universal tag for avian sexing (see also Griffiths et al. 1996, 1998; Huynen et al. 2002). Within the last decade, DNA sex-typing based on this and other molecular marker systems has become common practice in avian behavioral ecology (Komdeur et al. 1997; Millar et al. 1996; Westerdahl et al. 1997), conservation efforts (Double and Olsen 1997; Robertson et al. 2000), and related endeavors (Ellegren and Sheldon 1997).

In mammals, in which males are the heterogametic sex, analogous sex-typing methods have been developed based on Y chromosome markers (Fernando and Melnick 2001). For example, Sinclair et al. (1990) discovered a Y-specific probe that could be used in Southern blot analyses to identify each individual's sex in a wide range of mammal species. A remarkable early application of this approach involved humpback whales (*Megaptera novaeangliae*), which, like other baleen whales, lack obvious secondary sexual characteristics. A human Y chromosome sequence was employed as a hybridization probe in RFLP analyses to determine the gender of 72 free-ranging humpbacks from which skin biopsies had been collected by special dart (C. S. Baker et al. 1991). The sex of another individual was DNA-typed from sloughed skin collected from the whale's swimming path. More recently, cetaceans have also been sexed using PCR-based assays with primers for sex-specific *ZFY* and *ZFX* gene sequences (Bérubé and Pasbøll 1996). Wild brown bears (*Ursus arctos*) have similarly been sexed by Y-linked molecular markers, using DNA extracted from shed hairs (Taberlet et al. 1993, 1997).

One important application for sex-linked molecular markers is in estimating sex ratios at early developmental stages for comparison against adult sex ratios. An example involves the Japanese frog *Rana rugosa*, in which females are the heterogametic sex. Newly fertilized eggs of this species, sampled in the field throughout the summer months, were assayed for gender using a PCR-amplified sex-specific marker (Sakisaka et al. 2000). The researchers discovered a male-biased primary sex ratio (i.e., the sex ratio at or near conception) early in the reproductive season but a female-biased sex ratio later on, a result that they interpreted as indicative that adults might somehow be able to influence sex ratios among their progeny.

Another application for sex-linked molecular markers is in establishing the mode of sex determination in taxa for which it is otherwise unknown. An example involves gastropod mollusks in the genera *Busycon* and *Busycotypus*, which formerly were suspected to be protandrous (male-first) hermaphrodites. However, the fortuitous discovery of a sex-associated microsatellite marker with a transmission pattern analogous to that of X-linked genes in mammals strongly indicates that these whelks normally have separate sexes (Avise et al. 2004). This sex-linked microsatellite marker was then used to estimate near-primary sex ratios in brooded cohorts of tiny whelk embryos, for which it was otherwise impossible to assign gender.

Remarkably, a few plants possess sex chromosome systems somewhat analogous to those of vertebrates and various other animals. For example, in the perennial dioecious weed *Rumex acetosa*, females are XX and males are XY_1Y_2, so it seemed puzzling that sex ratios of flowering adults were female-biased. Surveys of DNA markers from both Y chromosomes resolved the quandary by permitting the ascertainment of sex and the estimation of sex ratios in seeds (Korpelainen 2002). It turned out that the primary sex ratio was about 1:1 in the total seed pool, and that the female-biased adult sex ratio resulted in part from higher male mortality during development.

Genetic Parentage

Molecular procedures for genetic assessment of parentage are similar in principle to those used to assess genetic identity versus non-identity, but with the added complication that rules of Mendelian transmission genetics must be taken into account when comparing the genotypes of sexually produced progeny against those of putative parents. Parentage analyses often address some version of the following question: Are the adults who are associated behaviorally or spatially with particular young the true biological parents of the offspring in question? If the answer proves to be no, a genetic exclusion has been achieved (Box 5.4). Whether the actual mothers and fathers also can be specified depends on the size and genetic composition of the pool of candidate parents and on the level of genetic variability monitored. Sometimes one biological parent is known from independent evidence and the problem simplifies to one of paternity (or maternity) exclusion or inclusion. In other cases, neither parent is known with certainty prior to the molecular study.

Parentage analyses utilize cumulative information from multiple polymorphic loci, scored either collectively, as complex DNA banding patterns on a gel (often in minisatellite DNA fingerprinting), or one at a time (e.g., by assays of allozymes or nuclear RFLPs), with the data tallied as discrete Mendelian genotypes accumulated across loci. In recent years, microsatellite assays have largely supplanted earlier methods of parentage assessment, especially in analyses of vertebrate animals, in which (unlike in many plants, for example) allozyme variation often tends to be insufficient to provide high exclusionary power. Even a few hypervariable microsatellite loci in a population often display sufficient genetic variation to yield combined exclusion probabilities well above 0.99, thereby offering exquisite information on biological paternity and maternity.

Various types of interpretive logic regarding parentage can be introduced by the following empirical examples, each a classic from the early literature:

1. *Maternity and paternity both uncertain, exclusions attempted.* T. W. Quinn et al. (1987, 1989) compared goslings within each of several broods of snow geese (*Chen caerulescens*) against their adult male and female nest attendants (putative parents) using genetic markers from multiple

---

BOX 5.4 Genetic Exclusions and Parentage Analyses

Using Mendelian molecular markers, estimates of genetic maternity or genetic paternity can be achieved by excluding as parents all adults whose genotypes are incompatible with those of the juveniles under consideration. Associated with such "genetic exclusions" are statistical probabilities that are a joint function of the variability of the genetic markers employed and the biological nature of the particular parentage problem (e.g., perhaps one of the two parents is known from independent evidence, such as pregnancy).

Exclusion probabilities may be either *specific* or *average*. Consider a neutral autosomal locus with two equally frequent alleles (*A* and *B*). In a large population at Hardy–Weinberg equilibrium, about 25% of all individuals would be homozygous *AA*, and another 25% homozygous *BB*. Suppose that molecular markers show that an *AA* mother has an *AA* offspring. All adult males in the population with genotype *BB* can be excluded as the youngster's biological father (barring mutation), so the specific exclusion probability in this case is 0.25. An average exclusion probability, by contrast, is the mean probability (or the expected proportion) of excluded parents for randomly chosen juveniles. A mean exclusion probability may be higher than some specific exclusion values and lower than others because it is calculated by combining all specific exclusion probabilities weighted by the expected frequency of each parent–offspring pair in the population. Biologists are often particularly interested in average exclusion probabilities because they indicate the strength of available genetic markers for parentage exclusions (values above 0.99 typically are sought) and because these are useful for comparing statistical power across published studies.

The first formulae for calculating average exclusion probabilities were published early in the twentieth century (e.g., Weiner et al. 1930), but later methods generalized and extended the underlying models, of which two are most common: the "one-parent-known" case, in which independent evidence provides secure knowledge of either the mother or the father (Dodds et al. 1996; Jamieson 1965; Jamieson and Taylor 1997; Weir 1996), and the "unknown parentage" case, in which neither parent is certain from extrinsic evidence (Crawford et al. 1993; Garber and Morris 1983). Formulae for single-locus mean exclusion probabilities under these respective models are as follows, where p_i is the frequency of the ith codominant allele at an autosomal locus:

$$P_E = 1 - 2\sum_{i=1}^{n} p_i^2 + \sum_{i=1}^{n} p_i^3 + 2\sum_{i=1}^{n} p_i^4 - 3\sum_{i=1}^{n} p_i^5 - 2\left(\sum_{i=1}^{n} p_i^2\right)^2 + 3\sum_{i=1}^{n} p_i^2 \sum_{i=1}^{n} p_i^3 \quad \textbf{(5.1)}$$

and

$$P_E = 1 - 4\sum_{i=1}^{n} p_i^2 + 2\left(\sum_{i=1}^{n} p_i^2\right)^2 + 4\sum_{i=1}^{n} p_i^3 - 3\sum_{i=1}^{n} p_i^4 \quad \textbf{(5.2)}$$

Mean exclusion values calculated for individual loci can then be combined across K independent marker loci into a total average exclusion probability for a given study (Boyd 1954):

$$P_{TE} = 1 - \prod_{j=1}^{K} (1 - P_{E(j)}) \quad \textbf{(5.3)}$$

The models described above typically assume that samples were taken from a randomly mating deme, but if biological phenomena such as population structure, philopatry, or inbreeding apply, these equations can artificially inflate exclusion probabilities relative to their true values (Double et al. 1997).

Many other statistical tasks associated with various nuances of parentage assessment have been developed (see review in Jones and Ardren 2003). For specified biological settings, maximum likelihood approaches are available to categorically assign individuals to their parents (Coltman et al. 1998b) or to fractionally assign parentage to multiple non-excluded adults (Smouse and Meagher 1994). Computer programs used to implement these or other methods include CERVUS (Marshall et al. 1998), FAMOZ (Marshall et al. 1998), KINSHIP (Goodnight and Queller 1999), PAPA (Duchesne et al. 2002), PARENTE (Cercueil et al. 2002), PATRI (Signorovitch and Nielsen 2002), and PROBMAX (Danzmann 1997). Some statistical approaches are targeted toward quite specific problems. For example, in many fishes and other species with hundreds or thousands of offspring in a clutch, methods of statistical correction (e.g., for finite marker data or incomplete sampling of candidate individuals in the population) have been devised to refine estimates of multiple mating and the mean number of reproductive adults contributing to a half-sib progeny array (DeWoody et al. 2000a; Fiumera et al. 2001; Jones 2001; Neff and Pitcher 2002), the proportion of next-generation offspring sired by a focal male (Fiumera et al. 2002a; Neff et al. 2000a,b, 2002), and the proportion of broods with at least two contributing members of each adult sex (Neff et al. 2002).

scnRFLP loci. From their data (Table 5.5), the following observations and deductions were made. Two goslings in family 2 (numbers 7 and 8) proved to be homozygous at some loci (*E* and *M*) for alleles not present in the female attendant. Such cases excluded the putative mother and were interpreted to reveal instances of intraspecific brood parasitism (IBP), or "egg-dumping" (Petrie and Møller 1991), whereby other females (not assayed) must have contributed eggs to the nest. Other goslings (e.g., number 6 in family 2) proved to be homozygous (locus *M*) for alleles not present in the male attendant. Such cases excluded the putative father and were interpreted to reveal likely instances of extra-pair fertilization (EPF) by other males. Some heterozygous loci (e.g., *J* in gosling 5, family 2) exhibited one allele not observed in either nest attendant and a second allele present in both attendants. Such loci exclude one of the putative parents, but do not alone determine which attendant is disallowed. Finally, some loci (e.g., *C* in gosling 4, family 1) were homozygous for alleles not observed in either nest attendant, thus excluding both. Overall, the genetic markers revealed that otherwise cryptic IBP and EPF behavioral events must be relatively common in snow goose populations (see also Lank et al. 1989).

2. *Maternity known, paternity to be decided among a few candidate males.* Burke et al. (1989) applied multi-locus DNA fingerprint assays to the dunnock sparrow (*Prunella modularis*), a species with a mating system in which two males often mate with a single female and defend her territory (Davies 1992). In the DNA fingerprints, those bands in progeny that could not have been inherited from the known mother were identified as paternally derived. Then, the true father was determined by comparing bands from the fingerprints of candidate sires against these paternal alleles in progeny. Figure 5.6 shows DNA fingerprints from one known mother (M), her four offspring (D–G), and two candidate sires (Pα and Pβ). In this family, the genetic data demonstrate that progeny G was sired by Pα, whereas D, E, and F were fathered by Pβ. Thus, molecular data confirmed that individual dunnock broods can be multiply sired.

3. *One parent or two?* Many plants and invertebrate animals are hermaphroditic; that is, an individual produces both male and female gametes. Such individuals may self-fertilize (in which case offspring have a single parent), or matings with other individuals may be facultative or compulsory (producing two-parent progeny). For wild-caught females whose mating habits are in question, genetic examination of progeny can reveal whether some of them carry alleles that are not present in the mother and, hence, derive from cross-fertilizations. Furthermore, comparisons of population genotypic frequencies against Hardy–Weinberg expectations can aid in deciding whether cross-fertilization or self-fertilization predominates at the population level (because selfing is an intense form of inbreeding whose continuance leads to pronounced

heterozygote deficits). In examples of these approaches, allozyme data were employed to show that cross-fertilization is the prevailing mode of reproduction in several freshwater species of hermaphroditic snails in the genera *Bulinus* and *Biomphalaria* (Rollinson 1986; Vrijenhoek and Graven 1992; Woodruff et al. 1985), that intermediate levels of self-fertilization characterize the Florida tree snail *Liguus fasciatus* (Hillis 1989) and the coral *Goniastrea favulus* (Stoddart et al. 1988), and that self-fertilization predominates in populations of the sea anemone *Epiactis prolifera* (Bucklin et al. 1984). In allozyme studies of 19 species of terrestrial slugs in the families Limacidae and Arionidae, most of the taxa were shown to be predominant outcrossers (Foltz et al. 1982, 1984).

TABLE 5.5   Diploid genotypes of nest attendants and goslings in three snow goose families

	RFLP locus					
	A	*B*	*C*	*D*	*E*	*F*
Family 1						
Male attendant	2,2	2,2	2,3	1,2	1,1	1,1
Female attendant	2,2	2,2	2,2	1,1	1,1	1,1
Gosling 1	2,2	2,2	2,2	1,2	1,1	1,1
Gosling 2	2,2	2,2	2,2	1,1	1,1	1,1
Gosling 3	2,2	2,2	2,2	1,2	1,1	1,1
Gosling 4	**2,3**[a]	2,2	**1,1**[d]	1,1	1,1	1,1
Family 2						
Male attendant	1,2	2,2	2,4	1,1	1,1	1,1
Female attendant	2,2	2,2	1,2	1,1	2,2	1,1
Gosling 5	2,2	2,2	1,2	1,1	1,2	1,1
Gosling 6	2,2	2,2	2,4	1,1	1,2	1,1
Gosling 7	1,2	2,2	2,4	1,1	**1,1**[b]	1,1
Gosling 8	**2,3**[a]	2,2	2,2	1,1	**1,1**[b]	1,1
Gosling 9	1,2	2,2	2,2	1,1	1,2	1,1
Family 3						
Male attendant	2,2	2,2	3,3	1,2	1,1	1,1
Female attendant	2,2	1,2	1,1	1,1	1,1	1,1
Gosling 10	2,2	1,2	1,3	1,2	1,1	1,1
Gosling 11	2,2	1,2	1,3	1,2	1,1	1,1
Gosling 12	2,2	2,2	1,3	1,2	1,1	1,1
Gosling 13	2,2	2,2	1,3	1,2	1,1	1,1

Source: After T. W. Quinn et al. 1987.

Notes: Letters are loci; numbers are allelic designations. Some goslings in families 1 and 2 (boldface) show genetic evidence of EPF or IBP (see text).
[a] Excludes one unspecified parent; [b]excludes putative mother; [c]excludes putative father; [d]excludes both putative parents.

4. *Self-fertilization or parthenogenesis?* As mentioned earlier, another form of uniparental reproduction is parthenogenesis, which can be experimentally distinguished from self-fertilization by examining diploid genotypes among offspring of a heterozygous parent. Fixed heterozygosity among progeny is inconsistent with expectations of Mendelian segregation under self-fertilization, but it is a hallmark of ameiotic parthenogenesis. Using allozyme markers in this context, Hoffman (1983) documented that a laboratory population of a slug species (*Deroceras laeve*) formerly suspected of self-fertilization actually reproduced parthenogenetically. Further discussion of parentage in the context of parthenogenetic reproduction will be deferred to Chapter 7.

			RFLP locus				
G	H	I	J	K	L	M	N
1,4	2,2	1,2	1,2	1,2	1,2	2,2	2,2
1,3	1,2	2,2	1,1	1,1	1,2	1,2	1,2
1,1	1,2	1,2	1,1	1,2	1,1	1,2	1,2
3,4	2,2	2,2	1,2	1,2	1,2	2,2	2,2
1,3	2,2	1,2	1,1	1,1	2,2	2,2	2,2
1,2[a]	1,2	1,2	1,1	1,1	1,1	**1,1**[c]	**1,1**[c]
1,1	1,2	1,1	1,1	1,1	1,2	1,1	1,2
1,2	1,2	2,2	1,1	1,1	1,1	2,2	1,2
1,1	2,2	1,2	**1,2**[a]	1,1	1,1	1,2	1,2
1,2	1,1	1,2	1,1	1,1	1,1	**2,2**[c]	1,1
1,1	1,2	**2,2**[c]	1,1	1,1	1,1	**1,1**[b]	1,1
2,2[c]	1,1	1,2	1,1	1,1	1,1	**1,1**[b]	2,2
1,2	2,2	1,2	1,1	1,1	1,2	1,2	1,2
1,1	1,2	1,2	1,1	1,1	1,1	1,1	1,2
1,1	2,2	1,1	1,1	1,2	1,2	1,2	1,2
1,1	1,2	1,2	1,1	1,2	1,1	1,2	1,2
1,1	1,2	1,2	1,1	1,1	1,2	1,1	1,2
1,1	2,2	1,2	1,1	1,2	1,1	1,2	1,2
1,1	2,2	1,2	1,1	1,1	1,2	1,1	2,2

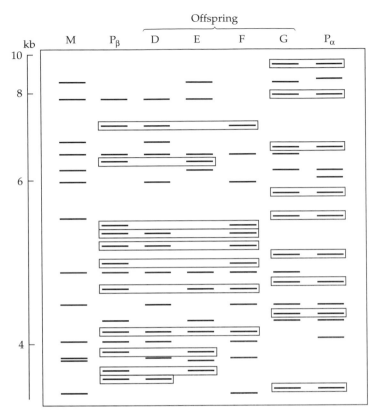

Figure 5.6 Parentage analysis by multi-locus DNA fingerprinting. Shown is a gel designed to assess whether each of four offspring (D–G) with known mother M was sired by P_α or P_β. Boxes encompass paternally derived bands in progeny that permitted choice between the two candidate sires. (After Burke et al. 1989.)

Behavioral and evolutionary contexts

Knowledge of biological parentage is important in many behavioral and ecological contexts. For example, matings are difficult to observe directly in nature for many species, but reproductive behaviors and patterns of gene flow (see Chapter 6) nonetheless can be deduced from molecular information on maternity and paternity. Proper interpretations of behavioral interactions between presumed family members depend on knowledge of genetic ties, including parentage. Even when matings can be readily observed, questions of genetic parentage remain of interest. In many birds and mammals, for example, copulations are known to occur outside the socially bonded pair, but until molecular markers were applied, the extent to which these matings resulted in illegitimate young remained uncertain, constituting a major deficiency in understanding of sexual selection and the ecology of mating systems (Fleischer 1996; Mock 1983; Trivers 1972). By revealing genetic parentage, molecular data provide direct assessments of realized reproductive success and therefore largely circumvent the danger of equating mating prowess or other components of reproduction with actual gene transfer across generations (Møller and Ninni 1998). Knowledge of biological parentage is also critical for correct interpretation of the transmission genetics or heritabilities of morphological and other phenotypic characteristics as deduced from field data on presumed parent–offspring relationships (Alatalo et al. 1984).

Typically, molecular data revealing genetic maternity and paternity within particular broods or clutches are accumulated across many families, such that the results collectively describe the "genetic mating system" (often quite different from the field-observed "social mating system") of a population or species. Figure 5.7 provides definitions of various genetic mating systems. It also summarizes their oft-suspected relationships to the intensity of sexual selection (resulting from differential abilities among individuals of the two genders to acquire mates) and the degree of elaboration in each sex of secondary sexual traits (those arising from sexual selection). For example, conventional wisdom holds that males in polygynous species are often under strong sexual selection and therefore display pronounced body adornments (e.g., large antlers in bull elk or flashy tails in peacocks) arising from intrasexual or intersexual competition for mates; whereas in polyandrous species, it is females who are likely to be under intense sexual selection and thus perhaps more ornamented with secondary sexual features.

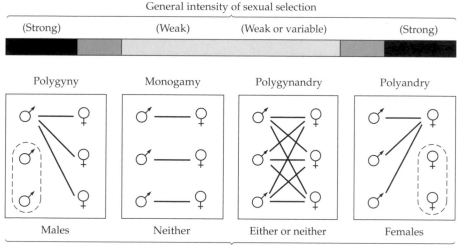

Figure 5.7   **Pictorial definitions of four possible genetic mating systems.** Lines connecting males and females indicate mating partners that produce offspring. Also shown are theoretical gradients in sexual selection intensities and degrees of gender dimorphism in secondary sexual traits often thought to be associated with these genetic mating systems. (From Avise et al. 2002.)

Parentage data provided by molecular markers afford fresh opportunities to critically test these and many other aspects of conventional sociobiological wisdom on the ecology and evolution of genetic mating systems.

Selected empirical examples by taxa

HUMANS.　In species with internal fertilization, or where close physical or behavioral associations connect mothers with offspring, maternity may be evident, whereas paternity remains uncertain. Such is often the case in humans; indeed, the great bulk of effort by molecular forensic laboratories involves identifying the father in cases of unclear or disputed paternity (Lewis and Cruse 1992).

　　Maternity in humans is far less often in doubt, but there are exceptions. Indeed, one of the first applications of DNA fingerprinting in human forensics involved maternity assessment in an immigration dispute (Jeffreys et al. 1985c). The case concerned a Ghanaian boy born in the United Kingdom who left to join his father and later returned alone to be reunited with his mother. However, immigration authorities suspected that an unrelated boy, or perhaps a son of one of the mother's several sisters living in Ghana, had been substituted for the mother's son. At the request of the family's solicitor, a DNA fingerprint analysis was conducted of the boy, his putative mother, and several of the child's undisputed siblings (the task was complicated because the boy's father was also uncertain). The first step was deducing paternal-specific bands in the boy's DNA fingerprint. These were DNA fragments present in at least one of the siblings but absent from the mother. The second step was subtracting these paternal-specific bands from the boy's DNA fingerprint. All 40 remaining fragments matched those present in the focal woman, indicating that she was indeed the child's biological mother. These genetic data were provided to the immigration authorities, who then dropped objections and granted the boy residence in the U.K.

OTHER PRIMATES.　Genetic parentage analyses have been conducted on many mammalian species, as will be illustrated here by a few classic examples from the primate literature.

　　Troops of rhesus macaques (*Macaca mulatta*) are characterized by strong dominance hierarchies (in both genders) whose behavioral underpinnings are postulated to have evolved in response to selective pressures favoring high-ranking individuals. Do males of higher social rank truly exhibit higher fitness through greater access to receptive females (Dewsbury 1982)? In a landmark genetic study, Duvall et al. (1976) coupled allozyme assays with behavioral observations to show that only 7 of 29 offspring (24%) produced over a 2-year study of a captive group actually were fathered by the alpha (top-ranking) male. Even low- and mid-ranking adolescent monkeys sired several offspring and, thus, clearly had access to ovulating females.

In another early study of a rhesus troop, Curie-Cohen et al. (1983) found that over an 8-year period, the dominant male sired only 13%–32% of the offspring, although he participated in 67% of observed copulations, whereas the second-ranking male sired 30%–48% of the offspring despite participation in only 14% of the matings. By genetically examining other groups of this same species, Smith (1981) found that reproductive success *was* significantly correlated with rank, although changes in rank appeared to follow, rather than precede, changes in reproductive success. Other seminal genetic findings on rhesus monkeys were observations that a reproductive advantage is enjoyed by sons of high-ranking mothers (Smith and Smith 1988) and that males and females of similar social rank tend to produce proportionate numbers of progeny (Small and Smith 1982). The overall picture from paternity analyses of this and other macaque species (e.g., Keane et al. 1997) is that social dominance and even copulation frequency inside a troop are at best imprecise predictors of male breeding success (although the possibility has not been eliminated that the lifetime genetic fitness of a socially dominant male is higher due to higher survival probabilities among his offspring, rather than his increased reproductive activity per se; Bernstein 1976).

Rhesus macaques typically live in large troops and are seasonal breeders, with females exhibiting synchronized fertile cycles during a well-defined period. Stern and Smith (1984) speculated that in such cases it is more difficult for males to monopolize females than in social systems in which troop size is smaller and estrus is temporally dispersed. A related macaque species (*M. fascicularis*) that breeds nonseasonally was employed to test this hypothesis. Genetic paternity analyses of 44 *M. fascicularis* offspring born over a 28-month period revealed no evidence, however, to support a positive association between a male's social rank and the number of offspring he sired (Shively and Smith 1985). A different pattern was uncovered in studies of the red howler monkey (*Alouatta seniculus*). In nine surveyed troops constituting single-male and multi-male harems, genetic evidence consistently implicated the dominant resident as the father of offspring conceived during his tenure (Pope 1990). This finding agreed with behavioral evidence because only top-ranking males were observed to mount females. Red howler troops are small and spatially cohesive, and this may facilitate behavioral monitoring by the alpha male. Pope (1990) further suggested that females avoid mating with subordinate males to avoid infanticide, because infants conceived during successful and attempted status changes by males are frequently killed.

Early molecular studies of paternity were also conducted on baboons, marmosets, lemurs, guenons, mandrills, gorillas, and chimpanzees, as well as several macaques and other monkeys (R. D. Martin et al. 1992). An important generality (results for red howler monkeys notwithstanding) is that social status and observed copulation frequency are often poor guides to male reproductive success. Thus, these traditional behavior-based methods of fitness estimation soon came to be appreciated as grossly inadequate predictors of successful progeny production in many primate societies.

More recent studies exemplify how microsatellite markers are further contributing to knowledge of primate parentage and social behavior. In the orangutan (*Pongo pygmaeus*), classic adult males are huge and display secondary sexual features including wide cheek pads ("flanges") and a large throat sac for emitting loud calls. Other males do not develop these features to nearly the same extent, but instead are "developmentally arrested" for up to 20 years after reaching sexual maturity. Thus, this species shows a pronounced "bimaturism" among males, which researchers had thought must be due to a social environment wherein the presence of a classic adult male hormonally suppresses the full maturation of subordinate males in his vicinity. A somewhat different perspective on the topic has recently come to light (Maggioncalda and Sapolsky 2002). Observations on the animals' behaviors and hormone levels did not square with the notion that subordinate males are abnormally stressed. Furthermore, microsatellite analyses revealed that about 50% of offspring in a Sumatran population were sired by unflanged males (Utami et al. 2002), who clearly force themselves upon (i.e., rape) females (a tactic used much less often by dominant males). These findings were interpreted to indicate that unflanged males are not pathological or debilitated specimens, but rather are employing a genetically successful "alternative reproductive tactic" to classic adult maleness.

Another recent paternity analysis, on savanna baboons (*Papio cynocephalus*), uncovered perhaps the first genetic evidence in any wild species that fathers can distinguish their own from other males' offspring in polygamous multi-male, multi-female assemblages. By combining results from microsatellite paternity analyses with 30 years of observational data on wild baboons in Kenya's Amboseli Basin, Buchan et al. (2003) showed that in resolving fights among juveniles, adult males were significantly more likely to support their own biological offspring than they were to intervene on behalf of unrelated young. This apparent knowledge of paternity by baboon males might be due to direct cues (such as a juvenile's appearance or smell), indirect cues (e.g., a male might assess his paternity probability based on his frequency of past copulations with mothers of particular offspring), or both. Whatever the mechanism of kinship recognition, nepotistic males tend to behave as if cognizant of their biological paternity. Among other ramifications, this genetic finding casts doubt on one conventional hypothesis for multiple mating by female baboons—that it confuses paternity within a troop and thereby serves either to enlist more adult males in offspring care or to reduce the risk of infanticide by unrelated adult males.

BIRDS. Parentage analyses via molecular markers have revolutionized thought in avian sociobiology by documenting that individual broods frequently contain progeny from at least one biological parent other than the attendant care-givers (Birkhead and Møller 1992; Ligon 1999; Westneat et al.

1990). Conventional wisdom was that Passeriformes (perching birds), in particular, are among the most monogamous of organisms (Gill 1990; Lack 1968). However, Gowaty and Karlin (1984; see also Gowaty and Bridges 1991a; Meek et al. 1994) were among the first to report genetic evidence for high frequencies of multiple paternity (ca. 8%–35% illegitimate young) in a purportedly monogamous passeriform, the eastern bluebird (*Sialia sialis*). Since then, in more than 150 molecular studies encompassing about 130 avian species and a total of more than 25,000 offspring (Griffith et al. 2002), many additional examples of multiple concurrent paternity have come to light, with foster young often found at high frequencies (Table 5.6).

TABLE 5.6   Representative frequencies of extra-pair offspring (EPOs) detected by molecular markers in various avian species

Species	No. broods assayed	% broods with EPO[a]	No. chicks assayed	% EPO chicks[a]	Reference
Bobolink, *Dolichonyx oryzivorous*	38	19	840	15	Bollinger and Gavin 1991
Corn bunting, *Miliaria calandra*	15	7	44	5	Hartley et al. 1993
Eastern kingbird, *Tyrannus tyrannus*	19	47	60	30	McKitrick 1990
Eurasian dotterel, *Charadrius morinellus*	22	5	44	5	Owens et al. 1995
Field sparrow, *Spizella pusilla*	17	41	52	19	Petter et al. 1990
Hooded warbler, *Wilsonia pusilla*	17	47	78	29	Stutchbury et al. 1994
House sparrow, *Passer domesticus*	183	26	536	14	Wetton and Parkin 1991
House wren, *Troglodytes aedon*	18	22	97	6	Price et al. 1989
Mallard duck, *Anas platyrhynchos*	46	17	298	3	Evarts and Williams 1987
Northern cardinal, *Cardinalis cardinalis*	16	19	37	14	Ritchison et al. 1994
Pied flycatcher, *Ficedula hypoleuca*	22	14	131	20	Gelter and Tegelström 1992
Red-cockaded woodpecker, *Picoides borealis*	28	4	48	1	Haig et al. 1994
White-crowned sparrow, *Zonotrichia leucophrys*	35	26	110	34	Sherman and Morton 1988
White-fronted bee-eater, *Merops bullockoides*	65	—	97	10	Wrege and Emlen 1987

Note: For details and more examples, see Gowaty 1996, and Westneat and Stewart (2003).

[a] Often minimum estimates because of limited exclusionary power in the markers employed; see Westneat et al. 1987 for discussion of this problem.

For example, an allozyme survey of the indigo bunting (*Passerina cyanea*) established that at least 37 of 257 offspring (14%) carried genotypes incompatible with the behaviorally suspected father (Westneat 1987). Statistical corrections that account for limited detection probabilities (because only a few polymorphic markers were employed) raised the estimated frequency of EPFs to as high as 42%. This latter estimate agreed with a DNA fingerprinting survey of another indigo bunting population, in which 22 of 63 nestlings (35%) were shown to have resulted from extra-pair fertilizations (Westneat 1990).

Not all genetic appraisals of passeriform species have produced evidence for extra-pair offspring (Griffith 2000). For example, among 176 offspring from 32 families of warblers in the genus *Phylloscopus*, no foster young were found by sensitive DNA fingerprint assays (Gyllensten et al. 1990). Likewise, nearly all 222 assayed juveniles of the bicolored wren (*Campylorhynchus griseus*) were produced by the primary mated pairs (Haydock et al. 1996), and the same proved true in microsatellite assays for nearly all of 139 scrub jay offspring (*Aphelocoma coerulescens*) produced by 34 assayed adult pairs (Quinn et al. 1999). At the other end of the continuum, the current record for highest frequency of extra-pair offspring in the nest may belong to the superb fairy-wren (*Malurus cyaneus*), in which molecular markers have documented that more than 70% of offspring are sired by males other than the putative father (Double and Cockburn 2000; Mulder et al. 1994).

Likewise, in many non-passeriform birds, molecular markers have often revealed multiple paternity or maternity within a clutch. An interesting example involved genetic documentation of communal nesting in the ostrich, *Struthio camelus*. From microsatellite assays of a population in Nairobi National Park, Kimwele and Graves (2003) discovered that only about 30% of all incubated eggs were genetically parented by both the resident territorial male and female. Instead, all surveyed males had fertilized at least some eggs in the clutches of neighboring males, and every surveyed female had laid eggs not only in her own nest, but also in neighboring ones. Some genetic reappraisals of non-passerines have found no evidence for EPFs or IBPs, however. For example, DNA fingerprints of black vultures (*Coragyps atratus*) confirmed that this species is genetically as well as socially monogamous (Decker et al. 1993), as did comparable molecular data for Leach's storm-petrel (*Oceanodroma leucorhoa*; Mauck et al. 1995), Cory's shearwater (*Calonectris diomedea*; Swatschek et al. 1994), the crested penguin (*Eudyptes pachyrhynchus*; McLean et al. 2000), and the endangered New Zealand takahe (*Porphyrio hochstetteri*; Lettink et al. 2002). The broader point is that molecular appraisals of numerous avian species evidence not only the power of marker-based parentage analyses, but also the shortcomings of traditional field observations as secure guides to actual mating behaviors and genetic mating systems.

Literature reviews have found no correlation between EPF rate and the nesting density or degree of coloniality in avian species (Westneat and Sherman 1997; Wink and Dyrcz 1999; but see also Møller and Birkhead 1993a). Thus, other factors must be involved. These factors have been the subject of many investigations that combine genetic assessments with behavioral or life history observations in the field (Petrie and Kempenaers 1998). In one such example, the frequency of foster nestlings proved to be significantly greater in the broods of eastern bluebird males who were in their first breeding season, who were paired with females who frequently strayed from home territory during their fertile periods, and who exhibited mate-guarding behavior (a counterintuitive result, unless it is supposed that these males sense a propensity for cheating by their mates and monitor them accordingly). Gowaty and Bridges (1991b) interpreted some of these trends as consistent with the postulate that female bluebirds actively pursue EPFs, rather than receiving them passively or by coercion from EPF-seeking males (as might be assumed in traditional mating system theory). In theory, EPFs might be selectively advantageous to a female for any of several reasons: they might generate higher genetic diversity among her progeny (Foerster et al. 2003); they might afford her more opportunities to obtain "good genes" for her progeny (Hamilton 1990; Møller and Alatalo 1999) or higher genetic compatibility with a male who will sire her offspring (Kempenaers et al. 1999; Tregenza and Wedell 2000); they might afford a female enhanced access to male resources or services (see reviews in Burley and Parker 1998; Kokko et al. 2003; Møller 1998); and they might provide "fertilization insurance" (as demonstrated experimentally in *Ficedula* flycatchers by Török et al. 2003; but see Olsson and Shine 1997 for a different outcome in a study involving lizards). Of course, extra-pair matings can have high costs as well, not least of which (for both sexes) is the danger of contracting sexually transmitted diseases (Kokko et al. 2002).

In DNA fingerprinting studies of the great reed warbler (*Acrocephalus arundinaceus*), Hasselquist et al. (1996) showed that females tend to obtain successful EPFs from neighboring males with larger song repertoires than their social mates, and that the offspring thereby produced also show higher survival. These results were interpreted to support the hypothesis that by engaging in EPFs, females in effect are seeking genetic benefits for their progeny. Additional evidence of this sort has added considerable strength to the notion that females (as well as males) in socially monogamous species do indeed obtain a variety of fitness benefits from extra-pair matings that underlie what is actually a genetically polygamous mating system (Gowaty 1996; Gray 1998; Ketterson et al. 1998).

Although EPFs may often benefit females (as well as cuckolding males), any males that get cuckolded would of course be disadvantaged by this phenomenon, leading to selection pressures on males not only for cuckoldry avoidance, but also for paternity assurance coupled to nestling investment (Møller and Cuervo 2000). Reed buntings (*Emberiza schoeniclus*) have exceptionally high extra-pair paternity, with 55% of 216 assayed young in 86% of 58 nests showing this phenomenon in one DNA fingerprinting study (Dixon et al. 1994). By combining these genetic results with field data on magnitudes of paternal investment in broods, these authors showed that males apparently can assess their likelihood of paternity and adjust their nestling provisioning rates accordingly. In similar studies of the zebra finch (*Taeniopygia guttata*)—another passeriform species with a high proportion (28%) of illegitimate young due to EPFs—males that appeared to be unattractive to females were found to accrue fitness gains by adopting a high parental investment (PI) strategy, whereas males that were attractive to females tended to benefit more by decreasing PI and increasing allocation to EPF behaviors (Burley et al. 1996).

Other detailed avian studies have integrated demographic or behavioral observations with genetic information on parentage. In both purple martins (*Progne subis*) and Bullock's orioles (*Icterus galbula*), DNA fingerprinting showed that older males achieved much higher reproductive success than did younger males, a result attributed to forced copulations by older males in the case of the martins (Morton et al. 1990) and to active female choice in the orioles (Richardson and Burke 1999). In a DNA fingerprinting study of polygynous red-winged blackbirds (*Agelaius phoeniceus*), the proportion of illegitimate chicks was found to be significantly greater in marshes with higher male densities, and the cuckolding males were often territorial neighbors (Gibbs et al. 1990). In a behavioral and DNA fingerprinting analysis of the blue tit (*Parus caeruleus*), Kempenaers et al. (1992) found that attractive males (those receiving many visits from neighboring females) were larger, survived better, and suffered less loss in paternity (had fewer extra-pair young in their own nests) than did unattractive males. These results were interpreted as supportive of a "genetic quality hypothesis" wherein females assess male quality and mate preferentially with superior individuals. In a series of studies on the barn swallow (*Hirundo rustica*), molecular markers coupled with experimental approaches revealed that males with longer and more symmetrical tail streamers tended to have increased paternity assurance within their own nests as well as more offspring in extra-pair broods, but that these fitness advantages via sexual selection were partially offset by natural selection against long-tailed individuals (Møller 1992; Møller and Tegelström 1997; Møller et al. 1998; Saino et al. 1997; H. G. Smith et al. 1991).

For the many birds that live in social groups, it is of interest to know which individuals actually participate in reproduction. The Galápagos hawk (*Buteo galapagoensis*) has an unusual social arrangement typically consisting of one adult female and up to eight oft-unrelated males. DNA fingerprinting assays of 66 hawks from ten breeding groups confirmed that the mating system is polyandrous ("cooperative polyandry" in this case), with males within a group exhibiting rather egalitarian reproductive success (Faaborg et al. 1995). Another form of grouping involves lekking behavior, wherein individuals assemble in a communal area for courtship display. Dominant males in such leks are often assumed to achieve a disproportionate share of successful matings. However, in a study that combined field observations with a microsatellite assessment of paternity in one such species, the buff-breasted sandpiper (*Tryngites subruficollis*), the variance in male reproductive success proved to be much lower than expected (Lanctot et al. 1997), due to multiple mating by females (sometimes with males off the leks as well) and to the use of alternative reproductive tactics by males (Lanctot et al. 1998).

In some birds, such as the acorn woodpecker (*Melanerpes formicivorus*) and stripe-backed wren (*Campylorhynchus nuchalis*), young often remain in their natal groups and appear to assist adult kin in rearing new broods. Under sociobiological theory, postponement of dispersal and breeding to assist in the rearing of others' progeny may be favored by natural selection if a juvenile's gain in inclusive genetic fitness from helping to produce close kin exceeds its expected gain in personal reproductive fitness had it dispersed (Brown 1987; Hamilton 1964). In the acorn woodpecker, DNA fingerprinting analyses revealed that helpers at the nest indeed are genetic offspring of the monogamous pair of parents that they assist (Dickinson et al. 1995). However, in the stripe-backed wren, DNA fingerprinting studies demonstrated a more direct avenue by which helper fitness can be enhanced. In some social groups, auxiliary males formerly thought to be nonreproductive actually sired some offspring (Rabenold et al. 1990). Such reproduction by subordinate male wrens may further help to explain their long tenure as helpers at the nest. In contrast, only dominant female wrens proved to be reproductively successful, a result interpreted as consistent with the proclivity of young females to compete for breeding sites outside the natal group.

In DNA analyses of another cooperatively breeding bird, the Arabian babbler (*Turdoides squamiceps*), one additional feature of reproduction by subordinates was uncovered. In this species, beta males that sired young proved to have significantly lower genetic similarity to the alpha male in their group than did those without offspring (Lundy et al. 1998). Thus, these more successful beta males probably were immigrants into the cooperative breeding groups, whereas the less successful beta males may have been stay-at-home young. Finally, in a similar microsatellite-based paternity analysis of cooperatively breeding carrion crows (*Corvus conone*), Baglione et al. (2003) found that although young male birds leave their natal groups to visit various others, they tend to settle and compete for matings in groups made up of individuals to whom they are moderately related (rather than unrelated). These results suggest that settlement patterns in this species are not just a passive consequence of random dispersal behavior, but instead register an active preference for association with kin (as might be predicted if kin selection plays an important evolutionary role in shaping cooperative reproductive alliances).

In addition to such species-focused studies, statistical summaries of data have identified several factors correlated with avian EPFs. For example, EPF frequencies often tend to be higher in species in which males are brightly plumaged (Møller and Birkhead 1994), have relatively large testes (Møller and Briskie 1995), and provide little or no offspring care (Møller and Birkhead 1993b), as well as in species in which females seem able to compensate for an absence of paternal care and also may receive important indirect fitness benefits from EPFs (Møller 2000). Reportedly, the incidence of EPFs is also higher in species with high molecular variation (a result interpreted as advantaging an EPF-seeking female by increasing genetic variety among her progeny; Petrie et al. 1998), in species with pronounced sexual dichromatism (Møller 1997), and in species that seem to have strong immunological defense mechanisms (a result interpreted as profiting EPF families by conferring offspring with superior resistance to virulent parasites; Møller 1997). Thus, such meta-analyses of the published literature have identified a number of empirical correlates and plausible causal interrelationships between avian genetic mating systems and various evolutionary and ecological phenomena such as sexual selection, sexual dimorphism, and individual behavioral tactics (Bennett and Owens 2002).

FISHES.   Molecular appraisals of parentage and genetic mating systems in fish populations did not become popular until the late 1990s (more than a decade after the molecular revolution in avian parentage began), but since then this field too has blossomed (see reviews in Avise 2001d; Avise et al. 2002). There are several reasons for special interest in fishes. First, unlike most birds (or mammals), fish typically have huge clutches that afford interesting statistical challenges (see Box 5.4) as well as novel biological opportunities for genetic parentage assessment. For example, de novo "clustered mutations" (which arise pre-meiotically in germ cell lineages and may enter the population not as singletons at a locus, but rather in clusters involving multiple siblings within a brood) are best sought and analyzed in species with large clutches (Jones et al. 1999a; Woodruff et al. 1996). Second, as indicated by a copious natural history literature, fishes collectively display diverse reproductive behaviors, ranging from pelagic group spawning to cooperative breeding to social monogamy, and this behavioral variety provides rich fodder for genetic assessments. Third, parental care in various fish species may be nonexistent, confined to one gender, biparental, or communal, and it can take such varied forms as oral or gill brooding, use of natural or constructed nests, open-water guarding of fry, or internal gestation by a pregnant mother or a pregnant father. Approximately 89 of 422 taxonomic families of bony fishes (21%) contain at least some species with parental care, and in nearly 70% of those families, the primary or exclusive parental custodian is the male (Blumer 1979, 1982). Paternal care of offspring is otherwise rather rare in the biological world (notable exceptions involve anuran amphibians; Clutton-Brock 1991; Wells 1977), and thus it affords a valuable mirror-image perspective on reproductive behaviors compared with the typical situation in most mammals, birds, and other groups, in which the female is normally the primary care-giver.

Molecular appraisals (almost always involving microsatellite markers) of fish parentage and mating systems usually have involved nest-tending species, in which dozens of tiny embryos in a nest are collected and individually genotyped, together with the nest-resident or "bourgeois" male (the suspected father) as well as other individuals in the vicinity. If the focal male was indeed the true sire, each offspring in his nest should carry one or the other of his alleles at each autosomal locus (barring de novo mutation), and maternity of the clutch can therefore be deduced by subtraction (using the logic illustrated in Figure 5.8). If some or all offspring in the nest were not sired by the resident male, this too should be evident (as a paternity exclusion) when they consistently fail to display that male's alleles. The most likely biological explanation for nonpaternity within a nest then

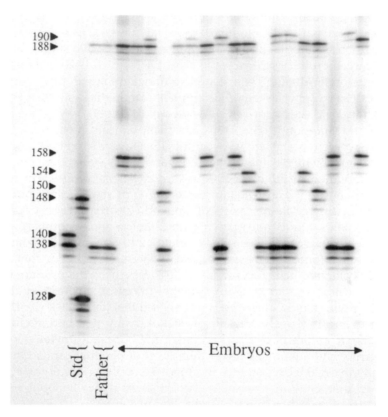

Figure 5.8 Genetic parentage analysis within a clutch tended by one known biological parent, in this case a pregnant male pipefish (*Syngnathus scovelli*). Shown is an autoradiograph of a microsatellite gel displaying banding patterns of standard controls (two leftmost lanes), of the pregnant male fish (third and fourth lanes from the left), and of each of 18 embryos taken from his brood pouch. Note that the father is heterozygous at this locus for alleles 188 and 138, and that each of his progeny displays one or the other of these alleles. Thus, the allele of maternal origin in each offspring is apparent by subtraction. Note also that four different maternal alleles are represented among these progeny, indicating that *at least* two dams (both presumably heterozygous) were involved. By combining such data and examining allelic associations across loci in many progeny (see DeWoody et al. 2000d), refined estimates of maternity within such a brood can be achieved. (From Jones and Avise 1997a.)

requires considered judgment that integrates the genetic findings with whatever natural history or other field information may be available in that particular instance.

One primary discovery from such appraisals is that a given fish nest often contains half-sib offspring from multiple (typically 2–8) dams (DeWoody and Avise 2001). Such multiple mating by the bourgeois male has been genetically documented in several species of *Lepomis* sunfishes (DeWoody et al. 1998, 2000b; Mackiewicz et al. 2002), *Etheostoma* darters (DeWoody et al. 2000c; Porter et al. 2002), *Spinachia* sticklebacks (Jones et al. 1998a), *Pomatoschistus* sand gobies (Jones et al. 2001a,b), and *Cottus* sculpins (Fiumera et al. 2002b). This pattern of multiple maternity within a nest is so prevalent that departures from it are of special interest. For example, genetic data demonstrated that nearly all nestmate embryos in a surveyed population of largemouth bass (*Micropterus salmoides*) were full sibs (not half-sibs or unrelated individuals). Thus, genetic monogamy apparently prevails in largemouth bass, a species that is also unusual among fishes in that both the sire and dam tend the offspring (DeWoody et al. 2000e).

A second common finding from genetic appraisals is occasional foster parentage wherein not all embryos within a nest were sired by the resident male. One documented route to such nonpaternity is cuckoldry, a phenomenon long studied in the bluegill sunfish, *Lepomis macrochirus*. In bluegill populations examined in Canada, three types of males exist: bourgeois males that mature at about 7 years of age and construct saucer-depression nests in colonies, attract and spawn with females, and vigorously defend nests and embryos; precocious sneaker males, 2 to 3 years of age, that often dart into a nest and release sperm; and older satellite males that mimic females in color and behavior, but also release sperm as the primary couple spawns (Gross and Charnov 1980). Genetic surveys (Colbourne et al. 1996; Neff 2001; Philipp and Gross 1994) have shown that about 20% of offspring in a bluegill colony are the result of cuckoldry by non-bourgeois males (Figure 5.9A). Studies have also suggested that bourgeois males can detect lost paternity and adaptively lower their level of parental care accordingly (Neff and Gross 2001). Cuckoldry has likewise been genetically documented in several other sunfish species, albeit at levels typically about an order of magnitude lower than in the bluegill (Figure 5.9B).

Another suspected route to nonpaternity by bourgeois males involves nest takeovers, often provisionally evidenced when few or none of the offspring in a given nest prove to have been sired by the resident male (see Figure 5.9B). Such nest piracies may be opportunistic responses by males to limited nest site availability, or perhaps a nest-holder captured at the time of sampling was merely a temporary visitor (e.g., there to cannibalize embryos). Yet another route to foster parentage by custodial males—egg thievery (wherein a few eggs are stolen from a neighbor's nest)—has been genetically as well as behaviorally documented in stickleback fishes (Jones et al. 1998a; Li and Owings 1978; Rico et al. 1992). Such egg-raiding behavior might seem highly counterintuitive, but one plausible explanation is that

(A)

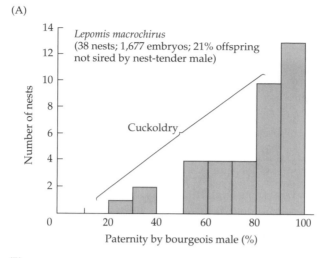

(B)

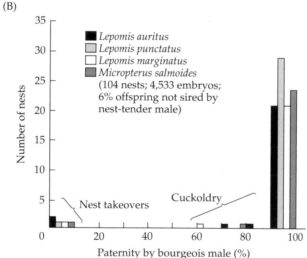

Figure 5.9   Molecular findings on genetic paternity in several species of North American sunfishes. Shown are percentages of progeny per nest that proved to have been sired by bourgeois males in (A) 38 nests of *Lepomis macrochirus* and (B) a total of 104 nests in four other species of Centrarchidae (see text for references). (From Jones and Avise 1997a.)

this behavior benefits the thief by seeding or "priming" his own nest with eggs, which are known to be effective in many fish species in eliciting spawning responses by additional females with whom the resident then mates (see review in Porter et al. 2002).

In one taxonomic family of fishes, Syngnathidae, male care of offspring has been taken to the extreme. In all of the 200+ living species of pipefishes and seahorses, females lay eggs on the ventral surface (usually an enclosed brood pouch) of a male, who then gestates the eggs for weeks before giving birth to dozens of offspring. Extensive molecular parentage analyses have been conducted on several syngnathid species in the genera *Syngnathus* (Jones and Avise 1997a,b; Jones et al. 1999b, 2001c), *Nerophis* (McCoy et al. 2001), and *Hippocampus* (Jones et al. 1998b, 2003). In sharp contrast to nest-tending fishes, no instances of cuckoldry by foreign males were detected (as might be expected for male-pregnant species with mostly internal fertilization). This complete assurance of paternity in the Syngnathidae in turn facilitates genetic analyses of maternity and mating systems. Several different outcomes among the surveyed species have been uncovered, ranging from genetic monogamy to polygynandry to polyandry. Furthermore, when these genetic findings were interpreted in the context of observed levels of sexual dimorphism and the presumed intensities and directions of sexual selection in the various species, results generally appeared compatible with conventional wisdom (as summarized in Figure 5.7) for taxonomic groups that include species with strong proclivities toward polyandry and "sex role reversal" (Jones and Avise 2001; Jones et al. 2000). For example, assayed syngnathid species that proved to have a polyandrous genetic mating system displayed greater sexual dimorphism and more pronounced secondary sexual characters in females than did monogamous species.

Another interesting finding to emerge from genetic parentage assessments in fishes is the first firm documentation in nature of filial cannibalism (eating one's own biological offspring). This phenomenon had been suspected from field observations that fish sometimes eat embryos from their own nests, but with genetic discoveries of widespread foster parentage, the possibility was raised that perhaps most cannibalism events were directed toward non-relatives rather than kin. DeWoody et al. (2001) critically tested this proposition by genotyping freshly eaten embryos dissected from the stomachs of wild-caught adult male sunfish and darters. Each of several dozen such embryos did indeed prove to have been consumed by its own biological father.

PLANTS.   Fatherhood in plants results from the spread of pollen, as mediated, for example, by insect pollinators or wind. Questions concerning pollen sources can be addressed by the same general types of molecular parentage analyses as described above for animals (Adams et al. 1992; Devlin and Ellstrand 1990). The task again is simplified when the mother is known (e.g., as the bearer of the seeds in question), but it can remain difficult when the pool of potential pollen donors is large. Paternity and the mating system may be addressed with regard to seeds within a fruit, fruits within a plant in a given season, or the lifelong seed set of an individual.

Many plant species are hermaphroditic (meaning that a given individual can produce both male and female gametes). Not all such specimens can self-fertilize, however, for several reasons: male and female flowers in a monoecious individual may mature at different times or be spatially separated on the plant; stamens and stigma within a perfect flower (a flower possessing both male and female parts) may be positioned such that mechanical pollen transfer is unlikely, or self-incompatibility genes may be present. These "self-sterility" genes are known to carry multiple alleles that appear to have been selected to prevent the possible deleterious effects of the intense inbreeding that self-fertilization entails. Operationally, when a maternal parent and a pollen grain share an allele at a self-sterility locus, sporophytic tissue discriminates against gametophytic tissue—for example, by inhibiting growth of the pollen tube down the style. Nevertheless, self-fertilization clearly has evolved independently from outcrossing on numerous occasions (Stebbins 1970; Wyatt 1988), probably at least 150 times in the Onagraceae alone (Raven 1979).

Thus, one of the first genetic questions regarding a hermaphroditic species concerns the frequency with which self-fertilization (as opposed to outcrossing) takes place. When the female parent is known, the problem simplifies to one of paternity assessment, with the issue in this case being how often the individual plant that mothered an array of progeny or seeds can be excluded as the father of those genotypes. For example, any offspring that exhibits alleles not present in its known mother must have arisen through an outcrossing event. Several statistical models are available to quantify rates of selfing versus outcrossing (s and t, respectively, where $s + t = 1$) from genotypic information at one or more loci (Brown and Allard 1970; Ennos and Clegg 1982; Ritland and Jain 1981; Schoen 1988; Shaw et al. 1981). For example, a widely employed "mixed-mating" model (Brown 1989; Clegg 1980) assumes that the mating process can be divided into two distinct components: random mating (i.e., random independent draws of pollen from the total population) and self-fertilization. This model may be especially appropriate for wind-pollinated species. A variant of this model that is likely to be more applicable to many insect-pollinated species assumes that outcrossing events within a family are correlated because they may involve successive pollen draws from a single male parent (Schoen and Clegg 1984).

From allozymes and other genetic markers, "mating system parameters" (s and t) have been empirically estimated for dozens of plant species, ranging from small herbaceous forms (Galloway et al. 2003) to intermediate-sized succulents (Massey and Hamrick 1999; Nasser et al. 2001) to large trees (Ruter et al. 2000). In an early summary of the literature by Lande and Schemske (1985), the overall frequency distribution of outcrossing rates proved to be bimodal, with most species either predominantly selfing or predominantly outcrossing (Figure 5.10A). These authors interpreted this bimodality as consistent with a scenario in which outcrossing is selected for in historically large species with substantial inbreeding depression, whereas selfing is favored in species in which prior pollinator failure or population bottlenecks have reduced the level of inbreeding depression via purging of deleterious recessive alleles. Empirical evidence does exist for high variance among plant species in degree of inbreeding depression (Schemske and Lande 1985), with outcrossers often exhibiting the highest reductions in fitness under inbreeding (although this may partly reflect a bias in the early

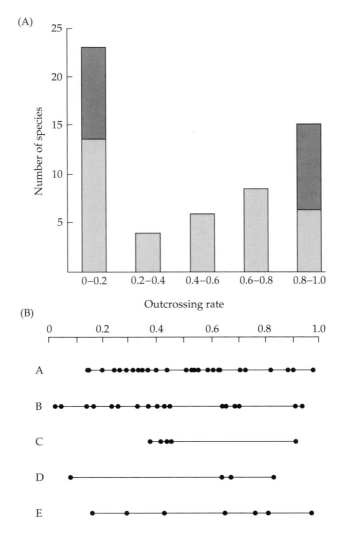

Figure 5.10   Outcrossing rates in plants. (A) Frequency distribution of mean outcrossing rates, as estimated from allozyme markers, for 55 hermaphroditic plant species. Lightly shaded portions of bars are animal-pollinated species; more darkly shaded portions are wind-pollinated species (Aide 1986). (After Schemske and Lande 1985.) (B) Inter-population variation in outcrossing rate within each of five plant species: A, *Lupinus succulentus*; B, *L. nanus*; C, *Clarkia exilis*; D, *C. tembloriensis*; and E, *Gilia achilleifolia*. Solid circles are population means; horizontal lines represent observed ranges among conspecific populations. (After Schemske and Lande 1985.)

literature, which included a disproportionate representation of selfing grasses and outcrossing trees; Aide 1986). Nevertheless, few hermaphroditic plant species are "fixed" for either pure outcrossing or pure selfing, and conspecific populations in some species show huge variation along the selfing–outcrossing continuum (Figure 5.10B).

In hermaphroditic species in which outcrossing has been established, or in any dioecious species, the next genetic question is, which plants were the pollen donors for particular outcrossed offspring? As illustrated in Box 5.5, molecular genetic markers again can supply the answer. The approach involves comparing the diploid genotype of each seed or progeny with that of its known mother, and thereby deducing (by subtraction) the haploid genotype of the fertilizing pollen. Candidate fathers are then screened for diploid genotype, and paternity is excluded for those whose genotypes do not match the deduced pollen contribution to the progeny. Sometimes all

BOX 5.5 Paternity Assignment

These data (taken from a much larger allozyme data set; see Ellstrand 1984) illustrate paternity assignment for five progeny from a known mother. The body of the table consists of observed diploid genotypes at each of six loci in the wild radish, *Raphanus sativus*.

	Allozyme locus					
	LAP	PGI	PGM1	PGM2	6PGD	IDH
Known mother	1,2	1,1	1,1	1,2	1,3	1,1
Potential fathers						
A	2,2	1,2	2,3	2,2	3,3	1,1
B	2,2	2,3	1,3	1,1	1,3	1,1
C	1,2	1,2	1,3	1,1	1,2	1,2
D	1,5	1,1	1,2	1,3	1,3	1,1
E	2,3	2,2	1,2	1,2	1,1	1,3
F	2,2	1,3	2,2	1,2	1,3	1,1
G	1,1	1,2	1,2	1,2	3,3	1,1
H	1,1	1,2	1,2	1,2	1,3	2,2
I	1,2	1,1	1,1	1,2	1,3	1,1
J	1,2	2,3	1,2	1,2	3,3	1,2
K	2,2	1,2	1,3	2,2	3,3	1,1
L	1,2	1,1	1,1	2,3	1,3	1,1
M	2,5	1,1	1,2	2,3	1,2	1,3
N	1,1	1,1	1,2	1,1	1,1	1,1
O	1,3	1,2	1,2	1,2	3,3	1,1

							Deduced paternity	
Offspring							Gamete	Assignment
P	2,2	1,2	1,3	1,2	1,1	1,2	223-12	C
Q	2,2	1,2	1,3	1,2	2,3	1,1	223-21	C
R	1,2	1,2	1,3	1,1	1,2	1,1	-23121	C
S	1,2	1,1	1,2	2,3	1,1	1,3	-12313	M
T	2,2	1,1	1,1	1,3	2,3	1,3	211323	M

Source: After Ellstrand 1984.

males except the true father can be excluded. When multiple candidates remain, procedures exist for assigning "fractional paternity" based on statistical probabilities of being the father. In the first large-scale application of these approaches, Ellstrand (1984) employed six highly polymorphic allozyme loci to establish paternity for 246 seeds from nine maternal plants in a closed population of the wild radish, *Raphanus sativus*. Multiple paternity was found for all assayed progeny arrays from a maternal plant and for at least 85% of all fruits, with the minimum paternal donor number averaging 2.3. The wild radish is a self-incompatible, insect-pollinated species. Subsequent work established that most multiply sired fruits were the consequence of a single insect vector having simultaneously deposited pollen from several plants (a phenomenon known as "pollen carryover"), and that a considerable fraction (up to 44%) of seed paternity for some plants involved immigrant pollen from sources at least 100 m away (Ellstrand and Marshall 1985; Marshall and Ellstrand 1985).

In another classic allozyme study, in this case of a small forest herb, *Chamaelirium luteum*, Meagher (1986) established paternity likelihoods for 575 seeds with known mothers. The distribution of inter-mate (pollen-flow) distances indicated that more nearby fertilizations had taken place than expected under random mating, but nonetheless some mating pairs were separated from one another by more than 30 m. A follow-up study of established seedlings (whose maternity was unknown) confirmed this pollen dispersal profile and also demonstrated that pollen and seed dispersal distances were similar (Meagher and Thompson 1987). Surprisingly, no relationship was found in this species between the size of the male plant (seemingly indicative of reproductive effort) and paternity success (Meagher 1991). Hamrick and Murawski (1990) conducted similar genetic paternity analyses on several tropical woody species and showed that a significant proportion of the pollen received by individuals came from relatively few pollen donors; many matings (30%–50%) appeared to take place between nearest neighbors, and about 10%–25% of matings involved long-distance pollen flow (greater than 1 km). Thus, the overall breeding structure appeared to have two components: a leptokurtic (i.e., peaked) pattern of pollen dispersal within populations, superimposed on a more even distribution of "background" pollen originating from outside the population.

Paternity analyses in plants are often referred to as providing direct estimates of gene flow (albeit across a single generation), as opposed to the indirect estimates of historical plus contemporary gene flow that can come from estimates of population genetic structure (see Chapter 6). In a common experimental setting, progeny within a focal population are monitored for paternal alleles that by genetic exclusion must have come from outside (rather than inside) the plot. Several such direct appraisals of paternal gene flow have documented instances (often in high frequencies) of immigrant pollen having arrived from rather distant sources. For example, in three species of fig trees (*Ficus*), molecular markers indicated that more than 90% of the pollen came from at least 1,000 meters away (Nason and Hamrick 1997). Other insect-pollinated trees in which molecular paternity analyses have often revealed long-distance pollen flow include *Calophyllum longifolium* (Stacy et al. 1996), *Pithecellobium elegans* (Chase et al. 1996), *Swietenia humilis* (White et al. 2002), and *Tachigali versicolor* (Loveless et al. 1998). Two wind-pollinated species for which frequent long-distance pollen flow has likewise been documented by molecular paternity analyses are *Quercus macrocarpa* (Dow and Ashley 1998) and *Pinus flexilis* (Schuster and Mitton 2000).

In native species of conservation concern, as well as in crop plants, genetic determinations of pollen sources often carry management or economic ramifications. For example, in a tropical tree of conservation interest, *Symphonia globulifera*, genetic paternity analyses demonstrated that a few large pasture specimens contributed disproportionately to the population's overall gene pool, thus producing a relative genetic bottleneck that would otherwise not have been apparent (Aldrich and Hamrick 1998). A more applied example involves commercial pine-seed orchards that provide a significant fraction of the zygotes used to establish tree plantations in the southeastern United States. One such seed orchard for the loblolly pine (*Pinus taeda*) in South Carolina was composed of grafted ramets of 50 loblolly clones that had been chosen and maintained for phenotypically desirable traits. Using allozyme markers, Friedman and Adams (1985) discovered that at least 36% of seeds from this orchard were fertilized by outside pollen, despite a surrounding 100-meter-wide buffer zone positioned explicitly to prevent such genetic contamination by non-selected males. Similarly, in a population of cultivated cucumbers (*Cucurbita pepo*), Kirkpatrick and Wilson (1988) showed by molecular markers that approximately 5% of progeny were fathered by native cucumbers (*C. texana*), an outcome illustrative of the potentials for appreciable genetic exchange that are known to exist between many cultivated crops and their wild relatives (Ellstrand 2003).

Selected empirical examples by topic

CONCURRENT MULTIPLE PATERNITY.   Molecular assays of individual litters, broods, and clutches have demonstrated concurrent multiple paternity for a wide variety of species in nature. Apart from the numerous birds and the nest-tending fishes mentioned above, these include many species of mammals (e.g., Birdsall and Nash 1973; Gomendio et al. 1998; Hoogland and Foltz 1982; Taggart et al. 1998; Tegelström et al. 1991; Xia and Millar 1991), snakes and lizards (Garner et al. 2002; Gibbs and Weatherhead 2001; Gibson and Falls 1975; Olsson and Madsen 2001), alligators (Davis et al. 2001), aquatic and terrestrial turtles (Bollmer et al. 1999; Palmer et al. 1998; Pearse and Avise 2001; Pearse et al. 2002), amphibians (Halliday 1998; Tennessen and Zamudio 2003; Tilley and Hausman 1976), female-pregnant fishes (Chesser et al. 1984; Soucy and Travis 2003; Travis et al. 1990; Trexler et al. 1997; Zane et al. 1999), ascidians (Bishop et al. 2000), mollusks (Avise et al. 2004; Baur 1998; Emery et al. 2001; Gaffney and McGee 1992; Mulvey and Vrijenhoek 1981; Oppliger et al. 2003), platyhelminth flatworms (Pongratz and Michiels 2003), and diverse arthropods and related groups (Baragona and Haig-Ladewig 2000; Brockman et al. 2000; Curach and Sunnucks 1999; Heath et al. 1990; Martyniuk and Jaenike 1982; Milkman and Zeitler 1974; Nelson and Hedgecock 1977; Parker 1970; Sassaman 1978; Walker et al. 2002; see also below). A fascinating natural history tour through this promiscuous biological world is provided by Birkhead (2000).

A few of these studies involved socially monogamous species and thus their results were somewhat surprising. Many others involved socially polygynous species and thus confirmed suspicions that multiple copulations or inseminations could indeed result in multiple successful fertilizations of a progeny cohort. For example, female Belding's ground squirrels (*Spermophilus beldingi*) are known to mate with several different males. Allozyme data established that an estimated 78% of litters were multiply sired, usually by two or three males (Hanken and Sherman 1981). In a similar study of an insect, the willow leaf beetle (*Plagiodera versicolora*), more than 50% of wild-caught females produced egg clutches with multiple sires (McCauley and O'Donnell 1984). On the other hand, not all molecular genetic analyses have uncovered evidence for multiple paternity within clutches. For example, using allozyme assays, Foltz (1981) demonstrated a high degree of genetic monogamy in the old-field mouse (*Peromyscus polionotus*), as did Ribble (1991) in DNA fingerprinting assays of the California mouse (*P. californicus*).

ALTERNATIVE REPRODUCTIVE TACTICS.   Alternative reproductive tactics (ARTs) are different behavioral modes employed by conspecific males (or females; Henson and Warner 1997) to achieve successful reproduction (Gross 1996; Taborsky 2001). They may be hard-wired genetic polymorphisms, or they may reflect behavioral or other phenotypic switches related to environmental conditions (e.g., hormone levels during development), but in either case they co-occur as distinctive reproductive strategies within a population or species. Examples were introduced above in discussions of flanged versus unflanged orangutans and bourgeois, sneaker, and satellite males in bluegill sunfish. Another example involves salmon, males of which may spawn either as full-sized anadromous adults after returning from the sea or as dwarf precocious parr that have remained in fresh water. Parentage analyses based on molecular markers have provided unprecedented information on individuals' relative reproductive success in populations displaying ARTs. For example, analyses of several populations of Atlantic salmon (*Salmo salar*) have shown that parr fertilize widely varying proportions (5%–90%) of eggs at different spawning sites (Garant et al. 2001; Garcia-Vasquez et al. 2001; Jordan and Youngson 1992; Moran et al. 1996; Thomaz et al. 1997).

In the spotted hyena (*Crocuta crocuta*), genetic parentage resulting from alternative male and female reproductive tactics was evaluated by microsatellite profiling of 236 offspring in 171 litters from three clans (East et al. 2002). Despite polyandrous mating and high frequencies of multiple paternity (35% of litters), female choice and sperm competition (see below) appeared to counter or trump pre-copulatory male tactics. This was evidenced by the finding that male hyenas rarely sired offspring with females whom they attempted to manipulate through monopolization or harassment, whereas males who invested energy and time in fostering amicable relationships with females proved to have sired most of the offspring.

Another example of parentage dissected by genetic markers involves side-blotched lizards (*Uta stansburiana*). Males in this species have three ARTs, each of which trumps, but is also susceptible to, one other tactic, much as in the children's game of rock–paper–scissors. One form of male has a blue throat, is territorial, and guards its mate. Another form has an orange throat and is hyper-territorial and polygynous, avidly mating with multiple females. A third form is yellow-throated and does not regularly defend a territory, but instead gains access to territories of defender males by mimicking a female and then sneaking copulations with resident females. Genetic parentage analyses coupled with field observations (Sinervo and Clobert 2003; Sinervo and Lively 1996; Zamudio and Sinervo 2000) have shown that the mate-guarding strategy of blue-throated males usually enables them to avoid cuckoldry by yellow-throated males, but leaves them vulnerable to cuckoldry by more aggressive orange-throated males. However, by virtue of their hyper-aggressive behavior, orange-throated males often obtain territories so large that they are unable to defend their females against yellow-throated sneaker males. So, the reproductive tactic of yellow-throated males (rock) can smash that of orange-throated males (scissors), which can snip that of blue-throated males (paper), which can cover that of yellow-throated males.

SPERM STORAGE.   Following a copulation event, the reproductive tract of females in many species is physiologically capable of storing viable sperm for varying periods of time (Birkhead and Møller 1993a; Howarth 1974; Smith 1984): typically a few days in mammals, weeks in many insects and birds, months in some salamanders, and up to several years in some snakes and turtles. Traditional evidence for this conclusion came from direct observations of live sperm (typically in special female storage organs referred to as spermathecae) and from the fact that captive females isolated from males for some period of time may continue to produce offspring (although the possibility that these progeny are parthenogenetic is not eliminated by this observation alone).

In recent years, molecular-based parentage analyses have added to our understanding of sperm storage phenomena. One illustration of the approach, involving a natural population of painted turtles (*Chrysemys picta*), may also provide the current record for the longest period of female sperm storage genetically verified in any species. Pearse et al. (2001a) used microsatellite markers to deduce paternity in successive clutches of physically tagged females. Exclusion probabilities were sufficiently high that unique-sire genotypes could be identified, and these genotypes sometimes were evidenced in the offspring of clutches that particular females had laid across as many as 3 successive years. By hard criteria, the possibility that a female had re-mated each year with the same male could not be eliminated entirely, but this explanation was deemed highly implausible given the asocial nature of this species, its high dispersal capability, and the high local densities of males. Thus, almost certainly, long-term female sperm storage and utilization had been documented by the genetic evidence.

SPERM AND POLLEN COMPETITION.   The widespread occurrences of genetic polygyny, concurrent multiple paternity, ARTs, and extended female sperm storage in many species all indicate that sperm from two or more males are often placed in direct competition for fertilization of eggs within a female's reproductive cycle. Several morphological characteristics and reproductive behaviors of males have been interpreted as adaptations to meet the genetic challenges resulting from this supposed competition with another male's sperm (Parker 1970). For example, in many worms, insects, spiders, snakes, and mammals, a male secretes a plug that serves temporarily as a "chastity belt" to block a female's reproductive tract from subsequent inseminations. In many damselflies and dragonflies (Odonata), males have a recurved penis that physically scoops out old sperm (from other males) from a female's reproductive tract during mating, thus helping to account for the genetic observation that last-mating males tend to sire disproportionate numbers of progeny (C. G. Cooper et al. 1996; Hooper and Siva-Jothy 1996). Other widespread male behaviors that have been interpreted as providing paternity assurance in the face of potential sperm competition include prolonged copulation (up to a week in some butterflies), multiple copulations with the same female, and post-copulatory mate guarding (Parker 1984).

From a female's perspective, mechanisms to prevent competition among sperm from different males are not necessarily desirable, which can lead to intersexual reproductive conflicts of interest (Eberhard 1998; Knowlton and Greenwell 1984). Growing evidence also suggests that the reproductive tracts of females may often play a more active role than previously supposed in post-copulatory choice of fertilizing sperm (Birkhead and Møller 1993b; Mack et al. 2003). These and related topics have made "sperm competition" one of the hottest topics in molecular ecology and evolution over the last two decades (Baker and Bellis 1995; Birkhead and Møller 1992, 1998; Smith 1984).

In individual clutches of multiply inseminated females, molecular markers can be employed to determine which among the competing males' sperm have achieved the fertilizations. Is the first-mating male at a reproductive advantage, or does the last-mating male achieve the highest fertilization success? Or is there no mating-order effect, the probability of fertilization instead merely being proportional to the number of sperm deposited by each male (the "raffle" scenario)? These questions have been addressed using genetic markers for numerous animal species (see Birkhead and Møller 1992; Smith 1984 for pioneering reviews). In insects, it often

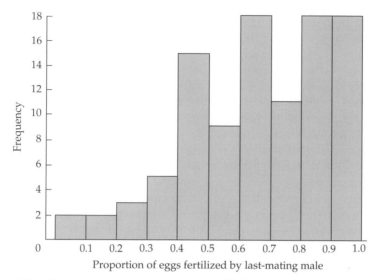

Figure 5.11   Outcomes of sperm competition in insects, typically as determined by genetic paternity analyses based on molecular markers. Shown is the frequency distribution (across more than 100 species) of the proportion of eggs fertilized by the second of two males to have mated sequentially with doubly inseminated females. (After Simmons and Siva-Jothy 1998.)

proved to be the case (but not always; Laidlaw and Page 1984) that the last male to mate with a female sired most of the offspring (Figure 5.11). For example, in the bushcricket *Poecilimon veluchianus*, Achmann et al. (1992) showed by DNA fingerprinting that the last male to mate achieved more than 90% of the fertilizations. Mating in this species involves transfer of a large spermatophore to a female, who often copulates with several males and may eat some of the spermatophores after copulation. The genetic findings appeared to eliminate the possibility that nourishment gained by a female from the spermatophore "gift" of an early-mating male reflected a paternal investment strategy enhancing that male's fitness.

The term "sperm displacement" conventionally was employed to describe the enhanced reproductive success exhibited by last-mating males. In an insect, the locust *Locusta migratoria*, an active "sperm flushing" process has been observed that probably contributes to the phenomenon (Parker 1984). In the dunnock sparrow, males peck at the cloaca of a female before copulating with her, apparently causing her to eject sperm from previous matings (Birkhead and Møller 1992). In other cases, mechanisms of sperm displacement appear less active. In chickens and ducks, for example, semen from different inseminations is stored in separate layers in the female reproductive tract, with the most recent contribution remaining on top and therefore perhaps most likely to fertilize the next available egg (McKinney et al. 1984). For such instances, more neutral terms, such as "sperm predominance" (Gromko et al. 1984) or "sperm precedence," may be preferred. In some birds, both raffle competition and sperm precedence are known to operate, but over different time scales. If inseminations occur more than about 4 hours apart, then last-male sperm precedence tends to operate, but a sperm raffle characterizes the process when two males inseminate a female in rapid succession (Birkhead and Møller 1992).

In a small proportion of insect species (see Figure 5.11) and in various other animals, *first*-mating males appear to have the fertilization advantage. For example, in the intertidal copepod *Tigriopus californicus*, allozyme studies showed that virtually all of a female's progeny are fathered by her first mate (Burton 1985). In this species, a male often clasps a female for a period of several days before her sexual maturation. In light of the genetic observations, Burton interpreted this prolonged clasping behavior by males as a pre-copulatory mate-guarding strategy to ensure that a potential mate has not been inseminated previously.

In the relatively asocial ground squirrel *Spermophilus tridecemlineatus*, synchronously breeding females are scattered spatially at low densities. As a consequence of this natural history, the mating system probably conforms to what has been labeled "scramble-competition polygyny." Indeed, behavioral observations suggest that the strongest phenotypic correlate of male mating success is male mobility during the breeding season, presumably because traveling males are more likely to encounter females in estrus (Schwagmeyer 1988). Using allozyme markers, Foltz and Schwagmeyer (1989) discovered that in wild populations of this species, the first male to copulate with a multiply mated female sired on average about 75% of the resulting progeny. These results were interpreted to indicate that a mating advantage for first males during pre-copulatory scramble competition translates into a genetic advantage during the ensuing post-copulatory sperm competition.

A remarkable example of first-male fertilization advantage was reported for the spotted sandpiper (*Actitis macularia*). In this polyandrous avian species with strong tendencies toward behavioral sex role reversal (including nest-tending by males), territorial females pair with, defend, and lay clutches for several males. Molecular studies based on DNA fingerprinting showed that males pairing early in the mating season cuckold their females' later mates by means of sperm storage in the females' reproductive tracts (Oring et al. 1992). Thus, not only does an early-pairing male have a greater confidence of paternity, but he thereby also appropriates the reproductive efforts of subsequent males toward enhancement of his own fitness.

The intriguing idea of sperm sharing was advanced for some species of hermaphroditic freshwater snails (Monteiro et al. 1984). According to this suggestion, a snail might pass on sperm from a previous mate to another partner, such that the transmitting individual acts mechanically as a male but achieves no genetic contribution to progeny. However, an empirical test of this hypothesis based on allozyme markers failed to support the sperm-sharing hypothesis (Rollinson et al. 1989). Instead, hermaphroditic snails proved capable of passing on their own sperm while still producing eggs fertilized by sperm received from an earlier mating. A variety of other issues regarding sperm competition in hermaphroditic species are reviewed by Michiels (1998).

In plants, opportunities also exist for competition among male gametes from different donors, as, for example, via differing rates of pollen tube growth through stigmatic tissue toward the egg (Snow 1990). Thus, pollen competition in plants is the analogue of sperm competition in animals (Delph and Havens 1998). Using allozyme markers to establish paternity, Marshall and Ellstrand (1985) demonstrated that most of the seeds in multiply sired fruits of the wild radish (*Raphanus sativus*) resulted from the first in a series of sequential pollen donors. Further study revealed that gametophyte competition among several pollen donors was more pronounced than that among male gametophytes from a single pollen source (Marshall and Ellstrand 1986). In the morning glory *Ipomoea purpurea*, similar allozyme analyses also revealed a strong fertilization advantage for first-pollinating males, even when pollen donations from a second source occurred immediately after the first (Epperson and Clegg 1987). In paternity studies of the herbaceous plant *Hibiscus moscheutos*, allozyme markers revealed that individuals with fast-growing pollen tubes sired a disproportionate number of seeds following mixed experimental pollinations (Snow and Spria 1991). More examples of pollen competition, in the context of barriers to interspecific hybridization, will be provided in Chapter 7.

MATERNITY ANALYSIS. In many taxonomic groups, such as mammals, maternity is usually more evident than paternity from direct behavioral observations, but in some cases the biological mother of particular offspring nonetheless remains in doubt. Tamarin et al. (1983) used an ingenious method for maternity assignment in small mammals. They injected pregnant or lactating females with unique combinations of gamma-emitting radionuclides (e.g., ^{58}Co ^{85}Sr ^{65}Zn), which were transferred to progeny via placenta or mother's milk. The isotopic profiles of young were determined spectrophotometrically and matched against those of prospective mothers to establish maternity (assuming that mothers nurse only their own offspring). Sheridan and Tamarin (1986) combined this method of maternity assignment with protein electrophoretic analyses to assess parentage in 40 offspring from a natural population of meadow voles (*Microtus pennsylvanicus*). Knowledge of maternity facilitated paternity analyses and led to the conclusion that about 38% of the adult males in the population bred successfully in the surveyed time period, fathering at most two litters each.

Each spring, pregnant females of the Mexican free-tailed bat (*Tadarida brasiliensis*) migrate to caves in the American Southwest and form colonies often containing several million individuals. Most females produce single pups, which within hours of birth are deposited on the cave ceilings or walls in dense creches. Lactating females return to the creches and nurse pups twice each day. Traditional thought was that nursing must be indiscriminate, such that mothers act "as one large dairy herd delivering milk passively to the first aggressive customers" (Davis et al. 1962), but McCracken (1984) challenged this view with protein electrophoretic evidence indicating that nursing was selective along genetic lines. This conclusion stemmed from comparisons of observed allozyme genotypes in female–pup nursing pairs with the expected frequencies of such genotypic combinations if nursing were random. A highly significant deficit of maternal genetic exclusions (relative to expectations from population genotype frequencies) indicated selective nursing by females of their own pups (or at least those of close kin). McCracken estimated that only 17% of the assayed females were nursing pups that could not be their offspring. A DNA fingerprinting analysis of maternity roosts in another bat species, *Myotis lucifugus*, likewise led to the conclusion that females preferentially suckle their own young (Watt and Fenton 1995).

VandeBerg et al. (1990) employed protein electrophoretic markers to validate pedigrees in captive squirrel monkeys (genus *Saimiri*). Among 89 progeny for which parentage had been inferred from behavioral observations, assignments for seven individuals proved incorrect, and retrospective examination of colony records in conjunction with further genetic typing permitted a correction of pedigree records. Five of the errors had involved cases of mistaken paternity, but two involved mistaken maternity. These latter cases apparently were the consequence of infant swapping between dams shortly after birth, an "allomaternal" behavior that previously had gone unrecognized.

Far more commonly, questions about maternity arise in oviparous animals such as birds, fishes, and insects, in which prolonged care of eggs outside the female's body opens possibilities for intraspecific brood parasitism or other means of egg or progeny mixing. Indeed, as described above, paternity in fishes is normally more field-evident than maternity (due to the prevalence of male parental care), so genetic maternity is typically one of the prime foci of molecular parentage analyses. In birds, traditional methods for inferring IBP include monitoring nests for supernormal clutch sizes, noticing the appearance of eggs deposited outside the normal laying sequence of the resident female, or detecting intra-clutch differences in the physical appearance of eggs in those species in which inter-clutch differences in egg patterning are pronounced. Molecular approaches provide more direct maternity assessments. For example, in wild zebra finches, a DNA fingerprinting analysis of 92 offspring from 25 families revealed that about 11% of offspring and 36% of broods resulted from IBP, and that the mean number of parasitic eggs per clutch was greater than one (Birkhead et al. 1990). In house wrens (*Troglodytes aedon*), a similar study based on allozymes led to the conclusion that about 30% of chicks were produced by females other than the nest attendant (Price et al. 1989).

Genetic markers have also been used to address issues concerning interspecific brood parasitism, a phenomenon in which females of one species surreptitiously lay their eggs in nests of other species. At molecular issue in this case is not how often this behavior occurs in nature (this is often evident from direct field observations, because eggs and young of the species involved are usually visually distinguishable), but rather how often brood parasitic behaviors have arisen in evolution. By mapping the phenomenon of brood parasitism (as opposed to personal nesting) onto an mtDNA-based molecular phylogeny for 15 species of cuckoos, Aragon et al. (1999) concluded that the phenomenon had a polyphyletic origin in the order Cuculiformes, having arisen separately in at least three well-defined clades. In one European species of interspecific brood parasite, the common cuckoo (*Cuculus canorus*), similar genetic analyses further showed that different gentes ("races" with different egg-color patterns) represent distinctive matrilines that nonetheless are closely similar to one another in their overall genetic makeup (Gibbs et al. 2000a; Marchetti et al. 1998).

POPULATION SIZE.   Genetic parentage analyses are also informative in terms of estimating local population size in at least two ways. First, by pinpointing which adults have actually sired and dammed progeny, molecular markers can offer better assessments of variance in reproductive success and of effective population size (see Box 2.2) than can mere census counts of potentially breeding adults (e.g., Hoelzel et al. 1999). Such knowledge can be important, for example, in assessing the magnitude of inbreeding in small captive or managed populations (Pope 1996).

A second means by which molecular parentage analyses can provide information about population numbers was introduced by Jones and Avise (1997b). In wildlife biology, a traditional approach is to use physical traps in mark–recapture protocols to estimate the contemporary size of a deme (Seber 1982). Under the oft-used Lincoln–Peterson statistic, for example, the number of individuals in a population is estimated as:

$$n = \frac{(n_1 + 1)(n_2 + 1)}{(m_2 + 1) - 1}$$

where n_1 is the number of animals captured and physically marked in an initial sample, n_2 is the number of animals caught later, and m_2 is the number of recaptured (marked) animals in the second sample (Pollock et al. 1990). The parentage analysis approach is a genetic analogue of this traditional method, in which the initial "marks" are, for example, the deduced genotypes of males (n_1) who sired progeny in assayed clutches. Genotypes of adult males from the population can then be considered the second sample (n_2), and those males that perfectly match deduced paternal genotypes in the clutches are considered "recaptures" (m_2). By plugging these genetically deduced parameter values into the Lincoln–Peterson equation, the current size of the adult breeding population can be estimated.

Pearse et al. (2001b) explored several variations on this theme. For example, in a population that is monitored over multiple breeding seasons, both marks and recaptures could come from the genetically deduced paternal (or maternal) genotypes in successive clutches. This method also has the advantage that there is never a need to physically trap (or even observe) the alternate sex, because polymorphic genes provide the marks and breeding individuals of one sex in effect provide both the captures and the recaptures of the opposite gender (via mating). Also, the resulting estimate of n for a given population refers explicitly to successful breeders (as opposed to all individuals), and thus may be of special interest in many ecological circumstances.

SUMMARY

1. Qualitative molecular markers from highly polymorphic loci provide powerful tools for assessing genetic identity versus non-identity and biological parentage (maternity and paternity).

2. In human forensics, DNA fingerprinting, first by VNTR loci and now by STR loci, provided a late-twentieth-century analogue of traditional fingerprinting. DNA fingerprinting has found wide application in civil litigation and criminal cases. Conservative procedures for calculating probabilities of a genotypic match can serve to ameliorate any potential biases against the defense.

3. For plants and animals that are known or suspected to reproduce asexually (clonally) as well as sexually, several types of polymorphic molecular markers have been used to assess reproductive mode in particular populations, to describe spatial distributions of particular genets (clonal descendants from a single zygote), and to estimate evolutionary ages of clonal lineages. Some clones have proved to be unexpectedly ancient, but these are the exception, not the rule.

4. In many microorganisms, including various bacteria, fungi, and protozoans, molecular markers have revealed unexpectedly strong proclivities for clonal reproduction in addition to mechanisms for occasional recombinational exchange of genetic material. These findings are of medical as well as academic interest because they can influence strategies for diagnosis of disease agents and for development of vaccines and curative drugs.

5. Molecular markers have found application in identifying genetic chimeras in nature, as well as in ascertaining gender in dioecious species, in which these features are not necessarily obvious from an inspection of external phenotypes alone.

6. Molecular assessments of genetic parentage can identify an individual's sire and dam (or at least exclude most candidate parents) when maternity or paternity are uncertain from other evidence. Methods of empirical analysis are influenced by the nature of the particular parentage problem, the size of the pool of candidate parents, and numbers of offspring in a clutch.

7. Individual clutches or broods in many vertebrate and invertebrate species have often proved upon molecular analysis to include varying proportions of foster young resulting from extra-pair fertilization (EPF) or sometimes intraspecific brood parasitism (IBP). Through such analyses, the distinction between social mating systems and genetic mating systems has become widely appreciated.

8. Topics that have been informed through genetic parentage analyses include alternative reproductive tactics, sperm storage, sperm (and pollen) competition, estimation of effective and census population size, and sociobiological patterns. In general, a powerful approach is to combine genetic parentage data with behavioral or other independent observations and interpret outcomes in the context of relevant ecological and evolutionary theory on mating systems, sexual selection, sexual dimorphism, and behavior.

6

Kinship and Intraspecific Genealogy

Community of descent is the hidden bond which naturalists have been unconsciously seeking.

C. Darwin (1859)

Clonal identity and parentage, the subject of Chapter 5, are extreme examples of close kinship. In this chapter we shall be concerned with applying molecular markers to reveal genetic relatedness within and among broader groups of extended intraspecific kin. Questions of genetic relatedness arise in virtually all discussions of social species in which particular morphologies and behaviors might have evolved as predicted under theories of inclusive fitness and kin selection (Box 6.1). Interest in kinship also arises for any species whose populations are spatially structured, perhaps along family lines. At increasingly greater depths in time, *all* conspecific individuals are related to one another through an extended pedigree that constitutes the composite intraspecific genealogy of a species.

Close Kinship and Family Structure

Molecular assessments of close kinship require qualitative genetic markers with known transmission properties, such as allozymes or microsatellites. However, compared with the rather straightforward situation in paternity and maternity analysis (in which genetic pathways connecting individuals extend across only one generation), appraisals of extended kinship are complicated by the fact that multiple generations and potential transmission pathways link more distant relatives. Thus, even when fairly large numbers of loci are assayed, the focus in

BOX 6.1 Within-Group Genetic Relatedness, Inclusive Fitness, and Kin Selection

Genetic Relatedness

Discussions of close kinship often require a quantitative measure of genetic relatedness (r). An intuitive interpretation of a coefficient of relatedness is provided by an answer to the following question: What is the probability that an allele carried by the focal individual is also possessed by the relative in question? In other words, what is the expected proportion of alleles shared by these individuals' genomes? In principle, $r = 0.50$ for full siblings and parent–offspring pairs (see Figure 6.1A); $r = 0.25$ for half-sibs or for an individual and its uncles, aunts, grandparents, and grandchildren; $r = 0.125$ for first cousins; and $r = 0.0$ for non-relatives. Generally, for any known pedigree, true values of r can be determined by direct pathway analysis of gene transmission routes (Cannings and Thompson 1981; Michod and Anderson 1979).

In nature, however, pedigrees usually are unknown, so several statistical methods have been developed and tailored for estimating coefficients of relatedness from polymorphic genetic markers, such as those provided by multi-locus allozymes (Crozier et al. 1984; Pamilo and Crozier 1982; Queller and Goodnight 1989), DNA fingerprints (Reeve et al. 1992), or microsatellites (Blouin et al. 1996; Henshaw et al. 2001; Queller et al. 1993; Strassman et al. 1996; Van de Casteele et al. 2001). Some of these approaches entail estimating average relatedness in assemblages of individuals, as implemented by computer programs such as *Relatedness* (Queller and Goodnight 1989). For example, Pamilo (1984a) derived an estimate for r that can be expressed in terms of heterozygosities observed at a locus (h_{obs}) and those expected under Hardy–Weinberg equilibrium (h_{exp}) within a colony m with N individuals, in comparison to heterozygosities observed (H_{obs}) and expected (H_{exp}) within a broader population composed of c colonies:

$$r = \frac{H_{exp} - \frac{1}{c}\Sigma h_{exp} - \frac{1}{c}\Sigma[(1/N-1)]\,[h_{exp} - \frac{1}{2}h_{obs}]}{H_{exp} - \frac{1}{2}H_{obs}} \tag{6.1}$$

This coefficient of relatedness may also be interpreted as a genotypic correlation among group members in a subdivided population (see Pamilo 1984a for derivations and discussion). Other approaches entail estimates of relatedness between specific pairs of individuals (Epstein et al. 2000; Lynch and Ritland 1999; Ritland 1996; Wang 2002), as implemented by computer programs such as *Kinship*, which uses a maximum likelihood statistical framework (Goodnight and Queller 1999).

Inclusive Fitness and Kin Selection

Classic genetic fitness is defined as the average direct reproductive success of an individual possessing a specified genotype in comparison to that of other individuals in the population. Inclusive fitness, which entails a broader view of the transmission of genetic material across generations, incorporates the individual's personal or classic fitness as well as the probability that its genes may be passed on through relatives (Queller 1989, 1996). These latter transmission probabilities are influenced by the coefficients of relatedness involved. Concepts of inclusive fitness have been advanced as an explanation for the evolution of "self-sacrificial" behaviors, wherein alleles influencing such altruism may have spread in certain populations under the influence of kin selection. For example, under the proverbial example of altruistic behavior, an individual's alleles would tend to increase in frequency if his or her personal fitness was completely sacrificed for a comparable gain in fitness by more than two full sibs, four half-sibs, or eight first cousins.

In general, according to Hamilton's (1964) rule, a behavior is favored by kin selection whenever

$$\Delta w_x + \Sigma r \Delta w_y > 0 \qquad \qquad (6.2)$$

where Δw_x is the change the behavior causes in the individual's fitness, Δw_y is the change the behavior causes in the relative's fitness, and r is the genetic relatedness of the individuals involved. Under Hamilton's rule, an allele will tend to increase in frequency if the ratio of the cost C that it entails (loss in expected personal reproduction through self-sacrificial behavior) to the benefit B that it receives (through increased reproduction by relatives) is less than r:

$$C/B < r \qquad \qquad (6.3)$$

genetic studies of broader kinship often shifts from attempts to enumerate relationships among particular individuals (but see below) to a concern with patterns of mean genetic relatedness within and among groups.

The concepts and reasoning that are involved in kinship assessment can be introduced by the following example (from Avise and Shapiro 1986): Juveniles of the serranid reef fish *Anthias squamipinnis* occur in social aggregations ranging in size from a few individuals to more than a hundred. Although eggs and larvae of this species are pelagic, drifting in the open ocean, Shapiro (1983) raised the intriguing hypothesis that juvenile aggregations might consist of close genetic relatives (predominantly siblings from a single spawn) that had stayed together through the pelagic phase and settled jointly. If so, then kin selection would have to be considered as a potential factor influencing behaviors within social aggregations, and furthermore, marine biologists would have to reevaluate the conventional wisdom that products of separate spawns are mixed thoroughly during the pelagic phase. To test the Shapiro hypothesis, genotypes were surveyed at each of three polymorphic allozyme loci in eight discrete social aggregations of juvenile *A. squamipinnis* from a single reef in the Red Sea. Allele frequencies are presented in Table 6.1.

TABLE 6.1   Allele frequencies observed at three allozyme loci in eight local aggregations of the reef fish *Anthias squamipinnis*

Allele		A	B	C	D	E	F	G	H	Total
					Aggregation					
Ldh locus	2n =	124	140	128	158	152	140	134	106	1,082
a		0.613	0.679	0.680	0.620	0.605	0.643	0.649	0.613	0.638
b		0.379	0.314	0.320	0.380	0.395	0.350	0.351	0.377	0.358
c		0.008	—	—	—	—	0.007	—	0.010	0.003
d		—	0.007	—	—	—	—	—	—	0.001
Pgm locus	2n =	124	140	128	162	156	140	162	106	1,118
a		0.944	0.936	0.937	0.938	0.904	0.957	0.926	0.943	0.935
b		0.048	0.043	0.016	0.050	0.070	0.022	0.049	0.028	0.042
c		0.008	0.021	0.031	0.012	0.026	0.021	0.025	0.019	0.020
d		—	—	0.008	—	—	—	—	—	0.001
e		—	—	0.008	—	—	—	—	0.010	0.002
Sod locus	2n =	124	140	128	162	156	140	162	106	1,118
a		0.758	0.707	0.734	0.729	0.679	0.722	0.698	0.792	0.723
b		0.113	0.150	0.180	0.148	0.211	0.164	0.191	0.142	0.164
c		0.089	0.100	0.070	0.080	0.071	0.057	0.087	0.047	0.076
d		0.040	0.036	0.008	0.037	0.026	0.036	0.012	0.019	0.027
e		—	—	0.008	—	0.013	0.021	0.012	—	0.007
f		—	0.007	—	—	—	—	—	—	0.001
g		—	—	—	0.006	—	—	—	—	0.001

Source: After Avise and Shapiro 1986.
Note: Estimates of within-aggregation genetic relatedness (based on the application of Equation 6.1 to data from the three loci) are $r = -0.01$, 0.01, and 0.02, providing strong evidence against the proposition that aggregations consist of close kin (see text).

The hypothesis of close genetic affinity of individuals within *A. squamipinnis* aggregations yields several genetic predictions that were not borne out by the allozyme data. If juveniles within an aggregation were strictly full sibs, they should (1) exhibit at most four alleles at a locus (instead, several aggregations displayed more than four alleles at *Pgm* and *Sod*); (2) exhibit alleles in frequencies consistent with transmission from two parents (instead, allele frequencies departed dramatically from these expectations, and best statistical fits to the empirical data required a *minimum* of 11–16 equally contributing parents per aggregation); (3) share any rare alleles observed in the overall study (instead, all rare alleles present in more than one individual were distributed across aggregations); (4) sometimes show strong heterozygote excesses relative to expectations based on within-group allele frequencies, as, for example, in aggregations that resulted from a mating between alternate homozygotes or between a homozygote and a heterozygote (but no such genetic patterns were uncovered); and (5) exhibit genotypic relatedness values (see Box 6.1) close to $r = 0.5$ (instead, estimates of r were invariably near zero). In addition, if *A. squamipinnis* aggregations existed primarily as full-sib assemblages, there should be large between-group variances in allele frequencies (but observed variances were close to zero), and there should be large differences among aggregations in single-locus and multi-locus genotypic frequencies (but no such differences were observed). Overall, these several lines of reasoning applied to the genetic data demonstrated that the assayed aggregations of *A. squamipinnis* represented nearly random samples of progeny from many matings. This finding casts considerable doubt on the hypothesis that larval kin routinely remain together throughout the pelagic phase.

In general, estimates of average genetic relatedness within groups can be quantified from such molecular data by approaches such as those outlined in Box 6.1. Estimated values of r then provide a genealogical backdrop against which to evaluate hypotheses regarding social behaviors and the possible influence of kin selection.

Eusocial colonies

Species considered "eusocial" consist of groups of individuals possessing the following traits (Wilson 1971): cooperation in care of young; reproductive division of labor, with more-or-less sterile individuals working on behalf of reproductives; and overlapping generations of colony workers. Eusociality has long intrigued biologists because its evolution entails a transition to an extreme form of altruism: sacrifice of direct reproduction by worker individuals.

HYMENOPTERAN INSECTS. The epitome of eusociality has been reached in colonial hymenopterans, which include nearly all ants and the highly social bees and wasps. In these species, female workers (who are sterile) provide exaggerated examples of "helpers at the nest." A colony may operate so efficiently and with such a high apparent level of cooperation and individual self-sacrifice that it has sometimes been referred to as a super-organism (Queller and Strassmann 2002).

Hamilton (1964) first proposed that the evolution of such extreme reproductive altruism might have been facilitated by the altered and asymmetric genetic relationship among parents and siblings stemming from haplodiploid sex determination. In hymenopterans, males develop from unfertilized eggs and hence are haploid for a single set of chromosomes inherited from their mother. Females develop from fertilized eggs and are diploid. Thus, daughters of a monogamous female share three-quarters of their alleles (all alleles from their father and half from their mother, such that $r = 0.75$; Figure 6.1B), a situation that contrasts with expected relationships in normal diploid species, in which full sibs share only half of their alleles ($r = 0.50$; Figure 6.1A). Thus,

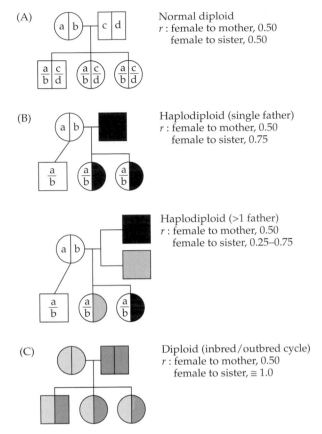

Figure 6.1 Transmission genetics and genetics relatedness in two-generation pedigrees. Shown are mean coefficients of genetic relatedness (r; see Box 6.1) expected among females under three reproductive modes: (A) normal diploid inheritance, (B) haplodiploidy, as in eusocial hymenopteran insects (with and without multiple insemination of queens), and (C) diploidy under an extreme inbreeding–outbreeding cycle. Circles (females) and squares (males) bisected by vertical lines represent diploid individuals; males without such lines are haploid. In the case of haplodiploidy with more than one father (polyandry), r depends on the number of sires and their proportionate contributions to a colony's worker population: mean $r = 0.5 (0.5 + \Sigma f_i^2)$, where f_i is the proportionate contribution by the ith male. Reproduction by multiple queens within a colony (polygyny) would serve further to lower the mean r.

according to Hamilton's insight, an ancestral hymenopteran with any behavioral predisposition toward rearing sisters, even at the risk of producing fewer young herself, might have increased her inclusive fitness by adopting this behavioral strategy. In other words, if the genetic ties within a generation are closer than those between generations, an individual might profit under the currency of inclusive fitness by investing in a parent's reproductive success rather than her own.

However, this prediction weakens considerably when additional factors germane to haplodiploid species are considered: that full-sister females are related to their brothers by only $r = 0.25$ (rather than $r = 0.50$ under normal diploidy); that reproductive females (queens) in some species mate with multiple males; that multiple queens may be present within a colony (this is especially true in many ant species); and that these queens are sometimes unrelated (Queller et al. 2000). Thus, depending upon sex ratio, numbers of parents, parental genetic relationships to one another, and relative contributions of different parents to the colony, genetic relatedness among colony members can be dramatically lower than idealized by Hamilton (Wade 1982; Wilson 1971; see Figure 6.1B). All else being equal, lower genetic relatedness within a colony will tend to decrease any inclusive fitness benefits that an individual can expect to receive through self-sacrificial cooperative behaviors, and it also greatly increases opportunities for conflicts of interest among individuals within a colony (Arévalo et al. 1998; Sundström and Boomsma 2001).

Empirical surveys using molecular markers have been conducted on numerous social hymenopterans to assess mean within-colony genetic relatedness (Table 6.2). For some species, such as *Rhytidoponera* ants, mean relatedness among colony workers has proved to be near $r = 0.75$, the value expected when a once-inseminated queen founds a monogyne (single-queen) nest. However, within-colony estimates of genetic relatedness for other hymenopteran species have proved to be significantly lower (Bourke and Franks 1995), often far below the critical threshold of $r = 0.50$, and therefore in a zone where the supposed genetic advantage of reproductive altruism by workers would seem to be absent. These molecular findings, coupled with field observations, have revealed that queens sometimes are multiply inseminated (but see Strassmann 2001) and that multi-queen ("polygyne") colonies are common (Cole 1983; Hölldobler and Wilson 1990; Ross and Carpenter 1991). These findings raised a conundrum about the evolution of hymenopteran eusociality: If same-generation individuals within a nest are often less closely related to one another than they would be to their own sons or daughters, then if kin selection is a powerful force, "what prevents evolution from leading to a more competitive state in which the workers (who have ovaries) try to take over reproduction?" (Hölldobler and Wilson 1990).

One possibility is that a colony is divided into cliques of individuals who can distinguish close kin from more distant kin and direct altruistic behaviors accordingly (Strassmann 1996). Experiments indicate that kin discrimination is possible in some hymenopteran species, but not others (Hölldobler and Wilson 1990). For example, worker honeybees (*Apis mellifera*) are able to assess their relatedness to other individuals and preferentially rear queens from larvae that are most closely related to themselves (Visscher 1986; but see also Gilley 2003), whereas workers in the wasp species *Polistes carolina* and *Parachartergus colobopterus* and in the ant *Rhytidoponera confusa* appear to lack such abilities (Crosland 1988; Strassmann et al. 1997, 2000a). In another ant

TABLE 6.2   Representative examples of coefficients of genetic relatedness (*r*) estimated among females within colonies of various eusocial hymenopteran insects

Species	Comparison	r	Colonies and queen matings[a]	Reference
Ants				
Camponotus lingiperda	Workers	0.08[b]	Polygyne	Gertsch et al. 1995
Formica aquilonia	Workers	0.09	Polygyne	Pamilo 1982
Formica polyctena	Workers	0.19–0.30	Polygyne	Pamilo 1982
Formica sanguinea	Workers	0.31–0.42	Polygyne; queens multiply-mated	Pamilo and Varvio-Aho 1979
Formica transkaucasica	Workers	0.33	Polygyne; queens singly-mated	Pamilo 1981, 1982
Myrmecia pilosula	Workers	0.17	Polygyne	Craig and Crozier 1979
Myrmica rubra	Workers	0.02–0.54	Polygyne	Pearson 1983
Nothomyrmecia macrops	Workers	0.17	Polygyne, occasionally	Ward and Taylor 1981
Rhytidoponera chalybaea	Workers	0.76	Monogyne; queen singly-mated	Ward 1983
Rhytidoponera confusa	Workers	0.70	Monogyne; queen singly-mated	Ward 1983
Solenopsis invicta	Workers	0.01–0.08	Polygyne; queens singly-mated	Ross and Fletcher 1985
Wasps				
Agelaia multipicta	Workers	0.27	Polygyne	West-Eberhard 1990
Cerceris antipodes	Females[c]	0.25–0.64	Polygyne	McCorquodale 1988
Microstigmus comes	Females[c]	0.60–0.70	Monogyne; often singly-mated	Ross and Matthews 1989a,b
Parachartergus colobopterus	Females[c]	0.11	Polygyne, probably	Queller et al. 1988
Polybia occidentalis	Females[c]	0.34	Polygyne	Queller et al. 1988
Polybia sericea	Females[c]	0.28	Polygyne; often singly-mated	Queller et al. 1988
Bees				
Apis mellifera	Workers	0.25-0.34	Highly polyandrous	Laidlaw and Page 1984

[a] Note how high relatedness within a nest depends on colonies being monogyne and possessing a queen who was singly-mated.
[b] Based on microsatellite data. Other estimates came from protein-electrophoretic analyses and represent mean values. Additional examples can be found in Crozier and Pamilo 1996.
[c] Reproductive and non-reproductive females not distinguished.

species (*Formica fusca*), however, workers *have* been found to display favoritism toward their own kin when rearing eggs and larvae in polygyne colonies (Hannonen and Sundström 2003).

Another suggestion is that high frequencies of polygyne colonies and multiple mating by queens represent derived behaviors, rather than the ancestral conditions under which eusociality evolved. Under this hypothesis, eusociality tends to arise through kin selection when populations are highly structured along family lines, whereas subsequent maintenance and elaboration into advanced eusociality can occur even when within-colony relatedness decreases. Eusocial colonies, once formed, may operate so smoothly and successfully that the inclusive fitness of workers remains higher than if workers became egg-layers, such that evolutionary reversion to a less eusocial condition is simply not feasible. In ants, it is difficult to test the hypothesis that polygyne colonies and multiple mating by queens are derived conditions because most species are strongly eusocial. In a primitively eusocial bee, *Lasioglossum zephyrum*, a high molecular genetic estimate of intra-colony relatedness ($r = 0.70$) indicated that kin selection may operate in this species (Crozier et al. 1987), but later analyses based on microsatellite markers in another *Lasioglossum* species (*malachurum*) showed that about one-third of nests had been taken over by unrelated queens prior to worker emergence (Paxton et al. 2002). Furthermore, in several wasp species that also have primitive or incipient eusociality, *r* values within a nest sometimes are only moderate to low (Strassmann et al. 1989, 1994). Although these wasps may not necessarily provide valid representations of ancestral behavioral conditions, the genetic findings do demonstrate that low within-colony relatedness is not confined to the most advanced hymenopteran societies.

Finally, various ecological–genetic hypotheses have been advanced to explain the conundrum of low genetic relatedness within some hymenopteran colonies. For example, high genetic diversity among nestmates might diminish susceptibility to infectious parasites (Shykoff and Schmid-Hempel 1991a,b) or permit the colony to perform better in some environments (Cole and Wiernasz 1999). Or caste determination might have a partial genetic basis that is conceivably allowed fuller expression by multiple mating or the formation of polygyne colonies (Crozier and Page 1985). To the extent that these or other strong adaptive benefits attend colonial living, the requirement of close kinship for eusociality should be somewhat relaxed. Another possibility is that collaborating queens fare proportionately better than individual queens in competition for limited nest sites (Herbers 1986). Under this hypothesis, concepts of inclusive fitness remain in partial effect if co-foundresses are genetic relatives, as sometimes (but not always) appears to be the case. In various hymenopteran species, molecular genetic appraisals of co-founding queens have revealed mean relatedness values ranging from $r \approx 0.00$ to $r \approx 0.70$ (Metcalf and Whitt 1977; Ross and Fletcher 1985; Schwartz 1987; Stille et al. 1991; Strassmann et al. 1989).

OTHER ARTHROPODS.   Highly eusocial systems (or behavioral components thereof) have been discovered in several other taxonomic groups, and these cases are valuable for the similarities and contrasts they provide with the eusocial hymenopterans. One remarkable example involves marine shrimp in the genus *Synalpheus*, in which individuals often live together by the hundreds within a large sponge. Field observations coupled with data from allozyme markers have shown that these colonies are eusocial and that each typically consists of full-sib animals (Duffy 1996; Duffy et al. 2002). *Synalpheus* shrimp are diploid. So too are termites, another group in which eusociality is well developed (Wilson 1975). These shrimp and termites demonstrate that eusociality is not invariably coupled to haplodiploid sex determination, a conclusion also evidenced by the fact that a few other arthropod groups (some mites, thrips, whiteflies, scale insects, and beetles) are haplodiploid but do not exhibit eusociality (Wilson 1975).

Several termite species possess sex-linked multi-chromosome translocation complexes that serve to elevate genetic relatedness both between sisters and between brothers (Syren and Luykx 1977; Lacy 1980), but this odd genetic system also lowers genetic relatedness between male and female siblings and thus is difficult to rationalize as being a prime causal factor in the evolution of termite eusociality (Andersson 1984; Leinaas 1983). Cyclic inbreeding–outbreeding is another proposed model that might promote eusociality by altering genetic relatedness within and among groups in such a way as to promote kin selection (Bartz 1979; see also Pamilo 1984b; Williams and Williams 1957). When male and female mates are unrelated but each is a product of intense inbreeding, their offspring can be nearly identical genetically, but only 50% like either parent (Figure 6.1C). When such conditions hold, any genes that behaviorally dispose siblings to stay together and assist their parents in rearing young might be favored for inclusive fitness reasons similar to those described above for the haplodiploid hymenopterans (see, however, Crozier and Luykx 1985). Termites possess several natural history features that favor social interactions and might set the stage for such a breeding cycle, such as living in protected and contained nests conducive to multi-generation inbreeding and passing symbiotic intestinal flagellates from old to young individuals by anal feeding (an arrangement that necessitates close social behavior; Wilson 1971).

In accounting for the evolution and maintenance of eusociality, an emerging sociobiological view is that haplodiploidy per se may seldom be the deciding factor after all, but instead is (at best) merely one of several elements in a broader kin-selection framework of cost–benefit fitness considerations. Queller and Strassmann (1998) describe several biological characteristics that consistently earmark two types of eusocial arthropods: "fortress defenders," such as social aphids (Stern and Foster 1996), social beetles, and termites, which live inside a nest or protected site (a valuable resource that is both possible and necessary to defend as a group); and "life insurers," such as ants, bees, and wasps, which forage in the open but nonetheless benefit from group behaviors because overlapping adult life spans are often needed to successfully care for young within the nest. In each case, the proposed benefits of sociality to an individual who cooperates closely with grouped kin (even if they are not extremely close relatives) presumably outweigh the high risks of go-it-alone personal reproduction.

NAKED MOLE-RATS.   Another noteworthy example of eusociality involves a colonial vertebrate, the naked mole-rat (*Heterocephalus glaber*) (Jarvis 1981; Sherman et al. 1991). Brood care and other duties in this underground rodent species are performed cooperatively by mostly non-reproductive workers or helpers, who represent offspring from previous litters. The helpers assist the queen in rearing progeny that are fathered by a few select males within the burrow system. Using DNA fingerprint assays of colony members, Reeve et al. (1990) documented high band-sharing coefficients (0.88–0.99) comparable in magnitude to estimates for highly inbred mice or monozygotic twins in cows and humans. From these molecular data, they estimated mean within-colony genetic relatedness at $r = 0.81$, and accordingly suggested that a great majority of matings within a colony must be among siblings or between parents and offspring. Intense within-colony inbreeding is consistent with a strong role for kin selection in the evolution of eusociality in naked mole-rats. However, ecological and life history considerations are also important, as are phylogenetic constraints, as evidenced by the fact that colonial and eusocial behaviors are displayed to widely varying degrees among different mole-rat species (Allard and Honeycutt 1992; Burda et al. 2000; Honeycutt 1992). For example, microsatellite assays of a more outbred eusocial species (*Cryptomys damarensis*) yielded an estimate of mean within-colony relatedness of only $r = 0.46$ (Burland et al. 2002). This finding suggests that even "normal" levels of family kinship within a colony can be sufficient for the evolution, or at least the retention, of eusociality in these mammals. It also suggests that while intense inbreeding and pronounced geographic population structure have been observed in mole-rats (Faulkes et al. 1997), these phenomena may first and foremost be responses to severe constraints on dispersal, especially given the predator-rich environments inhabited by these poor-sighted and rather defenseless animals (Braude 2000).

Non-eusocial groups

Most group-living species exhibit far less social organization and subdivision of labor than do eusocial arthropods and mole-rats, but genetic relatedness among group members remains of interest. A seminal compendium on known or suspected kinship in group-living animal species was provided by Wilson (1975). Traditionally, such genealogical understanding came from difficult and labor-intensive field observations of mating and dispersal (Fletcher and Michener 1987), but in the last three decades molecular markers have assisted greatly in these evaluations.

Eastern tent caterpillars (*Malacosoma americanum*), for example, are characterized by cooperative nest building as well as cooperative foraging along pheromone trails. Adult moths of this diploid species lay egg masses from which first-instar larvae emerge to feed on leaves at the tips of tree branches. Later, the caterpillars move to central locations in a tree to initiate tent (nest) construction. In a temporal genetic study using allozymes, Costa and Ross (1993) found that mean genetic relatedness within colonies of newly emerged larvae (from a single egg mass) was $r = 0.49$, not significantly different from the expected value of $r = 0.50$ for full siblings. However, during the ensuing 8 weeks, relatedness values declined to between $r = 0.38$ and $r = 0.25$. This temporal reduction in intra-colony relatedness represented an erosion of the initial simple family structure, apparently due to frequent exchanges of individuals among colonies after foragers encountered pheromone trails of non-siblings. The results indicated that immigrants are not overtly discriminated against, but rather can be accepted into a colony. Subsequent observations suggested an adaptive explanation (Costa and Ross 2003): The increased group size that results from acceptance of immigrants was found to enhance mean fitness by promoting larval growth and enhancing the final larval weights attained (which are highly correlated with adult reproductive success). Thus, in tent caterpillars, individual fitness benefits stemming from augmented group size apparently more than offset the dilution of biological relatedness in these genetically heterogeneous social groups.

Similar genetic studies were conducted on day-roosting colonies of *Phyllostomus hastatus* bats in Trinidad. These colonies are subdivided into compact clusters of adult females that remain highly stable over several years and are attended by a single adult male, who from allozyme evidence sires most of the babies born to females within the "harem" (McCracken and Bradbury 1981). Stable groups of adult females are fundamental units of social structure in this species. It was hypothesized that harem females are matrilineal relatives, such that kin selection might be a plausible factor underlying their social or cooperative behavior. However, based on allozyme assays in conjunction with field observations, the females within each harem proved to be random samples from the total adult population, and hence were unrelated (McCracken and Bradbury 1977, 1981). These results indicated that juveniles are not recruited into parental social units and, therefore, that contemporary kin selection cannot explain the maintenance of behavioral cohesiveness in these highly social mammals.

Conversely, several ground-dwelling squirrels in the family Sciuridae do have varying degrees of social organization built around matrilineal kinship (Michener 1983). For example, black-tailed prairie dogs (*Cynomys ludovicianus*) live in social groups (coteries) that typically consist of one or two adult males born outside the group, plus several adult females and young that are closely related. Females show strong tendencies to remain in their natal coteries for life (Hoogland and Foltz 1982). Genetic analyses based on pedigree and allozyme data documented that, despite this known matrilineal population structure, colonies are outbred due to coterie switching by males and social avoidance of father–daughter matings (Foltz and Hoogland 1981, 1983; see also Chesser 1983; Dobson et al. 1997, 1998).

The mound-building mouse (*Mus spicilegus*) constructs large earthen mounds containing nesting and food storage chambers. Each mound typically houses 3–10 animals. Garza et al. (1997) scored molecular markers at four autosomal and four X-linked microsatellite loci in individuals inhabiting 40 mounds in Bulgaria. Genetic results showed that at least two males and two females often had parented offspring in a mound, that parents of different sibships within mounds were more closely related than if they had been chosen at random from the population, and that adult females accounted for this excess relatedness. These genetic findings were interpreted to indicate that the mechanisms by which individuals congregate to build mounds are kin-based and that the evolution of communal nesting in this species could be due in part to kin selection.

Many cetacean species (whales and dolphins) live in social groups called pods, which have been the subject of several studies employing molecular markers (see reviews in Hoelzel 1991a, 1994, 1998). For example, in the long-finned pilot whale (*Globicephala melas*), social groups typically consist of 50–200 animals. Their herding instincts have been exploited by native peoples to drive entire pods into shallow bays for slaughter. Analyses of DNA fingerprints from tissues taken from such harvests in the Faroe Islands revealed that adult males are not closely related to adult females within a pod, and furthermore, that 90% of fetuses had not been sired by a resident male (Amos et al. 1991a,b, 1993). From these data and behavioral observations, the authors concluded that social groups in the pilot whale are built around matrilineal kinship, with most inter-pod genetic exchange mediated by males. As deduced by mtDNA analyses as well, a tendency toward matrifocal organization of structured groups (either locally or associated with particular migratory pathways) is a recurring theme in several species of dolphins (e.g., Hoelzel et al. 1998a; Pichler et al. 1998) and whales (Baker and Palumbi 1996; Baker et al. 1998; Brown-Gladden et al. 1997; Hoelzel 1991b, 1998; Hoelzel and Dover 1991b; Hoelzel et al. 1998b; O'Corry-Crowe et al. 1997; Palsbøll et al. 1997a,b). One ramification of this structure is that populations can show significantly greater genetic subdivision in mitochondrial than in nuclear gene markers, as has been demonstrated, for example, in sperm whales (*Physeter catodon*) on a global scale (Lyrholm et al. 1999).

Most reptiles are relatively asocial animals, but genetic and behavioral analyses of a large Australian lizard, *Egernia saxatilis*, have documented what is perhaps the first firm evidence for long-term "nuclear family" structure in a reptilian species. Parentage and kinship assessments based on microsatellite markers revealed tendencies toward multi-year monogamy and group stability, with up to three annual cohorts of full-sib offspring living with their parents (O'Connor and Shine 2003). Overall, 85% of the surveyed juveniles lived in social groups, and 65% lived in family groups with at least one of their biological parents (39% with both parents). An entirely different outcome was reported in an avian species that had been suspected of spending extended periods of time in family groups. In a DNA fingerprinting study of the long-eared owl (*Asio otus*), most birds in communal winter roosts proved not to be close kin, as evidenced by the fact that mean genetic relatedness within roosts was not significantly higher than that between roosts (Galeotti et al. 1997).

Questions concerning kinship also arise in grouped plants. The white-bark pine (*Pinus albicaulis*) frequently displays a multi-stem form. Allozyme analyses have demonstrated that stems within a clump are genetically distinct individuals (genets), but are nonetheless more similar to one another than to individuals in other clumps (Furnier et al. 1987). This family structure appears to be a direct result of seed-caching behavior by birds, especially Clark's nutcrackers (*Nucifraga columbiana*), which often deposit multiple seeds from related cones at particular locations.

The limber pine (*Pinus flexilis*) is another species that exhibits a multi-trunk growth form, perhaps registering similar seed-caching behavior by birds or perhaps registering the presence of multiple ramets from a single genetic individual. From allozyme analyses, nearly 20% of multi-trunk clusters proved to be composed of two to four genetically different individuals, and mean genetic relatedness within these clusters was $r = 0.19$, or slightly less than expected for half-sibs (Schuster and Mitton 1991). The authors note that such grouping of distinct but related genets opens the possibility of kin selection, a phenomenon seldom considered in plants. Occasional fusions or grafts among adjacent woody trunks are also observed in limber pines, and the authors found that fused genets were related significantly more closely than genets that were unfused. However, it remains uncertain whether such fusion behavior might have evolved in part under kin selection via possible adaptive advantages to the participants (such as joint translocation of water and nutrients, or added physical stability).

Kin recognition

The spatial co-occurrence of close kin in virtually all species raises additional questions about whether individuals can somehow assess their genetic relatedness to others and perhaps adjust competitive, cooperative, altruistic, or other behaviors accordingly (Waldman 1988; Wilson 1987). In studying such issues, ethologists traditionally monitored interactions among organisms supposed to exhibit varying levels of genetic relatedness as gauged by behavioral observations or by pedigree records in captive settings (Fletcher and Michener 1987; Hepper 1991). However, these conventional lines of evidence for kinship are less than fully reliable, and in any event are unavailable for many species. Molecular markers are now routinely employed to assist with relatedness assessments, several examples of which have already been mentioned.

Another classic example involved a free-living population of Belding's ground squirrels (*Spermophilus beldingi*) in California, for which Holmes and Sherman (1982) employed protein electrophoretic techniques to distinguish full siblings from maternal half-sibs resulting from multiple mating. Subsequent behavioral monitoring indicated that full sisters fought significantly less often and aided each other more than did half-sisters. Such nepotism (favoritism shown kin) must require an ability by ground squirrels to judge relatedness. Additional experiments indicated that the proximate cues by which this is accomplished in *S. beldingi* involve physical association during rearing as well as "phenotypic matching," whereby an individual behaves as if it had compared phenotypic traits (genetically determined) against itself or a nestmate template (Holmes and Sherman 1982).

Another postulated advantage of kin recognition involves behavioral avoidance of close inbreeding (Hoogland 1982). Like many amphibians, the American toad (*Bufo americanus*) exhibits site fidelity to natal ponds for breeding, and thus individuals are likely to encounter siblings as potential mates (Waldman 1991). Can siblings recognize close kin and avoid incestuous mating? Waldman et al. (1992) monitored mtDNA genotypes in 86 amplexed pairs of toads and found significantly fewer matings between possible siblings (with shared haplotypes) than expected from haplotype frequencies in the local population, which led the authors to suggest that "siblings recognize and avoid mating with one another." They further suggested that the proximate cues involved might include advertisement vocalizations by males, because resemblance among male calls proved to be positively correlated with genetic relatedness as gauged by band similarities in DNA fingerprints. Thus, females could potentially employ male vocalizations (or other genetically based clues such as odors) in kinship assessment.

Genetic relationships of specific individuals

Most of the cases described above entailed estimates of *average* relatedness within and among colonies or social groups, but another approach is to attempt assessments of genetic kinship among particular individuals. In one classic example, DNA fingerprinting assays were applied to African lions that, based on prior field observations, were thought to exist as matriarchal groups (Gilbert et al. 1991; Packer et al. 1991). A lion pride typically consists of 2–9 adult females, their dependent young, and 2–6 adult males, originally from outside the group, that have formed a coalition. Incoming males collaborate to evict resident males and often kill resident dependent juveniles. From analyses of minisatellite DNA fingerprints gathered from nearly 200 animals, the following conclusions were reached (Figure 6.2): female companions within prides proved invariably to be closely related; male coalition partners were either closely related (in some larger coalitions) or genetically unrelated (mostly in some smaller coalitions involving two or three males); and mating partners usually were unrelated. Furthermore, genetic parentage analyses revealed that resident males sired all cubs conceived during their tenure, and that the variance in male reproductive success increased greatly as coalition size increased. From these molecular observations, the authors concluded that lion prides are indeed matrilineal, and that a coalition male is likely to act as a non-reproductive "helper" only if the coalition that he entered includes closely related males.

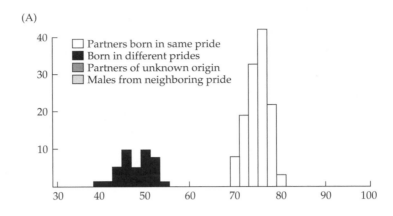

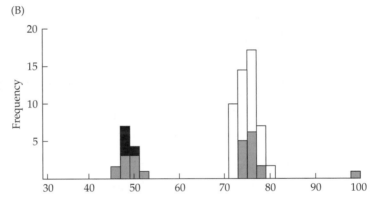

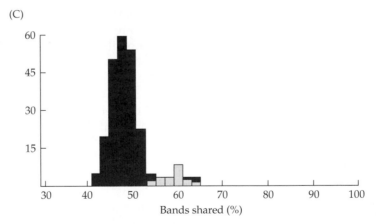

Figure 6.2 Frequency distributions of minisatellite band sharing in Serengeti lions. Percentages of bands shared are indicated between (A) females born into the same versus different prides; (B) male coalition partners known to have been born into the same versus different (or in some cases unknown) prides; and (C) coalition males and resident females. (After Packer et al. 1991.)

These molecular studies on lions benefited from the fact that long-term field observations and pedigrees were sometimes available to calibrate the extent of minisatellite band sharing against known or suspected kinship. Such analyses also revealed, however, that the relationship between magnitude of DNA band sharing and kinship for pairs of individuals can be nonlinear, population-specific, and can display a large variance. One technical reason may be that complex multi-locus banding profiles in minisatellite fingerprints are notoriously difficult to score (Baker et al. 1992; Prodöhl et al. 1992; van Pijlen et al. 1991). However, another potential complication in assessing kinship, which applies to all types of molecular assays, is biological: DNA profiles reflect kinship attributable not only to contemporary pedigrees, but also to earlier demographic histories that may have included such factors as population bottlenecks or inbreeding, which can leave lasting genetic signatures (Hoelzel et al. 2002a,b). Thus, kinship is a contextual concept, with empirical molecular estimates properly interpreted vis-à-vis some stated (or sometimes unstated) baseline that may include deep as well as shallow population history.

A case in point involves the dwarf fox (*Urocyon littoralis*), which colonized the Channel Islands off Southern California within the last 20,000 years. All assayed foxes from a small isolated island (San Nicolas) exhibited *identical* bands in DNA fingerprints, and several other island populations showed greatly enhanced levels of band sharing (75%–95%) relative to foxes from different islands (16%–56%) and relative to values (10%–30%) typifying outbred populations in many other vertebrate species (Gilbert et al. 1990). Thus, the astonishingly high kinship coefficients registered among individuals on San Nicholas may well reflect a history of population bottleneck(s) more than nonrandom mating (inbreeding) per se within the island population.

Molecular estimates of pairwise kinship among individuals inevitably have a large sampling variance when small numbers of loci are employed. Although it is usually quite feasible with small or modest numbers of molecular markers to reliably distinguish full sibs ($r = 0.50$) from half-sibs ($r = 0.25$) or nonrelatives ($r = 0.00$, in principle), finer meaningful distinctions (i.e., within the range of $r = 0.00–0.25$) remain problematic. Yet there are numerous biological settings in which ethologists and other researchers would welcome secure genetic knowledge of precise kinship between interacting individuals, the estimation of which is a current "holy grail" for molecular ecology. Thus, it will be extremely interesting to monitor developments in kinship assessment in this new era of massive genomic screening. With genetic information now obtainable in principle (at least in model organisms) from legions of qualitative genotypic markers such as microsatellites and SNPs, unprecedented opportunities should arise for refining pairwise estimates of individual relationships based on scores or even hundreds of independent loci (Glaubitz et al. 2003). Such applications are still in their infancy at the time of this writing, but might someday become hugely important in studies addressing such diverse topics as social behaviors in nature (e.g., Morin et al. 1994a), mating systems (Heg and van Treuren 1998), phenotypic heritabilities (Ritland 2000), and spatial population genetics (see below).

Geographic Population Structure and Gene Flow

Populations of nearly all species, social or otherwise, exhibit at least some degree of genetic differentiation across geography (Ehrlich and Raven 1969), if for no other reasons than because siblings usually begin life near one another and their parents and because mating partners seldom represent random draws from throughout a species' geographic range (Turner et al. 1982). In an influential study of such "population structure" on a microgeographic scale, Selander (1970) employed allozyme markers to demonstrate fine-scale spatial clustering of genotypes of house mice (*Mus musculus*) within and among barns on a farm. The spatial variation in this case was apparently due to tribal family structure in these mice and genetic drift in this small population.

Population genetic structure sometimes exists even in seemingly improbable settings. For example, mosquitofish (*Gambusia*) are abundant and highly dispersive creatures, yet extensive sampling revealed statistically significant differences in allozyme frequencies along a few hundred meters of shoreline (Kennedy et al. 1985, 1986), as well as significant temporal variation at particular locales over periods as short as a few weeks (McClenaghan et al. 1985). At broader geographic and longer temporal scales, mosquitofish populations have shown additional differentiation often hierarchically arranged at several levels: across ponds and streams within a local area, reservoirs within a river drainage, drainages within a region, and regional collections of drainages that house deep genetic differences associated with species-level separations perhaps dating to the Pleistocene (Scribner and Avise 1993a; M. H. Smith et al. 1989; Wooten et al. 1988). Various molecular markers have similarly been employed to assess geographic population structure due to genetic drift, various forms of selection, spatial habitat structure, isolation by distance, social organization, and other ecological and evolutionary factors in many hundreds of animal species at a wide variety of spatial and temporal scales.

Populations of most plant species also vary in genetic composition, sometimes over microspatial areas of a few kilometers or even meters (Levin 1979). For example, due in part to a self-fertilization reproductive mode and limited gene flow, large populations of wild wheat (*Triticum dicoccoides*) showed pronounced genetic structure over distances of less than 5 km (Golenberg 1989). In the grasses *Agrostis tenuis* and *Anthoxanthum odoratum*, sharp clinal variation was detected in several genetic characters across meter-wide ecotones between pastures and lead–zinc mines, as a result of strong disruptive selection for heavy metal tolerance and flowering time

(Antonovics and Bradshaw 1970; McNeilly and Antonovics 1968). In many plant species, gene flow via pollen and seed dispersal is sufficiently limited that estimates of neighborhood size (the population within which mating is random) often include less than a few hundred individuals occupying areas less than 50 m^2 (Bos et al. 1986; Calahan and Gliddon 1985; Fenster 1991; Levin and Kerster 1971, 1974; Smyth and Hamrick 1987). As with animal populations, additional genetic structure normally is to be expected over greater spatial and temporal scales.

A continuing challenge is to describe population genetic architectures within species (Box 6.2) and to identify and order the biological forces responsible. Broadly speaking, these forces may involve migration or gene flow (Box 6.3), random genetic drift, various modes of natural selection, mutational divergence, and the opportunity for genetic recombination mediated by organismal behaviors and mating systems. Finer considerations require partitioning these general categories into biological factors relevant to each organismal group. For example, numerous ecological and life history factors are predicted to influence population genetic structure in plants (Table 6.3). Comparative summaries of the allozyme literature for more than a hundred plant taxa revealed that magnitudes of genetic differentiation are indeed roughly associated with such factors as a species' breeding system, reproductive mode, pollination mechanism, floral morphology, life cycle, life form, and successional stage (Hamrick and Godt 1989; Loveless and Hamrick 1984).

In animals, a comparative summary of allozyme analyses on more than 300 species (Table 6.4) led Ward et al. (1992) to conclude that mobility tends to be especially well correlated with relative magnitudes of population structure. For example, vagile organisms such as insects and birds often show significantly less population structure than do relatively sedentary creatures such as some reptiles and amphibians. In another meta-review of population genetic structure with a similar outcome, Bohonak (1999) found that F_{ST} values (as gauged by molecular assays) were negatively correlated with dispersal potential (usually inferred from morphological traits of propagules) in 19 of 20 animal groups examined. In the sections that follow, a few specific cases will highlight how particular ecological and evolutionary factors can impinge on population genetic structure as revealed by molecular markers. Where possible, attempts will be made to draw meaningful parallels between results for taxonomically distinct groups such as plants and animals.

Autogamous mating systems

PLANTS.   In a paradigmatic series of studies employing allozyme markers, Allard and colleagues documented that the mating system can assume dramatic influence, especially in conjunction with natural selection, in shaping the multi-locus genetic architectures of plant species. The slender wild oat (*Avena barbata*) is a predominantly self-fertilizing species that was introduced to California from its native range in the Mediterranean during the

| TABLE 6.3 | Ecological and life history factors and their predicted effects on plant population genetic structure |

Factor	Genetic heterozygosity within populations	Genetic structure within populations	Genetic structure among populations
Breeding system			
Self-fertilizing	Low	High	High
Mixed mating	Moderate	Moderate	Moderate
Outcrossing	High	Low	Low
Floral morphology			
Monoecious	Depends on % selfing	Depends on % selfing	Depends on % selfing
Dioecious	High	Low	Low
Reproductive mode			
Apomictic	Depends on other factors	Depends on other factors	Potentially high
Sexual	Potentially high	Depends on other factors	Depends on other factors
Pollination mechanism			
Sedentary animal	Potentially low	Potentially high	High
Dispersive animal	High	Low	Low
Wind	High	Low	Low
Seed dispersal			
Limited	?	Potentially high	High
Long range	High	Low	Low
Seed dormancy			
Absent	Depends on other factors	Depends on other factors	Depends on other factors
Present	Increases potential	Reduces potential	Reduces potential
Phenology			
Asynchronous	No prediction	Increases potential	Increases potential
Synchronous	No prediction	Reduces potential	Reduces potential
Life form			
Annual	Reduced?	Increases potential	Increases potential
Long-lived	Increased?	Reduces potential	Reduces potential
Timing of reproduction			
Monocarpic[a]	No prediction	Increases potential	Increases potential
Polycarpic	No prediction	Reduces potential	Reduces potential
Successional stage			
Early	?	Depends on other factors	Increases potential
Late	?	Depends on other factors	Reduces potential
Geographic range			
Narrow endemic	Low	Low	High
Widespread	Potentially high	Depends on other factors	Depends on other factors
Population size, density			
High	High	Depends on other factors	Depends on other factors
Low	Low	Depends on other factors	Depends on other factors

Source: After Loveless and Hamrick 1984.
Note: Predictions remain qualified because categories may be interrelated and other confounding variables may pertain.
[a] Having only one fruiting period during the life cycle.

BOX 6.2 Statistical Description of Population Structure

Many approaches to the statistical description of population structure are available (e.g., Sokal and Oden 1978a,b; Weir 1990, 1996). One important example involves "F statistics," introduced by Wright (1951) as a way to describe genetic population structure in diploid organisms in terms of three allelic correlations (F_{IS}, F_{IT}, and F_{ST}) that are theoretically interrelated as follows:

$$(1 - F_{IT}) = (1 - F_{ST})(1 - F_{IS}) \tag{6.4}$$

Wright defined F_{IS} as the correlation between homologous alleles within individuals with reference to the local population and F_{IT} as the corresponding allelic correlation with reference to the total population. F_{IS} and F_{IT} are often called fixation indices (F_I). Their estimated values also may be interpreted as describing departures from expected Hardy–Weinberg (H–W) genotypic frequencies within local populations and within the total population, respectively (Nei 1973, 1977):

$$F_I = 1 - (h_{obs}/h_{exp}) \tag{6.5}$$

where h_{obs} and h_{exp} are observed and expected frequencies of heterozygotes at a locus (see Box 2.1). Thus, positive values for fixation indices indicate positive correlations among uniting gametes (heterozygote deficits), probably due to inbreeding (F_{IS}, local level) or population subdivision (F_{IT}, broader scale).

F_{ST} (or G_{ST}) can be interpreted as the variance (V_p) of allele frequencies among populations, standardized relative to the maximum value possible given an observed mean allele frequency (p):

$$F_{ST} = V_p/p(1 - p) \tag{6.6}$$

Equivalently, it can also be interpreted as the proportion of genetic variation distributed among (as opposed to within) subdivided populations, as calculated by

$$F_{ST} = (h_T - h_S)/h_T \tag{6.7}$$

where h_S is the mean expected heterozygosity at a locus (under H–W) within subpopulations, and h_T is the overall expected heterozygosity given allele frequencies in the total population. Values of F_{ST} can range from 0.0 (subpopulations genetically identical) to 1.0 (subpopulations fixed for different alleles).

The F_{ST} statistic is commonly used as a measure of population subdivision (despite several shortcomings; Neigel 2002), and as shown in Box 6.3, it also provides one approach for estimating inter-population gene flow in models that assume selective neutrality for allozyme alleles. An analogous measure (R_{ST}) is considered more appropriate for estimating population structure at microsatellite loci with many alleles (Slatkin 1995). Weir and Cockerham (1984) describe various nuances in calculating F-statistics from empirical data, such as how to handle multiple loci and unequal sample sizes across subpopulations.

TABLE 6.4   Comparative summary of population structures for 321 animal species surveyed by multi-locus protein electrophoresis

Taxonomic group	Population differences[a] (with SE)	Number of species
Vertebrates		
Mammals	0.242 ± 0.030	57
Birds	0.076 ± 0.020	16
Reptiles	0.258 ± 0.050	22
Amphibians	0.315 ± 0.040	33
Fishes	0.135 ± 0.040	79
TOTAL	0.202 ± 0.015	207
Invertebrates		
Insects	0.097 ± 0.015	46
Crustaceans	0.169 ± 0.061	19
Mollusks	0.263 ± 0.036	44
Others	0.060 ± 0.021	5
TOTAL	0.171 ± 0.020	114

Source: After Ward et al. 1992.
[a] Shown are proportions of total genetic variation within species due to genetic differences between geographic populations, as reflected in the "coefficient of gene differentiation," $(H_T - H_S)/H_T$, where H_S and H_T are mean heterozygosities estimated within local populations and within the entire species, respectively (Nei 1973).

BOX 6.3 Genetic Exchange among Populations

Gene flow is the transfer of genetic material between populations resulting from movements of individuals or their gametes. Usually, gene flow is expressed as a migration rate m, defined as the proportion of alleles in a population that is of migrant origin each generation. Gene flow is notoriously difficult to monitor directly, but it is commonly inferred from spatial distributions of genetic markers by several statistical approaches. Most of these approaches are based on equilibrium expectations derived from neutrality theory as applied to idealized models of population structure. Examples include the "island model," wherein a species is subdivided into equal-sized populations (demes or islands of size N), all of which exchange alleles with equal probability; and the "stepping-stone" model, wherein gene flow occurs between adjacent demes only. Allele frequencies in finite populations are also influenced by random genetic drift, which is a function of effective population size (see Box 2.2). Thus, the influences of drift and gene flow are difficult to tease apart, and most statistical procedures estimate only the product Nm, which can be interpreted as the absolute number of individuals exchanged between populations per generation. Also, Nm is of particular interest because under neutrality theory, the level of divergence among populations that are at equilibrium between gene flow and genetic drift is a function of migrant numbers rather than of the proportions of individuals exchanged. The most common approaches to estimating Nm and gene flow from molecular data are as follows:

1. *From F-statistics* (Cockerham and Weir 1993). Wright (1951) showed that for neutral alleles in an island model, equilibrium expectations are

$$F_{ST} \cong 1/(1 + 4Nm)$$

or, equivalently,

$$Nm \cong (1 - F_{ST})/4F_{ST} \tag{6.8}$$

Nei (1973) defined a related measure of between-population heterogeneity (gene diversity, or G_{ST}) that bears the same relationship to Nm and also is employed widely. Takahata and Palumbi (1985) suggested modifications of these basic statistics for extra-nuclear haploid genomes such as mtDNA, and Lynch and Crease (1990) proposed an analogue of the F_{ST} or G_{ST} indices (N_{ST}) that is applicable to data at the nucleotide level.

2. *From private alleles.* Private alleles are those found in only one population. For a variety of simulated populations, Slatkin (1985a) showed by computer analyses that the natural logarithm of the average frequency of private alleles $[p(1)]$ is related to the natural logarithm of Nm according to

$$\ln p(1) = -0.505 \ln (Nm) - 2.44$$

or, equivalently,

$$Nm = e^{-[(\ln p(1) + 2.44)/0.505]} \tag{6.9}$$

This result proved insensitive to most changes in parameters of the model, except that a correction for Nm due to differences in the mean number of individuals sampled per population was recommended (Barton and Slatkin 1986). The rationale underlying Slatkin's method is that private alleles are likely to attain high frequency only when Nm is low. In practice, when sufficient genetic information is available, the F_{ST} and private allele methods are expected to yield comparable estimates of gene flow under a wide variety of population conditions (Slatkin and Barton 1989).

3. *From allelic phylogenies.* Unlike the two approaches described above, which can be applied to phylogenetically unordered alleles (such as those provided by allozymes), this method requires knowledge of the phylogeny of non-recombining segments of DNA (such as mtDNA haplotypes). Given the correct gene tree and knowledge of the geographic populations in which the allelic clades are found, a parsimony criterion is applied to estimate the minimum number of migration events (s) consistent with the phylogeny. Slatkin and Maddison (1989) showed that the distribution of this minimum number is a simple function of Nm, which therefore can be estimated from empirical data by comparison with tabulated results from their computer-simulated populations.

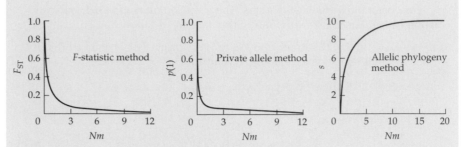

The graphs shown here plot theoretical relationships of Nm to each of these three empirical estimation parameters. Note in each case that the curvilinear relationship is detrimental to precise estimation of gene flow in large portions of the parameter space. For example, major sections of the curve relating Nm to F_{ST} are nearly flat, such that small differences in one variable (as might arise even from chance sampling error) translate into huge differences in the other (Templeton 1998). For this and other reasons (Bossart and Prowell 1998), conventional approaches to gene flow estimation, although popular, do have some serious limitations (see also Whitlock and McCauley 1999).

Thus, none of the procedures described above should be interpreted as providing precise estimates of genetic exchange among demes. Rather, each offers a qualitative (albeit numerical) impression as to whether populations are characterized by high, moderate, or greatly restricted gene flow. Roughly speaking, the average exchange of one individual per generation ($Nm \cong 1$) between populations, irrespective of deme size, is marginally sufficient in theory to prevent dramatic genetic differentiation by genetic drift alone (Allendorf 1983). The value $Nm = 1$ corresponds to mean $F_{ST} = 0.20$ (Equation 6.8) or $p(1) \cong 0.085$ (Equation 6.9). Thus, as a rule of thumb, "high-gene-flow" species are expected to exhibit estimates of F_{ST} and $p(1)$ that are lower than these values, whereas "low-gene-flow" species should display much higher values. Additional discussion of the strengths and limitations of these gene flow estimation procedures can be found in Neigel (1997), Slatkin (1985b, 1987), Templeton and Georgiadis (1996), and Waples (1998).

Neigel et al. (1991; Neigel and Avise 1993) introduced a philosophically different approach to gene flow analysis that yields estimates of single-generation dispersal distances (rather than Nm). The method is based on expected spatial distributions of lineages of various evolutionary ages in a gene tree, assuming an evolutionary clock for the molecule and assuming that lineage dispersal has occurred via a multi-generation "random walk" process from specifiable centers of origin for each clade. As applied to an empirical mtDNA gene tree for continent-wide populations of the deer mouse (*Peromyscus maniculatus*) (Lansman et al. 1983), this method yielded estimates of single-generation dispersal ($\cong 200$ m) that agreed well with direct mark–recapture data for this species. This approach is most suitable for low-dispersal species and for rapidly mutating, non-recombining genetic markers such as mtDNA. In such situations, mutations that delineate new descendant lineages may be dispersed at rates sufficiently low to prevent the attainment of the equilibrium between genetic drift and gene flow that many of the earlier models assume.

Several other statistical approaches can also be applied to molecular data to address geographic population structure (Epperson 2003). These methods include spatial autocorrelation analysis (Bertorelle and Barbujani 1995; Epperson 1993), principal component analysis (Cavalli-Sforza 1997), and multidimensional scaling (Lessa 1990). They are typically applied to genetic identity or distance summaries of allele frequency data, often from multiple loci.

Spanish period about 400 years ago, and then again during the Mission period some 250 years ago. Within California, it has since achieved a remarkable population genetic structure characterized by a great predominance of two apparently coadapted multi-locus gametic types (Allard et al. 1972; Clegg and Allard 1972). One genotype, labeled "1,2,2,2,1,B,H" (numbers and letters refer to alleles at each of five allozyme loci and two morphology genes), is characteristic of semiarid grasslands and oak savannas bordering the central Sacramento–San Joaquin Valley, whereas the complementary gametic type (2,1,1,1,2,b,h) is more common in the strip of coastal ranges and higher foothills of the Sierra Nevada mountains. These associations between genotype and environment (notably xeric versus mesic soils) also are maintained over microgeographic scales in transitional ecotones (Hamrick and Allard 1972), notwithstanding continued production of recombinant genotypes through occasional outcrossing (the estimated outcrossing rate is $t \cong 0.02$; Clegg and Allard 1973). The authors concluded that genetic variability in *A. barbata* in California has been genomically organized and spatially structured in less than 400 years by intense natural selection operating in conjunction with the severe constraints on recombination afforded by the mating system (Allard 1975). These studies added great empirical force to theoretical arguments (Box 6.4) that any factors such as linkage or inbreeding that tend to restrict genetic recombination can facilitate the operation of natural selection in molding coadapted multi-locus gene complexes (Clegg et al. 1972; Jain and Allard 1966).

"Autogamy" refers to the process of self-fertilization or self-pollination. In an early review of the literature on spatial genetic variation in plants, Heywood (1991) concluded that autogamous species often display levels of local genetic differentiation that far surpass those of predominantly outcrossing species. For example, in the highly autogamous annual plant *Plectritis brachystemon* (outcrossing rate 2%), most of the allozyme diversity proved to be partitioned among populations, whereas in a predominantly allogamous and sympatric congener, *P. congesta* (outcrossing rate 70%), most of the allozyme diversity was intra-populational (Layton and Ganders 1984). Another conclusion was that the spatial genetic architectures observed for predominantly autogamous species often are similar to those of taxa that reproduce by a mixture of apomixis (see Chapter 5) and sexual outcrossing (Heywood 1991). Especially when such species

BOX 6.4 Self-Fertilization, Restricted Genetic Recombination, and Multi-locus Organization

Self-fertilization is normally the most intense form of inbreeding. Among the consequences of predominant selfing are a severe restriction on intergenic recombination and an associated enhanced opportunity for the maintenance of coadapted gene complexes. These consequences can be illustrated qualitatively as follows: Assume that at four bi-allelic loci, the haploid genotypic combinations 1,1,1,1 and 2,2,2,2 (numbers refer to alleles at each locus) confer high viability on their diploid bearers and that the other 14 multi-locus haplotypes (1,2,1,1; 2,1,2,1; etc.) yield adaptively inferior individuals. During the course of a generation, the favored genotypes will tend to increase in frequency under natural selection from early to late stages in the life cycle. By contrast, under random outcrossing, recombination at gametogenesis will tend to undo the effects of selection in maintaining favored multi-gene complexes across generations. This process of decay in gametic phase disequilibrium (D) is inhibited when loci are physically linked on a chromosome because intergenic recombination is then restricted (see Box 5.3). By limiting *effective* recombination (even among unlinked loci), inbreeding likewise retards the rate at which D converges on zero. For two loci under a model of mixed selfing and random mating, this rate of disequilibrium decay is given by

$$1 - 0.5\{0.5(1 + \lambda + s) + [(0.5\,(1 + \lambda + s))^2 - 2s\lambda]^{\,0.5}\} \qquad \textbf{(6.10)}$$

where s is the selfing probability and λ is the amount of linkage (the recombination fraction is $c = 0.5$ for $\lambda = 0$) (Weir and Cockerham 1973). For example, with $s = 0.98$ and $\lambda = 0.00$ (or with $s = 0.00$ and $\lambda = 0.98$), the rate of decay of D is 1.0% per generation. The important point is that linkage and selfing enter this equation similarly and, therefore, retard disequilibrium decay in similar fashion.

One key difference between the effects of linkage and inbreeding in restricting recombination is that the former acts locally among physically linked loci, whereas the latter acts globally within the genome, restricting recombination between all loci regardless of chromosomal location. Thus, in theory and apparently in practice, inbreeding in conjunction with selection can serve to organize entire genomes into integrated multi-locus systems (Allard 1975).

occur in low-density situations, the effects of genetic drift, selection on multi-locus genotypes, and restricted recombination engendered by the mating system all collaborate to produce striking spatial variation in genotypic frequencies.

ANIMALS.   The hermaphroditic snail *Rumina decollata* likewise reproduces by facultative self-fertilization. Analogous population genetic analyses of this species provided a classic demonstration of the potential for adaptive convergence in the genetic architectures of selfing animals and plants

(Selander and Kaufman 1975a; Selander et al. 1974). This terrestrial snail is native to the Mediterranean region, where its populations exist as a complex of highly inbred or mildly outcrossed strains characterized by different suites of allozyme alleles and morphological markers. In surveyed areas of southern France, two strains predominate: a dark form that occupies protected mesic environments under logs or rocks, and a light form associated with open xeric habitat. These two strains showed fixed differences at 13 of 26 allozyme loci. Further analyses revealed that occasional outcrossing between the two strains releases extensive recombinational variation that otherwise is expressed as between-strain genetic differences. Nevertheless, these two strains tend to retain their separate identities and habitat correlations in nature.

These findings suggest that, as in *Avena barbata*, the strong multi-locus associations and pronounced population genetic structures in this snail species probably stem from natural selection (as well as stochastic population factors; Selander 1975) operating in conjunction with self-fertilization to reduce effective recombination (Selander and Hudson 1976). As a further note of interest, *R. decollata* was introduced to the eastern United States before 1822 and subsequently expanded across much of the continent. All assayed populations in North America proved to be identical in allozyme composition, and thus probably derived from a single strain introduced from Europe (Selander and Kaufman 1973b). These results indicate that despite an absence of appreciable genetic variation, some inbred strains of *R. decollata* can enjoy great ecological success (wide distributions and high population numbers), at least in the short term.

Gametic and zygotic dispersal

POLLEN AND SEEDS.   In outcrossing plants, mobile male gametes (pollen) are transferred to sedentary eggs in the female flower by wind or by animal vectors such as insects, birds, or mammals, some of which may be capable of moving long distances. Zygotes (seeds) are then dispersed from the maternal parent by animals (when seeds attach to the vector's body or pass through its digestive system), gravity, or wind. The extent to which various gametic and zygotic dispersal mechanisms influence gene flow and genetic structures of plant populations has been the subject of numerous investigations using molecular markers, mostly from the nucleus but also sometimes from the chloroplast genome (McCauley 1995).

Most tropical trees are predominantly outcrossing, with pollination and seed dispersal usually animal-mediated (Hamrick and Murawski 1990). Hamrick and Loveless (1989) compared population genetic structures in 14 outcrossing species of trees and shrubs in Panama. The authors first rank-ordered the species with regard to predicted magnitudes of gene flow based primarily on pollen and seed dispersal mechanisms. Then, using allozyme markers, they found that empirical estimates of Nm (see Box 6.3) were sig-

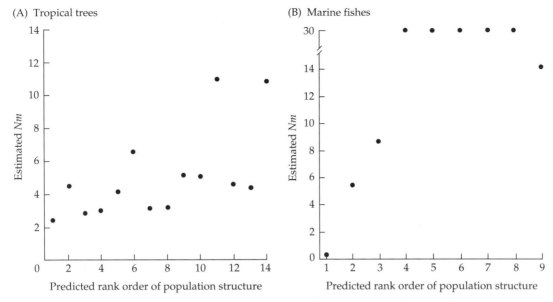

(A) Tropical trees

(B) Marine fishes

x-axis (both): Predicted rank order of population structure

y-axis (both): Estimated Nm

Figure 6.3   Relationship between gene flow and natural history dispersal potential. The relationship is shown for (A) fourteen species of tropical trees and shrubs (data from Hamrick and Loveless 1989) and (B) nine species of marine shore fishes (Waples 1987). Species are plotted in rank order according to the predicted magnitude of population structure based on suspected dispersal capabilities of pollen and seeds (plants) or larvae (fishes). Estimates of genetic exchange (Nm) came from allozyme data using Wright's F_{ST} approach (see Box 6.3) and, for both plants and fish, were significantly correlated with predicted dispersal rank ($P < 0.01$).

nificantly correlated with this ranking (Figure 6.3A). Thus, geographic population structure tended to be more pronounced in species with weakly flying pollinators and anemic mechanisms of seed dispersal. The few wind-pollinated species examined also had relatively strong population structures, consistent with earlier suggestions that wind may not be a particularly effective agent for pollen flow in low-density populations in tropical forests. Nonetheless, the correlations between reproductive biology and genetic structure were moderate at best, explaining considerably less than 50% of the overall variance in outcomes. Furthermore, over the spatial scales of several kilometers monitored, all species exhibited Nm greater than 1.0. This finding raises the more fundamental point that gene flow appeared to be moderate to high in most of these tropical forest trees.

Temperate zone trees, more of which are wind-pollinated, also commonly yield moderate to high estimates of gene flow. For example, among populations of pitch pine (*Pinus rigida*) throughout the species' range in the eastern United States, mean F_{ST} across allozyme loci was only 0.024 (Guries and Ledig 1982), a value associated with $Nm = 10.2$. Three other wind-pollinated pines that have been documented to exhibit low population differentiation across broad and continuous ranges are *P. banksiana* ($Nm = 6.7$;

Dancik and Yeh 1983), *P. contorta* (Nm = 8.1; Wheeler and Guries 1982), and *P. ponderosa* (Nm = 16.4; Hamrick et al. 1989). On the other hand, as might be expected, far greater spatial structure and more limited gene flow tend to be displayed by pine species whose populations are distributed as scattered isolates, such as *P. torreyana* (Nm = 0; Ledig and Conkle 1983), *P. halapensis* (Nm = 0.6; Scheller et al. 1985), and *P. muricata* (Nm = 1.0; Millar 1983).

Hamrick et al. (1992) reviewed published allozyme data on conspecific populations within more than 300 woody plant species. The overall mean estimate for the inter-population component of genetic diversity was G_{ST} = 0.085 (Nm = 2.7). However, only 16% of the heterogeneity in genetic structure across species could be accounted for by differences in seven life history and ecological traits considered. The authors concluded that other influences, including the idiosyncratic evolutionary histories of species, must have played important roles in determining how genetic diversity is partitioned. Another review of the allozyme literature for nearly 450 species of herbaceous as well as woody plants added breeding system and life form to the list of life history traits considered (Hamrick and Godt 1989). Across this wider taxonomic scale of comparison, the most important predictors of high population structure were selfing (as opposed to outcrossing) and annual (as opposed to perennial) life form. On average, G_{ST} = 0.357 (Nm = 0.45) for 146 annual species, and G_{ST} = 0.510 (Nm = 0.24) for 78 selfing species; whereas 131 long-lived woody perennials showed G_{ST} = 0.076 (Nm = 3.0) and 134 outcrossed wind-pollinated species showed G_{ST} = 0.099 (Nm = 2.3). Thus, plant species that self-fertilize and have short lives tend to exhibit considerably greater spatial structure than those with contrasting features.

MARINE GAMETES AND LARVAE.   Many marine invertebrates and fishes shed gametes and larvae into the water column in a manner analogous to the atmospheric release of pollen and seeds by many land plants. The time spent in this planktonic phase varies widely among species. For example, larvae of the sessile polychaete *Spirorbis borealis* remain free-living for only a few hours and are competent to settle immediately upon release from their parents, whereas polychaete larvae of the genus *Phyllochaetopterus* can delay settlement and metamorphosis for more than a year (Scheltema 1986). More typically, the larvae of benthic marine invertebrates are pelagic for a few days to weeks. Among marine fishes, larvae often drift in the plankton for weeks or months (Victor 1986), but the remarkable leptocephalus larvae of some eels may remain pelagic for 3 years or more (Castle 1984). Some marine invertebrates and fishes produce non-planktonic eggs or larvae. These less dispersive propagules may be demersal (found on or near the bottom), or they may be brooded by parents in oral cavities (as in marine catfishes), abdominal pouches (pipefishes and seahorses), or other storage sites (many invertebrates).

Are these widely varying potentials for gene flow via gametic and larval dispersal correlated with population genetic structure in marine species? Evidence suggests that they are, at least to some extent. Thus, for invertebrates, several genetic studies have reported a correspondence between increased potential for larval dispersal and diminished genetic differentiation among geographic populations (Ayre et al. 1997; Berger 1973; Crisp 1978; Gooch 1975; Liu et al. 1991). For example, a non-planktonic egg-casing snail, *Nucella canaliculata* (Sanford et al. 2003), and a larval-brooding snail, *Littorina saxatilis*, showed pronounced population structure in molecular markers that contrasted with the less structured pattern observed in a free-spawning marine snail, *L. littorea* (Janson 1987). In sea urchins of the genus *Heliocidaris*, one species (*H. tuberculata*) with a several-week planktonic larval stage showed little differentiation in mtDNA genotypes between populations separated by 1,000 km of open ocean, whereas populations of a congener (*H. erythrogramma*) with only a 3- to 4-day planktonic larval duration were strongly partitioned over comparable geographic scales (McMillan et al. 1992). In a solitary coral species that broods its larvae (*Balanophyllia elegans*), allozyme population structure along the California coast proved to be substantially greater than that in a co-distributed solitary coral species (*Paracyathus stearnsii*) with planktonic larvae (Hellberg 1996). Likewise, in comparative allozyme surveys of nine coral species in the genera *Acropora*, *Pocillopora*, *Seriatopora*, and *Stylophora*, population genetic structure along Australia's Great Barrier Reef usually (but not invariably) proved to be somewhat greater in brooding species than in broadcast spawners (Ayre and Hughes 2000).

Among the vertebrates, Pacific damselfishes with pelagic larvae showed allozyme uniformity over huge areas, whereas one assayed species that lacks a pelagic larval phase (*Acanthochromis polyacanthus*) was highly structured genetically (Ehrlich 1975; Planes and Doherty 1997). Another marine fish that lacks a pelagic phase, the black surfperch (*Embiotoca jacksoni*), likewise shows strong geographic population structure, as evidenced in this case by mtDNA (Bernardi 2000; Doherty et al. 1995). Waples (1987) assessed allozyme differentiation in several species of marine shore fishes sampled along the same geographic transect in the eastern Pacific and reported a strong negative correlation with dispersal capability as inferred from planktonic larval durations (Figure 6.3B): The species with the lowest potential for dispersal (a livebearer with no pelagic larval stage, *Embiotoca jacksoni*) exhibited the highest spatial genetic structure, whereas the species with the highest dispersal potential (a fish associated with drifting kelp and characterized by an extended larval duration, *Medialuna californiensis*) exhibited no detectable spatial genetic differentiation. Such results also appear to be generally consistent with the long-noted tendency for marine species with dispersive larvae to rapidly colonize oceanic islands and to exhibit broader geographic ranges than those with sedentary larvae (Jablonski 1986; Thorson 1961; but see Thresher and Brothers 1985 for exceptions).

Population genetic structures in North Atlantic eels have attracted particular interest because of the extraordinary catadromous life histories of these species (see review in Avise 2003b). Juvenile eels (*Anguilla rostrata* in the Americas, *A. anguilla* in Europe) inhabit coastal and inland waters for most of their lives, but during sexual maturation they migrate to the western tropical mid-Atlantic Ocean, where spawning takes place. Conventional wisdom (reviewed by Williams and Koehn 1984) was that conspecific larvae produced from each suspected mass spawn passively disperse via ocean currents to continental margins, perhaps settling at locales randomly oriented with respect to the homesteads of their parents. If mating is indeed quasi-panmictic and larval dispersal is passive, then all continental populations could represent nearly random draws from the species' gene pool, and accordingly would lack appreciable spatial genetic structure. Molecular data for *A. rostrata* and *A. anguilla* collected throughout their respective continental ranges are roughly consistent with this scenario. Several studies of *A. rostrata* from across eastern North America have documented a near or total absence of spatial structure in mtDNA and in polymorphic allozymes and microsatellite loci (Avise et al. 1986; Koehn and Williams 1978; Mank and Avise 2003; Williams et al. 1973; Wirth and Bernatchez 2003). For *A. anguilla* sampled across Europe, population genetic structure also appears slight (albeit statistically significant) at microsatellite loci ($F_{ST} = 0.002$) and in mtDNA (Lintas et al. 1998; Maes et al. 2002; Wirth and Bernatchez 2001).

In contrast, American and European eels are clearly distinct genetically, confirming the much-debated presence of at least two largely independent gene pools in the North Atlantic (see review in Avise 2003b). Additional genetic analysis also revealed the possible low-frequency presence of hybrids between *A. rostrata* and *A. anguilla* in Iceland (Avise et al. 1990b). This island is longitudinally intermediate to North America and Europe and is thousands of kilometers from where the zygotes presumably arose, so these genetic findings raise the intriguing possibility that hybrid larvae, if they truly exist (more definitive genetic data are needed), might have intermediate migratory behavior.

In general, long-duration planktonic larvae (as well as highly mobile adults in many marine taxa) afford opportunities for extensive gene flow, and such potential appears to have been realized in diverse species of marine invertebrates and vertebrates, as evidenced by a paucity of allozyme or mtDNA differentiation over vast areas. This is true, for example, among populations of several sea urchin species in the genera *Echinothrix* and *Strongylocentrotus* in long transects across parts of the Pacific Ocean (Lessios et al. 1998; Palumbi and Wilson 1990); among populations of rock lobster (*Jasus edwardsii*) across 4,600 km of Australasian habitat (Ovenden et al. 1992); in tiger prawns (*Penaeus monodon*) throughout the southwestern Indian Ocean (Forbes et al. 1999); in abyssal mussels (*Bathymodiolus thermophilus*) from hydrothermal vents scattered across the eastern Pacific (Craddock et al. 1995); within each of several species of Caribbean reef fishes from locales as much as 1,000 km apart (Lacson 1992; Shulman and Bermingham 1995); in walleye pollack (*Theragra chalcogramma*) sampled throughout the Bering Sea (Shields and Gust 1995); among damselfish (*Stegastes fasciolatus*) populations throughout the 2,500-km Hawaiian archipelago (Shaklee 1984); among milkfish (*Chanos chanos*) populations from

Pacific locales as far apart as 10,000 km (Winans 1980); among populations of orange roughy (*Hoplostethus atlanticus*) on a global scale (Smith 1986); and within each of several widespread or cosmopolitan species of pelagic billfishes and tunas (see reviews in Graves 1996, 1998).

This relative dearth of population structure in many (though certainly not all) marine species over large areas becomes especially evident when comparisons are made with related freshwater species surveyed similarly. For example, among open-water copepods, higher genetic patchiness has been reported among pond populations of *Diaptomis leptopus* in Quebec than in a pelagic marine species (*Calanus finmarchicus*) sampled across the North Atlantic Ocean (Bucklin and Kocher 1996; Bucklin et al. 1998; Kann and Wishner 1996). In fishes, as a general rule, species that inhabit fresh water or are anadromous (migrate from the sea to spawn in streams or lakes) often display higher population genetic structure than counterpart taxa that inhabit the sea or are catadromous, even when the latter have been assayed on much larger spatial scales (see reviews in Avise 2000a; Ward et al. 1994). Such outcomes are hardly surprising, given the contrasting physical structures of freshwater versus marine environments and the oft-differing lifestyles of creatures inhabiting these realms.

On the other hand, many marine species with pelagic larvae *have* exhibited dramatic population differentiation over mesogeographic or macrogeographic scales (Avise 1987, 2000a; Burton 1983, 1986; Hedgecock 1986; Palumbi 1996b). In the stomatopod *Haptosquilla pulchella*, for example, sharp genetic breaks in mtDNA separate regional populations in the Indo-West Pacific (Barber et al. 2002). In the horseshoe crab (*Limulus polyphemus*), a dramatic distinction in mtDNA was reported between two continuously distributed populations on the Gulf of Mexico and Atlantic coasts of the southern United States (Saunders et al. 1986), despite the presence in this species of trilobite larvae that are "specialized for dispersal" (Rudloe 1979). Analogous spatial genetic patterns have been observed in this same U.S. geographic region for several other maritime (coastal) invertebrate and vertebrate species (see below). Such cases appear to register long-standing separations between regional population assemblages, an aspect of historical population structure that will be deferred to the upcoming section on phylogeography.

Another form of population structure in marine species can arise in contemporary time and at microspatial scales. Sometimes termed "chaotic patchiness," this type of genetic structure may be ephemeral and is probably attributable to idiosyncrasies among local sites in histories of larval recruitment, perhaps coupled in some cases with strong ecological selective pressures favoring particular genotypes. From extensive molecular genetic evidence, particularly on Pacific oysters (*Crassostrea gigas*), Hedgecock and colleagues (1992) have persuasively argued that reproduction in high-fecundity species in the sea is a "sweepstakes" process entailing huge variances across families in contributions to the local gene pool, owing to the difficulties marine animals face in matching their reproductive activities with ephemeral oceanographic conditions conducive to gamete maturation, spawning, fertilization, larval development, and larval settlement. Various studies using molecular markers have supported several empirical predictions about sweepstakes reproduction: the occurrence of short temporal variation that can equal or surpass fine-scale spatial variation in population genetic structure (Hedgecock 1994a,b); the presence of larval cohorts that can differ considerably from one another in genetic composition; and the occurrence of significantly less genetic diversity within each cohort than in the total adult population (Li and Hedgecock 1998).

Chaotic patchiness may characterize many marine species, such as a rock-pool-inhabiting copepod (*Tigriopus californicus*) that displays strong local (as well as regional) differentiation in allozymes and mtDNA, notwithstanding a life cycle that includes free-swimming adults and larvae (Burton 1998; Burton and Feldman 1981; Burton and Lee 1994; Burton et al. 1979). Other marine invertebrates that display local genetic heterogeneity despite the presence of pelagic larvae, perhaps due in part to sweepstakes reproduction, include American lobsters (*Homarus americanus*) (Tracey et al. 1975), intertidal limpets in the genus *Siphonaria* (Johnson and Black 1982), *Echinometra* and *Strongylocentrotus* sea urchins (Edmands et al. 1996; Watts et al. 1990), and many others (Palumbi 1995). From an early compilation of examples documented by molecular markers, Burton (1983) concluded that although invertebrate species with planktonic larvae tend to show less spatial heterogeneity than those with non-motile larvae, "relationships between length of planktonic larval life and the geographic boundaries of panmictic populations are not strongly supported. In particular, substantial differentiation has been observed in several species that appear to have high dispersal capabilities."

There are additional reasons why a high dispersal potential of gametes or larvae may not always translate into spatial population genetic homogeneity and attendant high estimates of gene flow (Hedgecock 1986). First, actual levels of gene flow may be lower than presumed because of physical impediments to larval movement. Second, larvae may not always be the passive dispersal agents commonly assumed, but rather in some species may adopt active migration behaviors and settlement choices. Crisp (1976) and Woodin (1986) reviewed the evidence for discriminatory larval settlement in benthic intertidal and infaunal marine invertebrates, respectively, and found that larvae often fall far short of their dispersal potential. For example, in the shrimp *Alphaeus immaculatus*, a species in which adults live symbiotically with sea anemones but produce free-swimming larvae that can remain pelagic for an extended time, a detectable proportion of surveyed recruits nonetheless settled on anemone colonies within a few meters of their parents (Knowlton and Keller 1986).

Recent studies based on compositional analyses of trace elements in fishes' bodies (Swearer et al. 1999), mark–recapture experiments (G. P. Jones et al. 1999), theoretical treatments of diffusion processes in the sea (Cowen et al. 2000), and possible effects of oceanic currents on gene flow (Barber et al. 2000) have all demonstrated that contemporary generation-to-generation connections among populations of many species with planktonic larvae may be far lower than traditionally supposed. At least over time frames of perhaps hundreds to thousands of generations, most larval recruitment may be at a sufficiently local scale to permit substantial genetic differentiation among conspecific populations. One final example involves the cleaner goby (*Elacatinus evelynae*), which, despite an extended pelagic duration of 21 days, showed strong geographic structure in mtDNA across the Caribbean (Taylor and Hellberg 2003).

Conversely, diversifying natural selection acting on particular loci via differential survival or mating success might sometimes convey a false impression of low gene flow among highly connected populations. For example, in the blue mussel (*Mytilus edulis*), allele frequencies at a leucine aminopeptidase (*Lap*) allozyme locus are significantly heterogeneous spatially, but are strongly correlated with environmental salinity. Physiological and biochemical studies have indicated that these alleles function differentially in relieving osmotic stress in environments of varying salinity via their influence on the free amino acid pools and volumes of cells (Hilbish et al. 1982). Thus, frequencies of these non-neutral *Lap* alleles probably say more about environmental conditions than about the gene flow regime of the species (Boyer 1974; Koehn 1978; Theisen 1978). At other polymorphic allozyme loci, these same mussel populations exhibited large, moderate, and small inter-population variances in allele frequencies (Koehn et al. 1976), such that estimates of gene flow under assumptions of neutrality differed considerably across genes.

A sobering example of how different genetic markers can sometimes paint contrasting pictures of gene flow involves populations of the American oyster (*Crassostrea virginica*) from the Gulf of Mexico and Atlantic coasts of the southeastern United States. Surveys of polymorphic allozymes revealed a near uniformity of allele frequencies throughout this range (Figure 6.4A), a result understandably attributed to high gene flow resulting from "the rather long planktonic stage of larval development, since this species has the ability to disperse zygotes over great distances when facilitated by tidal cycles and oceanic currents" (Buroker 1983). However, mtDNA genotypes revealed a dramatic genetic "break," involving cumulative and nearly fixed mutational differences that cleanly distinguished most Atlantic from Gulf oyster populations (Reeb and Avise 1990). Subsequent surveys of nuclear DNA markers tended to support the dramatic Atlantic/Gulf mtDNA dichotomy (Karl and Avise 1992; Hare and Avise 1996, 1998; Figure 6.4B) and thus seem to eliminate differences in dispersal of male versus female gametes as a likely explanation for the contrasting population structures registered by allozymes and mtDNA. One possibility is that some of the allozyme loci surveyed may be under uniform balancing selection and thus do not register the population subdivision that seems clearly evidenced by multiple DNA markers in the nucleus and cytoplasm (Karl and Avise 1992). This suggestion may also be consistent with the long-

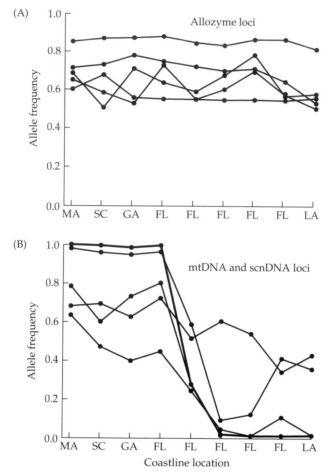

Figure 6.4   Allele frequencies in oyster populations along a coastal transect from Massachusetts through South Carolina, Georgia, Florida, and Louisiana. Shown are frequencies of the most common alleles at (A) five polymorphic allozyme loci: *Est1*, *Lap1*, *6Pgd*, *Pgi*, and *Pgm* (data from Buroker 1983), and (B) five loci assayed at the DNA level: mtDNA (heavy line; data from Reeb and Avise 1990) and four anonymous single-copy nuclear genes. (After Karl and Avise 1992.)

standing observation that allozyme heterozygosities in mollusks are strongly associated with presumed fitness components such as metabolic efficiency and growth rate (Garton et al. 1984; Zouros et al. 1980; see also Hare et al. 1996). Whether this explanation or its converse (that allozymes faithfully register high gene flow in oysters, but mtDNA and some nDNA markers differ between the Atlantic and Gulf coasts because of diversifying selection) is correct, the conclusion is that natural selection probably has acted on at least some of the genetic markers. This finding underlines the ever-present need for caution in inferring population structure and gene flow under an assumption of selective neutrality for all molecular markers.

Direct estimates of dispersal distances

In some cases, molecular markers that are rare or unique have been employed to monitor gene dispersal from known point sources. Such "genetic branding" approaches (Ferris and Berg 1986) are similar in concept to traditional labeling and tracking studies based on non-heritable markers such as physical tags, fluorescent dyes, radiotracers, morphological characters, and parasite loads (see review in Levin 1990), but with at least two added benefits: Microscopic species and even gametes (as well as large organisms traditionally studied by physical tags) can be analyzed by molecular markers, and the genetic tags can be monitored across multiple generations.

PLANTS.   Schaal (1980) monitored effective pollen flow in the Texas bluebonnet (*Lupinus texensis*) by assaying for heterozygotes among F_1 progeny in an experimental population in which an allozyme-marked pollen donor was surrounded by plants of an alternate genotype. She found that gene dispersal via pollen was quite restricted (a few meters), with most movement of pollen genes being to neighboring plants. Nonetheless, this gene movement was significantly greater than might have been inferred from direct observations of inter-plant flight distances by the pollinators (bees), apparently because of "pollen carryover," wherein some fraction of the pollen a bee deposits on a stigma comes from flowers that it visited prior to its most recent stop. Such studies of allelic movement and gene flow that involve experimental manipulations of genetic markers represent another application of genetic parentage analysis in natural plant populations (see Chapter 5).

ANIMALS.   Grosberg (1991) monitored effective sperm dispersal in the marine ascidian *Botryllus schlosseri* by placing allozyme-marked colonies on the pilings of a harbor dock and later assaying brooded embryos within nearby natural colonies for presence of the introduced allele (signaling fertilization by dispersed sperm). From a rapid decline in fertilization success observed beyond about 50 cm, Grosberg concluded that effective sperm dispersal in this species is extremely limited. In a conceptually similar experiment, Burton and Swisher (1984) introduced copepods (*Tigriopus californicus*) carrying rare allozyme alleles into several natural tide pool populations on rocky California headlands. What ensued was an initial decline in the frequency of the introduced alleles, followed within 6 weeks, however, by the spread of the marker alleles to non-recipient pools on the same rocky outcrop. Within 8 months, all populations on an outcrop were nearly homogeneous in allele frequency, indicating the cumulative effects of inter-pool gene flow over these small spatial scales. Similarly, Dillon (1988) introduced freshwater snails (*Goniobasis proxima*) carrying unique allozyme markers into isolated streams in the southern Appalachians of Virginia. Few introductions were successful (two of twelve), but from those that were, the introduced alleles spread at a rate of about 15–20 meters per year upstream and 5–10 meters per year downstream.

These genetic findings confirmed a suspected natural history feature of this species: its behavioral tendency to crawl against the current.

Nonetheless, several limitations attend attempts to monitor dispersal directly by genetic markers. First is the problem of generating or finding unique alleles that are not also present in the natural population to which the markers are to be introduced. A second difficulty is the possibility that the genetic markers themselves, or perhaps the artificial breeding programs often employed to generate them, might impair or otherwise modify the dispersal capabilities of the propagules in comparison to the natural state. A third possibility is that the dynamics of introduced alleles might be influenced by natural selection acting on the markers themselves (or their genetic backgrounds), perhaps more so than by the dispersal and gene flow regimes that usually are of primary interest in such studies. Finally, a great logistic challenge is the problem of monitoring long-distance movement of potentially vagile markers due to the large spatial scales involved and the inevitable effects of dilution from a point source (Jones et al. 1981). Undetected long-distance gene flow, even if rare, can have a significant long-term homogenizing influence on the genetic structure of a species.

Vagility, philopatry, and dispersal scale

Particularly in animals with non-dispersive gametes and larvae, adult vagility and habitat "grain" (the geographic scale at which organisms "perceive" environmental patchiness; Levins 1968) are major influences on gene flow and population structure. To a land snail, for example, parking lots and streets may be formidable barriers to dispersal, and allozyme analyses of *Helix aspersa* proved that colonies within and among city blocks do indeed exhibit significant population genetic structure (Selander and Kaufman 1975b). At the other extreme, many marine turtles are oceanic travelers that swim many thousands of kilometers during a lifetime, and nDNA assays of green turtles (*Chelonia mydas*) have documented that their population genetic structure throughout an entire ocean basin is roughly comparable in magnitude to that among *Helix* populations on adjacent city blocks (mean $F_{ST} \cong$ 0.13; Karl et al. 1992).

Although the spatial scale of potential gene flow is clearly influenced by an animal's mobility, apparent population genetic structure is not inextricably tied to organismal vagility for several reasons, including the presence of physical or ecological barriers that overmatch dispersal capability; behavioral considerations such as social alliances, habitat choice, or philopatry (site faithfulness) with regard to reproduction; gender-biased dispersal and gene flow patterns (which especially can affect molecular markers with sex-limited transmission); influences of natural selection on particular genetic markers (or loci linked to them); and historical demographic events that may have removed populations from expectations of evolutionary equilibrium (e.g., between gene flow and genetic drift). These theoretical possibilities will be empirically illustrated, in turn.

PHYSICAL DISPERSAL BARRIERS.   Bluegill sunfish (*Lepomis macrochirus*) are active swimmers, abundant throughout their freshwater range in North America. An allozyme survey of 2,560 specimens divided equally among 64 localities (eight sites per reservoir, four reservoirs in each of two adjacent river drainages) revealed that about 90% of the total allele frequency variance occurred between reservoirs in a drainage, whereas within reservoirs (which ranged in size up to more than 100,000 acres) allele frequencies seldom were significantly heterogeneous (Avise and Felley 1979). Clearly, the subdivided structure of the physical environment (reservoirs separated by dams) had imposed a corresponding genetic structure on these otherwise highly mobile fish.

Gyllensten (1985) reviewed allozyme literature on geographic population structure within each of 19 fish species characterized by lifestyle: strictly freshwater, anadromous, and marine. The average percentages of total intraspecific gene diversity that were distributed among locales (as opposed to within them) increased dramatically in the following order by habitat: marine taxa (1.6%), anadromous species (3.7%), and freshwater species (29.4%). Thus, differences in spatial distributions of genetic variability generally coincided with qualitative differences in the occurrence of obvious geographic barriers to movement. Few trends in population genetics are without exception, however, and molecular analyses of several marine and anadromous species have sometimes documented levels of range-wide population structure that are quite comparable to those typifying many freshwater fish species (e.g., Avise et al. 1987b; Bowen and Avise 1990).

In a flightless water strider (*Aquarius remigis*) that migrates by rowing on water surfaces, an allozyme survey by Preziosi and Fairbairn (1992) revealed that whereas populations distributed along a given stream are nearly undifferentiated ($F_{ST} = 0.01$), those inhabiting different streams in a watershed are highly structured ($F_{ST} = 0.46$). By contrast, a water strider species (*Limnoporus canaliculatus*) with functional wings exhibited nearly homogeneous allele frequencies throughout several Atlantic seaboard states (Zera 1981). An mtDNA survey has also been conducted on open-ocean water striders, or sea-skaters (*Halobates* spp.), one of the few insect groups to have invaded the marine environment. Although the data are not extensive, they suggest that population genetic structure in these species may be partitioned primarily on the spatial scale of large oceanic regions (Andersen et al. 2000). Collectively, these available results on aquatic Hemiptera suggest that inherent dispersal capacities, in conjunction with the physical nature of the environment, exert a huge influence on species' population genetic structures.

On the other hand, genetic comparisons of population structure in five species of carabid beetles revealed no correlation with degree of flight-wing development (ranging from vestigial to fully winged). A positive correlation was noted, however, between F_{ST} values and the elevations of the collecting sites (Liebherr 1988), suggesting in this case that habitat fragmentation (of highland sites) is more important than dispersal capability alone in molding population genetic structures in these beetles.

Numerous other species likewise occupy discontinuous habitats and may show significant population genetic structures related to environmental patchiness, which often overrides their normal dispersal capabilities. For example, troglobitic (obligate cave-dwelling) crickets in the genera *Hadenoecus* and *Euhadenoecus* were shown to exhibit greater allozyme population structure in their isolated pockets of habitat (different cave systems) than did their epigean (surface-dwelling) counterparts (Caccone and Sbordoni 1987). Similarly, for mice on islands, even narrow ocean channels must be huge hurdles to dispersal, and small island populations of *Peromyscus* do indeed often show less within-population variability and greater between-island genetic differences than do their mainland counterparts over comparable geographic scales (Ashley and Wills 1987, 1989; Avise et al. 1974; Selander et al. 1971). For any habitat specialist, suitable environments may be scattered. To pick one more setting as a final example, granite outcrops are scattered across the southeastern United States like small islands in a matrix of mesophytic forest. They house several endemic species, such as the beetle *Collops georgianus*, whose populations proved to display far more pronounced genetic structure among outcrops ($F_{ST} = 0.19$) than within them ($F_{ST} = 0.01$) (King 1987). Similar population genetic patterns have been documented by molecular markers in a variety of plant species endemic to these isolated patches of rock (Wyatt et al. 1992; Wyatt 1997).

PHILOPATRY TO NATAL SITE.   Each reproductive season, female marine turtles typically migrate hundreds or thousands of kilometers from foraging grounds to nesting locales, where they deposit eggs on sandy beaches. For example, green turtles (*Chelonia mydas*) that nest on Ascension (a small, isolated island on the mid-Atlantic oceanic ridge) otherwise inhabit feeding pastures along the coast of Brazil, some 2,000 km distant. From repeated captures of physically tagged adults, it was long known that green turtles exhibit strong nest site fidelity; that is, Ascension females nest on Ascension and nowhere else, Costa Rican and Venezuelan nesters are faithful to their respective rookeries, and so on. What remained unknown was whether the site to which a female is fidelic as an adult was also her natal rookery. If female "natal homing" prevails, most rookeries should exhibit clear genetic differences from one another with regard to matrilines (and hence mtDNA), even if appreciable inter-rookery exchange of nuclear genes occurs via the mating system and male-mediated gene flow (Karl et al. 1992). In the first genetic surveys of green turtle rookeries around the world (Bowen et al. 1992; Meylan et al. 1990), a fundamental split in mtDNA genealogy was found to distinguish all surveyed specimens in the Atlantic–Mediterranean from those in the Indian–Pacific, and pronounced genetic substructure also proved to characterize rookeries within each ocean basin (Figure 6.5). Indeed, distinct mtDNA haplotypes completely (or nearly) distinguished many pairs of nesting colonies within an ocean basin, a finding indicative of a strong propensity for natal homing by nesting females.

Following these pioneering studies, similar population genetic surveys have been conducted on most of the world's seven or eight marine turtle species (see reviews in Avise 2000a; Bowen and Avise 1996; Bowen and Karl 1997). These surveys include further molecular analyses of green turtles (e.g., Encalada et al. 1996) as well as hawksbills (*Eretmochelys imbricata*; Broderick et al. 1994; Bass et al. 1996), loggerheads (*Caretta caretta*; Bowen et al. 1993b, 1994; Encalada et al. 1997), and ridleys (*Lepidochelys* species; Bowen et al. 1998). The empirical findings, often qualitatively paralleling those described above, exemplify how even some of the world's most highly mobile species nonetheless can display dramatic matrilineal population structures, due in this case to both geographic constraints (e.g., physical barriers between oceans) and inherent natal homing behaviors (to particular rookeries within oceanic basins).

Whales too are impressive mariners, normally traveling many thousands of kilometers seasonally. Several analyses of mtDNA (and nDNA) from skin biopsies of humpback whales (*Megaptera novaeangliae*) sampled globally have found genetic differences between various groups, including those previously reported to show distinct migration routes within an ocean basin between summer feeding grounds in subpolar or temperate environs and winter breeding areas in the tropics (Baker et al. 1990, 1993, 1994, 1998; Larsen et al. 1996; Palsbøll et al. 1995, 1997b). Such spatial partitioning of matrilineal genotypes appears due in large part to female-directed fidelity to specific migratory destinations. Several other cetacean species have similarly been shown to be subdivided into matrilineal groups through which cultural traditions are passed (Whitehead 1998). These results illustrate how social behaviors can be another factor promoting population genetic structure in highly mobile marine animals.

Salmon are active and powerful swimmers, but also are notorious for suspected natal homing propensity, in this case by both sexes. In anadromous forms of these species, juveniles spawned in freshwater streams migrate to the sea before returning to their natal stream years later as adults to complete the life cycle. Numerous surveys of nuclear genes (e.g., via allozymes, microsatellites) and mtDNA from both Atlantic and Pacific species have revealed significant genetic differences among spawning populations at various microspatial, mesospatial, and macrogeographic scales (some early studies were by Billington and Hebert 1991; Ferguson 1989; Gyllensten and Wilson 1987a; Ryman 1983; and Ståhl 1987). Small or modest allele frequency shifts often characterize spawning populations within and among nearby drainages (e.g., Banks et al. 2000; Laikre et al. 2002; J. L. Nielsen et al. 1997; Scribner et al. 1998; G. M. Wilson et al. 1987), or even

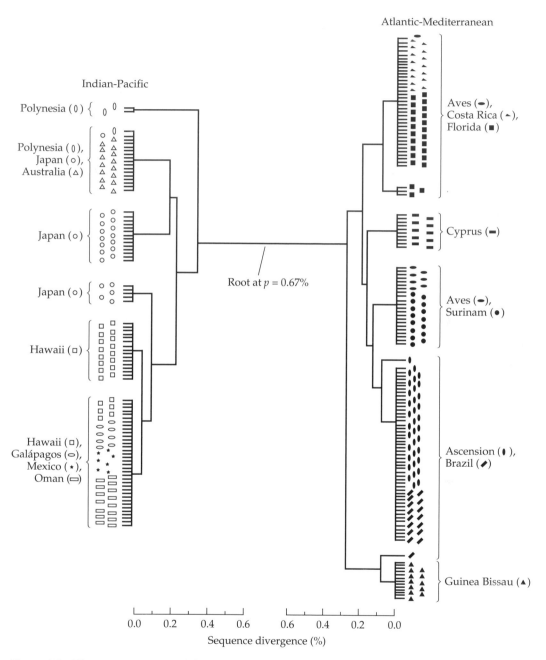

Figure 6.5   Phenogram summarizing relationships among 226 sampled nests of the green turtle. To conserve space, sequence divergence (*p*) axes on the bottom are presented as mirror images centered around the root leading to two distinct clonal assemblages (Atlantic-Mediterranean versus Indian-Pacific ocean basins). (After Bowen et al. 1992.)

populations within a drainage that spawn at different times of the year (Fillatre et al. 2003). Additional and sometimes fixed genetic differences often distinguish conspecifics from more distant regions (e.g., Bermingham et al. 1991; Bernatchez and Osinov 1995; Davidson et al. 1989). The relative paucity of local differentiation probably is due to occasional "homing mistakes" that underlie gene flow between otherwise isolated spawning areas, as well as the fact that most high-latitude streams currently utilized by salmon were colonized within the last few thousand years following glacial retreats. Hence, nearby populations are often tightly connected historically, notwithstanding strong propensities for natal homing. The relatively greater extent of genetic differentiation between regional assemblages presumably is due to longer-term population separations related to historical biogeography (in collaboration with ongoing natal homing trends).

Avian taxa present a special enigma with regard to magnitude and pattern of population structure. On the one hand, most birds have high dispersal potential (because of flight, and often a migratory propensity), and many species have broad geographic distributions, suggesting that gene flow may be high and population differentiation minimal. On the other hand, many species exhibit strong tendencies toward nest site philopatry and are composed of recognized subspecies that display obvious geographic variation in body size, song, or plumage (Newton 2003), suggesting that inter-population gene flow might be low. Barrowclough (1983) reviewed the early allozyme literature on geographic variation among populations within 57 vertebrate species. The surveyed birds (primarily temperate zone passerines) exhibited minor geographic differentiation (mean F_{ST} = 0.02), compared with mean F_{ST} estimates of 0.11, 0.23, 0.30, and 0.38 for fishes, mammals, reptiles, and amphibians, respectively. These results were attributed in part to the inherent high vagility of birds compared with more sedentary animals such as salamanders and rodents (but see Barrowclough 1983 for a different interpretation).

Subsequent research employing more discriminatory mtDNA assays (Avise and Zink 1988) revealed that avian species exhibit a wide variety of population genetic structures (see reviews in Avise 2000a; Crochet 2000). Some species, such as the downy woodpecker (*Picoides pubescens*; Avise and Ball 1991; Ball and Avise 1992), pintail duck (*Anas acuta*; Cronin et al. 1996), greenfinch (*Carduelis chloris*; Merilä et al. 1997), red knot (*Calidris canutus*; Baker and Marshall 1997; A. J. Baker et al. 1994), and bluethroat (*Luscinia svecica*; Questiau et al. 1998), show minimal or restrained mtDNA differentiation across their respective ranges (sometimes continent-wide). Other avian species, such as the seaside sparrow (*Ammodramus maritimus*; Avise and Nelson 1989), Canada goose (*Branta canadensis*; Van Wagner and Baker 1990), fox sparrow (*Passerella iliaca*; Zink 1991), LeConte's thrasher (*Toxostoma lecontei*; Zink et al. 1997), Adelaide's warbler (*Dendroica adelaidae*; Lovette et al. 1998), flightless brown kiwi (*Apteryx australis*; Baker et al. 1995), and many others (Zink et al. 1995; Zink 1997) show deep historical subdivisions in intraspecific mtDNA genealogy. Superimposed on these deep partitions may be shallower within-region genetic separations presumably reflective of more recent or contemporary ecological factors (e.g., Hare and Shields 1992; Zink 1991).

When an avian species is subdivided into *highly distinct* mtDNA clades, there typically have proved to be only about two to six such principal historical units per species, and they often tend to align quite well with biogeographic or morphological partitions (Avise and Walker 1998). The pronounced spatial structure in some avian species also permits genetic analysis of migratory behaviors. For example, the dunlin sandpiper (*Calidris alpina*)—a long-distance migrant that nests at high northern latitudes—proved by mtDNA evidence to be historically subdivided into several primary matrilineal groups, each with a different regional breeding distribution (Wenink et al. 1993, 1996). Genetic analyses of dunlins captured on migratory and wintering grounds then helped to reconstruct migration pathways (Wennerberg 2001), revealing, for example, that birds from different breeding regions sometimes occur in mixed assemblages at these other times of year (Wenink and Baker 1996). Similar analyses of several other species of migrant shorebirds have been conducted using nuclear markers (Haig et al. 1997).

GENDER-BIASED DISPERSAL AND GENE FLOW.   An animal's degree of faithfulness to its natal site or social group often is gender-dependent. For example, as described earlier, lion prides consist of related females, and most inter-pride movement is by males (Packer et al. 1991; Schaller 1972). The same general pattern holds in various cetacean species (see above) and in some bats (Petri et al. 1997) and primates (Rosenblum et al. 1997), which often tend to display pronounced population structure in maternally inherited mtDNA. Conversely, scrub jays (*Aphelocoma coerulescens*) live in extended family groups built around patrilineal kinship. Young jays often remain at the nest as helpers, accession of territories is patrilineal, and most dispersal (albeit over rather short distances) is by females (Woolfenden and Fitzpatrick 1984). Population genetic structure in this species across Florida, as estimated from microsatellite data, is high ($G_{ST} = 0.48$; $Nm = 0.27$), presumably due to these cohesive social behaviors and also to a patchy distribution of suitable habitats (McDonald et al. 1999).

In general, most mammalian species with asymmetric philopatry exhibit male-biased dispersal, whereas most such avian species exhibit female-biased dispersal (P. J. Greenwood 1980; Greenwood and Harvey 1982). Asymmetric dispersal by gender may mean that a species can exhibit qualitatively distinct patterns of geographic population structure at genes with biparental transmission (most nuclear loci) versus those with uniparental transmission (e.g., the Y chromosome of mammals, the W chromosome of birds, or mtDNA). Although several Y- and W-linked genes were identified long ago in various mammalian and avian species, respectively (Casanova et al. 1985; Ellis et al. 1990; Page et al. 1985; Rabenold et al. 1991; Rasheed et al. 1991; Whisenant et al. 1991), they usually have proved to be insufficiently variable at the intraspecific level to be of special genealogical utility (Dorit et al. 1995; Maynard Smith 1990; but see Bishop et al. 1985). Thus, most molecular analyses of gender-biased dispersal have relied instead on mtDNA data (indicative of matrilineal history) interpreted in conjunction with population genetic information from autosomal markers such as allozymes or microsatellites (indicative of biparental histories).

A case in point involves macaque (*Macaca*) monkeys, in which mirror-image patterns of geographic variation have been reported in nuclear-encoded allozymes versus mtDNA. Male macaques typically leave their natal group before reaching sexual maturity, whereas females remain for life. Melnick and Hoelzer (1992) reviewed the literature on molecular variation in several macaque species (*M. fascicularis*, *M. mulatta*, *M. nemestrina*, and *M. sinica*) and reported patterns of geographic population structure that are consistent with these gender-specific behaviors. For example, in the nuclear genome of *M. mulatta*, only 9% of total intraspecific diversity proved attributable to variation among geographic locales, whereas 91% of overall diversity in the mitochondrial genome occurred between populations. Thus, spatial genetic patterns registered by these two genomes are "intimately linked to the asymmetrical dispersal patterns of males and females and the maternal inheritance of mtDNA" (Melnick and Hoelzer 1992). Furthermore, as shown by subsequent DNA sequence analyses, bifurcations in macaque mtDNA gene trees typically predate Y chromosome divergences at the same phylogenetic nodes, as might be expected for these female-philopatric animals (Tosi et al. 2003).

A similar genetic pattern of extreme sex-biased dispersal has been reported in a communally breeding, nonmigratory bat (*Myotis bechsteinii*). Based on a comparison of mitochondrial and nuclear microsatellites, almost complete separation was uncovered in mtDNA markers due to near-absolute female philopatry, despite extensive male dispersal that had produced only a weak (albeit statistically significant) population genetic structure at nuclear loci (Kerth et al. 2002).

Molecular studies of at least one avian species have reported exactly the opposite population genetic pattern, suggesting sex-biased dispersal in favor of females (rather than males). In red grouse (*Lagopus lagopus*) from northeastern Scotland, molecular analyses of 14 populations revealed significant spatial structure in nuclear microsatellite markers but not in mtDNA (Piertney et al. 2000). Although at first thought these findings might seem contradictory (because female-mediated gene flow would move nuclear as well as mitochondrial markers), the authors identified theoretical models under which these outcomes are plausible, provided that specifiable differences in the dispersal and ecology of males versus females are such that local effective population sizes of the nuclear genome (more so than the mitochondrial genome) are severely reduced (see Piertney et al. 2000).

Another possible example of distinctive genetic signatures resulting from gender-based differences in behavior involves the green turtle (*Chelonia mydas*). As already mentioned, most rookeries within an ocean basin are strongly isolated with regard to mtDNA lineages (mean inferred

$Nm \cong 0.3$), indicating a strong propensity for natal homing by females (Bowen et al. 1992). However, these same rookeries proved to be somewhat less differentiated at assayed nuclear loci (mean $Nm = 1.7$; Karl et al. 1992), perhaps because of occasional male-mediated gene flow. Green turtles are known to mate at sea, often on feeding grounds or other locales spatially removed from the nesting sites. Thus, inter-rookery matings could provide an avenue for nuclear gene exchange that largely is closed to mtDNA because of female natal homing. Other marine organisms that have shown matrifocal genetic arrangements and contrasting population structures in cytoplasmic versus nuclear loci include not only various cetaceans (as already mentioned) but also some species of pinnipeds (sea lions and allies). The southern elephant seal (*Mirounga leonina*), for example, displays significantly greater population structure in mtDNA markers than in nDNA markers across the southern oceans (Hoelzel et al. 2001; Slade et al. 1998).

In interpreting such molecular contrasts, one potential complication is the theoretical fourfold lower effective population size for uniparentally than for biparentally transmitted genes. Thus, all else being equal, mtDNA is more subject to genetic drift effects, which also can promote the emergence of salient population structure. This factor can be taken into account in data analyses, as illustrated by Wilmer et al. (1999) when they documented higher population subdivision in the Australian ghost bat (*Macroderma gigas*) for mtDNA than for nuclear markers even after factoring in the expected difference in N_e for these two sets of genes. Nonetheless, whenever possible, it is also desirable to gauge dispersal by the two sexes not only indirectly via assessment of population genetic structure, but also from more direct observational evidence (as has been done for southern elephant seals; Fabiani et al. 2003). As described next, comparisons between indirect (genetic) and direct contemporary appraisals of dispersal in the field are often far more informative than either form of evidence interpreted in isolation.

Many waterfowl provide exceptions to the prevalent pattern of male-biased philopatry in birds. In the lesser snow goose (*Chen caerulescens*), as in some other migratory waterfowl, pair formation occurs on wintering grounds, where birds from different nesting areas often gather in mixed assemblages. Then a mated pair normally returns to the female's natal or prior nesting area. Among all avian species for which direct banding returns are available (Cooke et al. 1975), according to P. J. Greenwood (1980), "the lesser snow goose is the best documented example of male biased natal and breeding dispersal." This natural history pattern suggests considerable inter-colony gene flow mediated by males, an expectation consistent with results of both allozyme (Cooke et al. 1988) and nRFLP studies (Quinn 1988; Quinn and White 1987). This behavior also suggests that colonies should be isolated with regard to matriarchal lineages, but surprisingly, this has not proved to be the case. In an mtDNA survey of 160 geese from colonies across the breeding range (from Russia to the eastern Canadian Arctic), no significant differences were observed in the spatial frequencies of two major mtDNA clades, a result indicative of considerable population connectivity and gene flow involving females (Avise et al. 1992b; Quinn 1992). One likely explanation is that the entire current range of the snow goose was colonized recently from expansion out of Pleistocene refugia, where separation between the two mtDNA clades may have been initiated. A related possibility is ongoing gene flow, either via occasional lapses in philopatry by females (a phenomenon that has been documented by direct banding returns) or via episodic pulses of mass movement of individuals during periods of colony perturbation (also suspected from field observations). Whatever the process, snow goose colonies must have been in recent matrilineal contact notwithstanding the propensity for natal philopatry by females.

From these comparisons of banding and genetic data for snow geese, two important object lessons emerged: that direct behavioral or marking studies on contemporary populations can in some cases provide a misleading picture of the geographic distributions of genetic traits because they fail to reveal the important evolutionary aspects of population connectivity revealed in genes; and conversely, that geographic distributions of genetic markers can in some cases provide a misleading picture of contemporary dispersal and gene flow because they retain a record of evolutionary events and demographic parameters that may differ from those of the present. Thus, a full appreciation of geographic population structure in any species requires an integration of evolutionary (genetic) and contemporary (behavioral) perspectives.

It is also true, however, that some waterfowl populations have shown striking matrilineal differentiation. In the spectacled eider (*Somateria fisheri*), mtDNA markers revealed much higher regional population structure than did sex-linked and autosomal microsatellite loci (Scribner et al. 2001b). From these genetic data, the authors estimated that per generation rates of interregional gene flow were almost 35 times greater for males than for females (1.28×10^{-2} and 3.67×10^{-4}, respectively). Male-biased dispersal and gene flow have also been genetically deduced in some passeriform species, such as the yellow warbler (*Dendroica petechia*; Gibbs et al. 2000b) and red-bellied quelea (*Quelea quelea*; Dallimer et al. 2002).

Invertebrates also have be the subject of critical molecular analyses of gender-asymmetric dispersal. Africanized "killer" bees are aggressive forms of *Apis mellifera* that spread rapidly in the New World following the introduction of African honeybees into Brazil in the late 1950s. Two competing hypotheses were advanced for their mode of spread and the composition of their colonies. Perhaps queens are sedentary, such that most of the geographic expansion in aggressive behavior has resulted from gene flow mediated by drones. Under this hypothesis, males might travel considerable distances and mate with the docile honeybees of European ancestry that formerly constituted domesticated hives in the Americas. Alternatively, perhaps gene flow has resulted from colony swarming, a mechanism of maternal migration wherein a queen and some of her workers leave a hive and fly elsewhere to establish a new colony. Under this hypothesis, hybridization with domesticated European bees is not required.

Molecular analyses illuminated the issue by first demonstrating the involvement of colony swarming: Surveyed colonies of Africanized bees in the Neotropics often proved to carry African-type (as opposed to European-type) mtDNA (Hall and Muralidharan 1989; Hall and Smith 1991; D. R. Smith et al. 1989). Furthermore, allozymes and other nDNA markers showed that African and European honeybees had hybridized in the Neotropics, at least occasionally, and that this hybridization led to introgression of nuclear genes as part of the Africanization process, albeit to an argued degree (Hall 1990; Lobo et al. 1989; Rinderer et al. 1991; Sheppard et al. 1991). More recently, an intensive molecular investigation into the Africanization process in Mexico's Yucatán Peninsula has been reported (Clarke et al. 2002). Bees of African ancestry first arrived there in 1986. Based on analyses of mitochondrial and nuclear microsatellite markers that distinguish African from European forms, the genetic composition of Yucatán populations changed dramatically in the ensuing 15 years. By 1989, substantial paternal gene flow from invading Africanized drones had occurred, but maternal gene flow was negligible. By 1998, however, a radical shift had occurred, such that African nuclear alleles (65%) and African-derived mtDNA (61%) both predominated in the formerly European colonies.

Dispersal can also be sex-biased in many plants, notably due to the fact that pollen tends to be far more dispersive than seeds. One net consequence in such species is a greater opportunity for the spread of nuclear alleles than of maternally transmitted cytoplasmic alleles. Using cpDNA markers, often in conjunction with those provided by nuclear DNA or allozyme loci, such possibilities have been investigated and sometimes (not invariably) documented in several plant species (Grivet and Petit 2002; Latta and Mitton 1997; McCauley et al. 1996; McCauley 1998; Oddou-Muratorio et al. 2001).

Non-neutrality of some molecular markers

Lewontin and Krakauer (1973) pointed out that one expected signature of natural selection on genetic markers is the appearance of significant heterogeneity across loci in allele frequency variances among geographic populations. In theory, genetic drift, gene flow, and the breeding structure of a species should affect all neutral autosomal loci in a similar fashion, so different population genetic patterns across loci might signify either that allele frequencies at geographically variable loci are under diversifying selection (despite high gene flow as evidenced by geographically uniform genes), or that allele frequencies at geographically uniform loci are under stabilizing or equilibrium selection (despite low gene flow as evidenced by heterogeneous allele frequencies at geographically variable loci). Lewontin and Krakauer applied this reasoning to suggest that natural selection had acted on at least some human blood group polymorphisms (Cavalli-Sforza 1966), which on a global scale showed allele frequency variances spanning a wide range ($F_{ST} = 0.03$ to $F_{ST} = 0.38$). The "Lewontin–Krakauer" test subsequently was criticized on the grounds that its statistical methods seriously underestimated variances in gene frequencies expected under the null (neutral) theory (Nei and Maruyama 1975; Robertson 1975; see also Lewontin and Krakauer 1975). Nevertheless, it remains true that different loci within a species can sometimes paint very different pictures of population structure and gene flow, and that some of these patterns can be strongly suggestive of various departures from selective neutrality.

A noteworthy early example involved the deer mouse (*Peromyscus maniculatus*). In allozyme surveys of populations from across North America, F_{ST} values at six polymorphic loci ranged from 0.04 (inferred $Nm = 6.0$) to 0.38 ($Nm = 0.4$) (Avise et al. 1979c). Especially remarkable was the observation that surveyed populations from central Mexico to northern Canada and from the Pacific coast to the Atlantic all exhibited roughly similar frequencies ($F_{ST} = 0.05$) of the same two electromorphs at the aspartate aminotransferase (*Got-1* or *Aat-1*) locus. Subsequent screening by varied electrophoretic techniques and other discriminatory assays failed to reveal any appreciable "hidden protein variation" within these two *Aat-1* electromorph classes (Aquadro and Avise 1982b). Yet this relative geographic near-homogeneity at *Aat-1* contrasts sharply with the extreme geographic heterogeneity exhibited by this species in morphology, ecology, karyotype, and mtDNA sequence (Baker 1968; Blair 1950; Bowers et al. 1973; Lansman et al. 1983). For example, the number of acrocentric chromosomes ranges from 4 to 20 across populations (Bowers et al. 1973), and regional populations often show deep historical subdivisions involving cumulative and fixed differences in mtDNA (Lansman et al. 1983). It is difficult to escape the conclusion that *Aat-1* provides a serious underestimate of the overall magnitude of population genetic structure in this species. One theoretical possibility is that geographically uniform selection somehow balances *Aat-1* allele frequencies despite severe historical and contemporary restrictions on gene flow apparently registered by numerous other genetic traits.

The converse of this situation may apply to a classically studied allozyme polymorphism in *Drosophila melanogaster*. The main biochemical function of alcohol dehydrogenase (ADH) is to metabolize ethanol, which is abundant in fermented fruits in the flies' natural environment. Several studies have shown that the Adh^F allele has significantly higher enzymatic activity than Adh^S, but is less heat-resistant, and that these and other biochemical and physiological attributes translate into fitness differences between *Adh* genotypes under particular experimental regimes (Sampsell and Sims 1982; van Delden 1982). In natural populations, frequencies of these two *Adh* alleles often vary locally (e.g., inside versus outside wine cellars; Hickey and McLean 1980) and also show strong latitudinal clines, with Adh^F more common with increasing latitude in both the Northern and Southern hemispheres (Oakeshott et al. 1982). Such evidence for diversifying selection on *Adh* implies that prima facie estimates of gene flow based on this polymorphism alone could be misleadingly low. Based on several other genetic traits, Singh and Rhomberg (1987) concluded that gene flow in *D. melanogaster* is sufficiently high ($Nm \cong 1–3$), even on continental scales, to theoretically homogenize nuclear genes in the absence of selection (but see Begun and Aquadro 1993; Hale and Singh 1991). On the other hand, further molecular analyses of *D. melanogaster* and related species have uncovered a rich heterogeneity of population genetic signatures, suggesting that natural selection, genetic drift, mutation rate, recombination rate, and other evolutionary factors must have all contributed (often interactively) to the observed patterns (Aquadro 1992).

Based on similar arguments from comparative geographic patterns, natural selection on at least some molecular markers has been implicated for various genetic polymorphisms in many other species as well (e.g., Ayala et al. 1974). A recent example involving humans entailed calculating genetic differentiation at more than 330 short tandem repeat (STR) loci in Africans and Europeans (Kayser et al. 2003). For about a dozen loci that displayed unusually large genotypic differences, additional linked loci were then genotyped, and they too showed significant genetic divergences between these populations. The authors concluded that these loci displaying aberrant genetic distances from the genomic norm probably earmark chromosomal regions that have been under unusually intense diversifying selection related to environmental circumstances.

When such findings are coupled with further lines of evidence for balancing, directional, or diversifying selection on particular proteins (see Chapter 2) or DNA sequences (e.g., Hughes and Nei 1988; Kreitman 1991; MacRae and Anderson 1988; Nei and Hughes 1991), it becomes clear that interpretations of geographic population structure under the assumption of strict neutrality are made with some peril. At the very least, conclusions about genomically pervasive forces shaping population structure in any species should be based on information from multiple independent loci.

Historical demographic events

For reasons of mathematical tractability, many theoretical models in population genetics yield only equilibrium expectations between counteracting evolutionary forces, such as the diversifying influence of genetic drift in small populations versus the homogenizing influence of gene flow under an island model or a stepping-stone model (see Box 6.3). Seldom is it feasible to formally consider the idiosyncratic histories of particular species or to treat non-equilibrium situations. Yet demographic histories and phylogenies of real species *are* highly idiosyncratic and are likely to produce population genetic signatures that depart in various ways from theoretical equilibrium expectations. Hence, empirical genetic structures of natural populations are notoriously challenging to interpret.

For example, in comparative analyses of three anadromous fish species along the same coastline transect in the southeastern United States, Bowen and Avise (1990) were led to consider several historical demographic and biogeographic factors that might have produced the observed differences in population structure as registered by mtDNA. All three species showed significant differences in haplotype frequency between the Atlantic and the Gulf of Mexico, but in magnitudes and patterns that differed greatly among taxa. Populations of the black sea bass (*Centropristis striata*) showed little within-region polymorphism and a clear phylogenetic distinction between the Atlantic and the Gulf; menhaden (*Brevoortia tyrannus* and *B. patronus*) showed extensive within-region polymorphism and a paraphyletic relationship of Atlantic to Gulf populations; and sturgeon (*Acipenser oxyrhynchus*) exhibited extremely low mtDNA variation within and between regions. Based on the magnitude of mtDNA variation observed in regional populations of these three species, estimates of evolutionary effective population size varied by more than four orders of magnitude—from $N_{F(e)} = 50$ (Gulf of Mexico sturgeon) to $N_{F(e)} = 800,000$ (Atlantic menhaden)—and their rank order was correlated with present-day census sizes. These differences in $N_{F(e)}$, which presumably reflect the idiosyncratic demographic histories of the three species, may help to explain some of their distinctive phylogenetic features, including the clean distinction between Atlantic and Gulf forms of the sea bass versus the paraphyletic pattern in menhaden (assuming that regional populations in both of these species were separated by similar historical vicariant events). However, even grossly different effective population sizes in the biogeographic context of shared vicariance cannot explain all the contrasting features of population genetic structure in these three co-distributed fish species. Thus, for menhaden and sturgeon (but not sea bass), recent gene flow between the Atlantic and Gulf is strongly implicated by the shared presence in these two regions of several nearly identical mtDNA haplotypes.

Whether these particular inferences are correct or not, they serve to introduce some of the historical demographic considerations and non-equilibrium environmental conditions that *must* have affected genetic structures in real populations. In interpreting empirical data on population structure, deciding how far to pursue idiosyncratic demographic explanations is a difficult challenge, particularly because these explanations can seldom be tested critically in controlled or replicated settings (however, see Fos et al. 1990; Scribner and Avise 1994a; Wade and McCauley 1984), and because competing scenarios might also be compatible with the data. Nonetheless, cognizance of the limitations of equilibrium theory, and of the potential effect of historical demographic factors on population genetic structures, represents an important step toward greater realism.

Population assignments

Most of the molecular assessments of geographic structure and gene flow described above employed sample allele frequencies from composite assemblages of individuals—"populations"—that had been defined a priori, typically by subjective spatial and phenotypic criteria. Any such population, real or not, will of course have some quantifiable genetic relationship to others, but this summary characterization may obscure much that is of biological interest. In other words, an undesirable element of circular reasoning is introduced into traditional assessments of spatial structure that take particular populations as "givens" at the outset of the analysis. Although this may seldom be a fatal difficulty in practice, it is desirable in many situations to treat individual organisms (whose genetic reality or coherence is seldom in dispute) as basic units of genealogical analysis.

An early example of this approach involved summarizing genotypic data from 30 microsatellite loci into an evolutionary phenogram whose external nodes were the 148 individual humans examined (Figure 6.6). Despite the small variation in allele frequencies between regionally defined populations around the world, branches connecting individuals into a neighbor-joining tree (based on percentages of alleles shared across loci) proved to reflect these people's geographic origins "with remarkable accuracy" (Bowcock et al. 1994). Other informative examples of this sort, in which individual organisms were treated as fundamental units in population structure analysis (based on genotypic data from multiple nuclear loci), include molecular analyses of *Apis* honeybees (Estoup et al. 1995), *Heterocephalus* mole-rats (O'Riain et al 1996), and *Odocoileus* deer (Blanchong et al. 2002).

A general goal in such studies is to classify particular individuals into populations (Davies et al. 1999; Guinand et al. 2002). One appropriate biological context is when a number of suspected source populations may have contributed individuals to a sample of interest, in which case a mixed-stock analysis can be conducted (Waser and Strobeck 1998). Allele frequencies in candidate source populations are estimated at a series of unlinked loci, and the statistical likelihood (based on multi-locus genotype) that each individual of unknown origin came from each potential source is calculated (Letcher and King 1999; Rannala and Mountain 1997; Smouse et al. 1990). In a recent modification of this approach, Pritchard et al. (2000) introduced a Bayesian clustering method that attempts to assign multi-locus genotypes of individuals to specific populations while simultaneously estimating population allele frequencies. This method can be applied even in situations in which the source populations are not explicitly specified at the outset. The authors successfully applied this approach to humans and to an endangered avian species, *Turdus helleri*. Similar kinds of applications in individual-based population assignment also arise routinely in the context of human and wildlife molecular forensics (Campbell et al. 2003; Foreman et al. 1997; Paetkau et al. 1995; Roeder et al. 1998; see Chapter 9).

As described in the next section, one context in which individuals *are* routinely treated as fundamental units of population genetic analysis is in the field of mtDNA-based phylogeography. In this special case, historical relationships registered in individuals' mtDNA sequences reflect the matrilineal component of population structure.

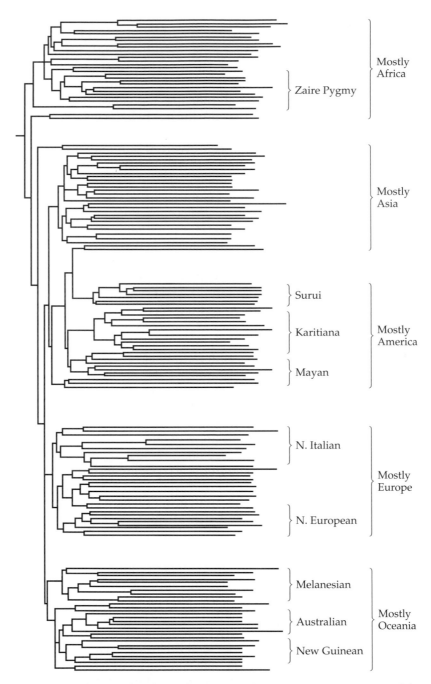

Figure 6.6   Neighbor-joining tree for 148 people. The tree was constructed from pairwise genetic distances at 30 microsatellite loci. The 148 subjects were treated in this analysis as individuals. Note the generally good agreement of genetic clusters with geographic origins. (After Bowcock et al. 1994.)

Phylogeography

Phylogeography is a field of study concerned with principles and processes governing the geographic distributions of genealogical lineages, especially those within and among closely related species. In other words, the discipline focuses explicitly on historical or phylogenetic components of population structure (including how these may have been influenced by genetic drift, gene flow, natural selection, or any other evolutionary forces). In broad terms, phylogeography's most important contributions to biology have been to emphasize non-equilibrium aspects of population structure and microevolution, clarify the tight connections that inevitably exist between population demography and historical genealogy (Box 6.5), and build conceptual and empirical bridges between the formerly separate fields of traditional population genetics and phylogenetic biology (Figure 6.7). The field of phylogeography was reviewed in a recent textbook (Avise 2000a) that included approximately 1,500 references to the literature and that in many respects represents a companion volume to this current edition of *Molecular Markers*. Thus, only an introductory qualitative treatment of phylogeography, involving a few select examples, will be presented here.

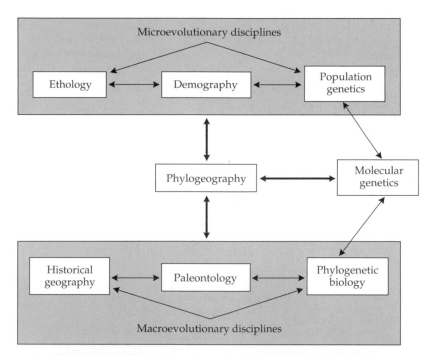

Figure 6.7 Phylogeography serves as a bridging discipline between several traditionally separate fields of study in the micro- and macroevolutionary sciences. (After Avise 2000a.)

BOX 6.5 Branching Processes and Coalescent Theory

Branching process theory and coalescent theory are formal mathematical disciplines that address inherent connections between population demography and genealogy at the intraspecific level (Griffiths and Tavaré 1997; Herbots 1997; Taib 1997). As such, they provide conceptual frameworks for interpreting many of the findings of empirical phylogeography.

The notion that genealogy and demography are intimately related can be introduced by the following scenario. Imagine that females in a population produce daughters according to a Poisson distribution with a mean (and hence variance) equal to 1.0. Under this model, the expectation that a female contributes zero daughters to the next generation (or, the expected frequency of daughterless mothers) is $e^{-1} = 0.368$ (e is the base of the natural logarithms), and her probabilities of producing n offspring ($n \geq 1$) are given by $e^{-1} (1/n!)$. Thus, the chances that a female contributes 1, 2, 3, 4, and 5 or more daughters are 0.368, 0.184, 0.061, 0.015, and 0.004, respectively. These probabilities apply across a single organismal generation. Mathematical "generating functions" (available for several theoretical distributions of family size, including the Poisson) can then be employed to recursively calculate the probability that a matriline goes extinct across multiple non-overlapping generations. Application of the Poisson generating function, for example, yields cumulative probabilities of matriline extinction that increase asymptotically from 0.368 in the first generation to 0.981 by generation 100. In other words, due to the turnover of lineages that inevitably accompanies reproduction, a few fortunate matrilines may proliferate at the expense of many others that die out along the way.

Thus, with respect to matrilineal genealogy, individuals in any extant population invariably trace back, or "coalesce," to common ancestors at various depths of times in the past. In other words, the individuals alive at any moment are historically connected to one another in a hierarchically branched intraspecific genealogy, as in this hypothetical illustration:

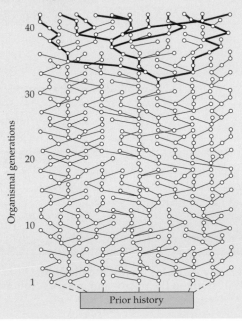

Indeed, the only situation in which this would not be true is if each female in every generation replaced herself with exactly one daughter, in which case there would be no lineage sorting, no hierarchical branching structure (all matrilines would exist as a series of parallel lines of descent through time), and no coalescent. Of course, families in all real populations show variances in contributions to the progeny pool; the larger that variance, the more rapid the pace of lineage sorting and the more shallow the resulting coalescent point (all else being equal). In a roughly stable population with N_F females and a Poisson distribution of family sizes, the expected mean time (in generations, G) to common matrilineal ancestry for random pairs of individuals is $G \cong N_F$, and the expectation for the coalescent point of the entire suite of lineages is $G \cong 2N_F$ (Nei 1987).

The same logic of branching processes and coalescence applies to patrilines (the lineages traversed, for example, by the Y chromosome of mammals). It also applies in principle to "gene genealogies" at any autosomal locus, except that coalescent depths under neutrality are expected to be about fourfold greater (a twofold effect for biparental as opposed to uniparental inheritance, and another twofold effect for diploid versus haploid inheritance). Although it is beyond the scope of the current discussion, coalescent theory and related approaches have also been extended to populations that are historically dynamic in size (Harvey and Steers 1999), receive outside gene flow (e.g., Beerli and Felsenstein 1999, 2001; Nee et al. 1996a; Rogers and Harpending 1992; Wakeley and Hey 1997), and are geographically structured into metapopulations (Bahlo and Griffiths 2000; Hey and Machado 2003; Hudson 1998; Nei and Takahata 1993; Pannell 2003; Wakeley and Aliacar 2001). In general, this new theoretical framework has promoted recognition of the close relationships between genealogy and population demography that are highly germane to interpreting intraspecific gene trees estimated by molecular markers, most notably from mtDNA.

History and background

The introduction of mtDNA analyses to population genetics in the late 1970s prompted a revolutionary shift toward historical, genealogical perspectives on intraspecific population structure. Because mtDNA sequences evolve rapidly and show non-recombinational inheritance, they typically provide haplotype data that can be ordered phylogenetically within a species, yielding an intraspecific phylogeny (gene genealogy) interpretable as the matriarchal component of an organismal pedigree. Mitochondrial transmission in animal species constitutes the female analogue of male surname transmission in many human societies (Avise 1989b): Both sons and daughters inherit their mother's mtDNA genotype, which only daughters normally transmit to the next generation. Thus, mtDNA lineages reflect mutationally interrelated "female family names" of a species, and their historical dynamics can be interpreted according to the types of theoretical models long used by demographers to analyze surname distributions in human societies (Lasker 1985; Lotka 1931; Box 6.5). A thumbnail sketch of the history of phylogeography is presented in Box 6.6.

BOX 6.6 Brief Chronology of Some Key Developments in the History of Phylogeographic Analysis

1974 Brown and Vinograd demonstrate how to generate restriction site maps for animal mtDNAs.

1975 Watterson describes basic properties of gene genealogies, marking the beginnings of modern coalescent theory.

Brown and Wright introduce mtDNA analysis to the study of the origins and evolution of parthenogenetic taxa.

1977 Upholt develops the first statistical method to estimate mtDNA sequence divergence from restriction digest data.

1979 Brown, George, and Wilson document rapid mtDNA evolution.

Avise, Lansman, and colleagues present the first substantive reports of mtDNA phylogeographic variation in nature.

1980 Brown provides an initial report on human mtDNA variation.

1983 Tajima and also Hudson initiate statistical treatments of the distinction between a gene tree and a population tree.

1986 Bermingham and Avise initiate comparative phylogeographic appraisals of mtDNA for multiple co-distributed species.

1987 Avise and colleagues coin the word "phylogeography," define the field, and introduce several phylogeographic hypotheses.

Cann and colleagues describe global variation in human mtDNA.

1989 Slatkin and Maddison introduce a method for estimating inter-population gene flow from allelic phylogenies.

1990 Avise and Ball introduce principles of genealogical concordance as a component of phylogeographic assessment.

1992 Avise summarizes the first extensive compilation of phylogeographic patterns for a regional fauna.

1996 Edited volumes by Avise and Hamrick and by Smith and Wayne summarize conservation roles for phylogeographic data.

1998 A special issue of the journal *Molecular Ecology* is devoted to phylogeography.

Templeton reviews statistical roles of "nested clade analysis" in phylogeography (Templeton 1993, 1994, 1996; for a critical appraisal, see Knowles and Maddison 2002).

2000 The first textbook on phylogeography is published, by Avise.

2001 *Molecular Ecology* introduces a continuing subsection entitled "Phylogeography, Speciation, and Hybridization."

Source: Avise 1998b.

One of the earliest phylogeographic studies based on mtDNA data still serves as a useful illustration of the types of population structure that are frequently revealed (Avise et al. 1979b). The southeastern pocket gopher (*Geomys pinetis*) is a fossorial rodent that inhabits a three-state area in the southern United States. Analysis of 87 individuals from across this range by six restriction enzymes revealed 23 different mtDNA haplotypes, whose phylogenetic relationships and distributions are summarized in Figure 6.8. Clearly, most mtDNA genotypes in these gophers were localized geographically, appearing at only one or a few adjacent collection sites. Furthermore, genetically related clones tended to be geographically contiguous or overlapping, and a major gap in the matriarchal phylogeny cleanly distinguished all eastern from all western populations. This principal phylogeographic gap was also registered in the nuclear genome by shifts in frequencies of distinctive karyotypes and protein electrophoretic alleles.

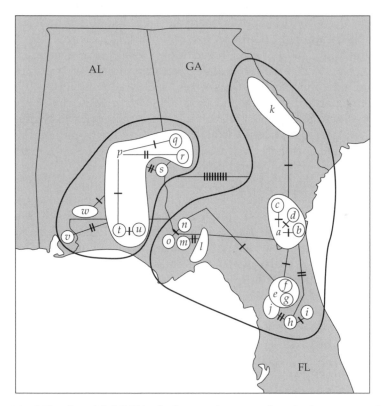

Figure 6.8   Mitochondrial DNA phylogeny for 87 pocket gophers across the species range in Alabama, Georgia, and Florida. Lowercase letters represent different mtDNA genotypes, which are connected by branches into a parsimony network that is superimposed over the geographic sources of the collections. Slashes across network branches reflect the number of inferred mutational steps along a pathway. Heavier lines encompass two distinct mtDNA clades that differ by at least nine mutational steps. (After Avise et al. 1979b.)

Population subdivision characterized by localized genealogical structure or significant mtDNA phylogenetic gaps across a species' range soon were likewise reported in a huge number of animal species: mammals ranging from voles and mice to whales (early studies by Carr et al. 1986; Cronin et al. 1991a; Cronin 1992; MacNeil and Strobeck 1987; Plante et al. 1989; Prinsloo and Robinson 1992; Riddle and Honeycutt 1990; Wada et al. 1991), birds ranging from sparrows to geese (Avise and Nelson 1989; Shields and Wilson 1987b; Van Wagner and Baker 1990; Zink 1991), reptiles ranging from geckos to tortoises (Densmore et al. 1989a; Lamb and Avise 1992; Lamb et al. 1989; Moritz 1991), amphibians (e.g., Wallis and Arntzen 1989), freshwater and marine fishes (Avise 1987; Bermingham and Avise 1986; Crosetti et al. 1993), insects (Hale and Singh 1987, 1991; Harrison et al. 1987), crustaceans (Saunders et al. 1986), echinoderms (Arndt and Smith 1998a; Williams and Benzie 1998), mollusks (Murray et al. 1991; O'Foighil and Smith 1996; Quesada et al. 1995, 1998), and many others. A few species proved to exhibit little or no mtDNA phylogeographic structure across broad ranges, but these were the exception rather than the rule. Examples included some large, mobile mammals (Lehman and Wayne 1991; Lehman et al. 1991), some birds (Ball et al. 1988; Tegelström 1987a), several marine fishes (Arnason et al. 1992; Avise 1987), some migratory insects (Brower and Boyce 1991), and miscellaneous other species, such as a nematode (*Ostertagia ostertagi*) that parasitizes cattle and probably was spread widely by livestock transport (Blouin et al. 1992). It soon became apparent for a wide array of species that differences in organismal vagility and environmental fragmentation (past and present) had exerted major influences on patterns of mtDNA phylogeographic population structure.

One common finding is that regional assemblages of conspecific populations often are distinguished by deep genealogical separations compared with the shallow separations in mtDNA genealogy normally observed within each assemblage. Such highly distinctive matrilineal clades within a species are sometimes provisionally referred to as "evolutionarily significant units" (ESUs; see Chapter 9) or as salient "intraspecific phylogroups" (Avise and Walker 1999). Furthermore, most species display only a small number of such ESUs (typically about 1–8), and they are usually spatially oriented in ways that make considerable sense in terms of geographic history (such as known or suspected Pleistocene refugia and subsequent dispersal routes) or taxonomy (e.g., they may agree well with traditionally described subspecies). Several examples are provided below, and many more are summarized by Avise (2000a).

Presumably, the localization of closely related mitochondrial genotypes and clades in most species reflects contemporary restraints on gene flow (at least via females), and many of the deeper genetic breaks (distinguishing provisional ESUs) register much longer-term historical population separations. Such observations quickly prompted a deeper appreciation of distinctions between contemporary gene flow and historical population connectivity in a genealogical sense (Avise 1989a; Larson et al. 1984; Slatkin 1987). What follows are a few illustrations (chosen to make particular points) of how mtDNA analyses have added a phylogenetic dimension to perspectives on intraspecific population structure.

Case studies on particular populations or species

GREEN TURTLES ON ASCENSION ISLAND.   Ascension Island, a tiny (8-km diameter) island situated on the mid-Atlantic ridge halfway between Brazil and Liberia, is a major rookery for green turtles (*Chelonia mydas*). From direct tagging studies, it was known that females that nest on Ascension otherwise inhabit shallow-water feeding pastures along the South American coastline. Thus, for each nesting episode (every 2 to 3 years for an individual), females embark on a 5,000-km migration to Ascension Island and back, a several-months-long odyssey requiring navigational feats and endurance that nearly defy human comprehension. How might Ascension turtles have established such an unlikely migratory circuit, particularly since suitable nesting beaches along the South American coast are utilized by other green turtles? Carr and Coleman (1974) proposed a historical biogeographic scenario involving plate tectonics and natal homing. Under their hypothesis, the ancestors of Ascension Island green turtles nested on islands adjacent to South America in the late Cretaceous, soon after the equatorial Atlantic Ocean opened. Over the past 70 million years, these volcanic islands have been displaced from South America by seafloor spreading (at a rate of about 2 cm per year). A population-specific instinct to migrate to present-day Ascension Island thus might have evolved over tens of millions of years of genetic isolation (at least with regard to matrilines) from other green turtle rookeries in the Atlantic.

Bowen et al. (1989) critically tested the Carr–Coleman hypothesis by comparing mtDNA genotypes of Ascension Island nesters with those of other green turtles. They identified fixed or nearly fixed mtDNA differences between Ascension and many Atlantic rookeries, a finding consistent with severe restrictions on contemporary inter-rookery gene flow by females and, thus, supportive of the natal homing aspects of the Carr–Coleman hypothesis. However, the magnitude of mtDNA sequence divergence from several other Atlantic rookeries was tiny ($p < 0.002$; see Figure 6.5), indicating that any current genetic separation of the Ascension colony was initiated very recently, probably within the last 100,000 years at most. Indeed, the time elapsed may have been much less than this, because the predominant Ascension haplotype proved indistinguishable in available assays from a genotype characterizing a Brazilian rookery (Bowen et al. 1992). In any event, these genetic results clearly were incompatible with the temporal aspects of the Carr–Coleman scenario. Instead, the colonization of Ascension by green turtles, or at least extensive matrilineal gene flow into the population, was evolutionary recent.

SMALL MAMMALS OF AMAZONIA. To explain why the Amazon basin contains the world's richest biota, several hypotheses have been advanced: the refugial model, which posits that populations were sundered when their habitats were disjoined during cyclic expansions and contractions of forests during alternating wet and dry episodes of the Pleistocene (Cracraft and Prum 1988; Haffer 1969); ecological models, which posit that diversification was driven by selection pressures associated with high ecological and environmental heterogeneity in the region (Tuomisto et al. 1995); and the riverine barrier model, which suggests that large rivers promoted genetic divergence in terrestrial organisms by blocking inter-regional gene flow (Ayres and Clutton-Brock 1992).

The riverine barrier hypothesis was put to phylogeographic test in mtDNA analyses of more than a dozen small mammal species across large portions of Amazonia (da Silva and Patton 1998; Patton et al. 1994a,b, 1997; Peres et al. 1996). Salient phylogeographic partitions were uncovered within several species, but these genetic breaks did not correspond with the current positions of major rivers. Instead, highly divergent clades typically were observed in upstream versus downstream regions, in positions generally demarcated by geological arches associated with Andean uplifts of the mid to late Tertiary. For at least some taxonomic groups, these observations prompted a new hypothesis for Amazonian phylogeography: that these ancient, quasi-isolated paleobasins may have been historical centers of diversification (da Silva and Patton 1998).

Lessa et al. (2003) tested a prediction of the refugial model: that organisms originally isolated in Pleistocene refugia should have experienced substantial population growth when climates ameliorated and new habitats opened. Using coalescent theory (see Box 6.5) as applied to mtDNA sequence data for several small mammal species in western Amazonia, these authors uncovered no evidence for demographic expansions following the late Pleistocene. By contrast, pronounced and oft-concordant genetic footprints of recent population expansions were found in similar mtDNA analyses of several mammals occupying high latitudes in North America (Lessa et al. 2003). These results illustrate how historical demographic responses to climatic changes can be genetically tracked, and they also suggest that such responses may have varied across latitudinal gradients of biodiversity.

BROWN BEARS AND ALLIES. Several genetic surveys of mtDNA in brown bears (*Ursus arctos*) have collectively spanned most of this species' Holarctic range (Cronin et al. 1991b; Leonard et al. 2000; Matsuhashi et al. 2001; Paetkau et al. 1998; Taberlet and Bouvet 1994; Talbot and Shields 1996a,b; Waits et al. 1998). Results indicate the presence of about 5–6 provisional ESUs or phylogroups, each confined to a distinct region in North America, Europe, or Asia (Figure 6.9). Most likely, the species was subdivided historically into several regional assemblages whose matrilines gradually accumulated the evident mtDNA differences.

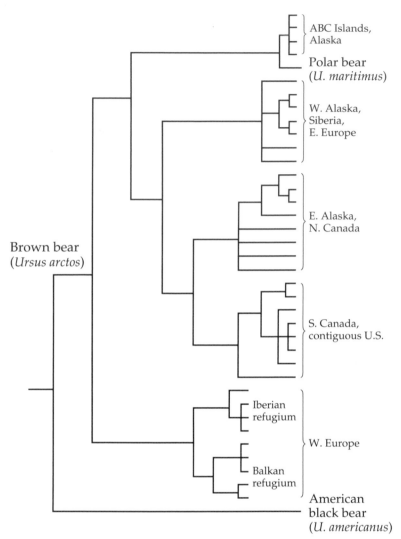

Figure 6.9   Global mtDNA phylogeography in the brown bear. This depiction uses the American black bear as the outgroup, and also shows the matrilineal position of the polar bear. It is a simplified summary compiled from several references cited in the text. (After Avise 2004d.)

Another interesting feature of these data is the position of the polar bear (*Ursus maritimus*) within this phylogeny (see Figure 6.9). In terms of matriarchal ancestry as registered by mtDNA, polar bears appear to be closely allied to some brown bears in the "ABC Islands" of southeastern Alaska, thus making this clade a tiny subset of the broader lineage diversity within brown bears globally (Shields et al. 2000). In other words, brown bear matrilines appear to be paraphyletic with respect to those of polar bears. One possibility for this unexpected pattern is that introgressive hybridization recently transferred some mtDNA lineages from brown bears to polar bears (these two species can produce fertile offspring in captivity). Another possibility involves historical lineage sorting. Perhaps polar bears arose within the past few tens of thousands of years from coastal populations of brown bears, to which their matrilines now appear most closely related. If so, then polar bears possess a suite of derived morphological characteristics that may have evolved rapidly in response to the special selective conditions of the Arctic, a suggestion that has some support from fossil and other evidence (Talbot and Shields 1996a). Finally, the direction of evolution might have been exactly the reverse: ABC Islands brown bears may have arisen recently from a few polar bears that moved south.

RED-WINGED BLACKBIRDS.   An absence of dramatic phylogeographic population structure can be just as interesting and informative as its presence. One widely distributed species that failed to display ancient subdivisions in matrilineal phylogeny (based on evidence from mtDNA restriction site assays) is the red-winged blackbird (*Agelaius phoeniceus*). A total of 34 different mtDNA haplotypes were observed among 127 specimens collected from across North America, but all of these haplotypes were closely related, and they were not obviously partitioned geographically (Ball et al. 1988). Indeed, almost all of the haplotypes were related in a "starburst" pattern, with the most common haplotype (nearly ubiquitously distributed) at its core (Figure 6.10A). These findings indicate that redwing populations throughout most of the continent have been in strong and recent genetic contact. To a first approximation, the entire species can be considered a single, tight-knit evolutionary unit.

Furthermore, the data could be grouped into a frequency histogram of pairwise mtDNA genetic distances among sampled individuals, and this in turn could be converted (assuming a conventional molecular clock) into a distribution of estimated times to shared matrilineal ancestry (Figure 6.10B). Such histograms, termed "mismatch distributions," bear somewhat predictable relationships under coalescent theory (see Box 6.5) to historical population demography and evolutionary effective population size (Fu 1994a,b), in this case of females ($N_{F(e)}$). For red-winged blackbirds, a reasonably good fit of the data to coalescent expectations was obtained by assuming $N_{F(e)} \cong 40,000$ individuals. Furthermore, mild departures of this mismatch distribution from the theoretical expectation for a single population of this size were in a direction suggestive of a recent population expansion (Rogers and Harpending 1992). These findings make considerable biological sense because *A. phoeniceus* must have expanded its range across much of the continent within the last 18,000 years, following the retreat of the most recent Pleistocene glaciers.

GLACIAL REFUGIA FOR HIGH-LATITUDE FISHES.   Phylogeographic appraisals have been conducted on several species of freshwater fishes inhabiting high latitudes of North America and Eurasia. In many cases, the genetic foot-

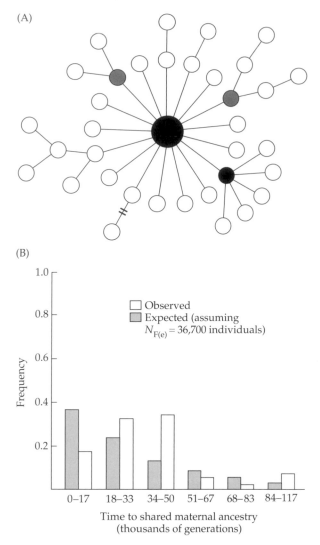

Figure 6.10 Mitochondrial DNA patterns in a continent-wide restriction site survey of red-winged blackbirds. (A) Starburst phylogeny, with the most common and widespread haplotypes shown in black, hypothetical haplotypes (not observed) indicated by gray shading, and other haplotypes (all rare) indicated by open circles. (B) Mismatch frequency distribution, showing inferred times to shared maternal ancestry and estimated evolutionary effective population size for this species.

prints of Quaternary refugia have been evident in the contemporary spatial distributions of mtDNA phylogroups or clades (see review in Bernatchez and Wilson 1998). Some of these surveys have been Holarctic in scale (e.g., Brunner et al. 2001). Typically, differentiated matrilineal clades in these fishes appear to be regionally organized in ways that reflect postglacial dispersal and sometimes secondary overlaps of distinctive phylogroups that had

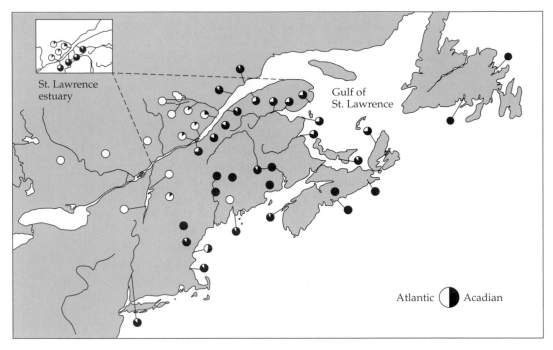

Figure 6.11 Population frequencies of the two major mtDNA clades across the range of the rainbow smelt (*Osmerus mordax*). (After Bernatchez 1997.)

accumulated genetic differences in allopatry. When secondary sympatry has been achieved, molecular markers also permit examination of reproductive compatibility between the divergent forms.

One such example involves the rainbow smelt (*Osmerus mordax*) of northeastern North America. Bernatchez (1997) mtDNA-genotyped a total of 1,290 smelt from 49 populations across the species' native range, and uncovered two highly divergent clades whose spatial distributions are summarized in Figure 6.11. Eastern populations were largely dominated by one mtDNA clade and western populations by the other. Furthermore, this genealogical dichotomy proved to be largely independent of life history forms, which include lake-dwelling and anadromous fish (see also Taylor and Bentzen 1993). Most likely, the so-called Atlantic and Acadian races had survived in glacial refugia along the Atlantic coastal plain and in the Grand Banks area, respectively. Based on paleogeographic as well as this genetic evidence, Bernatchez (1997) further postulated that the Atlantic race then colonized northern regions about 5,000 years prior to the Acadian race, with both clades eventually coming into secondary contact in the St. Lawrence River estuary, where a suspected evolution of reproductive isolating mechanisms between the two races then ensued. All of these interpretations depart dramatically from the conventional (pre-molecular) biogeographic wisdom that all rainbow smelt populations originated from one coastal refugium.

In Europe, an emerging view is that various fishes (as well as other biotas; Bilton et al. 1998; Stewart and Lister 2001) survived the last glaciation not only in isolated refugia of the Mediterranean region (Hewitt 2000; Taberlet and Cheddadi 2002; Taberlet et al. 1998), but also farther north (Bernatchez 2001; Hänfling et al. 2002; Kotlík and Berrebi 2001). An interesting example involves *Barbus* freshwater fishes, especially in the Black Sea area. Today, the Black Sea is an inland body of salt water, fed by numerous large rivers draining most of Europe and connected to the Atlantic Ocean via the Mediterranean. Toward the end of the Pleistocene, however, it had become a giant freshwater lake as inflow of Mediterranean seawater was interrupted during the last glaciation. Then, about 7,500 years ago, marine conditions were reestablished in the Black Sea basin when catastrophic flooding by Mediterranean waters occurred. Kotlík et al. (2004) used mtDNA sequences to test whether *Barbus* in rivers surrounding the Black Sea might all trace to a recent common ancestor that could have inhabited the Black Sea basin during its freshwater phase. Results showed instead that highly divergent lineages now occupy different river drainages entering the Black Sea, indicating that multiple refugial populations probably survived throughout the late Pleistocene in the vicinity of this ancient lake.

LACERTID LIZARDS ON ISLANDS.   Molecular phylogeographic patterns can also serve as genealogical backdrops for interpreting evolutionary histories of organismal phenotypes. This exercise can be thought of as a microevolutionary analogue of "phylogenetic character mapping" (PCM) at intermediate and higher taxonomic levels (discussed in Chapter 8).

An illustration of PCM in the context of intraspecific phylogeography involves the Canary Island lizard (*Gallotia galloti*). A molecular genealogy for this species, based on mtDNA and other genetic data (Thorpe et al. 1993, 1994), indicated the presence of two distinct lineages whose colonization histories could be rigorously hypothesized as a colonization of La Palma Island from North Tenerife and separate sequential colonizations of Gomera and Hierro islands from South Tenerife (Figure 6.12). Thorpe (1996) then used these historical inferences to interpret the distributions of 30 variable morphological features. Nine of these 30 characteristics (30%) were significantly associated with the molecular phylogeny. For example, blue spots on the foreleg and hindleg are present in the southern lineage, but absent in the northern lineage and in congeneric outgroup species, indicating that these features (thought to be employed in sexual communication) are synapomorphies that probably arose once in a common ancestor of the lizards on Gomera and Hierro islands. Examples of phenotypic characters not clearly associated with phylogeny were dorsal yellow bars and gracile heads, both of which tend to be present in animals inhabiting wet, lush areas irrespective of the molecular lineage to which they belong. Perhaps these characters are strongly selected for under these ecological circumstances, or perhaps there are direct dietary effects on phenotype (especially with regard to head and jaw size).

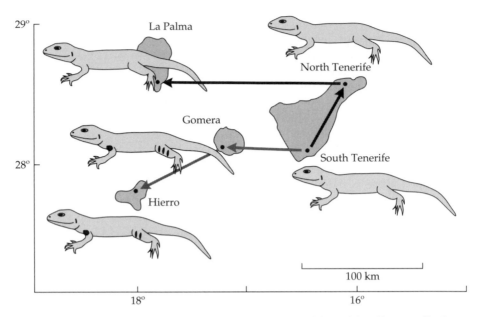

Figure 6.12 Evolutionary history of the Canary Island lizard (*Gallotia galloti*). Arrows indicate the colonization sequence of two major evolutionary lineages as deduced from molecular genetic evidence. Note the pronounced spots on the foreleg and hindleg, which appear to be synapomorphies unique to the southern lineage. (After Thorpe 1996.)

As in any such PCM exercise, this analysis of island lizards assumes that the molecular markers employed are reliable indicators of genomic history. To the extent that this might not be entirely true, interpretations of the histories and causal factors impinging on phenotypes could, of course, be compromised.

MÜLLERIAN MIMICRY BUTTERFLIES.   Many invertebrates show mtDNA phylogeographic patterns quite like those of vertebrates: that is, genealogical separations at varying evolutionary depths, and often major genetic breaks between regional population arrays or phylogroups. A case in point is the tropical butterfly species *Heliconius erato*, traditionally described as being composed of more than a dozen allopatric races, each displaying a unique wing coloration pattern. These wing patterns not only vary geographically across northern South America, but they do so in parallel with wing-color races of a related species (*H. melpomene*). Both species are unpalatable to predators, so these butterflies collectively provide a classic example of Müllerian mimicry. It has long been of interest to calibrate the evolutionary rates and processes by which the different wing-color forms have arisen.

Toward that end, Brower (1994) assayed mtDNA sequences in *H. erato* across much of its range. More than a dozen different haplotypes were detected, but these haplotypes grouped into two highly distinct clades that appear to be confined to opposite sides of the Andes Mountains, and which by molecular clock considerations separated about 1.5 to 2.0 million years ago. Within each phylogroup, by contrast, mtDNA sequence differences were small to negligible. Yet nearly identical wing-color patterns were observed within and between the two mtDNA phylogroups. Overall, the phylogeographic backdrop provided by mtDNA suggests that *H. erato* experienced a rather ancient population sundering event related to the gene flow barrier that the Andes provide, and that since that time there have been multiple instances of rapid and often convergent evolution in wing coloration patterns (Brower 1994).

EUROPEAN TREES.   Until fairly recently, phylogeographic studies in plants lagged far behind those in animals, due mostly to the perceived poor suitability of plant cytoplasmic genomes for such tasks. However, with the advent of better laboratory methods and the larger data sets to which they permit access, chloroplast (cp) DNA has become a powerful workhorse for botanical phylogeographic analyses at the intraspecific level (Petit et al. 2001, 2003a,b; Schaal et al. 1998, 2003; D. E. Soltis et al. 1992).

Some of the earliest and most extensive work involved detailed analyses of cpDNA variation in eight species of European white oaks (*Quercus*) sampled from more than 2,600 sites in 37 European countries (see reviews in Petit and Verdramin 2004; Petit et al. 2003b). Genetic footprints from the chloroplast genome revealed several primary and secondary Pleistocene refugia where genetic differentiation must have been initiated, as well as specific postglacial colonization routes from those isolated southern pockets. Polymorphic genetic markers from cpDNA have also helped to reveal hybridization patterns between various European oak species (Bacilieri et al. 1996; Belahbib et al 2001; Petit et al. 2002).

Such analyses based on cpDNA sequences were then extended to 22 widespread species of trees and shrubs (Petit et al. 2003b). This massive phylogeographic survey not only helped to identify primary glacial refugia in Europe, but also demonstrated that the most genetically diverse populations now occur at intermediate (rather than southern) latitudes, probably as a consequence of genetic admixture of divergent lineages that had expanded outward from their ancestral homes. Thus, Pleistocene refugia in Europe were historical wellsprings of genetic diversity in plants, but modern admixture zones are the current melting pots.

FREE-LIVING MICROBES.   Even some of the world's smallest creatures have attracted phylogeographic attention. Some of this work has been motivated by the "ubiquitous dispersal" hypothesis (see Finlay 2002), which posits that by virtue of their numerical abundance and ease of dispersal, most free-living species of common microbial eukaryotes with body sizes less than about 2 mm, as well as superabundant free-living prokaryotes that are much smaller, probably lack appreciable population structure across huge (even global) geographic scales. This suggestion, based mostly on theoretical considerations and indirect evidence such as morphology-based taxonomy and microbial community structure, has been controversial (Coleman 2002).

In recent years, the ubiquitous dispersal hypothesis has been put to preliminary empirical genetic tests. In partial support of this notion is a finding by Darling et al. (2000) that Arctic and Antarctic subpolar populations within each of two planktonic foraminiferan species (*Globigerina bulloides* and *Turborotalita quinqueloba*) share at least one identical genotype at an otherwise variable rRNA gene, suggesting that trans-tropical gene flow has occurred recently. Likewise, an identical rDNA genotype has been reported in some flagellated protists worldwide, including at such disparate sites as a shallow fjord in Denmark and hydrothermal deep-sea vents in the eastern Pacific (Atkins et al. 2000). On the other hand, various green algal protists in several recognized genera, such as *Pandorina* and *Volvulina,* have proved upon genetic examination to consist of numerous sexually isolated groups (syngens) that are otherwise morphologically nearly identical (Coleman 2000). Furthermore, within some of these surveyed syngens, genetic distances among rDNA sequences appear to increase with geographic distance between collecting sites, suggesting considerable biogeographic population structure (Coleman 2001). Likewise, in rDNA surveys of *Sulfolobus* microbes (Archaea) that live in isolated geothermal environments, significant genetic differentiation has been documented among various populations scattered around the world, thus contradicting predictions of the unrestricted dispersal hypothesis (Whitaker et al. 2003).

One phenomenon that has complicated biogeographic reconstructions in some microbes, notably bacteria (Parker and Spoerke 1998; Qian et al. 2003; Spratt and Maiden 1999), is horizontal gene transfer (see Chapter 8), which can create genomes with mosaic evolutionary histories and conflicting phylogeographic patterns across loci (Parker 2002). Other challenging difficulties include identifying particular taxa or clades to begin with (because the morphological evidence is often inadequate) and sampling them extensively enough from across vast regions of Earth to critically test the ubiquitous dispersal hypothesis using molecular markers (see John et al. 2003 for an example).

HUMAN POPULATIONS.   Not surprisingly, more attention has been paid to phylogeography in *Homo sapiens* than in any other species. Among many early studies (e.g., Ballinger et al. 1992; Cann et al. 1984; Denaro et al. 1981; DiRienzo and Wilson 1991; Excoffier 1990; Hasegawa and Horai 1991; Johnson et al. 1983; Merriwether et al. 1991; Stoneking et al. 1986; Torroni et al. 1992; Vigilant et al. 1991; Ward et al. 1991; Whittam et al. 1986), two captured the essence of the situation and stand out as having had major historical and conceptual impacts.

First, an early glimpse of global mtDNA diversity came from RFLP analyses of 21 people of diverse racial and geographic origins (Brown 1980). Genetic differentiation proved to be quite limited (mean sequence divergence $p = 0.004$). Vastly more mitochondrial data have accumulated since Brown's original study, but current estimates of mtDNA sequence diversity remain nearly identical to that preliminary appraisal. Thus, the overall picture for human matrilines remains one of fairly shallow evolutionary separations relative to those reported among conspecific populations in most other species. In this regard, the mtDNA results also parallel long-standing findings from the nuclear genome that human populations and races are remarkably similar in molecular makeup, notwithstanding obvious phenotypic differences in traits such as hair texture and skin color (Boyce and Mascie-Taylor 1996; Nei and Livshits 1990; Nei and Roychoudhury 1982). For example, from an early summary of the protein electrophoretic literature, Nei (1985) concluded that "net gene differences between the three races of man, Caucasoid, Negroid, and Mongoloid, are much smaller than the differences between individuals of the same races, but this small amount of gene differences corresponds to a divergence time of 50,000 to 100,000 years."

Brown (1980) also included a provocative statement in his original study: that the observed magnitude of mtDNA diversity "could have been generated from a single mating pair that existed $180–360 \times 10^3$ years ago, suggesting the possibility that present-day humans evolved from a small mitochondrially monomorphic population that existed at that time." This statement implied that the coalescence of extant human matrilines might trace to a single female (dubbed "Eve" by the popular press) within the last few hundred thousand years, and furthermore, that the data indicated a severe bottleneck in absolute human numbers (the "Garden of Eden" scenario). The latter conclusion was soon challenged with results of models and computer simulations of population lineage sorting as a function of historical population demography. From such gene tree theory, Avise et al. (1984a) concluded that "Eve could have belonged to a population of many thousands or tens of thousands of females, the remainder of whom left no descendants to the present day, due simply to the stochastic lineage extinction associated with reproduction." Several other authors likewise pointed out that simply because the genealogy of mtDNA (or any other locus) is observed to coalesce does not necessarily imply an extreme bottleneck in absolute population numbers at the coalescent point (Ayala 1995; Hartl and Clark 1989; Latorre et al. 1986; Wilson et al. 1985). Later analyses of various nuclear genes in humans bolstered the notion that *Homo sapiens* may never have experienced an acute bottleneck: "Genetic variation at most loci examined in human populations indicates that the [effective] population size has been $\cong 10^4$ for the past one Myr ... [and] population size has never dropped to a few individuals, even in a single generation" (Takahata 1993).

The second of the hallmark phylogeographic studies on humans extended the mtDNA survey to 147 people from around the world and produced a parsimony tree whose root traced to the African continent (Cann et al. 1987). These findings led to the "out of Africa" hypothesis, stating that maternal lineages ancestral to modern humans originated in Africa and spread within the last few hundred thousand years to the rest of the world, replacing those of other archaic populations. This conclusion also provoked initial controversy. One criticism came from some paleontologists who, on the basis of fossil or other evidence, favored a multi-regional origin for humans far preceding the apparent evolutionary date of the mtDNA spread (e.g., Wolpoff 1989; Wolpoff et al. 1984; but see also Stringer and Andrews 1988; Wilson et al. 1991). Another criticism came from some geneticists who found that the postulated African root of the molecular phylogeny was not strongly supported (nor refuted) when additional tree-building analyses were applied to the mtDNA data (Hedges et al. 1992c; Maddison 1991; Templeton 1992). What remained unchallenged, however, was another argument for an ancestral African homeland for mtDNA: that extant African populations house by far the highest level of mtDNA polymorphism and, indeed, are paraphyletic with respect to populations on other continents.

Being based on mtDNA evidence, these early discussions about human origins referred to the matrilineal component of our ancestry. To illuminate patrilineal ancestry, and perhaps to reveal "Adam" (the "father of us all"; Gibbons 1991), analogous molecular studies were then conducted on the human Y chromosome. Genealogical analyses of several such sequences uncovered modest genetic variation in our species (Dorit et al. 1995; Whitfield et al. 1995), with results interpreted to indicate a relatively recent coalescent event for human patrilines in Africa (Hammer 1995; Ke et al. 2001).

At first thought, it might be supposed that knowledge of the matrilineal and patrilineal components of human ancestry would complete the story, but this is far from true. The vast majority of any sexual species' genetic heritage involves nuclear loci whose alleles have been transmitted via both genders across the generations. Due to the vagaries of Mendelian inheritance (random segregation and independent assortment), nuclear genealogies inevitably differ somewhat from gene to unlinked gene, as well as from the uniparental transmission pathways traversed by mtDNA and the Y chromosome (Avise and Wollenberg 1997). So attempts have been made to add nuclear gene genealogies (other than those on the Y) to analyses of human origins. Among published examples are studies of the X-linked *ZFX* gene (Huang et al. 1998) and of autosomal genes encoding apolipoproteins (Rapacz et al. 1991; Xiong et al. 1991), β-globin (Fullerton et al. 1994; Harding et al. 1997), and others (Ayala 1996; Tishkoff et al. 1996; Wainscoat et al. 1986). Takahata et al. (2001) used DNA sequence data from ten X-chromosomal regions and five autosomal regions to deduce ancestral haplotypes at each locus, and the analyses offered substantial support for African (rather than Asian) human genetic origins during the Pleistocene. Findings from most loci surveyed to date are also generally consistent with the presence of tight genealogical connections among human populations worldwide (Takahata 1995).

Thus, most available nuclear data also tend to support a relatively recent out-of-Africa expansion scenario for our species (e.g., Armour et al. 1996; Goldstein et al. 1995b; Nei and Takezaki 1996; Reich and Goldstein 1998), while not necessarily eliminating the possibility that some relatively small fraction of genes may also have had early diversification centers elsewhere, such as in Asia (Giles and Ambrose 1986; Takahata et al. 2001; Xiong et al. 1991). One plausible scenario is that humans with modern anatomical features appeared first in Africa and then spread throughout the world, not completely replacing archaic populations, but rather interbreeding with them to some extent (Li and Sadler 1992). To address such issues comprehensively will require secure knowledge of many more nuclear gene genealogies in peoples from around the world.

Genealogical concordance

The literature on mtDNA variation in animals (and cpDNA variation in plants) has demonstrated that nearly all species are likely to be genetically structured across geography, at least to some degree. Given that geographic variation can vary tremendously in magnitude and will have been affected by forces that operated at a wide range of ecological and evolutionary time frames, it is important to recognize not only the proper spatial scale, but also the proper temporal scale in each case.

One empirical way to do so (and perhaps the only way from molecular genetic evidence) is by appealing to "genealogical concordance" principles (Avise and Ball 1990), which in general provide a conceptual framework for empirically distinguishing historically deep (ancient) from shallow (recent) population structures, based on levels of agreement between independent genetic characters or data sets. For heuristic purposes, genealogical concordance has four distinct aspects. These aspects are diagrammed in Figure 6.13 and will be described in turn with a few examples provided of each.

ACROSS MULTIPLE SEQUENCE CHARACTERS WITHIN A GENE.   Almost by definition, any deep phylogenetic split deduced in a gene genealogy will have been registered concordantly by multiple independent sequence changes within the molecule. For example, the evolutionary separation between eastern and western matrilines in southeastern pocket gophers (see Figure 6.8) was deemed to be relatively ancient precisely because at least nine independent restriction profiles agreed perfectly in delineating these matrilineal clades, whereas in these same molecular assays, haplotypes within either the eastern or the western clade usually differed by no more than two such changes. Numerous species have likewise proved to consist of geographic sets of populations that differ from other such groups by many more mutational steps (i.e., display higher sequence divergence) than typically occur within a geographic region.

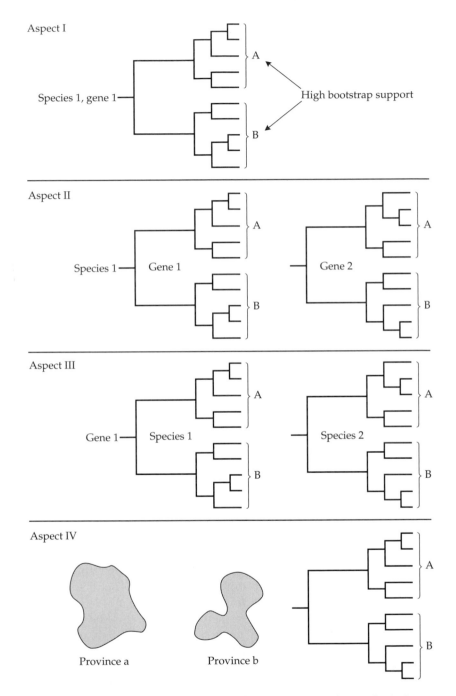

Figure 6.13   Schematic description of four distinct aspects of genealogical concordance. A and B are distinctive phylogroups in a gene tree. (After Avise 2000a.)

The quantitative significance of genealogical concordance aspect I is that bootstrapping or related statistical criteria permit the recognition of putative clades in a gene tree only when multiple characters consistently distinguish what therefore become well-supported clades. In theory, at least three or four diagnostic genetic characters, uncompromised by homoplasy, are required for robust statistical recognition of a putative gene tree clade. Empirically, many more nucleotide substitutions than that often cleanly distinguish regional sets of populations in phylogeographic surveys of mtDNA. The biological significance of aspect I is that appreciable evolutionary time must normally have elapsed for multiple independent mutations to accumulate between lineages within a non-recombining gene tree. Furthermore, under neutrality theory and molecular clock concepts, the greater the magnitude of sequence divergence, the greater the time elapsed (all else being equal).

ACROSS GENE GENEALOGIES WITHIN A SPECIES.   In principle, phylogeographic breaks in a gene tree can arise not only from long-term vicariant separations, but also from isolation by distance in continuously distributed species (Irwin 2002; Neigel and Avise 1993). To distinguish gene-idiosyncratic or spatially haphazard genealogical breaks (due to isolation by distance, or perhaps to gene-specific selection) from ancient vicariance-induced genealogical breaks (whose effects are more likely to be genomically extensive), aspect II of genealogical concordance should be addressed.

Suppose that within a given species, gene genealogies have been estimated not only for mtDNA or cpDNA, but also for each of multiple unlinked nuclear loci within each of which inter-allelic recombination had been rare or absent over the time frame under scrutiny. Suppose further that each of those gene trees displayed a deep genealogical split (as defined by aspect I of genealogical concordance), and that those splits agreed well or perfectly with respect to the particular populations distinguished. This is what is meant by aspect II of genealogical concordance. Its biological significance is that these concordant partitions across independent gene trees within an organismal pedigree almost certainly register a fundamental (i.e., genomically pervasive) phylogenetic split at the population level. In other words, extant populations that concordantly occupy different major branches in multiple gene trees probably separated from one another long ago.

As discussed in earlier chapters, several technical as well as biological complications have conspired to inhibit molecular appraisals of nuclear gene trees at intraspecific levels, but a few informative cases do exist (Hare 2001). One of the earliest and most interesting involved the killifish *Fundulus heteroclitus*, an inhabitant of salt marshes along the eastern seaboard of the United States. Near the midpoint of this coastline, two common alleles at a lactate dehydrogenase (LDH) nuclear gene proved to

(A)

(B)

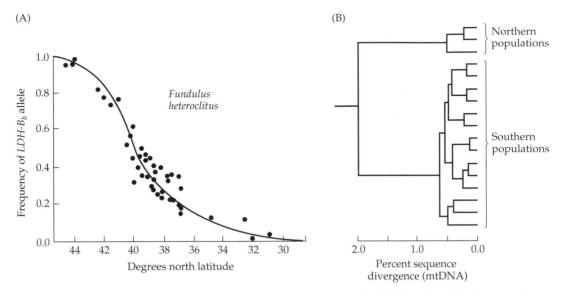

Figure 6.14 Molecular geographic patterns in the killifish *Fundulus heteroclitus*.
(A) Latitudinal cline in population frequencies of the *b* allele in an LDH nuclear polymorphism. (B) Phenogram of mtDNA haplotypes in the same populations, showing a deep phylogenetic distinction between northern and southern areas (a similar phylogeographic pattern was observed in a gene tree of LDH haplotypes). (After Powers et al. 1991a.)

exhibit a pronounced clinal shift in frequency (Figure 6.14A). Detailed laboratory studies revealed kinetic and biochemical differences between these LDH alleles that predicted significant differences among individuals in metabolism, oxygen transport, swimming performance, developmental rate, and relative fitness (Powers et al. 1991a; Schulte et al. 1997). The nature of these differences was such that latitudinal shifts in environmental temperature were posited as directly responsible for the clinal allelic structure (Mitton and Koehn 1975; Powers et al. 1986). Does contemporary adaptation to local ecological conditions provide the entire story for the genetic architecture of these killifish populations?

Researchers then generated an mtDNA gene tree (González-Villaseñor and Powers 1990) as well as a sequence-based LDH gene tree (Bernardi et al. 1993) for killifish populations sampled along the same coastal transect. These trees demonstrated a pronounced phylogenetic subdivision of *F. heteroclitus* into northern versus southern matrilineal clades (Figure 6.14). Thus, northern and southern populations were probably isolated from each other during the Pleistocene and now hybridize secondarily along the mid-Atlantic coast in such a way as to contribute to the clinal structure observed in LDH allele frequencies and in some other nuclear genes. This example demonstrates two points: that genealogical concordance across independent loci (aspect II) can provide empirical support for historical vicariance at the population level, and that phylogenetic and selective mechanisms need not be opposing influences and may in some cases act in concert to achieve an observed population structure (Powers et al. 1991b). Although the differences in LDH allele frequency are mediated to a significant degree by environmental selection, the historical context in which this selection had taken place added an important dimension to knowledge on contemporary population genetic structure in killifish.

Only a small number of studies to date have explicitly searched for intraspecific genealogical concordance across multiple unlinked loci (introns and/or exons). In a fungal taxon, *Coccidioides immitis*, Koufopanou et al. (1997) showed strong phylogeographic agreement across five loci that partitioned California from non-California populations (which might therefore be cryptic species). In the European grasshopper *Chorthippus parallelus*, concordant patterns in allozymes, nuclear DNA sequences, and other characters generally distinguish parapatric subspecies (Cooper and Hewitt 1993; Cooper et al. 1995). In the tide pool copepod *Tigriopus californicus*, Burton (1998) found general agreement between nDNA and mtDNA with regard to a deep phylogeographic partition. On the other hand, Palumbi and Baker (1994) found sharply contrasting phylogeographic structures for nuclear intron sequences and mtDNA in humpback whales (*Megaptera novaeangliae*). The discrepancy in this case probably reflects either differences in genetic drift in mitochon-drial versus nuclear genes (due to their expected difference in effective population size) or asymmetric dispersal by sex, in which males transferred nuclear DNA (but not mtDNA) to offspring they sired in foreign matrilineal groups.

In phylogeographic surveys of the European rabbit (*Oryctolagus cuniculus*) across the Iberian Peninsula, a well-defined genealogical break in mtDNA (Branco et al. 2002) is spatially concordant with patterns in nuclear protein and immunological polymorphisms, but is not registered by major shifts in microsatellite allele frequencies (Queney et al. 2001). The authors concluded that microsatellite loci mutate so rapidly as to produce extensive homoplasy, which obscured what apparently was a relatively ancient (2-million-year-old) population separation. In general, rapidly evolving microsatellite loci may be better suited for revealing recent or shallow population structures than for confirming deep historical structures that may be registered in other classes of molecular markers (Gibbs et al. 2000b; Mank and Avise 2003).

Because nuclear gene trees are difficult to obtain at the intraspecific level, a surrogate approach that falls within a broader conceptual framework of concordance aspect II is to examine other kinds of nuclear genetic evidence for possible phylogeographic agreement with a cytoplasmic gene tree. For example, in the North American sharp-tailed sparrow (*Ammodramus caudacutus*), a rather deep split in matrilineal genealogy (as registered in mtDNA) distinguished an assemblage of birds currently inhabiting the continental interior as well as the northern Atlantic coast from a southern group that occurs along the Atlantic coast from southern Maine to Virginia (Rising and Avise 1993). Populations belonging to these two mtDNA clades also proved to be concordantly recognizable in multivariate analyses of morphology, song, and flight displays (Greenlaw 1993; Montagna 1942). Such concordance among independent lines of (presumably genetic) evidence indicates that the split in the mtDNA gene tree also reflects a rather deep phylogenetic distinction in organismal phylogeny. Indeed, these two sets of populations were later elevated to the status of separate taxonomic species.

In other empirical cases, phylogeographic breaks in an mtDNA gene tree do not seem to coincide with sudden changes in other organismal traits (e.g., Bond et al. 2001; Irwin et al. 2001a,b; Puorto et al. 2001). As discussed earlier in this chapter, possible reasons for such discrepancies are many and must be scrutinized on a case-by-case basis.

ACROSS CO-DISTRIBUTED SPECIES.   Suppose now that several species with similar geographic ranges have been genetically surveyed, that several or all of them display deep phylogeographic structure (as evidenced by aspects I or II of genealogical concordance, as described above), and that the phylogeographic partitions are at least roughly similar in spatial placement and perhaps temporal depth. Aspect III of genealogical concordance would then have been documented. The biological significance of aspect III is that it strongly implicates historical biogeographic factors as having shaped the genetic architectures of multiple species in similar fashion. Studies conducted under this multi-species orientation exemplify what has been termed the "regional" (Avise 1996), "landscape" (Templeton and Georgiadis 1996), or "comparative" (Bermingham and Moritz 1998) approach to phylogeography.

The first extensive phylogeographic appraisals of a regional biota involved numerous freshwater and maritime species in the southeastern United States (see reviews in Avise 1992, 1996, 2000a). In both of these environmental realms, a remarkable degree of aspect III phylogeographic concordance was evidenced in molecular genetic surveys. For example, within each of several freshwater fishes, including *Micropterus salmoides* bass, *Amia calva* bowfins, *Gambusia* mosquitofish, and each of four species of *Lepomis* sunfish, deep phylogeographic partitions in the mtDNA molecule typically distinguished populations from most river drainages entering the Gulf of Mexico from those inhabiting most watersheds of the Atlantic Ocean (Figures 6.15 and 6.16). Some of these species have also been surveyed for

Figure 6.15   Relationships among mtDNA haplotypes in seven species of fresh- ▶ water fish surveyed across the southeastern United States. Data are all plotted on the same scale of estimated sequence divergence. Data for *Lepomis* and *Amia* are from Bermingham and Avise (1986) and Avise et al. (1984b), those for *Gambusia* are from Scribner and Avise (1993a), and those for *Micropterus* are from Nedbal and Philipp (1994). (For some of these taxa, populations were considered conspecific at the time of the original assays but have since been taxonomically subdivided into eastern and western sister species. For example, the largemouth bass was split into *M. salmoides* and *M. floridanus*; Kassler et al. 2002.) (After Avise 2000a.)

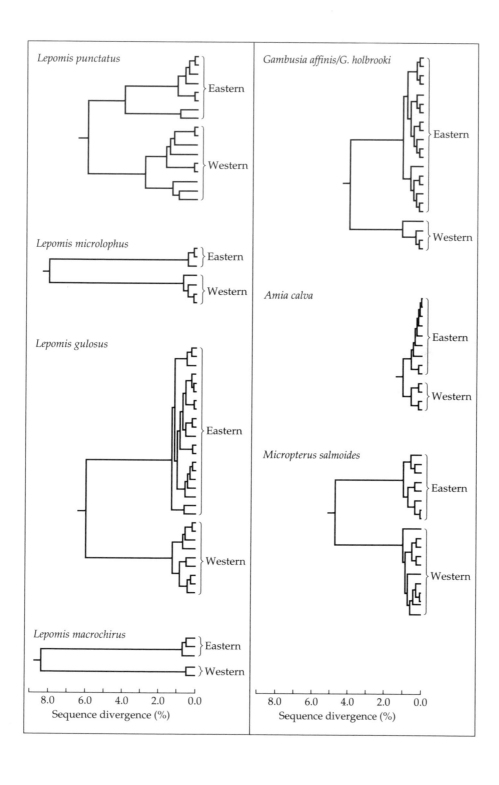

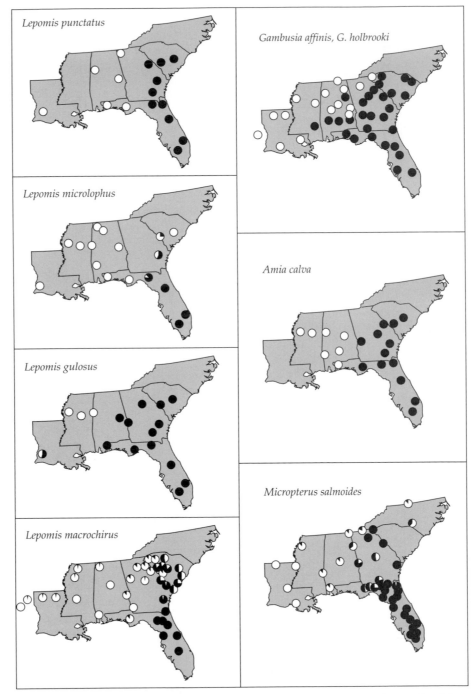

Figure 6.16 **Geographic distributions of major mtDNA clades** within each of the seven freshwater fish species described in Figure 6.15. For each species, pie diagrams summarize observed frequencies of the two fundamental clades at various geographic sites. (After Avise 2000a.)

nuclear allozyme markers, for which comparable genetic breaks between these "eastern" and "western" forms (as well as hybridization between them in secondary contact zones) have been documented (Avise and Smith 1974; Philipp et al. 1983b; Scribner and Avise 1993a; Wooten and Lydeard 1990). Similar (albeit more complicated) phylogeographic patterns in mtDNA also characterize several freshwater turtle species surveyed throughout this same geographic area (Roman et al. 1999; Walker et al. 1995, 1997, 1998a,b; see review in Walker and Avise 1998).

Likewise, in the maritime realm of the southeastern United States, major and concordant genetic subdivisions have been reported in a wide variety of vertebrate and invertebrate species (Figure 6.17). Thus, apparently deep historical partitions, as registered in mtDNA or various nuclear

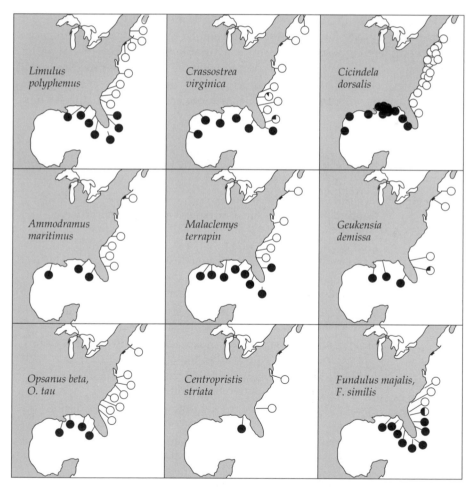

Figure 6.17   Geographic distributions of primary genetic subdivisions observed within each of nine maritime taxa of the southeastern United States. Pie diagrams follow the format described in Figure 6.16. (After Avise 2000a.)

assays or both, tend to characterize most surveyed Atlantic versus most Gulf populations of *Limulus* horseshoe crabs (Saunders et al. 1986), *Crassostrea* oysters (Hare and Avise 1996, 1998; Reeb and Avise 1990), *Cicindela* beetles (Vogler and DeSalle 1993, 1994), *Ammodramus* sparrows (Avise and Nelson 1989), *Malaclemys* terrapins (Lamb and Avise 1992), *Geukensia* mussels (Sarver et al. 1992), *Opsanus* toadfish (Avise et al. 1987b), *Centropristis* sea bass (Bowen and Avise 1990), and *Fundulus* killifish (Duggins et al. 1995), among others. Although this phylogeographic pattern is far from universal among maritime species that have been genetically analyzed from this region (Avise 2000a; Gold and Richardson 1998), it nonetheless is prevalent enough to suggest that historical biogeographic factors (see below) have affected the genetic architecture of a significant fraction of this regional biota.

This general kind of multi-species concordance in phylogeographic pattern is not unique to faunas in the southeastern United States. Several comparative molecular surveys are now available for other regional biotas around the world. These surveys sometimes have (and sometimes have not) documented aspect III concordance to varying degrees. For example, concordant intraspecific phylogeographic patterns across multiple species have

TABLE 6.5   Examples and outcomes of comparative phylogeography studies

Organisms	Outcome and region
Marine vertebrates and invertebrates	Concordant partitions but apparently varying temporal depths between geminate taxa across the Isthmus of Panama
Marine invertebrates, mostly	Four distinct phylogeographic patterns, each observed in several species, documenting colonization histories of trans-Arctic taxa
Amphipods and other crustaceans	Concordant subdivision of each of six species into two units, one inhabiting the Black Sea and the other the Caspian Sea region
Butterflies and vertebrates	Considerable congruence of phylogeographic patterns in diverse invertebrate and vertebrate taxa of Amazonia
Butterfly fishes (*Chaetodon*)	Striking phylogeographic concordance between two species groups in the Indo-West Pacific marine realm
Darter fishes	No appreciable phylogeographic concordance among five species in highlands of the south-central United States
Freshwater fishes	Species-idiosyncratic patterns suggesting perhaps three distinct waves of invasion into Central America from South America
Freshwater and anadromous fishes	Consistently deep phylogeographic partitions in species from non-glaciated regions compared with more northern taxa
Herpetofauna	No appreciable concordance in phylogeographic patterns across species in the North American desert southwest
Lizards	Ancient and concordant fragmentation of clades on either side of Wallace's Biogeographic Line in southeastern Asia

been identified for elements of both the herpetofauna and the avifauna in rainforest remnants of northeastern Australia (Joseph and Moritz 1994; Joseph et al. 1995; Moritz and Faith 1998; Schneider et al. 1998) and for disjunct conspecific populations of eight marine fish species in the northern Gulf of California versus the outer Pacific coast (Bernardi et al. 2003). Table 6.5 summarizes several additional examples of comparative molecular phylogeography as applied to regional faunas and floras.

BETWEEN GENEALOGICAL AND OTHER BIOGEOGRAPHIC INFORMATION.   A final aspect of genealogical concordance is between molecular genetic data and traditional biogeographic evidence based on nonmolecular data. Such concordance may apply to individual species if, for example, particular populations that are cleanly demarcated in a molecular genealogy correspond to those also recognized (perhaps in taxonomic summaries) from morphological, behavioral, geological, or other more traditional lines of evidence. Many such examples are reviewed by Avise (2000a). Or, aspect IV concordance can broadly refer to agreement between molecular phylogeographic patterns for a regional biota and comparable patterns registered in more conventional biogeographic appraisals.

Primary reference or review
Bermingham et al. 1997; Bermingham and Lessios 1993; Knowlton et al. 1993
Cunningham and Collins 1998
Cristescu et al. 2003
Hall and Harvey 2002
McMillan and Palumbi 1995
Turner et al. 1996
Bermingham and Martin 1998
Bernatchez and Wilson 1998
Lamb et al. 1989, 1992
Schulte et al. 2003

(continued)

TABLE 6.5 (*continued*)   **Examples and outcomes of comparative phylogeography studies**

Organisms	*Outcome and region*
Pelagic seabirds and marine turtles	Similar phylogeographic partitions of rookeries on a global scale, but contrasting patterns within ocean basins
Birds	In 7 of 13 species, rather consistent distinctions between populations on alternate sides of Beringia
Birds	No appreciable concordance in phylogeographic patterns in species with transcontinental ranges
Birds	Species-idiosyncratic phylogeographic patterns in species of the Caribbean islands
Birds	Concordant genetic evidence from two African species for three biogeographic regions in area of Cameroon, western Africa
Mammals, birds, reptiles, amphibians	Otherwise cryptic but often concordant breaks distinguishing phylogeographic units in diverse Baja California species
Vertebrates, arthropods, plants	Little concordance in phylogeographic patterns, but strong similarities in postglacial colonization routes across Europe
Cats (*Leopardus*)	Remarkably concordant phylogeographic partitions in ocelot and margay cats across Central and South America
Bats and other mammals	Limited population structure of mtDNA lineages in Neotropical bats contrasts with strong structure in small non-volant mammals
Monkeys, toads	Concordant areas of endemism documented by molecular markers for macaques (*Macaca*) and toads (*Bufo*) on Sulawesi
Lemurs (primates)	Madagascar's landscape features that acted as phylogeographic barriers revealed by mtDNA sequences of several lemur species
Various plants	Strong concordance in major molecular phylogeographic clades across several plant species in the Pacific Northwest
Plants and animals	Congruent patterns of genetic diversity and biodiversity hotspots revealed for diverse biotas in the California Floristic Province

Note: In each case, multiple co-distributed species were surveyed, using molecular markers (typically mtDNA in animals, cpDNA in plants), for genealogical patterns on a regional scale.

For example, in the maritime realm of the southeastern United States, biogeographers have long recognized the existence of two distinct faunal assemblages (temperate versus subtropical) that meet along the east-central Floridian coastline in the general area of Cape Canaveral (Briggs 1958, 1974). In other words, the southern range limit of many temperate species occurs in this transition zone, as does the northern range limit of many species adapted to warmer waters. For several other species that *are* continuously distributed across this transition zone, molecular data have revealed otherwise cryptic genealogical breaks in this same region (see Figures 6.4 and 6.17). Thus, there is a general spatial agreement between major biogeographic provinces, as defined by traditional faunal lists, and major phylogeographic subdivisions often registered within species.

Primary reference or review
Avise et al. 2000
Zink et al. 1995
Zink 1996
Bermingham et al. 1996
Smith et al. 2000
Riddle et al. 2000
Taberlet et al. 1998
Eizirik et al. 1998
Ditchfield 2000
Evans et al. 2003
Pastorini et al. 2003
Soltis et al. 1997
Calsbeek et al. 2003

These findings, if they can be generalized, indicate that biogeographic provinces and the boundaries between them may often have been shaped by a combination of historical vicariance and contemporary selection. In the southeastern United States, Pleistocene or earlier events probably separated populations periodically into Gulf versus Atlantic zones, where adaptations to local conditions also arose. Populations in one or the other region then sometimes went extinct, in which case only their sister taxa in the other region remained present for observation today (thus contributing to the distinctness of faunal provinces based on species lists). For species whose populations survive in both regions, genetic footprints of the sundering events are often retained in extant genomes. Today, these distinctive forms now characterize the Gulf and Atlantic regions, and they often meet and mix in boundary zones, which therefore exist where they do in part for historical reasons and in part for reasons related to contemporary selection gradients and associated gene flow barriers.

A similar spatial agreement (between traditionally recognized biogeographic provinces and concentrations of intraspecific phylogroups) exists for freshwater fishes of the southeastern United States. Based on the known geographic ranges of 241 fish species in the area, Swift et al. (1985) estimated and then clustered faunal similarity coefficients for the region's approximately 30 major river drainages. The deepest split in this faunal similarity phenogram cleanly distinguished Gulf coast drainages from those along the Atlantic coast and in peninsular Florida, thereby defining two major biotic provinces that agree quite closely with intraspecific phylogeographic patterns often registered in molecular appraisals (Figure 6.18). Again, historical as well as contemporary factors must have operated to shape these concordant regional features of the biotic and genetic landscape.

Another example of aspect IV concerns European biotas. Extensive molecular appraisals of many animal and plant species across Europe have revealed both species-specific idiosyncrasies and generalized trends in phylogeographic patterns (see reviews in Hewitt 1999, 2000; Petit et al. 2003b; Petit and Verdramin 2004; Taberlet et al. 1998). Among the latter, most noticeable are genetic as well as traditional biogeographic data documenting that a small number of Pleistocene refugia (notably in the Balkans, the Iberian Peninsula, and the Apennine or Italian Peninsula) were major centers of historical genetic isolation and were the primary foci from which postglacial recolonizations of Europe often took place along specifiable and rather consistent routes.

Genealogical discordance

The flip side of genealogical concordance is the lack of phylogeographic agreement across various genetic characters or data sets. Genealogical discordance likewise has four distinct aspects, and these also can be highly informative when uncovered in particular instances.

Discordance aspect I occurs when different sequence characters within a gene suggest conflicting or overlapping clades in the gene tree. If the locus was historically free of inter-allelic recombination (as is normally true for mtDNA), then homoplasy (evolutionary "noise" involving convergences, parallelisms, or reversals in some character states) must be responsible. If the locus is housed in the nucleus, and if the nucleotide differences that display overt phylogenetic discordance are clustered in distinct portions of a gene sequence, then inter-allelic recombination may have been responsible (see Chapter 4). Alleles that arose via intragenic recombination consist of amalgamated stretches of sequence that truly had different genealogical histories within the locus.

Aspect II discordance involves disagreement across genes in phylogeographic signatures within a given species. Some degree of phylogenetic dis-

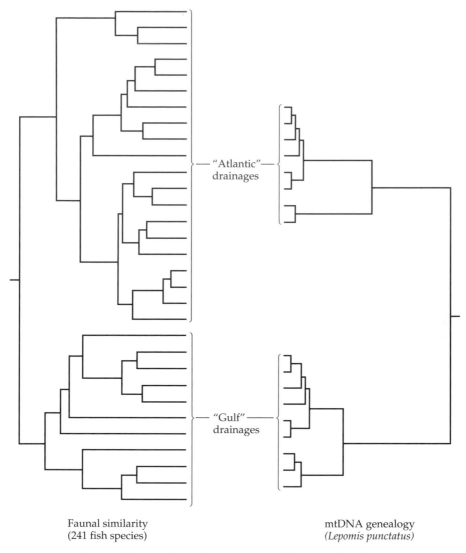

Figure 6.18   Aspect IV genealogical concordance illustrated by fishes of the southeastern United States. *Left*, phenogram summarizing faunal relationships among 31 river drainages, based on faunal similarity coefficients from compilations of range data from 241 fish species (data from Swift et al. 1985). *Right*, geographic distributions of the two major branches in an intraspecific matrilineal genealogy for the spotted sunfish, *Lepomis punctatus* (data from Bermingham and Avise 1986). (After Avise 2000a.)

Faunal similarity
(241 fish species)

mtDNA genealogy
(Lepomis punctatus)

"Atlantic"
drainages

"Gulf"
drainages

cordance across gene trees is an inevitable consequence of Mendelian inheritance and the vagaries of lineage sorting at unlinked loci through a sexual pedigree. As described earlier, however, additional biological factors can also produce genealogical heterogeneity among loci. Two such factors apply with special force to comparisons between nuclear and mtDNA (or cpDNA) genealogies: the expected fourfold difference (all else being equal) in effective population sizes for genomes housed in the nucleus versus the cytoplasm; and behavioral differences between the genders, which can produce contrasting phylogeographic patterns at cytoplasmic versus nuclear loci if, for example, one sex or its gametes are far more dispersive than the other (see above). Finally, dramatic heterogeneity among gene trees within an extended pedigree can characterize any sets of molecular markers if some but not all of them have been under intense forms of balancing or diversifying selection.

Aspects III and IV of phylogeographic discordance simply suggest that whatever ecological or evolutionary forces may have been at play in a given instance have operated mostly in a species-idiosyncratic fashion, rather than in a generic way that might otherwise have concordantly shaped the genetic architectures of a regional biota. Even when each species proves to have a unique or peculiar phylogeographic history, this finding can be of considerable interest, for example, in developing conservation plans for particular rare or endangered species (see Chapter 9).

Microtemporal Phylogeny

Most nucleic acids evolve far too slowly to permit direct detection of significant de novo sequence evolution over yearly or decade-long scales. One valiant attempt to describe such microtemporal changes involved comparisons of mtDNA sequences in modern versus earlier populations of a kangaroo rat (*Dipodomys panamintinus*) sampled at three locales in California (Thomas et al. 1990). Sequences from extant specimens were compared with PCR-amplified sequences from dried museum skins prepared in 1911, 1917, and 1937. Results indicated temporal stability, with the three populations showing identical genetic relationships in the early- and late-twentieth-century collections. However, even if a dramatic population genetic change had been observed, it presumably would have entailed frequency shifts of preexisting haplotypes (as can occur rapidly by random genetic drift, differential reproduction, or migration of foreign lineages into the site), rather than the in situ origin and spread of de novo mutations over such a short period of time. A recent genetic survey of white-footed mice (*Peromyscus leucopus*) in the Chicago area compared mtDNA haplotypes of nineteenth-century museum skins with those present in modern samples, and it did indeed reveal such allele frequency changes at the population level (Pergams et al. 2003).

Exceptional molecular systems do exist that mutate so rapidly as to permit de novo sequence evolution to be documented and monitored in contemporary time (Jenkins et al. 2002). These systems are RNA viruses such as HIVs, the human immunodeficiency retroviruses responsible for AIDS (acquired immune deficiency syndrome). The mean rate of synonymous nucleotide substitution for HIV genomes is approximately 10^{-2} per site per year (W.-H. Li et al. 1988), or about a million times greater than typical rates in the nuclear genomes of most higher organisms (see Chapter 4). High mutation rates in particular regions of RNA viruses, combined with occasional inter-strain recombination, underlie the astonishingly rapid changes often observed in viral pathogenicity and antigenicity (Coffin et al. 1986; Gallo 1987).

The HIV viruses come in two distinct classes (HIV-1 and HIV-2) that emerged in Africa (Diamond 1992), probably within the past few decades following natural (or perhaps unnatural; see Marx et al. 2001; Poinar et al. 2001a; Weiss 2001) transfers from related "simian" viruses (SIVs) that infect wild primates. Judging from recent phylogenetic reconstructions based on nucleotide sequences (Figure 6.19A), HIV-1 and HIV-2 originated when distinctive SIVs from chimpanzees and sooty mangabeys, respectively, jumped into humans, perhaps on more than one occasion each (Bushman 2002; Gao et al. 1999; Hahn et al. 2000; Korber et al. 2000; Lemey et al. 2003; T. Zhu et al. 1998).

Beginning in the 1980s, phylogenetic analyses of HIV sequences had already helped to document the histories of the viral lineages that spread the AIDS pandemic to millions of people worldwide (Desai et al 1986; Gallo 1987; Yokoyama and Gojobori 1987). For example, sequences analyzed from 15 HIV isolates from the United States, Haiti, and Zaire were the basis for the phylogenetic appraisal summarized in Figure 6.19B. Results provided some of the earliest genetic support for HIV's African origins and the timing of the virus's subsequent expansion to Haiti and the United States. The most remarkable aspect of this viral phylogeny is its short time frame; various branching events date only to the 1960s and 1970s. Based on similar phylogenetic analyses, the whole HIV-1 pandemic traces to a common ancestral viral sequence from about 1931 (Korber et al. 2000).

Historical reconstructions based on molecular data have likewise been accomplished for other disease-causing RNA viruses, a case in point being the dengue virus (*Flavivirus* sp.). Dengue is an emerging tropical disease now affecting more than 50 million people. A phylogeny based on nucleotide sequences indicates that the virus arose approximately 1,000 years ago, that its transfer from monkeys to humans led to sustained human transmission beginning about one to three centuries ago, and that current global diversity in the virus involves four or five primary lineages (Twiddy et al. 2003).

Another consequence of rapid sequence evolution in RNA viruses is that different people (and sometimes even the same individual at different times; Holmes et al. 1992) may carry recognizable variants of the virus, a finding with forensic ramifications. For example, in comparisons of HIV-1 sequences from a Florida dentist, seven of his infected patients, and 35 other local HIV-1 carriers, it was shown that the dentist's particular viral strain was genealogically allied to those of five of his clients (Ou et al. 1992). These molecular findings were interpreted to provide the first genetic confirmation of HIV transmission (unintentional) from an infected health care professional to clients. Another such case of HIV transmission, documented by molecular markers, involved criminal intent (Metzker et al. 2002).

(A)

(B)

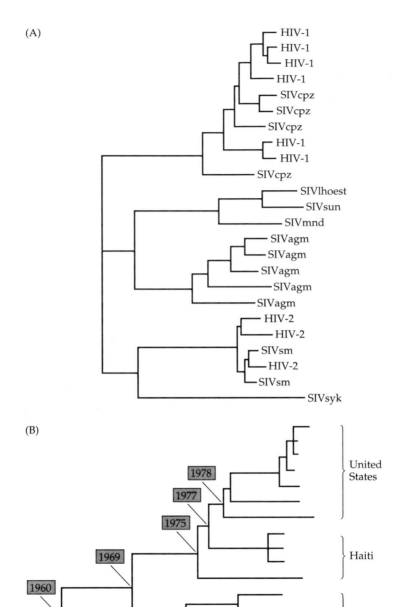

◀ **Figure 6.19   Molecular microphylogenies for HIV isolates.** (A) Phylogeny based on protein sequences from the *Pol* gene in various HIV strains and in SIVs from several primate species. Note the close relationship of HIV-1 to SIVcpz (from chimpanzees, *Pan troglodytes*) and of HIV-2 to SIVsm (from sooty mangabeys, *Cercocebus atys*). Other primates whose SIVs were analyzed include *Cercopithecus lhoesti* (lhoest), *C. solatus* (sun), *C. albogularis* (syk), *Mandrillus sphinx* (mnd), and *Chlorocebus* sp. (agm). (After Hahn et al. 2000.) (B) Phylogeny, based on several sequenced regions of HIV genomes, indicating one early route in the spread of AIDS from Africa to the New World. (After W.-H. Li et al. 1988.)

SUMMARY

1. All conspecific individuals are genetically related through a time-extended pedigree (mating partners and parent–offspring links) that constitutes the full intraspecific genealogy of a species. Molecular markers can help to recover various components of this extended pedigree.

2. Molecular approaches for assessment of kinship within a species normally require highly polymorphic, qualitative genetic markers with known transmission patterns. However, the complexity of potential transmission pathways between relatives more distant than parents and offspring, coupled with the relatively narrow range of potential kinship coefficients (0.00–0.25), means that distinguishing precise categories of genetic relationship for specific pairs of individuals can be accomplished only in rather special cases. On the other hand, assessments of *mean* genetic relatedness within groups are readily conducted.

3. In eusocial species, such as many haplodiploid hymenopterans, estimates of mean intra-colony genetic relatedness sometimes have proved to be high, but numerous exceptions present a conundrum for some sociobiological theories on the evolution of reproductive altruism.

4. Within non-eusocial groups, a variety of mean genetic relatedness values have been observed using molecular markers, and these values are often interpretable in terms of the suspected behaviors and natural histories of the particular species assayed. Genetic markers have also helped to address questions regarding mechanisms and genetic consequences of kin recognition.

5. Populations of almost all species are genetically structured across geography. These genetic architectures have been characterized for numerous species using molecular markers, and they clearly can be influenced by ecological and evolutionary factors operating over a wide variety of spatial and temporal scales. Among these influences are mating systems and gene flow regimes, which in turn can be influenced by the species-specific dispersal capabilities of gametes and zygotes, by the behaviors of organisms (including their vagility and social cohesiveness) and by the physical structure of the environment.

6. The contemporary genetic architecture of any species will also have been influenced by biogeographic and demographic factors of the past. In large measure, historical perspectives on population genetic structure were stimulated by the extended genealogical reconstructions made possible by molecular assays for non-recombining haplotypes in cytoplasmic genomes (especially animal mtDNA). A relatively new discipline termed phylogeography has enriched

biogeographic analyses and provided an empirical and conceptual bridge between the formerly independent fields of traditional population genetics and phylogenetic biology.

7. Some viruses evolve so rapidly that genetic changes can be directly observed across time frames of years or decades. For example, molecular phylogenetic appraisals have revealed important details about the origin and spread of HIV viruses within the last century.

7

Speciation and Hybridization

Without gene flow, it is inevitable that there will be speciation.

M. H. Wolpoff (1989)

Numerous "species concepts" have been advanced for sexually reproducing organisms (Mayden 1997), the most historically influential of which are summarized in Box 7.1. Most of these concepts entail the perception of conspecific populations as a field for gene recombination—that is, as an extended reproductive community within which genetic exchange potentially takes place. For example, under the popular biological species concept (BSC) championed by Dobzhansky (1937), species are envisioned as "groups of actually or potentially interbreeding natural populations which are reproductively isolated from other such groups" (Mayr 1963). Many authors have expressed sentiments on the BSC similar to those of Ayala (1976b):

> Among cladogenetic processes, the most decisive one is speciation—the process by which one species splits into two or more. ... Species are, therefore, independent evolutionary units. Adaptive changes occurring in an individual or population may be extended to all members of the species by natural selection; they cannot, however, be passed on to different species.

Thus, under the BSC and related concepts, species are perceived as biological and evolutionary entities that are more meaningful and perhaps less arbitrary than other taxonomic categories such as subspecies, genera, or orders (Dobzhansky 1970; Howard and Berlocher 1998). Nonetheless, several complications attend the application of biological (or other) species concepts.

One difficulty of the BSC involves the discretionary judgments that are often required about the specific status of closely related forms in allopatry (and also of extant forms to their evolutionary ancestors). Inevitably, reproductive isolating barriers, or

BOX 7.1   Representative Species Concepts and Definitions

1. *Biological species concept* (BSC) (Dobzhansky 1937): "Species are systems of populations: the gene exchange between these systems is limited or prevented by a reproductive isolating mechanism or perhaps by a combination of several such mechanisms."

   Comment: Unquestionably the most influential concept for sexually reproducing species, the BSC remains popular today.

2. *Evolutionary species concept* (ESC) (Simpson 1951): "a lineage (ancestral–descendant sequence of populations) evolving separately from others and with its own unitary evolutionary role and tendencies."

   Comment: Applicable both to living and extinct groups, and to sexual and asexual organisms. However, this concept is vague operationally in what is meant by "unitary evolutionary role and tendencies."

3. *Phylogenetic species concept* (PSC) (Cracraft 1983): a monophyletic group composed of "the smallest diagnosable cluster of individual organisms within which there is a parental pattern of ancestry and descent."

   Comment: Explicitly avoids all reference to reproductive isolation and focuses instead on phylogenetic histories of populations. A serious problem involves how monophyly is to be recognized and how to distinguish histories of traits (e.g., gene trees) from histories of organisms (pedigrees).

4. *Recognition species concept* (RSC) (Paterson 1985): the most inclusive population of biparental organisms which share a common fertilization system.

   Comment: Similar to the BSC in viewing conspecific populations as a field for gene recombination. However, this concept shifts the focus away from isolating mechanisms as barriers to gene exchange between species and toward the

"RIBs" (Box 7.2), develop between geographically separated populations as an ancillary by-product of genomic divergence, but the time frames involved and the magnitudes of differentiation are matters for study in particular instances. The "acid test" for biological species status—whether populations retain separate identities in sympatry—often has not been carried out in nature. A second practical difficulty involves the issue of how much genetic exchange disqualifies populations from status as separate biological species. Thus, the study of speciation conceptually links the topic of gene flow (see Chapter 6) with that of introgressive hybridization. Under the BSC, there are no black-and-white solutions to either of these difficulties because genetic divergence and speciation are gradual processes that in many cases can yield gray outcomes at particular points in evolutionary time, and because levels of genetic exchange can vary along a continuum from nil to extensive (Dobzhansky 1976).

positive role of reproduction-facilitating mechanisms among members of a species. Although reproductive barriers can arise as a by-product of speciation, under the RSC they are not viewed as an active part of the speciation process.

5. *Cohesion species concept* (CSC) (Templeton 1989): "the most inclusive population of individuals having the potential for cohesion through intrinsic cohesion mechanisms."

   Comment: Attempts to incorporate strengths of the BSC, ESC, and RSC and avoid their weaknesses. The major classes of cohesion mechanisms are genetic exchangeability (factors that define the limits of spread of new genetic variants through gene flow) and demographic exchangeability (factors that define the fundamental niche and the limits of spread of new genetic variants through genetic drift and natural selection).

6. *Concordance principles* (CP) (Avise and Ball 1990): a suggested means of recognizing species by the evidence of concordant phylogenetic partitions at multiple independent genetic attributes.

   Comment: Attempts to incorporate strengths of the BSC and PSC and avoid their weaknesses. This approach accepts the basic premise of the BSC, with the understanding that the reproductive barriers are to be interpreted as intrinsic as opposed to extrinsic (purely geographic) factors. When phylogenetic concordance is exhibited across genetic characters solely because of extrinsic barriers to reproduction, subspecies status is suggested.

These and other species concepts all have limitations related to the inevitable ambiguities of cleanly demarcating continuously evolving lineages (Hey 2001; Hey et al. 2003). Thus, for taxonomy, conservation, and other purposes, it may be wiser to accept (rather than bemoan) such uncertainty as simply being inherent in the nature of evolutionary processes.

Another challenge in applying the BSC involves a need to distinguish the evolutionary origins of RIBs from their genetic consequences. Normally, reproductive barriers under the BSC are considered *intrinsic* biological factors rather than purely *extrinsic* limits to reproduction resulting from geographic separation alone. However, this distinction blurs when syntopic populations (those occupying the same general habitat) are isolated via preferences for different microhabitats, particularly when these ecological proclivities are coupled with differences in mate choice (Diehl and Bush 1989). In such situations, one substantive as well as semantic issue is whether speciation may have occurred sympatrically, versus allopatrically followed by secondary range overlap. Another issue is whether certain types of RIBs arise in direct response to selection pressures favoring homotypic matings (see Box 7.2) or whether they reflect non-selected byproducts of genomic differentiation that occurred for other reasons.

BOX 7.2 Classification of Reproductive Isolating Barriers (RIBs)

1. Prezygotic Barriers

   a. Ecological or habitat isolation: Populations occupy different habitats in the same general region, and most matings take place within these microhabitat types.

   b. Temporal isolation: Matings take place at different times (e.g., seasonally or diurnally).

   c. Ethological isolation: Individuals from different populations meet, but do not mate.

   d. Mechanical isolation: Inter-population matings occur, but no transfer of male gametes takes place.

   e. Gametic mortality or incompatibility: Transfer of male gametes occurs, but eggs are not fertilized.

2. Postzygotic Barriers

   a. F_1 inviability: F_1 hybrids have reduced viability.

   b. F_1 sterility: F_1 hybrids have reduced fertility.

   c. Hybrid breakdown: F_2, backcross, or later-generation hybrids have reduced viability or fertility.

One rationale for distinguishing between prezygotic and postzygotic RIBs is that, in principle, only the former are directly selectable. Under the "reinforcement" scenario of Dobzhansky (1940; Blair 1955), natural selection can act to superimpose prezygotic RIBs over preexisting postzygotic RIBs that may have arisen, for example, in former allopatry (Liou and Price 1994). As stated by Dobzhansky (1951), "Assume that incipient species, A and B, are in contact in a certain territory. Mutations arise in either or both species which make their carriers less likely to mate with the other species. The nonmutant individuals of A which cross to B will produce a progeny which is adaptively inferior to the pure species. Since the mutants breed only or mostly within the species, their progeny will be adaptively superior to that of the non-mutants. Consequently, natural selection will favour the spread and establishment of the mutant condition."

Notwithstanding its conceptual appeal, Dobzhanksy's suggestion has proved difficult to verify observationally or experimentally (see review in Butlin 1989). Koopman (1950) and Thoday and Gibson (1962) provide widely quoted examples of selective reinforcement of prezygotic RIBs, but other such experimental studies have produced equivocal outcomes (e.g., Spiess and Wilke 1984). It is true that "reproductive character displacement" (greater interspecific mate discrimination in sympatry than in allopatry) is quite common in nature (Noor 1999). However, the extent to which it reflects direct selection against hybrids as opposed to other evolutionary mechanisms (such as spatially varying intensities of sexual selection without hybrid dysfunction) remains uncertain (Day 2000; Turelli et al. 2001).

Associated with the speciation process, under any definition, is the conversion of genetic variability within a species to between-species genetic differences. However, because RIBs retain primacy in demarcating species under the BSC, no arbitrary magnitude of molecular genetic divergence can provide an infallible metric to establish specific status, especially among allopatric forms. Furthermore, as noted by Patton and Smith (1989), almost "all mechanisms of speciation that are currently advocated by evolutionary biologists … will result in paraphyletic taxa as long as reproductive isolation forms the basis for species definition" (see below). How, then, can molecular markers inform speciation studies? First, molecular patterns provide distinctive genetic signatures (Figure 7.1) that often can be related to demographic events during speciation or to the geographic settings in which speciation took place (Barraclough and Vogler 2000; Harrison 1998; Neigel and Avise 1986; Templeton 1980a). Second, estimates of genetic differentiation between populations at various stages of RIB acquisition are useful in assessing temporal aspects of the speciation process (Coyne 1992). Finally, molecular markers are invaluable for assessing the magnitude and pattern of genetic exchange among related forms, and thereby can help to elucidate the intensity and nature of RIBs.

The Speciation Process

What follows are examples of the diversity of questions in speciation theory that have been empirically addressed and answered, at least partially, through use of molecular genetic markers.

How much genetic change accompanies speciation?

TRADITIONAL PERSPECTIVES.   One long-standing view of species differences, stated clearly by Morgan early in the last century (1919), is that species differ from each other "not by a single Mendelian difference, but by a number of small differences." A counterproposal frequently expressed was that new species, or even genera, might arise by single mutations of a special kind—"macromutations" or "systemic mutations" (deVries 1910; Goldschmidt 1940)—that suddenly transform one kind of organism into another. Although such suggestions for saltational speciation are untenable in their original formulation, more recent theories have stressed plausible routes by which species can arise rapidly, perhaps in some cases with minimal molecular genetic divergence overall. Examples of such avenues to sudden speciation in plants and animals are summarized in Box 7.3. But apart from these "special cases" (which nonetheless may be fairly common), can species arise quickly and with little genetic alteration?

One class of arguments for sudden speciation came from paleontology. Based on a reinterpretation of "gaps" in the fossil record, Eldredge and Gould (1972) proposed that diagrams of the Tree of Life, in which divergence is plotted on one axis and time on the other, are best represented as "rectangular"

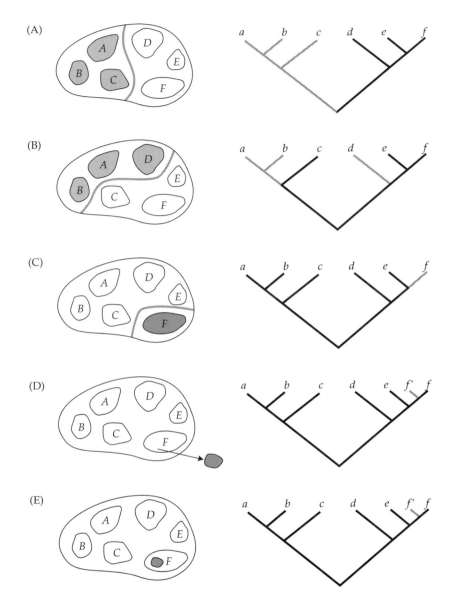

Figure 7.1 Five modes of speciation and corresponding gene trees. Shown on the right are distributions of allelic lineages in the two daughter species (solid versus gray lines). For simplicity, each population is represented as monomorphic, and the gene genealogy in each case is {[(*a*,*b*)(*c*)][(*d*)(*e*,*f*)]}. In reality, most populations are likely to be polymorphic and hence, upon separation, are expected to evolve through intermediate states of polyphyly and paraphyly in the gene tree (see Figure 4.13). (A) Speciation by geographic subdivision, with the physical partition congruent with an existing phylogenetic discontinuity. (B) Speciation by subdivision, with the partition not congruent with an existing phylogenetic discontinuity. (C) Speciation in a peripheral population. (D) Speciation via colonization of a new habitat by propagule(s) from a single source population. (E) Local sympatric speciation. (After Harrison 1991.)

BOX 7.3 Sudden Speciation

Several known pathways to rapid speciation entail little or no change in genetic composition at the allelic level (beyond the rearrangement or sorting of genetic variation from the ancestral forms).

1. *Polyploidization.* The origin of stable polyploids usually is associated with hybridization between populations or species that differ in chromosomal constitution. If the hybrid is sterile only because its parental chromosomes are too dissimilar to pair properly during meiosis, this difficulty is removed by the doubling of chromosomes that produces a polyploid hybrid. Furthermore, such a polyploid species spontaneously exhibits reproductive isolation from its progenitors because any cross with the parental species produces progeny with unbalanced (odd-numbered) chromosome sets. For example, a cross between a tetraploid and a diploid progenitor produces mostly sterile triploids.

   Examples: The treefrog *Hyla versicolor* is a tetraploid that, on the basis of allozyme and immunological comparisons, is believed to have arisen recently from hybridization between distinct eastern and western populations of its cryptic diploid relative *H. chrysoscelis* (Maxson et al. 1977; Ralin 1976). Polyploidy is relatively uncommon in animals, however, and is confined primarily to forms that reproduce asexually (see the section on hybridization in this chapter). On the other hand, at least 70%–80% of angiosperm plant species may be of demonstrably recent polyploid origin (Lewis 1980), and most plants have probably had at least one polyploidization event somewhere in their evolutionary history.

   Especially for recent polyploids, molecular genetic data often provide definitive evidence identifying the ancestral parental species. For example, allozyme analyses established that two tetraploid goatsbeards (*Tragopogon mirus* and *T. miscellus*) arose from recent crosses between the diploids *T. dubius* and *T. porrifolius* and the diploids *T. dubius* and *T. pratensis*, respectively (Roose and Gottlieb 1976). These allopolyploids (polyploids arising from combinations of genetically distinct chromosome sets) additively expressed all examined protein electrophoretic alleles inherited from their progenitors. Similar molecular analyses involving allozymes or cpDNA have demonstrated that some polyploid forms are of polyphyletic (i.e., multi-hybridization in this case) origin, including *Asplenium* ferns (Werth et al. 1985), *Glycine* soybeans (Doyle et al. 1990), *Heuchera* alumroots (Soltis et al. 1989), *Plagiomnium* bryophytes (Wyatt et al. 1988), and *Senecio* composites (Ashton and Abbott 1992). An especially ingenious application of genetic markers documented the complex cytological pathway leading to a new species of tetraploid fern, *Asplenium plenum*. Using allozyme markers, Gastony (1986) showed that *A. plenum* must have arisen through a cross between a triploid *A. curtissii* (which had produced an unreduced spore) and a diploid *A. abscissum* (which had produced a normal haploid spore). The nearly sterile triploid *A. curtissii* itself was shown to have arisen through a cross between a tetraploid species, *A. verecundum*, and diploid *A. abscissum*. New autopolyploid taxa (polyploids that arise by a multiplication of one basic set of chromosomes) also have been described through molecular assays (Rieseberg and Doyle 1989).

(continued)

In the modern era of massive DNA sequencing, genomic dissections of poly-ploidization phenomena have become highly sophisticated. For example, by com-puter-searching the entire *Arabidopsis* genomic sequence, Bowers et al. (2003) not only identified numerous specific chromosomal segments that had been duplicat-ed by a probable polyploidization event after *Arabidopsis* diverged from most di-cots, but also analyzed genomic patterns of subsequent loss or "diploidization" (restoration of the diploid condition) for many of these chromosomal regions.

2. *Chromosomal rearrangements.* Closely related taxa differing in a variety of struc-tural chromosomal features—including translocations, inversions, or chromo-some number—may exhibit reproductive isolation for at least three reasons. First, some structural differences themselves may cause difficulties in chromo-some pairing and proper disjunction during meiosis in hybrids, resulting in partial or complete sterility. Second, some gene rearrangements may diminish fitness in hybrids through disruptions of gene expression patterns resulting from position effects. For either of these reasons, formation of a new species via major chromosomal rearrangements probably entails a relatively quick transi-tion through an "underdominant" (fitness-diminished) heterozygous phase to a condition of population homozygosity for the new karyotype (Spirito 1998). Third, structural rearrangements in specific chromosomal regions have the ef-fect of reducing recombination (even if not fitness) when in heterozygous con-dition, and this reduction per se can also act as a partial barrier to gene ex-change in genomic regions that differ karyotypically (Navarro and Barton 2003a).

   Examples: White (1978a) compiled evidence that chromosomal rearrangements often are involved in the speciation process for animals (see also Sites and Moritz 1987), as did Grant (1981) for plants. When chromosomal rearrange-ments have conferred reproductive isolation recently, *allelic* differentiation be-tween the descendant species may still be minimal. Some examples in which the reported magnitude of allozyme divergence between chromosomally differ-entiated forms is about the same as that between populations within a species involved subterranean *Thomomys* and *Spalax* rodents (Nevo and Shaw 1972; Nevo et al. 1974; Patton and Smith 1981) and *Sceloporus* lizards (Sites and Greenbaum 1983). In some of these cases, however, the chromosomal differ-ences are not complete barriers to reproduction.

3. *Changes in the mating system.* Many plant species exhibit self-incompatibility, whereby pollen fail to fertilize ova from the same individual. The mechanisms may involve alleles at a self-incompatibility locus that is known to be highly polymorphic within some species (Ioerger et al. 1990) or a physical barrier, such as a difference in the lengths of styles and stamens (heterostyly), that inhibits self-pollination. A switch in mating system, for example, from self-incompati-bility to self-compatibility (autogamy) as mediated by a change from heterosty-ly to homostyly can precipitate a rapid speciation event with little change in overall genic composition. Other alterations of the breeding system, such as the timing of reproduction, similarly can generate reproductive isolation rapidly.

   Examples: In many plant groups, closely related taxa exhibiting contrasting re-productive modes suggest that "the evolution of floral syndromes, and their influence on mating patterns, is intimately associated with the development of re-productive isolation and speciation" (Barrett 1989). For example, self-compatible *Stephanomeria malheurensis* apparently arose from a self-incompatible progenitor, *S. exigua coronaria*, and also differs from it by chromosomal rearrangements that are the principal cause of hybrid sterility (Stebbins 1989). High allozyme similarities (Gottlieb 1973b) suggest that the process took place recently, such that the derivative species was extracted from the repertoire of genetic polymorphisms present in the ancestor (Gottlieb 1981). Such evolution of self-fertilization probably favors the establishment of chromosomal rearrangements that contribute to reproductive isolation of the selfing derivative (Barrett 1989).

branching patterns (Stanley 1975) reflecting evolution through "punctuated equilibria." According to this view, a new species arises rapidly and, once formed, represents a well-buffered homeostatic system, resistant to within-lineage change (anagenesis) until speciation is triggered again, perhaps by an alteration in ontogenetic (developmental) pattern (Gould 1977). A second class of arguments for sudden speciation came from molecular genetics: The molecular events responsible might involve changes in gene regulation, perhaps mediated by relatively few control elements that could have a highly disproportionate influence on organismal evolution, including the erection of RIBs (Britten and Davidson 1969, 1971; Krieber and Rose 1986; McDonald 1989, 1990; Rose and Doolittle 1983; Wilson 1976). A third class of arguments involved demographic and population genetic considerations. Mayr (1954) suggested that "founder effects" in small geographically isolated populations might produce "genetic revolutions" leading to new species. Carson (1968) advanced a "founder-flush" model in which rapid population expansion and relaxed selection following a severe founder event facilitated the appearance and survival of novel recombinant genotypes leading to a new species (see also Slatkin 1996). Templeton (1980b) introduced a "transilience model" in which speciation involves a fast shift to a new adaptive peak under conditions in which founder events cause rapid but temporary inbreeding without severely depleting genetic variability. Carson and Templeton (1984) compared these models, and Provine (1989) discussed their histories.

The generality of such quick-speciation scenarios proved difficult to document, however. This is in part because speciation is a highly variable and eclectic process, probably differing greatly in mean tempo and mode in different kinds of organisms (such as mobile marine fishes versus sedentary insects). Furthermore, even exceptionally rapid speciation is normally an extended temporal process, seldom directly observable from start to finish during a human lifetime (see below). Finally, many of the genetic and demographic events proposed to be associated with speciation often occur at the population level as well *without* producing new species (e.g., Rundle et al. 1998).

On the other hand, many authors viewed speciation as a rather unexceptional continuation of the same microevolutionary processes that generate geographic population structure, albeit with the added factor of RIB acquisition (see early reviews in Barton and Charlesworth 1984; Charlesworth et al. 1982). This view was termed "phyletic gradualism" by Eldredge and Gould (1972). However, Sewall Wright (1931) and some others who interpreted speciation mostly as a continuation of microevolution (Provine 1986) nonetheless emphasized that episodic shifts in evolution could result from genetic drift (in conjunction with selection) facilitating rapid leaps across "fitness peaks" in "adaptive landscapes." Paleontologists likewise were long aware that evolutionary changes (at least in morphology) often occur in fits and starts, rather than at a steady pace (Simpson 1944). Thus, the crucial distinction is not whether evolutionary change is gradual or episodic, but whether or not speciation as a process is somehow uncoupled from processes of intraspecific population differentiation (as Gould proposed in 1980). In earlier approaches to addressing these issues, many nonmolecular assessments of the magnitude and pattern of genetic differentiation associated with species formation were made.

These traditional approaches often involved the study of phenotypes in later-generation crosses between related species that could be hybridized. One method was to measure the variance among F_2 hybrids for particular behavioral or morphological characters. Frequently, such variances proved to greatly exceed those in either the parental or the F_1 populations, and few F_2 hybrids fell into the parental classes (DeWinter 1992; Lamb and Avise 1987; Rick and Smith 1953). Such results appeared attributable to recombination-derived variation, and they indicated that for the assayed characters, the parental species must differ in multiple genes, each with small effect (although only the minimum number of such polygenes can be estimated by this approach; Lande 1981). Another traditional method of assessing genetic differences between species involved chromosomal mapping of prezygotic or postzygotic RIB genes through searches for consistent patterns of co-segregation in experimental backcross progeny (see reviews in Charlesworth et al. 1987; Richie and Phillips 1998; Wu and Hollocher 1998). For example, in sibling species of *Drosophila*, partial hybrid sterility and inviability proved attributable to differences at several (mostly anonymous) loci on each chromosome (Dobzhansky 1970, 1974; Orr 1987, 1992; Wu and Davis 1993), with X-linked genes typically having the greatest effects (Coyne and Orr 1989a). Unfortunately, such studies could only be conducted on model experimental species with well-known genetic systems.

There were at least two other serious limitations to these classic Mendelian approaches. First, they could be applied only to hybridizable taxa. Second, patterns of allelic assortment could be inferred only for loci distinguishing the parental species; genes that were identical in the parents escaped detection. But to determine the *proportion* of genes distinguishing species, both divergent and non-divergent loci must be monitored. Thus, following the introduction of allozyme methods in the mid-1960s, many researchers revisited the issue of genetic differentiation during speciation, under the rationale that these molecular assays permitted, for the first time, examination of a large sample of gene products presumably unbiased with regard to magnitude of divergence. Early reviews of this effort were provided by Ayala (1975), Avise (1976), and Gottlieb (1977).

CLASSICAL MOLECULAR EVIDENCE. A classic survey of allozyme differentiation accompanying RIB acquisition involved the *Drosophila willistoni* complex (Ayala et al. 1975b), which includes populations at several stages of the speciation process, as gauged by reproductive relationships and geographic distributions. This complex, which is distributed widely in northern South America, Central America, and the Caribbean, provides a paradigm of gradual speciation involving geographic populations that are fully compatible reproductively; different "subspecies" that are allopatric and exhibit incipient reproductive isolation in the form of postzygotic RIBs (hybrid male sterility, in this case, in laboratory crosses); "semispecies" that overlap in geographic distribution and show both postzygotic RIBs and prezygotic RIBs (homotypic mating preferences), the latter presumably having evolved under the influence of natural selection after sympatry was secondarily achieved between subspecies (see Box 7.2); sibling species that show complete reproductive isolation but remain nearly identical morphologically; and non-sibling species that are phenotypically distinct and presumably diverged at earlier times. Frequency distributions of genetic similarities across 36 allozyme loci are summarized in Figure 7.2. Between subspecies (or semispecies), nearly 15% of these genes showed substantial or fixed allele frequency differences involving detected replacement substitutions. Results were interpreted to indicate that a substantial degree of genetic differentiation occurs during the first stage of speciation (Ayala et al. 1975b). Subsequent analyses of more pairs of closely related *Drosophila* taxa were reviewed by Coyne and Orr (1989b, 1997). These studies examined allozyme divergence in a cross section of populations at various stages of the speciation process, as defined by geographic distributions and experimentally determined levels of prezygotic and postzygotic reproductive isolation. Results demonstrated that even partial reproductive isolation is often associated with large genetic distances (Table 7.1).

Other noteworthy early studies demonstrating moderate to large allozyme distances between populations at various stages of speciation involved *Lepomis* sunfishes (Figure 7.2), *Peromyscus* mice (Zimmerman et al. 1978), and *Helianthus* sunflowers (Wain 1983). In a sense, these and the *Drosophila* studies merely affirm what was emphasized in Chapter 6; that is, that considerable genetic differentiation among geographic populations can accumulate prior to the completion of intrinsic reproductive isolation.

Among the vertebrates, perhaps the record for magnitude of genetic differentiation among forms that had been considered conspecific involves the salamander *Ensatina eschscholtzii*. This complex of morphologically differentiated populations encircles the Central Valley of California in ringlike fashion, with adjacent populations usually capable of genetic exchange (Jackman and Wake 1994; Wake and Yanev 1986; Wake et al. 1986). Remarkably, various populations in this "ring species" show allozyme distances up to $D = 0.77$ (values

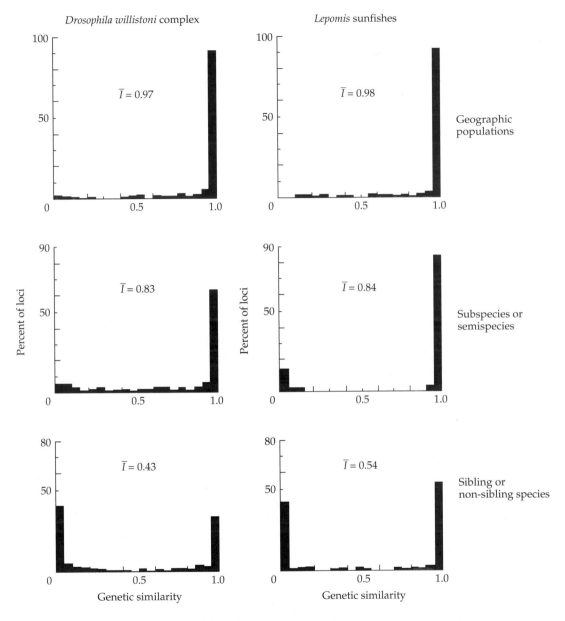

Figure 7.2   Distributions of allozyme loci with respect to genetic similarity (Nei's 1972 measure) in some of the first multi-locus comparisons among populations at various stages of evolutionary divergence. Shown are results from the *Drosophila willistoni* complex of fruit flies (data from Ayala et al. 1975b) and *Lepomis* sunfishes (data from Avise and Smith 1977).

TABLE 7.1   Means and standard errors of genetic distance (Nei's *D*, allozymes) characterizing *Drosophila* taxa at indicated levels of prezygotic and postzygotic reproductive isolation

Reproductive isolation index[a,b]	Number of comparisons	Mean genetic distance (SE)	
		Prezygotic[a]	Postzygotic[b]
0.00	13	0.122 (0.046)	0.138 (0.058)
0.25	8	0.370 (0.078)	0.251 (0.083)
0.50	21	0.257 (0.080)	0.249 (0.032)
0.75	29	0.578 (0.098)	0.722 (0.198)
1.00	13	0.523 (0.089)	0.991 (0.127)

Source: After Coyne and Orr 1989b.
[a] Prezygotic isolation index = 1 − [(frequency heterotypic matings) / (frequency homotypic matings)].
[b] The postzygotic isolation index is a measure of hybrid inviability and hybrid sterility, scaled from zero to one.

more typically associated with highly divergent congeneric species; see Figure 1.2). Huge genetic distances ($p > 0.12$) are apparent in this taxon's mtDNA genome as well (Moritz et al. 1992a). Wake et al. (1989) interpreted the results to evidence "several stages of speciation in what appears to be a continuous process of gradual allopatric, adaptive divergence," implying that speciation in these salamanders must be extremely slow. On the other hand, Frost and Hillis (1990) argued that *E. eschscholtzii* should instead be considered an assemblage of several highly distinct species that separated in allopatry long ago. This example illustrates the kinds of *taxonomic* uncertainties (not to mention the dangers of circular reasoning) that can arise in attempts to describe "the amount of genetic differentiation during the speciation process."

At the opposite extreme, some animals and plants considered distinct taxonomic species show very small allozyme distances (i.e., $D < 0.05$), well within the range of values normally associated with conspecific populations. Early examples were reported in herbaceous plants (Ganders 1989; Witter and Carr 1988), insects (Harrison 1979; Simon 1979), snakes (Gartside et al. 1977), birds (Thorpe 1982), and mammals (Apfelbaum and Reig 1989; Hafner et al. 1987). Presumably, the paucity of protein electromorphs distinguishing such biological species indicates that insufficient time has elapsed for accumulation of greater de novo mutational differences. Indeed, the time-dependent aspect of allozyme divergence permitted reassessments of speciation dates. For example, two minnow species in California that had been placed in different genera (*Hesperoleucus* and *Lavinia*) proved to exhibit an allozyme distance of only $D \cong$ 0.05, suggesting a far more recent separation than their generic assignments had implied (Avise et al. 1975). In the plant genera *Clarkia, Erythronium, Gaura,* and *Lycopersicon,* particular progenitor–derivative species pairs were reinterpreted to be of relatively recent origin when they were found to exhibit unexpectedly low allozyme distances (Gottlieb 1974; Gottlieb and Pilz 1976; Pleasants and Wendel 1989; Rick et al. 1976). Conversely, the self-pollinating plant *Clarkia franciscana* was formerly thought to have evolved from *C. rubicunda* by rapid and recent reorganization of chromosomes, but the two species proved to share few or no alleles at 75% of allozyme loci, indicating that they had separated much longer ago than formerly supposed (Gottlieb 1973a).

Overall, with regard to observed magnitudes of genetic divergence and inferred ecological or evolutionary times associated with speciation, data from the allozyme era documented a wide spectrum of outcomes. The same general message has emerged from post-allozyme molecular analyses (Harrison 1991), such as DNA hybridization (e.g., Caccone et al. 1987) and DNA sequencing (see Figure 1.3). Numerous examples appear throughout Chapters 6–8. Despite the heterogeneity in genetic patterns (which provides rich fodder for analyzing and comparing various speciational processes), the molecular revolution has made abundantly clear at least one consistent point: Even closely related taxonomic species normally differ in many genetic features, not just one or a few. Consider, for example, two vertebrate species estimated to differ by a net sequence divergence of 1% (this would be considered a small genetic distance, of the approximate magnitude differentiating humans and chimpanzees, for example). If each of these genomes contained 3 billion bp, then a total of about 30,000,000 nucleotide substitutions would distinguish these two species. Although only a small fraction of these genetic changes might be directly involved in RIB formation, they would all provide potential molecular markers for analyzing a plethora of issues relating to speciation and hybridization.

IDENTIFICATION AND ANALYSIS OF SPECIATION GENES. A somewhat different research tack in the post-allozyme era is to focus analyses more directly on "speciation genes" (Coyne and Orr 1998; Orr 1992), an approach with at least two distinct aspects. The first is to employ large banks of molecular markers to identify the approximate number and genomic positions of quantitative trait loci, or QTLs (see Box 1.2), underlying morphological, behavioral, or other features that distinguish particular species of interest (Albertson et al. 2003; Streelman et al. 2003; Via and Hawthorne 1998). The second is a "candidate gene" approach, wherein specific loci already known or suspected to play an important role in RIB formation are analyzed in depth, phylogenetically or functionally or both.

Orr (2001) reviewed the literature on QTL-mapped genetic differences for more than 50 traits between closely related pairs of animal and plant species (Table 7.2). Many of these phenotypes—such as differing floral traits in related flower species, courtship songs in crickets (K. L. Shaw 1996, 2000), and genital morphologies and pheromone hydrocarbons in fruit flies (Coyne 1996; Coyne and Charlesworth 1997; J. Liu et al. 1996)—contributed directly to RIBs themselves (Bradshaw et al. 1995, 1998). The number of genes provisionally identified ranged from 1 to nearly 20 per trait.

Sister species compared	Phenotypic character	Minimum number of genes
Drosophila fruit flies	Adult toxin resistance	5
	Larval toxin resistance	3
	Oviposition site preference	2
	Fine larval hairs	1
	Various posterior lobe features	4 to 19 each
	Sex comb tooth number	2
	Testis length	7
	Cyst length	3
	Tibia length	5
	Male pheromone	5
	Female pheromone	5
	Various counts of bristle number	1 to 6 each
	Fifth sternite	1
	Anal plate area	3
	Cuticular hydrocarbon profile	1
	Male courtship song	2
Nasonia wasps	Wing size	2
Laupala crickets	Song pulse rate	8
Mimulus monkeyflowers	Concentrations of various pigments	1 to 3 each
	Lateral petal width	8
	Various corolla features	4 to 10 each
	Petal reflexing	4
	Nectar volume	3
	Stamen lengths in various species comparisons	3 to 7 each
	Pistil lengths in various species comparisons	1 to 13 each
	Bud growth rate	8
	Anther–stigma separation in various comparisons	2 to 5 each

TABLE 7.2 Genetic analyses, via QTL mapping, of species' differences in numerous phenotypic traits

Source: Modified from a review by Orr (2001).

However, several cautionary points should be made about such genetic appraisals. First, the statistical power to detect significant associations varied considerably across studies, so not all results are directly comparable. Second, such tallies alone do not accurately describe the distribution of the magnitude of phenotypic effects across loci, a particular weakness being in the identification of genes with small or modest effect. Thus, the number of polygenes contributing to a trait is normally underestimated by this QTL approach, and a reporting bias exists toward the notion that most identified genes have substantial effects on phenotype. Third, critical attention in QTL mapping should be devoted to evaluating epistasis and dominance—that is, non-additive allelic interactions on phenotype among and within loci, respectively (Turelli and Orr 2000)—but unfortunately these phenomena are often neglected (but see Kim and Rieseberg 2001, for a nice exception). Finally, such QTL studies can be applied only to hybridizable species pairs, and normally to phenotypic traits that differ cleanly between them. The net effect of these and other qualifications about QTL mapping, plus the wide variety of observed outcomes reported to date, led Orr (2001) to a cautious conclusion: "Such results do not encourage the idea that the genetics of species differences shows any regularities."

Nevertheless, considering the total number of phenotypic differences often distinguishing congeners and the minimum tallies of responsible genes per trait in several taxa studied to date by QTL mapping, the composite number of genetic changes between even closely related species must normally be large. For example, Kim and Rieseberg (1999) identified 56 QTL loci contributing to differences in 15 morphological traits between closely allied species of *Helianthus* sunflowers. Furthermore, the number of genes deduced to contribute directly to prezygotic and postzygotic RIBs seems to be fairly substantial in available studies (see Table 7.2; Civetta et al. 2002; Hollocher and Wu 1996; Suwamura et al. 2000). On the other hand, in many animal and plant species, the number of genetic changes distinguishing long-separated allopatric populations may also be large, so an important future task will be to conduct similar kinds of QTL mapping experiments on conspecific populations and compare the results to the interspecific outcomes. Another way to look at this challenge is to appreciate that results from available QTL mapping studies provide tallies of accumulated differences between species, but that some or all of these genetic changes might have either predated or postdated completion of the speciation process itself (the same cautionary note applies to nearly all molecular analyses of genetic differences between extant species). Furthermore, although QTL analyses identify chromosomal regions contributing to phenotypic differences, they do not by themselves identify the actual genes that are mechanistically responsible.

The second approach, known as the "candidate gene" method (Haag and True 2001), focuses even more directly on particular loci suspected to play an immediate role in providing RIBs between sister species. These analyses can be phylogenetic, functional, or both. A phylogenetic appraisal is illustrated by studies of *Odysseus* (*OdsH*), a gene known to be responsible for hybrid male sterility in fruit flies (Perez et al. 1993). Phylogenetic analyses of sequence polymorphisms in *Drosophila mauritiana* and *D. simulans* showed that these closely related species are reciprocally monophyletic in the *OdsH* gene tree, but not invariably so at other loci not directly involved in RIB formation (Ting et al. 2000). The authors concluded that RIB-causing genes such as *OdsH* faithfully track the evolutionary history of reproductive isolation (i.e., speciation per se), whereas non-RIB loci are expected to display a much wider variety of phylogenetic patterns due to evolutionary factors such as retention of ancestral polymorphisms or post-speciation movement of genes via hybridization (see also Wu 2001).

Another example involved a detailed molecular dissection of a gene (*Nup96*) in *Drosophila* that encodes a nuclear pore protein (Presgraves et al. 2003). Found on chromosome 3, *Nup96* interacts with one or more unknown genes on the X chromosome. Within either *D. simulans* or *D. melanogaster*, the protein products of these genes apparently interact well together, but in hybrids between them, the genes interact epistatically to reduce viability severely. The authors demonstrate that this nuclear pore protein evolved by positive natural selection in both species' lineages such that the hybrid inviability is merely a by-product of adaptive intraspecific protein evolution.

Another nice empirical example of this sort involved experimental analyses of functional coadaptation between two cellular proteins (cytochrome *c* and cytochrome *c* oxidase) in a marine copepod, *Tigriopus californicus*. These two proteins interact during a final step of the mitochondrial electron transport system (ETS), which plays a central role in cellular energy production. In laboratory assays, Rawson and Burton (2002) discovered that cytochrome *c* variants isolated from each of two genetically divergent copepod populations had significantly higher reactivity with cytochrome *c* oxidase molecules derived from their own population than with those from the alien population. These results indicate the presence of positive epistatic interactions between coadapted ETS proteins. It is also known that inter-population crosses in *T. californicus* yield later-generation hybrids with reduced performance in a wide variety of fitness-related traits (e.g., Edmands 1999). Taken together, these studies suggest that any hybridization between genetically divergent copepods would disrupt the coadapted ETS complex, thereby contributing to functional incompatibilities in cellular respiration that underlie partial reproductive isolation.

Functional as well as population genetic analyses of speciation genes are especially well illustrated by studies of "gamete recognition" loci (Palumbi 1998; Snell 1990), a subcategory within the broader set of genes responsible for prezygotic RIBs between many closely related species (Howard et al. 1998). Most marine invertebrates release their gametes into the water, so sperm and eggs of each species must find and recognize each other for successful fertilization. The cellular mechanisms involved have proved to be diverse. In echinoderms, for example, gametic attachment and fusion are mediated by sperm bindin proteins that interact with carbohydrates attached to proteins on the egg surface (Palumbi 1999; Vacquier et al. 1995), whereas in mollusks, a lysin protein mediates how well a sperm can burrow through an egg's chorion layer (Vacquier and Lee 1993). In addition to functional studies of these genes' modes of action, population genetic analyses of DNA sequences have shown that characteristic regions in the bindin and lysin molecules evolve rapidly both within and among related species (Lee et al. 1995; Metz and Palumbi 1996; Swanson et al. 2001a). Observed rates and patterns of amino acid substitution also indicate that these gamete recognition loci are often under positive diversifying selection. One hypothesis is that sperm in general should be under strong selection for rapid egg entry because, in the open ocean, any sperm cell is likely to encounter at most only one egg and must take advantage of the opportunity; whereas eggs are under strong selection for defense against untoward sperm advances because a typical egg may encounter swarms of sperm cells, only one of which is genetically required (Palumbi 1998). One net result is that eggs and sperm may be engaged in a coevolutionary "arms race" of offensive and defensive tactics that promote rapid molecular evolution in gamete recognition genes (Rice 1998). Another evolutionary consequence is an opportunity for rapid reproductive character displacement at gametic recognition loci as a barrier to detrimental hybridization between species (Geyer and Palumbi 2003).

In species with internal fertilization, it has been found that proteins intimately associated with reproduction also often evolve extremely rapidly, probably for similar kinds of selective reasons. Examples of genes showing rapid sequence evolution in replacement sites include the zona pellucida genes whose glycoprotein products coat the eggs of mammals (Swanson et al. 2001b) and genes for accessory gland proteins that occur in the seminal fluid of *Drosophila* males (Begun et al. 2000; Swanson et al. 2001c).

Do founder-induced speciations leave definitive genetic signatures?

All the sudden modes of speciation described in Box 7.3 no doubt are initiated by very small numbers of individuals who first acquire the relevant chromosomal or reproductive alterations. Apart from such situations, do founder events underlie speciations in many other animal and plant groups? If so, such speciations might entail significant shifts in frequencies of ancestral polymorphisms, but at the outset probably little de novo sequence evolution, and the ancestral species would normally be paraphyletic with respect to the derivative for at least some evolutionary time following their separation. A severe and prolonged population bottleneck accompanying speciation should also greatly diminish genetic variability in the neospecies.

A remarkable radiation of drosophilid flies has occurred in the Hawaiian archipelago, which is home to about 800 species endemic to the islands, compared with about 2,000 species in the remainder of the world (Carson and Kaneshiro 1976; Wheeler 1986). Founder-induced speciation models figured prominently in early discussions of the prolific speciation among Hawaiian *Drosophila* (Giddings et al. 1989), in which species formation was postulated to follow the colonization of new islands, perhaps by one or a small number of gravid females. However, molecular genetic data seemed to be equivocal about these scenarios. Some sister species, such as *D. silvestris* and *D. heteroneura*, did indeed prove to exhibit high allozyme similarities suggestive of recent speciation (Sene and Carson 1977). On the other hand, many recently derived Hawaiian species proved to be no less variable genetically than typical continental *Drosophila*, a result used by Barton and Charlesworth (1984) to dispute the founder model, but defended by Carson and Templeton (1984) as consistent with the founder-flush and transilience models of speciation. *D. silvestris* and *D. heteroneura* also showed relatively high genotypic and nucleotide diversities in mtDNA (DeSalle et al. 1986a,b), further suggesting to Barton (1989) that founder-induced speciation was not involved.

In the Australian fish *Galaxias truttaceus*, a landlocked form of which constitutes an incipient species separated from coastal ancestors within the last 3,000–7,000 years (based on geological evidence), surveys of both allozymes and mtDNA variation have been conducted (Ovenden and White 1990). Of the total of 58 mtDNA haplotypes observed in coastal populations, only two characterized the landlocked forms. Heterozygosities at allozyme loci nonetheless were nearly identical in the landlocked and coastal populations. These genetic results were interpreted to indicate that a severe but transitory population bottleneck accompanied the transition to lacustrine habitat (because in principle, such bottlenecks might affect the genotypic diversity of mtDNA more than that of nDNA).

In terms of gene genealogy, a founder-induced speciation should initially produce a paraphyletic relationship between the ancestral and descendant species (see Figure 7.1C–E). Many examples of paraphyly in mtDNA or scnDNA gene trees have been reported for related species (Avise 2000a; Powell 1991). Indeed, a recent literature survey of more than 2,200 species identified more than 500 cases (ca. 23%) in which a paraphyletic relationship between related taxonomic species had been statistically documented (by bootstrap criteria) in mtDNA gene trees (Table 7.3). Four empirical examples are illustrated in Figure 7.3. For example, the deer mouse (*Peromyscus maniculatus*) occupies most of North America and exhibits a paraphyletic relationship to the old-field mouse (*P. polionotus*), a species confined to the southeastern United States. Similarly, the mallard duck (*Anas platyrhynchos*), with broad Holarctic distribution, appears paraphyletic in mtDNA genealogy to the American black duck (*Anas rubripes*), which inhabits eastern North America only (Avise et al. 1990a). As detailed in Chapter 6, perhaps the most unexpected and remarkable of such examples involves the brown bear (*Ursus arctos*), which appears paraphyletic in matrilineal pattern to the polar bear (*Ursus maritimus*) despite these species' grossly different phenotypes (see Figure 6.9).

However, for the species depicted in Figure 7.3 and others like them, the mere appearance of genealogical paraphyly in a gene tree, or even in a composite organismal phylogeny, is insufficient for concluding that founder-induced speciations necessarily were involved, for several *biological* reasons (i.e., apart from "bad taxonomy," mistaken gene trees, or other artifactual causes). First, paraphyly is expected even in the absence of severe population bottlenecks when a derivative, geographically restricted species emerges via gradual allopatric divergence (Figure 7.1C). Second, under most geographic modes of speciation entailing even moderate or large populations, paraphyly in gene trees is a fully anticipated stage preceding reciprocal monophyly and often following genealogical polyphyly (see Chapter 4). Finally, the appearance of paraphyly also can result from secondary introgressive hybridization that has transferred some allelic lineages from one species to another (see examples in Freeland and Boag 1999; Shaw 2002; Sota and Vogler 2001; and the section on hybridization below). The latter possibility has been invoked, for example, to account in part for the paraphyly or polyphyly of the mallard duck to the black duck and other related species with which it often hybridizes extensively (McCracken et al. 2001; Rhymer et al. 1994).

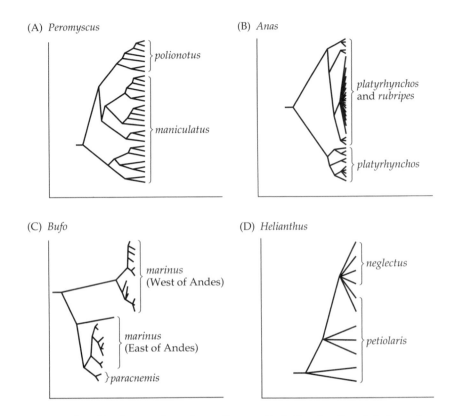

(A) *Peromyscus*

polionotus

maniculatus

(B) *Anas*

platyrhynchos
and *rubripes*

platyrhynchos

(C) *Bufo*

marinus
(West of Andes)

marinus
(East of Andes)

paracnemis

(D) *Helianthus*

neglectus

petiolaris

Figure 7.3   Empirical examples of genetic paraphyly for closely related species.
(A, B) mtDNA gene trees for *Peromyscus* mice and *Anas* ducks (see text for references).
(C) mtDNA gene tree for *Bufo* toads (Slade and Moritz 1998). (D) cpDNA and nuclear
rDNA gene trees for *Helianthus* sunflowers (Rieseberg and Brouillet 1994). (After
Avise 2000a.)

The influences of founder events on patterns of genetic differentiation
among species and on magnitudes of genic variability within species are in theory also functions of the size and duration of each population bottleneck and
the subsequent rate of population growth when variation recovers (Nei et al.
1975). All else being equal, uniparentally inherited cytoplasmic genes might
register founder effects more clearly than autosomal nuclear loci because of
their expected fourfold lower effective population size (Palumbi et al. 2001).
However, this is merely a baseline expectation from which departures can arise
for a variety of biological reasons (Hoelzer 1997; Moore 1995, 1997), even apart
from the high inherent stochasticity with regard to which lineages happen to
survive in a bottlenecked (or other) population to contribute to genetic diversity in descendants (Edwards and Beerli 2000). Furthermore, firm genetic inferences about historical demographic events accompanying speciations can also
be confounded by non-speciational population bottlenecks that may have predated and/or postdated erection of RIBs themselves.

TABLE 7.3   Instances of statistically documented paraphyly (including "polyphyly") uncovered in a literature survey of mtDNA gene trees for congeneric animal species

Taxonomic group	Number of studies	Number of genera	Number of species	% of species paraphyletic
Mammals	139	102	469	17.0
Birds	74	87	331	16.7
Reptiles	56	45	147	22.4
Amphibians	35	26	137	21.3
Fishes	100	99	371	24.3
Arthropods	143	126	702	26.5
Other invertebrates	37	41	162	38.6
Total	584	526	2319	23.1

Source: After Funk and Omland 2003.

Theoretical objections to the founder-induced speciation model have emphasized the low likelihood that small populations can successfully traverse major adaptive peaks (Barton 1989, 1996; Barton and Charlesworth 1984) or, in general, that they are more predisposed to speciation than large populations (Orr and Orr 1996). For example, one recent theory posits that reproductive isolation is often driven by conflicts of interest between the sexes and so might evolve most rapidly in large, dense populations, in which this type of selection should be most effective (Gavrilets 2000; Martin and Hosken 2003). However, not everyone accepts such theoretical objections to speciation in small populations (Hollocher 1996), and the original observation that motivated founder-induced speciation scenarios—namely, that insular populations or those at the periphery of a species range often show unusually high phenotypic divergence (e.g., Berry 1996)—still holds. In theory, rapid founder-induced speciations should be reproducible in appropriate experimental settings, especially in organisms with short generation times, but such "population cage" tests (mostly in dipteran flies) have yielded equivocal results at best (Moya et al. 1995). In summarizing this overall state of affairs, Howard (1998) concluded, "The question of whether small founder populations play an important role in genetic divergence and speciation is still open, although there is probably less enthusiasm for the role of founder events in speciation now than existed a decade ago."

What other kinds of phylogenetic signatures do past speciations provide?

Several other approaches to translating molecular observations on extant species into plausible inferences about the nature and tempo of speciations past have been suggested (Harvey et al. 1994; Kirkpatrick and Slatkin 1993; Klicka and Zink 1999; Losos and Adler 1995; Nee et al. 1994a; Rogers 1994). Typically, these approaches employ phylogenetic methods to assess the shapes of evolutionary trees and thereby address whether cladogenesis across time departs significantly from specified null models of lineage diversification.

"Lineage-through-time" analyses (Barraclough and Nee 2001; Nee 2001) can serve to illustrate the general conceptual orientation of several such approaches (Figure 7.4). The lineage-through-time model views speciations and extinctions in a supraspecific phylogeny as analogous to births and deaths of individuals in a population or gains and losses of lineages in an intraspecific gene tree. It assumes that a phylogram is available (e.g., from molecular data) for the *extant* species under analysis. It then asks whether the changing number of total lineages in the tree, when graphed as a lineages-through-time plot (log scale), might indicate a constant uniform rate of speciation in all branches throughout the tree, in which case the plot could show a straight line with slope equal to the per lineage speciation rate; recent accelerated speciation in the tree, in which case the plot could be concave upward; or recent decelerated speciation, in which case the plot could be concave downward. These expectations are somewhat equivocal, however (Kubo and Isawa 1995). For example, a concave downward curve might also be interpreted to register a recent increase in the extinction rate (Nee et al. 1994b).

Notwithstanding such caveats, this and alternative statistical phylogenetic approaches (e.g., Wollenberg et al. 1996) have been employed, provisionally, to infer nonrandom patterns of cladogenesis in large evolutionary clades ranging from *Cicindela* tiger beetles (Barraclough et al. 1999) to *Dendroica* warblers (Lovette and Bermingham 1999). For example, statistical analyses of the shapes of molecule-based phylogenies provided evidence for a significant temporal clustering of *ancient* cladogenetic events in a "species flock" of *Sebastes* rockfishes in the North Pacific (Johns and Avise 1998b), which provided an interesting comparison to *recent* explosive cladogenesis in a species flock of African freshwater cichlid fishes (to be described later in this chapter).

Are speciation rates and divergence rates correlated?

One intriguing possibility is that speciation events themselves might accelerate evolutionary differentiation within clades. If so, then magnitudes of divergence between extant species could be proportional to numbers of speciation events in their evolutionary histories, rather than to elapsed times since common ancestry. With regard to morphological divergence, this is indeed a logical consequence of the original model of punctuated equilibrium (Eldredge and Gould 1972), which posited stasis for organismal lineages except during speciation events. To test this possibility at the genetic level, Avise and Ayala (1975) introduced a conceptual approach that involves comparing pairwise genetic distances within clades that have experienced different rates of speciation. If genetic divergence is proportional to time, then mean genetic distances among extant species should be similar in rapidly speciating (species-rich) and slowly speciating (species-poor) clades of similar evolutionary age, whereas if genetic divergence is a function of the number of speciation events, mean genetic distance among extant forms should be obviously greater in the species-rich clade (Figure 7.5).

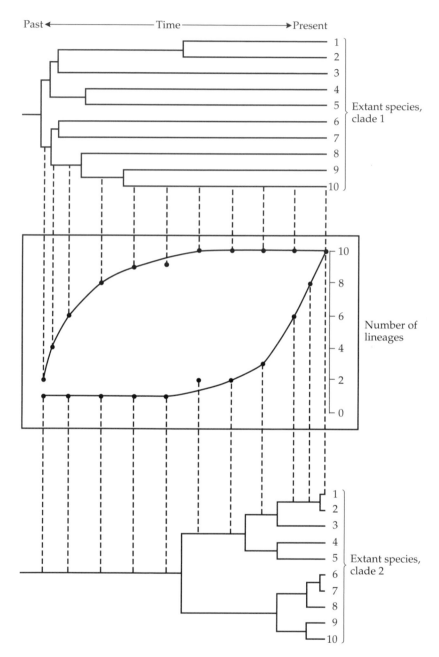

Figure 7.4 Simplified illustration of the use of "lineages-through-time" plots (center) for inferring past cladogenetic patterns from the shapes of phylogenetic trees (top and bottom). (In actual cases, the clades to be examined would be much larger, and each lineage-through-time plot would be on a logarithmic scale; Barraclough and Nee 2001.) The method involves tallying and graphing the number of species lineages at successive temporal points to assess possible departures from constant rates of speciation or extinction through time (see text).

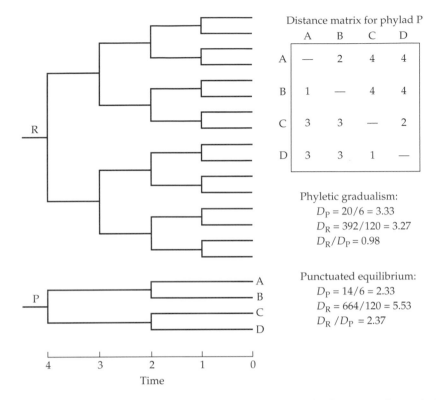

Distance matrix for phylad P

	A	B	C	D
A	—	2	4	4
B	1	—	4	4
C	3	3	—	2
D	3	3	1	—

Phyletic gradualism:
$D_P = 20/6 = 3.33$
$D_R = 392/120 = 3.27$
$D_R/D_P = 0.98$

Punctuated equilibrium:
$D_P = 14/6 = 2.33$
$D_R = 664/120 = 5.53$
$D_R/D_P = 2.37$

Figure 7.5 Explanation of models underlying a test for whether rates of genetic divergence and rates of speciation are correlated. "R" and "P" are species-rich and species-poor clades of comparable evolutionary age. The distance matrix in the upper right applies to clade P (the larger matrix for R is not presented) and shows expected distances between species pairs when differentiation is either time-dependent ("phyletic gradualism," above diagonal) or speciation-dependent ("punctuated equilibrium," below diagonal). At the lower right are shown expected ratios of *mean* distances (*D* values) for extant species within R and P under these competing models. Under phyletic gradualism, $D_R/D_P \cong 1.0$, whereas under punctuated equilibrium, $D_R/D_P \gg 1.0$. (After Avise and Ayala 1975.)

One set of empirical tests utilizing this approach involved the speciose North American minnows (approximately 200 species, more than 100 then recognized within *Notropis* alone) and the relatively depauperate sunfishes (approximately 30 species, 11 within *Lepomis*). Based on fossil evidence, these groups both appeared to have Miocene origins on the continent. A key assumption underlying the test was that the minnows had speciated more rapidly than the sunfishes (rather than having experienced lower rates of extinction). Allozyme comparisons involving more than 80 species (Avise 1977a; Avise and Ayala 1976) revealed nearly identical mean genetic distances among species within the two groups (e.g., $D = 0.62$ and $D = 0.63$ in *Notropis* and *Lepomis*, respectively), as well as similar mean heterozygosities per species ($H = 0.052$ and $H = 0.049$; Avise 1977b). These results were provisionally interpreted as inconsistent with punctuated equilibrium with regard to allozyme evolution in these fishes.

Gould and Eldredge (1977) justifiably questioned the relevance of these findings to the broader debate between punctuated equilibrium and phyletic gradualism, pointing out that the controversy referred primarily to patterns of morphological divergence, whereas allozymes were neutral molecular characters and, hence, irrelevant to the discussion. Douglas and Avise (1982) responded by reexamining these fishes for quantitative morphological differences and showed by multivariate statistical analyses that the overall magnitudes of *phenotypic* divergence were closely similar in the minnows and the sunfishes, a result again inconsistent with some of the predictions of punctuated equilibrium.

Nonetheless, these particular tests had several acknowledged weaknesses. For example, minnow and sunfish taxa originally were recognized on the basis of qualitative morphological appraisals, so the quantitative reassessment by Douglas and Avise (1982) probably included elements of "circular" reasoning in this supposed test of "rectangular" evolution! Another concern, voiced by Mayden (1986), was that the minnows examined might not be monophyletic. However, if the North American minnows do not constitute a clade, then the actual intervening number of speciation events would be even greater than assumed, in which case the minnows should have displayed even larger mean genetic distances under the rectangular evolution model. So non-monophyly should have biased the outcome toward (rather than against) the predictions of punctuated equilibrium. In any event, it is true that such tests ideally should involve sister clades so that, by definition, the evolutionary times over which the differential species proliferations took place would be identical. However, it is difficult to find sister clades that are differentially branchy or "unbalanced" (Heard 1996) on opposite sides of a basal node.

Despite the importance of this topic, relatively few explicit tests of this sort involving molecular data from extant species have been attempted (but see Barraclough and Savolainen 2001; Lemen and Freeman 1981, 1989; McCune and Lovejoy 1998; Ricklefs 1980). Using published data from DNA hybridization and mtDNA sequences to contrast numerous independent pairs of tropical and temperate avian taxa, Bromham and Cardillo (2003) tested and provisionally rejected a hypothesized link (Rohde 1992) between rates of speciation and rates of molecular evolution along global latitudinal gradients. Mindell et al. (1990) reported a different outcome when they summarized available literature on allozyme genetic distances within 111 vertebrate genera and noted a positive correlation between genetic divergence and species richness, which they attributed to the accelerating influence of speciation on molecular differentiation. However, among taxa this diverse, one cannot be certain that other uncontrolled variables do not also correlate with, and possibly influence, molecular rates. Similarly, in a recent review of 56 published molecular phylogenies, Webster et al. (2003) identified a significant association between number of speciations and rate of genetic evolution in 30%–50% of the evolutionary trees. Results were interpreted to suggest that molecular clocks are not gradual, but rather are punctuated according to speciation events. On the other hand, at least half of those analyzed phylogenetic trees showed no significant correlation between rate of molecular evolution and apparent rate of speciation, so any association between the two must be either non-universal or rather subtle.

Can speciation occur sympatrically?

A long-standing issue in evolutionary biology concerns how often new species arise sympatrically, most likely under the influence of diversifying selection on resource utilization or mate choice (Levine 1953; Maynard Smith 1966; Schluter 1996a, 2000; Tauber and Tauber 1989). Some biologists question whether sympatric speciation plays any significant role in evolution (e.g., Coyne 1992; Felsenstein 1981b; Futuyma and Mayer 1980; Mayr 1963), whereas others perceive it as a prevalent mode in at least some taxonomic groups, such as insects (e.g., Berlocher and Feder 2002; Bush 1975; Dieckmann and Doebeli 1999; Higashi et al. 1999; Kondrashov and Mina 1986). Here, several examples will be provided in which molecular markers have shed some light on the possibility of sympatric speciation.

HOST SHIFTS OR HABITAT SWITCHING IN INSECTS.   Changes in host usage by phytophagous or zoophagous parasites might give quick rise to new host races or neospecies that are reproductively isolated from their sympatric progenitors (Bush 1975; Wood et al. 1999). Host-specific parasites that faithfully complete their entire life cycle on a particular host species are especially strong candidates for sympatric speciation (or at least for local syntopic speciation). In such cases, any rare shift in host utilization almost by definition causes a simultaneous shift in mating options, such that ecological differentiation and prezygotic isolation go hand in hand (Hawthorne and Via 2001; Kondrashov and Kondrashov 1999). For any such speciation event that took place recently and involved a small number of founders initiating the host shift, population genetic signatures should include a reduction of variation in the neospecies and a paraphyletic relationship of the progenitor species to the derivative. About a dozen additional phylogeographic signatures of various types of sympatric speciation events have been suggested (Via 2001), but only seldom do these alone permit firm elimination of all competing hypotheses (Berlocher 1998).

The classic example of host switching involves frugivorous *Rhagoletis* flies, in which several host-specific forms, such as an apple race and a hawthorn race of *R. polmonella*, are postulated to be undergoing sympatric speciation in modern times (Bush 1969; Linn et al. 2003). Available genetic data are not inconsistent with this possibility. Collections of flies from apple and hawthorn trees in local sympatry show small but significant allozyme differentiation (Feder et al. 1988, 1997; McPheron et al. 1988), and such differences between paired samples extend across the eastern United States and Canada (Feder et al. 1990a,b). Furthermore, Berlocher and Bush (1982; see also Berlocher 2000; Smith and Bush 1997) concluded from molecular data that the phylogeny for many *Rhagoletis* flies and their relatives differs from that of their hosts, a result consistent with host-switching aspects of sympatric speciation scenarios (see below). A recent phylogeo-graphic analysis based on nuclear and mitochondrial loci added a twist to sym-patric speciation scenarios for *Rhagoletis:* An ancestral, hawthorn-infesting popu-lation apparently was subdivided about 1.5 million years ago into Mexican and North American isolates that diverged in adaptive diapause traits; subsequent gene flow from Mexico may then have aided North American flies in adapting to a variety of plants with different fruiting times, thereby helping to spawn new species, perhaps sympatrically (Feder et al. 2003).

Molecular markers have likewise indicated genetic differences between several other insect "races" suspected of undergoing host-shift speciations, such as gallmaker flies (*Eurosta;* Brown et al. 1996; Waring et al. 1990), pea aphids (*Acyrthosiphon;* Via 1999), yucca moths (*Prodoxus;* Groman and Pellmyr 2000), and seed-parasitic moths (*Greya;* Brown et al. 1997). About 25%–40% of all ani-mal species are thought to be phytophagous specialists (Berlocher and Feder 2002), so even if sympatric speciation is confined to a modest fraction of such or-ganisms, it could be highly important to speciation theory and biodiversity.

FLOCKS OF FISHES.   Within each of several isolated lakes or drainages scattered around the world, particular groups of related fishes constitute "species flocks" (Echelle and Kornfield 1984) that *might* have diversified through sympatric speciation. The numbers of named species in such fish flocks range from just a few, as in salmonid complexes in high-latitude lakes of the Northern Hemi-sphere, to many hundreds, as in cichlids in some of the Rift Valley lakes of east-ern Africa (Fryer and Iles 1972; Greenwood 1981). As gauged by molecular ge-netic evidence, some of these flocks are evolutionarily old (e.g., *Cottus* sculpins in Russia's Lake Baikal; Grachev et al. 1992), whereas others are extremely young (e.g., cichlids in Africa's Lake Victoria; Stiassny and Meyer 1999).

Especially for the youngest flocks, one preliminary question is whether dif-ferences among sympatric morphotypes indeed reflect the presence of distinct species (i.e., reproductively isolated gene pools), rather than intraspecific pheno-typic polymorphisms, perhaps due to ontogenetic switches triggered by ecologi-cal conditions. Molecular markers are well suited to address this question. For example, allozyme markers were used to examine *Cichlasoma* fishes in an isolat-ed basin near Coahuila, Mexico, where three trophic morphs described as sepa-rate species occur: a snail-eating form with molariform teeth and a short gut, a detritus-feeding or algae-eating form with papilliform teeth and a long gut, and a fish-eating form with a fusiform body. At all 27 monomorphic and polymor-phic loci assayed, these morphs proved to be indistinguishable (Kornfield and Koehn 1975; Kornfield et al. 1982), a result interpreted by Sage and Selander (1975) to indicate that trophic radiation in these cichlids was achieved by ecolog-ical polymorphism, not speciation. Aquarium experiments confirmed this con-clusion when it was shown that distinct morphotypes could be generated among progeny within a brood by altering the rearing conditions (Meyer 1987).

A similar report of dramatic trophic polymorphism due to phenotypic plasticity involved *Ilyodon* fishes. In some Mexican streams, sharply dichoto-mous trophic morphs, formerly considered distinct species, were shown to be indistinguishable at several polymorphic allozyme loci, and pooled genotypic frequencies conformed to Hardy–Weinberg expectations for randomly mating populations (Grudzien and Turner 1984). These data strongly suggested that the trophic types are conspecific (Turner and Grosse 1980).

Within many salmonid species, coexisting forms often exhibit contrasting life histories: Anadromous individuals hatched in streams or lakes migrate to sea before returning to fresh water to spawn, whereas non-anadromous individuals spend their entire lives in fresh water. Some landlocked populations, lacking present-day access to the ocean, include both stream-resident and lake-migratory individuals. Do these various life history types at a given site constitute separate gene pools? Allozyme studies of several such complexes in species such as sockeye salmon (*Oncorhynchus nerka;* Foote et al. 1989), rainbow trout (*O. mykiss;* Allendorf and Utter 1979), cutthroat trout (*O. clarki;* Campton and Utter 1987), brown trout (*Salmo sutta;* Hindar et al. 1991), and Atlantic salmon (*S. salar;* Ståhl 1987) revealed high genetic similarities between migratory and resident fish spawning in the same area, but significant differences between spawning populations in disjunct geographic regions (Ferguson 1989; Ryman 1983; see also Chapter 6). In other studies, small but statistically significant genetic differences were demonstrated among nearby or sympatric salmonids that differed in lifestyle (Baby et al. 1991; Birt et al. 1991; Krueger and May 1987; Skaala and Nævdal 1989; Vuorinen and Berg 1989; see review in Schluter 1996b). Some of the impediments to gene flow probably reflect habitat structure per se, but in some cases innate (gene-based) differences in microhabitat preferences or spawning times apparently augment these external barriers (Allendorf 1996).

Whether or not migratory and resident salmonid populations at a particular locale are fully isolated reproductively at the present time, the small genetic distances often involved and the polyphyletic appearance of particular life history patterns across a species' range suggest that the isolations are evolutionarily ephemeral. Clearly, lifestyle switches can be rapid and common, so any contemporary genetic separations are probably of recent origin (as indicated also by the fact that most of the high-latitude locales under consideration were covered by glacial ice as recently as 10,000 years ago). Furthermore, rearing and tagging studies of brown trout have shown that freshwater-resident individuals can develop from anadromous parents, and vice versa (Skrochowska 1969), indicating a considerable element of phenotypic plasticity in lifestyle. In this case, the freshwater-resident lifestyle is plausibly associated with a slow growth rate of parr (Hindar et al. 1991), which itself is influenced by both genetic and environmental factors.

Molecular analyses of several other flocks of fishes *have* revealed significant differences in allozymes or mtDNA among sympatric forms, thus confirming the presence of multiple biological species. Examples include representatives of the atherinids of central Mexico (Echelle and Echelle 1984), cyprinodontids of eastern Mexico (Humphries 1984), various cichlids in African Rift Valley lakes (Sage et al. 1984; Sturmbauer and Meyer 1992), and the now-extinct cyprinids of Lake Lanao in the Philippines (Kornfield and Carpenter 1984). However, despite significant allelic differences at particular loci, overall genetic distances among species remained remarkably small in several of these cases, suggesting that the evolutionary separations were very recent.

A follow-up question concerning such validated species flocks is whether the speciations took place allopatrically or sympatrically. In the Allagash basin of eastern Canada and northern Maine, coexisting dwarf and normal-sized lake whitefish (*Coregonus clupeaformis*) represent independent gene pools, albeit displaying only a small allozyme distance (Kirkpatrick and Selander 1979). If sympatric speciation was involved, then populations of dwarf and normal-sized fish in the Allagash should be one another's closest relatives. Contrary to this expectation, mtDNA findings indicated that the Allagash populations stem from a secondary overlap of two monophyletic groups that probably evolved in separate refugia during the most recent Pleistocene glaciation (Bernatchez and Dodson 1990, 1991, 1994). Western populations outside the Allagash basin belong to one mtDNA clade, eastern populations to another, and only in the Allagash basin do these clades overlap and appear alternately fixed in dwarf and normal whitefish. However, subsequent genetic analyses of dimorphic whitefishes (dwarf versus normal and limnetic versus benthic) at additional Canadian locales (Bernatchez et al. 1996) revealed instances of both "sympatric divergence and multiple allopatric divergence/secondary contact events on a small geographic scale" (Pigeon et al. 1997).

The most famous fish flocks occur in several African Rift Valley lakes. Some of the genetic findings for these flocks have been almost as stunning as the biological radiations themselves, which can involve hundreds of now-sympatric cichlid species. In particular, early comparisons of mtDNA sequences, including a normally highly variable portion of the control region, revealed little phylogenetic differentiation among morphologically diverse representatives of Lake Victoria's flock of more than 500 species, but a large genetic distinction (more than 50 assayed base substitutions) from cichlids in nearby Lake Malawi (Meyer et al. 1990). Results indicated a recent monophyletic origin for many or most Lake Victoria cichlids, conflicting with a traditional notion that the taxonomic complex is "a super-flock comprised of several lineages whose members cut across the boundaries imposed by the present-day lake shores" (P. H. Greenwood 1980). Nagl et al. (2000) expanded the mtDNA analyses and concluded that more than one founding cichlid entered Lake Victoria. Verheyen et al. (2003) extended the mtDNA analyses yet again, and concluded that two seeding lineages colonized Lake Victoria from Lake Kivu to the west. Based on analyses of nuclear AFLP markers, Seehausen et al. (2003) suggested that the broader species flock of which Lake Victoria cichlids are a part may not be strictly monophyletic and, at the outset, might have entailed hybrid swarms.

The Lake Victoria basin was nearly dry about 15,000 years ago, so one long-discussed possibility was that explosive sympatric speciation post-dated this desiccation event (Fryer 1997, 2001; Johnson et al. 1996; Seehausen 2002). However, the analysis by Verheyen et al. (2003) indicates that the major lineage diversification took place about 100,000 years ago. Regardless of the exact dates, the fact remains that speciation rates in the Lake Victoria assemblage have been dramatically high. Lakes Malawi and Tanganyika also house large cichlid flocks, but these flocks are phylogenetically more diverse (Danley and Kocher 2001; Rüber et al. 1999). These other Rift Valley lakes may have supported rapid speciations as well, albeit longer ago (McCune 1997).

The genetic and geological evidence make it tempting to conclude that at least some of the prolific speciation in Africa's Rift Valley lakes has been sympatric. However, micro-allopatric speciation is hard to eliminate. Small bodies of water inside a lake basin might have retained isolated fish populations during dry periods, or multiple closely related progenitors might have invaded a re-formed lake from outside refugia. To address this question more critically, researchers have also examined smaller cichlid flocks in volcanic crater lakes. These lakes are conical, bowl-like structures with steep walls, so they probably have always lacked *internal* physical barriers to fish dispersal. Yet, genetic analyses have shown that mini-flocks of endemic cichlids in two West African crater lakes (Lake Barombi Mbo, with eleven species, and Lake Bermin, with nine) are each of recent monophyletic origin, a finding consistent with sympatric speciation (Schliewen et al. 1994). For other cichlid mini-flocks in crater lakes in Central America, phenotypic and mtDNA analyses indicate that sexual selection has contributed there to assortative mating and sympatric speciation (A. B. Wilson et al. 2000).

ECOLOGICAL SPECIATION.   Sympatric and micro-allopatric speciation will probably remain difficult to dichotomize sharply, but an emerging perspective conceptualizes these processes as "ecological speciation" (Funk et al. 2002; Orr and Smith 1998; Schluter 1996b; Streelman and Danley 2003; Via 2002) or "adaptive speciation" (Dieckmann et al. 2001). This approach is illustrated by multifaceted research on a northern high-latitude fish, the threespine stickleback (*Gasterosteus aculeatus*) (Bell and Foster 1994). In this species, related ecotypes that differ in morphology and behavior (skeletal armor, feeding adaptations, and benthic versus limnetic lifestyle) often co-occur in particular bodies of water. Molecular markers interpreted in conjunction with experimental studies have shown that multiple unlinked genes underlie the ecotypic differences (Peichel et al. 2001); that strong assortative mating between ecotypes in various locations has arisen rapidly and repeatedly, both in sympatry and in allopatry (Rundle et al. 2000; Thompson et al. 1997); that hybrids are often viable and fertile; and that distinct ecotypes nonetheless often persist within a lake, their differences reinforced by ecological character displacement. In general, these studies suggest that divergent ecological selection pressures can play huge roles in promoting rapid adaptive radiations and local "speciations" (Schluter 2001).

Several authors have suggested that sexual selection, by driving changes in mate recognition, is at least as potent a force as natural selection in promoting ecological speciation (Panhuis et al. 2001; West-Eberhard 1983). By definition, sexual selection operates on features associated with mating success, so any phenotypic divergences driven by this phenomenon (within or among geographic populations) are also prime candidates for becoming prezygotic RIBs

(Questiau 1999). Consistent with this notion, speciation rates have been found to be correlated with apparent intensities of sexual selection in birds (Barraclough et al. 1995; Mitra et al. 1996; Møller and Cuervo 1998). Molecular markers have helped in analyzing several other taxonomic groups, such as Rift Valley cichlids (described above) and columbine plants with diverse floral spurs (Hodges and Arnold 1995), in which rapid speciation appears to have been associated with strong sexual selection on "key innovations" that promoted assortative mating and speciation.

Sympatric speciation is easiest to envision when assortative mating *and* disruptive selection operate on the same phenotypic characters (Doebeli 1996; Kawecki 1996, 1997; Kondrashov 1986). Data from molecular markers have assisted in addressing one such case involving male-pregnant seahorses (Jones et al. 2003). Microsatellite analyses of genetic parentage demonstrated that seahorses mate assortatively by body size in nature. Using a quantitative genetic model, this empirical mating preference was shown to be strong enough, when coupled with even modest disruptive natural selection, to produce rapid evolutionary divergence for body size and thereby curtail gene flow between populations of large and small fish in sympatry. This finding is potentially relevant to speciation modes in these fishes because several instances are known from around the world in which closely related species of dwarf and normal-sized seahorses co-occur.

In many of the host-switching insects, fish species flocks, and other sympatric species assemblages in which the RIBs are primarily prezygotic, reproductive isolation may be rather fragile (i.e., easily lost). Accordingly, these ecotypic species may be evolutionarily ephemeral not only due to routine extinction processes, but also because the lineages may re-amalgamate via introgressive hybridization if the ecological or behavioral barriers dissolve, for whatever reason. Thus, even if rapid ecological speciation proves to be a common phenomenon in contemporary studies of various organisms, its broader significance in generating new lineages that regularly withstand the longer tests of evolutionary time would remain debatable. In any event, the possibility that ecological or behavioral barriers can dissipate, causing gene pools of ecotypic species to merge through hybridization, has become of practical conservation concern for the cichlid fishes in Lake Victoria. There, human-caused eutrophication of the lake's otherwise clear waters threatens to compromise the visual mate-choice cues that otherwise help to maintain each species' current genetic integrity (Seehausen et al. 1997).

What are the temporal durations of speciation processes?

The instances highlighted in previous sections and in Box 7.3 demonstrate unequivocally that speciation can and sometimes does occur on short ecological time scales (within a matter of decades or less, at the extreme). However, such examples might be highly misrepresentative of speciation processes in general, perhaps having attracted special research attention precisely because they are unusual, and also because they offer exceptional experimental or observational tractability. Almost certainly, most speciations are associated with and promoted by geographic population separations (e.g., Coyne and Price 2000), and as such should be interpreted as extended temporal processes rather than as point events in evolution. What are the approximate durations of geographic speciation in absolute time?

Avise and Walker (1998) introduced a way to address this issue by identifying temporal ceilings and floors on speciation durations, using molecular evidence and the following logic. The temporal ceilings are estimates of separation times between extant sister species, which place an upper bound on how long ago genetic separation leading to speciation might have been initiated. The temporal floors are estimates of separation times between recognizable geographic clades or molecular "phylogroups" at the intraspecific level, which indicate minimum evolutionary times that have transpired without speciation having gone to apparent completion (at least according to current taxonomic assignments). Thus, the approximate mean duration of geographic speciation must lie somewhere (perhaps at the midpoint?) between these temporal floors and ceilings. The sidereal dates of these floors and ceilings are provisionally estimated using molecular clocks as applied to empirical genetic distances observed between relevant extant taxa.

The first application of this methodology involved birds. Klicka and Zink (1997) already had compiled genetic distances (based on mtDNA comparisons) between extant sister species of North American songbirds, from which they concluded that most of these speciations had been initiated in the Pliocene or early Pleistocene, rather than in the middle to last Pleistocene, as had been conventional wisdom (e.g., Brodkorb 1971; Rand 1948). Likewise, Avise and Walker (1998) then compiled a summary of the mtDNA literature on avian intraspecific phylogroups, most of which (76%) dated by molecular evidence to various separation times during the Pleistocene. Overall, the midpoint between inferred separation dates for sister species and those for intraspecific phylogroups suggested that the mean duration of a typical avian speciation might be approximately 2 million years. Avise et al. (1998) then repeated this entire exercise for other major vertebrate assemblages (mammals, fishes, and reptiles and amphibians). The composite evidence again suggested that vertebrate speciation durations were often on the order of 1–3 million years.

Johnson and Cicero (2004) later reexamined these estimates for birds, using additional molecular data for more taxa as well as several taxonomic realignments that had accumulated in the interim. In particular, some of the presumptive "sister species" in the Klicka and Zink (1997) study had proved not to be closest relatives when more taxa were examined, and some of the "intraspecific phylogroups" in the Avise and Walker (1998) compilation had been raised to full species status. These revisions, taken at face value and in conjunction with the newer data, tended to translate into accordingly shorter estimates of avian speciation durations, which might typically thus last "only" hundreds of thousands of years. Results again show that species concepts and taxonomic protocols can impinge considerably on perceptions of "speciation" processes.

Regardless of which temporal estimates are most valid, it remains clear from molecular evidence that allopatric speciation, which undoubtedly predominates in most vertebrates and many other animal groups as well, is a rather slow process. Thus, the temporal durations of classic geographic speciation typically extend far beyond the time scales of various sympatric and ecological speciation events.

How prevalent is co-speciation?

Parasites (or other symbionts) may sometimes be tied so closely to specific host species that any speciation events in the host taxa would simultaneously confer reproductive isolation upon the symbionts they sponsor. When repeated time and again during evolution, this "co-speciation" phenomenon should result in near-perfect topological matches between the phylogenies of host taxa and their associates. Conversely, if the utilizers occasionally or frequently shift among host species (as occurs, for example, in some of the phytophagous insects described earlier), then the phylogenies of hosts and their associates could be conspicuously uncoupled, the host shifts in effect producing reticulations of guest lineages across host lineages (e.g., Baverstock et al. 1985; Page 1993, 2003). Thus, there has been considerable empirical interest in jointly assessing host and guest phylogenies using molecular markers.

First, however, a cautionary note: Lack of congruence between the phylogenies of coevolving species could also result from asynchronous speciations and lineage sorting processes (Charleston 1998; Page 1994), such that surviving lineages in the parasite or symbiont trace to phylogenetic splits either predating or postdating nodes in the host phylogeny (Figure 7.6). Patterns of lineage sorting in a guest phylogeny embedded within a host phylogeny bear some analogy to patterns that can distinguish a gene tree from a species tree (see Chapter 4) and are also somewhat reminiscent of distinctions between orthology and paralogy in phylogenetic analyses (Fitch 1970; see Chapter 1).

Remarkable examples of coincident phylogenies are provided by aphids (Aphididae) and their bacterial symbionts in the genus *Buchnera* (Clark et al. 2000; Moran et al. 1993, 1998; Tamas et al. 2002). These insects feed on plant sap, but depend upon *Buchnera* bacteria living within their cells for nutrients not present in their phloem diet. For 100 million years or more, aphids and their bacterial endosymbionts probably have been tightly interdependent mutualists, so perhaps it is not too surprising that their phylogenies (estimated from molecular data) proved to be almost perfectly coincident (Figure 7.7). This example extends further: Like fleas on fleas, tiny plasmids are symbiotically associated with the *Buchnera* bacteria, and molecular phylogenies of assayed plasmids have proved to be congruent with those of their hosts (Funk et al. 2000). These impressive juxtapositions of multiple symbiont phylogenies might best be described as coevolution rather than co-speciation, however, because neither the bacteria nor their plasmids (unlike their aphid hosts) are sexual reproducers in the normal sense to which biological species concepts typically apply.

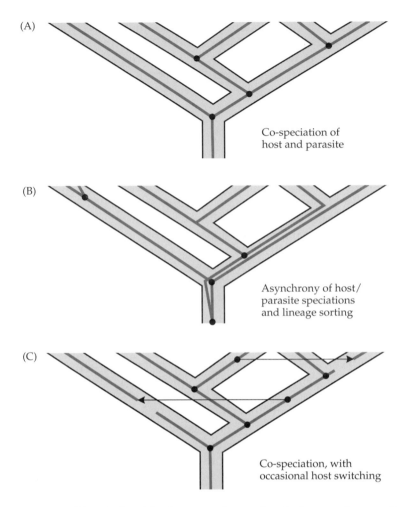

(A)

Co-speciation of
host and parasite

(B)

Asynchrony of host/
parasite speciations
and lineage sorting

(C)

Co-speciation, with
occasional host switching

Figure 7.6   Possible relationships between the species phylogeny of a parasite and that of its host taxon under models of (A) co-speciation, wherein host and parasite phylogenies are perfectly concordant; (B) asynchronous speciation, wherein non-concordant lineage sorting characterizes host and parasite lineages; and (C) host switching by the parasite. The parasite phylogeny is depicted by the inner branches; black dots indicate speciations. The thick outer branches depict the host phylgeny.

As judged by similar molecular phylogenetic analyses (of mitochondrial genes, nuclear genes, or both) in numerous other taxonomic groups, perfect congruence between the cladogenetic histories of interacting species appears to be the exception, not the rule. For example, the phylogeny of *Schistosoma* trematodes bears little resemblance to the phylogeny of the snails they parasitize (Morgan et al. 2002), and the same can be said for *Puccinia* rust fungi and their plant hosts (Roy 2001), *Wolbachia* bacteria and their insect hosts (Kittayapong et al. 2003; Shoemaker et al. 2002), herbivorous *Ophraella* beetles and the plants they utilize

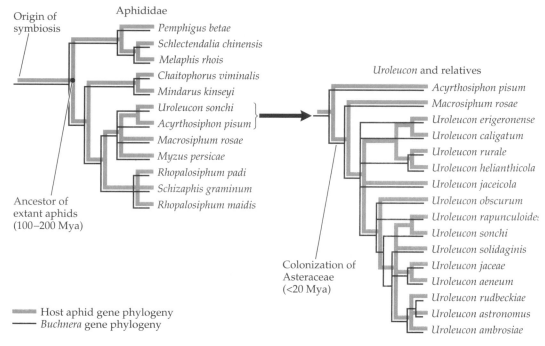

Figure 7.7   Near-perfect phylogenetic congruence between aphids and their endosymbiotic *Buchnera* bacteria. The diagram on the left is at the taxonomic family level (Aphididae), and that on the right is for a subset of aphid hosts mostly in the genus *Uroleucon*. (After Moran 2001.)

(Funk et al. 1995), mushroom-eating *Drosophila* and their nematode parasites (Perlman et al. 2003), and brood parasitic finches and their avian host species (Klein and Payne 1998). Another common outcome, illustrated by various avian taxa and their lice (Johnson et al. 2003; Page et al. 1998; Paterson et al. 2000) and by marine sponges and their bacterial symbionts (Erpenbeck et al. 2002), is partial correspondence between the phylogenies of host taxa and their associates. Such instances indicate a tendency for co-speciation, but with the patterns confounded by occasional host shifts or other complicating factors (see above) that have partially uncoupled historical coevolution between the interacting taxa.

Another kind of species interaction that has been hypothesized to promote congruent phylogenies (as well as sympatric speciation) is "social parasitism," as illustrated by *Polistes* wasps. In these colonial hymenopterans, a social parasite uses workers of another social insect species (usually a close evolutionary relative) to rear its own progeny. However, phylogenetic analyses based on allozymes and on rDNA sequences from the mitochondrial genome revealed that all social parasites in the genus are monophyletic and recently evolved relative to their host species (Carpenter et al. 1993; Choudhary et al. 1994). These results are more consistent with the idea that speciations in these social parasites occurred allopatrically and independently of the evolution of social parasitism.

Can morphologically cryptic species be diagnosed?

Molecular markers can be of great utility in diagnosing closely related species, even where morphological or other traditional markers fail or are ambiguous. In an illustrative early example, several sibling species of *Drosophila* proved readily separable (see Figure 7.2) with a battery of allozyme markers despite their near-identity in morphological appearance (Ayala and Powell 1972b; Ayala et al. 1970; Hubby and Throckmorton 1968). Such molecular markers can find application in diagnosing the species composition of field collections where this is otherwise uncertain. For example, Pascual et al. (1997) used a battery of allozyme, mtDNA, and RAPD markers to distinguish three sibling species of *Drosophila* (*subobscura*, *athabasca*, and *azteca*) and thereby illuminate the geographic distributions of these flies following their colonization of North America's west coast.

Sibling species of *Trachyphloeus* weevils (Jermiin et al. 1991) and *Chthamalus* barnacles (Hedgecock 1979) likewise have been readily distinguished by allozymes, as have various *Alpheus* and *Penaeus* shrimp by mtDNA sequences (Mathews et al. 2002; Palumbi and Benzie 1991). A polychaete worm (*Capitella capitata*) used as an indicator of marine pollution and environmental disturbance was once thought to be a single cosmopolitan species, but allozyme analyses indicated the presence of at least six sibling species (Grassle and Grassle 1976). In corals assigned to *Montastrea annularis*, some morphological variation suspected to be due to phenotypic plasticity proved instead, upon allozyme analysis, to involve distinct sympatric species (Knowlton et al. 1992; Lopez et al. 1999). In sea anemones (genus *Anthopleura*), allozyme analyses confirmed the presence in sympatry of two closely related species that differ in having solitary versus clonal reproductive lifestyles (McFadden et al. 1997). In general, allozymes and other molecular markers have identified numerous sibling species in the sea (Knowlton 1993) and elsewhere (e.g., Gómez et al. 2002; Jarman and Elliott 2000; McGovern and Hellberg 2003; Trewick 2000). Such studies can inform ecology and management as well as systematics. For example, molecular genetic analyses of sibling species of *Carcinus* crabs have helped to identify cryptic marine invasions and the geographic sources of colonizing taxa (Geller et al. 1997). Molecular genetic identifications of scale insects (*Quadraspidiotus* sp.) enabled researchers to determine which particular pest species had been attracted to artificial pheromone traps (Frey and Frey 1995).

Various classes of molecular markers can likewise be used to distinguish closely related plant species (e.g., Whitty et al. 1994). An interesting but atypical example involved use of species-specific "transposon signatures" to distinguish species and subspecies within the *Zea*/*Tripsacum* complex of maize-like plants (Purugganan and Wessler 1995). Transposable elements (TEs; see Box 1.3) are abundant components of prokaryotic and eukaryotic genomes, in which they reside at specific positions often characteristic of particular populations or species. The approach employed by Purugganan and Wessler was to amplify particular DNA regions containing TEs and then digest them with restriction enzymes. The resulting digestion profiles provided diagnostic molecular signatures for different *Zea* taxa.

Zooxanthellae are unicellular endosymbiotic algae that live within the tissues of diverse marine invertebrates such as sea anemones, corals, and gorgonians. Due to a paucity of distinguishing morphological characters, most zooxanthellae had been placed in the genus *Symbiodinium*, but the numbers of species, their evolutionary relationships, and their particular host associations were poorly known. Rowan and Powers (1991) assayed nuclear rDNAs from zooxanthellae isolated from 22 invertebrate taxa and identified several genetically distinct forms, some of which resided in congeneric hosts. Conversely, some zooxanthellae that were genetically similar came from divergent hosts of ordinal or greater taxonomic separation. Results suggested that many cryptic species of zooxanthellae exist and that symbioses can arise by symbiont shifts among even distant host taxa. The conclusion that *Symbiodinium* is a highly diverse and ancient evolutionary assemblage was further supported by the finding that the collective genetic diversity of its rDNA sequences rivals that between taxonomic orders of non-symbiotic dinoflagellates (Rowan and Powers 1992). Furthermore, this diversity appears to be organized into about half a dozen highly divergent evolutionary clades, according to evidence from cpDNA and nuclear rDNA sequences (LaJeunesse 2001; Pochon et al. 2001; Santos et al. 2002).

Fig-pollinating wasps in the family Agonidae have refined mutualistic interactions with *Ficus* trees: The fig trees depend on female wasps for pollination, and the wasps depend on fig inflorescences for oviposition sites and nurseries for their young (Janzen 1979; Wiebes 1979). These tightly integrated fig–wasp symbioses have long provided model systems for research into the ecology and evolution of animal–plant mutualisms, with a common assumption being that one species of pollinator wasp specializes on each species of host fig. However, recent molecular analyses have cast doubt on this latter assumption by documenting substantial mtDNA sequence differences among wasps within a majority of fig species surveyed, thus strongly suggesting the presence of many formerly cryptic wasp species (Molbo et al. 2003). By undermining the prevalent notion of a strict one-to-one specificity between figs and their pollinators, these genetic findings will necessitate various reinterpretations of conventional wisdom about this mutualistic complex (Machado et al. 2001; see also Weiblen and Bush 2002).

Among the vertebrates, many problematic species have been distinguished using molecular characters (e.g., Belfiore et al. 2003). Morphologically cryptic fish species, such as several closely related species of *Thunnus* tunas (Bartlett and Davidson 1991), can be distinguished by allozyme or mtDNA markers (see early reviews in Powers 1991; Shaklee 1983; Shaklee et al. 1982). In *Gastrotheca* frogs, immunological and protein electrophoretic evidence revealed that at least six species in two different groups formerly had masqueraded under the name *G. riobambae* (Duellman and Hillis 1987; Scanlan et al. 1980). Remarkable examples of morphological stasis despite extensive speciation are provided by the lungless plethodontid salamanders, in which multiple fixed allozyme differences proved to be common among populations formerly thought to be conspecific (Larson 1984, 1989). For example, the slimy salamander of the eastern United States had been considered a single species (*Plethodon glutinosus*), but dramatic and often sympatric differences in protein markers (Highton 1984; Maha et al. 1989) revealed the presence of at least 16 groups that probably warrant recognition as species or semispecies (Highton et al. 1989). Among avian taxa, sibling species in the flycatcher genera *Empidonax* and *Contopus* are difficult to distinguish morphologically, but species and clades proved to exhibit diagnostic molecular markers (Johnson and Cicero 2002; Zink and Johnson 1984). In general, birds are among the best known of organisms at the species level, but in the early 1990s a shrike named *Lanarius liberatus* became the first "new" avian species described solely on the basis of molecular data (Hughes 1992; E. F. G. Smith et al. 1991).

In many taxonomic groups, adults of different species are distinguishable morphologically, whereas their juveniles or larvae are not. By comparing molecular characters in unknown larvae against known adults, species assignments can often be made. In some pioneering examples of this approach, researchers used protein electrophoresis or DNA restriction mapping to make species assignments for various larval marine organisms ranging from oysters (Hu et al. 1992) to fishes (Graves et al. 1988, 1989; Morgan 1975; Sidell et al. 1978; Smith and Benson 1980; Smith and Crossland 1977). The introduction of PCR-based methods dramatically improved scientists' abilities to identify small larval forms or eggs, as, for example, when Silberman and Walsh (1992) amplified and digested nuclear 28S rRNA genes from phyllosome larvae of three species of spiny lobsters (*Panulirus argus*, *P. guttatus*, and *P. laevicauda*) and thereby revealed species-diagnostic DNA banding patterns. In another interesting early application of PCR-based larval identification, Olson et al. (1991) amplified and sequenced 16S rDNAs from sea cucumbers (Echinodermata) and thereby assigned a collection of bright red pentacula larvae to *Cucumaria frondosa*. This assignment came as a surprise—based on coloration and morphological considerations, these larvae had been thought to belong to a distantly related sea cucumber, *Psolus fabricii*. In recent years, PCR-assisted DNA sequencing has become almost routine for taxonomic identifications of marine larvae (Coffroth and Mulawka 1995) and in systematic assessments of small sea creatures such as copepods (Bucklin et al. 1999).

PCR-based methods and additional molecular techniques have also further revolutionized scientists' abilities to discriminate among species of tiny or microbial forms, including mites (Fenton et al. 1995), soil nematodes (Floyd et al. 2002), mosses (A. J. Shaw 2000), fungi (Bidochka et al. 1997; Fell et al. 1992), foraminiferans (Bauch et al. 2003), viruses (e.g., Allander et al. 2001), and bacteria (Laguerre et al. 1994). For example, Wimpee et al. (1991) employed two highly conserved regions of the *luxA* gene as PCR primers to develop species-specific probes that can identify four major groups of marine luminous bacteria from field isolates.

Schmidt et al. (1991) used molecular assays of bulk genomic DNA to address the species composition of picoplankton collections from the central Pacific Ocean. They cloned mixed populations of DNA into phage, screened these clones for the presence of 16S rRNA genes, and sequenced the isolates. When compared against an established information base of rRNA gene sequences, these data allowed identification of 15 distinctive bacterial sequences that could be related to cyanobacteria and proteobacteria. Similar molecular approaches have been employed to characterize the approximate phylogenetic positions of previously unknown microbial taxa from a wide variety of soils and aqueous environments (Angert et al. 1998; Blank et al. 2002; Dawson and Pace 2002; de Souza et al. 2001; Giovannoni et al. 1990; Weller and Ward 1989; Weller et al. 1991). Although such molecular analyses are still quite some way from providing complete species-level descriptions in the prokaryotic world (Curtis et al. 2002), they can be applied to all bacterial forms (whether they can be artificially cultured or not), and they are revolutionizing scientific capabilities to delve into microbial ecology and diversity (Pace 1997; Seidler and Fredrickson 1995; Wilson 2003).

In industrial and epidemiological areas also, molecular markers routinely play many key roles in distinguishing otherwise cryptic lineages. For example, Salama et al. (1991) employed subspecies-specific rRNA gene probes to identify *Lactococcus lactis cremoris*, a bacterium whose few available strains are relied upon by the dairy industry for the manufacture of cheddar cheese that is free of fermented and fruity flavors. Similarly, Regnery et al. (1991) employed amplified DNA sequences from the citrate synthase and 190-kDa antigen genes as substrates for molecular assays that proved to distinguish various rickettsial species causing spotted fever. In another application with epidemiological relevance, PCR-based assays were used to identify sequences from the introduced West Nile virus in wild birds and mosquitoes from the northeastern United States, thereby helping to assess the origins and spread of the West Nile disease outbreak in North America (Anderson et al. 1999).

As transmitting agents for many human diseases, mosquitoes began to receive molecular genetic attention more than two decades ago. Among the early efforts was the identification by Miles (1978) and Finnerty and Collins (1988) of diagnostic allozyme and RFLP markers for mosquitoes in the *Anopheles gambiae* complex, which includes some of the primary vectors for African malaria. In the United States, another *Anopheles* mosquito (formerly considered *A. quadrimaculatus*) proved upon molecular and cytological examination to consist of at least four cryptic species. Numerous molecular markers, including allozymes (Narang et al. 1989a,b,c), rDNAs (Mitchell et al. 1993), and mtDNA (Kim and Narang 1990), showed concordant genetic partitions that also agreed with reproductive boundaries as revealed in experimental crosses. Indeed, many molecular characters were diagnostic, enabling construction of dichotomous keys to species identification (Table 7.4). Other pathbreaking molecular studies on mosquitoes discriminated morphologically cryptic and sometimes sympatric forms of *Aedes*, including behavioral types in the *A. aegypti* complex, the primary vector to humans of dengue fever and yellow fever viruses (Munstermann 1988; Powell et al. 1980; Tabachnick and Powell 1978; Tabachnick et al. 1979). Examples of more recent developments in molecular genetic diagnosis and systematics of various mosquito taxa can be found in Krzywinski et al. (2001), Lehmann et al. (2003), Miller et al. (1996), Munstermann (1995), and

---

TABLE 7.4 Diagnostic allozyme loci and dichotomous biochemical key to four sibling species in the *Anopheles quadrimaculatus* complex of mosquito sibling species

Diagnostic loci[a] for the indicated species pairs

A:B	**A:C**	**A:D**	**B:C**	**B:D**	**C:D**
Idh-1	*Acon-1*	*Acon-1*	*Acon-1*	*Acon-1*	*Got-1*
Idh-2	*Idh-2*	*Idh-2*	*Idh-1*	*Idh-1*	*Had-1*
Est-2	*Had-1*	*Got-1*	*Had-1*	*Got-1*	*Had-3*
Est-5	*Had-3*	*Got-2*	*Had-3*	*Got-2*	*Pep-4*
Est-7	*Pep-2*	*Pep-2*	*Got-2*	*Pep-2*	*Pgi-1*
Had-1	*Got-2*	*Pep-4*	*Pep-2*	*Pep-4*	*Me-1*
6Pgd-1	*Pgi-1*	*Me-1*	*Pgi-1*	*Me-1*	*Est-2*
	Est-2	*Mpi-1*	*Est-4*	*Est-2*	*Mpi-1*
	Est-6		*Est-5*	*Est-7*	
	Mpi-1		*Est-6*	*Mpi-1*	
	6-Pgd-1		*Est-7*		
	Xdh-3		*Mpi-1*		
	Ao-1		*Xdh-3*		

Biochemical key[b]

1. *Mpi-1* slow (62 allele, rarely with 52 as heterozygote)...............Species D
   Mpi-1 faster (78 or greater)..Go to 2

2. *Idh-1* slow (86) and *Idh-2* fast (162)........................…........... Species B
   Idh-1 faster (≥100, sometimes with 86 as heterozygote);
       Idh-2 fast or slower (100, 132, 162)....................................Go to 3

3. *Had-3* slow (45); *Pgi* slow (95)..…...........Species C
   Had-3 faster (100, sometimes with 45 as heterozygote);
       Pgi faster (100, rarely with 95 as heterozygote)........…...............Species A

Source: After Narang et al. 1989b.
[a] Those providing correct identification with probability >99%.
[b] In this key (one of many that could be generated), numbers indicate electromorph gel mobilities in an unknown sample relative to the electromorph in a standard strain.

Walton et al. (2001). The entire nuclear genome of *Anopheles gambiae* was recently published (Holt et al. 2002).

Hebert et al. (2003; see also Tautz et al. 2003) raised the prospect that routine biodiversity assessments of the future may involve "DNA barcodes" more than conventional taxonomic appraisals based on morphology. They suggested that traditional systematic expertise is collapsing rapidly for many taxonomic groups, and that "the sole prospect for a sustainable identification capability lies in the construction of systems that employ DNA sequences as taxon barcodes." This Orwellian specter is perhaps more sad than exciting.

Should a phylogenetic species concept replace the BSC?

Throughout the twentieth century and continuing today, the biological species concept (BSC) has been the major theoretical framework orienting research on the origins of biological diversity. However, a serious recent challenge to the BSC has come from some systematists, who argue that it lacks a sufficient phylogenetic perspective and, hence, provides an inappropriate guide to the origins and products of evolutionary diversification (de Queiroz and Donoghue 1988; Donoghue 1985; Eldredge and Cracraft 1980; Mishler and Donoghue 1982; Nelson and Platnick 1981; see reviews in Hull 1997; Wheeler and Meier 2000). Modern critics of the BSC argue that "reproductive isolation should not be part of species concepts" (McKitrick and Zink 1988) and that "as a working concept, the biological species concept is worse than merely unhelpful and non-operational—it can be misleading" (Frost and Hillis 1990). These criticisms have led to a call for replacement of the BSC with a phylogenetic species concept, or PSC (see Box 7.1), under which a species is defined as a monophyletic group composed of "the smallest diagnosable cluster of individual organisms within which there is a parental pattern of ancestry and descent" (Cracraft 1983).

One key motivation for this suggestion is that *biological* speciations, as described above, often result initially in paraphyletic taxa (an anathema to cladists) at the species level. In principle, this "problem" could be remedied under the PSC by elevating all diagnosable populations to full species rank (Omland et al. 1999; Voelker 1999), or, alternatively, by synonymizing paraphyletic taxa and the subset taxa nested therein. From the perspective of the BSC, however, neither of these alternatives is desirable, in part because they would either ignore the genetic and reproductive distinctness of the nested lineage or neglect what might be high gene flow within the paraphyletic lineage (Funk and Omland 2003; Olmstead 1995; Sosef 1997; Wiens and Penkrot 2002). Thus, the PSC can also be justifiably accused of being biologically unhelpful, if not misleading (Johnson et al. 1999), even in the strict genealogical context it otherwise intends to inform (see also Wiens 1999).

Because molecular data provide unprecedented power for phylogeny estimation, it might be supposed that molecular evolutionists would be among the strongest advocates for the PSC, but this has not necessarily been the case (Avise 2000a,b; Avise and Wollenberg 1997). One serious difficulty with existing PSC proposals concerns the nature of the evidence required to justifiably diagnose a monophyletic group warranting species recognition. Molecular technologies have made it abundantly clear that multitudinous derived traits often can be employed to subdivide named species into diagnosable subunits (see Chapter 6). Indeed, most individuals and family units within sexually reproducing species can be distinguished from one another with high-resolution molecular assays. If each individual or kinship unit is genetically unique, then to group multiple individuals into phylogenetic "species" would require that distinctions below some arbitrary threshold be ignored (unless each specimen is to be considered a unique species). The evolutionary significance of any such threshold surely must be questionable. For these and other reasons, Avise and

Ball (1990) suggested that if the broader framework of the PSC were to contribute to a significant advance in systematic practice (as they believed that it could), a shift from issues of diagnostics to issues of magnitudes and patterns of phylogenetic differentiation, and of the historical and reproductive reasons for such patterns, would be required. Toward that end, they introduced the notion of "genealogical concordance" principles that might be employed to combine desirable elements of both the PSC and the BSC.

Within any sexual organismal pedigree, allelic phylogenies can differ greatly from locus to locus (Ball et al. 1990; Baum and Shaw 1995; Maddison 1997), if for no other reasons than the Mendelian nature of meiotic segregation and syngamy and the inevitable vagaries of lineage sorting within and among gene trees. An array of individuals phylogenetically grouped by one locus may differ from an array of individuals grouped by another locus, unless some overriding evolutionary force has concordantly shaped the phylogenetic structures of multiple quasi-independent genes. One such force expected to promote genealogical concordance across loci (aspect II; see Chapter 6) is intrinsic reproductive isolation (the focal point of the BSC). Through time, due to processes of lineage turnover, biological species isolated from one another by intrinsic RIBs inevitably tend to evolve toward a status of reciprocal monophyly in particular gene genealogies. Furthermore, through time, the genealogical tracings of independent loci almost inevitably sort in such a way as to partition these species concordantly. Thus, various aspects of genealogical concordance per se become deciding criteria by which biologically meaningful genetic partitions can be distinguished from partitions that are "trivial" or gene-idiosyncratic with respect to organismal phylogeny.

However, for populations that are geographically isolated for sufficient lengths of time relative to effective population sizes, genealogical concordance across loci also can arise from purely extrinsic barriers to reproduction. As emphasized in Chapter 6, dramatic phylogenetic partitions are routinely observed among populations considered conspecific under the BSC. It might be argued that such populations also warrant formal taxonomic recognition (albeit not necessarily at the species level) on the grounds that they represent significant biotic partitions of relevance to such areas as biogeographic reconstruction and conservation biology.

From consideration of these and additional factors, Avise and Ball (1990) suggested the following conceptual framework for biological taxonomy, based on genealogical concordance principles. The biological and taxonomic category "species" should continue to refer to groups of actually or potentially interbreeding populations isolated by intrinsic RIBs from other such groups. In other words, a retention of the basic philosophical framework of the BSC is warranted, in no small part because RIBs are a powerful evolutionary force in generating significant historical partitions in organismal phylogenies (i.e., in generating salient "genotypic clusters"; Mallet 1995). Within such units, "subspecies" warranting formal recognition could then be conceptualized as groups of actually or potentially interbreeding populations (normally mostly allopatric) that are genealogically highly distinctive from, but reproductively compatible with, other such groups. Importantly, the empirical evidence for genealogical distinction must come, in principle, from concordant genetic partitions across multiple, independent, genetically based molecular (or phenotypic; Wilson and Brown 1953) traits. This phylogenetic approach to taxonomy near and below the species level represents a novel compromise between the BSC and the PSC, and is a clear conceptual outgrowth from molecular genetic and coalescence-based perspectives on microevolutionary processes.

Hybridization and Introgression

The term "hybridization" is as difficult to define as is speciation, and for similar reasons. In the early literature of systematics, a "hybrid" was deemed to be an offspring resulting from a cross between species, whereas the term "intergrade" was reserved for any product of a cross between recognizable conspecific populations or subspecies. But as we have seen, this distinction can be rather subjective, so "hybridization" is now usually employed in a broad sense to include crosses between genetically differentiated forms regardless of their current taxonomic status. "Introgression" refers to gene movement between species (or sometimes between well-marked genetic populations) mediated by hybridization and backcrossing.

Frequencies and geographic settings of hybridization

Hybridization and introgression are common phenomena in many plant and animal groups. More than 30 years ago, Knobloch (1972) compiled a list of nearly 24,000 reported instances of interspecific or intergeneric plant hybridization (despite the availability of detailed studies on only a small fraction of the botanical world). Introgression is more challenging to assess, but Rieseberg and Wendel (1993) provided a compilation of 155 noteworthy cases of plant introgression, many of which include molecular documentation. Hybridization is especially common in outcrossing perennials (Ellstrand et al. 1996). Similarly, hybridization and introgression have been uncovered in numerous animal taxa (Dowling and Secor 1997; Harrison 1993). For example, Schwartz (1981) compiled a list of nearly 4,000 references dealing with natural and artificial hybridization in fishes, many cases of which have been verified and characterized further using molecular markers (Avise 2001c; Campton 1987; Verspoor and Hammar 1991). Among the vertebrates, fishes with external fertilization appear most prone to hybridization, but the phenomenon is widespread.

Both the frequency of hybridization and the extent of introgression can vary along a continuum from nil to extensive, and molecular markers are invaluable for empirically assessing where a given situation falls. Near one extreme, hybridization may be confined primarily to the production of F_1 hybrids, which may be abundant or rare. For example, analyses based on nuclear and mtDNA markers revealed that hybrids between brook trout (*Salvelinus fontinalis*) and bull trout (*S. confluentus*) in Montana are mostly nearly sterile F_1 individuals, but they are also common (at some locales) and arise from crosses in both directions with respect to sex (Leary et al. 1993). By contrast, hybrids between blue whales (*Balaenoptera musculus*) and fin whales (*B. physalus*) are rare, but similar molecular marker analyses proved that three phenotypically anomalous individuals were F$_1$ hybrids and that one hybrid female also carried a backcross fetus with a blue whale father (Árnason et al. 1991; Spilliaert et al. 1991). At the opposite extreme, introgressive hybridization can be so extensive that populations merge into one panmictic gene pool. This situation is exemplified by hybrid swarms between genetically distinct subspecies of bluegill sunfish (*Lepomis macrochirus macrochirus* and *L. m. purpurescens*) in Georgia and the Carolinas (Avise and Smith 1974; Avise et al. 1984b) and between cutthroat trout subspecies (*Oncorhynchus clarki lewisi* and *O. c. bouvieri*) in Montana (Forbes and Allendorf 1991). In both cases, these taxa normally inhabit different geographic regions, but can hybridize extensively when they meet.

Reports of extensive introgressive hybridization between well-marked taxa occasionally appear. One case in point involves two crayfish species in northern Wisconsin and Michigan: the native *Orconectes propinquus* and an introduced congener, *O. rusticus*. Based on cytonuclear molecular analyses, female *rusticus* often mate with male *propinquus*, producing F$_1$ hybrids that mediate extensive introgression (Perry et al. 2001). One net result has been the near-elimination of genetically pure *O. propinquus* in one Wisconsin lake. A similar case in vertebrate animals involves spotted bass (*Micropterus punctulatus*) and smallmouth bass (*M. dolomieui*) in Lake Chatuge in northern Georgia. In the late 1970s, spotted bass were introduced into the lake, which formerly was inhabited only by smallmouths. Within about 10 years, only a small percentage of genetically pure smallmouth bass remained, as judged by species-diagnostic nuclear and mitochondrial markers (Avise et al. 1997). Furthermore, this demographic shift had been accompanied by extensive introgression, such that more than 95% of the remaining smallmouth bass alleles in Lake Chatuge had become "genetically assimilated" into the gene pool of fish with hybrid ancestry. Such an outcome, sometimes referred to as "genetic swamping," can be interpreted as a local genetic extinction of a population via hybridization and introgression. Several additional examples of this phenomenon are known (see review in Rhymer and Simberloff 1996).

In many taxonomic groups, organisms separated for long periods of evolutionary time nonetheless may retain the anatomical and physiological capacity for hybrid production. Using micro-complement fixation assays, Wilson and colleagues (1974a, 1977; Prager and Wilson 1975) compared immunological distances in numerous pairs of mammal species, bird species, and frog species that were known to be capable of generating viable hybrids in captivity or in the wild. The genetic distances were then translated into estimates of absolute divergence times for the species involved, using molecular clocks calibrated specifically for each taxonomic group (Figure 7.8). Results indicated that the hybridizable frog species had separated from one another, on average, more than 20 million years ago, as had the hybridizable birds assayed, whereas mean separation time for the hybridizable mammal species was only about 2–3 mya. The dramatically faster pace at which mammals had lost the potential for interspecific hybridization was provisionally attributed to a faster pace of chromosomal evolution or a

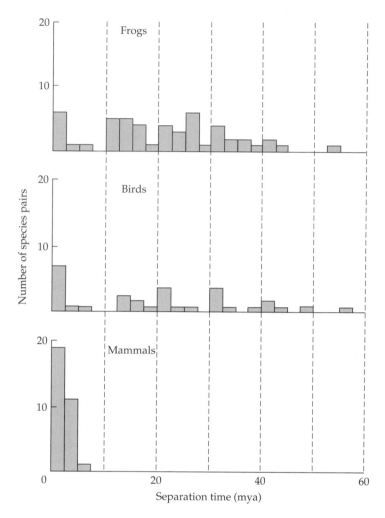

Figure 7.8 Evolutionary separation times, as estimated from albumin immunological distances (ID) for more than a hundred pairs of vertebrate species capable of producing viable hybrids. Molecular clocks used to generate these times were calibrated at about 1.7 ID units per million years in frogs and mammals (Prager and Wilson 1975) and about 0.6 ID units per million years in birds (Prager et al. 1974). (After Prager and Wilson 1975; Wilson et al. 1974a.)

higher evolutionary rate in their regulatory genes (Prager and Wilson 1975; Wilson et al. 1974a,b). Regardless of the explanation, many organisms clearly retain the physiological capacity to hybridize over very long periods of evolutionary time. How often such potential is realized in nature is another issue, of course, and one that can be powerfully addressed using molecular markers.

Frequently, as in the basses mentioned above, hybridization follows human-mediated transplantations (Scribner et al. 2001a). In the early 1980s, the pupfish *Cyprinodon variegatus* was introduced to the Pecos River in Texas, where it then hybridized with an endemic species, *C. pecosensis*. Protein electrophoretic data revealed that within 5 years, panmictic admixtures of the two pupfishes occupied approximately 430 river kilometers, or roughly one-half of the historic range of the endemic species (Echelle and Connor 1989; Echelle et al. 1987, 1997). A similar study of land snails involved Bahamian *Cerion casablancae* that were introduced in 1915 to the range of *C. incanum* on Bahia Honda Key, Florida. Introgressive hybridization ensued, and analyses of allozymes and morphology later in the century revealed that the snails had become panmictic on Bahia Honda, that no pure *C. casablancae* remained, and that there had been a 30% reduction in frequency of the introduced genome (Woodruff and Gould 1987). Another notable example of hybridization precipitated by artificial transplantations involves salmonid fishes in the western United States. There, repeated introductions of millions of hatchery-reared rainbow trout to endemic cutthroat trout habitats, and of cutthroat trout from one locale to another, were followed by extensive genetic introgression that has been thoroughly documented using molecular markers (Allendorf and Leary 1988; Busack and Gall 1981; Gyllensten et al. 1985b; Kanda et al. 2002a; Leary et al. 1984).

With respect to geography, natural hybridization may occur sporadically between broadly sympatric species or be confined to particular contact areas. Hybrid zones are regions in which genetically distinct populations meet and produce progeny of mixed ancestry (Barton and Hewitt 1989; Harrison 1990), and they are often spatially linear (Hewitt 1989) or mosaic (Harrison and Rand 1989; Rand and Harrison 1989). They can also move through time (Barton and Hewitt 1981), sometimes rapidly, as has been documented with the help of molecular markers in plants (Martin and Cruzan 1999), butterflies (Dasmahapatra et al. 2002), fishes (Childs et al. 1996), birds (Rohwer et al. 2001), and mammals (Hafner et al. 1998), among others. Hybrid zones typically represent secondary overlaps of formerly allopatric or parapatric (abutting) taxa, and they are often evidenced by a general concordance across loci in allelic clines that transect the presumed contact zone (e.g., Dessauer 2000). (In theory, however, concordant clines could also be generated by intense diversifying selection within a continuously distributed population; Endler 1977.) Secondary hybrid zones may be persistent or ephemeral. Persistent hybrid zones are usually hypothesized to register either "bounded hybrid superiority," wherein hybrids have superior fitness in areas of presumed ecological transition, or "dynamic equilibrium" (also known as genetic "tension"; Barton and Hewitt 1985; Key 1968), wherein the hybrid zone is maintained through a balance between continued dispersal of parental types into the area and hybrid inferiority (Moore and Buchanan 1985).

Hybrid zones are marvelous settings in which to apply molecular markers for several reasons (Hewitt 1988). First, the populations or species involved are genetically differentiated (by definition), such that multiple markers for characterizing each hybrid gene pool normally can be uncovered. Second, because each hybrid zone involves an amalgamation of independently evolved genomes, exaggerated effects of intergenomic interactions can be anticipated. These effects magnify the impact of such processes as recombination and natural selection, making these evolutionary forces easier to study. Third, various sexual asymmetries frequently are involved in hybrid zones, and powerful approaches now exist for dissecting these factors by utilizing joint data from cytoplasmic and nuclear markers, as described next (see also Box 7.4).

Sexual asymmetries in hybrid zones

The power of molecular markers in dissecting hybridization phenomena can be introduced by considering a study in which cytonuclear analyses (i.e., joint examination of nuclear and cytoplasmic markers) were applied to a hybrid population between *Hyla cinerea* and *H. gratiosa* in ponds near Auburn, Alabama. These genetically distinct treefrog species are distributed widely and sympatrically throughout the southeastern United States, but judging from morphological evidence they hybridize at least sporadically, and have done so extensively at the Auburn site across several decades. One reason for particular interest in the Auburn population stems from behavioral observations suggesting the potential for a sexual bias in the direction of interspecific matings. During the breeding season, *H. gratiosa* males call from the water surface, whereas *H. cinerea* males call from perches along the shoreline (Figure 7.9A). In the evenings, gravid females of both species approach the ponds from surrounding woods and become amplexed (mated). Thus, one hypothesis is that interspecific matings might primarily involve *H. cinerea* males with *H. gratiosa* females, rather than the reverse, because *H. gratiosa* females must "hop a gauntlet" of *H. cinerea* males before reaching conspecific partners.

Lamb and Avise (1986) employed five species-diagnostic allozyme loci plus mtDNA to characterize 305 individuals from this hybrid population. The allozyme loci were chosen because they exhibited fixed allelic differences between the species, thus allowing provisional assignment of each individual at the Auburn site to one of the following six categories: pure *cinerea*, pure *gratiosa*, F_1 hybrid, progeny from a backcross to *cinerea*, progeny from a backcross to *gratiosa*, or later-generation hybrid. For example, an F_1 hybrid should be heterozygous at all marker loci, and a *cinerea* backcross progeny would probably appear heterozygous at some loci and homozygous for *cinerea* alleles at others. [Probabilities of misclassifying an individual can be calculated from basic Mendelian considerations, and they are low whenever multiple diagnostic markers are used. For example, a true first-generation *cinerea* backcross progeny would be mistaken for a pure *cinerea* with probability $k = (0.5)^n$, where n is the number of fixed marker loci, so in this case, $k = 0.03$.] The mtDNA genotypes then allowed assignment of the female (and hence male) parent for each allozyme-characterized treefrog at the Auburn site.

For this hybrid population, the molecular data revealed a striking genetic architecture that generally proved consistent with the suspected mating behaviors of the parental species (Table 7.5; Figure 7.9B). Thus, all 20 F_1 hybrids carried *gratiosa*-type mtDNA, showing that they had *gratiosa* mothers. Furthermore, 52 of 53 individuals identified as backcross progeny to *gratiosa* possessed *gratiosa*-type mtDNA (as predicted, because their mothers were either F_1 hybrids or pure *gra-*

(A)

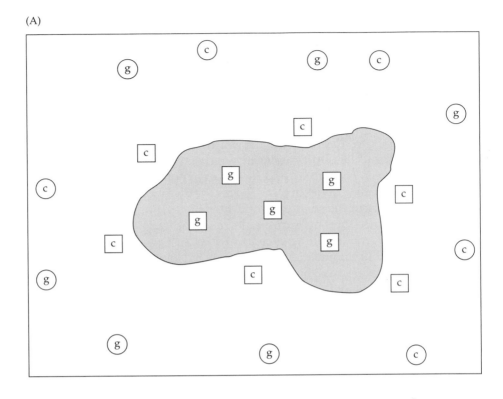

(B)

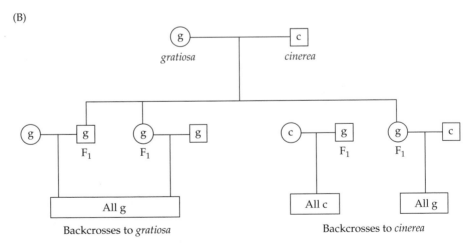

Figure 7.9   Biological setting of a hybrid *Hyla* population. (A) Diagrammatic aerial view of the Auburn pond, showing the typical spatial positions of male and female frogs before the mating process. (B) Expected pedigree involved in production of F_1 hybrids and various backcross classes, under the assumption that the hybridization events typically entailed matings of male *H. cinerea* with female *H. gratiosa*. In both (A) and (B), each letter refers to the species origin of the mtDNA genotype ("c", *cinerea*; "g", *gratiosa*), and squares and circles indicate males and females, respectively.

TABLE 7.5   Genetic architecture of a hybrid population involving the tree frogs
Hyla cinerea and *H. gratiosa*[a]

Allozyme category	gratiosa-type mtDNA		cinerea-type mtDNA	
	Observed	Expected	Observed	Expected
Pure *H. gratiosa*	103	—	0	—
Pure *H. cinerea*	0	—	60	—
F_1 hybrid	20	20	0	0
H. cinerea backcross	22	29	36	29
H. gratiosa backcross	52	53	1	0
Later-generation hybrids	9	Some[b]	2	Some[b]

Source: After Lamb and Avise 1986.
[a] Shown are numbers of frogs in each hybrid or non-hybrid category as identified by multi-locus allozyme genotype, as well as the female parent species for those individuals as identified by mtDNA markers. Also shown are expected numbers based on the behaviorally motivated hypothesis (see text) that interspecific crosses are in the direction *H. cinerea* male × *H. gratiosa* female, and that F_1 hybrids of both sexes (who thus have *H. gratiosa* mtDNA) have contributed equally to a given backcross category.
[b] Both *cinerea*-type and *gratiosa*-type mtDNA genotypes are expected among later-generation hybrids, but relative frequencies are dependent on additional factors and, thus, are hard to predict.

tiosa). Furthermore, among the progeny of backcrosses to *cinerea*, individuals carrying either *gratiosa*-type or *cinerea*-type mtDNA were both well represented (also as predicted, because the mtDNA genome transmitted in a given mating would depend on whether the F_1 hybrid parent was a male or female; see Figure 7.9B). Nevertheless, asymmetric mating alone cannot explain all aspects of the data, because individuals with pure *cinerea* and pure *gratiosa* genotypes remained present in high frequency (Table 7.5). Additional factors may involve selection against hybrids or continued migration of parental species into the area. In formal models that allowed variation in parental immigration rates and included tendencies for positive assortative mating between conspecifics, Asmussen et al. (1989) found an excellent fit to the empirical cytonuclear data when, at equilibrium, about 32% of the inhabitants of the hybrid zone were pure-species immigrants in each generation. However, the possibility of selection against hybrids was not formally modeled.

How much of this pronounced genetic structure in the *Hyla* population would have been uncovered from a traditional morphological assessment alone? Lamb and Avise (1987) applied multivariate analyses to numerous phenotypic characters in these same treefrog individuals and compared results against those obtained from the molecular genetic assessments. Although pure *gratiosa* and pure *cinerea* specimens (as classified by molecular genotype) could be distinguished cleanly by discriminate analyses of morphological characters, various hybrid classes proved less recognizable. Thus, by morphology, 18% of true F_1 hybrids were indistinguishable from pure parental species, 27% of backcrosses in either direction were not distinguished from F_1 hybrids, 50% of *gratiosa* backcross progeny were misidentified as pure *gratiosa*, and 56% of *cinerea* backcross progeny were misidentified as pure *cinerea*. By contrast, expected mis-classification rates based on the molecular genotypes surveyed were invariably less that 4% (based on straightforward Mendelian considerations). Furthermore, the pronounced asymmetry in mating behavior that apparently exerted profound influence on the genetic architecture of this hybrid population would have remained *completely undetected* by morphological assessment alone.

More hybrid zone asymmetries

Numerous molecular genetic analyses of hybridization and introgression have appeared in the past three decades, with early and recent landmark reviews or compilations provided by Barton and Hewitt (1985; Barton 2001; Hewitt 2001),

---

BOX 7.4 Cytonuclear Disequilibria in Hybrid Zones

Cytonuclear disequilibria (CD) are nonrandom associations within a population between genotypes at nuclear and cytoplasmic loci (Arnold 1993; Clark 1994). Consider a population whose individuals have been scored at a diploid autosomal gene and at a haploid cytoplasmic locus (mtDNA in animals; cpDNA or mtDNA in plants), and assume further that each locus has two alleles. Six different cytonuclear genotypes are possible. It is convenient to organize the data into a three-by-two table as follows (wherein each of the six cells in the table refers to the frequency of a cytonuclear genotype):

| Cytoplasm | Nuclear genotype | | | Total |
	AA	Aa	aa	
M	u_1	v_1	w_1	x
m	u_2	v_2	w_2	y
Total	u	v	w	1.0

Using such tables, Asmussen et al. (1987) introduced the following four formal measures of genotypic and allelic cytonuclear disequilibria (D):

Genotypic disequilibria

D_1 = freq. (AA/M) – freq. (AA) freq. $(M) = u_1 - ux$

D_2 = freq. (Aa/M) – freq. (Aa) freq. $(M) = v_1 - vx$

D_3 = freq. (aa/M) – freq. (aa) freq. $(M) = w_1 - wx$

(Note: $D_1 + D_2 + D_3 = 0$)

Allelic disequilibrium

D = freq. (A/M) – freq. (A) freq. $(M) = u_1 + 1/2v_1 - (u + 1/2v)\,x$

(Note: $D = D_1 + 1/2D_2$)

As shown in the following diagram (after Avise 2001c), various phenomena in hybrid zones can leave characteristic CD signatures when the cytoplasmic genome is maternally inherited (Arnold 1993).

Three-by-two table	Cytonuclear signature	Likely explanation
(A) <table><tr><td>+++</td><td>0</td><td>0</td></tr><tr><td>0</td><td>0</td><td>+++</td></tr></table>	$D_1 = -D_3 = D \neq 0$ $D_2 = 0$	Absence of hybridization
(B) <table><tr><td>obs=exp</td><td>obs=exp</td><td>obs=exp</td></tr><tr><td>obs=exp</td><td>obs=exp</td><td>obs=exp</td></tr></table>	$D_1 = D_2 = D_3 = 0$ $D = 0$	Random-mating hybrid swarm
(C) <table><tr><td>++</td><td>obs=exp</td><td>0</td></tr><tr><td>0</td><td>obs=exp</td><td>++</td></tr></table>	$D_1 = -D_3 = D \neq 0$ $D_2 = 0$	Hybridization without introgression; no sex-based directionality to interspecific matings
(D) <table><tr><td>++</td><td>++</td><td>0</td></tr><tr><td>0</td><td>--</td><td>++</td></tr></table>	$D_1, D_2, D_3, D \neq 0$	Hybridization without introgression; sex-based directionality to interspecific matings
(E) <table><tr><td>++</td><td>obs=exp</td><td>--</td></tr><tr><td>--</td><td>obs=exp</td><td>++</td></tr></table>	$D_1 = -D_3 = D \neq 0$ $D_2 = 0$	Introgression into both species; no sex-based directionality to interspecific matings
(F) <table><tr><td>++</td><td>++</td><td>--</td></tr><tr><td>0</td><td>0</td><td>++</td></tr></table>	$D_1, D_2, D_3, D \neq 0$	Possible introgression into both species; sex-based directionality to interspecific matings
(G) <table><tr><td>obs=exp</td><td>++</td><td>--</td></tr><tr><td>obs=exp</td><td>--</td><td>++</td></tr></table>	$D_2 = -D_3 = 2D \neq 0$ $D_1 = 0$	Sex-based directionality to interspecific matings; hybrids preferentially backcross to less discriminating species

In these tables, plus signs indicate excesses and minus signs indicate deficits (relative to random-mating expectations) in the observed frequencies of particular cytonuclear genotypic classes. These various CD signatures are consistent with (but not proof of) several possible hybrid zone phenomena described in the right-hand column.

For example, in a well-mixed hybrid swarm (case B above), observed (obs) frequencies of all six cytonuclear genotypes are in statistical accord (Asmussen and Basten 1994) with expectations (exp) based on products of the marginal frequencies of the single-locus genotypes, and all cytonuclear disequilibria are zero. In cases C and D above, the CD signatures suggest that hybridization was confined to the F$_1$ generation, perhaps due to hybrid sterility or other mechanisms of reproductive isolation. In case C, the interspecific matings occurred with equal likelihood in either direction with respect to gender, whereas in case D there is evidence for a pronounced asymmetry such that females of only one species and males of only the other were primarily involved.

Cytonuclear disequilibrium theory has been extended to other kinds of population genetic settings as well, such as nuclear–dicytoplasmic plant systems involving both mitochondrial and chloroplast genomes (Schnabel and Asmussen 1989), paternal as well as maternal cytoplasmic inheritance (Asmussen and Orive 2000), the estimation of gene flow via pollen versus seeds (Goodisman et al. 2000; Orive and Asmussen 2000), haplodiploid species and X-linked genes (Goodisman and Asmussen 1997), apomictic species (Overath and Asmussen 2000a), and tetraploid species (Overath and Asmussen 2000b).

Arnold (1992, 1997), Harrison (1993), and Richie and Butlin (2001), among others. Several classic studies will be encapsulated here to illustrate the diversity of issues addressed, with special attention devoted to additional categories of genetic asymmetry that frequently attend natural hybridization and introgression. Some types of asymmetries reflect differential compatibilities of introgressed alleles on heterologous genomic backgrounds, a phenomenon that sometimes is revealed by significant contrasts across unlinked loci in the steepness, width, or placement of clines across a hybrid zone (Barton 1983). Other asymmetries may reflect differences between the sexes in genetic fitness or behavior (as in the *Hyla* treefrog example above), which are often revealed by contrasting patterns in cytoplasmic and nuclear markers.

DIFFERENTIAL INTROGRESSION AND mtDNA CAPTURE ACROSS A HYBRID ZONE. A classic hybrid zone, which has been examined using a variety of molecular markers in studies spanning three decades (Fel-Clair et al. 1998; Selander et al. 1969), involves the house mice *Mus musculus* and *M. domesticus*. These forms, sometimes considered subspecies, meet and hybridize along a narrow line bisecting central Europe (Figure 7.10). In one early analysis, Hunt and Selander (1973) surveyed diagnostic allozyme markers in nearly 2,700 mice from the contact zone in Denmark. They discovered free interbreeding within the hybrid zone, as indicated by agreement of genotype frequencies with random-mating expectations; an asymmetry of introgression adjacent to the zone, with extensive introgression of some *domesticus* alleles into *musculus*, but little gene movement in the opposite direction; and a marked increase in the width of the hybrid zone in western Denmark as compared with the east, where 90% of the transition in genic characters occurred across a distance of only 20 km. The different slopes and spatial patterns in the allelic clines that were observed across loci were interpreted as evidence of different selective values for various alleles on foreign genetic backgrounds. In other words, "selection against introgression of the genes studied (or chromosomal segments that they mark) is presumed to involve reduced fitness in backcross generations caused by disruption of coadapted parental gene complexes" (Hunt and Selander 1973).

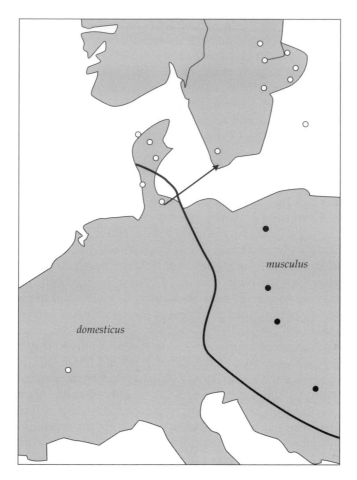

Figure 7.10   Distributions of two species of house mice in central Europe. The heavy line indicates the position of a hybrid zone at the contact between *Mus musculus* (to the north and east, including Scandinavia) and *M. domesticus* (to the south and west). Open and solid circles indicate mtDNA genotypes normally characteristic of *M. domesticus* and *M. musculus*, respectively. The arrow indicates the postulated route of colonization of Scandinavia by female *M. domesticus*. Note that the mtDNA distributions are strikingly discordant with the ranges of the two species as defined by morphology and nuclear genes (see text). (After Gyllensten and Wilson 1987b.)

A molecular study of the 20-km-wide hybrid zone in southern Germany revealed that about 98% of the house mice there had backcross genotypes (Sage et al. 1986). Furthermore, these hybrids were unusually susceptible to parasitic pinworms, other nematodes, and tapeworms, leading to the conclusion that the hybrid zone acts as a low-fitness genetic sink that interferes with gene flow between the species. (In plant hybrid zones as well, parasites or herbivores often are more abundant than in non-hybrid populations, presumably because hybrid individuals tend to have less effective defenses; Strauss 1994.) In the *Mus* hybrid zone at various European sites, subsequent studies of Y-linked and X-linked markers, mtDNA haplotypes, and chromosomal features revealed sharp clines further indicative of genomic incompatibilities and severe restrictions on introgression at both nuclear and cytoplasmic loci (Boissinot and Boursot 1997; Fel-Clair et al. 1996; Tucker et al. 1992; Vanlerberghe et al. 1986, 1988).

Mus musculus and *M. domesticus* normally differ sharply in mtDNA composition (Ferris et al. 1983a,b), but an unexpected pattern emerged in parts of Scandinavia, where mice that by evidence from nuclear DNA and morphology appeared to be pure *musculus* nevertheless carried exclusively *domesticus*-type mtDNA (see Figure 7.10). Gyllensten and Wilson (1987b) proposed that the "foreign" mtDNA originated from a small population of female *domesticus* that colonized Scandinavia from a southern source (Prager et al. 1993), perhaps in association with the spread of farming from northern Germany to Sweden some 4,000 years ago. Continued backcrossing to *musculus* males might thereby have introduced mtDNA from *domesticus* into populations that retained a predominant *musculus* nuclear genetic background. If this interpretation is correct, it provides an example of the phenomenon of "mitochondrial capture," wherein mtDNA genotypes characteristic of one species sometimes occur against a predominant nuclear background of another species because of past introgression between them.

Many such cases of interspecific cytoplasmic capture (mtDNA in animals, usually cpDNA in plants) have been reported (Harrison 1989). Representative examples are listed in Table 7.6. These studies (most of which provisionally eliminated lineage sorting from a polymorphic ancestor as an alternative explanation) document a widespread occurrence of historical gene exchange between species. They also highlight the considerable potential risk of misinterpreting data from any single gene in studies of phylogeography and phylogenetics. In general, cytoplasmic capture is usually easier to document than is the capture of particular nuclear genes because whole linked blocks of non-recombinant markers in the mitochondrial (or chloroplast) genome jointly signify the event.

CHROMOSOMAL ALTERATIONS AS INTROGRESSION FILTERS. A recent theoretical model focuses on structural rearrangements in chromosomes as important contributors to differential introgression as well as speciation (Navarro and Barton 2003a; Noor et al. 2001; Rieseberg 2001). The model is based on the well-supported notion that genetic recombination is reduced in chromosomal regions heterozygous for structural alterations, which thus act as partial reproductive barriers. One net result can be the emergence between karyotypically distinct populations of a semipermeable reproductive sieve, in which gene flow continues unabated in non-rearranged chromosomes, but is partially curtailed in the rearranged chromosomal segments.

One prediction of this model is that positively selected genetic changes should accumulate preferentially in rearranged as opposed to non-rearranged chromosomal regions, a finding that found support in the following empirical test. Humans and chimpanzees differ in major chromosomal rearrangements

TABLE 7.6 Known or suspected examples of cytoplasmic genomic "capture" reportedly due either to modern or ancient introgressive hybridization between related species

Genus	Common name	Reference
Animal mtDNA		
Caledia	Grasshoppers	Marchant 1988
Clethrionomys	Voles	Tegelström et al. 1988
Coregonus	Whitefishes	Lu et al. 2001
Drosophila	Fruit flies	Solignac and Monnerot 1986
Gila	Chub fishes	Gerber et al. 2001
Gryllus	Crickets	Harrison et al. 1987, 1997
Hyla	Treefrogs	Lamb and Avise 1986
Lepus	Hares	Thulin et al. 1997; Alves et al. 2003
Luxilus	Cyprinid fishes	Duvernell and Aspinwall 1995
Micropterus	Black bass	Avise et al. 1997
Mus	House mice	Ferris et al. 1983a
Mytilus	Mussels	Rawson and Hilbish 1998
Notropis, Luxilis	Minnows	Dowling et al. 1989, 1997; Dowling and Hoeh 1991
Odocoileus	Deer	Carr et al. 1986; Cathey et al. 1998
Oreochromis	Cichlid fishes	Rognon and Guyomard 2003
Phrynosoma	Horned lizards	Reeder and Montanucci 2001
Rana	Frogs	Spolsky and Uzzell 1984, 1986
Salvelinus	Charr and trout	Wilson and Bernatchez 1998; Redenbach and Taylor 2002
Stercorarius	Skuas (birds)	Andersson 1999
Tamias	Chipmunks	Good et al. 2003
Thomomys	Pocket gophers	Ruedi et al. 1997
Zosterops	White-eyes (birds)	Degnan 1993
Plant cpDNA		
Argyroxiphium	Silverswords	Baldwin et al. 1990
Brassica	Cabbage and allies	Palmer et al. 1983
Dubautia	Silverswords	Baldwin et al. 1990
Eucalyptus	Australian trees	Jackson 1999
Gossypium	Cottons	Wendel et al. 1991; Wendel and Albert 1992
Helianthus	Sunflowers	Rieseberg et al. 1990b
Heuchera	Heucheras	Soltis et al. 1991; Soltis and Kuzoff 1995
Persea	Avocados	Furnier et al. 1990
Pinus	Pines	Latta and Mitton 1999
Pisum	Peas	Palmer et al. 1985
Populus	Poplars	Smith and Sytsma 1990; Martinsen et al. 2001
Quercus	Oaks	Whittemore and Schaal 1991; Dumolin-Lapegue et al. 1999a
Salix	Willows	Brunsfeld et al. 1992
Tellima	(Perennial herb)	Soltis et al. 1991
Zea	Teosintes, maize	Doebley 1989

involving 10 of the 22 human autosomes. While searching existing databases for 115 protein-coding autosomal genes from these two primate species, Navarro and Barton (2003b) discovered that protein evolution was significantly faster in the rearranged than in the co-linear chromosomes. The authors interpreted these results as consistent with the possibility that full reproductive isolation between humans and chimpanzees evolved gradually with the accumulation of chromosomal rearrangements, and that during this time, genes in co-linear chromosomal regions may have continued to flow back and forth even as the reproductive barriers moved slowly toward completion. If this model can be generalized, it suggests that chromosomal rearrangements may be important factors contributing to the inter-locus variation in introgression patterns commonly observed in parapatric or secondary contact zones.

Differential levels of hybrid-mediated genetic exchange across loci have also been suggested through an explicit gene tree approach. Machado and Hey (2003; see also Machado et al. 2002) used molecular sequence data to estimate gene genealogies, separately, for each of 16 loci in several closely related species of *Drosophila*. Nine of these gene trees are pictured in Figure 7.11, and they illustrate a remarkable heterogeneity of outcomes. Genealogies of five X-linked loci (A–E in Figure 7.11) each supported traditional thought

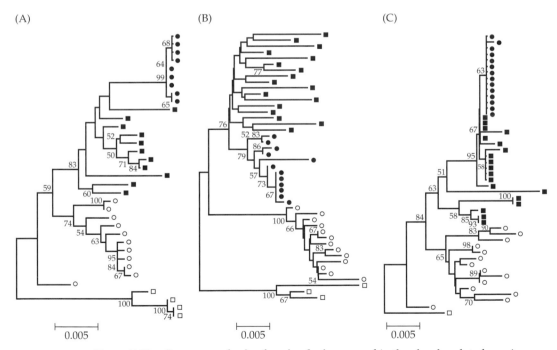

Figure 7.11 Gene genealogies for nine loci surveyed in the closely related species *Drosophila pseudoobscura* (solid squares), *D. persimilis* (open circles), *D. bogotana* (solid circles), and *D. miranda* (open squares). Numbers on each tree indicate levels of bootstrap support for various branches. The scale bar represents nucleotide divergence per base pair. (After Machado and Hey 2003.)

(D)

(E)

(F)

(G)

(H)

(I)

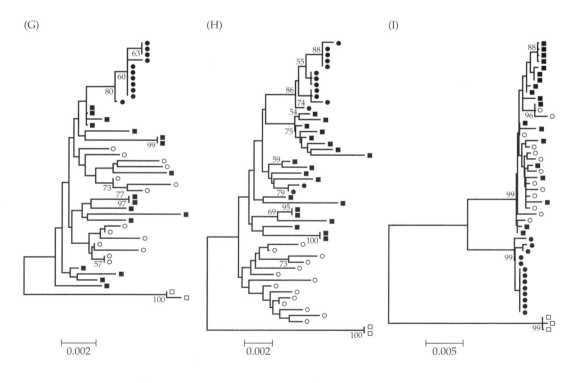

about the phylogeny of a portion of this complex: [(*pseudoobscura*, *bogotana*), *persimilis*]. A mitochondrial gene tree (I) showed an entirely different pattern suggestive of recent gene flow between *pseudoobscura* and *persimilis*. Loci located on nuclear chromosomes 2, 3, and 4 showed other patterns. For example, two trees (F and H) showed no monophyly for any single species or pair of species, and one tree (G) showed monophyly of *bogotana* sequences, but lineage paraphyly for *pseudoobscura* and *persimilis*. These findings are important for several reasons. They provide one of the first available empirical assessments of multiple gene trees within and among closely related species. They show conclusively that composite genomes can have highly mosaic histories (and, hence, that estimating species phylogenies from only one or a few gene trees can have serious pitfalls). Finally, they prove that gene genealogies from unlinked loci can differ rather strikingly in topology, in this case probably due to introgressive gene flow of some portions of the genome, but not others (although differential lineage sorting from polymorphic ancestors might also explain some of the gene tree heterogeneity).

The following sections provide additional examples of how differential introgression can occur, especially via asymmetries stemming from different behaviors or different fitnesses of males and females. Especially intriguing are the patterns of differential genetic exchange frequently observed between nuclear and cytoplasmic loci. Although patterns of variation for cytoplasmic and nuclear markers are highly concordant in some hybrid zones (e.g., Baker et al. 1989; Nelson et al. 1987; Szymura et al. 1985), in others (e.g., the *Hyla* treefrogs discussed above) they may show pronounced discordances for a variety of reasons.

HALDANE'S RULE.   The heterogametic sex is the gender that carries unlike sex chromosomes. In humans, for example, males normally are heterogametic for sex chromosomes conventionally labeled X and Y, whereas in birds, females are heterogametic for Z and W. An empirical generality first noticed by Haldane (1922) is that "when in the F_1 offspring of two different animal races one sex is absent, rare, or sterile, that sex is the heterozygous [heterogametic] sex." Thus, in species with heterogametic males (such as mammals and fruit flies), male hybrids tend to show more severe reductions in viability or fertility, whereas in species in which females are heterogametic (such as birds and butterflies), female hybrids more often show decreased fitness (Coyne and Orr 1989a; Orr 1997; Presgraves 1997). In principle, such asymmetries could influence gene flow across hybrid zones.

In birds and butterflies, introgression of nuclear DNA might occur via fertile male hybrids even if mtDNA introgression is blocked due to female sterility. One possible example involves a hybrid zone between two European flycatchers, *Ficedula albicollis* and *F. hypoleuca*. In accord with Haldane's rule, male hybrids in these birds are known to be more fertile than females (Gelter et al. 1992; Sætre et al. 1999), and hybrids of both sexes are also known to be less fit genetically than pure parentals. Following earlier allozyme and mtDNA analyses by Tegelström and Gelter (1990), Sætre et al. (2001) compared microsatellite and mtDNA markers and found that mtDNA gene flow was somewhat lower than nuclear gene flow in the hybrid zone, albeit not significantly so. In several other avian hybrid zones, mtDNA introgression *has* proved to be significantly diminished relative to that for at least some nuclear loci (Bensch et al. 2002; Brumfield et al. 2001; Helbig et al. 2001; Sattler and Braun 2000). Similarly, in a butterfly hybrid zone between *Anartia fatima* and *A. amathea* in Panama, reduced fitness of female hybrids apparently has placed restrictions on levels and patterns of backcrossing (Davies et al. 1997).

Also in accord with Haldane's rule are voles (small mammals), in which interspecific crosses produce fertile female hybrids but sterile males (Tegelström et al. 1988). In populations of *Clethrionomys glareolus* and *C. rutilus* in northern Scandinavia, Tegelström (1987b) observed a pronounced discrepancy between species boundary and mtDNA phylogeny, leading him to conclude that *rutilus*-type mtDNA had introgressed into *glareolus*, perhaps following a limited hybridization episode dating to a postglacial colonization of the region some 10,000 years ago. This historical scenario is strongly reminiscent of the suspected case of mtDNA capture (discussed above) in Scandinavian populations of *Mus* mice.

In *Drosophila* hybrids, the heterogametic males often show partial or complete sterility, sometimes in one direction of a cross only. In studies spanning three decades, the genetic basis of hybrid sterility has been dissected using chromosomal and molecular markers, and the loci responsible have been mapped to sex chromosomes and various autosomes (Coyne and Berry 1994; Dobzhansky 1974; Kaluthinal and Singh 1998; Orr and Irving 2001; Vigneault and Zouros 1986). The homogametic female hybrids, by contrast, often remain fertile and thus provide what some researchers have interpreted as potential bridges for interspecific exchange of mtDNA via introgression (Powell 1983, 1991). In *D. mauritiana*, for example, some individuals carry an mtDNA genotype also found in nearby populations of *D. simulans*, an observation interpreted by Solignac and Monnerot (1986) to indicate recent introgression of *simulans*-type mtDNA into *D. mauritiana*. This hypothesis gained support from population cage studies in which the predicted takeover by *D. simulans* mtDNA was documented experimentally over a few generations of introgressive hybridization (Aubert and Solignac 1990). On the other hand, DeSalle and Giddings (1986) reported that the mtDNA phylogeny for several closely related species of Hawaiian *Drosophila* matches the species phylogeny quite well, despite postulated historical introgression that sometimes has complicated phylogenetic reconstructions based on nuclear genes.

DIFFERENTIAL MATING BEHAVIORS.   The *Hyla* example discussed earlier provides a powerful illustration of how a behavioral asymmetry has influenced the genetic architecture of a hybrid zone. Another example involves a contact zone in France between hybridizing newts, *Triturus cristatus* and *T. marmoratus* (Arntzen and Wallis 1991). Allozymes again were employed to characterize the hybrid status of individuals, and mtDNA genotypes were used to identify the female parents. All F_1 hybrids possessed *cristatus*-type mtDNA, perhaps due to a strong asymmetry in mate choice. An absence of mtDNA introgression in areas where *T. cristatus* replaced *T. marmoratus* is also consistent with this interpretation.

Sequence data from the Y chromosome have been used in collaboration with data from mtDNA (and sometimes autosomal genes) to deduce sex-biased patterns of dispersal and introgression in mammals. Cases in point involve hybridization between species of macaque (*Macaca*) monkeys, for which Y chromosome introgression has been reported in various settings in the absence of interspecific movement of mtDNA (Evans et al. 2001; Tosi et al. 2002, 2003). Results probably reflect sex-biased dispersal in these primates, wherein males typically emigrate from their natal troops and engage elsewhere in matings with resident females (evidently including those of other species, on occasion).

Sunfishes in the genus *Lepomis* are renowned for their propensity to hybridize, both in artificial ponds and in nature. As stated by Breder (1936), "There is probably no group of fishes, North American at least, in which there would seem to be a concatenation of reproductive and other events so well arranged as to lead to extensive hybridizing; i.e., the species are numerous; there is less geographic separation than usual; spawning occurs at about the same temperature threshold; spawning sites are limited and similar for most species; nests are exchanged among species." From observations of diminished hybrid fertility in both sexes, Hubbs and Hubbs (1933; see also Hubbs 1955) concluded that natural hybridization probably was limited to the F_1 generation. However, later work with experimental populations revealed that "a number of different kinds of hybrid sunfishes … are not sterile, are fully capable of producing abundant F_2 and F_3 generations, and can be successfully backcrossed to parent species and even outcrossed to non-parental species" (Childers 1967). Nonetheless, about a dozen recognized species within the genus normally exhibit large genetic distances at protein-coding loci (see Figure 7.2) and, for the most part, retain distinctive morphological identities throughout their ranges. Thus, questions remained about the magnitude of introgression between *Lepomis* in nature. In one early study, Avise and Saunders (1984) characterized a total of 277 sunfish from two locations in northern Georgia for species-diagnostic allozymes and mtDNA. The genetic data revealed a low frequency (5%) of interspecific hybrids (all of which appeared to be F_1 individuals), involvement of five sympatric species in the production of these hybrids, and no evidence for introgression at these study locales. Furthermore, most of these hybrids were between parental species that differed greatly in abundance and had mothers that were from the less common of the hybridizing species. These data suggest a density-dependent mating pattern in which a paucity of conspecific spawning stimuli and mates for females of the rarer species might be key factors increasing the likelihood of interspecific hybridization.

Another molecular analysis of natural sunfish hybridization involved bluegill (*Lepomis macrochirus*) and pumpkinseed sunfish (*L. gibbosus*) in a southern Canadian lake. Among 44 phenotypically intermediate individuals examined for allozymes and mtDNA, all proved to be F_1 hybrids with pumpkinseed mothers (Konkle and Philipp 1992). As described in Chapter 5, some bluegill males (but presumably not pumpkinseed males) display specialized fertilization-thievery tactics that they normally employ with considerable success in intraspecific spawns (Gross 1979; Gross and Charnov 1980). The gender asymmetry of interspecific hybridization is consistent with the postulate that male bluegills also cuckold heterospecific pumpkinseed males (albeit to no avail in terms of the bluegills' ultimate genetic fitness).

DIFFERENTIAL GAMETIC EXCHANGE.   Especially in plants, a pronounced uncoupling of male and female components of gene flow across species is possible due to the two distinct avenues for gene movement: pollen and seeds (Paige et al. 1991). When genetic transfer is mediated solely by pollen, maternally inherited genetic markers (e.g., cpDNA in most angiosperms) cannot introgress, whereas seed migration into a foreign population might lead to introgression of both nuclear and cytoplasmic genes when the resulting plants are fertilized by pollen from resident individuals. Clearly, alternative modes of gene transfer in plants can leave different signatures on cytonuclear associations in hybrid zones (Asmussen and Schnabel 1991).

In a series of observational and experimental studies involving allozymes, cpDNA, and other molecular markers, Arnold et al. (1990a,b, 1991, 1992; see reviews in Arnold 2000; Arnold and Bennett 1993; Johnston et al. 2001) have documented many ecological and genetic aspects of introgression between Louisiana irises (genus *Iris*), a classic example of a plant taxonomic complex engaged in hybridization (Anderson 1949). For example, the authors found asymmetric gene flow between *Iris fulva* and *I. hexagona* at particular locales (Figure 7.12): Many individuals of hybrid ancestry (as evidenced by recombinant nuclear genotypes) nonetheless retained cpDNA of *I. hexagona*, indicating that they were products of pollen transfer from *I. fulva* onto *I. hexagona* flowers. In such cases, introgression of nDNA markers occurred in the absence of cpDNA transfer. Results appear to be consistent with *Iris* natural history, which includes pollen movement by mobile bumblebees and hummingbirds, but only short-distance seed dispersal.

Might long-distance pollen movement lead to the exceptionally wide areas of introgression that have been postulated for some plant species complexes? In the southeastern United States, three parapatric species of buckeye trees (*Aesculus sylvatica*, *A. flava*, and *A. pavia*) appear to have experienced introgression across a region at least 200 km wide, as inferred from patterns of morphology, geographic distribution, and meiotic irregularities associated with decreased germination of pollen from putative interspecific hybrids. Allozyme data gathered for these species are also consistent with introgression scenarios, and they further raise the possibility that long-distance gene movement beyond the hybrid zone (as recognizable by morphology) has taken place (dePamphilis and Wyatt 1990). For example, one suspected hybrid region appears to be highly asymmetric, with alleles characteristic of coastal-plain *A. pavia* also found in Piedmont populations, where *A. sylvatica* normally occurs. The authors hypothesized that hummingbirds (important pollination agents for buckeyes) may have effected long-distance pollen flow during their northward spring migration, thus accounting for the width and asymmetry of the hybrid zone (dePamphilis and Wyatt 1989).

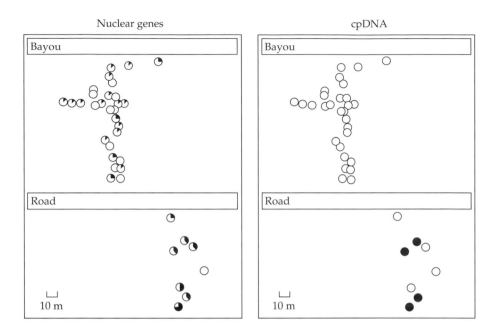

Figure 7.12   Asymmetric introgression between *Iris fulva* and *I. hexagona*, probably resulting from pollen flow. Each circle represents a single plant. *Left*, relative proportion of *I. fulva* (shaded) and *I. hexagona* (unshaded) nuclear markers. *Right*, similar representation for maternally transmitted chloroplast DNA markers. Note especially the population between the road and the bayou, in which multi-locus nuclear genotypes suggest the presence of advanced-generation hybrids or backcrosses, but cpDNA markers suggest the absence of seed dispersal from *I. fulva*. (After Arnold 1992.)

Once pollen has arrived on a heterospecific style, additional challenges await it that could lead to asymmetric barriers to successful hybridization. Experimental studies on Louisiana irises illustrate what can happen. In pollen competition experiments in which pollen grains from two or more species were placed on individual stigmas, genetic marker analyses of the resulting progeny revealed that homospecific pollen tubes typically out-competed (or out-paced) heterospecific pollen tubes in their race toward the ovary (Carney et al. 1996; Emms et al. 1996). The primary exceptions occurred only when foreign pollen grains were given a significant head start in these egg fertilization contests. Similar experiments on "interspecific pollen competition" in *Helianthus* sunflowers again revealed reproductive barriers between species, but no evidence for differential pollen tube growth rates in this case (Rieseberg et al. 1995b). Another example, but with yet a different pattern, involves *Eucalyptus* trees. In interspecific crosses, *E. nitens* pollen tubes grow slowly and never reach full length in the larger *E. globulus* styles, whereas *globulus* pollen tubes grow rapidly in *nitens* styles and enter the ovary (Gore et al. 1990). Hybridization between several species of *Eucalyptus* occurs rather commonly in nature (Griffin et al. 1988; Jackson et al. 1999), and the unilateral cross-incompatibility mediated by this asymmetry might play an important role in structuring cytonuclear associations—for example, those involving cpDNA, which is maternally inherited in these trees (McKinnon et al. 2001).

Of course, a variety of other selective mechanisms operating at prezygotic or postzygotic stages also could lead to asymmetric introgression in *Eucalyptus* species, and in other organisms (Gore et al. 1990; Potts and Reid 1985). One such postzygotic mechanism is cytoplasmic male sterility (CMS), wherein hybrid males tend to show greatly reduced fertility. CMS is common in arthropods, for example, where it is often attributable at least in part to the presence of cytoplasmically housed *Wolbachia* bacteria (and other intracellular microbial taxa; Stouthamer et al. 1993) that usually are transferred to hybrid progeny through females, but not males (O'Neill et al. 1997; Werren 1998). Thus, this asymmetry could leave a genetic footprint on hybridizing populations (Giordano et al. 1997). In one genetic test of this possibility, Mandel et al. (2001) used molecular markers to assay for presence versus absence of *Wolbachia* in a hybrid zone of field crickets (*Gryllus*). In this case, however, results indicated that *Wolbachia* were unlikely to have been involved in the unidirectional incompatibility of hybrid crosses between *G. firmus* and *G. pennsylvanicus*.

CMS is also a widespread phenomenon in plants (Edwardson 1970), in which it often involves pollen abortion in hybrid progeny due to an interaction of cytoplasmically transmitted mutations (usually in mtDNA) with a foreign nuclear background (Budar et al. 2003; Hanson 1991). Male sterility in first-generation hybrids or backcross progeny could have the effect of attenuating the introgression of nuclear genes, while nonetheless leaving an open avenue for cytoplasmic transfer via fertile females. CMS is known to occur, for example, in hybrids between some *Helianthus* species that have figured prominently in discussions of inter-taxon exchange of cpDNA, as described next.

RETICULATE EVOLUTION AND cpDNA CAPTURE.   Due to a strong propensity for introgressive hybridization in many plant taxa, botanists have been especially concerned with the possibility of widespread reticulate evolution (Grant 1981; Rieseberg and Morefield 1995; Stebbins 1950), wherein the phylogeny of a particular taxonomic group might be anastomotic or "netlike" rather than strictly dichotomous and branched. Molecular analyses based on phylogenetic contrasts between nDNA and cpDNA markers have provided considerable support for this phenomenon (see Table 7.6). The usual evidence consists of a gross incongruity between the phylogeny of a particular gene (typically cpDNA) and the consensus species phylogeny from other sources of information, including nuclear markers. One powerful explanation for such incongruities is ancient cytoplasmic capture via introgressive hybridization. However, caution should be exercised before accepting poorly documented cases because other evolutionary processes (such as idiosyncratic sorting of polymorphic lineages that had been retained across temporally close speciation events; Moran and Kornfield 1993) can also generate topological discordance between a gene tree and a species tree (see Chapters 4 and 8; see also Rieseberg et al. 1996b).

In *Helianthus* sunflowers, gross discrepancies between phylogenies estimated from morphology, chromosomal variation, experimental crossing success, and various molecular markers have shown conclusively that both recent and relatively ancient episodes of inter-taxon gene exchange have produced a reticulate pattern of relationships among these species (Rieseberg 1991; Rieseberg et al. 1988, 1991). For example, a cpDNA-based phylogeny for numerous *Helianthus* taxa contrasts dramatically (at the indicated positions in the phylogeny presented in Figure 7.13) with suspected relationships based on morphological characters and on nuclear rDNA sequences. Thus, each of five species (*H. anomalus, H. annuus, H. debilis, H. neglectus,* and *H. petiolaris*) possesses some highly distinct cpDNA genotypes otherwise characteristic of different clades within the genus. Furthermore, some species appear to have captured cytoplasms of other species on multiple occasions. For example, *H. petiolaris* acquired the cytoplasm of *H. annuus* at least three times, as evidenced by the geographic sites of introgression and by the particular cpDNA genotypes of *H. annuus* that were exhibited (Rieseberg and Soltis 1991). Another

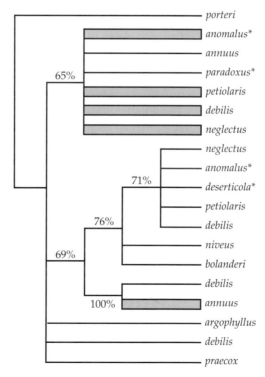

Figure 7.13   Early evidence for cytoplasmic introgression in *Helianthus* sunflowers. Shown is a parsimony tree based on cpDNA data; shaded branches indicate areas of overt conflict with a morphological classification. These discrepancies are indicative of interspecific cpDNA transfer mediated by hybridization. Numbers are levels of bootstrap support for putative clades. Asterisks specify taxa that are stabilized hybrid derivatives (see text). (After Rieseberg and Soltis 1991.)

well-documented example of apparent cytoplasmic capture and reticulate evolution involves *Gossypium* cottons (Wendel and Albert 1992; Wendel et al. 1991), a phylogeny for which is shown in Figure 7.14. Rieseberg (1995) concludes that such situations are rather common in plants and that "detailed surveys of the current literature might yield over 100 potential examples."

In general, like animal mtDNA, the cpDNA molecule may be especially helpful in revealing cases of reticulation because its clonal haploid transmission allows particular ancestral sources to be identified without the complication of recombination (including recombination at the intragenic scale), which can lead to a mosaic ancestry for the nuclear genome. Apart from this detection bias, cytoplasmic DNA might tend to introgress *more* readily than nuclear DNA if genes contributing to reproductive isolation are housed primarily in the nucleus (Barton and Jones 1983). Indeed, in a detailed genetic dissection of three natural hybrid zones between the sunflowers *Helianthus petiolaris* and *H. annuus*, based on 88 marker loci distributed across 17 chromosomes, Rieseberg et al. (1996a, 1999) found introgression to be significantly reduced relative to neutral expectations for 26 chromosomal segments, suggesting that each of these nuclear segments contained factors contributing to RIBs. On the other hand, an opposite argument can be made: Because cytoplasmic DNA is haploid and is therefore potentially exposed to selection in all individuals, and because it houses non-recombining genes, several of whose protein products must functionally interact with those of nuclear genes, cytoplasmic introgression might be severely impeded relative to that of typical nuclear DNA. If so, then the reticulation events revealed most clearly by cytoplasmic captures might be only the tip of the iceberg of interlineage gene exchange (see Chapter 8).

In summarizing this section, it is abundantly clear that hybridization phenomena are best analyzed through multiple lines of evidence involving several types of molecular (and other) markers. Barriers to reproduction between closely related taxa are seldom absolute, and they often appear differentially "semipermeable" to cytoplasmic and various nuclear alleles. Thus, a rich and varied fabric of gene genealogies (seldom evident from traditional morphological assessment alone) characterizes many hybrid zones, revealing varying degrees of reticulation among the phylogenetic branches connecting related species.

More hybrid zone phenomena

CONSISTENCY OF OUTCOMES?   Hybrid zone contacts tend to be rather idiosyncratic evolutionary happenings, so it is seldom possible to examine perfect replicates in nature and thereby critically assess any repeatability in outcomes. Even when multiple transects along a linear or mosaic hybrid zone are examined, uncontrolled key variables, such as genetic background or ecological setting, may differ. An alternative is to replicate hybrid zones experimentally, initiating them with characterized genetic stocks under controlled ecological conditions and then monitoring the temporal course of hybridization and introgression (Emms and Arnold 1997; Hodges et al. 1996).

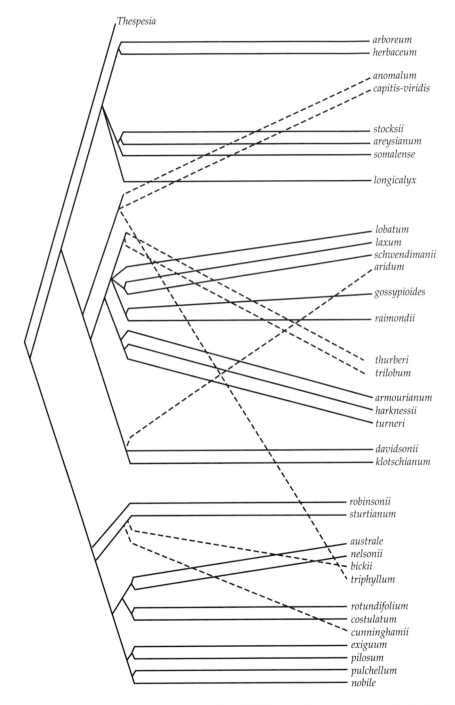

Figure 7.14   Heuristic representation of a cpDNA-based phylogeny for 34 diploid species of *Gossypium*. Dashed lines crossing solid lines indicate probable instances of cytoplasmic capture and, hence, reticulate evolution. (After Wendel and Albert 1992.)

An example of this approach involved two species of mosquitofish (*Gambusia affinis* and *G. holbrooki*) that, as judged by allozyme and mtDNA markers, hybridize naturally across a broad portion of the southeastern United States (Scribner and Avise 1993a). Scribner and Avise (1994a) examined replicated sets of experimental pond and pool populations, each initiated with specified equal numbers of adult male and female *affinis* and *holbrooki*. These captive populations were then monitored periodically for changes in cytonuclear genetic composition over a 2-year period. The dynamics of hybridization and introgression proved to be remarkably consistent among replicates. In each case, there was an initial flush of hybridization, followed by backcrossing and a rapid decline in the frequencies of *affinis* nuclear and cytoplasmic alleles. Similar outcomes across 2 years were also observed in experimental *Gambusia* populations set up in more complex habitats inside the Biosphere 2 facility in Arizona (Scribner and Avise 1994b). Overall, the genetic analyses documented extensive introgressive hybridization accompanied by strong directional selection that promoted rapid and consistent evolutionary changes favoring *holbrooki* alleles (Figure 7.15). The results were also relevant to behavioral, demographic, and life history aspects of natural hybridization between these small fishes (Scribner and Avise 1993b).

HYBRID FITNESS.   One traditional notion is that interspecific hybridization is costly to the participants, typically yielding progeny with diminished fitness and resulting in hybrid zones that act as genetic sinks (Barton 1980). In recent years, there has been a resurgence of interest (Arnold 1997; Dowling and Secor 1997; Rieseberg 1995) in an old suggestion (Anderson and Stebbins 1954; Lewontin and Birch 1966) that, by virtue of their possessing novel recombinant genotypes, some hybrid populations might also be sources of adaptive evolution and lineage diversification.

One prediction of the latter view is that the fitnesses of hybrid organisms should sometimes surpass those of their parents. Indeed, studies employing molecular markers (often in conjunction with other information used to identify hybrid classes and to estimate various components of reproductive fitness) have shown that particular recombinant genotypes in at least some ecological settings do outperform the non-hybrid genotypes of parentals. Examples have been documented in a wide variety of animal and plant taxa ranging from *Sceloporus* lizards (Reed and Sites 1995; Reed et al. 1995a,b) to *Mercenaria* clams (Bert and Arnold 1995) to *Artemesia* sagebrushes (Freeman et al. 1995, 1999; Graham et al. 1995; Wang et al. 1997) and *Iris* plants (Arnold et al. 1999; Burke et al. 1998). In a review of 37 research studies on this topic, Arnold and Hodges (1995) reported that hybrids possessed the highest fitness values in 5 cases (13%), were equivalent to the most fit parental class in 15 cases (40%), were intermediate to the two parents in 7 cases (20%), and were the least fit in 10 cases (27%). Even if they represent only a minority of outcomes (Burke and Arnold 2001), such instances of increased fitness of recombinant genotypes in hybrid zones are certainly intriguing.

Figure 7.15 Temporal genetic changes in four sets of experimental *Gambusia* hybrid ▶ populations monitored over 2 years. Shown are observed frequencies of non-hybrid specimens (pure *holbrooki* or pure *affinis*), probable F$_1$ hybrids, backcrosses in either direction, and later-generation hybrids, as gauged by species-diagnostic mtDNA and nuclear (allozyme) markers. Numbers above histogram bars are observed percentages of specimens with *holbrooki*-type mtDNA. Pools and ponds were located at the Savannah River Ecology Lab (SREL). (After Avise 2001c.)

There are two basic ways in which natural hybridization might contribute significantly to longer-term evolution: by transferring adaptations from one taxon to another and by promoting the foundation of new evolutionary lineages (i.e., new species). The first type of creative potential is probably evidenced by at least some of the many remarkable instances of interspecific gene capture (discussed above) mediated by introgressive hybridization. Such cases provide prima facie evidence that transferred genes (which often involve whole cytoplasmic genomes) clearly can function *at least* adequately in a heterologous genetic background, despite what must have been earlier long-term independent evolution in separate species before the introgression events. The second type of creative potential for hybridization—generating new species—seems even more remarkable, but this phenomenon is also well documented, as described next.

Speciation by hybridization

Differentiated genomes brought together through hybridization can sometimes produce a new species. One such mechanism, allopolyploidization, already has been described (see Box 7.3). Surveys of molecular markers have revealed other hybridization-mediated routes to speciation as well (Abbott 1992).

DIPLOID OR HOMOPLOID SPECIATION.   In the plant literature, a traditional proposal has been that species isolated from each other by a chromosomal sterility barrier might give rise via hybridization to new fertile diploid species that are at least partially reproductively isolated from both parents (Grant 1963; Stebbins 1950). Such "recombinational speciation" (Grant 1981), without change in chromosome number, was verified by the experimental synthesis of new hybrid species under artificial conditions (see review in Rieseberg et al. 1990a). However, questions remained about the prevalence of this speciation mode in nature. In various plant taxa, numerous candidates for hybrid species were identified in early studies of morphology, ecology, and geographic ranges, but full confirmation of hybrid ancestry for particular species awaited the application of molecular markers (Rieseberg 1997).

A diploid annual plant native to southern California, *Stephanomeria diegensis*, was suspected to have originated by stabilization of recombinant genotypes from a natural cross between two divergent diploid relatives (*S. exigua* and *S.*

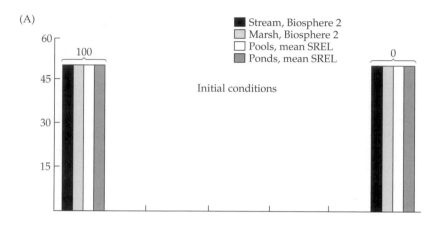

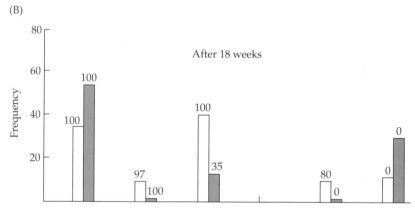

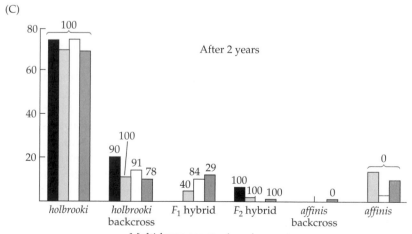

virgata) with the same chromosome number. Gallez and Gottlieb (1982) demonstrated that *S. diegensis* indeed displays an additive profile of allozyme alleles characteristic of its presumed relatives, a finding consistent with the plants' intermediate morphology and karyotype and supportive of its postulated hybrid origin. Similarly, findings based on more than a hundred nuclear RAPD markers are consistent with a hybrid origin of the diploid shrub *Encilia virginensis* from two other diploid congeners, *E. actoni* and *E. frutescens* (Allan et al. 1997). Another plant species of putative hybrid origin, *Iris nelsoni*, proved upon molecular analysis to possess a combination of nuclear genes characteristic of three species—*I. fulva*, *I. hexagona*, and *I. brevicaudis*—that all appear to have been involved in its formation (Arnold et al. 1990b, 1991). However, not all molecular genetic reappraisals have confirmed the suspected hybrid origins of problematic plant taxa. The diploid annual *Lasthenia burkei* proved not to possess a combination of allozyme alleles present in *L. conjugens* and *L. fremontii*, thus disputing earlier hypotheses that *L. burkei* is a stabilized hybrid derivative of those two species (Crawford and Ornduff 1989).

Rieseberg et al. (1990a) noted that, by hard criteria, apparent additivity of alleles in the nuclear genome is insufficient to confirm the hybrid origin of a diploid species because it does not exclude the possibility that the taxon in question is ancestral to its putative parents. One solution to this problem is to use additional markers (such as those from cpDNA) to establish the evolutionary polarity of relationships. Applying this philosophy to allozymes and cpDNA markers in diploid sunflowers, Rieseberg et al. (1990a, 1995c) concluded that three problematic taxa, appropriately named *Helianthus paradoxus*, *H. anomalus*, and *H. neglectus*, had dissimilar pathways of origin. The last species in this list appears to be a recent non-hybrid derivative of *H. petiolaris*, but the first two species did prove to be hybrids having arisen from crosses between *H. annuus* and *H. petiolaris*. The analysis of *H. anomalus* also illustrates the exceptional refinement of some of these genomic dissections of hybrid speciation. Rieseberg et al. (1995c) employed approximately 200 mapped RAPD markers to identify the precise genomic linkage blocks in *H. anomalus* that had stemmed (perhaps on multiple independent occasions; Schwarzbach and Rieseberg 2002) from the two parental species that produced this hybrid taxon (Figure 7.16).

Rieseberg et al. (1996c) also experimentally synthesized three independent hybrid lineages (by crossing *H. annuus* with *H. petiolaris*), then compared the genomic compositions of these synthetic hybrids with that of natural *H. anomalus*. Their most important discovery was that patterns of introgression of the two parental genomes were significantly correlated across all of these hybrids. In other words, particular blocks of introgressed loci were nonrandomly similar in all cases, thus strongly indicating that natural selection, rather than chance, often governed the detailed genomic composition of the hybrid species. Further genomic and phenotypic comparisons of ancient and synthetic hybrids between *Helianthus* species have also shown how complementary gene action following hybridization may produce extreme phenotypes that can facilitate adaptive evolution and ecological divergence (Rieseberg et al. 2003).

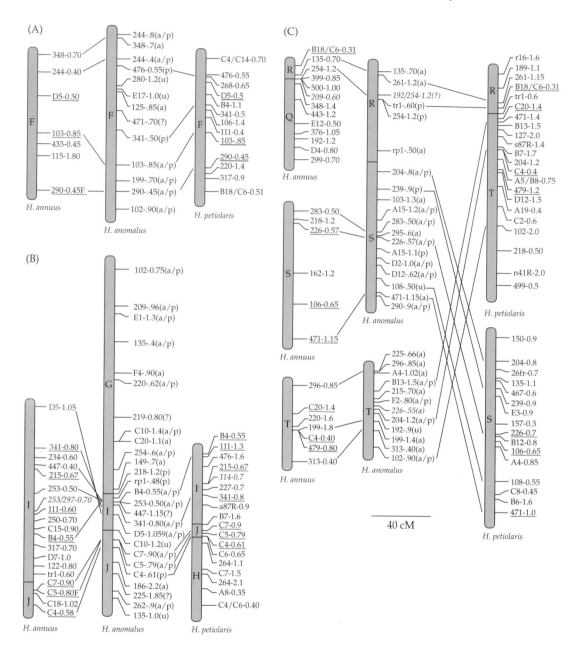

Figure 7.16   Genetic linkage maps for nearly 200 RAPD markers in an ancient hybrid species, *Helianthus anomalus*, and in the two parental species from which it was derived. Linkage groups are arranged into three structural sets: (A) complete co-linearity, (B) inversions, and (C) inter-chromosomal translocations. For *H. anomalus*, letters in parentheses after each marker indicate parental origin (p, *petiolaris*; a, *annuus*; a/p, either *annuus* or *petiolaris*; u, unique to *anomalus*). Lines between linkage groups connect loci that are known to be homologous between the hybrid and its parental species. Underlines denote loci that are homologous between *H. annuus* and *H. petiolaris*. (After Rieseberg et al. 1995c.)

Less attention has been devoted to the possibility of homoploid speciation via hybridization in animals. On the basis of morphological and allozyme data, Highton et al. (1989) suggested that the salamander *Plethodon teyahalee* had a hybrid origin from past interbreeding between *P. glutinosus* and *P. jordani*. Several other such cases have been postulated (but seldom well documented) in mollusks, crustaceans, insects, and various vertebrates (see review in Dowling and Secor 1997). Among fishes in particular, several examples of stabilized hybrid forms have been suspected from morphological or distributional considerations. For example, the "Zuni sucker" in the Little Colorado drainage of the western United States was suggested to be a hybridization-derived intermediate between *Catostomus discobolus* of the Colorado drainage and *C. plebeius* of the Rio Grande, and the "white shiner" in the Roanoke and adjacent drainages of the eastern United States was proposed to have arisen from hybridization between nearby *Luxilus cornutus* and *L. cerasinus*. However, in the case of the Zuni sucker, molecular reevaluations did not support the hybrid origin scenario (Crabtree and Buth 1987), and in the case of the white shiner, molecular data were equivocal due to the difficulty of distinguishing among the possibilities of ancestral polymorphism, convergence, and past hybridization, all of which could account for the observed allozyme distributions (Meagher and Dowling 1991).

Molecular reappraisals *have* provided support for postulated hybrid origins of a few recognized fish species. By virtue of its mosaic genetic structure, as registered in mitochondrial and nuclear gene assays, one cichlid species (*Neolamprologus marunguensis*) in Africa's Lake Tanganyika appears to have arisen via introgressive hybridization between two ancient and genetically distinct species (Salzburger et al. 2002). In the case of the cyprinid *Gila seminuda* of the Virgin River in the western United States, an intermediate morphology led to suspicion that this species also was a hybrid, derived in this case from crosses between the roundtail chub (*G. robusta*) and bonytail chub (*G. elegans*). *G. seminuda* proved to be polymorphic for allozyme alleles at two loci otherwise diagnostic for the putative parental taxa; as judged by mtDNA, the matriarchal lineage retained by *G. seminuda* derived from *G. elegans* (DeMarais et al. 1992). Phylogenetic analyses of additional *Gila* species (Dowling and DeMarais 1993) identified phylogenetic conflicts between nuclear and mtDNA markers, a finding further suggestive of past episodic introgression. As noted by the authors, such stabilized hybrid derivatives might be relatively common in some groups of fishes, but remain unrecognized due to a lack of detailed molecular studies. As also noted by the authors, the formal taxonomic status of such introgressed forms will probably remain a point of contention, especially when the population in question is currently isolated from its parental species by extrinsic (geographic) barriers to reproduction. Should the hybrid form be considered a distinct population, a subspecies, or a species? This question is not merely academic, but is relevant to the implementation of conservation programs (see Chapter 9).

ORIGINS OF UNISEXUAL BIOTYPES.   Unisexual "biotypes" that reproduce by parthenogenesis, gynogenesis, or hybridogenesis (see Figure 5.4) are not biological species in the usual sense applied to sexually reproducing taxa, but they are nonetheless isolated genetically from their sexual relatives (as well as from other unisexual lineages) and are typically afforded formal taxonomic recognition. Parthenogenetic forms are found in a variety of taxonomic groups, and many of them have been examined for evolutionary origins using molecular markers. For example, using mtDNA sequence data, Johnson and Bragg (1999) identified the female of the sexual species involved in hybrid-mediated geneses of asexual *Campeloma* snails in the southeastern United States, and they also showed that the parthenogenetic taxa were of polyphyletic origin. Similarly, Delmotte et al. (2003) used nuclear and cytoplasmic markers to show that many parthenogenetic aphid strains arose through multiple hybridization events between *Rhopalosiphum padi* and an unknown sibling species. Among the vertebrates, essentially all of the 70 known unisexual taxa also arose through hybridization events between related sexual species, and molecular markers have likewise been highly informative in revealing the mode of origin and parentage of these all-female biotypes.

In most cases, the particular bisexual progenitors of various unisexual vertebrate taxa were suspected from earlier comparisons of morphology, karyotype, or geographic range, but molecular surveys of allozymes and mtDNA have confirmed those hybrid origins in several instances and, for the first time, revealed the sexual directions of the original crosses. For example, a gynogenetic livebearing fish, *Poecilia formosa*, in northeastern Mexico exhibits nearly fixed heterozygosity at numerous protein and allozyme loci that distinguish or are polymorphic in the sexual species *P. latipinna* and *P. mexicana* (Abramoff et al. 1968; Balsano et al. 1972; Turner 1982). These molecular findings confirmed earlier evidence from morphology and geography that *P. formosa* arose via hybridization between these sexual species. The gynogen *P. formosa* also carries the mtDNA of *P. mexicana*, which is highly divergent from that of *P. latipinna*, indicating that the direction of the initial hybrid cross(es) was *P. mexicana* female × *P. latipinna* male (Avise et al. 1991). Similar molecular inspections have allowed unambiguous determination of the sexual progenitors for more than 25 unisexual biotypes (examples are given in Table 7.7).

A few unisexuals carry genomic contributions from more than two sexual ancestors. The gynogenetic fish *Poeciliopsis monacha-lucida-viriosa*, for example, includes *P. viriosa* nuclear genes apparently introgressed as a result of occasional matings of *P. monacha-lucida* females with *P. viriosa* males, rather than with males of their usual sexual host, *P. lucida* (Vrijenhoek and Schultz 1974). As judged by allozymes and other evidence, several triploid parthenogenetic lizards in the genus *Cnemidophorus* also carry genes from three sexual progenitors (Dessauer and Cole 1989; Good and Wright 1984), probably as a result of multiple hybridization events involving these species.

In many cases, molecular analyses have further pinpointed the geographic and genetic sources of particular unisexual biotypes. For example, based on mtDNA comparisons, the matrilineal components of nine unisexual biotypes in the *sexlineatus* group of *Cnemidophorus* lizards all appear to stem from females within one of the four nominate geographic subspecies of *C. inornatus*: *C. i. arizonae* (Densmore et al. 1989a). Similar molecular inspections have traced the maternal ancestry of five triploid unisexual strains in the *Poeciliopsis monacha-lucida* complex of fishes to a bisexual species, *P. monacha*, from the Río Fuerte in northwestern Mexico (Quattro et al. 1992b).

Typically, the bisexual relatives of unisexual species have proved highly distinct in their mtDNA genotypes, whereas the mtDNAs of the unisexuals are closely related to or indistinguishable from those of one of the sexual progenitors. Thus, an emerging generalization is that most extant unisexual biotypes originated through asymmetric hybridization events occurring in one direction only (e.g., *A* female × *B* male versus *B* female × *A* male). Whether this phenomenon reflects some asymmetric mechanistic constraint on the origin of unisexuals or merely the survival of a limited subset of lineages from crosses in both directions generally remains unclear. However, in the case of the *Poeciliopsis* hybridogens, laboratory crosses of *P. monacha* females × *P. lucida* males sometimes result in the spontaneous production of viable hybridogenetic lineages, whereas the reciprocal matings do not (Schultz 1973). This direction of crossing is consistent with the molecularly inferred origins of natural hybridogenetic strains, all of which possess *monacha*-type mtDNA (Quattro et al. 1991). Furthermore, these extant natural hybridogens have arisen multiple times through separate hybridization events, as gauged by their links to several different branches in the mtDNA phylogeny of their maternal ancestor *P. monacha* (Figure 7.17).

One exception to such straightforward hybrid origins involves the hybridogenetic frog *Rana esculenta* of Europe, in which individuals exhibit mtDNA genotypes normally characteristic of either *R. lessonae* or *R. ridibunda* (Spolsky

TABLE 7.7   Species parentage determined for unisexual vertebrates of hybrid origin

Unisexual biotype	Ploidy level	Reproductive mode[a]
Cnemidophorus (lizards)		
uniparens	3n	P
tesselatus	2n	P
velox	3n	P
laredoensis	2n	P
Heteronotia binoei (lizard)	3n	P
Menidia clarkhubbsi (fish)	2n	G
Phoxinus eos-neogaeus (fish)	2n, 3n	G
Poecilia formosa (fish)	2n	G
Poeciliopsis (fishes)		
monacha-lucida	2n	H
monacha-occidentalis	2n	H

Note: See Avise et al. 1992c for an extended list.

[a] P = parthenogenetic; G = gynogenetic; H = hybridogenetic (see Figure 5.4).

[b] The bisexual parental species were identified in each case from allozymes, morphology, karyotype, geographic range, or other information. The female parent species was identified by mtDNA comparisons.

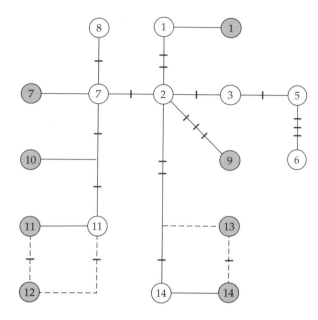

Figure 7.17   Relationships among mtDNA haplotypes in the sexual fish *Poeciliopsis monacha* (open circles) and its unisexual derivative, *P. monacha-lucida* (shaded circles). Slashes are inferred mutations along the parsimony network; dashed lines indicate alternative network pathways. (After Quattro et al. 1991.)

Bisexual parental species[b]		Reference
Male	Female	
		Moritz et al. 1992b
burti	*inornatus* (2)	Densmore et al. 1989a
septemvittatus	*marmoratus*	Brown and Wright 1979;
		Densmore et al. 1989b
inornatus (2)	*burti* or *costatus*	Moritz et al. 1989
sexlineatus	*gularis*	Wright et al. 1983
binoei	sp. "CA6"	Moritz 1991
beryllina	*peninsulae*	A. A. Echelle et al. 1989
eos	*neogaeus*	Goddard et al. 1989
latipinna	*mexicana*	Avise et al. 1991
lucida	*monacha*	Quattro et al. 1991
occidentalis	*monacha*	Quattro et al. 1992a

and Uzzell 1986). The hybridogen *R. esculenta* is unique among assayed "asexual" biotypes in consisting of high frequencies of both males and females. From behavioral considerations, the initial hybridizations producing *R. esculenta* were postulated to involve male *R. lessonae* × female *R. ridibunda*. Once the hybridogen was formed, occasional matings of male *R. esculenta* with female *R. lessonae* may have introduced *lessonae*-type mtDNA secondarily into *R. esculenta*. Furthermore, females belonging to such *R. esculenta* lineages appear to have served as a natural bridge for interspecific transfer of *lessonae* mtDNA into certain *R. ridibunda* populations via matings with *R. ridibunda* males (Spolsky and Uzzell 1984). Such crosses apparently produced "*R. ridibunda*" frogs with normal nuclear genomes (because *R. lessonae* chromosomes are excluded during meiosis), but with *lessonae*-type mtDNA.

Another complex scenario surrounds the hypothesized maternal ancestry of the triploid salamander *Ambystoma 2-laterale-jeffersonianum*, which, according to allozyme evidence, contains nuclear genomes of the bisexual species *A. laterale* and *A. jeffersonianum*, but reportedly carries mtDNA from *A. texanum* (Kraus and Miyamoto 1990). The authors favor an explanation in which an original *A. laterale-texanum* hybrid female produced an ovum with primarily *A. laterale* nuclear chromosomes, but the female-determining sex chromosome (W) and the mtDNA of *A. texanum*. If such a female were fertilized by a male *A. laterale*, female progeny with two *A. laterale* nuclear genomes and the mtDNA of *A. texanum* would result. Subsequent hybridization with male *A. jeffersonianum* could then produce the observed *A. 2-laterale-jeffersonianum* biotypes carrying *texanum*-type mtDNA. Although this scenario is speculative, its mere feasibility suggests that distinct reticulate histories could characterize different genomic elements in some hybridogenetic taxa.

Another question about unisexual vertebrates addressed by molecular markers concerns mechanistic modes of polyploid formation. More than 60% of known unisexual biotypes are polyploid (Vrijenhoek et al. 1989), and two competing hypotheses have been advanced to account for the evolutionary origins of these forms (Figure 7.18). Under the "primary hybrid origin" hypothesis, a disruption of meiotic processes in an F_1 interspecific hybrid leads to the production of unreduced diploid eggs whose subsequent fertilization by sperm leads to a triploid condition (Schultz 1969). Alternatively, under the "spontaneous origin" hypothesis, parthenogenetic triploids might have arisen when unreduced oocytes from a diploid non-hybrid were fertilized by sperm from a second bisexual species (Cuellar 1974, 1977). As diagrammed in Figure 7.18, joint comparisons of mitochondrial and nuclear markers permit an empirical test of these competing possibilities. If a unisexual biotype arose spontaneously from bisexual ancestors and hybridization was involved only secondarily, the paired homospecific nuclear genomes should derive from the maternal parent and, thus, should be coupled with mtDNA derived from the same species. Conversely, under a model of primary hybrid origin, the paired homospecific nuclear genomes could be coupled with the mtDNA type from either of the sexual ancestors, depending on whether the nuclear genome was duplicated or added (see below).

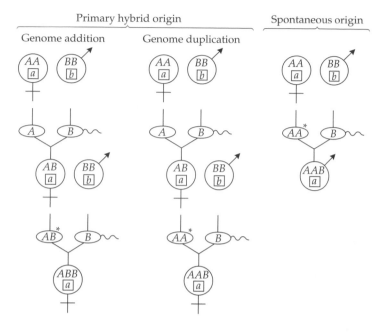

Figure 7.18   Competing scenarios for the origin of triploid unisexual taxa. Each uppercase letter represents one nuclear gene set (*A* or *B*) from the respective parental species, and lowercase letters in boxes similarly refer to maternally transmitted mtDNA genomes. Smaller ovals represent sperm (tailed) and eggs (non-tailed); asterisks indicate eggs that are unreduced (i.e., diploid). In the genome duplication scenario, suppression of reduction occurs during an equational division such that the *AB* hybrid produces *AA* (or *BB*) ova. (After Avise et al. 1992c.)

Cytonuclear genetic analyses for several unisexual taxa have provided support for the primary hybrid origin hypothesis. For example, the triploid parthenogen *Cnemidophorus flagellicaudus* possesses the mtDNA of *C. inornatus*, but two homospecific nuclear genomes from *C. burti*, and a similar cytonuclear pattern was observed for eight of ten parthenogenetic *Cnemidophorus* biotypes examined (Densmore et al. 1989a; Moritz et al. 1989). Similarly, the triploid gynogenetic fish *Poeciliopsis monacha-2 lucida* possesses the mtDNA of *P. monacha*, but two nuclear genomes from *P. lucida* (Quattro et al. 1992b). These results appear to refute the "spontaneous origin" scenario (unless the diploid non-hybrid that produced unreduced gametes was a male, in which case all bets are off).

Assuming correctness of the primary hybrid origin scenario, two further cytogenetic pathways to triploidy can be distinguished (see Figure 7.18). Under the "genomic addition" scenario (Schultz 1969), interspecific F_1 hybrids produce unreduced ova (*AB*) that then unite with a haploid gamete from one of the sexual ancestors to produce allotriploid backcross biotypes *AAB* or *ABB*. Under the "genomic duplication" scenario (Cimino 1972), suppression of an equational cellular division in an F_1 hybrid could produce unreduced *AA* or *BB* ova, which following a backcross to species *A* or *B* would produce *AAB* or *ABB* offspring (autopolyploid *AAA* or *BBB* progeny could also result from this process, but no self-sustaining populations of autopolyploid unisexual vertebrates are known). An important distinction between these pathways is the predicted level of heterozygosity in the homospecific nuclear genomes. Heterozygosity should be extremely low under the genome duplication pathway (the only variation being derived from post-formational mutations), whereas normal heterozygosity is predicted under the genomic addition pathway. At least one test of these scenarios has been done: In triploid *Poeciliopsis* gynogens, all assayed strains proved to be heterozygous for homospecific nuclear markers at one or more allozyme loci, a result that probably excludes the genome duplication hypothesis for these fishes (Quattro et al. 1992b).

Molecular markers have also contributed to the study of androgenesis, a sort of male analogue of gynogenesis, but in this case involving the development of an individual solely under the influence of its paternally derived chromosomes (i.e., without instructions from the mother's genetic material). This rare process has been demonstrated experimentally in the lab (see Giorgi 1992) and studied in nature as well. Using a combination of markers from allozymes, chromosomes, and mtDNA, Mantovani and colleagues (1991, 2001; Mantovani and Scali 1992) identified wild hybridogenetic strains of Italian stick insects (*Bacillus rossius-grandii*) that had arisen from hybridization between *B. rossius* females and *B. grandii* males. Hybridogenetic males are infertile, whereas females reproduce clonally or hemiclonally. The most surprising discovery, however, was that when female *B. rossius-grandii* were crossed with some sexual males, up to 20% of the offspring had the nuclear genetic makeup solely of their father. Such androgenetic individuals proved to be diploid (via fusion of two sperm heads) and fertile.

Overall, such detailed understanding of the evolutionary genetics of "unisexual" taxa was unimaginable prior to the application of molecular markers. As recently as 1978, in referring to a hybrid-derived parthenogenetic grasshopper, a leading student of the speciation process lamented that "we are never likely to know which species was the female parent" (White 1978b). Since then, however, molecular-based parentage determinations for taxa with parthenogenetic or related reproductive modes have become routine, and indeed, are viewed as merely a starting point for more refined genetic analyses of these species' evolutionary origins and pathways.

SUMMARY

1. Various speciation patterns are expected to leave characteristic phylogenetic signatures on the genomes of recently separated species. By revealing such signatures, molecular markers have prompted reexamination of several long-standing questions in speciation theory: How much genetic change accompanies speciation? What specific genes are involved? Do most speciations entail severe population bottlenecks? Are rates of speciation correlated with rates of genomic evolution? What are the temporal durations of alternative speciation modes? How and how often do natural selection and sexual selection play direct (or indirect) roles

in the speciation process? How often do speciation events occur in sympatry? How prevalent is co-speciation? These and related questions have been provisionally answered for several studied groups.

2. Molecular markers have practical utility in distinguishing closely related taxa, including sibling species that may have gone unrecognized by morphological or other appraisals. Molecular diagnoses sometimes involve sibling species of medical or economic importance.

3. Molecular approaches have enriched the traditional biological species concept (BSC) by adding an explicit phylogenetic perspective to discussions of population relationships and histories. The fundamental distinction between gene trees and species trees has led to the development and elaboration of genealogical concordance principles for the recognition of subspecies and species.

4. Molecular markers provide powerful means for identifying hybrid organisms and for characterizing patterns of introgression. Degrees of hybridization and introgression have proved to vary along a continuum, from instances of sporadic production of F_1 individuals only to extensive introgression leading to genetic mergers between formerly separate taxa.

5. Through cytonuclear analyses, several sources of genetic asymmetry in hybrid zones have been theoretically appreciated and empirically documented. Differential patterns of introgression can result from inter-locus variation in selection intensities against alleles on heterologous genetic backgrounds; from Haldane's rule, whereby gender-specific fitness differences characterize hybrid organisms; from differential mating behaviors by the sexes engaged in hybridization; or from other sources of differential gametic exchange.

6. Contrasts between nuclear phylogenies and mitochondrial or chloroplast gene trees have identified many instances of "cytoplasmic capture" mediated by past introgressive hybridization. This and other lines of genetic evidence suggest that reticulate evolution has been fairly common, especially in several plant groups.

7. Genetic markers have revealed that several plant and animal taxa are of hybrid origin. Especially impressive have been detailed molecular genetic characterizations of the genomic contributions from the parental species involved and of the precise cytological pathways leading to the production of hybridization-derived unisexual biotypes.

8

Species Phylogenies and Macroevolution

All the organic beings which have ever lived on this Earth may be descended from some one primordial form.

C. Darwin (1859)

Study of the gene at the most fundamental level will soon tell us more about the phylogenetic relationships of organisms than we have managed to learn in all the 173 years since Lamarck.

R. K. Selander (1982)

We proceed now to the traditional provenance of molecular phylogenetics: estimation of evolutionary relationships among species and higher taxa. After speciation has been completed and reproductive barriers are in place, the genomes of separated taxa are free to diverge further, and many of their DNA sequences typically do so in more or less time-dependent fashion (see Chapter 4). Thus, overall magnitudes of genetic distance provide at least crude guides to evolutionary times since species last shared ancestors. Furthermore, particular molecular markers, considered individually and, especially, in combination, provide powerful characters for delineating clades.

Many molecular methods (protein electrophoresis, immunological assays, DNA restriction analysis, DNA hybridization, and others) have been used to estimate phylogenies, but in recent years direct nucleotide sequencing has revolutionized the field. Molecular phylogenetic analyses have been conducted on hundreds of taxonomic groups and at temporal scales collectively ranging from Holocene and Pleistocene separations to pre-Paleozoic divergences. No attempt will be made here to treat this vast literature exhaustively. Instead, diverse technical and conceptual approaches will be described in order to highlight exciting outcomes in higher-level molecular phylogenetics.

Rationales for Phylogeny Estimation

Oddly, recovery of phylogeny per se is seldom the final goal of a phylogenetic analysis. Rather, molecular (or other) estimates of phylogeny are desired because of their value as a historical backdrop for interpreting ecological (Harvey et al. 1995, 1996; Losos 1996) and evolutionary processes (Nee et al. 1996b). These appraisals may include assessments of evolutionary rates and patterns in organismal phenotypes (such as particular morphological, physiological, or behavioral traits), biogeographic configurations of taxa, frequencies of horizontal genetic transmission and reticulate evolution, and countless other biological topics.

Phylogenetic character mapping

Closely related taxa often tend to be more similar in phenotype than distant taxa, although many exceptions exist due to variable evolutionary rates and homoplasy (evolutionary convergences, parallelisms, and reversals). Phylogenetic hypotheses (explicit or implicit) underlie virtually all conclusions in comparative evolution (Harvey and Purvis 1991). For example, the inference that powered flight in mammals is a derived rather than an ancestral condition rests upon the restricted phylogenetic position of bats (Chiroptera) within a group (Mammalia) whose ancestral forms unquestionably were terrestrial. At issue in this case is whether flight evolved once or more than once (convergently) in bat evolution, a question whose answer depends on whether the members of Chiroptera are monophyletic or polyphyletic.

Whenever a phylogeny is known with reasonable assurance (e.g., from secure molecular evidence), the evolutionary origin(s) and directions of change in morphological, behavioral, or other organismal features can be illuminated by superimposing trait occurrences on the tree. These traits may be alternative states of composite attributes (such as wings) or more narrowly defined characters (ultimately, the genes or nucleotides actually responsible for a given character), and evolutionary interpretations must be adjusted accordingly (Zink 2002). For example, the broad attribute "flight" is clearly polyphyletic in animals, whereas more specific characteristics often associated with flight (such as feathers in birds, echolocation in bats, or compound eyes in some insects) might each have arisen only once or a few times. However, in this chapter I will use "character" in a generic sense. The shorthand phrase "phylogenetic character mapping" (PCM) will therefore refer to any attempt to match character states with their associated species on a cladogram, the purpose being to reveal the evolutionary origins and histories of those traits. The cladogram itself is estimated using data that are different from and independent of the character states to be mapped. With the advent of molecular approaches, such independent appraisals of phylogeny have become commonplace.

An important task for biologists is to understand the history as well as the mechanistic operation of adaptive (and other) organismal features (Autumn et al. 2002; Frumhoff and Reeve 1994; Givnish and Sytsma 1997), including evolutionarily labile and continuously variable ones such as body size, age of senescence, or metabolic rate. Quantitative traits of this sort pose at least two special challenges for PCM. First, the statistical non-independence of character values, which is inherent in the fact that multiple related taxa have partially overlapping phylogenetic histories (partially shared tree branches), must be accommodated (Martins and Hansen 1997; Richman and Price 1992). If it is erroneously assumed, for example, that all body size comparisons in a set of taxa under consideration are phylogenetically independent, then in effect, the degrees of freedom would be overestimated in a statistical analysis of evolutionary trends. To circumvent such errors, especially when analyzing rapidly evolving quantitative traits, accommodation methods such as "independent contrasts" have been devised (Box 8.1).

Second, most complex traits are polygenic (almost by definition), and various genes and alleles underlying particular phenotypes can wax and wane along the branches of phylogenetic trees (Figure 8.1). The final phenotype may be an outcome of rheostatic or threshold genetic (or environmental) controls whose particular mechanisms of operation remain unknown for most phenotypic attributes and characters. All of this can complicate the interpretation of PCM. So too can the fact that a molecular phylogeny itself is seldom a perfect estimate of organismal history (as will become apparent from some of the case studies to come). For this and other reasons, evolutionary conclusions from PCM should always be viewed as working hypotheses subject to revision with new evidence.

Despite these many challenges, PCM is a popular and informative endeavor in molecular evolution. Most of the examples considered in this chapter will entail slowly evolving qualitative characters (Maddison 1994) that are relatively straightforward to map onto a molecular phylogeny.

ANATOMICAL FEATURES.   Bats are distinctive flight-specialized mammals, so it might seem almost certain that they originated only once in evolution (Figure 8.2A). Under the traditional view, microbats (small, nocturnal, echolocating species; suborder Microchiroptera) and megabats (large diurnal forms; suborder Megachiroptera) are monophyletic sister taxa. Nonetheless, on the basis of neuroanatomical features, an alternative "diphyletic" hypothesis (Pettigrew 1986, 1991) was proposed, according to which megabats are phylogenetically closer to primates than to microbats. If so, then wings and powered flight could have evolved on at least two separate occasions in mammals (Figure 8.2B): once in an ancestor of Microchiroptera and again in a lineage leading to Megachiroptera after its separation from Primates. Thus, a dilemma exists: Either the shared mor-

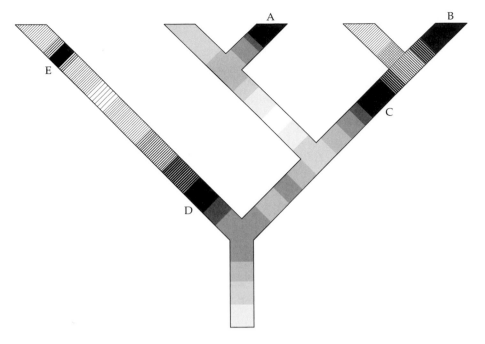

Figure 8.1 Concept of gradients and thresholds in PCM. Consider a polygenic trait that varies as a function of appropriate alleles at many unlinked loci. The diagram shows how genetic changes at these loci can produce gradual evolutionary shifts in trait expression (intensities of shading). Furthermore, different suites of underlying alleles may be responsible for a given phenotypic outcome. A quantitative phenotype may also require some threshold number or pattern of alleles for expression. Assume, for example, that black in the diagram falls above the required threshold (indicating presence of the trait) and non-black falls below the threshold (indicating trait absence). "Trait presence" would then have arisen polyphyletically (at positions A–E) due to shifts in levels of polygenic support. Thus, a quantitative trait could have complex mixtures of both homologous and homoplasious genetic elements.

phologies associated with powered flight evolved convergently in megabats and microbats, or shared neuroanatomical traits evolved convergently in primates and megabats.

Several phylogenetic analyses of both nuclear and mtDNA sequences have led to firm rejection of the "flying primate" hypothesis for megabats and have generally supported a monophyletic scenario for Chiroptera (Adkins and Honeycutt 1991; Bailey et al. 1992; Bennett et al. 1988; Mindell et al. 1991; Van den Bussche et al. 1998). Thus, when interpreted against the backdrop of molecular phylogeny, the original suite of anatomical features associated with bat flight probably arose just once. Subsequent molecular analyses added an interesting twist, however, by demonstrating that microbats are paraphyletic with respect to megabats (Figure 8.2C). This finding implies that microbats' unique echolocation abilities were lost secondarily

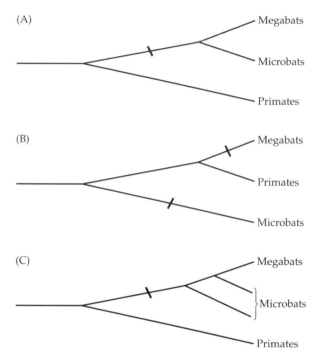

Figure 8.2   Alternative scenarios for the phylogenetic relationships of microbats, megabats, and primates. Slashes across tree branches indicate hypothesized origins of powered flight.

in the megabats, or else that they were gained independently in different microbat lines (Springer et al. 2001; Teeling et al. 2000, 2002).

King crabs of Alaska (genera *Lithodes* and *Paralithodes*) are well-known decapod crustaceans that look like large versions of "typical crabs," showing a strongly calcified exoskeleton and a reduced abdomen that folds up under the body. By contrast, hermit crabs (approximately 800 species in more than 80 genera) have a long, decalcified abdomen that each animal coils into a vacant gastropod shell, the hermit's adopted home. Thus, at least superficially, the morphology of hermit crabs is in between that of true crabs and that of the other major groups of decapod crustaceans: lobsters and shrimp. Nonetheless, researchers have long suspected close genealogical ties between hermits and king crabs for several reasons (Gould 1992): the abdomen of king crabs, although reduced, is asymmetrical, as in hermit crabs; some pairs of legs are reduced in both king and hermit crabs, whereas all ten legs are fully developed in typical crabs; larval king crabs and hermits are remarkably alike in form; and carcinization (evolution of crab-like features) appears to have been a recurring theme in hermit crab evolution under ecological circumstances in which shells of gastropod snails are in limited supply (as in the deep sea).

BOX 8.1 Independent Contrasts

For phenotypic characters displaying only a few discrete states that change rarely (such as "wing presence" and "wing absence"), it is usually possible to trace trait evolution and deduce ancestral states simply by superimposing observed distributions of character states on a known or suspected phylogeny. However, for phenotypes that vary continuously (such as body size or life span), or those that may be evolutionarily labile, or those that may themselves show phylogenetic correlations, such exercises are more challenging.

The method of independent contrasts (Felsenstein 1985b) is one of several statistical procedures that can be used when evaluating the evolution and coevolution of continuously varying traits (Bennett and Owens 2002; Garland et al. 1992; Harvey et al. 1996; Martins 1995, 1996). These methods entail a known phylogeny (often estimated from molecular data; Gittleman et al. 1996a), a range of phenotypes scored in present-day taxa, and a theoretical model for phenotypic evolution. For example, Felsenstein's original method (several elaborations exist; e.g., McPeek 1995; Pagel 1992, 1994) assumes that the phenotypes in question have evolved as if by random Brownian motion, and it statistically corrects for phylogenetic non-independence of character comparisons by transforming measured data for N extant species into a set of $N - 1$ standardized independent contrasts. In the diagram below (after Martins and Hansen 1996), four contrasts are phylogenetically independent (have no overlapping tree branches), and two of them involve internal nodes.

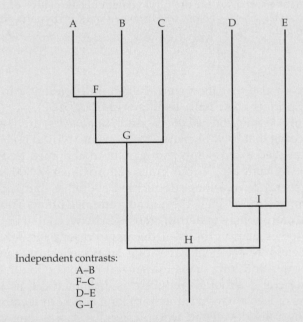

Independent contrasts:
A–B
F–C
D–E
G–I

Ancestor states (or their probabilities) at internal nodes can often be reconstructed based on tree topology and on character measurements in contemporary species (e.g., Schluter et al. 1997). The assignments are often self-evident, e.g. when all species stemming from a node share an identical chracter state, or when outgroup comparisons securely polarize the qualitative states examined. For highly variable or quantitative characters, where assignments may be far less obvious, the state in the ancestor can be estimated by appropriately averaging character values across extant members of that clade (Bennett and Owens 2002).

Examples of continuously varying traits whose evolutionary histories have been analyzed by independent contrasts or related procedures include senescence and several other quantitative life history features in mammals (Gaillard et al. 1994; Gittleman et al. 1996b), musculoskeletal functions in salamanders (Lauder and Reilly 1996), and patterns of song evolution and courtship displays in birds (Irwin 1996). A computer software package by Purvis and Rambaut (1995) is available for "comparative analyses by independent contrasts" (CAIC).

Phylogenetic analyses of mitochondrial rRNA gene sequences appear to have clinched the case for close genetic links between hermit and king crabs (Figure 8.3). Apparently, king crabs arose from a genealogical subset of hermit crabs and, indeed, are nested within the hermit genus *Pagurus*. Furthermore, based on molecular clock considerations and extrapolation from fossil and geographic evidence, this split from hermit ancestors was estimated to have occurred about 13–25 million years ago, thus placing an upper bound on the time that transpired during evolutionary loss of the shell-living habit and the complete carcinization of king crabs (Cunningham et al. 1992). Evolutionary changes in the timing of events during organismal development (heterochrony) probably account for these dramatic morphological shifts.

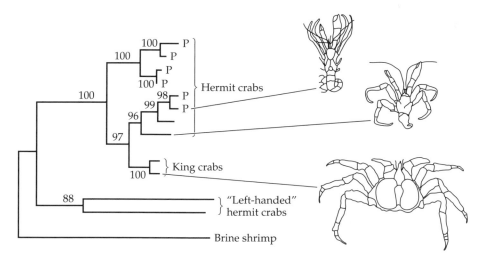

Figure 8.3 Molecular phylogeny for several species of king crabs and hermit crabs, oriented using the brine shrimp (*Artemia*) as outgroup. "P" designates species traditionally placed in the genus *Pagurus*. Numbers indicate percentages of parsimony bootstrap support for various clades. (After Cunningham et al. 1992.)

Living cetaceans traditionally were divided into two distinct suborders: Odontoceti (echolocating toothed whales and dolphins) and Mysticeti (filter-feeding baleen whales). Surprisingly, molecular phylogenies derived from sequence analyses of several mitochondrial and nuclear genes have shown that carnivorous sperm whales are not most closely related to other toothed cetaceans, but rather to baleen whales (Hasegawa et al. 1997; Milinkovitch et al. 1994a,b, 1996). Cetaceans as a whole proved to constitute a clade, but the relationship of Odontoceti to Mysticeti seems to be one of paraphyly, not reciprocal monophyly (Figure 8.4). These findings prompted major reinterpretations of evolutionary transformations among several cetacean features (Milinkovitch 1995). For example, PCM analysis indicates that baleen whales

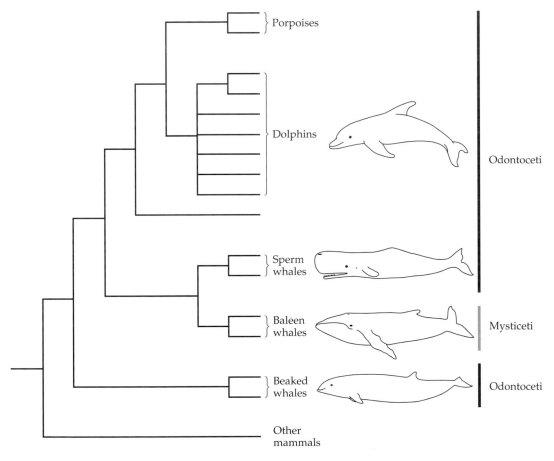

Figure 8.4   Molecular phylogeny for representative cetaceans based on mitochondrial rRNA gene sequences. Levels of bootstrap support were near 100% for most of the putative clades identified in the parsimony and neighbor-joining analyses. (After Milinkovitch et al. 1993.)

probably lost the morphological apparatus and capacity for echolocation secondarily (alternatively, echolocation abilities could have been gained independently by sperm whales and the other toothed cetaceans).

Molecular analyses also have prompted a reexamination of the phylogenetic position of whales and dolphins within Mammalia (Novacek 1992). An evolutionary connection of cetaceans to ungulates (hoofed animals) was first suggested more than a century ago, and this notion has been amply confirmed by molecular (Goodman et al. 1985; Irwin et al. 1991; Milinkovitch 1992; Miyamoto and Goodman 1986; Southern et al. 1988) as well as paleontological and phenotypic evidence (see review in Milinkovitch 1995). More controversial has been the question of whether cetaceans are phylogenetically closer to the odd-toed Perissodactyla (a taxonomic order including the horse, tapir, and rhinoceros) or the even-toed Artiodactyla (including the pig, camel, and deer). Most mitochondrial and nuclear DNA evidence favors the Artiodactyla connection (Gatesy 1997; Graur and Higgins 1994). Indeed, a recent discovery that whales share several uniquely derived SINEs (short interspersed DNA elements) with ruminants and hippopotomuses (Shimamura et al. 1997) suggests that cetaceans are embedded well within the artiodactyl lineage (Milinkovitch and Thewissen 1997; Nikaido et al. 1999).

Crossopterygia ("lobe-finned" fishes) are probable evolutionary links between Actinopterygia ("ray-finned" fishes) and tetrapods (land vertebrates). In 1938, scientists were thrilled by the discovery of a living coelacanth (*Latimeria chalumnae*), a member of a taxonomic subset of lobe-finned fishes that formerly was thought to have gone extinct about 65 million years ago. Hopes were high that detailed studies of this "living fossil" would resolve three long-standing competing hypotheses for the early branching history in tetrapod phylogeny: lungfishes (another group of lobe-finned fishes) as sister group to coelacanths plus tetrapods (Figure 8.5A); tetrapods as sister group to lungfishes plus coelacanths (Figure 8.5B); and coelacanths as sister group to lungfishes plus tetrapods (Figure 8.5C). Initial analyses of mitochondrial genes (notably cytochrome *b* and 12S rRNA) provided support for the lungfish + tetrapod clade (Hedges et al. 1993; Meyer and Dolven 1992), and this phylogenetic arrangement was used to interpret the histories of 22 morphological traits (presence versus absence of a glottis and internal nostrils, pelvic girdles joined versus unjoined, etc.) early in vertebrate evolution (Meyer and Wilson 1990). However, subsequent analyses of nuclear sequences from 28S rDNA gave significant support to the lungfish + coelacanth clade (Zardoya and Meyer 1996a). The contradictory nature of these and other molecular findings (Gorr et al. 1991; Normark et al. 1991; Sharp et al. 1991; Stock et al. 1991) next prompted Zardoya and Meyer (1996b, 1997) to sequence the complete mtDNA genomes of an extant lungfish (*Protopterus dolloi*) and a coelacanth. These expanded data still were phylogenetically inconclusive (Zardoya et al. 1998), however, in part because of strong heterogeneity in molecular evolutionary rates across loci. Curole and Kocher (1999) used these data and other examples from the literature to urge extreme caution in

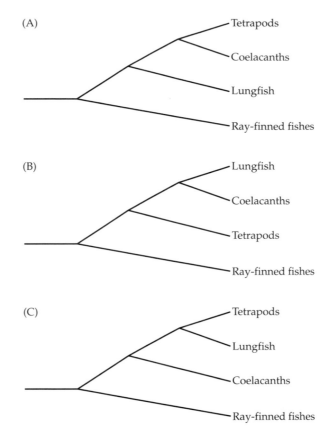

Figure 8.5 Alternative hypotheses for the phylogenetic root of tetrapods. (After Meyer and Wilson 1990.)

extrapolating from the phylogenies of particular genes (or even of whole mtDNA genomes) to the phylogenies of organisms. Nonetheless, the balance of current evidence seems to favor scenario C in Figure 8.5 (Meyer and Zardoya 2003).

Since it was first described in 1869, the giant panda (*Ailuropoda melanoleu-ca*) has been a phylogenetic enigma. It generally looks like a bear (family Ursidae), but also has many non-bearlike traits: flattened teeth and other adaptations associated with a bamboo diet; an opposable "thumb" (really a modified wrist bone); lack of hibernation; a bleating voice like a sheep; and a karyotype of only 42 chromosomes, compared with bears' 74. More than 40 morphological treatises reached no consensus on whether giant pandas are phylogenetically allied to bears, raccoons, or neither, but molecular appraisals apparently solved the mystery (O'Brien 1987). Based on diverse data from protein electrophoresis (Goldman et al. 1989), protein immunology, DNA hybridization (O'Brien et al. 1985a; Sarich 1973), and DNA sequencing

(Hashimoto et al. 1993; Slattery and O'Brien 1995), the giant panda lineage originated about 20 mya as an early offshoot of the Ursidae clade (Figure 8.6).

This molecular phylogeny also prompted a reexamination of the giant panda's bizarre karyotype. Using refined laboratory methods that reveal details of chromosomal banding patterns, the differences between the giant panda's 42 chromosomes and bears' 74 chromosomes were shown to be mechanistically superficial, apparently attributable to simple centromeric fusions along the giant panda lineage (O'Brien et al. 1985a). Overall, these PCM studies on pandas illustrate two broader points: that phylogenies are

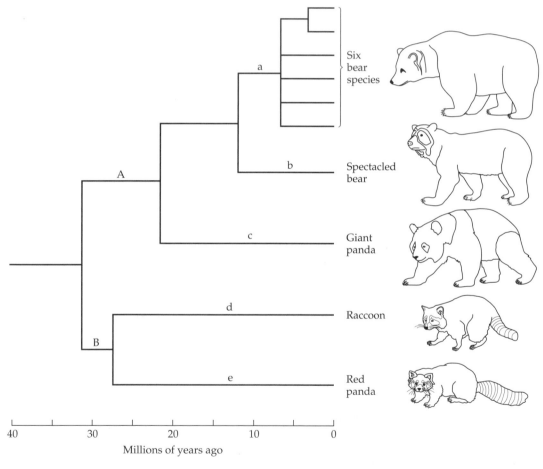

Figure 8.6   Consensus molecular phylogeny showing the genealogical position of the giant panda relative to bears (Ursidae, node A) and raccoons (Procyonidae, node B). Lowercase letters represent suggested subfamily designations. For recent molecular data and refinements of thought concerning phylogenetic relationships of the *red* panda (another phylogenetic enigma), see Flynn et al. (2000). (After O'Brien 1987.)

most convincing when supported concordantly by multiple lines of molecular evidence; and that consensus trees can offer an informative backdrop for interpreting the evolutionary histories of problematic molecular-level and cellular-level characters, just as they can for morphological traits.

BEHAVIORAL, PHYSIOLOGICAL, AND LIFE HISTORY FEATURES. Molecular phylogenies can also provide a useful backdrop for interpreting the evolutionary histories of behavioral and other classes of organismal characteristics. A case in point involves interspecific brood parasitism, or "egg-dumping," by females into other species' nests with the purpose of duping what often then become foster parents. In the Americas, several cowbird species are dedicated brood parasites: *Molothrus rufoaxillaris* specializes on one host species; *M. aeneus* and *Scaphidura oryzivora* parasitize confamilial genera only; and *M. ater* and *M. bonariensis* utilize a wide variety of host taxa. PCM analyses based on mtDNA cytochrome *b* gene sequences from nearly three dozen species of Icteridae (cowbirds, plus other blackbirds that are non-parasitic) indicate that all the brood parasites listed above form a clade within which the single-host specialist (*M. rufoaxillaris*) branched off earliest. Furthermore, the two generalist species (*M. ater* and *M. bonariensis*) constitute a terminal subclade (Lanyon 1992; Lanyon and Omland 1999). These results suggest that host specificity is the ancestral condition and that host generalization is derived. This conclusion challenged an earlier notion that across evolutionary time brood parasites might increasingly specialize on fewer host taxa (because at least some host species eventually evolve defensive mechanisms against the brood parasite).

Researchers have used molecular PCM approaches to chart the direction of behavioral evolution in *Lasioglossum* sweat bees, a hymenopteran group containing both solitary and eusocial members with a diversity of nest architectures. Danforth et al. (2003) employed DNA sequences from two nuclear genes and one mitochondrial gene to estimate a phylogeny for 42 *Lasioglossum* taxa, onto which they then mapped social behaviors. Salient findings were that eusociality had a single origin within the group, but that multiple (about six) reversals to solitary nesting must have occurred within the eusocial clade. Results supported the view that eusociality may be hard to evolve, but somewhat easier to lose (Figure 8.7). In an earlier study of eight species in the subgenus *Evylaeus*, Packer (1991) had employed multi-locus allozyme data to estimate a cladogram onto which he then mapped architectural features of the bees' nests. From this analysis, one notable characteristic—an extended opening of brood cells during development—was inferred to have originated on two separate evolutionary occasions among the species considered.

Figure 8.7 **DNA sequence-based phylogeny for sweat bees** (*Lasioglossum*) onto ▶ which have been mapped nesting behaviors. Numbers on branches indicate levels of statistical support (Bayesian posterior probabilities) for various groups. (After Danforth et al. 2003.)

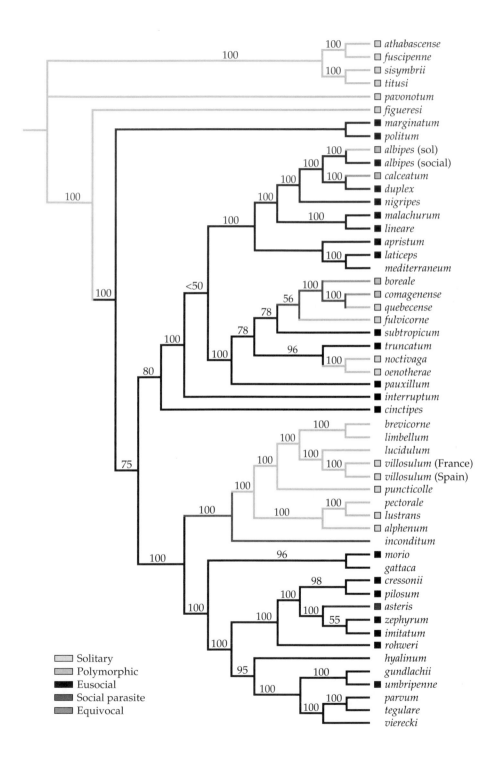

Solitary
Polymorphic
Eusocial
Social parasite
Equivocal

Figure 8.8   Coarse-focus and finer-focus PCM analyses of nesting habits in birds. ▶
(A) Phylogenetic distribution of "safe" nests in major avian groups. (After Owens
and Bennett 1995.) (B) A closer view of the phylogenetic distribution of nesting
modes in 17 swallow species plus an outgroup. (After Winkler and Sheldon 1993.)
In each case, the molecular phylogenies were based on DNA hybridization data.

A PCM analysis of nesting habits in birds further illustrates how
"behaviors" and "extended morphological phenotypes" can be interrelated.
Birds' nests can be broadly categorized as "safe" (e.g., in burrows or tree
cavities) or "open," and PCM analyses indicate that safe nesting habits have
arisen multiple times in avian evolution (Figure 8.8A). The swallow family
Hirundinidae has been of special interest because its approximately 90
species collectively display perhaps the greatest diversity of nest construction modes known in any avian taxonomic family. Some swallow species
burrow into cliffs, some use adopted natural cavities, and others construct
mud nests with designs ranging from open cups to roofed and eaved
abodes. From maps of these nesting modes superimposed on a molecular
phylogeny for representative swallows (Figure 8.8B), the following were
deduced: burrowing probably was the primitive nesting mode in the group,
predating cavity adoption and mud nesting; obligate cavity adoption is confined mostly to a New World clade; mud-nest construction originated once
in the family and diversified primarily in Africa; and mud nests "evolved"
higher complexity through time, from simple cups to fully enclosed abodes.

Endothermy, the ability to maintain elevated body temperatures by
metabolic means, is rare among fishes, the only documented examples
being within large teleosts of the suborder Scombroidei (including tunas,
mackerels, and billfishes). Did endothermy evolve once or multiple times
within this assemblage? The character state endothermy, when mapped
onto a phylogeny estimated from cytochrome b mtDNA sequences, indicates that this physiological adaptation arose independently at least three
times within the Scombroidei (Block et al. 1993). Diverse physiological and
anatomical pathways are involved in these phenotypic convergences
(Figure 8.9).

Ascidians (sea squirts) are thought to be primitive chordates, as gauged
in part by the presence of a notochord in their tadpole-like larvae.
Phylogenetic analyses based on rDNA sequences (Field et al. 1988) and on
molecular features of muscle actins (Kusakabe et al. 1992) bolstered this
view by placing ascidians closer to vertebrates than to invertebrates. Two
distinctive reproductive/developmental modes are exhibited among the
2,300 known species of Ascidiacea: solitary forms, and communal forms that
can asexually generate colonies by budding, strobilation, or regeneration.
Under an orthodox classification based on characteristics of the branchial
sac and gonads, ascidians were divided into the orders Enterogona and
Pleurogona, irrespective of solitary versus colonial lifestyle. If this taxonomy reflects phylogeny, then one or both lifestyles probably evolved multiple

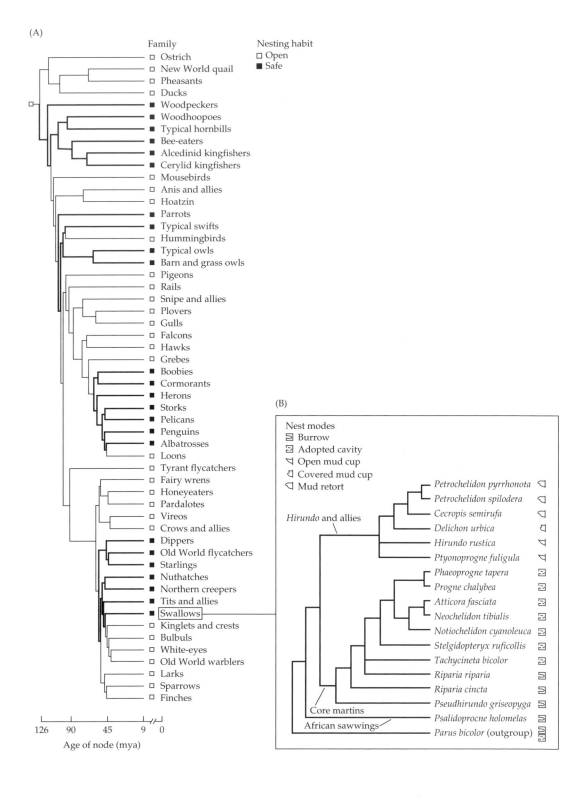

(A)

Family

Nesting habit
□ Open
■ Safe

- □ Ostrich
- □ New World quail
- □ Pheasants
- □ Ducks
- ■ Woodpeckers
- ■ Woodhoopoes
- ■ Typical hornbills
- ■ Bee-eaters
- ■ Alcedinid kingfishers
- ■ Cerylid kingfishers
- □ Mousebirds
- □ Anis and allies
- □ Hoatzin
- ■ Parrots
- ■ Typical swifts
- □ Hummingbirds
- ■ Typical owls
- ■ Barn and grass owls
- □ Pigeons
- □ Rails
- □ Snipe and allies
- □ Plovers
- □ Gulls
- □ Falcons
- □ Hawks
- □ Grebes
- ■ Boobies
- ■ Cormorants
- ■ Herons
- ■ Storks
- ■ Pelicans
- ■ Penguins
- ■ Albatrosses
- □ Loons
- □ Tyrant flycatchers
- □ Fairy wrens
- □ Honeyeaters
- □ Pardalotes
- □ Vireos
- □ Crows and allies
- ■ Dippers
- ■ Old World flycatchers
- ■ Starlings
- ■ Nuthatches
- ■ Northern creepers
- ■ Tits and allies
- ■ Swallows
- □ Kinglets and crests
- □ Bulbuls
- □ White-eyes
- □ Old World warblers
- □ Larks
- □ Sparrows
- □ Finches

126   90   45   9   0

Age of node (mya)

(B)

Nest modes
- ⊟ Burrow
- ⊟ Adopted cavity
- ◁ Open mud cup
- ◁ Covered mud cup
- ◁ Mud retort

Hirundo and allies

- *Petrochelidon pyrrhonota* ◁
- *Petrochelidon spilodera* ◁
- *Cecropis semirufa* ◁
- *Delichon urbica* ◁
- *Hirundo rustica* ◁
- *Ptyonoprogne fuligula* ◁
- *Phaeoprogne tapera* ⊟
- *Progne chalybea* ⊟
- *Atticora fasciata* ⊟
- *Neochelidon tibialis* ⊟
- *Notiochelidon cyanoleuca* ⊟
- *Stelgidopteryx ruficollis* ⊟
- *Tachycineta bicolor* ⊟
- *Riparia riparia* ⊟
- *Riparia cincta* ⊟
- *Pseudhirundo griseopyga* ⊟
- *Psalidoprocne holomelas* ⊟
- *Parus bicolor* (outgroup) ⊟⊟

Core martins

African sawwings

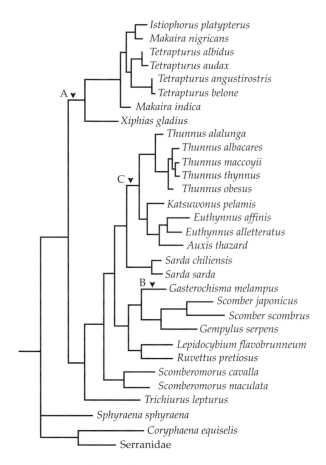

Figure 8.9 Distribution of endothermy in the phylogeny of Scombroidei marine fishes. The phylogeny was estimated from mtDNA sequences. Letters indicate three separate origins of endothermy, each with a different physiological basis: A, modification of the superior rectus muscle into a thermogenic organ; B, modification of the lateral rectus muscle into a thermogenic organ; and C, use of vascular countercurrent heat exchangers in the muscles, viscera, and brain. (After Block et al. 1993; see also Block and Finnerty 1994.)

times. Alternatively, if the distinct lifestyles register a basal phylogenetic split, then the branchial sac and gonadal characters would be homoplasious. To distinguish between these hypotheses, Wada et al. (1992) phylogenetically analyzed ascidian 18S rDNAs. They showed that ascidian species fell into two distinct clades coinciding with orthodox taxonomy. Thus, solitary and colonial lifestyles probably were gained or lost independently after the phylogenetic split between Enterogona and Pleurogona.

Behavioral and morphological traits can be thoroughly intertwined, a point further illustrated by male tail lengths in *Xiphophorus* platyfishes and swordtails. These fishes had raised an evolutionary question: Might female mating preferences for males with swordlike tails predate the evolutionary origin(s) of such tails? If so, then sexual selection based on mate choice could be driven in large part by this "preexisting bias." Reinterpreting original molecular studies by Meyer et al. (1994) and Basolo (1995), Schluter et al. (1997) showed through PCM analyses that the ancestral condition of *Xiphophorus* males (and hence the issue of preexisting female mating bias) could not be decided definitively, due to rapid evolutionary transitions between sword-carrying and swordlessness within the genus (Figure 8.10). This study illustrates how ancestral states can be deduced by PCM, but it also highlights the statistical or probabilistic nature of the enterprise (D. R. Maddison 1995), especially for rapidly evolving traits.

Plants also have morphologies and behaviors that can be subjected to PCM via molecular markers. Ever since Darwin (1877) speculated on the matter, androdioecy (in which male and hermaphroditic flowers occur on separate plants) has been viewed as an intermediate step in the evolution of dioecy (separate sexes) from monoecy (hermaphroditism). However, in the Datiscaceae, which includes one androdioecious species (*Datisca glomerata*)

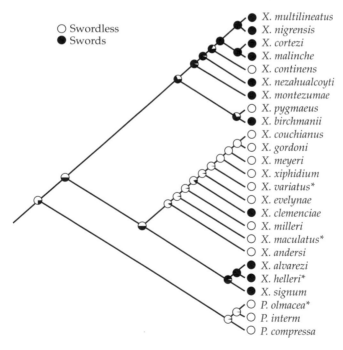

Figure 8.10   Evolution of males' tail conditions (swords versus swordless) in 25 species of *Xiphophorus* (*X*) and *Priapella* (*P*) fishes. Asterisks indicate species in which female preference for males with swords has been experimentally documented (Basolo 1995). Pie diagrams indicate the relative likelihoods from PCM analysis of particular ancestral tail conditions at the nodes. (After Schluter et al. 1997; based on the molecular phylogeny from Meyer et al. 1994.)

plus several dioecious ones, androdioecy was postulated to represent a derived condition (Liston et al. 1990). To test this hypothesis, Rieseberg et al. (1992) utilized cpDNA restriction site data to construct a molecular phylogeny for Datiscaceae and outgroup taxa. Results suggested that *D. glomerata* occupies a derived position relative to dioecious members of the family and, hence, that androdioecy in this case evolved from dioecy rather than from monoecy. However, later phylogenetic analyses of *rbcL* and 18S rDNA sequences disagreed as to whether dioecy or monoecy was ancestral to *Datisca*, so the issue of evolutionary direction may not yet be fully settled (Swensen et al. 1998).

Darwin (1875) was also well aware of carnivorous plants, which possess whole suites of anatomical and physiological features associated with attracting, trapping, killing, and digesting animals and absorbing their products. Given this complexity, it might be supposed that the evolution of carnivory would be difficult and rare in the plant world. Nonetheless, when this lifestyle was recently mapped onto a phylogeny based on *rbcL* gene sequences for several dozen taxonomic plant families, carnivory was revealed to be polyphyletic, and even some subcategories, such as "pitcher traps," may have arisen three or more times independently (Albert et al. 1992). On the other hand, "snap-traps" proved to be monophyletic (Cameron et al. 2002), as did various other detailed components of carnivory that clustered phylogenetically. Thus, overall, carnivory in plants displays a mixture of homologous and analogous elements. As phrased by Albert et al. (1992), "form is not a reliable indictor of phylogenetic relationships among carnivorous plants at highly inclusive levels (such as trapping mechanism), whereas it appears to be at less inclusive ones (such as glandular anatomy)."

Even microbes and their metabolic capabilities have been the subject of PCM. In one interesting example, DeLong et al. (1993) used small-subunit rRNA gene sequences to estimate the phylogeny of bacteria that employ iron-rich magnetosomes inside their cells to orient to Earth's geomagnetic field as they swim. Findings from this PCM exercise indicated a polyphyletic origin for magnetotaxis: magnetosomes based on iron oxides evolved independently from those based on iron sulfides.

The studies described above merely introduce the logic and scope of molecular PCM approaches. Several additional examples are described in Table 8.1.

Biogeographic assessment

A second rationale for phylogeny estimation is its use in biogeographic reconstruction. Just as phylogeographic relationships among conspecific populations have been revealed through molecular analyses (see Chapter 6), so too have phylogeographic relationships among species and higher taxa.

VICARIANCE VERSUS DISPERSAL.   The potential geographic range of any taxon is of course limited by the suitability of environmental conditions.

Within that zone of ecological tolerance, realized geographic distributions are additionally influenced by historical factors. The "success" of countless human-introduced plants and animals around the world is testimony to the fact that not all habitats suitable for a species are occupied naturally by that species (or, perhaps, that humans have disturbed native habitats so much that such introductions often succeed). Thus, whether a species occurs in a particular area is a function not only of ecology, but also of historical demographic and dispersal patterns, which themselves are influenced by the proximity and spatial relationships of habitable environments. The study of such historical factors is the focus of molecular biogeography.

When related taxa show disjunct geographic ranges, two competing hypotheses often are advanced to account for these spatial arrangements (Box 8.2). Under dispersalist scenarios, such a taxonomic group came to occupy its current range through active or passive dispersal across a preexisting geophysical or ecological barrier. Alternatively, under vicariance models, the more or less continuous ranges of ancestral taxa were sundered by geophysical or ecological events. Such "vicariant events" might have been the uplift of a mountain range sundering lowland species, a continental breakup that partitioned terrestrial organisms, or subdivision of an ocean basin by the rise of an isthmus separating marine taxa. Dispersal and vicariance models both subscribe to the proposition that speciation is predominantly allopatric. However, an important prediction of vicariance biogeography (not shared by dispersalist scenarios) is that the cladogram for a taxonomic group should match the historical "area cladogram" of environments occupied (Box 8.2).

Critical tests of specific vicariance scenarios traditionally involved phylogenetic appraisals based on nonmolecular characters (e.g., Cracraft 1986), but since the 1980s molecule-based phylogenies have played ever-increasing roles. For example, Bermingham et al. (1992) used data from mtDNA restriction sites to test a vicariance model for the evolution of North American birds in the black-throated green warbler complex. It had been proposed that episodic glacial advances during the Pleistocene repeatedly fragmented the ranges of forest-dwelling birds into eastern and western populations in such a way that subsequent speciations produced a series of western endemics, each linked phylogenetically to the widespread eastern form (*Dendroica virens* in this case), but at different evolutionary depths (Mengel 1964). Molecular data proved not to be fully consistent with this scenario, however, suggesting instead that some western warblers in the black-throated green complex budded off from one another (perhaps via inter-montane isolations) rather than directly from *D. virens* (Bermingham et al. 1992).

A similar Pleistocene scenario constituted conventional wisdom for anuran evolution in southwestern Australia, where several western endemics were hypothesized to have arisen following multiple invasions from eastern source stocks. However, immunological studies of albumins from more than 20 frog species led to rejection of the multiple-invasion scenario in favor of a model that includes speciation events within southwest-

TABLE 8.1	Molecule-based studies illustrating the wide diversity of evolutionary time frames, taxa, and organismal phenotypes that have been the subject of phylogenetic character mapping (PCM)

Basis for molecular phylogeny	*Phenotype mapped*
Multiple gene sequences	Photosynthesis in prokaryotes
Complete genomic sequences	Circadian clock genes in prokaryotes
Chloroplast gene sequences	Endosperm in flowering plants
Nuclear and cpDNA gene sequences	Nectar spurs in columbine plants
rRNA gene sequences	Position of the ovary in herbaceous plants
rDNA sequences and AFLPs	Flower pollination by hummingbirds
Nuclear and mtDNA gene sequences	Fruiting body structures in fungi
Nuclear and mtDNA gene sequences	Symbiosis of termites and fungi
A large variety of gene sequences	Venom composition in cone snails
mtDNA gene sequences	Mutualism between ants and aphids
Nuclear gene sequences	Agriculture in ants
Nuclear gene sequences	Wing features in stick insects
Nuclear and mtDNA gene sequences	Social behavior in thrip insects
mtDNA gene sequences	Courtship songs in lacewing insects
mtDNA and nuclear gene sequences	Features of microsatellite loci in wasps
Ribosomal gene sequences	Compound eye in arthropods
mtDNA gene sequences	Eusociality in marine shrimp
mtDNA gene sequences	Egg-mimic structures in male fish
mtDNA gene sequences	Brood pouches in male-pregnant fish
mtDNA gene sequences	Placentas in live-bearing fish
mtDNA gene sequences	Sexual isolation behaviors in fish
mtDNA gene sequences	Müllerian mimicry in frogs
mtDNA sequences and DNA hybridization	Flightlessness in birds
Mitochondrial gene sequences	Polygynous mating in birds
Mitochondrial gene sequences	Plumage features in orioles
Allozymes	Dietary habits in Galápagos finches
Ribonuclease gene sequences	Ruminant digestion in mammals
Mitochondrial gene sequences	Human domestication of horses
Mitochondrial gene sequences	Coat colors in wild mice

ern Australia (Maxson and Roberts 1984; Roberts and Maxson 1985). Furthermore, based on molecular evidence, many of these speciation events appeared to predate the Pleistocene significantly.

Considerable molecular attention has been devoted to the possible roles of Pleistocene ice ages in vicariant population separations and speciation in plants (e.g., Comes and Kadereit 1998) and animals (Hewitt 1996), both in temperate regions (Knowles 2001) and in the tropics (Moritz et al. 2000; Patton et al. 1994a,b). Based on molecular phylogenies for literally hundreds of species (Avise 2000a; see Chapter 6), it is now abundantly clear that many conspecific populations were isolated and began differentiating in Quaternary refugia, sometimes achieving the status of taxonomic species by

Reference
Raymond et al. 2002
Dvornyk et al. 2003
Geeta 2003
Hodges and Arnold 1994
Kuzoff et al. 1999
Beardsley et al. 2003
Hibbett and Binder 2002
Aanen et al. 2002
Olivera 2002
Shingleton and Stern 2002
Mueller et al. 1998; Currie et al. 2003
Whiting et al. 2003
Morris et al. 2002
Wells and Henry 1998
Zhu et al. 2000
Oakley and Cunningham 2002
Duffy et al. 2000
Porter et al. 2002; Porterfield et al. 1999
Wilson et al. 2001, 2003
Reznick et al. 2002
Mendelson 2003
Symula et al. 2001
Cubo and Arthur 2001
Searcy et al. 1999
Omland and Lanyon 2000
Yang and Patton 1981; Schluter et al. 1997
Jermann et al. 1995; Schluter et al. 1997
Vila et al. 2001
Nachman et al. 2003

the present, and sometimes not (Avise and Walker 1998; Klicka and Zink 1997). Thus, for contemporary distributions of closely related taxa, a recurring phylogeographic theme is vicariant separation(s) during the Pleistocene and earlier, followed by post-Pleistocene dispersal to the current configurations (Bernatchez and Wilson 1998; Hewitt 1999, 2000; Jaarola and Tegelström 1995, 1996; Taberlet et al. 1998).

Vicariance versus dispersal questions also apply in the marine realm. For example, closely related pairs of more than 50 tropical shoreline-restricted fish taxa have disjunct distributions in the Pacific Basin, being separated by at least 5,000 km (the distance between the closest islands in the central versus eastern Pacific). Long-distance dispersal, probably via planktonic larvae

BOX 8.2 Vicariance versus Dispersal in Biogeography

Disjunct distributions of related taxa are biogeographically intriguing. Traditionally, disjunct ranges were interpreted to evidence dispersal across pre-existing geographic or ecological barriers, usually from a biogeographic "center of origin" where a given taxonomic group presumably originated (Darlington 1957, 1965). Dispersalist explanations sometimes became quite strained, however, as, for example, in accounting for global arrangements of relatively sedentary creatures such as the flightless ratites (ostriches, rheas, emus, and their presumed relatives). Vicariance hypotheses challenge dispersalist ones by proposing that disjunct ranges result from historical sunderings of ancestral taxa by geophysical events, without the need to invoke improbable long-distance dispersal or "sweepstakes" (rare and lucky) colonizations (Simpson 1940).

Vicariance biogeography as a formal discipline (Humphries and Parenti 1986; Nelson and Platnick 1981; Nelson and Rosen 1981; Rosen 1978) grew out of two developments in the 1960s and 1970s (Wiley 1988): a growing appreciation, based on the study of plate tectonics and other geologic processes, that the Earth's features were not fixed, but rather were historically dynamic; and the growth of cladistics and parsimony (see Chapter 4), which armed researchers with new conceptual outlooks and analytical tools for phylogeny reconstruction.

Under strict vicariance hypotheses (in contrast to dispersalist expectations), cladistic relationships among related disjunct taxa should mirror faithfully the historical relationships among the geographic regions occupied. Indeed, one major goal of vicariance biogeography is to answer questions about the biological history of the real estates themselves (Wiley 1988): for example, "Is Cuba more closely related to Hispaniola than to Jamaica?" This is accomplished through comparative searches for congruent patterns ("generalized tracks") in organismal phylogeography, while also realizing that most biotic communities will have had both "branching" (vicariant) and "reticulate" (secondary dispersal) events in their histories (Enghoff 1995; Ronquist 1997).

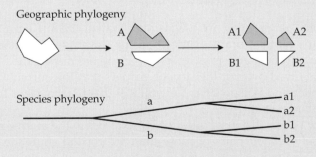

traveling west to east with equatorial currents, may have connected popula-tions in these areas fairly recently. Alternatively, fish taxa inhabiting the east-ern Pacific might be vicariant relics of former worldwide ancestors in the Tethys Sea, an ocean that separated Laurasia from Gondwana following the breakup of the supercontinent Pangaea more than 90 mya, during the Mesozoic (McCoy and Heck 1976).

For several such fish groups on opposite sides of the Pacific, as well as some invertebrates with similar distributions, molecular data have clearly eliminated the ancient vicariance hypothesis by demonstrating relatively small genetic distances that are more consistent with recent (< 3 mya) or sometimes ongoing genetic contacts across the Pacific basin (Baldwin et al. 1998; Palumbi 1994; Palumbi et al. 1997; Rosenblatt and Waples 1986). Indeed, even on a circumglobal scale, phylogenetic analyses of mtDNA in several taxonomic assemblages of oft-cryptic marine species and genera have yielded estimated evolutionary separation dates falling within the last 0–20 million years (Bermingham et al. 1997; Bowen and Grant 1997; Bowen et al. 2001; Colborn et al. 2001; see Aoyama et al. 2001, for a possible excep-tion involving *Anguilla* eels). Of course, even if most of these marine con-geners separated rather recently, that does not negate the likelihood that ancient vicariant events (perhaps related to the Tethys Sea) had genetic effects as well. Some populations sundered in ancient times may since have evolved into what today are recognized as higher taxa (often with second-ary changes in geographic distributions resulting in range overlaps of some of the constituent species).

Thus, one important advantage of molecular approaches is that tempo-ral issues (based on molecular clock considerations) can be examined in addition to cladistic assessments per se. In another example, extensive molecular data sets have been employed in conjunction with other biogeo-graphic evidence to assess the evolutionary times of separation between a wide variety of fishes, amphibians, reptiles, birds, and mammals in the Caribbean region (see review in Hedges 1996). Geologic evidence indicates that the Greater Antilles islands formed in close proximity to North and South America about 110–130 mya, but via plate tectonics had begun sepa-rating from these mainlands by the late Cretaceous (80 mya). Much of the present West Indian biota might reflect these ancient proto-Antillean vicari-ant separations. Alternatively, post-vicariant dispersal might account for the presence of related taxa on the various islands.

Hedges' (1996) analyses proved inconsistent with the hypothesis of ancient vicariance for most of this biota in two regards (Figure 8.11). First, in phylogenetically independent comparisons of numerous pairs of related island taxa, or island versus mainland forms, genetic distances showed a large variance, suggesting widely differing colonization dates (as predicted under most dispersalist scenarios). Second, most of the estimated diver-gence dates, based on molecular clocks, fell well within the Cenozoic, long

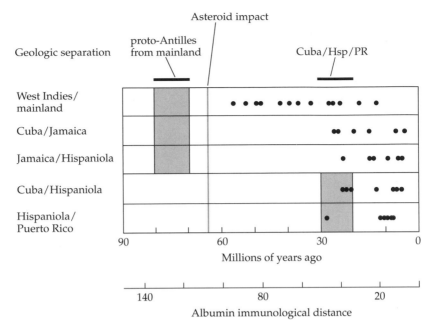

Figure 8.11   Empirical molecular tests of dispersalist versus vicariance hypotheses for the origins of terrestrial vertebrates in the Caribbean region. Shown are immunological distances (in albumin proteins) and associated estimates of evolutionary separation times (under an evolutionary clock) between various island and mainland species (each comparison is indicated by a dot). Shaded areas indicate dates of faunal separation predicted under vicariance scenarios based on geologic events. (After Hedges et al. 1992b.)

after the postulated vicariant geographic events that otherwise might have been involved. Closer phylogenetic inspections further revealed that South America was the original source for most island lineages of amphibians and reptiles (nonvolant animals that presumably arrived by flotsam drifting on oceanic currents), whereas North America and Central America were the dispersal sources for most birds and bats (volant animals that simply flew to the islands from these more proximate continental regions).

Many more phylogeographic insights have been gleaned through these and other molecular comparisons of Caribbean faunas, both vertebrate and invertebrate (Burnell and Hedges 1990; Hedges 1989; Hedges et al. 1991; Klein and Brown 1995; Losos et al. 1998; Seutin et al. 1994; Shulman and Bermingham 1995). For example, the six species of *Anolis* lizards on Jamaica constitute a clade that radiated from a colonizer that probably arrived about 14 mya (Hedges and Burnell 1990). Similarly, the nine species of Jamaican land crabs (unique among all the world's crabs in providing active brood care for larvae and juveniles) apparently arose in situ from a common ancestor that inhabited the island about 4 mya (Schubart et al. 1998).

A remarkable historical instance of long-distance dispersal was intimated by molecular findings for two spatially disjunct species of annual plants: the diploid *Senecio flavus* of the Saharo-Arabian and Namibian deserts in Africa and the tetraploid *S. mohavensis* of the Mojave and Sonoran deserts in North America. In allozyme assays, these two species proved to be remarkably similar (Nei's $I \cong 0.95$), a genetic result that not only affirmed their traditional congeneric status based on their gross morphology, but also implied that recent intercontinental dispersal accounts for these species' current distributions (Liston et al. 1989). How this "sweepstakes dispersal" (see Box 8.2) occurred remains a mystery, but one possibility is that the plants' sticky seeds were transported by migrating or lost birds.

Much deeper along the temporal scale, considerable attention has been devoted to possible phylogeographic consequences of the Mesozoic breakup of the world's supercontinents and the subsequent movements of landmasses by plate tectonics. For example, characiform fishes consist of about 1,200 living species in 16 families, nearly all restricted to fresh waters of Africa and South America. By assaying slowly evolving rDNA sequences, Ortí and Meyer (1997) identified three ancient trans-Atlantic clades, each of which apparently was sundered by the 90-mya vicariant disjunction of these continents. This ancient Gondwanaland breakup has likewise been inferred to have left deep and concordant molecular phylogeographic footprints on two other large taxonomic assemblages of circumtropical freshwater fishes, Cichlidae and Aplocheiloidei (Figure 8.12), as well as on circumtropical birds in the orders Psittaciformes (parrots) and Piciformes (including barbets and toucans) (Miyaki et al. 1998; Sibley and Ahlquist 1986, 1990).

Overall, however, it is probably unwise to dichotomize dispersal and vicariance too sharply (except for heuristic purposes) because both processes play key roles in many instances. This point is forcefully illustrated by the global phylogeography of ratite birds (ostriches, rheas, emus, kiwis, and their allies). Molecular and other evidence have often been interpreted as consistent with ancient lineage separations and Cretaceous vicariance events (the continental breakup of Gondwanaland) as accounting for the current distribution of these flightless birds on southern landmasses (Cooper 1997; Lee et al. 1997; Prager et al. 1976; Sibley and Ahlquist 1981; Stapel et al. 1984; see Härlid et al. 1998 for a dissenting view). Complete mtDNA genomic sequences for extant ratite species recently appear to have confirmed the general outlines of this view (Haddrath and Baker 2001). However, molecular phylogeographic findings have also demonstrated at least two instances of dispersal in ratite evolution: a secondary colonization of New Zealand by kiwis (Cooper et al. 1992, 2001), and a colonization of Africa by an ostrich ancestor that probably arose in South America about 60 mya before dispersing to Africa via the Northern Hemisphere (Haddrath and Baker 2001; van Tuinen et al. 1998).

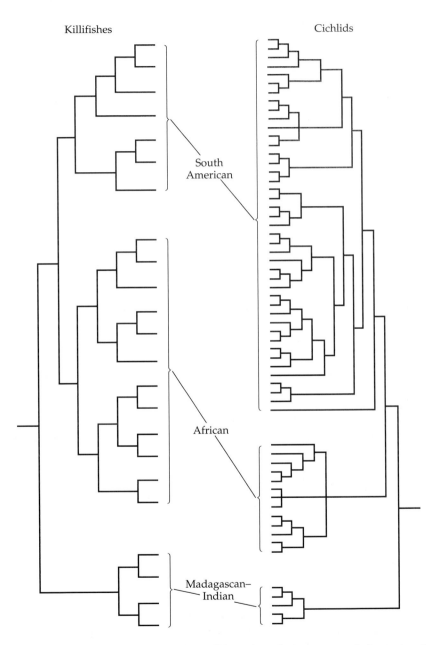

Figure 8.12   Maximum parsimony trees for two speciose groups of circumtropi-cal freshwater fishes: Aplocheiloidei (Murphy and Collier 1997) and Cichlidae (Farias et al. 1999; see also Streelman et al. 1998; Zardoya et al. 1996). Trees were derived from mtDNA sequences. (Modified from Avise 2000a.)

COMMON ANCESTRY VERSUS CONVERGENCE. In molecular PCM analyses, a typical goal is to distinguish between common ancestry and convergent evolution as the source of particular phenotypic features shared between extant taxa. These issues often come into sharp focus in a geographic context. For example, many Australian marsupials bear general morphological or behavioral resemblance to particular placental mammals elsewhere in the world. Examples of such convergences include kangaroos and deer, the Tasmanian wolf and placental carnivores, and bandicoots and rabbits. Such cases of "ecological equivalence" are "known" to register evolutionary convergence because all of the marsupials retain detailed and almost certain signatures of common ancestry, most notably a pouch (marsupium). Furthermore, the suspected monophyly of marsupials, nearly all of which live in the Southern Hemisphere, makes biogeographic sense. Australia became isolated from other landmasses during the early and middle Tertiary period (ca. 30–60 mya), and the marsupials there adaptively radiated to fill many ecological niches occupied by placental mammals elsewhere. Many cladistic details of this evolutionary radiation of marsupials have been worked out using a panoply of molecular methods, including protein immunology (Baverstock et al. 1987; Kirsch 1977), allozyme analyses (Sinclair 2001), DNA hybridization (Kirsch et al. 1990; Springer and Kirsch 1989, 1991; Springer et al. 1990), microsatellite analyses (Pope et al. 2000), and direct sequencing of nuclear and mitochondrial genes (Belov et al. 2002; Colgan 1999; Janke et al. 2002; Springer et al. 1998).

One of the most dramatic findings in early molecular phylogenetics involved what may have been an analogous evolutionary radiation of Australian songbirds. European ornithologists described and named the Australian endemics only after having classified much of the Old World avifauna. Many of Australia's species appeared to fit neatly into taxa previously established. For example, Australian warbler-like birds were placed into Sylviidae (Eurasian warblers), Australian flycatchers into Muscicapidae (Afro-Eurasian flycatchers), treecreepers into Certhiidae (Eurasian-American creepers), and sitellas into Sittidae (Holarctic nuthatches). More than a century later, however, molecular analyses (based initially on DNA hybridization) of hundreds of avian species worldwide indicated that these traditional assignments were incorrect (Sibley 1991; Sibley and Ahlquist 1986, 1990).

Instead, based on the molecular evidence, many Australian songbirds seemed to stem from a common ancestor on the continent, leading Sibley and colleagues to conclude that oscine songbirds of the world (those with a complex syrinx or voice box) constitute two major phylogenetic lines: suborder Passerida, which evolved in Africa, Eurasia, and North America, and suborder Corvida, which originated in the Australian region (Figure 8.13). These findings were remarkable. They overturned conventional taxonomic wisdom for birds, indicated rampant convergence of general phenotype

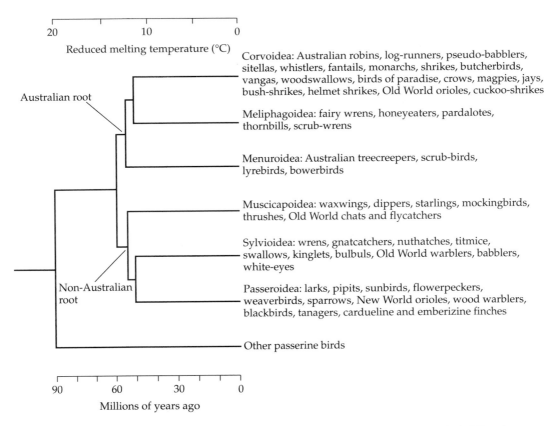

Figure 8.13   Phylogeny of the oscine songbirds based on DNA hybridization data. Two major historical groups were postulated: one (Corvida) tracing to an Australian root (some of whose members, such as crows and jays, secondarily radiated elsewhere in the world) and the other (Passerida) tracing to a non-Australian origin. (After Sibley and Ahlquist 1986.)

between various members of Corvida and Passerida, and suggested that the phylogeographic radiation of a major segment of the Australian avifauna had roughly paralleled that of mammals native to Australia. The general thrust of these phylogenetic suggestions has been supported in subsequent analyses of other types of molecular data (e.g., Ericson et al. 2000, 2002; Lovette and Bermingham 2000), albeit not without particular points of contention (e.g., Barker et al. 2002).

More surprises about the early (e.g., ordinal-level) evolution of avian lineages are emerging from extensive DNA sequence analyses. Much of this molecular effort was prompted by a radical reinterpretation of the avian fossil record by Feduccia (1995, 1996), who hypothesized that all modern birds might trace back to one or a few fortunate lineages that survived the mass extinction at the Cretaceous/Tertiary (K/T) boundary 65 mya. However, recent molecular dating studies appear not to support this bottleneck scenario (Cooper and Penny 1997). Instead, they indicate that many avian lineages now recognized as different taxonomic orders originated in the early to mid-Cretaceous and survived to traverse the K/T boundary (Hedges et al. 1996; van Tuinen and Hedges 2001; van Tuinen et al. 2000; Waddell et al. 1999). These findings also support theories that ancient continental breakups were probably an important driving force in early avian evolution (Ericson et al. 2002; Paton et al. 2002).

Other molecular discoveries have prompted major revisions of thought concerning deep evolutionary branches in the global phylogeny of placental mammals (Waddell et al. 1999). In particular, analyses of many mitochondrial and nuclear genes plus protein sequences indicate that an ancient and previously unrecognized lineage of placental mammals originated and diversified on the African continent (Nikaido et al. 2003; Springer et al. 1997; Stanhope et al. 1998b; van Dijk et al. 2001). This "Afrotheria" clade, which includes such unlikely evolutionary cousins as elephants, hyraxes, aardvarks, golden moles, and tiny elephant shrews (named for their long snouts), appears to be one of about four deep molecular branches of placental mammals. Another branch includes rodents, rabbits, and primates; a third includes bats, hoofed animals, and most carnivores; and the fourth branch is composed of armadillos, sloths, and their allies (Madsen et al. 2001; Murphy et al. 2001). Furthermore, based on molecular clock considerations and earliest fossil appearances, several deep mammalian lineages (and probably many others nested within them) are now believed to have separated well before the end of the Cretaceous (Hedges 2001; Hedges et al. 1996; Kumar and Hedges 1998; Springer et al. 2003). These findings challenge earlier hypotheses that the primary diversification of mammalian lineages occurred after the K/T mass extinction, and they also support theories that continental configurations have been a key driving force in early mammalian evolution (Delsuc et al. 2002).

RECENT ISLANDS, ANCIENT INHABITANTS.   Another spectacular evolutionary radiation studied via molecular markers is that of Drosophilidae flies in the Hawaiian Islands, the native home to an estimated 800 or more species (a remarkable count because the whole archipelago accounts for only 0.01% of Earth's total land area). These flies traditionally are divided into two groups—the drosophiloids and scaptomyzoids—that were postulated to derive from one or two founder populations of unknown continental origin (Throckmorton 1975). From molecular analyses, it now appears likely that all Hawaiian *Drosophila* and *Scaptomyza* form one large monophyletic group (Kwiatowski and Ayala 1999). Geologically, the most ancient of the major islands above water today is only 5 million years old. Was the incredible proliferation of the drosophilid clade in the Hawaiian archipelago truly accomplished within such a short evolutionary time span?

From immunological comparisons of larval hemolymph proteins, Beverley and Wilson (1985) provided some of the first evidence that various lineage separations in these Hawaiian flies vastly predate the volcanic emergence of present-day islands, a contention that soon gained further support from data on nuclear and mitochondrial gene sequences (DeSalle 1992a,b; Thomas and Hunt 1991). Apparently, extant Hawaiian drosophilids stem from a colonist that landed on the archipelago about 30 mya (Kambysellis and Craddock 1997). This paradox has a resolution: The current land archipelago contains merely the latest in a succession of islands dating back more than 70 million years, remnants of which now exist as drowned seamounts or low atolls northwest of the current chain. Thus, according to the molecular findings, many speciations probably occurred tens of millions of years ago on islands no longer in existence as flies island-hopped to newly arisen terrain. Similar speciation processes are continuing today, as evidenced by molecular and other data on particular *Drosophila* subgroups, such as the picture-winged flies (Carson 1976, 1992; Piano et al. 1997). On the other hand, molecular analyses of about 20 other taxonomic groups on the Hawaiian Islands (mostly various native birds and plants) have typically yielded estimates of divergence times falling within the past 5 million years (see review in Price and Clague 2002).

Analogous questions apply to biotas on another volcanic island chain that has figured prominently in evolutionary studies. All of the present-day Galápagos Islands are less than 3 million years old, an age considered by some researchers to place an outer bound on the maximum time over which the exuberant evolution on the archipelago must have taken place (Hickman and Lipps 1985). Another possibility, however, is that many speciation events transpired on former islands before they sank beneath the sea (Christie et al. 1992). Molecular phylogenetic data have helped to decide between these competing possibilities. For example, small genetic distances within the 14-species clade of Darwin's finches (Geospizinae) are consistent with the hypothesis of an in situ evolutionary radiation within the 3-million-year time frame of the modern islands (Grant and Grant 2003; Petren et al. 1999; Polans 1983; Sato et al. 1999, 2001; Yang and Patton 1981). Similarly, molecular findings on the Galápagos tortoises (*Geochelone nigra*) indicate that morphologically diverse populations of this species are monophyletic and just a few million years old, having originated from the Chaco tortoise (*G. chilensis*) of mainland South America (Caccone et al. 1999, 2002).

On the other hand, larger genetic distances suggestive of separation dates of greater antiquity were reported between the marine iguana (*Amblyrhynchus cristatus*) and land iguanas (*Conolophus pallidus* and *C. subcristatus*) of the Galápagos (Rassman 1997; Wyles and Sarich 1983). Perhaps these genera were phylogenetically separated on now-drowned islands (as their relatively ancient yet monophyletic molecular status might suggest), but the possibility that they trace to two or more separate invasions of the islands cannot be excluded. Similar multi-colonization scenarios have been advanced to account for large genetic distances observed among some other lizards and rodents native to the Galápagos (Lopez et al. 1992; Patton and Hafner 1983; Wright 1983). The broader message is that extensive taxonomic sampling and secure molecular phylogenies are often necessary to draw definitive conclusions about the origins and subsequent evolutionary histories of island biotas (Emerson 2002).

Academic pursuit of genealogical roots

In truth, many studies in molecular phylogeny probably are initiated by sheer intellectual curiosity about a particular group's ancestry. Most systematists have a favorite taxon (be it fishes or fungi), the phylogenetic understanding of which can become an obsession. Although it would be presumptuous to choose particular examples in molecular systematics as being of special inherent interest, the Primate assemblage that includes humans stands out as worthy of mention, because no other topic in molecular phylogenetics and evolution has attracted so much attention (Donnelly and Tavaré 1997; Lewin 1999; Tashian and Lasker 1996).

Traditionally, *Homo sapiens* has been classified as the sole extant species in Hominidae, a taxonomic family belonging to the superfamily Hominoidea, which also includes the Asiatic apes [gibbons (*Hylobates*), siamangs (*Symphalangus*), and orangutans (*Pongo*)] and the African apes [gorillas (*Gorilla*) and chimpanzees (*Pan*)]. Our closest relatives outside the Hominoidea clearly are Old World monkeys (Cercopithecoidea). Within Hominoidea, conventional wisdom has been that humans' closest living relatives are the great apes of Africa (Pongidae).

Prior to the availability of molecular data, a popular paleontological scenario was that the lineage leading to humans split from a line leading to gorillas and chimpanzees about 15–30 mya (see Patterson 1987). In 1967, in a stunning report based on immunological assays (Figure 8.14), Sarich and Wilson challenged this belief by concluding that the phylogenetic lineage eventuating in *Homo sapiens* separated from that leading to the African apes only about 5 mya. Furthermore, the African apes might not form a distinct clade, because lineages leading to chimpanzee, gorilla, and human constituted an unresolved phylogenetic "trichotomy" in these immunological analyses.

Nearly 40 years and dozens of molecular studies later, both of these conclusions have been largely vindicated. It is now generally accepted that the proto-human lineage separated about 4–6 mya, and that humans, chimpanzees, and gorillas are related roughly equidistantly to one another (but see below). These conclusions grew from an early consensus of molecular information from protein immunology, protein electrophoresis, amino acid sequencing, DNA hybridization, restriction site analyses, and nucleotide sequencing from mtDNA as well as many nuclear genes and noncoding regions. Some of pioneering studies were by Bruce and Ayala (1979), Caccone and Powell (1989), Goodman et al. (1990), Hasegawa (1990), Miyamoto and Goodman (1990), Nei and Tajima (1985), Sibley et al. (1990), and Williams and Goodman (1989). Prior competing scenarios from mor-

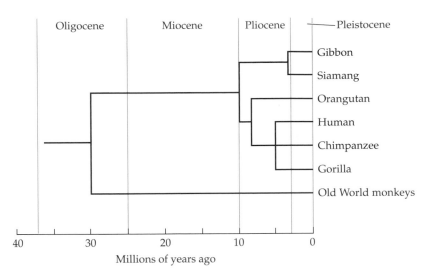

Figure 8.14   One of the first molecular-based estimates of the phylogenetic position of *Homo sapiens* within the primates. This phylogeny, based on immunological distances in albumins, revolutionized thought about human origins, and its major features generally have been confirmed with much additional molecular evidence. (After Sarich and Wilson 1967; see also Goodman 1962.)

phology and fossil evidence were also reinterpreted, at least partially, to accommodate these molecular revelations (Andrews 1987; Pilbeam 1984). Recent genome-scale comparisons of human and chimpanzee sequence data (nuclear and mitochondrial, respectively) can be found in Chen et al. (2001) and Ingman et al. (2000).

Much attention has been focused on resolving the human–chimpanzee–gorilla phylogenetic trichotomy. Early discussions centered on the suitability of various classes of molecular data as well as appropriate methods of statistical analysis (e.g., Kishino and Hasegawa 1989; Nei et al. 1985; Saitou and Nei 1986; Templeton 1983). A consensus gradually emerged favoring a human–chimpanzee clade, to which the gorilla represents a close sister lineage (Hasegawa 1990; Horai et al. 1992; Li and Graur 1991; Ruvolo et al. 1994; Sibley and Ahlquist 1984; Williams and Goodman 1989). This sentiment remains preeminent today, albeit with an added recognition that the temporal closeness of nodes in the phylogenetic "trichotomy" means that some (a minority) of gene trees might actually ally humans with gorillas rather than with chimpanzees (Pääbo 2003). In any event, from the extensive molecular analyses of great ape relationships, at least two salient points have emerged: humans are phylogenetically very close to our primate relatives, and the traditional placement of *Homo sapiens* in a monotypic family (Hominidae) reflects anthropocentric bias more than objective phylogenetic reality.

Some Special Topics in Phylogeny Estimation

Another way to organize thought about the plethora of phylogenetic studies is to focus on the diversity of molecular methodologies employed. This section will highlight several additional pioneering approaches that have had significant effects on phylogenetic analyses of particular taxa, or at particular temporal depths in the evolutionary hierarchy.

DNA hybridization and avian systematics

One of the first mega-analyses in molecular phylogenetics involved a decade-long series of DNA hybridization studies by Charles Sibley and Jon Ahlquist on more than 1,700 avian species representing nearly all of the 171 taxonomic families of birds. The collective result was what became known in ornithological circles as "The Tapestry," an extended estimate of avian phylogeny that in its printed version spanned 42 pages in the authors' summary tome (Sibley and Ahlquist 1990). This work was revolutionary not only for its unprecedented taxonomic sampling, but also for its many provocative conclusions. One of these was the proposed evolutionary radiation of the Australian Corvida, described above. Another concerned New World and Old World vultures, which Sibley and Ahlquist (1990) claimed were not closely related. This result was generally confirmed (albeit with differences in detail) in subsequent direct analyses of cytochrome *b* mtDNA sequences (Seibold and Helbig 1995; Wink 1995), thus implying that vultures and their carrion-feeding lifestyle are polyphyletic in origin.

Several other unorthodox conclusions from The Tapestry are described in Table 8.2. Although Sibley and Ahlquist's approach and conclusions were by no means met with universal approbation (e.g., Cracraft 1992; Cracraft and Mindell 1989; Sarich et al. 1989), their findings continue to warrant serious consideration for several reasons: the DNA hybridization approach is presumably powerful, at least in theory; most branching patterns in The Tapestry do agree well with traditional ornithological thought based on morphological or other evidence; and some (but not all) of even the most controversial aspects of The Tapestry have gained additional support from subsequent direct sequencing analyses of mitochondrial or nuclear genes (see Newton 2003; Sorenson et al. 2003 for additional examples).

Regardless of how these various phylogenetic hypotheses are eventually resolved, an important point regarding this immense effort by Sibley and Ahlquist should not be lost: Their study was among the first sweeping attempts to capitalize explicitly on a "common yardstick" rationale (see Chapter 1) in molecular phylogenetics. These researchers promulgated a standardized molecular metric (ΔT_m, presumably related to evolutionary time) that could serve as a potentially universal measure of the magnitude of evolutionary separation among all avian (as well as other) taxa, be they hummingbirds or ratites. Indeed, Sibley and Ahlquist went further by advo-

Provocative conclusion from DNA hybridization	Follow-up molecular support
Loons (Gaviiformes) and grebes (Podicipediformes) are not close evolutionary relatives.	Hedges and Sibley 1994
Game birds (Galliformes) and waterfowl (Anseriformes) are one another's closest living relatives, albeit anciently separated (ca. 90 mya)	Mindell et al. 1997; van Tuinen and Hedges 2001
New World barbets (in Capitonidae) are evolutionarily closer to toucans (Ramphastidae) than to Old World barbets.	Lanyon and Hall 1994
The traditional order Pelecaniformes is not monophyletic, but instead includes at least 3–4 highly distinct clades, often with closest living relatives elsewhere.	Hedges and Sibley 1994; Siegel-Causey 1997
The American wrentit (*Chamaea fasciata*) is related to Old World *Sylvia* warblers.	Shirihai et al. 2001
New World quails (Odontophoridae) are a sister clade to the pheasants rather than being embedded within the Phasianidae.	Kornegay et al. 1993

Note: Each example listed here has gained some support from subsequent molecular analyses often based on direct DNA sequencing. However, not all of Sibley and Ahlquist's conclusions have been upheld by subsequent DNA sequence data. For example, some of their proposed relationships among species of Gruiformes now appear to be questionable (Houde et al. 1997), as do particular details concerning relationships of some of the passerine species in the Australian corvid radiation (see text).

cating that these metrics be directly translated into a universal taxonomy based on "categorical equivalency." For example, they suggested that species should be classified at the family level when they exhibit a ΔT_m in the range of 9–11°C, and at the sub-ordinal level when $\Delta T_m = 18$–20°C (Table 8.3). Whether or not this translation, or perhaps others based directly on evolutionary time (see the section "Toward a Global Phylogeny ...", p. 460), eventually is adopted, the comparative molecular perspectives promoted by Sibley and Ahlquist will remain an important historical landmark along any path that eventually may lead to a universally standardized systematics.

Mitochondrial DNA and the higher systematics of animals

In addition to its great utility as a microevolutionary marker at the intraspecific level (see Chapter 6), mtDNA also is employed widely as an informative guide to phylogenetic relationships among higher animal taxa. For these latter purposes, sequence regions that evolve more slowly than average or unique structural alterations in the molecule are monitored.

TABLE 8.3 Suggested levels of genetic divergence (as measured by DNA hybridization) to be associated with indicated levels of taxonomic recognition in birds, as per recommendations of Sibley et al. (1988)

Taxonomic category	Suffix	ΔT_m range	Example
Class	—	31–33	Aves
Subclass	-ornithes	29–31	Neornithes
Infraclass	-aves	27–29	Neoaves
Parvclass	-ae	24.5–27	Galloanserae
Superorder	-morphae	22–24.5	Anserimorphae
Order	-iformes	20–22	Anseriformes
Suborder	-i	18–20	—
Infraorder	-ides	15.5–18	Anserides
Parvorder	-ida	13–15.5	—
Superfamily	-oidea	11–13	—
Family	-idae	9–11	Anatidae
Subfamily	-inae	7–9	Anatinae
Tribe	-ini	4.5–7	Anatini
Subtribe	-ina	2.2–4.5	—
Congeneric spp.	—	0–2.2	*Anas* (puddle ducks)

EXTENSIVE MtDNA SEQUENCES.   One common approach in higher-animal phylogenetics involves direct sequence comparisons of relatively slowly evolving mtDNA loci, or portions thereof, such as transversions or non-synonymous substitutions in protein-coding regions. Several examples already have been presented above (e.g., regarding bats, crabs, whales, coelacanths, and blackbirds; see also Table 8.1), and more will follow. Especially popular have been mitochondrial genes encoding rRNA subunits (e.g., Colgan et al. 2000; Sullivan et al. 1995), cytochrome oxidases (Hebert et al. 2003; Sena et al. 2002), and cytochrome *b* (Lydeard and Roe 1997). However, none of these or other mtDNA sequences are without shortcomings for phylogeny estimation (Meyer 1994b; Naylor and Brown 1998), and each has a restricted temporal window of resolution (Moore and DeFilippis 1997).

As sequencing procedures have become streamlined and automated, a growing approach is to conduct "brute force" analyses of whole (ca. 16 kb each) mtDNA genomes (e.g., Kumazawa and Nishida 1999; Mindell et al. 1999). In a remarkable example of this approach, Miya et al. (2003) gathered and phylogenetically analyzed complete mtDNA sequences from 100 species representing 74 taxonomic families in 26 orders of teleost fishes. For any species, such analyses exhaust the information content of the mtDNA genome, which nonetheless represents only a single "gene" from a phylogenetic viewpoint, and therefore only a tiny fraction of the hereditary history in any organismal phylogeny.

ECCENTRIC MtDNA MARKERS.   The DNA hybridization studies discussed earlier, by providing a numerical "average" genetic distance over a large but unspecified number of low-copy genes, represent one end of a continuum of empirical and philosophical approaches in molecular systematics. Near the other end of this spectrum are analyses that focus on much smaller numbers of molecular features that nonetheless could be of special cladistic relevance by virtue of supposed singularities of evolutionary origin.

One such potential suite of characters involves gene arrangements and other unusual compositional properties of mtDNA. The same ensemble of about 37 genes constitutes the mtDNA of most animal species, but gene orders vary somewhat (Hoffmann et al. 1992; Hyman et al. 1988; Lee and Kocher 1995; Macey et al. 1997; Quinn and Mindell 1996; Sankoff et al. 1992; Wolstenholme et al. 1985). For example, although identical mtDNA gene arrangements are displayed by most placental mammals, amphibians, and fishes, the molecule's tRNA clusters are reordered relative to the main vertebrate theme in several surveyed marsupials (Janke et al. 1994; Pääbo et al. 1991). Likewise, gene order in quail was shown to differ from the vertebrate norm, in this case by a transposition of five loci (Desjardins and Morais 1991). In another example, alternative mtDNA gene orders distinguish two primary groups of songbirds (oscines and suboscines) that also are well demarcated by morphological and other molecular evidence (Mindell et al. 1998). A review of mitochondrial genome organization in vertebrates is provided by Pereira (2000).

Distinctive mtDNA gene arrangements have also been employed as phylogenetic markers in fungi (Bruns and Palmer 1989; Bruns et al. 1989) and in a wide variety of invertebrate animal taxa (Arndt and Smith 1998b; Boore and Brown 2000; von Nickisch-Rosenegk et al. 2001). Particular gene orders have suggested, for example, that the following groups of Arthropoda are monophyletic: the phylum as a whole; Mandibulata (insects, myriapods, and crustaceans); and a subclade composed of insects plus crustaceans (Boore et al. 1995, 1998). Echinodermata is another invertebrate phylum for which mtDNA gene arrangements have added important clues to the cladogram. This assemblage is traditionally divided into five classes: Asteroidea (sea stars), Echinoidea (sea urchins), Ophiuroidea (brittle stars), Holothuroidea (sea cucumbers), and Crinoidea (crinoids). Fossil evidence suggests that these taxa split from one another early in the Paleozoic (albeit in controversial order; A. B. Smith 1992). In molecular studies reviewed by M. J. Smith et al. (1993), mtDNA gene arrangements in ophiuroids and asteroids proved to be similar, but contrasted sharply with the layout common to echinoids and holothuroids (crinoids were not assayed). Outgroup rooting (against vertebrates) suggested that the asteroid-ophiuroid condition is synapomorphic, thus defining a clade.

Because mtDNA gene rearrangements are evolutionarily rather rare, particular gene orders might therefore seem to provide ideal clade markers, but such is not invariably the case. In their survey of mtDNA genomes in 137 species representing 13 avian taxonomic orders, Mindell et al. (1998) discovered that one novel gene arrangement had originated independently in woodpeckers (Picidae), cuckoos (Cuculidae), songbirds (Passeriformes), and birds of prey (Falconiformes). The authors concluded that although mtDNA gene arrangements offer excellent power for clade delineation, it is also true that "gene order characters appear susceptible to some parallel evolution because of mechanistic constraints." Several of these kinds of mechanisms and constraints have been illuminated (Lavrov et al. 2002; Macey et al. 1998).

For a few invertebrates, another phylogenetically useful feature of the mtDNA genome is its linear (not circular) condition. Within the phylum Cnidaria (corals, anemones, jellyfishes, and allies), ascertainment of relationships among taxonomic groups has posed a classic phylogenetic challenge due to a paucity of phenotypic characters that are independent of dramatic differences in the life cycles of various forms. A surprising molecular discovery was that all surveyed members of the classes Cubozoa, Scyphozoa, and Hydrozoa possess linear mtDNA, as opposed to the circular mtDNA of Anthozoa, Ctenophora (a supposed outgroup), and most other surveyed metazoan phyla (Bridge et al. 1992). Thus, the derived (linear) condition appears to define a clade. This finding (through its incorporation into PCM exercises) implied that a benthic polyp stage, rather than a pelagic medusa form, probably came first in cnidarian evolution. The mtDNA genomes of various metazoan groups also show other types of evolutionary novelties, such as differences in tRNA secondary structures, sizes of rRNA genes, peculiar features of the control region, and presence of specific introns. These features also can provide phylogenetic signals (Beagley et al. 1998; Wolstenholme 1992).

One especially intriguing set of mtDNA features involves codon assignments in protein translation (reviewed in Knight et al. 2001a,b). Following the initial discovery of modified codon assignments in mammalian mtDNA (Barrell et al. 1979), it was found that genetic codes in mitochondria not only depart from the "universal" code of nuclear genomes, but also vary across taxa. Wolstenholme (1992) summarized available data on mtDNA codon assignments in 19 metazoan animals representing six phyla and superimposed the observed differences on a suspected phylogeny (Figure 8.15). Several of the patterns made phylogenetic sense. For example, in the mtDNAs of all assayed invertebrate phyla (except Cnidaria), AGA and AGG specify serine, whereas they are stop codons in vertebrate mtDNA. Thus, based on this apparent shared derived condition, vertebrates form a clade (see also Boore et al. 1999). On the other hand, evidence for homoplasy in codon assignments was also uncovered. The best example involved the nucleotide triplet AAA, which in two supposedly unrelated phyla (Echinodermata and Platyhelminthes) may have convergently evolved to specify asparagine, not lysine as in most other metazoan mtDNAs and in the universal code (Figure 8.15). Also, in an ancestor of echinoderms, an apparent reversion may have occurred to the presumed ancestral condition in which ATA specifies isoleucine rather than methionine. Finally, the AAA codon was later discovered to be missing altogether in hemichordates (Castresana et al. 1998).

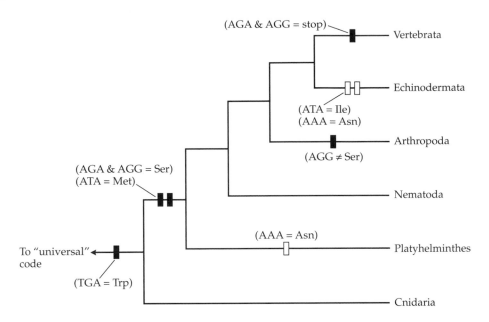

Figure 8.15 One evolutionary scenario for alterations of the genetic code in metazoan mtDNA. Observed departures from the "universal genetic code" are plotted on a provisional phylogeny for metazoan animals based on other evidence. Solid bars across branches indicate probable synapomorphies marking various clades; open bars indicate probable instances of evolutionary convergence or reversal. (After Wolstenholme 1992.)

Although studies of mtDNA genome arrangements have clearly been phylogenetically informative, they also have amplified a cautionary note regarding all attempts to define clades by idiosyncratic genetic markers considered individually. No matter how secure a synapomorphy might appear, the possibility of homoplasy seldom can be eliminated entirely. Thus, before definitive phylogenetic conclusions are drawn, confirmation from multiple independent sources of information should always be attempted.

Chloroplast DNA and the higher systematics of plants

What the mtDNA molecule has done for animal higher systematics, cpDNA has analogously done for plants. The chloroplast genome is well suited for plant phylogenetic analyses for several reasons: it is abundant in plant cells and essentially ubiquitous taxonomically; much background information is available to facilitate experimental and comparative work; it often houses distinctive structural features of cladistic utility; and its general pace of nucleotide substitution is moderate to slow (Clegg and Zurawski 1992). Again, researchers use two distinct phylogenetic approaches (Olmstead et al. 1990): monitoring of taxonomically idiosyncratic molecular features of cpDNA, the rationale being that these features should be especially power-ful for clade identification; and sequencing of specific cpDNA genes or regions. Sequences cumulatively provide vastly larger pools of genetic markers, but on the other hand, character states at individual nucleotide positions can be afflicted with higher probabilities of homoplasy.

EXTENSIVE cpDNA SEQUENCES. In 1993, a landmark paper with 42 authors offered a compendium appraisal of seed plant phylogeny based on a huge database of nucleotide sequences from the *rbcL* gene (Chase et al. 1993), which encodes the large polypeptide subunit of ribulose-1,5-biphosphate carboxylase. This chloroplast gene became a favored sequencing target for several reasons: much comparative information had accumulated for it early on (Baum 1994; Duvall et al. 1993; Gaut et al. 1992; Kim et al. 1992; Soltis et al. 1990); the locus is large (>1,400 bp) and provides many phylo-genetically informative characters; and its rate of sequence evolution proved appropriate for addressing questions about plant phylogeny, especially at intermediate and higher taxonomic levels. In subsequent years, sequences from additional species and several other genes, notably another plastid locus (*atpB*), a nuclear gene encoding 18S rRNA, and some slowly evolving mitochondrial genes, were added to the mix (Parkinson et al. 1999; Qui et al. 1999; Soltis et al. 1998, 1999).

Summary gene sequence phylogenies for hundreds of seed plant species are shown in Figure 8.16. Many of the clades had been identified in earlier studies, but seldom with the high levels of statistical support that the combined genetic data provided. The following are among many notable conclusions reached for angiosperms (flowering plants) (Figure 8.16A): eudicots form a large monophyletic group, as do angiosperms at a higher clade level; magnoliids, which include monocots, form a distinct clade that branched off early in angiosperm evolution; plants that engage in nitrogen-fixing symbioses with nodulating bacteria form a subclade within the eurosid I clade; and many "model species" that are used widely in genetic and evolutionary research, such as *Arabidopsis* (mustards), *Brassica* (cab-bages and allies), and *Gossypium* (cotton), lie within the rosid clade and, hence, encompass only a tiny fraction of total phylogenetic diversity in seed plants. Several molecular conclusions disagreed with conventional wisdom, as evidenced by the fact that some DNA-based clades within the core eudi-cots cut across subclass boundaries in previous classifications (Soltis et al. 1999). The molecular findings thus have given impetus to a revision of angiosperm ordinal classifications (Nyffeler 1999).

Gymnosperms (naked-seeded plants) have been phylogenetically ana-lyzed in similar fashion (Bowe et al. 2000; Chaw et al. 2000). Some primary conclusions are as follows (Figure 8.16B): extant gymnosperms are mono-phyletic; cycads are the most basal clade of gymnosperms, followed by *Ginkgo*; and members of the long-problematic order Gnetales are allied to conifers. Also based on molecular evidence, two other major groups of vas-cular plants—ferns and club mosses (or lycophytes)—probably branched off

(A)

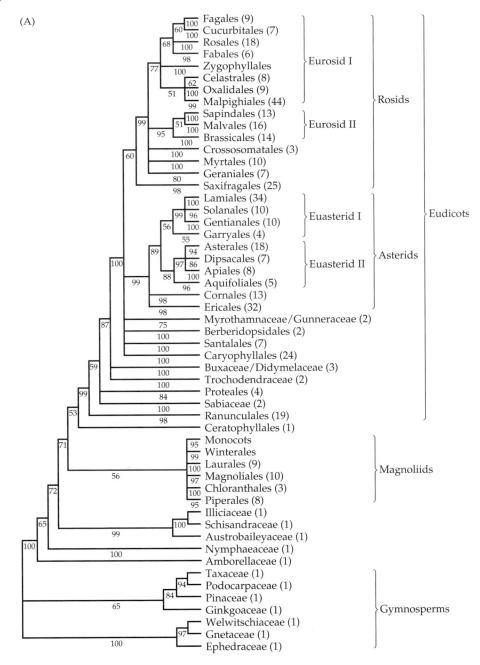

Figure 8.16   Consensus phylogenies for representative seed plants based on DNA sequence data from multiple cytoplasmic and nuclear loci. (A) Molecular phylogeny primarily for angiosperms. Numbers in parentheses are tallies of species examined (560 total, but not all are pictured). (After Soltis et al. 1999.) (B) Molecular phylogeny emphasizing relationships among gymnosperms. (After Chaw et al. 2000). In both diagrams, numbers on branches indicate levels of statistical support for various clades.

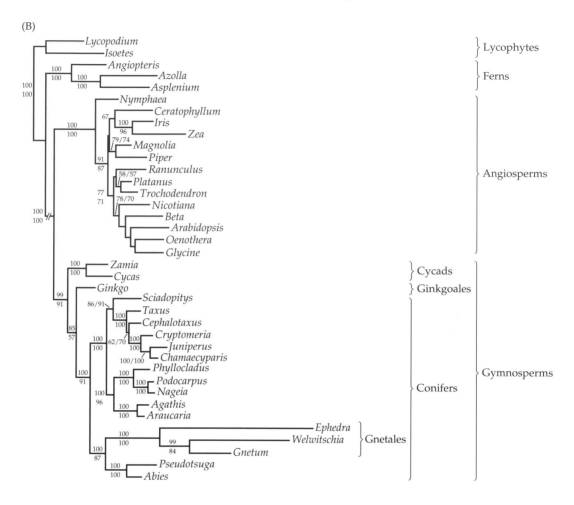

(B)

prior to the "seed plant" clade (gymnosperms plus angiosperms). Apart from these analyses of vascular plants, molecular data have also contributed to knowledge of the deeper phylogeny of land plants, which also include mosses, hornworts, and liverworts. Emerging results from this vast phylogenetic effort are summarized and periodically updated on the Deep Green Web site (ucjeps.berkeley.edu/TreeofLife/hyperbolic.php).

One advantage of sequence analyses is that general estimates of divergence times (as well as branching patterns) can be attempted, using appeals to molecular clock considerations (Olmstead et al. 1992; Wolfe et al. 1989). In taking advantage of this utility, Wikström et al. (2001) used molecular findings in conjunction with fossil-based evidence to estimate geologic times for more than 500 nodes in the angiosperm tree. For example, they estimated that the crown group of extant angiosperms arose in the Early to Middle Jurassic (about 179–158 mya), and that the eudicot lineage originated in the Late Jurassic to mid-Cretaceous (147–131 mya).

It has not gone unnoticed (Rieseberg and Brunsfeld 1992) that molecular conclusions in plant higher systematics have been based to a large degree on the same molecule (cpDNA) for which reticulate evolution due to introgression has been documented between several closely related plant species (see Chapter 7). Could this phenomenon also have led to discrepancies between cpDNA gene trees and organismal trees in deep plant phylogeny? In theory, yes, but probably seldom in practice, according to Clegg and Zurawski (1992): "It is reasonable to assume that the approximation to organismal history will improve as time increases, because the biases introduced by interspecific hybridization or intraspecific polymorphism will diminish with an increase in time scale." Nevertheless, plant phylogenies estimated from cpDNA sequences (or any other single-gene genealogies) always remain provisional pending corroboration from additional loci.

ECCENTRIC cpDNA MARKERS.   Phylogeny estimation in plants has also benefited from relatively unusual or idiosyncratic features of cpDNA. Such phylogenetic markers include inversions, losses of genes and introns, and losses of a large inverted repeat region of the molecule (Downie and Palmer 1992).

With some notable exceptions, such as those involving *Pisum* (Palmer et al. 1988b), *Trifolium* (Milligan et al. 1989), and conifers (Strauss et al. 1988), gene order is normally a conservative feature of cpDNA in vascular plants. This is illustrated by the fact that the arrangement of cpDNA genes in tobacco (*Nicotiana tabacum*) is similar to the presumed ancestral order also found in most other examined angiosperms, ferns, and *Ginkgo* (Palmer et al. 1988a). Furthermore, when gene order differences have been discovered, one or a few inversions usually appear to be responsible. For example, cpDNAs from bryophytes, mosses, and lycopsids differ from those of most vascular plants by a 30-kb inversion (Calie and Hughes 1987; Ohyama et al. 1986; Raubeson and Jansen 1992a), one of the few large structural alterations in cpDNA accepted over the hundreds of millions of years of evolution involved. This inversion appears to be a synapomorphy for vascular plants minus the lycopsids, and it provisionally identifies a basal evolutionary split among vascular plants.

One of the first and most comprehensive phylogenetic studies of a cpDNA rearrangement involved a 22-kb inversion found to be shared by 57 genera representing all tribes of Asteraceae (sunflowers), a taxonomic family with more than 20,000 species in 1,100 genera (Jansen and Palmer 1987). Absence of this inversion from the subtribe Barnadesiinae of the Mutisieae tribe, and from all families allied to Asteraceae, suggested that Barnadesiinae represents the most basal lineage of Asteraceae and that, contrary to earlier opinion, Mutisieae is not monophyletic. These conclusions were supported by congruent results obtained from phylogenetic analyses of cpDNA restriction sites (Jansen and Palmer 1988) and sequences (Kim et al. 1992; see review in Jansen et al. 1992).

Unlike the typical case for mtDNA in most higher animals, cpDNA genes often carry numerous introns (Shinozaki et al. 1986). Losses of entire introns (in contrast to oft-observed changes in intron length) seem to be relatively rare events, and hence are phylogenetically informative when they are observed. For example, the absence of an intron in the *rpl2* gene marked all examined members of Caryophyllales (Downie et al. 1991). Caution in the use of such markers was also indicated, however, because this particular intron apparently was lost independently in at least five unrelated dicot lineages (Downie et al. 1991). Apart from introns, evolutionary losses of particular cpDNA genes (perhaps often to the nucleus; see below) also have been employed as phylogenetic signals (Downie and Palmer 1992).

Most land plant cpDNAs possess a large (20–30-kb) inverted repeat (IR) region. A shared deletion of one copy of this repeat indicated probable monophyly for six tribes in the subfamily Papilionoideae (Lavin et al. 1990). Although this character state ("IR loss") is phylogenetically informative locally, it again could be a misleading guide to global phylogeny because IR losses seem to have occurred independently in more than one plant group (Doyle et al. 1992). Conifers, for example, also possess only one IR element (Lidholm et al. 1988; Raubeson and Jansen 1992b).

In general, convergent evolutionary gains of rare features are even less likely than convergent losses (although both types of events can be troubled by homoplasy). Hence, the shared possession of de novo genomic additions should be of special significance in clade delineation. The phylogenetic origin of embryophytes (land plants) has long intrigued botanists. One candidate sister group was Charophyceae (a particular group of green algae), a possibility consistent with early molecular findings on cpDNA. All previously examined algae, as well as eubacteria, lacked introns in their tRNA^{Ala} and tRNA^{Ile} genes, whereas all assayed embryophytes, as well as charophyceaens, possessed them. This observation was interpreted to indicate the evolutionary acquisition of a genetic novelty, perhaps marking a clade of land plants plus Charophyceae (Manhart and Palmer 1990). However, a later discovery was that mosses, hornworts, and all major vascular plant lineages share *three* mtDNA introns that are not possessed by green algae or other eukaryotes. This discovery was taken as refined evidence that even better candidates for the closest ancestors of land plants exist (Qui et al. 1998). In general, at the time of this writing, the whole topic of early plant (and eukaryote) origins is under intense discussion and possible revision (e.g., Cavalier-Smith 2003; Cavalier-Smith and Chao 2003; Nozaki et al. 2003).

Ribosomal gene sequences and deep phylogenies

Phylogenetic analyses of slowly evolving rDNA sequences have also been hugely important in resolving deep and intermediate branches in the Tree of Life. The main function of rRNAs is protein synthesis, so it is not surprising that genomes of all organisms contain sequences coding for these essential molecules (Wheelis et al. 1992). Starting mostly in the late 1980s, many researchers began to take advantage of the phylogenetic markers provided by these genes at meso- and macroevolutionary levels (e.g., Edman et al. 1988; Fuhrman et al. 1992; Gerbi 1985; Hamby and Zimmer 1992; Hedges et al. 1990; Hillis and Dixon 1989, 1991; Hori et al. 1985; Jorgensen and Cluster 1988; Kumazaki et al. 1983; Mindell and Honeycutt 1990; Nairn and Ferl 1988; Sogin et al. 1989). The number of such studies has grown explosively since then, a fact evidenced, for example, by the journal *Molecular Phylogenetics and Evolution*, in which 346 of 965 articles (36%) published during its first 11 years (1992–2002) included phylogenetic analyses of rRNA genes.

A few examples of this approach, as applied to early animal lineages, are listed in Table 8.4, and Figure 8.17 illustrates the impressive amount of historical information that can be recovered from even a single ribosomal gene (in this case, the sequence encoding the 18S rRNA subunit in diverse metazoan lineages). On the other hand, it is also understood that sequence data from individual genes (ribosomal or otherwise) can be misrepresentative or even "positively misleading" (Felsenstein 1978b) regarding ancient organismal relationships. One well-known danger is "long-branch attraction" (Hendy and Penny 1989), in which unrelated lineages falsely appear to constitute ancient clades because of backward, parallel, and convergent nucleotide substitutions that tend to accumulate over long evolutionary time and increase the ratio of phylogenetic noise to signal. Later sections of this chapter will suggest further reasons to retain a cautious view of phylogenies estimated from nucleotide sequences of just one or a few genes.

Genomic Mergers, DNA Transfers, and Life's Early History

From Greek antiquity to recent times, a common notion was that living organisms could be divided into two kingdoms: animals and plants. Following invention of the microscope and the discovery of microbes, a different view came into vogue, which subdivided all of life into prokaryotes (microorganisms lacking a membrane-bound nucleus) and eukaryotes (organisms consisting of cells with true nuclei). Later, in the 1960s, a proposal by Whittaker (1959) became popular, in which five primary kingdoms were recognized: prokaryotes, unicellular eukaryotes, fungi, plants, and animals.

A breakthrough occurred in 1977 when Carl Woese and colleagues analyzed 16S rRNA gene sequences and concluded that all living systems should be divided in a different fashion, along what appeared to be distinct phylogenetic lines of descent (Fox et al. 1977, 1980; Woese 1987; Woese and Fox 1977). Their scenario (Figure 8.18) proposed that all forms of life could be classified into three "domains" above the rank of kingdom: Eucarya (eukaryotic organisms, or at least the nuclear component of their cells); Bacteria (previously Eubacteria); and Archaea (formerly Archaebacteria), including methanogens and thermophilic forms. This view was generally supported by sequence analyses of additional rRNA genes (Gouy and Li

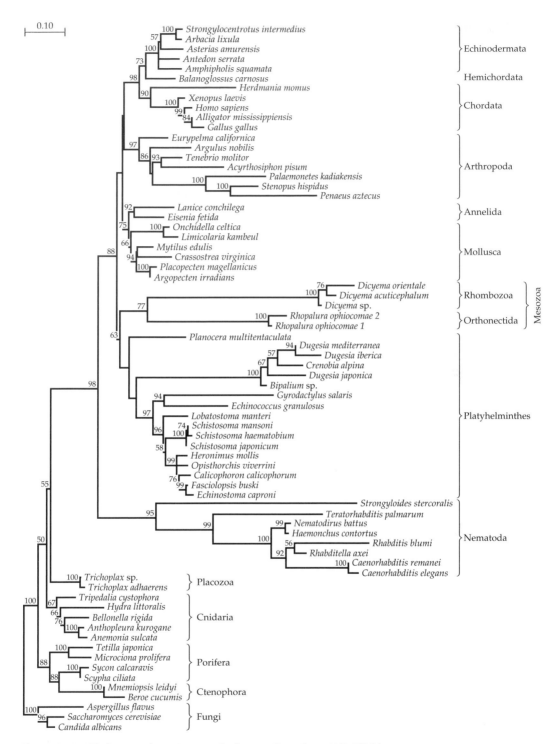

Figure 8.17 Estimate of metazoan phylogeny based on 18S rDNA sequences.
Numbers on branches indicate statistical support for putative clades. This diagram
is intended merely to illustrate rDNA approaches and should not be interpreted as
definitive regarding metazoan relationships. (From Winnepenninckx et al. 1998.)

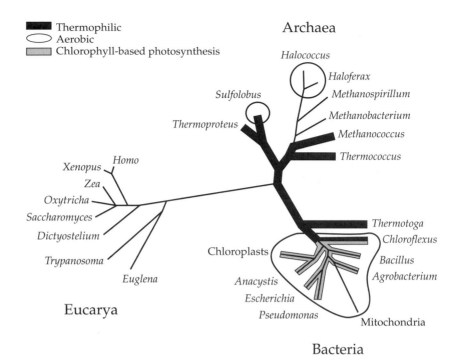

Figure 8.18   One of the first extensive reconstructions of deep phylogenetic topology in the Tree of Life, inferred by Woese and colleagues from sequences of small-subunit rRNA genes. Codings indicate phylogenetic distributions of three physiological traits as indicated by their occurrence within at least some members of the taxonomic groups pictured.

TABLE 8.4	Recent examples of phylogenetic assessments of various animal groups that included or focused on sequences from rDNA (usually the 18S nuclear-encoded subunit)
Group	**Reference**
Invertebrates	Field et al. 1988
Deuterostomes	Cameron et al. 2000
Deuterostomes, chordates	Winchell et al. 2002
Molting animals	Aguinaldo et al. 1997
Protostomes	Mallatt and Winchell 2002
Bilaterians	Peterson and Eernisse 2001
Ctenophora	Podar et al. 2001
Metazoans	Bromham et al. 1998; Giribet 2002
Basal groups	Medina et al. 2001
Reptiles	Hedges and Poling 1999
Jawless fishes	Mallatt and Sullivan 1998
Bony fishes	Obermiller and Pfeiler 2003
Bony fishes	Chen et al. 2002
Cartilaginous fishes	Douady et al. 2003

1989) and other loci (see below). The molecular differences separating extant members of these lineages appeared to be "of a more profound nature than the differences that separate typical kingdoms, such as animals and plants" (Woese et al. 1990).

One immediate question of interest concerned where the root might belong on the Eucarya–Bacteria–Archaea tree. This issue was addressed not only using rRNA genes (e.g., Hori and Osawa 1987), but also from a cladistic standpoint by examining distributions of shared derived features post-dating the separations of two primordially duplicated genes: those encoding elongation factors and ATPase subunits (Gogarten et al. 1989; Iwabe et al. 1989). All of these studies concluded that Archaea and Eucarya constitute sister lineages within a larger clade, such that the root of life lies somewhere within the Bacteria. This view is still favored today, but not without major qualifications relating to ancient intergroup mergers and lateral DNA transfers (Ribeiro and Golding 1998; see following sections).

Although some competing interpretations of Woese's data and the deep-branching structure of life were voiced early (see Day 1991; Lake 1991), there could no longer be any doubt that major phylogenetic lineages, previously unrecognized, exist among prokaryotic life. With regard to taxonomy, some authors nonetheless contended that the differences in levels of biological organization between prokaryotes and eukaryotes remain so profound that continued recognition of these two traditional assemblages is desirable (Mayr 1990, 1998). Other researchers retorted that a failure to formalize Archaea and Bacteria as higher taxa would perpetuate an artificial and flawed classification that disregards phylogeny (Woese 1998a; Woese et al. 1991). This debate raises a general question: Should classifications strictly reflect branching struc-

Authors' conclusions

Cnidarians are separate from other animal lineages
Echinoderms + hemichordates form a clade, as do chordates
Echinoderms + hemichordates form a clade; lancelets sister to chordates
Arthropods, nematodes, and other "molters" form a clade
Supported conclusions of molting study above; identified subclades
Annelids group with mollusks and other taxa with spiral cleavage
Comb jellies are monophyletic, related most closely to cnidarians
Early metazoan phylogeny evaluated against supposed Cambrian explosion
Bilateria + Cnidaria + Ctenophora + Metazoa form a clade
Crocodilians and turtles form a clade; squamates at base of reptile tree
Cyclostomes (lampreys and hagfishes) are monophyletic
Fishes with leptocephalus larvae are not demonstrably monophyletic
Main lineages of higher teleosts identified
Sharks form a clade; rays and skates are basal to elasmobranch lineage

tures in evolutionary trees, or should perceived grades of organismal resemblance somehow be incorporated as well? This is a subjective (albeit an operationally important) question whose answer depends on one's view of the nature of information that a formal classification should convey.

Regardless of their taxonomic implications, the original studies by Woese and colleagues stimulated molecular examinations of many more taxa and genes in efforts to clarify life's deep phylogeny (for updated reviews, see Olsen and Woese 1997; Wolf et al. 2002). So too did a proposal by Cavalier-Smith (1983) to formally recognize an assemblage (Archezoa) of single-celled eukaryotes hypothesized to be so old that they might have predated the endosymbiotic origins of mitochondria and chloroplasts (see next section). Subsequent molecular research soon intimated that several eukaryotic protists, such as diplomonads, entamoebas, euglenids, slime molds, and trichomonads, do indeed encompass incredible phylogenetic diversity (Knoll 1992; Sogin 1991), with many lineages probably tracing back independently across billion-year time frames (Knoll 1999). Recent molecular findings also indicate, however, that at least some of these "Archezoa" lost their organelle genomes secondarily, after the endosymbiotic mergers (Roger 1999). In any event, various lineages of eukaryotic protists are certainly extremely ancient, as illustrated by a recently published consensus phylogeny for eukaryotes (Figure 8.19). To appreciate the relative divergence scales involved, note in particular the placement of "animals" (nested inside the opisthokonts) within this broader genealogical framework.

From ancient endosymbioses to recent intergenomic transfers

ORIGINS OF EUKARYOTIC CELLS.   Traditionally, two hypotheses were advanced to account for the distinctive nuclear and cytoplasmic genomes (mtDNA and cpDNA) found in modern eukaryotic cells. One hypothesis (discussed in Cavalier-Smith 1975; Uzzell and Spolsky 1981) stipulated that organelle genomes arose autogenously within eukaryotes as fragments from the nuclear genome were incorporated into membrane-encased mitochondria or chloroplasts. The competing hypothesis stipulated that organelle genomes had exogenous origins, stemming from bacteria that invaded (or were engulfed by) proto-eukaryotic host cells bearing precursors of the nuclear genome (Margulis 1981, 1995). Beginning mostly in the 1980s, this "endosymbiont theory" received considerable molecular support from phylogenetic analyses of several genes and gene families (Bremer and Bremer 1989; Gray et al. 1989; Howe et al. 1992; Kishino et al. 1990; Lockhart et al. 1992; W. Martin et al. 1992; Schwartz and Dayhoff 1978; Van den Eynde et al. 1988; Van de Peer et al. 1990; Villanueva et al. 1985).

The type of molecular evidence supporting the endosymbiont theory is illustrated by cytoplasmic small-subunit rRNA genes, which show much closer phylogenetic affinities to rDNA sequences in Bacteria than they do to nuclear rDNA sequences in their own eukaryotic host cells (see Figure 8.18). Specifically, it was shown that representative mitochondrial rDNAs phyloge-

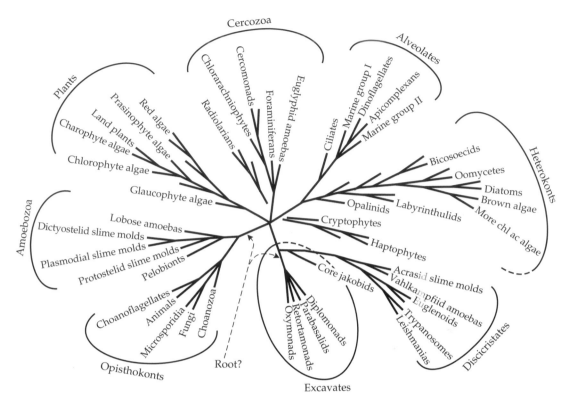

Figure 8.19   Deep phylogeny for eukaryotes. This diagram represents a (highly provisional) consensus picture from extensive molecular data (e.g., Baldauf et al. 2000; Hirt et al. 1999; Katz and Sogin 1999; Keeling et al. 2000) and ultrastructural morphology. (After Baldauf 2003.)

netically group with those of the α-subdivision of purple proteobacteria (Cedergren et al. 1988; Yang et al. 1985), and that chloroplast rDNAs group with those of photosynthetic cyanobacteria (Giovannovi et al. 1988). Likewise, analyses of several other genes (e.g., Cammarano et al. 1992; Morden et al. 1992; Pühler et al. 1989; Recipon et al. 1992) supported an alliance of mtDNA and cpDNA with Bacteria rather than Eucarya, and also confirmed the overall distinctness of Archaea (as first proposed by Woese and colleagues).

Alternative viewpoints persist about the finer details of the original or primary endosymbiotic events. For example, some authors argue for a polyphyletic origin of chloroplasts (Stiller et al. 2003), whereas most other researchers interpret available molecular and other evidence as indicative of ultimate monophyly for all plant plastids, albeit probably with some secondary symbiotic transfers and many genomic rearrangements at later times (Palmer 2003; Turmel et al. 2002). There seems to be less doubt that mitochondria arose singularly, near the base of eukaryotic evolution (Palmer 2003).

In any event, further data from more genes and protein sequences have evidenced an even more complicated and interesting phylogenetic history than is suggested in Figure 8.18. For example, eukaryotic nuclear genes encoding proteins involved in transcription and translation often cluster phylogenetically with Archaea, whereas those for metabolic proteins often tend to cluster with Bacteria (Rivera et al. 1998). Furthermore, the metabolic proteins of eukaryotes often group with those of α-proteobacteria, and those of plants often group with cyanobacteria. These results were again interpreted to reflect endosymbiotic origins of mitochondria and plastids from these respective microbial assemblages, but with the mergers followed by functional transfers of various symbiont genes to host nuclei (Golding and Gupta 1995; Gupta 1998; Lang et al. 1999; Margulis 1996).

GENETIC COMMERCE BETWEEN CELL ORGANELLES AND NUCLEI. Post-endosymbiotic transfers of cytoplasmic DNA to the nucleus had long been suspected from the observation that modern mitochondrial and chloroplast genomes house only a small subset of the genes required for their own replication and expression, with complementary functions being encoded by nuclear DNA (Gray 1992; Wallace 1986). Focused studies of specific DNA sequences (e.g., Kubo et al. 1999), as well as computer database searches (e.g., Blanchard and Schmidt 1995), have confirmed earlier suggestions that physical transfers (recent as well as ancient) of genetic material between organelle and nuclear genomes have been relatively common during evolution (Baldauf and Palmer 1990; Baldauf et al. 1990; Fox 1983; Gantt et al. 1991; Gellissen et al. 1983; Nugent and Palmer 1991; Stern and Palmer 1984). These events are sometimes referred to as intracellular gene transfers (IGT).

In animals, most inter-genomic DNA movements that eventuated in actual transfers of function probably occurred soon after the endosymbiotic mergers (as evidenced, for example, by the fact that nearly all of the dozens of fully sequenced mtDNA genomes of diverse metazoan animals contain exactly the same set of 13 protein-coding genes). Recent migrations of animal mtDNA sequences to the nucleus have proved to be pervasive as well (Collura and Stewart 1995; Mourier et al. 2001; M. F. Smith et al. 1992; Sorenson and Fleischer 1996; Williams and Knowlton 2001; Woischnik and Moraes 2002), but most such sequences trafficked to the nucleus are now functionless pseudogenes (Bensasson et al. 2001). Unless care is taken (e.g., by examining for stop codons), researchers sometimes can misinterpret these nucleus-housed pseudogenes as bona fide cytoplasmic mtDNA (e.g., when using PCR-based assays with mtDNA primers; Collura et al. 1996; Zhang and Hewitt 1996). If the true origin of a transferred sequence goes unrecognized, this can create interpretive errors in phylogenetic assessments (analogous to traditional problems of confusing paralogy with orthology in analyses of multi-gene families), and it can even lead to diagnostic mistakes in medical or other applications for "mtDNA" markers (Wallace et al. 1997).

In plants, many taxa possess a nearly identical set of about 40 different mtDNA genes, presumably retained from algal ancestors more than a billion years ago. Nonetheless, some angiosperms have repeatedly and recently lost mitochondrial genes to the nucleus (Adams et al. 2000, 2001, 2002), and some of these ongoing IGTs are functionally effective. For example, Adams et al. (1999) documented multiple shifts of mitochondrial *cox2* sequences to nuclei, followed by inactivation of the gene in one genome or the other. Chloroplast DNA sequences are also heavily involved in this type of inter-genomic commerce (Huang et al. 2003; Millen et al. 2001; Thorsness and Weber 1996).

The ancient lineage mergers discussed above, and undoubtedly others early in the history of life (Delwiche 1999; Hartman and Fedorov 2002; Margulis and Sagan 2002), plus subsequent inter-genomic gene traffic (Syvanen 2002; see also next section), mean that eukaryotic cells and their constituent genomes are phylogenetic "mosaics" (Ribeiro and Golding 1998) or "chimeras" (Katz 1999) housing mixtures of distinct evolutionary lines. Thus, "phylogeny" in the primordial biological world was probably anasto-motic or network-like (Figure 8.20), rather than mostly branched and hier-

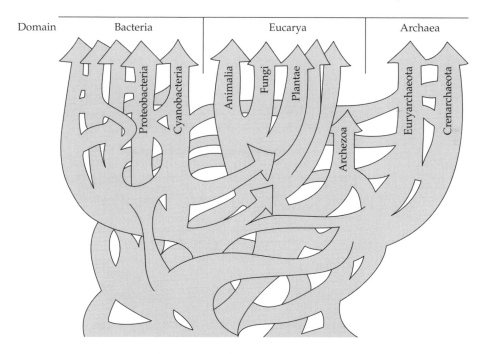

Figure 8.20 Schematic representation of genomic mergers and lateral transfer events of the kind that probably characterized early life. Included in the diagram are the acquisitions by Eucarya of mitochondrial and chloroplast genes from pro-teobacteria and cyanobacteria, respectively. (After Doolittle 1999.)

archical, as it was later to become (Woese 1998b, 2002). Such possibilities raise important issues about phylogeny estimation, and even about the meaning of deep organismal phylogeny. On the other hand, some "core" genes that were never exchanged may exist, and these genes may yet permit reconstruction of mainstream relationships among ancient life forms (Doolittle et al. 2003).

Ancient genomic mergers represent biological reality, so the mosaic nature of modern genomes is not merely an artifact of phylogenetic "sampling noise." Furthermore, the phylogenetic information in mosaic genomes can be turned to advantage in historical reconstructions. In one such example, Hedges et al. (2001) synthesized and integrated available molecular data regarding eukaryotic origins with evidence about physical conditions on primordial Earth (Figure 8.21). Their results suggested an early divergence (ca. 4.0 bya) between Archaea and the archaeal genes now present in eukaryotic cells; subsequent occurrences of at least two genetic mergers (at 2.7 and 1.8 bya) that transferred genes from Bacteria into eukaryotes; an early phylogenetic separation for *Giardia*, an unusual eukaryote with only tiny remnant mitochondria (Roger et al. 1998; Tovar et al. 2003); and the appearance of cyanobacteria (presumably the originators of oxygen-generating photosynthesis) immediately prior to the earliest undisputed evidence for the rise of oxygen in the Earth's primitive atmosphere (ca. 2.5 bya). These intriguing suggestions will merit further critical evaluation.

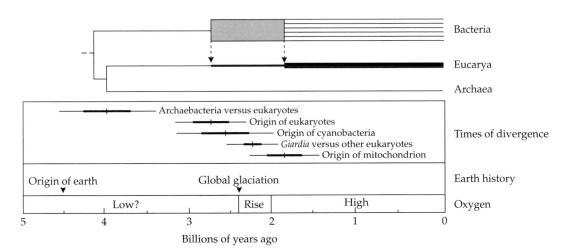

Figure 8.21 Postulated relationships between key phylogenetic events deep in the history of life and their temporal relationship to environmental conditions of primordial Earth and its atmosphere. (After Hedges et al. 2001.)

Horizontal gene transfer

Genetic transmission is overwhelmingly "vertical"; that is, genes are transmitted from parents to offspring. If this were not true, heredity would have little meaning, and phylogenetic trees built from independent characters would seldom exhibit the coherent structures that tend to characterize most well-studied taxonomic groups. As described above, however, remarkable instances of "horizontal gene transfer" (HGT) *have* occurred in evolution—for example, during the probable endosymbiotic mergers that led to eukaryotic cells.

Apart from these ancient endosymbioses, have HGT events shuttled particular nucleotide sequences between otherwise isolated species? The unequivocal short answer to this question is "yes." Of current scientific interest are the frequencies, mechanisms, and evolutionary consequences of such lateral gene movements (Gogarten 2003; Gogarten et al. 2002; Syvanen and Kado 2002). HGT must be distinguished from interspecies gene movement via introgressive hybridization (see Chapter 7), which is merely a special case of vertical heredity. For the purposes of the current discussion, HGTs between taxa must also be distinguished from IGTs between organelle and nuclear genomes strictly within a genetic lineage (as described above).

MOLECULAR CRITERIA FOR INFERRING HGT.   Provisional molecular support for HGT usually comes from one or another of the patterns described in Table 8.5. However, all of these patterns represent indirect or surrogate lines of evidence, rather than definitive documentations of HGT phenomena. This often leaves ample room for alternative interpretations, and several authors have justifiably criticized particular empirical claims for HGT (Eisen 2000; Koski et al. 2001; Ragan 2001).

For example, one oft-used class of evidence for HGT between taxa is a gross discrepancy between the apparent phylogeny for a given segment of DNA and an overwhelming consensus phylogeny for those taxa based on other data (Gogarten 1995; M. W. Smith et al. 1992). In such cases, an HGT event is postulated to have produced the aberrant phylogeny for the "odd-man-out" sequence. However, several evolutionary processes other than HGT can also lead to apparent incongruence between the phylogeny of a particular gene sequence and the broader phylogeny of the genome (Figure 8.22): shared retention of ancestral states by the taxa in question; pronounced heterogeneity in molecular evolutionary rates across lineages; convergent evolution; introgressive hybridization; mistaken assumptions of orthology for loci that actually are paralogous; and idiosyncratic gene loss in separate lineages. Thus, extreme caution is indicated in deducing HGT events from "discordant" phylogenetic signatures alone.

Indeed, some of the earliest reports of HGT were soundly criticized (by Lawrence and Hartl 1992; Leunissen and de Jong 1986; Shatters and Kahn 1989; Steffens et al. 1983) for failure to eliminate competing possibilities.

TABLE 8.5   Four molecular criteria by which putative instances of horizontal gene transfer often are inferred

Criterion	Description
1. Incongruent trees	Phylogenetic trees for specific DNA sequences or proteins show discordant relationships with known or suspected phylogenetic trees for the taxa in which they are housed. Caution is indicated, however, because incongruent trees can also result from several other evolutionary forces (see Figure 8.22). Nonetheless, satisfaction of this criterion often provides the strongest available line of evidence for HGT events.
2. Unusual nucleotide composition	Shifts in nucleotide composition between neighboring sequences (such as an increase in frequency of GC over AT base pairs) could indicate that foreign genes or coding sequences exist as "islands" in the genome. Natural selection or mutation bias in the absence of HGT could mimic this outcome, however. Also, the signal of a bona fide ancient HGT event will normally attenuate over time, making older HGT events difficult to detect by this criterion.
3. Unusual species distributions of genes	In a particular species, the presence of a gene that otherwise is found in distant relatives but not in close relatives could signal an HGT event from the distant taxon. Alternative explanations should be eliminated, such as the possibility of gene loss in the intervening lineages. Also, this criterion clearly cannot work for genes that are taxonomically universal.
4. Unexpected homology patterns	"BLAST" searches (Altschul et al. 1997) of computer databases might indicate that a given gene sequence in the species of interest shows significant similarity (putative homology) to a gene or genes otherwise known only from distant taxonomic groups. This method provides a "quick and dirty" screen for possible HGT events, but for many reasons is error-prone.

Source: After Brown 2003, and Eisen 2000.

Two studies that were challenged involved suspected movements of a superoxide dismutase gene from a fish into its bacterial symbionts (Bannister and Parker 1985; Martin and Fridovich 1981) and of a glutamine synthetase gene from a plant into its bacterial symbionts (Carlson and Chelm 1986). However, many other early reports of HGT were less easily dismissed (e.g., Bork and Doolittle 1992; Doolittle et al. 1990; Heinemann 1991; Heinemann and Sprague 1989; Kidwell 1992; Mazodier and Davies 1991; Smith and Doolittle 1992; M. W. Smith et al. 1992; Sprague 1991; Zambryski et al. 1989), and compelling evidence has gradually accumulated for such events in a wide diversity of eukaryotic as well as prokaryotic taxa.

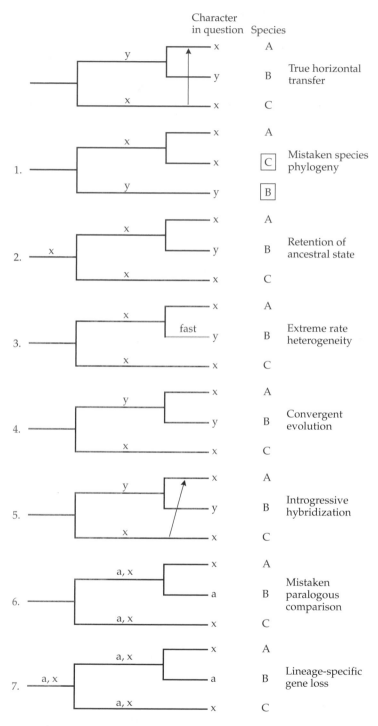

Figure 8.22 **Seven competing hypotheses** to account for an apparent character-state discordance that otherwise might be attributable to a horizontal gene transfer event.

PROKARYOTE–PROKARYOTE HGT.   Horizontal genetic exchanges between prokaryotic species can occur via several routes, including plasmid exchange (even between distantly related taxa), transduction (viral-mediated transfers), and transformation (uptake and incorporation of DNA from the environment). Such processes have long been known, but their full impact remains a matter of contention. In one key study, Lawrence and Ochman (1998) reported that *Escherichia coli* has experienced no less than 230 HGT events during the last 70 million years, and that 755 of its 4,288 protein-coding genes (17.6%) are of foreign origin (presumably introduced mostly by plasmids, phages, and transposable elements). The authors considered these to be minimum estimates because evidence for older insertions probably blurs with time, and because many sequence insertions may be evolutionarily transient. Of course, all molecular evidence for these events was necessarily indirect, involving various criteria (described in Table 8.5) that are potentially subject to challenge.

Similar molecular analyses soon suggested that HGT events had been fairly common in evolution, moving pieces of DNA among a wide variety of prokaryotic species (Figure 8.23), including some apparent transfers between Bacteria and Archaea (e.g., Forterre et al. 2000; Nelson et al. 1999). For example, Rest and Mindell (2003) inferred at least four lateral gene transfers of retroids (elements bearing reverse transcriptases) from Bacteria to Archaea (the latter were previously not known to possess such elements). The evidence presented for putative HGT events is sometimes quite compelling, and some authors now view lateral gene movement in prokaryotes to be so prevalent as to require fundamental reorientations of thought about the nature of bacterial taxa (Lawrence 2002) and the basis of innovative bacterial evolution (Jain et al. 2002, 2003; Koonin et al. 2001; Ochman et al. 2000). Others reserve judgment pending more definitive molecular evidence for ubiquitous HGT in prokaryotic evolution (Eisen 2000; Ragan 2001). Still other researchers, while acknowledging that HGT events may be fairly common in bacterial evolution, suggest that extensive DNA sequence comparisons of many genes nonetheless will permit robust phylogenetic reconstructions of a single primary tree topology for microbial lineages (e.g., Daubin et al. 2003; Lerat et al. 2003).

PROKARYOTE–EUKARYOTE HGT.   Lateral gene transfers from prokaryotes to eukaryotes are also well known (Klotz and Loewen 2003). One of the first-discovered examples involved *Agrobacterium tumefaciens*, the bacterial agent of "crown gall" disease in plants. This bacterium infects wounded sites on a tree, resulting in tumor-like growth. During the process, it also transfers some of its genetic material (T-DNA, carried on its plasmids) into the plant's nuclear genome. This type of HGT is so effective that purposefully engineered strains of *A. tumefaciens* are now employed routinely by biotechnologists as transformation vectors for introducing specific transgenes into commercially valuable plants (see Avise 2004c). Another interesting example of prokaryote-to-eukaryote HGT involves glycosyl hydro-

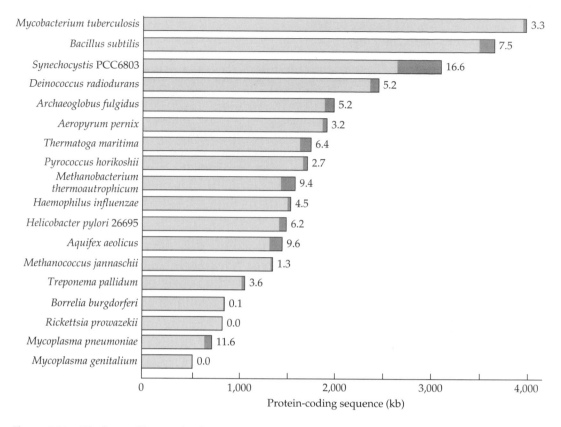

Figure 8.23 Horizontally acquired DNA in 18 sequenced bacterial genomes. The light gray portion of each horizontal bar indicates the fraction of the DNA sequence thought to be native to that species. The darker gray portion indicates the suspected fraction of the DNA sequence (also shown numerically as a percentage) that is thought to be of foreign origin. (After Ochman et al. 2000.)

lase genes that apparently jumped from bacteria to fungi in the rumens of cattle, thereby providing the fungi with a useful capacity to degrade the cellulose and other plant polysaccharides in that environment (Garcia-Vallvé et al. 2000).

It is a remarkable irony that if HGT events between prokaryotes and microbial eukaryotes have indeed been common in evolution, they might actually call into question HGT's original poster-child scenario: the endosymbiosis theory. To some authors (e.g., Andersson et al. 2003), HGT is so pervasive, and "fusion" theories so implausible mechanistically (Rotte and Martin 2001), as to demand a rigorous reexamination and a possible reinterpretation of the evolutionary origins of eukaryotic cells. Perhaps, they suggest, these nucleated cells arose from "routine" HGT events between microbes, rather than from wholesale genomic amalgamations all at once.

The frequency of natural HGTs from prokaryotes to metazoan animals, including humans, is also under current debate. One provocative claim emerged from an initial analysis of the draft human genomic sequence (Lander et al. 2001), in which it was suggested that at least 223 genes from prokaryotes had been imported into the human nuclear genome. The evidence consisted primarily of unexpected homology patterns (criterion 4 in Table 8.5); also, many of those genes seemed not to be the result of transfers from the human mitochondrial genome. However, subsequent analyses of expanded data sets and reinterpretations of the molecular evidence challenged these HGT claims on several grounds (Genereux and Logsdon 2003; Salzberg et al. 2001; Stanhope et al. 2001)—for example, by documenting that many sequences purportedly imported from bacteria actually have closer phylogenetic ties to other eukaryotic genes.

EUKARYOTE–EUKARYOTE HGT.   One of the earliest convincing cases of HGT between eukaryotic species involved "P elements" in fruit flies (Daniels et al. 1990). These transposable elements have a patchy phylogenetic occurrence confined mostly to *Drosophila* and related dipteran genera (Perkins and Howells 1992). A remarkable molecular discovery was that P-element sequences in *D. melanogaster* are nearly identical to those in *D. willistoni*, despite a suspected evolutionary separation of these host species of tens of millions of years. Furthermore, close relatives of *melanogaster* appear to lack *P* elements entirely, whereas these elements are widespread in species of the *willistoni* group. The compelling conclusion was that an HGT event must have moved proliferative *P* elements from the *willistoni* complex into *melanogaster*, probably within the last century (Kidwell 1992). A semiparasitic mite (*Proctolaelaps regalis*) may have been the mediating vector (Houck et al. 1991).

Other pioneering studies recognized HGT events among as well as within various animal, plant, fungal, and microbial taxa (Calvi et al. 1991; Flavell 1992; Mizrokhi and Mazo 1990; Simmons 1992). Such findings were merely the tip of the iceberg, however, as many such lateral genetic transfers by mobile elements are now well documented (Jordan et al. 1999; Kapitonov and Jurka 2003; Kidwell 1993; McDonald 1998; Rosewich and Kistler 2000). For example, Cho et al. (1998; see also Cho and Palmer 1999) estimated that mobile self-splicing introns had invaded *cox1* genes by cross-species HGT on at least 1,000 independent occasions during angiosperm evolution. The extent and genomic consequences of such lateral movements of nucleic acids may be profound, as emphasized, for example, by Brosius (1999) who concluded that "genomes were forged by massive bombardments with retroelements and retrosequences" (mobile sequences that insert into genomes via reverse transcription).

Furthermore, once a mobile element invades a lineage, it often replicates dramatically therein. Incredibly, at least 50%, and probably much more, of the human genome (like the genomes of most other eukaryotes) consists of remnants of retrotransposable elements and retroviruses that invaded the host genome, proliferated, and whose copies are now in various stages of expansion or decay (Promislow et al. 1999). To cite just one example, molecular phylogenetic analyses indicate that one family of retroviral sequences invaded the proto-human genome anciently, stayed quiescent for eons, and then underwent a burst of transpositional activity approximately 6 million years ago, coinciding in time with the separation of proto-human and proto-chimpanzee lineages (Jordan and McDonald 2002).

Most HGT events between eukaryotes probably involve mobile genetic elements, but several recently documented instances of HGT between unrelated plants seem harder to explain by this route. These cases involve mitochondrial housekeeping genes (encoding ribosomal and respiratory proteins) that appear to have transferred during evolution between unrelated species, yielding outcomes that include gene duplications, recaptures of functional genes formerly lost by a lineage, and appearances of chimeric loci whose sequences are essentially half-monocot and half-dicot (Bergthorsson et al. 2003). Perhaps vectoring agents such as viruses, bacteria, fungi, insects, or pollen were involved, or perhaps there has been transformational uptake of plant DNA from the soil. Whatever the explanation, the authors suggest that such HGT events between higher plants may be reasonably frequent on an evolutionary time scale of millions of years. With regard to higher animals, another potential route for HGT—via food ingestion—has been the subject of discussion and some limited experimentation (Doolittle 1998; Schubbert et al. 1997).

For any HGT event between eukaryotes to be "successful," the foreign DNA must somehow enter germ line cells and then be passed to successive generations. Although such horizontal DNA transfer is far too infrequent to overturn conventional genetic wisdom about the overwhelming predominance of vertical transmission in eukaryotic evolution, it nonetheless is proving to be at least an occasional contributor to the taxonomic and evolutionary distributions of particular DNA sequences in multicellular organisms (Bushman 2002; Syvanen and Kado 2002).

Relationships between retroviruses and transposable elements

Many suspected instances of HGT involve transposable elements (TEs; see Box 1.3), classes of DNA sequences that seem to be predisposed to such inter-taxon movements by virtue of their inherent proclivity to shift from one chromosomal site to another (albeit typically within cell lineages). The mechanisms by which TEs occasionally escape the confines of a host lineage to colonize other taxa are poorly understood, but several possible routes exist, such as by hitchhiking on viruses or parasites. Once inside host cells, some TEs (class I retrotransposable elements) can transpose proliferatively by reverse transcription of RNA intermediates, whereas others (class II elements) merely jump from spot to spot by DNA-to-DNA transposition mechanisms (Finnegan 1989). Most TEs have characteristic structures that include gene sequences coding for enzymes involved in the transposition process, usually flanked by terminal repeat sequences of varying lengths. In the retrotransposable elements (RTEs), one of these genes encodes reverse transcriptase (RT), which catalyzes the transcription of RNA to DNA.

Particularly intriguing are the biological and structural similarities between RTEs and retroviruses (RVs). Retroviruses are small, single-stranded RNA viruses that resemble RTEs in several ways, including the production of a reverse transcriptase and the presence of long terminal repeats (LTRs) flanking the coding region. However, like other viruses and unlike RTEs, retroviruses can encase themselves in a protective envelope that facilitates independent infectious transport across the cells of the same or different organisms. These observations raised an interesting evolutionary question (Doolittle et al. 1989; Finnegan 1983): Might RTEs represent degenerate retroviruses that secondarily lost much of their facility for autonomous intercellular transport? Or, alternatively, did retroviruses evolve from ancestral RTEs by secondary acquisition of these capabilities?

In addition to its presence in both RTEs and RVs, a gene for reverse transcriptase is found in several other genetic elements, including the hepadnaviruses of animals and the caulimoviruses of plants. The RT gene also exhibits structural similarities (suggestive of shared ancestry) to the RNA-directed RNA polymerases of some other viruses. Xiong and Eickbush (1990) took advantage of available nucleotide sequences from reverse transcriptase genes and RNA polymerase genes to estimate a molecular phylogeny for more than 80 RT-containing genetic elements (Figure 8.24). Their retroelement phylogeny consisted of two primary branches: one leading to non-LTR retrotransposons, the other leading to LTR retrotransposons, retroviruses, caulimoviruses, and hepadnaviruses. Most members within each of these five named assemblages grouped together in terms of RT phylogeny. The authors concluded that retroviruses probably evolved from retrotransposable elements rather than vice versa (as evidenced by RVs' restricted position in the broader phylogeny of RT-housing elements).

However, other researchers have questioned whether conventional phylogenetic concepts really apply to different assemblages of viruses and other mobile genetic elements (e.g., Bushman 2002). Their contention is that different groups may actually be independent in origin, or contain idiosyncratic amalgamations of genes that themselves have been subject to convergent functional evolution as well as possible sequence exchanges among different ancestors. Thus, the presumptive phylogeny in Figure 8.24 is somewhat controversial with regard to overall retroelement history. Nonetheless, it does serve to illustrate how molecular appraisals are stimulating novel ideas about phylogenetic relationships among even some of the simplest "life forms."

Further Topics in Molecular Phylogenetics

Toward a global phylogeny and universal systematics

Figure 8.25 illustrates how molecular phylogenies can be estimated for particular taxa (the fungus *Fusarium oxysporum*, in this case) at levels spanning

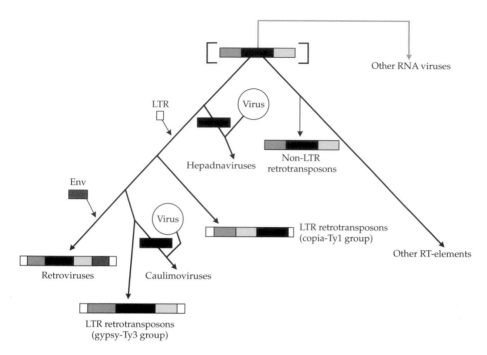

Figure 8.24 Possible phylogeny for retroelements based on RT-like sequences, with structural features of the elements superimposed. Boxes represent stretches of coding sequence: black areas, RT-like region; leftmost dark gray areas, *gag* gene region; lighter gray areas, integrase region; rightmost dark gray area in retroviruses, envelope gene; white terminal areas, LTRs. A hypothesized ancestral structure is shown in parentheses. (After Xiong and Eickbush 1990.)

the full evolutionary spectrum: from intraspecific genealogy to distant relationships early in the history of life. Within the next decade or two, as many more genes and taxa are surveyed and molecular data are assembled and integrated with information from traditional systematics and paleontology (Tudge 2000), it should become possible to reconstruct much of the (Super)Tree of Life (Beninda-Emonds et al. 2002; Pennisi 2001; Sugden et al. 2003). This synthesis will stand as one of the great achievements in the history of biology, not only for its magnitude of effort, but also for its seminal importance in such diverse areas as conservation biology (Vázquez and Gittleman 1998), the study of adaptations (Mooers and Heard 1997), and many other kinds of evolutionary hypothesis testing (Beninda-Emonds et al. 1999). Indeed, a properly reconstructed Tree of Life will be indispensable as a historical "road map" for orienting evolutionary knowledge and guiding virtually all research in comparative biology (Avise 2003b).

Several points should be made about this ongoing enterprise. First, although a *complete* Tree of Life would include micro-genealogies for each of the world's tens of millions of species, the hierarchically nested and pyram-

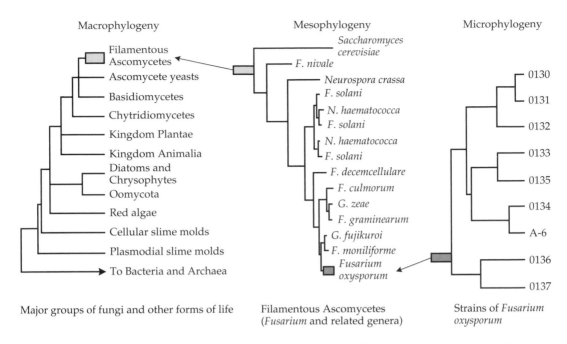

Major groups of fungi and other forms of life · Filamentous Ascomycetes (*Fusarium* and related genera) · Strains of *Fusarium oxysporum*

Figure 8.25 Early empirical example of a tiered phylogenetic assessment based on molecular data. Shown from right to left are relationships among strains of the wilt fungus, *Fusarium oxysporum*, from mtDNA restriction analyses; and, based on small-subunit rRNA gene sequences, the estimated position of this species within the filamentous ascomycetes, and the position of these fungi within the broader phylogenetic hierarchy of life. For additional and updated details about deep molecular relationships among fungal and related lineages, see Baldauf and Palmer (1993), Heckman et al. (2001), and Figure 8.19. (Compiled from diagrams and information in Bowman et al. 1992; Bruns et al. 1991; Gaudet et al. 1989; and Jacobson and Gordon 1990.)

idal nature of phylogeny means that fewer and fewer appraisals will be required at successively deeper (more inclusive) nodes. Thus, it should be feasible to achieve a consensus phylogeny for many, if not all, major and intermediate branches in the Tree of Life, plus detailed snapshots of "twig" relationships among select populations and species in the Tree's current outer canopy.

Second, interest in reconstructing the Tree of Life has spawned the development of various analytical phylogenetic techniques, such as "matrix representation with parsimony" (Baum 1992; Ragan 1992), that can knit disparate small phylogenies into "supertrees." Such procedures (Figure 8.26) are necessary for two primary reasons: different types of genetic data are differentially informative at varying temporal depths in a tree, due largely to variation in molecular evolutionary rates; and the number of possible branching orders in multi-taxon supertrees is much larger than available computer algorithms can search exhaustively for optimality criteria. The

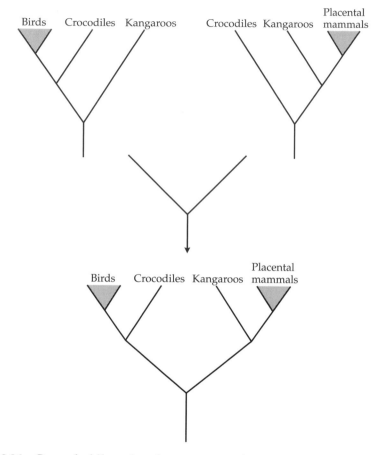

Figure 8.26   **General philosophy of supertree amalgamation from smaller, over-lapping phylogenies.**

solution is to compartmentalize the problem by estimating manageable phylogenies for subsets of data and taxa, then amalgamate these into composite pictures using regions of overlap and suitable representative information from the separate trees (Purvis 1995; Sanderson et al. 1998, 2003).

Third, the reconstructed Tree of Life ideally will include dated nodes. Purely cladistic appraisals aimed at resolving branch topology are important, especially for applications such as phylogenetic character mapping, but they risk diverting attention from the equally important topics of branch lengths and divergence times. For example, the fungal cladograms depicted in Figure 8.25 (like many such tree diagrams in the current literature) were based on procedures designed primarily to recover branch topology, but knowledge of absolute divergence times would enrich the representations greatly. Phylograms (trees with branch lengths and, preferably, dated nodes) are far more difficult to estimate than mere cladograms, in part because paleontology-based estimates usually postdate true lineage divergence times (due to missing fossils), whereas molecular-based estimates often predate true divergence times (due to statistical biases in the estimation procedures; Rodriguez-Trelles et al. 2002). However, a rapprochement between these two kinds of dating exercises is likely to be forthcoming as suitable correction factors are implemented and as both of these sources of temporal information about evolution improve with more data (Benton and Ayala 2003).

Nonetheless, even rough estimates of evolutionary time can greatly expand the information conveyed by a supertree. Empirical molecular studies supporting this contention (including some described in detail in earlier sections of this chapter) have explicitly addressed temporal issues for some of life's earliest lineages (Doolittle et al. 1996; Hasegawa and Fitch 1996; Heckman et al. 2001; Wang et al. 1999), as well as for metazoan phyla (Ayala et al. 1998), major clades of animals (Knoll and Carroll 1999; Lynch 1999), seed plants (Wikström et al. 2001), higher vertebrate taxa (Nei et al. 2001), amphibians (Bossuyt and Milinkovitch 2001), carnivorous mammals (Beninda-Emonds and Gittleman 2000; O'Brien et al. 1999), and many smaller taxonomic groups, such as the world's squirrels (Mercer and Roth 2003). These and numerous other branches in the Tree of Life now have provisionally dated nodes stemming from various combinations of molecular and paleontological data.

A fourth point concerns an underlying assumption of the Tree of Life: that it is based primarily on histories of vertical rather than horizontal genetic transmission. But if lateral DNA transfer between branches has been evolutionarily common, as appears to be true particularly in the prokaryotic and early eukaryotic worlds, then some portions of the Tree of Life may be extensively reticulate, rather than strictly non-anastomotic. Especially in such cases, an organismal genome is a genuine mosaic of evolutionary pasts, and the challenge is to disentangle and interpret the histories of its separate parts. For this and other reasons relating to distinctions between phylogenies of specific genes and a phylogeny of species (see Chapter 4), treelike depictions of organismal history should be conceptualized as prevailing genomic trends at best, with interesting and oft-specifiable exceptions.

One final point concerns nomenclatural issues. Assuming that the Tree of Life, or at least major portions of it, can be reliably approximated using molecular and other data, excellent opportunities will be afforded to develop more informative taxonomies. Unfortunately, current biological classifications are fundamentally flawed (Ereshefsky 2001) because they fail to standardize ranking criteria across different kinds of organisms. No direct comparability now exists between a genus or a family of mammals and their counterpart taxa in fishes, much less in invertebrates, plants, or microbes. One suggested response to this problem is to abandon the Linnaean hierarchical system altogether and erect instead a rankless nomenclature (de Queiroz and Gauthier 1992, 1994). Another possibility is to retain a Linnaean system (or some other nomenclatural analogue) that makes classifications explicitly phylogenetic and also *standardized*.

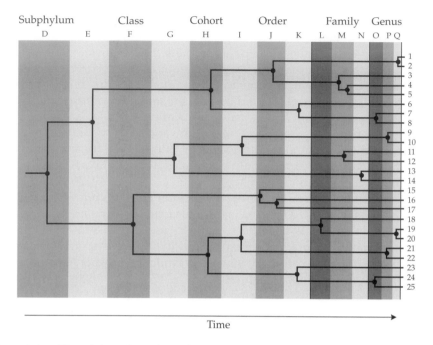

Figure 8.27 Pictorial explanation of "temporal banding" as applied to a hypothetical phylogeny for 25 extant species. Under this proposed scheme, clades are assigned taxonomic ranks defined by specified windows of evolutionary time. (After Avise and Johns 1999.)

Hennig (1966) proposed one means for creating such a phylogenetic system: Let the categorical rank of each taxon denote its geologic age. In practice, the ranks could be traditional Linnaean categories and subcategories (a full list is given in Mayr and Ashlock 1991), or a different recording system, such as alphanumeric code, could be used. The basic idea, elaborated by Avise and Johns (1999), is that "temporal bands" superimposed on phylograms would provide deciding criteria for assignment of taxonomic rank (Figure 8.27). The exact boundaries and widths of these bands are in principle arbitrary, but once agreed upon and ratified by the systematics community, would be universally applied. All extant species that last shared a common ancestor during a specified window of time would be united into a taxonomic family, for example, and those tracing to a common ancestor within successively deeper windows of time would be placed in a superfamily, suborder, and so forth. To retain manageability, only the deepest clades within a window would be given formal taxonomic recognition (i.e., every named taxon would be a clade, but not every clade would be a formal taxon). This also means that all evolutionary lineages traversing a given temporal band (without coalescing inside that band with other such traversing lineages) would be afforded taxonomic distinctions at that categorical level.

A primary advantage of this hypothetical scheme, apart from its explicit focus on phylogeny, is that it adopts absolute evolutionary time as classification's universal common denominator (rather than questionable second-order surrogates, such as magnitudes of genetic divergence or oft-incomparable levels of phenotypic divergence among disparate organismal groups). Thus, nomenclatures per se would become far more informative and phylogenetically meaningful. The actual classifications would probably differ considerably from those in present use. Many anciently separated metazoan lineages might warrant elevation to higher categorical ranks, for example, and various rearrangements would be entailed at lower taxonomic levels as well (Figure 8.28).

As this book testifies, numerous twigs, branches, and limbs in the Tree of Life have been analyzed to varying extents using molecular genetic markers, and initial attempts have been made to root and assemble them into more comprehensive phylogenetic pictures. In some ways, molecular phylogeneticists are at preliminary stages of biotic description analogous to those provided by European naturalists in the 1700s and 1800s, during their explorations of a newly discovered world. At that time too, vast quantities of systematically relevant biological data were being gathered rapidly, and an important challenge was to catalog, interpret, and synthesize the findings into a broad comparative framework.

Molecular paleontology

Molecular appraisals normally are directed toward extant organisms, with phylogenetic inferences representing extrapolations to mutational changes and cladistic events of the past. A long-standing dream of molecular evolutionists has been to assess extinct biota more directly, through recovery of biological macromolecules from fossil material.

In 1980, Prager and co-workers reported a phylogenetic signal retained in the serum albumin proteins of a 40,000-year-old mammoth (*Mammuthus primigenius*) whose carcass had been preserved in the frozen soil of eastern Siberia. In immunological tests, rabbits injected with ground mammoth muscle produced antibodies that reacted strongly with albumins from extant Indian and African elephants, weakly with sea cows (in a related taxonomic order, Sirenia), and still more weakly or not at all with other mammalian albumins. Using similar assays, Lowenstein et al. (1981) showed that albumins from the extinct Tasmanian wolf (*Thylacinus cynocephalus*) produced phylogenetically informative levels of immunological reaction against albumins from other extant Australian marsupials. The preserved tissue in that study was dried muscle from museum specimens collected in the late nineteenth and early twentieth centuries. Apart from a few such examples involving fortuitously well-preserved or recent tissues, most other attempts to extract genetic information from fossil proteins met with little success (Hare 1980; Wyckoff 1972).

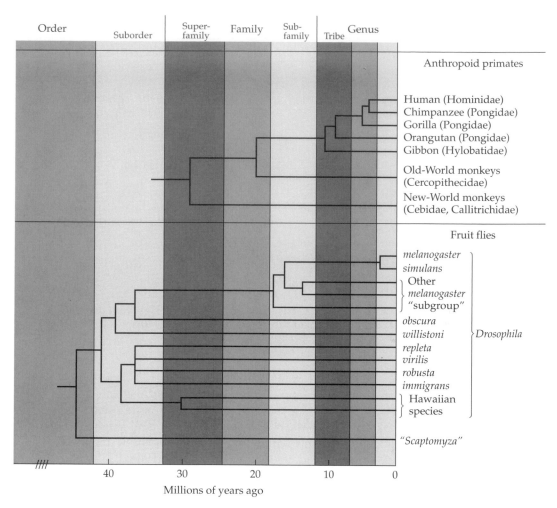

Figure 8.28   Examples of taxonomic disparity (with respect to both clade structure and evolutionary time) in existing classifications of primates and fruit flies. Under traditional taxonomic assignments (families in parentheses), Pongidae is paraphyletic to Hominidae; also, several families of anthropoid primates share common ancestors far more recently than did many fruit fly species placed within the single genus *Drosophila*. Also shown is one way in which these disparities could, in principle, be rectified using time-standardized taxonomic ranks (listed across the top) under a temporal-banding framework. (After Avise and Johns 1999.)

Early studies of ancient DNA (Pääbo 1989) fared somewhat better. In the first successful retrieval of phylogenetically informative DNA sequences from museum material, Higuchi et al. (1984, 1987) recovered short mtDNA segments from a 140-year-old study skin (salt-preserved) of the extinct quagga (*Equus quagga*), an African species with an enigmatic mixture of

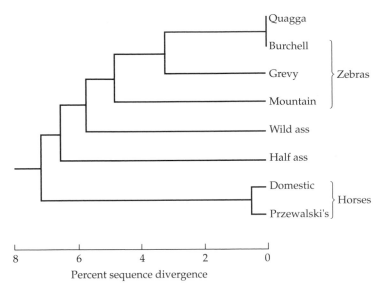

Figure 8.29 Phylogenetic tree based on mtDNA sequences from the extinct quagga and extant members of the genus *Equus*. (After Pääbo et al. 1989.)

horselike and zebralike features. Fragments of DNA isolated from dried muscle and connective tissue were cloned into a lambda phage vector, and sequences totaling 229 base pairs were obtained and compared against those of extant species in the family Equidae. Phylogenetic analyses of these and additional molecular data (George and Ryder 1986; Pääbo and Wilson 1988) showed that the quagga had been closely related to the Burchell zebra, *Equus burchelli* (Figure 8.29).

In 1985, Pääbo reported the isolation and biological cloning of nuclear DNA pieces from an Egyptian mummy 2,400 years old. One year later, Doran et al. (1986) reported the successful extraction of DNA from human brain tissue 8,000 years old, which had been buried in a swamp in central Florida. Notwithstanding these and a few other apparent success stories, traditional isolation procedures seldom yielded ancient DNA sequences in a sufficient state of preservation to be of practical utility for phylogenetic comparisons.

Much excitement, therefore, attended early PCR-based attempts to recover ancient DNA from museum materials and fossil templates (Pääbo et al. 1989). Especially in the late 1980s and early 1990s, many researchers (see Brown 1992) claimed to have sequenced various PCR-amplified fragments of "fossil DNA" from remains of plants and animals that died as long as many tens of millions of years ago (e.g., Cano et al. 1992, 1993; DeSalle et al. 1992; Golenberg et al. 1990; P. S. Soltis et al. 1992a; Woodward et al. 1994). Alas, most such reports now seem highly implausible; almost certainly, researchers mistakenly amplified modern DNA sequences that had contaminated the fossil material (Austin et al. 1997; Hofreiter et al. 2001; Poinar 2002; Wayne et al. 1999).

When an organism dies, its DNA is normally degraded quickly, initially by nucleases in the cells themselves and later by relentless exogenous forces, including background radiation as well as oxidative and hydrolytic processes (Lindahl 1993a,b). Low temperatures, high salt concentrations, and desiccation can slow some of the degradation, but the physical chemistry of DNA is such that even under superb preservation conditions, native sequences hundreds of nucleotides long seldom (if ever) survive as suitable templates for PCR for more than about a million years. Indeed, even under the best of natural circumstances, the oldest fossil material from which suitable DNA fragments can be recovered with current technology is now believed to be about 50–100 millennia.

Retrieval of ancient DNA from well-preserved fossils less than 100,000 years old sometimes *is* possible, but the PCR assays are fraught with the danger of sample contamination from pervasive modern DNA (Cooper and Poinar 2000). Thus, all claims of authenticity for fossil DNA must be viewed with skepticism unless stringent criteria have been met (Hofreiter et al. 2001). These criteria may include the use of appropriate controls (e.g., mock extractions and PCR reactions without template), a demonstration that identical PCR products emerge from multiple extracts, quantification of the number of template DNA molecules (PCR amplifications are problematic when less than 1,000 molecules initiate the process), observation of an inverse relationship between DNA fragment length and amplification efficiency (short sequences amplify more readily), and reproducibility of outcomes by independent investigators. Furthermore, the laboratory itself must be routinely bleached and UV-irradiated to destroy contaminating modern DNA, and it must be physically separated from other work areas. At the very least, researchers must use protective clothing and face shields.

One helpful approach to ensuring the authenticity of ancient DNA also serves as a preliminary molecular screen for precious fossils and museum specimens. Most amino acids can exist in two mirror-image isomeric forms (L and D) that rotate plane-polarized light in opposite directions. In living tissues, however, the L-forms greatly predominate due to the action of specialized racemase enzymes. After an organism dies and racemase activity halts, some of the L-isomers gradually convert to D-isomers at predictable rates. This chemical process is the basis for conventional "racemization" dating of fossils, but it can also be used to address whether fossil tissue samples might be well enough preserved to contain endogenous DNA (Poinar et al. 1996). Starting with tiny bits of museum tissue, the extent of racemization can be measured, and the samples thereby screened for potential suitability as sources of bona fide ancient DNA.

The danger of sample contamination is especially acute when dealing with ancient hominid remains (Stoneking 1995). Nonetheless, a developing success story in molecular paleontology has involved the retrieval of DNA from fossil bones of several Neanderthal specimens (Krings et al. 1997, 2000;

Ovchinnikov et al. 2000). *Homo sapiens neanderthalensis* was a morphologically distinct hominid that appeared in Europe and western Asia at least 200,000 years ago and persisted until about 30,000 years ago, thus overlapping temporally with modern humans (*H. sapiens sapiens*), including Cro-Magnons. Long-standing debates have centered on reproductive and phylogenetic relationships between these taxonomic subspecies. Using Neanderthal fossils from widely separated sites in Europe, mtDNA sequences were recovered and found to fall essentially outside the range of mtDNA sequence variety observed among humans worldwide today. (The fact that Neanderthal sequences were distinctive and phylogenetically unified was further testimony to their own authenticity.) The magnitude of sequence divergence suggests that the matrilineal separation between Neanderthals and the ancestors of modern humans occurred more than half a million years ago, a date considerably older than estimates of the matrilineal coalescent time for all extant humans (ca. 200,000 years bp). Furthermore, molecular analyses of 24,000-year-old fossils of anatomically modern humans of the Cro-Magnon type indicated that the mtDNA sequences of these individuals fell well within the range of sequence variation in living humans (Caramelli et al. 2003), although in this case contamination with modern DNA was almost impossible to eliminate as an explanation. If these latter molecular findings involving Cro-Magnons are valid, the genetic distinction between *H. s. sapiens* and *H. s. neanderthalensis* assumes even greater potential biological significance.

PCR-based sequence comparisons between deceased and modern DNAs have similarly been achieved for several other animal groups. These studies typically involve mtDNA because targeted fragments of these cytoplasmic genomes occur in much higher abundance than single-copy nuclear genes, and because suitable PCR primers are often available. For example, Thomas et al. (1989) used mtDNA sequences from the extinct Tasmanian wolf to confirm that this species was related more closely to other Australian marsupials than to carnivorous marsupials in South America (see also Krajewski et al. 1997). Janczewski et al. (1992) characterized mitochondrial and nuclear sequences from 14,000-year-old bones of the sabre-toothed cat (*Smilodon fatalis*) from tar pits in Los Angeles, thereby uncovering the phylogenetic position of this extinct species within the evolutionary radiation of Felidae.

Other extinct animals from which phylogenetically informative DNA sequences have been recovered include the mastodon (*Mammut americanum;* Yang et al. 1996), woolly mammoth (*Mammuthus primigenius;* Noro et al. 1998), blue antelope (*Hippotragus leucophaeus;* Robinson et al. 1996), Steller's sea cow (*Hydrodamalis gigas;* Ozawa et al. 1997), ground sloth (*Mylodon darwinii;* Höss et al. 1996), cave bear (*Ursus spelaeus;* Hänni et al. 1994; Hofreiter et al. 2002; Loreille et al. 2001; Orlando et al. 2002), pig-footed bandicoot (*Chaeropus ecaudatus;* Westerman et al. 1999), a bovid from the Balearic

Islands (*Myotragus balearicus;* Lalueza-Fox et al. 2000), and various flightless (*Thambetochen, Ptaiochen*) and flighted (*Anas*) waterfowl from Hawaii (Cooper et al. 1996; Sorenson et al. 1999b). In 2001, Cooper and colleagues were the first to publish *complete* mtDNA genomic sequences from fossil remains (two species of extinct moas from New Zealand). Two years later, two research groups (Bunce et al. 2003; Huynen et al. 2003) were perhaps the first to use a sex-linked molecular marker to identify the gender of individuals from fossil remnants of extinct species (also moas).

In some instances, mtDNA sequences have been amplified from multiple fossil specimens or geographic sites and compared with modern samples, permitting direct temporal assessments of population genetic or phylogeographic patterns. For example, in the brown bear (*Ursus arctos*), distinctive matrilines currently restricted to different North American regions apparently co-occurred in a Beringian population 36,000 years ago, as judged by comparisons of fossil and extant mtDNAs (Leonard et al. 2000; see also Barnes et al. 2002). In a similar study of horses (*Equus caballus*) based on several late Pleistocene fossils (12,000 to 28,000 years old), genetic diversity was found to be high, but the mtDNA lineages fell mostly in one distinct portion of the broader matrilineal tree for extant horse populations (Vila et al. 2001). Lambert et al. (2002) recovered mtDNA sequences from 7,000-year-old subfossil bones of Adélie penguins (*Pygoscelis adeliae*) to assess population genetic changes in a hypervariable region of the molecule, as did Hadly et al. (1998) for mtDNA cytochrome *b* sequences using 2,400-year-old remains of pocket gophers (*Thomomys talpoides*). Across a more recent time frame, Guinand et al. (2002) used archived scales from lake trout (*Salvelinus namaycush*) that resided in the Upper Great Lakes during the 1940s and 1950s as a source of microsatellite DNA to compare against allele frequencies in extant populations. Modern collections showed reduced genetic variation, probably due to severe population declines of this species during the mid-twentieth century.

Studies of ancient DNA have even been used to examine ecological conditions of the past. For example, by scrutinizing animal mtDNAs and plant cpDNAs recovered from 12,000-year-old packrat middens in a Chilean desert and comparing these fossil sequences against those of extant species, Kuch et al. (2002) identified a number of animals (a vicuña, two rodents, and a bird) plus several plant species representing taxonomic families no longer found at the site today. Results suggest a diverse biota and a more humid climate at that location when the middens were deposited. Other amazing studies in molecular paleontology used cpDNA sequences from coprolites (fossil dung) to reconstruct the diet of an extinct herbivorous ground sloth (*Nothrotheriops shastensis*) that inhabited Nevada in the late Pleistocene (Hofreiter et al. 2000; Poinar et al. 1998). Finally, from ancient human feces 2,000 years old, both cpDNA and mtDNA sequences were recovered by PCR and used to deduce the diet of omnivorous Native Americans at a cave site in Texas (Poinar et al. 2001b).

SUMMARY

1. Phylogenetic hypotheses underlie virtually all conclusions in comparative organismal evolution. Phylogenetic character mapping (PCM) has become a popular means of making these hypotheses more explicit and testable. In PCM, particular organismal features are matched with their associated species on a cladogram, with the purpose of revealing the evolutionary histories of those traits. These independent appraisals of phylogeny are now based routinely on molecular markers.

2. Phylogenetic character mapping against a molecular backdrop has been accomplished for numerous anatomical, physiological, and behavioral features of plants, animals, and microbes. Some organismal traits have proved to be monophyletic, others polyphyletic. Concepts of gradients and thresholds in the phylogeny of quantitative traits have also been stimulated by PCM exercises.

3. Molecular markers are used widely in biogeographic assessment: for example, to test dispersalist versus vicariance explanations for the appearance of related taxa in disjunct geographic regions, to distinguish between common ancestry and convergence as an explanation for organismal similarities among different regional biotas, and to reconstruct biogeographic histories of island inhabitants.

4. Phylogenetic assessments above the species level have been based on many molecular approaches, including DNA hybridization, immunological methods, and restriction site analyses, but nucleotide sequencing has become the method of choice in recent years for examining slowly evolving nuclear and cytoplasmic genes in a macroevolutionary context. The phylogenetic content of the sequences themselves, plus that of eccentric molecular features such as alternative gene orders, presence versus absence of particular introns, or patterns of codon assignment, offer many special opportunities for clade delineation.

5. Lateral genetic transfers were probably common early in the history of life, the most notable examples being the endosymbioses that eventuated in the distinctive nuclear and cytoplasmic genomes of eukaryotic cells. Such reticulation events raise important questions about the frequency of DNA exchanges across lineages near the base of the Tree of Life, and even about the fundamental meaning of organismal "phylogeny." Nonetheless, molecular phylogenetic studies are revealing major features in the evolutionary histories of even the simplest and most ancient forms of life.

6. In addition to ancient inter-genomic exchanges of nucleic acids, horizontal gene transfers (HGTs) have proved to be far more common in subsequent evolution than formerly supposed. Through the various types of phylogenetic "signatures" or "footprints" that HGTs produce, many contemporary as well as historical reticulation events (often involving transposable elements) have been provisionally documented within and between prokaryotes and eukaryotes. Caution is called for in reading such historical signatures, however, because several non-HGT evolutionary processes can mimic their effects.

7. In the near future, prospects are great for developing a global or universal phylogeny for all of life. Molecular methods will play a huge role in that endeavor. Important questions have arisen about how to analyze, interpret, standardize, and taxonomically summarize the wealth of molecular genetic information that is now becoming available.

8. With the advent of PCR, molecular phylogenetic appraisals have been extended to ancient DNA sequences recovered from creatures no longer alive. In exceptional circumstances, "fossil DNAs" have even been extracted and phylogenetically analyzed at community-level scales and from well-preserved materials up to tens of thousands of years old.

9

Molecular Markers in Conservation Genetics

Modern biology has produced a genuinely new way of looking at the world ... to the degree that we come to understand other organisms, we will place a greater value on them, and on ourselves.

E. O. Wilson (1984)

Conservation genetics is a subdiscipline within the broader field of conservation biology (Meffe and Carroll 1997). It has sometimes been characterized primarily as the study of inbreeding effects and losses of adaptive genetic variation in small populations (Frankham 1995), but this is an unduly narrow characterization, as this chapter will attest. A brief history of major developments in the young but expanding field of conservation genetics is presented in Box 9.1.

In the final analysis, biodiversity *is* genetic diversity. As we have seen, this genetic diversity is genealogically arranged across diverse temporal scales: from family units, extended kin groups, and phylogeographic population structures within species to graded magnitudes of genetic divergence among species that became phylogenetically separated at various times in the evolutionary past. As we have also learned, visible phenotypes of organisms are not infallible guides to the way in which this genealogical diversity is arranged. Sadly, even as powerful molecular tools have become available to assess this genetic variety in exciting new ways, the marvelous biodiversity that has carpeted our planet is being lost at a pace that is nearly unprecedented in the history of life (Ehrlich and Ehrlich 1991). Biodiversity is in serious decline, with, for example, approximately 50% of vertebrate animal species and 12% of all plants now considered vulnerable to near-term extinction, mostly as a result of effects of habitat alteration associated with human population growth (Frankham et al. 2002). Earth recently entered the

BOX 9.1 Brief Chronology of Some Key Developments in the History of Conservation Genetics

Many other important contributions could also be listed, but the following provide representative examples and a general time frame of events.

1966   Lewontin and Hubby introduce protein electrophoretic techniques to population biology.

1973   The Endangered Species Act sets a legal precedent in the United States for identifying and conserving rare taxa.

1975   Frankel and Hawkes edit a volume focusing on management of crop genetic resources; Martin edits a volume on captive breeding of endangered species.

1979   Ralls and colleagues draw attention to the wide occurrence of inbreeding depression in captive populations (for later updates, see Ralls et al. 1988; Frankham 1995). Avise and colleagues, and independently Brown and colleagues, introduce mtDNA approaches to population biology.

1980   Soulé and Wilcox publish the first of several conservation books with an evolutionary genetic as well as an ecological orientation (see also Frankel and Soulé 1981; Soulé 1986, 1987; Soulé and Kohm 1989).

1982   Laerm and colleagues publish the first multifaceted genetic appraisal of the taxonomic status of a wild endangered species.

1983   Schonewald-Cox and colleagues edit the first major volume devoted explicitly to genetic perspectives in conservation. O'Brien and colleagues initiate a series of studies on inbreeding, heterozygosity, and population bottlenecks in wild felids. Mullis invents the polymerase chain reaction technique for in vitro amplification of DNA (see Mullis 1990).

1984   Templeton and Read initiate an influential series of studies on eliminating inbreeding depression in captive gazelles.

1985   The Society for Conservation Biology is formed. Jeffreys and colleagues introduce DNA fingerprinting methods.

1986   Ryder brings the phrase "evolutionarily significant unit" to wide attention in conservation biology.

1987   Ryman and Utter edit a volume on population genetics in fisheries management. The first issue of *Conservation Biology* (Blackwell) is published, complementing earlier journals such as *Biological Conservation* and *Journal of Wildlife Management*. Avise and colleagues introduce the term "phylogeography" and outline the field's major principles.

1988   Lande draws focus to genetic versus demographic concerns for small populations.

1989   The Captive Breeding Specialist Group begins a series of population viability analyses (PVAs) for endangered taxa (see review in Ellis and Seal 1995). The U.S. Fish and Wildlife Service opens a laboratory facility (in Ashford, Oregon) devoted explicitly to wildlife forensics. Microsatellites are introduced (by Tautz as well as Weber and May, and others) as a source of highly polymorphic nuclear markers. Avise promotes novel roles for molecular genetics in the recognition and conservation of endangered species.

1990 Hillis and Moritz edit a volume summarizing the many laboratory genetic approaches to molecular systematics.

1991 Vane-Wright and colleagues raise novel issues about phylogenetic diversity and conservation value (see also Forey et al. 1994; Humphries et al. 1995). Falk and Holsinger edit a volume on conservation genetics in rare plants.

1992 Avise empirically introduces a comparative perspective in conservation genetics for a regional biota. Hedrick and Miller address conservation biology notably from the vantage of genetic diversity and disease susceptibility. Groombridge edits a taxonomic and genetic inventory of global biodiversity as a backdrop for conservation efforts.

1993 Thornhill edits an important volume on the natural history and consequences of inbreeding and outbreeding.

1994 Avise publishes the first edition of this current textbook. Loeschcke and colleagues produce an edited volume on conservation genetics. Burke edits a special issue of *Molecular Ecology* devoted to conservation genetics. Baker and Palumbi provide a powerful application of molecular forensics in monitoring endangered species products.

1995 Ballou and colleagues edit a volume on genetic and demographic management issues for small populations.

1996 Avise and Hamrick, and, independently, Smith and Wayne, edit compendia of molecular studies in conservation genetics. Rhymer and Simberloff review the topic of genetic extinction via introgressive hybridization.

1997 Hanski and Gilpin edit a volume on the metapopulation concept, including issues of genetics, evolution, and extinction (see also Rhodes et al. 1996).

1998 Allendorf edits a special issue of the *Journal of Heredity* devoted to conservation genetics of marine organisms.

1999 Landweber and Dobson edit a volume on genetics and species extinctions. Wildt and Wemmer review and preview reproductive technologies (cloning, embryo transfer, etc.) in conservation biology.

2000 The journal *Conservation Genetics* (Kluwer) is launched. Avise publishes the first textbook on phylogeography, a field with a genealogical slant on genetic variation and conservation.

2002 Frankham and colleagues publish the first introductory "teaching textbook" on conservation genetics.

Source: After Avise 2004b.

sixth mass extinction episode in its history (Leakey and Lewin 1995; Wilson 1992), the only one caused by a living creature. Not since 65 million years ago, when a large asteroid slammed into the planet, has there been such a sudden negative impact on global biodiversity (Wilson 2002).

One goal of conservation biology is to preserve genetic diversity at any and all possible levels in the phylogenetic hierarchy—that is, to save as much as possible of the Tree of Life (Mace et al. 2003). Another goal is to promote the continuance of ecological and evolutionary processes that foster and sustain biodiversity (Bowen 1999; Crandall et al. 2000; Moritz 2002). Genetic drift, gene flow, natural selection, sexual selection, speciation, and hybridization are examples of natural and dynamic evolutionary processes that orchestrate how genetic diversity is arranged.

This concluding chapter addresses the following question: How can molecular markers contribute to assessments of genetic diversity and natural processes in ways that are serviceable to the field of conservation biology? The most general answer is simple: Molecular genetic tools help us to understand the nature of life. More specifically, molecular markers offer conservation applications in all of the topical areas that formed the organizational framework for this book, including assessments of genetic variation within populations, biological parentage, kinship, gender identification, population structure, phylogeography, wildlife forensics at various levels, speciation, hybridization, introgression, and phylogenetics. In this chapter we will revisit these topics in order, but now using illustrations that are especially germane to conservation efforts.

Within-Population Heterozygosity Issues

About 30% of all publications in the field of "conservation genetics" have focused on how best to preserve variability within rare or threatened populations (Avise 2004b). A common assumption underlying these studies is that higher mean heterozygosity (H, a measure of within-population genetic variation; see Chapter 2) enhances a population's survival probability over ecological or evolutionary time. Traditional approaches to heterozygosity assessment and management have been indirect. Management of H in captive populations (e.g., in zoos) often occurs de facto through breeding programs designed to avoid intense inbreeding, either by maintaining populations above some "minimum viable population size" or by exchanging breeding individuals among sites. For natural populations, some analogous management approaches have been to ensure adequate habitat such that local effective population sizes remain above levels at which inbreeding (and its associated fitness depression) becomes pronounced and to maintain habitat corridors that facilitate natural dispersal and gene flow among populations (Hobbs 1992; Simberloff and Cox 1987; but see Simberloff et al. 1992 for a critical appraisal of corridor programs). Thus, concerns about inbreeding depression in small populations, both captive and wild, have motivated much of the work in conservation genetics (Hedrick and Kalinowski 2000).

With the advent of molecular techniques, more direct estimates of heterozygosity were made possible. These estimates typically involve assays of multiple marker loci (such as allozymes or microsatellites). Such molecular heterozygosity estimates raise two major conservation-related issues: Is molecular variability reduced significantly in rare or threatened populations? If so, is this reduction a cause for serious concern about the future of those populations?

Molecular variability in rare and threatened species

In the mid-1800s, indiscriminate commercial harvests of northern elephant seals (*Mirounga angustirostris*) reduced this formerly abundant species to dangerously low levels. Fewer than 30 individuals survived through the 1890s (all on a single remote island west of Baja California), but following legislative protection by Mexico and the United States, the species rebounded and now numbers tens of thousands of individuals, distributed among several rookeries. Bonnell and Selander (1974) first surveyed 24 allozyme loci in 159 of these seals from five rookeries and observed absolutely no genetic variability, a striking finding given the high heterozygosities reported for most other species similarly assayed by protein electrophoretic methods. Several additional molecular analyses have since confirmed and extended these findings of an exceptional paucity of genetic variation in northern elephant seals (Hoelzel 1999; Hoelzel et al. 1993, 2002b). Results cannot be attributed to recent phylogenetic legacy or to some other peculiarity of marine Pinnipedia because, in identical molecular assays, the closely related southern elephant seal (*M. leonin*) displayed normal levels of genetic variation (Slade et al. 1998).

In recent decades, an isolated and endangered population of gray wolves (*Canis lupus*) on Isle Royale in Lake Superior declined from about 50 individuals to as few as a dozen. Molecular studies then revealed that approximately 50% of allozyme heterozygosity had been lost relative to mainland samples (Wayne et al. 1991b). Furthermore, only a single mtDNA genotype remained on the island. In terms of multi-locus nuclear DNA fingerprints, these Isle Royale wolves were about as similar genetically as were full-sibling wolves in a captive colony, suggesting that the island population had become severely inbred (Wayne et al. 1991b).

Hillis et al. (1991) employed allozyme markers to estimate genetic variability in the Florida tree snail (*Liguus fasciatus*), many of whose populations are threatened or already extinct. Among 34 genes monitored in 60 individuals, only one locus was polymorphic, and mean heterozygosity was only 0.002. Perhaps a population bottleneck accompanied or followed this species' colonization of Florida from Cuba, or perhaps the snail's habit of partial self-fertilization played a role in its loss of heterozygosity (at least within local populations). Surprisingly, the lack of appreciable variation at the allozyme level contrasts diametrically with this species' exuberant morphological variability, especially with regard to genetically based shell patterns. Results highlight the fact that mean heterozygosity (as registered by molecular markers) and magnitude of phenotypic variation (even that which is genetically encoded) are not necessarily similar.

Genetic variation in remnant populations of the Sonoran topminnow (*Poeciliopsis occidentalis*) in Arizona, where the species is endangered, was compared at 25 allozyme loci with genetic variation in populations from Sonora, Mexico, where the fish is widespread and abundant. The peripheral populations in Arizona exhibited significantly less variation than did the Mexican populations near the center of the species' distribution (Vrijenhoek et al. 1985). The molecular analysis also revealed three major genetic groups within this species' range. The authors concluded that these three units should be maintained as discrete entities in nature because most of the over-all genetic diversity in *P. occidentalis* is attributable to these inter-group differences. They also recommended that any restocking efforts in Arizona employ local populations whose mixing would increase within-population heterozygosity without eroding the genetic differentiation that characterizes the broader geographic assemblages.

Another endangered species assayed extensively for molecular genetic variability is the cheetah (*Acinonyx jubatus*). The South African subspecies of this large cat was first surveyed at 47 allozyme loci, all of which proved to be monomorphic, and at 155 abundant soluble proteins revealed by two-dimensional gel electrophoresis, at which heterozygosity also proved to be low ($H = 0.013$; O'Brien et al. 1983). Subsequent assays of more allozyme markers and of RFLPs at the major histocompatibility complex (MHC) gave further support to the notion that this population is extremely genetically depauperate (O'Brien et al. 1985b; Yuhki and O'Brien 1990). Additional confirmation came from the fact that these cats fail to acutely reject skin grafts from "unrelated" conspecifics. The low molecular genetic variation documented in cheetahs cannot be attributed to some inherent property characteristic of all cats, because other species of Felidae often exhibit normal to high levels of genic heterozygosity in these same kinds of assays (O'Brien et al. 1996). Later surveys of rapidly evolving molecular systems (mtDNA and VNTR nuclear loci) did uncover modest genetic variation in cheetahs across their broader range, but the overall magnitude remained low, leading Menotti-Raymond and O'Brien (1993) to conclude that the heterozygosity present today could be due to post-bottleneck mutational recovery over a time frame of roughly 6,000 to 20,000 years. O'Brien et al. (1987) proposed that the cheetah experienced at least two population bottlenecks: one approximately 10,000 years ago, prior to geographic isolation of the two recognized subspecies (which are highly similar genetically), and a second within the last century, which may have produced the exceptional genetic impoverishment of the South African form.

A similar scenario of bottleneck effects emerged regarding the Asiatic lion (*Panthera leo persica*), which now occurs as a remnant population in the Gir Forest Sanctuary in western India. Allozyme surveys (ca. 50 loci) detected absolutely no variation in a sample of 28 individuals from this subspecies, whereas the Serengeti population of the African subspecies had much higher genetic variation (Wildt et al. 1987). Similar results emerged from DNA fingerprinting methods and analyses of MHC loci (O'Brien et al. 1996). The relict group of lions in the Gir Forest is descended from a population that contracted to fewer than 20 animals in the first quarter of the twentieth century. The obvious interpretation is that this population reduction profoundly affected genomic variation. Analogous scenarios of population bottlenecks and inbreeding in felids were documented for an isolated population of lions in Africa's Ngorongoro Crater and an isolated population of cougars (*Puma concolor coryi*) in North America's Florida Everglades (O'Brien et al. 1996) (see below).

In a review of 38 endangered mammals, birds, fishes, insects, and plants, Frankham (1995) reported that 32 species (84%) displayed lower genetic diversity, as estimated by molecular markers (usually allozymes), than did closely related non-endangered species. A similar trend subsequently was reported in DNA-level appraisals (VNTRs, RAPDs, AFLPs, or STRs) of threatened species versus their more common relatives (Frankham et al. 2002). Examples include endangered populations of such diverse creatures as the beluga whale (*Delphinapterus leucas*; Patenaude et al. 1994), black robin (*Petroica traversi*; Ardern and Lambert 1997), and burying beetle (*Nicrophorus americanus*; Kozol et al. 1994), and the plants *Lysimachia minoricensis* (Calero et al. 1999) and *Cerastium fischerianum* (Maki and Horie 1999). The magnitudes of these reductions in H were not invariably great, but there are many examples of rare or threatened populations that reportedly show extremely low molecular variation (some of these are listed in Table 9.1). In most of these instances, results were provisionally attributed to effects of genetic drift attending historical bottlenecks in population size.

On the other hand, many rare or endangered species have proved *not* to be unusually constrained in genetic variation. Examples of threatened species that have displayed more or less normal levels of molecular variability include a federally protected spring-dwelling fish (*Gambusia nobilis*) endemic to the Chihuahuan desert (A. F. Echelle et al. 1989); Przewalski's horse (*Equus przewalskii*), which is extinct in the wild but survived by several hundred animals in zoos (Bowling and Ryder 1987); the endangered manatee (*Trichechus manatus*) in Florida (Garcia-Rodriguez et al. 1998; McClenaghan and O'Shea 1988); and Stephens' kangaroo rat (*Dipodomys stephensi*), whose historical range is restricted to interior coastal valleys of southern California (Metcalf et al. 2001). Additional examples include several endangered avian species (see review in Haig and Avise 1996): a flightless parrot (*Strigops habroptilus*) native to New Zealand (Triggs et al. 1989), the American wood stork (*Mycteria americana*; Stangel et al. 1990), and most populations of the red-cockaded woodpecker (*Picoides borealis*) in the southern United States (Stangel et al. 1992).

In theory, the demographic details of population bottlenecks (such as their size, duration, and periodicity) should exert important influences on the severity of expected reductions in neutral genetic variability (Luikart et al. 1998). For example, the loss in mean heterozygosity can be minimal if population size increases rapidly following a single bottleneck of short duration

TABLE 9.1	Examples of rare or endangered species with exceptionally low genetic variability, as documented by multi-locus molecular methods
Species	**Observation**
Plants	
Bensoniella oregona	Complete absence of allozyme variation (24 loci) within or among populations of this endemic herbaceous perennial in southwest Oregon and northwest California.
Eucalyptus phylacis	The last remnant stand of this Australian tree is so depauperate in genetic variation as to consist in effect of a single clone. Another closely related local endemic was moderately variable, however.
Harperocallis flava	This endemic to the Apalachicola lowlands of the Florida Panhandle was monomorphic at all 22 allozyme loci scored, in sharp contrast to high genetic variation in a related lily species widespread in that region.
Howellia aquaticus	Complete absence of allozyme variation (18 loci) within or among populations of this rare and endangered aquatic plant in the Pacific Northwest.
Pedicularis furbishiae	Complete absence of allozyme variation (22 loci) within or among populations of this endangered hemiparasitic lousewort in northern Maine.
Posidonia oceanica	A population of this seagrass in the northern Adriatic Sea appears to be in effect a single clone, as gauged by microsatellite assays.
Saxifraga cernua	Complete absence of variation in RAPD or allozyme markers within each of several glacial relict populations in the European Alps.
Trifolium reflexum	Complete absence of allozyme variation (14 loci) in the only known population of this rare native clover in Ohio; however, allozyme assays (20 loci) of an endangered congener, *T. stoloniferum*, did reveal moderate levels of genetic variation.
Animals	
Bison bison	Only one allozyme locus (among 24 tested) was polymorphic in a bison herd in South Dakota known to be descended from a small founder group; other bison herds show microsatellite heterozygosities that are correlated with numbers of founding animals.
Castor fiber	Scandinavian beavers, which went through a severe bottleneck in the 1800s due to overhunting, now show extremely low variation at DNA fingerprinting and MHC loci; nonetheless, the population recovered and expanded tremendously during the twentieth century.
Monachus schauinslandi	Populations of this critically endangered Hawaiian monk seal showed extremely low genetic variation in nuclear DNA fingerprints and mtDNA.
Mustela nigripes	Only one allozyme locus (among 46 tested) was polymorphic in the one known remaining population of the highly endangered black-footed ferret; microsatellite variation was also considerably reduced.
Perameles gunnii	Complete absence of allozyme variation (27 loci) within an endangered, isolated population of the eastern barred bandicoot in Australia (however, a widespread and dense population of the same species in Tasmania also lacked genetic variation at these same loci).
Strix occidentalis	Complete absence of allozyme variation (23 loci) in six populations of the endangered spotted owl from Oregon and California.

Reference
P. S. Soltis et al. 1992b
Rossetto et al. 1999
Godt et al. 1997
Lesica et al. 1988
Waller et al. 1987
Ruggiero et al. 2002
Bauert et al. 1998
Hickey et al. 1991
McClenaghan et al. 1990; Wilson and Strobeck 1999
Ellegren et al. 1993
Kretzmann et al. 1997
O'Brien et al. 1989; Wisely et al. 2002
Sherwin et al. 1991
Barrowclough and Gutiérrez 1990

(Nei et al. 1975). An empirical example of a severe population reduction that, for suspected demographic reasons, did not result in low heterozygosity involves the endangered one-horned rhinoceros (*Rhinoceros unicornis*). Prior to the fifteenth century, about half a million of these animals ranged across a broad area from northwestern Burma to northern Pakistan. Land clearing and human settlement then began to fragment and destroy rhino habitat, and by 1962 fewer than 80 animals remained, all in what is now Nepal's Royal Chitwan Park. Surprisingly, this herd proved to exhibit one of the highest allozyme heterozygosity values reported for any vertebrate: $H \cong 0.10$ (Dinerstein and McCracken 1990). One possibility is that loss of rhino habitat across the Indian subcontinent compressed surviving populations into the Chitwan area, thereby concentrating into a single locale considerable genetic variation that formerly had been distributed among regions.

The particular molecular markers employed can also dramatically influence estimates of genetic heterozygosity. Most of the early studies used allozymes (see Table 9.1), but multi-locus microsatellite appraisals have been employed increasingly in recent years to reassess within-population variation and relate it to population demography and fitness components potentially associated with inbreeding depression (e.g., Coltman et al. 1998a,b; Coulson et al. 1998; Hedrick et al. 2001; Pemberton et al 1999; Rossiter et al. 2001; Slate et al. 2000). However, STR loci have high mutation rates and tend to recover genetic variation quickly, so the molecular footprints of population bottlenecks on these loci should be less long-lasting than on allozyme loci.

Does reduced molecular variability matter?

The examples cited above indicate that genic heterozygosity is indeed reduced in many (though certainly not all) rare or threatened populations and species. Do these findings carry any special significance for conservation efforts? Although it is tempting to assume that a paucity of genetic variation jeopardizes a species' future, the goal of firmly documenting a causal link between molecular heterozygosity and population viability remains elusive (see Chapter 2). In general, there are several reasons for exercising caution in interpreting low molecular heterozygosities reported for rare species: most of the reductions in genetic variation presumably have been outcomes, rather than causes, of population bottlenecks; at least a few widespread and successful species also appear to have low H values as estimated by molecular methods; and in some endangered species (such as the northern elephant seal), low genetic variation has not seriously inhibited population recovery from dangerously low levels (at least to the present).

Another point is that the fitness costs of inbreeding (Box 9.2) are known to vary widely among species (and often even among conspecific populations; e.g., Kärkkäinen et al. 1996; Montgomery et al. 1997). Thus, some taxa are highly susceptible, but others relatively immune, to fitness depression effects from consanguineous matings (Frankham et al. 2002; Laikre and Ryman 1991; Price and Wasser 1979; Ralls et al. 1988). Furthermore, inbreeding depression

BOX 9.2 Inbreeding Depression

Inbreeding depression is the decrease in growth, survival, or fertility often observed following matings among relatives. The phenomenon is of special concern in conservation biology because inbreeding is likely to be severe in small populations. Genetically inbred populations have reduced heterozygosity (increased homozygosity) due to increased probabilities that individuals carry alleles that are identical by descent (stem from the same ancestral copy in earlier generations of a pedigree). This probability for an individual I is the inbreeding coefficient, which for known pedigrees can be calculated as

$$F_I = \Sigma (1/2)^i (1 + F_A)$$

where the summation is over all possible paths through all common ancestors, i is the number of individuals in each path, and A is the common ancestor in each path (for computational details, see Ballou 1983 and Boyce 1983).

Two competing hypotheses for the genetic basis of inbreeding depression have been debated for decades (see Charlesworth and Charlesworth 1987). Under the "dominance" scenario, lowered fitness under inbreeding results from particular loci being homozygous for otherwise rare deleterious recessive alleles, which in outbred populations are usually masked in expression (in heterozygotes) by their dominant counterparts. Under the competing "overdominance" or "heterozygous advantage" scenario, genome-wide heterozygosity per se is the critical influence on fitness. Recent literature seems to indicate considerable support for the dominance model, with overdominance (including epistasis) also contributing to inbreeding depression as a secondary factor (e.g., Carr and Dudash 2003).

The dominance and overdominance hypotheses make different predictions about the relative tolerance of populations to inbreeding (Lacy 1992). If deleterious recessive alleles cause inbreeding depression, then selection will be more likely to have removed most such alleles from populations that have long histories of inbreeding. All else being equal, such populations should rebound and for a while be resistant to further inbreeding effects. In other words, under the dominance hypothesis, populations that survive severe inbreeding may be temporarily "purged" of deleterious recessive alleles, and mean heterozygosity (as estimated, for example, by molecular markers) should therefore have little general predictive value of a population's genetic health. (However, the frequencies of genes of large and small effect and other population genetic factors can also have important effects on the extent to which inbreeding depression is purged; see Byers and Waller 1999; Crnokrak and Barrett 2002.) On the other hand, if inbreeding depression occurs because of a selective advantage to genome-wide heterozygosity (the overdominance hypothesis), then inbred (and homozygous) populations should show reduced fitness, and under future inbreeding might be expected to fare no better than would highly heterozygous populations.

Whatever the mechanistic explanation, different populations empirically exhibit widely varying fitness costs associated with inbreeding. For example, in a survey of captive populations of a variety of mammalian species, relative reduction in survival in crosses between first-degree relatives (such as full sibs) varied across more than two orders of magnitude (Ralls et al. 1988, as summarized by Hedrick and Miller 1992):

Species	Cost of inbreeding
Sumatran tiger (*Panthera tigris sumatrae*)	<0.01
Bush dog (*Speothos venaticus*)	0.06
Short bare-tailed opossum (*Monodelphis domestica*)	0.10
Gaur (*Bos gaurus*)	0.12
Pygmy hippopotamus (*Choeropsis liberiensis*)	0.33
Greater galago (*Galago c. crassicaudatus*)	0.34
Dorcas gazelle (*Gazella dorcas*)	0.37
Elephant shrew (*Elephantulus refuscens*)	0.41
Golden lion tamarin (*Leontopithecus r. rosalia*)	0.42
Brown lemur (*Lemur fulvus*)	0.90

Because inbreeding depression typically is weaker in captive than in natural populations, these values should perhaps be considered minimal estimates of what may occur in the wild. Many additional examples are discussed by Frankham et al. (2002), who also provide extended discussions of the causes and consequences of inbreeding depression in a conservation context.

seems to have a "stochastic nature" (Frankham et al. 2002), which is reflected in varying outcomes depending on the particular fitness components examined in a given species (Lacy et al. 1996), as well as in hit-or-miss expressions of the phenomenon in diverse organismal groups (ranging from birds and mammals to other vertebrates, invertebrates, and plants; Crnokrak and Roff 1999). For all of these reasons, caution is indicated in drawing firm universal conclusions about levels of molecular variation as they might relate to a population's susceptibility to extinction.

A further concern about interpreting the evolutionary significance of molecular variation is that published estimates based on any single class of markers (such as allozymes or microsatellites) may inadequately characterize genome-wide heterozygosity (Hedrick et al. 1986), including quantitative variability that may underlie morphological or physiological traits of potential adaptive significance (Pfrender et al. 2000). For example, in a recent literature review of microsatellite heterozygosity values, Coltman and Slate (2003) concluded that available estimates of variation at STR loci are only weakly correlated with phenotypic or life history variation, and that far larger sample sizes (typically more than 600 individuals per study) will have to be employed in the future to detect any statistically robust relationships that might exist with inbreeding depression. Years earlier, Carson (1990) already had gone even further in suggesting that "genetic variance available to natural selection may actually increase following a single severe bottleneck" and that "character change in adaptation and speciation may, in some instances, be promoted by founder events." These conclusions stemmed from observations and experiments with bottlenecked populations of fruit flies and house flies.

From these and additional considerations, the argument has been made that demographic, ecological, and behavioral considerations should often be of greater immediate importance than genetic (i.e., heterozygosity) issues in the formulation of conservation plans for endangered species (Lande 1988, 1999). For example, individuals in many species show decreased reproduction at low population densities because of a lack of the social interactions that are necessary for breeding, difficulties in finding a mate, or other density-dependent behavioral and ecological factors collectively known as "Allee effects" (Andrewartha and Birch 1954). Furthermore, when populations are few in number and small in size, the possibility of species extinction through "stochastic" demographic fluctuations (irrespective of levels of molecular heterozygosity) can be of paramount immediate concern (Gilpin and Soulé 1986; Hanski and Gilpin 1997).

On the other hand, some authors have forcefully argued that heterozygosity, as measured by molecular markers, *is* important to a population's health and continued survival and should be monitored accordingly in enlightened management programs (e.g., O'Brien et al. 1996; Vrijenhoek 1996). In case studies involving a variety of taxa, plausible arguments have been advanced for rather direct associations between observed molecular variability and the viability of an endangered taxon. For example, in Isle Royale's population of gray wolves, Wayne et al. (1991b) speculated that an observed behavioral difficulty in adult pair bonding might be due to a recognition-triggered instinct for incest avoidance (because molecular data suggested that these wolves were highly inbred). For the isolated and genetically uniform Gir Forest population of Asiatic lions, O'Brien and Evermann (1988) concluded that high frequencies of abnormal spermatozoa and diminished testosterone levels in males, relative to lions of the African Serengeti, were attributable to intense inbreeding (because similar damaging effects on sperm development have been observed in inbred mice and livestock). In the Sonoran topminnow, Quattro and Vrijenhoek (1989) experimentally monitored several fitness components (survival, growth, early fecundity, and developmental stability) in laboratory-reared progeny of fish collected from nature. All of these fitness traits proved to be positively correlated with mean allozyme heterozygosities in the populations from which the parents originated. In endangered plain pigeons (*Columba inornata wetmorei*) of Puerto Rico, four measures of reproductive fitness (total eggs, fertile eggs, number of hatchlings, and number of fledglings) were significantly correlated with genetic variation as measured in DNA fingerprints (Young et al. 1998). In the greater prairie chicken (*Tympanuchus cupido*), a wild population that had experienced an extreme demographic contraction not only lost heterozygosity as measured by STR loci, but also suffered a significant decline in hatching rates (Bouzat et al. 1998).

Perhaps the most intriguing case for a causal link between inbreeding, low molecular heterozygosity, and diminished genetic fitness involves cheetahs. As mentioned above, there is multifaceted molecular evidence for severely reduced heterozygosity in this species, including at MHC genes that encode cell surface antigens involved in the immune response. Several years ago, a disease (feline infectious peritonitis, or FIP, caused by a coronavirus) swept through several captive cheetah colonies and caused 50%–60% mortality over a 3-year period. This same virus in domestic cats (which have normal levels of MHC variation, as indicated by graft rejections and molecular assays) has an average mortality rate of only 1%. O'Brien and Evermann (1988) speculate that an FIP virus might have acclimated initially to one cheetah, then spread rapidly to other individuals who were genetically uniform in their immunological defenses. In general, enhanced susceptibility to infectious diseases or parasitic agents probably constitutes one of the most serious challenges faced by a population with low genetic variation (O'Brien and Evermann 1988).

Interestingly, emphasis on the special adaptive significance of immunorecognition genes led to a suggestion that captive breeding programs for endangered species be designed with the specific goal of maintaining diversity at MHC loci, because individuals heterozygous at all or most MHC loci would be protected against a wider variety of pathogens than homozygous specimens (Hughes 1991). Furthermore, according to Hughes (1991), "at most loci loss of diversity should not be a cause for concern, because the vast majority of genetic polymorphisms are selectively neutral." This suggestion was immediately criticized on several grounds (Gilpin and Wills 1991; Miller and Hedrick 1991), not least of which was its assumption that variability at genes other than the MHC is adaptively irrelevant. The critics argued that several other loci are known to contribute to disease resistance itself, and that polygenes underlying numerous other quantitative traits of potential adaptive relevance should not be cavalierly ignored (Vrijenhoek and Leberg 1991). Furthermore, selective breeding designed explicitly to maintain MHC diversity could have a counterproductive consequence: accelerated inbreeding, with concomitant accelerated loss of diversity elsewhere in the genome.

In conclusion, especially when large numbers of loci are monitored and multiple assays are performed (e.g., of allozymes, MHC loci, and microsatellites), molecular markers can provide quite reliable estimates of genome-wide heterozygosity, which in turn are theoretically interpretable in terms of historical effective population sizes (Box 9.3). Thus, molecular analyses can help to identify natural or captive populations that display severe genetic impoverishment from past population bottlenecks or inbreeding. Less clear (except perhaps in extreme cases, such as in cheetahs) is the extent to which molecular heterozygosity is a reliable gauge of a population's short-term survival and long-term adaptive potential. Thus, managing captive or natural populations for genetic heterozygosity per se should not come at the expense of neglecting important behavioral, ecological, or environmental factors. Usually, however, all such considerations are mutually reinforcing.

BOX 9.3 Effective Population Size in a Conservation Context

The concept of effective population size (see Box 2.2) is relevant to many conservation efforts. For example, in the early literature of conservation genetics, Frankel (1980) and Soulé (1980) suggested that a minimum effective population size of 50 individuals would be required to stem inbreeding depression, and Frankel (1980; Frankel and Soulé 1981) added that an effective population size of 500 would prevent the long-term erosion of variability by genetic drift. These specific management guidelines became known as the 50/500 rule, and at first they were taken quite literally in some conservation practices (Simberloff 1988). Later, Lande (1995) and Lynch and Lande (1998) concluded from further theoretical considerations that these numbers were about one or two orders of magnitude too small (see also Frankham et al. 2002). All such recommendations are assumption-laden and provide only crude guidelines for population management (Varvio et al. 1986), but they do illustrate the kind of attention that has been devoted to estimating N_e in a conservation context.

As described in Chapter 2, average molecular heterozygosity (H) and *long-term* effective population size (evolutionary N_e) are interrelated under models that assume selective neutrality for genetic variation. If average mutation rates for particular classes of molecular markers are known or assumed, and if the current standing crop of variation in those markers has been assayed in a given extant population, results can be translated into provisional estimates of evolutionary N_e for that population by several statistical procedures (Estoup and Angers 1998; Luikart and England 1999). In one application of this approach involving a population of Tanzanian leopards (*Panthera pardus*), Spong et al. (2000) assayed genetic variation at 17 microsatellite loci and then converted the resulting expected heterozygosity values (assuming Hardy–Weinberg equilibrium) to estimates of evolutionary effective population size using the formula

$$N_e = (1 \; / \; [1 - H_E]^2 - 1) \; / \; 8\mu$$

(see Lehmann et al. 1998). When a microsatellite mutation rate of $\mu = 2 \times 10^{-4}$ was assumed, these molecular data suggested that long-term N_e was about 40,000 for leopards in this region of Africa.

Molecular data from extant populations can also be used to estimate temporal historical demographies of populations. The general philosophy underlying this approach was introduced in the discussions of lineage sorting and coalescent theory in Chapters 2 and 6. One approach involves examining mismatch frequency distributions (i.e., pairwise genealogical distances between extant individuals as estimated from molecular markers). Different kinds of population histories are predicted to leave different types of footprints on these mismatch distributions. For example, a rapid population expansion in the past theoretically "makes a wave" (Rogers and Harpending 1992) in the distribution because many extant lineages will have traced back (coalesced) to that approximate time of evolutionary expansion, thereby making a peak or wave in the mismatch histogram coinciding with that historical period.

A related coalescent approach utilizes lineages-through-time plots analogous to the one presented in Figure 7.4, except that such lineages in the current context are those in intraspecific genealogies (derived from mtDNA or from genetic distance analyses of multiple nuclear loci), rather than branch lengths in species phylogenies. If the number of intraspecific lineages has neither increased nor decreased disproportionately in a population's recent evolutionary history, a lineages-through-time plot should be roughly linear. However, the plot should be concave upward or concave downward, respectively, if the population has experienced recent accelerated or decelerated growth (Nee et al. 1996a). By applying this type of lineage-based reasoning to their genetic data, Lehmann et al. (1998) were led to conclude that the Tanzanian leopard population mentioned above had been large and roughly stable in size over recent evolutionary time.

Such marker-based appraisals of the temporal dynamics of population demographic history now appear quite often in the literature (Lavery et al. 1996; Nee et al. 1996a,b; Rooney et al. 1999), but the results must be interpreted with great caution because they rest critically on several underlying assumptions: that the focal population has been genetically closed and spatially unstructured; that mutation rate estimates are reliable; that genealogies are estimated with considerable accuracy; and that the particular molecular markers employed are appropriate for the ecological or evolutionary time frame presumably covered by the analysis. Thus, marker-based estimates of historical population sizes and their temporal dynamics normally have major uncertainties and wide biological confidence limits (Hedrick 1999; Marjoram and Donnelly 1997).

Molecular markers can also be employed to estimate generation-by-generation N_e in modern or contemporary time. One such method requires that neutral allele frequencies be monitored across two or more generations. Then, effective population size for that time interval can be estimated by statistical procedures that relate N_e to any of several molecular outcomes that might be empirically observed: reductions in heterozygosity due to inbreeding; changes in allele frequencies due to genetic drift; or rates of decay of linkage disequilibrium among loci. Details of these and similar methods can be found in Frankham et al. (2002), Neigel (1996), and Schwartz et al. (1998). Apart from the conservation relevance of numerical abundance per se (because rarity is often an indicator of extinction vulnerability), there are other genetic reasons for interest in contemporary population size: advantageous mutations are far more likely to arise in large than in small populations; and natural selection's influence (relative to drift) is theoretically stronger in larger populations.

TABLE 9.2   Advantages and disadvantages of molecular markers, natural phenotypic markings, and physical tags in the identification of individual wildlife specimens

Tagging technique	Universal?	Heritable?	Permanent?
Molecular markers	Yes	Yes	Yes
Natural phenotypic markings	No	Yes or no	Yes or no
Human-applied physical tags	No	No	Yes or no

Source: Modified from Palsbøll 1999.

For example, management programs for rare or endangered species often promote heterozygosity preservation de facto through captive breeding programs designed to avoid close inbreeding or via habitat preservation that promotes larger effective population sizes in the wild. So, genetic heterozygosity issues are typically just one of several reinforcing elements in a nexus of management considerations for endangered taxa.

Genealogy at the Microevolutionary Scale

Although the heterozygosity and inbreeding issues discussed above have often dominated discussions in conservation genetics, molecular markers are perhaps even better suited for forensic, genealogical, and phylogenetic appraisals of organisms in various conservation contexts, at both micro- and macroevolutionary scales. This section begins to address various microevolutionary applications of molecular genetic techniques in conservation.

Tracking individuals in wildlife management

Wildlife movements traditionally are monitored through the use of physical devices attached by researchers (such as fin tags placed on fish, leg bands on birds, or radio collars on mammals) or by field observations of individuals that are distinguishable by natural phenotypic markings (such as variable color patterns or scars on whales). Protein and DNA molecules can provide specimen tags too, and they offer several advantages over traditional tracking methods (Table 9.2): all individuals in all species come ready-made with these natural labels; genetic tags are transmitted across generations under specifiable modes of inheritance; and modern laboratory techniques (notably PCR) make it routinely possible to obtain genotypes non-destructively and non-invasively (from shed hair, feathers, eggshells, feces, etc.), sometimes without the need to handle or even see the animals analyzed (Morin and Woodruff 1996; Taberlet and Luikart 1999). Another advantage is that population genetic variation in sexual species is normally so high that molecular markers from multiple hypervariable loci are expected to distinguish individuals with high probability (Paetkau and Strobeck 1994).

Suitable for remote field scanning?	Non-invasive?	Permit assessment of random match?
No	Yes or no	Yes
Yes	Yes	Yes or no
Yes or no	No	Yes or no

In genetic studies of humpback whales (*Megaptera novaeangliae*), Palsbøll et al. (1997b) "ground-truthed" the individual-diagnostic power of molecular markers by comparing identifications from genotypes at six polymorphic microsatellite loci with those from natural markings (pigmentation patterns and scar marks) known to distinguish particular specimens. More than 3,000 samples for molecular analysis came either from skin biopsies or from sloughed skin collected from free-swimming whales. The molecular findings not only confirmed the individual-diagnostic power of these genetic markers, but also supported earlier assumptions about the animals' annual migration routes and their patterns of site fidelity to specific summer feeding grounds.

Comparable studies have been conducted on large land animals. In a study of mountain lions (*Puma concolor*) in Yosemite Valley, California, Ernest et al. (2000) "tracked" individual animals by analyzing multi-locus genotypes from fecal samples collected along hiking trails, and also compared these genotypes with those from animal tissues sampled at necropsy. One of the largest molecular studies of this sort involved black bears (*Ursus americanus*) and brown bears (*U. arctos*) in western Canada, for which Woods et al. (1999) used "hair traps" (barbed wire attached to trees) to collect nearly 2,000 hair samples from the wild. Following PCR, these samples were genotyped at the mtDNA control region, at a Y chromosome segment, and at six autosomal microsatellite loci. Each genotypic match was deemed to be a repeat collection from the same individual, whereas distinct genotypes clearly came from different specimens. Through these genetic analyses, the authors were able to identify the species, sex, and individuality of each sample without the need to capture or even observe these free-ranging animals directly. Molecular studies on European bears have achieved similar success using PCR-amplified DNA from fecal samples (Kohn et al. 1995; Taberlet et al. 1997).

Apart from obviating the need to disturb (or be disturbed by) such large and difficult-to-observe animals, another rationale for this kind of individual tracking by non-invasive molecular markers is to estimate current population size. For example, based on microsatellite analyses of coyote (*Canis latrans*) fecal samples systematically collected within a 15-km^2 region of the Santa Monica Mountains near Los Angeles, California, Kohn et al. (1999) estimated a population size of $N = 38$ for these otherwise difficult-to-count animals.

Parentage and kinship

Multi-locus genotypes obtained by non-invasive sampling can also be used in molecular studies of genetic paternity and kinship. For example, Morin et al. (1994a,b) used hair samples, and Gerloff et al. (1995) used fecal samples, as DNA sources for PCR-based assessments of genetic relatedness in wild chimpanzees (*Pan troglodytes*). Such genetic appraisals often have conservation or management relevance. For example, microsatellite paternity analyses in one group of captive chimpanzees showed that the dominant male had sired most (but not all) offspring (Houlden et al. 1997). If these results are typical, they suggest that particular males might generally dominate reproduction in captivity, in which case N_e in each zoo population could be far lower than might otherwise have been assumed. By contrast, a similar marker-based study of wild chimpanzees revealed that slightly more than 50% of analyzed offspring had been fathered by males from outside the focal group (Gagneux et al. 1997). Thus, managers might wish to consider the occasional exchanges of male chimpanzees (or at least their sperm for artificial insemination) between zoos, both to diminish inbreeding within captive colonies and perhaps to mimic natural behavioral and genetic conditions more closely.

Longmire et al. (1992) used DNA fingerprints to assess paternity within a small captive flock of endangered whooping cranes (*Grus americana*) that had been maintained at the Patuxent Wildlife Center in Maryland since 1965. Because of this species' long generation time and low reproductive output, crane husbandry in captivity requires substantial human investment. Thus, in attempts to maximize the captive flock's efficiency in producing fertile eggs, Patuxent researchers sometimes artificially inseminated adult females with semen from several males. However, this procedure also made paternity uncertain, with the undesirable consequence that breeding plans based on maximizing N_e (i.e., avoiding inbreeding) were compromised. The molecular genetic study rectified this situation by providing a posteriori knowledge of biological parentage that in turn could be used as pedigree information in the design of subsequent matings within the flock. A later analysis of microsatellite DNA extended this approach to additional captive and reintroduced wild populations of whooping cranes (Jones et al. 2002).

Molecular data on parentage and kinship have been used to verify breeding records and correct "studbook" errors for several captive or reintroduced endangered species, such as the Waldrapp ibis (*Geronticus eremita*; Signer et al. 1994) and Arabian oryx (*Oryx leucoryx*; Marshall et al. 1999). Such analyses can also be used to help decide which specific individuals in a captive population of known pedigree should have breeding priority when the goal is to maximize population genetic variation (Haig et al. 1990; Hedrick and Miller 1992). For this purpose, two explicit breeding guidelines have been suggested and sometimes implemented: crosses should be designed to equalize expected genetic contributions from the original founder individuals (Lacy 1989); and reproduction by individuals who are deemed to have special genetic significance by virtue of their outlying genealogical positions in the pedigree should be emphasized (Geyer et al. 1989).

Marker-based studies related to parentage and kinship can often provide conservation-relevant information on managed non-captive populations as well. One example involves the Mauna Kea silversword (*Argyroxiphium sandwicense*), which, due to grazing by alien ungulates, had been reduced by the late 1970s to a single remnant population on the island of Hawaii. To improve the species' prospects for survival, beginning in 1973, several hundred plants were intermittently introduced to other sites, but it turned out that all of these outplants were probably first- or subsequent-generation offspring of only two maternal founders. An analysis of molecular markers from 90 RAPD loci indicated that detectable polymorphism had decreased by nearly 75% during this reintroduction program (Robichaux et al. 1997).

An interesting application of genetic parentage analysis involving both captive and wild stocks has been conducted on steelhead trout, *Oncorhynchus mykiss*. In 1996, at Forks Creek, Washington, hatchery-raised fish were released into the wild to supplement natural stocks that were in decline. Later, through parentage analyses based on microsatellite markers, McLean et al. (2003) showed that hatchery-raised females had produced only about one-tenth as many surviving adult progeny per capita as had wild females. The poor performance of the hatchery-reared fish may have been an inadvertent consequence of their domestication, or of their having spawned too early in the winter season, or both, but in any event, the results indicated the near-futility of this management scheme (at least at this site).

Perhaps the greatest value of molecular parentage and kinship analyses in conservation efforts lies in their ability to reveal otherwise unknown aspects of the reproductive biology and natural history of threatened (and other) species in the wild. A case in point involves the rare blue duck (*Hymenolaimus malacorhynchos*), which inhabits isolated mountain streams in New Zealand. From patterns of band sharing in DNA fingerprints, new insights emerged about this species' probable family-unit structure. Significantly higher genetic relatedness was documented within than between blue duck populations from different rivers, probably due to a social system that includes limited dispersal from natal sites and frequent matings among relatives (Triggs et al. 1992), as well as source–sink population dynamics (King et al. 2000). Another example involves mating patterns in marine turtles, about which little was formerly known because these animals mate at sea and individual specimens are difficult to distinguish visually. Molecular paternity analyses of several populations of these endangered species have revealed varied percentages (ca. 10%–60%) of clutches that were the result of successful multiple mating by females (Bollmer et al. 1999; Crim et al. 2002; Fitzsimmons 1998; Harry and Briscoe 1988; Kichler et al. 1999; Parker et al. 1996). Furthermore, maternity analyses based on mtDNA have revealed much new information about individuals' patterns of yearly as well as lifetime migration between nesting sites and foraging grounds (see below).

In one final example, the northern hairy-nosed wombat (*Lasiorhinus krefftii*) may be Australia's most endangered marsupial, but its fossorial and nocturnal habits make it difficult to observe in the wild. Genotypes at nine microsatellite loci were scored in more than two-thirds of the 85 known individuals of this species. The results unexpectedly showed that both males and females were significantly more closely related to their same-sex burrow companions than to random individuals from the population, whereas opposite-sex burrow companions were not closely related (Taylor et al. 1997). Results suggest that wombats may have dispersal mechanisms that lead to associations of same-sex relatives, but that these mechanisms do not lead to a high incidence of inbred matings because close relatives of opposite gender are not significantly associated in space.

Gender identification

Many endangered species in captive breeding programs are sexually non-dimorphic in their visible features, so sex-linked molecular markers can play a key role in identifying gender. In the critically endangered black stilt (*Himantopus novaezelandiae*) of New Zealand, knowledge gained through molecular sexing was used to avoid same-sex pairings in a recovery program that began with only about a dozen adult birds (Millar et al. 1997). Gender-specific molecular markers (notably on the female-specific W chromosome) likewise have been used to sex individuals in captive breeding programs for the critically endangered Taita thrush (*Turdus helleri*; Lens et al. 1998), Norfolk Island boobook owl (*Ninox novaeseelandiae undulata*; Double and Olsen 1997), and others. DNA from molted feathers was PCR-amplified and used to determine the sex of the last Spix's macaw (*Cyanopsitta spixii*) in the wild (Griffiths and Tiwari 1995). The individual turned out to be a male, so a captive female was released as a prospective mate. By establishing the gender of otherwise unknown individuals in any population, molecular markers can also provide information suitable for estimating effective population size, because the relative number of breeding adults of the two genders is an important determinant of N_e (see Box 2.2).

An interesting application of molecular sexing to a wild endangered population involved the Seychelles warbler (*Acrocephalus sechellensis*; Komdeur et al. 1997). Each breeding pair of this non-dimorphic species occupies the same territory for as long as 9 years, producing one clutch per year with a single egg. A daughter (but seldom a son) often remains with the parents for 2 to 3 years, supposedly helping to raise later offspring. Indeed, field studies have shown that these helpers do increase the survival chances of nestlings in high-quality parental territories, but not in poor-quality ones, where the helpers do more harm than good (probably by competing for limited resources). By molecular sexing of baby birds from many nests, researchers discovered that adult warblers on high-quality territories produced mostly stay-home daughters, whereas adult warblers on poor-quality territories produced mostly leave-home sons. The authors interpreted these genetic findings and field observations as evidence that parent Seychelles warblers must have some (unknown) means for adaptively adjusting the sex ratio of their progeny to environmental conditions.

Estimating historical population size

As mentioned in Chapter 2, most high-gene-flow species with large numbers of living individuals have much shallower intraspecific histories (i.e., show far less molecular genetic variation) than might have been expected given their current abundance, suggesting that such species probably experienced demographic contractions (and/or selective sweeps) in times past. Interestingly, molecular markers sometimes register an entirely different relationship between inferred historical population size and contemporary census size in species that today are rare or endangered. A case in point involves several cetacean species that in recent centuries have been the targets of commercial whaling.

Only about 10,000 humpback whales (*Megaptera novaeangliae*), 56,000 fin whales (*Balaenoptera physalus*), and 149,000 minke whales (*B. acutorostrata*) remain in the North Atlantic today. However, based on coalescent theory as applied to current observed standing crops of intraspecific mtDNA sequence diversity, Roman and Palumbi (2003) estimated that historical population sizes were much greater in these species: about 240,000 humpbacks, 360,000 fin whales, and 265,000 minke whales. These estimates are far higher than traditional ones (e.g., from whalers' logbooks), and they have generated considerable controversy (Lubick 2003). However, if these genetically based estimates of historical population size are valid even to a first approximation, scientists and whaling regulators may have to seriously revise their assumptions about the pre-exploitation state of large mammals in the oceans.

Dispersal and gene flow

Management strategies for natural populations can often benefit from knowledge about the movement patterns of organisms (and their gametes) from their natal sites and the magnitudes of realized gene flow (i.e., successful reproduction) that such dispersal entails. For example, an emerging discipline of "restoration genetics" (a synthesis of restoration ecology and population genetics) is concerned with issues such as the spatial extent of local adaptations, magnitudes of gene flow, and the potential risks of introducing foreign genotypes that might cause outbreeding depression or be unwise for other reasons (Hufford and Mazer 2003). Molecular markers can assist greatly in these analyses, and the findings often have conservation implications.

One case in point involves marine protected areas, or MPAs (National Research Council 2000a). A sad truth is that negative human impacts, including pollution and overharvesting, have radically degraded the world's oceans. This situation has triggered calls for efforts to protect and restore marine ecosystems (Lubchenko et al. 2003; National Research Council 1999, 2000a,b). The good news is that the establishment of local or regional "no-take" zones can be highly effective not only in increasing biodiversity and the abundances of many species within the boundaries of such reserves, but also in supplementing commercial and sport fisheries when the larvae of high-fecundity species naturally move from MPAs into nearby waters (Halpern 2003; Halpern and Warner 2003; Palumbi 2001). How large should MPAs be, and how should they be spaced? To a considerable degree, the answers depend on each species' dispersal and recruitment patterns, and in particular, on magnitudes of larval production and transport.

Using molecular markers applied to many marine taxa, scientists are examining genetic patterns in the sea and beginning to interpret results in the context of MPA design. For example, Palumbi (2003) used a combination of genetic findings and computer simulations to deduce that many marine fishes and invertebrate species probably have larval dispersal distances of about 25–150 km, and that genetic divergence under isolation-by-distance models will often be most evident when the demes compared are separated by about 2–5 times the mean larval dispersal range. These results are preliminary, but they do offer a hint at the appropriate spatial scales of demographic connectivity among MPAs (which may be distributed, for example, as "stepping stones" along a coast).

Population Structure and Phylogeography

Issues of kinship, dispersal, and contemporary gene flow grade into those of geographic population structure and intraspecific phylogeny. At these levels too, many applications of molecular markers have been employed in a conservation context. For example, especially in agronomically important plants, much effort has been devoted to collecting and storing "seed banks" from which future genetic withdrawals may prove invaluable in developing needed strains (Frankel and Hawkes 1975). Similar proposals have been made for generating DNA banks for endangered animal species (Ryder et al. 2000). The genetic diversity of such collections can be maximized through knowledge of how natural variation is partitioned within and among populations, a task for which molecular genetic markers are well suited (Schoen and Brown 1991). In general, by revealing how genetic variation is partitioned within any plant or animal species, molecular methods can help to characterize the intraspecific genetic resources that conservation biology seeks to preserve.

Genetics–demography connections

In some cases, molecular genetic findings on extended kinship have clear and immediate relevance for conservation strategies. For example, mtDNA analyses have shown that different rookeries of the endangered green turtle often are characterized by distinctive maternal lineages, indicating a strong propensity for natal homing by females of this highly migratory species (see Chapter 6). Irrespective of the level of inter-rookery gene flow mediated by males and the mating system, this matrilineal structure of rookeries indicates that green turtle colonies should be considered demographically independent of one another at the present time, because any colony that might be extirpated would be unlikely to be reestablished naturally by females hatched elsewhere, at least over ecological time frames (Figure 9.1). This genetics-based deduction is consistent with the observation that many rookeries exterminated by humans over the past four centuries (including those on Grand Cayman, Bermuda, and Alto Velo) have not yet been recolonized. So, the continuing decline of many sea turtle colonies through overharvest-

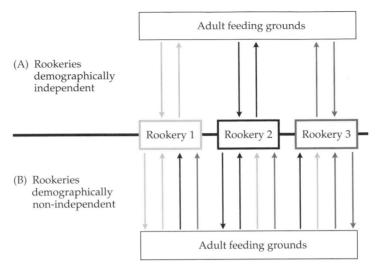

Figure 9.1 Alternative scenarios for the genetics and demography of green turtle rookeries. Coded arrows indicate possible migration pathways of females between natal and feeding grounds. (A) In the diagram above the heavy horizontal line, females are assumed to home to natal sites, in which case rookeries would be independent of one another both genetically (with respect to mtDNA) and demographically (with respect to female reproduction). (B) In the diagram below the horizontal line, females commonly move between rookeries, yielding genetic and demographic non-independence. Molecular data from mtDNA are mostly consistent with scenario (A). (After Meylan et al. 1990.)

ing is not likely to be ameliorated in the near term by natural recruitment from foreign rookeries, meaning that each remaining colony should be valued and protected individually (Bowen et al. 1992).

This argument can be generalized and extended, as shown in Figure 9.2. When both sexes in any species disperse widely from their natal sites to reproduce (lower right quadrant of the figure), high gene flow and little population genetic structure are anticipated in any class of neutral genetic markers, perhaps implying also that local populations are demographically well-connected. Conversely, when both sexes are sedentary (upper left quadrant), low gene flow and strong population genetic structure should be registered in any suitable cytoplasmic or nuclear marker, implying considerable demographic autonomy for each local population. When female dispersal is high and male dispersal low (upper right quadrant), a "Y-linked" gene may or may not show pronounced population structure, depending on whether the vagabond females carry zygotes or unfertilized eggs. If the dispersing females carry haploid gametes which are then fertilized by local males, strong differentiation in Y-linked allele frequencies might be anticipated, and the populations also would tend to be demographically independent from one another with regard to male reproduction. However, if dispersing females carry zygotes or juveniles between populations (e.g., via pregnancy

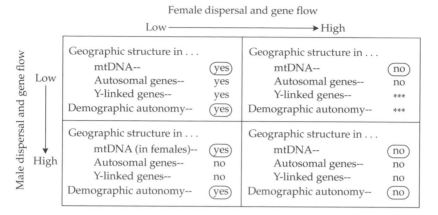

Figure 9.2 Relationships between population genetic structure and gender-specific dispersal and gene flow regimes. Population genetic structure is as might be registered in particular classes of neutral genetic markers with contrasting modes of inheritance. See the text for further explanation, especially for nuanced interpretations of the upper right quadrant where asterisks appear. (After Avise 1995.)

or social interactions), they could maintain demographic ties (as well as avenues for the exchange of Y chromosomes) between geographic locales to which adult males are strictly fidelic.

The most intriguing connection between population demography and matrilineal structure occurs when female dispersal is extremely low and male dispersal high (lower left quadrant in Figure 9.2). Then, as already mentioned, populations could be independent demographically even in the absence of significant spatial structure as registered in nuclear genes. In that case, gene flow estimates based solely on nuclear loci could, if taken at face value, provide a grossly misleading base for management decisions requiring a demographic perspective, such as how many distinct management units, or stocks, exist (see below), how they might respond to harvest, or how habitat corridors between refugia might couple otherwise separated populations.

On the other hand, there are certain conditions under which inferred patterns of female dispersal and matrilineal gene flow (taken at face value) could also be misleading in demography-based population management. For example, if strong geographic matrilineal structure resulted solely from density-dependent restrictions on female movement, populations that are overexploited nonetheless might recover quickly by foreign recruitment as density-based impediments to female dispersal are relaxed. This situation might apply with special force to high-fecundity species such as many marine fishes and invertebrates, in which immigration of even a few gravid females might quickly replenish a local population. Another misleading situation could involve species that conform to a source–sink demographic model (Pullium 1988), in which most geographic populations persist via continued recruitment from reproductively favorable areas. Matrilines would exhibit little spatial structure, suggesting at face value that local populations could recolonize quickly, yet extirpation of critical source populations could doom an entire regional assemblage.

There are additional reasons for caution in relying exclusively on mitochondrial (or other) molecular markers as phylogeographic guides to conservation strategies: the molecular assays themselves may fail to reveal some true population genetic structure; empirical patterns might be attributable to evolutionary forces other than historical gene flow and genetic drift (such as habitat-specific natural selection); and, perhaps most importantly, historical gene flow might have been high enough to homogenize frequencies of genetic lineages across locales ($Nm \gg 1$), yet nonetheless far too low (in terms of absolute numbers of exchanged individuals per generation) to imply population demographic unity in any sense relevant to management. Thus, in assessing whether populations exhibit sufficient demographic autonomy to qualify as quasi-independent entities in contemporary time, field observations and experiments on animal dispersal will remain necessary because they complement and extend the kinds of information on historical gene flow that often come from molecular markers.

Inherited versus acquired markers

In characterizing wildlife stocks—in fisheries management, for example—an important distinction is between heritable (genetic) and acquired features (Booke 1981; Ihssen et al. 1981). The latter include a variety of environmentally induced attributes often employed in population analysis, such as isotope ratios in tissues (Chamberlain et al. 2000; Hobson 1999; Marra et al. 1998), parasite loads (Caine 1986; T. P. Quinn et al. 1987), or various phenotypic characters, such as vertebral counts, that may be developmentally plastic and at least partially indicative of environmental exposures (Jockusch 1997). Acquired characters also include the physical tags, bands, or transmitters that researchers attach to animals to monitor their movements.

Acquired markers unquestionably serve an important role in population analysis because they reveal where individuals have spent various portions of their lives. For example, a popular exercise in recent years has been to reconstruct where particular birds have migrated not only through banding studies, but also by measuring the ratios of stable isotopes (notably those of carbon or hydrogen) in their bodies (Kelly and Finch 1998). These isotopes tend to differ predictably with environmental factors such as latitude (Rubenstein et al. 2002). However, physical tags and other acquired characteristics are seldom transmitted across generations, so they do not necessarily illuminate the reproductive phenomena that can also be highly germane to population management, nor can they reveal the principal sources of phylogeographic diversity within a species. By combining information from acquired characters with data from molecular genetic markers, more and different kinds of information about movement patterns are obtained than from either source of data considered alone (e.g., Clegg et al. 2003).

Population structure registered by

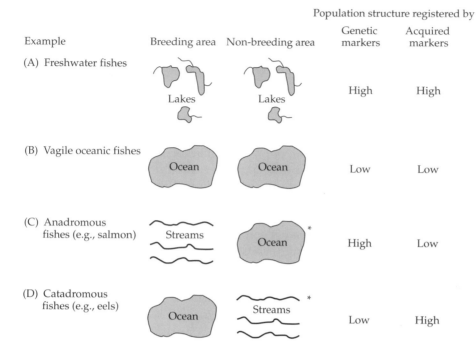

Figure 9.3   Four possible categories of relationship between magnitudes of apparent population structure as registered by distributions of inherited and of environmentally acquired markers. For categories (C) and (D) , it is assumed that the environmental markers were acquired during the oceanic and freshwater life stages, respectively (as indicated by the asterisks).

In general, four possible relationships can be envisioned between the apparent population structures registered by genetic and by environmentally induced tags (Figure 9.3). These two classes of markers may reveal high and concordant structures, as, for example, when populations inhabit separate environments that induce different phenotypes in individuals and also promote population genetic divergence. Such a concordant outcome might characterize a sedentary species, or perhaps one that is geographically strongly partitioned (such as many freshwater taxa) (Figure 9.3A). Alternatively, both genetic and acquired markers might reveal relative population homogeneity over broad areas, as in many vagile marine species (Figure 9.3B).

However, two types of discordant outcomes are also possible. First, strong population structure might be evidenced by genetic markers despite an absence of population differences in acquired characteristics. Such could be the case in a natal-homing anadromous species assayed at the freshwater adult life stage, provided that the acquired characters were gained during the oceanic phase of the life cycle (Figure 9.3C). Conversely, significant geographic structure in acquired characters might be evident despite an absence of genetic differentiation, as, for example, in a randomly mating catadromous species sampled during the freshwater portion of its life cycle (provided that the acquired characters were incorporated during the freshwater phase) (Figure 9.3D). In other words, a true population genetic pattern would probably not be registered accurately by any non-genetic marker acquired at a locality or life stage other than where reproduction occurs.

These distinctions between genetic and acquired markers can be important in population management and conservation applications (Swain and Foote 1999). In the case of anadromous salmon originating in different rivers, for example, diagnostic population markers that happened to coincide with political boundaries (state or federal) could be ideal for equitably apportioning oceanic catches to fishermen from the relevant jurisdictions. For these purposes, any marker (genetic or otherwise) would do. Thus, naturally acquired stream-specific parasite loads, as well as dyes or tags artificially applied to smolts (juvenile salmon in streams), would be perfectly suitable for these management purposes. However, the "populations" thus identified would have little or no evolutionary genetic significance if, for example, sufficient gene flow between stream populations occurs via lapses in natal homing. The point is that both inherited and acquired markers can find gainful employment for various management objectives, but only genetic markers have direct connection to population genetic and evolutionary issues.

Mixed-stock assessment

In fisheries management in particular, much attention has been devoted to "stock assessment" (Ovenden 1990; Ryman and Utter 1987; Shaklee 1983; Utter 1991), which can be viewed as a practical application of the principles and procedures of population structure analysis. Molecular markers can materially help in identifying such stocks (Shaklee and Bentzen 1998). Indeed, computerized databases now exist that summarize such information for fishes (Imsiridou et al. 2003). Molecular markers can also be invaluable in assessing the genetic composition of mixed stocks as a prelude, for example, to setting harvesting quotas or otherwise managing finite fish resources.

In many commercial and sport fisheries, amalgamations of native and introduced (hatchery-produced or otherwise transplanted) stocks are harvested. In several instances, molecular markers that distinguish genetic exotics from natives have been employed to monitor the fate ("success") of these introductions and to assess whether hybridization and introgression with native fish have taken place. The outcomes of such appraisals have varied. Among brown trout (*Salmo trutta*) in the Conwy River of North Wales, an introduction of fry from anadromous populations into landlocked populations resulted in considerable hybridization between these two salmonid forms, as gauged by allozyme markers (Hauser et al. 1991). Thus, a stocking program designed to bolster catches of trout in landlocked bodies of water had come at the risk of introgressive loss of unique genetic characteristics in the native landlocked form. For this same species in Spain, however, hatchery supplementation was a "failure": Protein electrophoretic analyses showed that genetically marked fry introduced from hatcheries into indigenous populations failed to reach sexual maturity and apparently did not contribute to the pool of catchable and reproductive fish (Moran et al. 1991). Similar molecular studies of the Japanese ayu (*Plecoglossus altivelis*) likewise demonstrated that introduced stocks contributed little to reproduction in a native river population of this fish species (Pastene et al. 1991).

Commercial fisheries may also involve exploitation of mixed native stocks, an example being the capture of anadromous salmon at sea. As alluded to above, a long-sought goal in salmon management has been to discover diagnostic markers that distinguish anadromous fish originating from different river drainages or regions (Beacham et al. 2000, 2001, 2002, 2003; Bermingham et al. 1991; Potvin and Bernatchez 2001). Using such markers, the proportionate contribution of each breeding population to a mixed oceanic fishery could be determined, and harvesting strategies or capture allocations might then be tailored to varying population sizes or other attributes of the respective reproductive stocks. However, seldom have native salmon from nearby rivers displayed fixed allele frequency differences that would make such stock assignments unambiguous within a management jurisdiction (Stewart et al. 2003; but see Baker et al. 2003). Rather, observed genetic differences often involve mere shifts (albeit statistically significant ones) in population allele frequencies at multiple polymorphic loci. Thus, even when particular source populations cluster together in terms of composite genetic distances (Figure 9.4), alleles typically are shared, such that precise genetic contributions from different stocks can be estimated only in a probabilistic or statistical framework (e.g., Millar 1987; Pella and Milner 1987; Xu et al. 1994).

One example of such a statistical mixed-stock analysis in a conservation context involved the endangered loggerhead turtle, *Caretta caretta*. In the Mediterranean Sea, a longline fishery for swordfish incidentally captures an estimated 20,000 juvenile loggerheads per year, of which at least 20% perish (see Bowen and Avise 1996). To assess which breeding populations are affected by this source of mortality, mtDNA cytochrome *b* sequences in longline captures were compared with those found in North Atlantic and Mediterranean loggerhead rookeries (Bowen 1995; Laurent et al. 1993, 1998). Based on a maximum likelihood estimate from a mixed-stock model, only about 50% of loggerheads in the longline bycatch were derived from Mediterranean nesting beaches; most of the rest had come from the western Atlantic.

Molecular markers have likewise been employed to track migratory movements of marine turtles in other contexts (Avise and Bowen 1994). Based on mtDNA assays, juvenile loggerheads taken near Charleston, South Carolina, were shown to have been hatched at various rookery sites as far south as Florida (Sears et al. 1995). Loggerheads collected in the Azores and Madeira (near Africa) proved to have originated from rookeries in the southern United States and Mexico (Bolten et al. 1998); and young specimens from

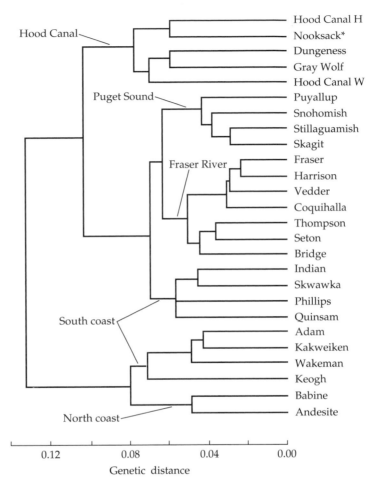

Figure 9.4   Genetic relationships among populations of pink salmon
(*Oncorhynchus gorbuscha*) from 26 streams in Washington and British Columbia.
Shown is a phenogram based on allele frequencies at more than 50 allozyme loci.
Reasonably strong genetic clustering is evident for five geographic regions
(although the Nooksack River, indicated by an asterisk, enters Puget Sound).
However, this clustering was due primarily to differing frequencies of shared alle-
les rather than fixed allelic differences. (After Shaklee et al. 1991.)

Baja California proved to have migrated across the Pacific from Japanese
rookeries (Bowen et al. 1995). For green turtles (*Chelonia mydas*) as well,
mixed-stock analyses of mtDNA data have been used to quantify the relative
contributions of various North Atlantic rookeries to two of that species'
major feeding grounds: off the northeastern coast of Nicaragua (Bass et al.
1998) and in the Bahamas (Figure 9.5).

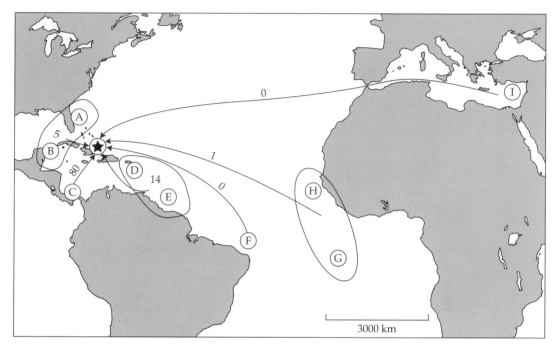

Figure 9.5   Genetic contributions of different green turtle rookeries (lettered circles) to a feeding-ground population (starred circle) on Great Inagua, Bahamas. Numbers indicate maximum likelihood estimates of percent contributions of different rookeries to this feeding population, based on a mixed-stock model using observed mtDNA haplotypes. (After Lahanas et al. 1998.)

Shallow versus deep population structures

A general point about the recognition of population units in conservation biology involves the distinction between shallow (recent or contemporary) and deep (ancient) population genetic structure. In other words, "statistically significant" population genetic structures are not all equal, but instead may be reflective of widely varying depths of evolutionary separation. Furthermore, populations of many species may be strongly isolated from one another at the present time (and at most time horizons in the past), but nonetheless remain tightly connected in a genealogical sense through recent or pulsed episodes of gene flow. Such considerations have given rise to a key distinction in conservation genetics between "management units" (MUs) and "evolutionarily significant units" (ESUs).

The conceptual distinction between MUs and ESUs (Box 9.4) is nicely illustrated by population genetic findings on the green sea turtle. Nesting rookeries of this endangered species display evident but "shallow" mtDNA population structure within a given ocean basin, due in large part to natal

BOX 9.4 MUs and ESUs

In many, if not most, species, organismal dispersal is far too low to promote appreciable demographic connections between geographic populations on a generation-by-generation basis. Limited dispersal may be due to inherent low vagility or to physical or ecological impediments to individual movements. With regard to population genetic evidence, the logic underlying the concept of a potential "management unit" (MU) is as follows: Any population that exchanges so few migrants with others as to be genetically distinct from them will normally also be independent of them demographically, at least at the present time. In the literature of commercial fisheries, MUs are often referred to as "stocks," toward which harvesting quotas or other management plans are directed (Avise 1987; Ovenden 1990; Ryman and Utter 1987). In principle, for any species, populations that are genuinely autonomous in contemporary time could qualify as distinct MUs (regardless of whether or not genetic markers may have helped to document such demographic autonomy).

When (as is often the case) empirical field evidence on the magnitude of demographic connection between natural populations is limited or absent, MUs can nonetheless be identified provisionally by significant differences in allele frequencies at neutral marker loci. Mitochondrial haplotypes are especially powerful for identifying potential MUs (Avise 1995; Moritz 1994) because of their typically fourfold smaller effective population size (compared with haplotypes at autosomal loci), and because of the special relevance of matrilines to population demography.

An "evolutionarily significant unit" (ESU), in principle, is one or a set of conspecific populations with a distinct long-term evolutionary history mostly

homing by females (see Chapter 6). Thus, despite the recency of the genetic separations, most rookeries are demographically autonomous with regard to reproduction and, accordingly, each should qualify as an MU. The magnitude of genetic divergence between rookeries in the Atlantic versus Pacific Ocean basins is far greater (see Figure 6.5), apparently due to much longer-term barriers to gene flow. Thus, these regional assemblages of rookeries should qualify as distinct ESUs. Similar genetic architectures have been found in global molecular surveys of several other (but not all; Dutton et al. 1999) species of marine turtles (Bowen and Avise 1996), and these findings are likewise highly informative in identifying provisional MUs and ESUs.

Thus, both demographic and evolutionary separations are relevant to conservation efforts and management strategies (Avise 1987; Dizon et al. 1992). Even shallow population genetic separations can be important because they indicate, for example, that magnitudes of contemporary recruitment from outside sources are probably insufficient to enable overexploited populations to recover quickly via natural immigration. Deep genetic separations within a species are also significant because they register major

separate from that of other such units (Ryder 1986). As such, ESUs are the primary sources of historical (and perhaps adaptational; Fraser and Bernatchez 2001) genetic diversity within a species and are thereby worthy of special consideration in conservation efforts (Avise 2000a; Bernatchez 1995; Moritz 1994). In practice, ESUs often conform closely to "intraspecific phylogroups," as defined by Avise and Walker (1999). They might also be appropriately interpreted as validated "subspecies" (the practical reason for not doing so being the likelihood of introducing taxonomic confusion: Most currently recognized subspecies were described from limited information in the pre-molecular era, and might or might not qualify as bona fide ESUs upon more detailed analysis).

Operational genetic criteria for recognizing ESUs have ranged from broad to detailed. A general suggestion is that ESUs contribute substantially to the overall genetic diversity within a species (Waples 1991b). A more explicit recommendation is that ESUs be identified as groups of populations "reciprocally monophyletic for mtDNA alleles and also differ[ing] significantly for the frequency of alleles at nuclear loci" (Moritz 1994). Any such empirical suggestion is arbitrary to some extent because there can be no clean line of demarcation along the continua of possible magnitudes of population genetic differentiation or temporal depths of population separation. Application of genealogical concordance principles (see Chapter 6) offers perhaps the best hope for critically evaluating evolutionary depths of population separation using molecular genetic markers. Notwithstanding the empirical challenges of genetically identifying potential MUs and ESUs in particular instances (e.g., Bowen 1998; Dimmick et al. 1999; Paetkau 1999), the concepts themselves remain among the most important, if not revisionary, perspectives to have emerged from phylogeographic appraisals of microevolution.

sources of evolutionary genetic diversity that can be especially important to conservation's broader goal of biodiversity preservation. What follows are a few more examples of how both shallow and deep population structures within a species can inform conservation plans for rare or endangered taxa.

Africa's highly endangered black rhinoceros (*Diceros bicornis*) and white rhinoceros (*Ceratotherium simum*) have been decimated by poachers who supply lucrative markets in ornamental dagger handles and supposed medicines made from rhino horns. Even if poaching were eliminated, genetic and demographic problems in the few remaining populations could also threaten these species' survival. Should all remaining conspecific rhinos be considered members of a single population for conservation purposes? Or should various recognized forms (including different taxonomic subspecies) be managed as separate population entities? The strategy of mixing and breeding conspecific rhinos from separate geographic sources might enhance each species' chance of survival by increasing effective population sizes and thereby forestalling inbreeding depression (and also reducing the risk of stochastic demographic extinctions). On the other hand, attempted amalgamations of well-differentiated genetic forms might lead to outbreeding depression or cause an overall erosion of genetic variation, most of which might occur among (rather than within) geographic regions.

To address such issues, several molecular studies have been conducted. Merenlender et al. (1989) reported only a small genetic distance ($D = 0.005$ in allozyme comparisons) between two white rhino subspecies. Likewise, Swart and Ferguson (1997) found only very limited allozyme divergence between black rhinoceros subspecies, and Ashley et al. (1990) and O'Ryan et al. (1994) similarly found low overall mtDNA sequence divergence ($p <$ 0.5%). These genetic results suggested that outbreeding depression in crosses between black rhino subspecies would probably not be a serious problem. A subsequent survey of the rapidly evolving mtDNA control region in two black rhino subspecies resulted in an estimate of about 2.6% sequence divergence, a value that the authors interpreted to indicate a separation time of perhaps a million years (Brown and Houlden 2000). Nonetheless, this sequence divergence in the mtDNA control region remains far lower than the directly comparable estimate of 14% sequence divergence between black rhinoceros and white rhinoceros.

In a complex of endangered desert pupfishes (genus *Cyprinodon*) in and near Death Valley, California, early allozyme surveys revealed little polymorphism within, but considerable genetic divergence among, most populations, which currently are confined to isolated springs and streams that are remnants of inland lakes and interconnected watercourses that existed in former pluvial times (Turner 1974). A different pattern of geographic structure within one of the desert pupfish species may be an exception that proves the rule. In *C. macularius* populations of the Salton Sea area of southern California, polymorphism within remnant colonies accounted for 70% of the total genetic variance, with differences among colonies contributing only 30% (Echelle et al. 1987; Turner 1983). The hydrologic history of this region suggests an explanation: These populations probably were in repeated contact due to historical cycles of flooding of the lake basin, perhaps most recently in the early twentieth century when water broke out of the Colorado River irrigation system (Turner 1983).

In general, fishes in desert basins of North America are declining at an alarming rate, with more than 20 taxa having gone extinct in the last few decades and many more at risk of the same fate. Meffe and Vrijenhoek (1988) considered management recommendations that might stem from the two different types of population structure exhibited by the remaining species, and these scenarios should apply to other biological settings as well (Vrijenhoek 1996). In the "Death Valley model" (patterned after the desert pupfishes), populations are small and isolated, such that most of the total genetic diversity within a species is likely to be partitioned among sites. In managing such populations, there is no need for concern about human-mediated interruptions of gene flow because natural genetic contact has been absent since the time when the ancestral watercourses desiccated. Indeed, precautions should be taken to avoid artificial gene flow among healthy populations that are naturally isolated and are probably in the process of evolutionary diversification. The main concern should be maintenance of high N_e within locales to alleviate both genetic and demographic dangers of small population size.

Under an alternative "stream hierarchy model" (patterned after the endangered *Poeciliopsis occidentalis* and certain members of the *Cyprinodon* complex), populations in small dendritic water systems exhibit varying degrees of connection and gene flow, such that a larger fraction of total genetic diversity occurs within colonies. Meffe and Vrijenhoek (1988) suggested that management programs in such situations should be aimed at preserving the genetic integrity of each species while at the same time maintaining genetic variability. Thus, moving conspecific individuals among sites within a river basin should pose no special difficulties because such movement probably occurs naturally. However, precautions should be taken to avoid artificial movements between separate drainages, especially when this involves mixing different species that might compete or hybridize.

Considerable attention has also been devoted to assessing molecular genetic variation within and among populations of endangered "big-river" desert fishes in the American Southwest (Dowling et al. 1996b; Garrigan et al. 2002), and then devising management recommendations accordingly (Dowling et al. 1996b; Hedrick et al. 2000). In a bold and proactive conservation plan based on genetic and demographic considerations for several species that inhabit the main stem of the Colorado River, Minckley et al. (2003) proposed that these native fishes should be bred and their progeny allowed to grow in isolated, protected, off-channel habitats from which competitor non-native fishes (otherwise abundant following human-mediated introductions) are excluded. Panmictic adult populations (the source of brood stock for each native species) would continue to reside in the main channel of the river, but these would be supplemented each generation with individuals taken from the isolated off-channel habitats. With respect to genetics and demography, the intent of this program would be to significantly increase effective population sizes in these critically endangered species while otherwise being faithful to their natural biology.

Molecular analyses of population genetic structure have likewise been conducted on numerous endangered plants (see reviews in Hamrick and Godt 1996; Rieseberg and Swenson 1996). To pick just one example, the meadowfoam *Limnanthes floccosa californica* is a geographically restricted annual that is endemic to vernal pools in Butte County, California. Protein electrophoretic analyses revealed that nearly all of the total genetic diversity (> 95%) was distributed among rather than within the 11 known populations, such that estimates of inter-population genetic exchange were very low ($Nm < 0.02$; Dole and Sun 1992). Based on these findings, the authors recommended a conservation plan that emphasizes preservation of as many of these MUs as possible.

Conservationists' ultimate aim is to preserve biodiversity, an important currency of which is genetic diversity at a variety of levels (Ehrlich and Wilson 1991). A widely held sentiment is that added conservation value or worth should attach to groups of organisms that are genealogically highly (rather than minimally) distinct from other such groups (Barrowclough 1992; Faith 1994; Vane-Wright et al. 1991). At the intraspecific level, this means that ESUs (often identified now by molecular markers) should also be assigned special conservation importance.

Consider the great apes, which have been the subject of several molecular surveys. In chimpanzees (*Pan troglodytes*), two highly divergent mtDNA lineages proved to distinguish a geographically disjunct population in western Africa from populations to the east (Goldberg and Ruvolo 1997; Morin et al. 1992, 1994a). Similarly, in gorillas (*Gorilla gorilla*), relatively deep genetic subdivisions between eastern and western forms in Africa have been reported (Garner and Ryder 1996; Ruvolo et al. 1994; Saltonstall et al. 1998). In orangutans (*Pongo pygmaeus*), marked genetic divergence between populations in Sumatra and Borneo has led to suggestions that two taxonomic species be recognized (Warren et al. 2001; Xu and Arnason 1996; Zhi et al. 1996). Table 9.3 describes similar examples of other rare or endangered taxa in which deep phylogeographic splits have been documented (typically in mtDNA sequences).

Rather ancient genealogical subdivisions sometimes appear even in the most unlikely of situations. The African elephant (*Loxodonta africana*) is one of the most conspicuous and supposedly well known species on Earth, but recent DNA sequence analyses of four independent nuclear genes (Roca et al. 2001) uncovered a deep genealogical partition between forest-dwelling and savanna-dwelling forms (Figure 9.6). This general pattern was supported and extended by findings of salient genetic distinctions between these and other geographic populations of African elephants at microsatellite loci and in mtDNA sequences (Comstock et al. 2002; Eggert et al. 2002). These molecular data, interpreted together with morphological, ecological, and reproductive evidence, led to a recent taxonomic revision in which two African species are now recognized (*L. africana* and *L. cyclotis*). At least two major mtDNA clades also exist in the Asian elephant (*Elephas maximus*), but these currently are recognized as conspecific ESUs rather than species (Fernando et al. 2003b; Fleischer et al. 2001). These examples also illustrate how molecular studies of "intraspecific" phylogeography can grade into issues of species-level taxonomy relevant to conservation (see below).

Lessons from intraspecific phylogeography

Ehrlich (1992) lamented that time is running out for saving biological diversity: "The sort of intensive, species-focused research that I and my colleagues have carried out [on checkerspot butterflies] … appears to have a very limited future in conservation biology. Instead, if a substantial portion of remaining biodiversity is to be conserved, detailed studies of single species must be replaced with 'quick and dirty' methods of evaluating entire ecosystems, designing reserves to protect them, and determining whether those reserves are working." Willers (1992) went further: "To dwell endless-

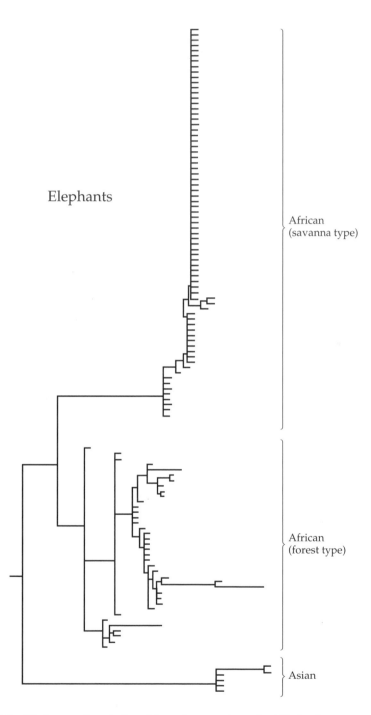

Elephants

African
(savanna type)

African
(forest type)

Asian

Figure 9.6   Phylogenetic relationships of approximately 120 elephants sampled
from multiple geographic locations and habitat types. (Relationships are based on
nuclear DNA sequences.) (After Roca et al. 2001.)

TABLE 9.3 Examples of rare or endangered species in which molecular markers have identified relatively deep phylogeographic separations between at least some populations that conventionally have been considered conspecific

Threatened species	Major phylogeographic distinction
Lasmigona sp. (bivalve mollusk)	Northern versus southern populations in eastern U.S.
Cicindela dorsalis (tiger beetle)	Atlantic Coast versus Gulf of Mexico populations
Latimeria chalumnae (coelacanth)	Indonesian versus African ocean waters
Litoria pearsoniana (hylid frog)	Opposite sides of Brisbane River Valley, Australia
Lachesis muta (bushmaster snake)	Central versus South America
Gnypetoscincus queenslandiae (rainforest skink)	Northern versus southern populations in Queensland
Gopherus polyphemus (gopher tortoise)	Eastern versus western regions of southeastern U.S.
Xerobates agassizi (desert tortoise)	East versus west of the Colorado River in western U.S.
Apteryx australis (kiwi bird)	North Island versus South Island, New Zealand
Strix occidentalis (spotted owl)	Northern versus southern populations in western U.S.
Lanius ludovicianus (loggerhead shrike)	San Clemente Island versus southern California mainland
Burramys parvus (pygmy possum)	Northern, central, and southern Australian Alps
Dasyurus maculatus (marsupial tiger quoll)	Tasmania versus mainland Australia
Perameles gunnii (barred bandicoot)	Mainland Australia versus Tasmania
Petrogale xanthopus (rock wallaby)	Southern Australia versus Queensland
Cervus eldi (Eld's deer)	Three distinctive and disjunct units in south Asia
Lycaon pictus (wild dog)	Southern versus eastern Africa
Eumetopias jubatus (Stellar sea lion)	Gulf of Alaska northward versus southeast Alaska, Oregon
Trichechus manatus (West Indian manatee)	Three or four major phylogeographic units in species range

Reference
King et al. 1999
Vogler and DeSalle 1993, 1994
Holder et al. 1999
McGuigan et al. 1998
Zamudio and Greene 1997
Cunningham and Moritz 1998
Osentoski and Lamb 1995
Lamb et al. 1989
Baker et al. 1995
Barrowclough et al. 1999; Haig et al. 2001
Mundy et al. 1997
Osborne et al. 2000
Firestone et al. 1999
Robinson 1995
Pope et al. 1996
Balakrishnan et al. 2003
Girman et al. 1993; Roy et al. 1994
Bickham et al. 1996
Garcia-Rodriguez et al. 1998

ly on the tasks of obtaining more and ever more data for the expressed purpose of managing a biological reserve is to suggest that enough knowledge is just around the corner. This is not so." In a sense, these authors are correct. The severe conservation challenges facing society cannot be solved solely by detailed genetic and ecological studies of particular taxa or regions. A lack of political and social will to implement existing scientific understanding is a far greater impediment to conservation efforts than is a lack of detailed scientific information.

On the other hand, there are important lessons to be gained from detailed phylogeographic studies of the sort described in this book. Most importantly, it has become abundantly clear that most species should not be viewed as undifferentiated monotypic entities for conservation (or other) purposes. Instead, they typically consist of geographically variable populations with hierarchical and sometimes deep genealogical structures. From this recognition have come two rather unorthodox but fairly generalizable management guidelines for conserving intraspecific phylogeographic diversity.

LIMIT UNNECESSARY TRANSPLANTATIONS. Most biologists recognize that introductions of exotic species can cause irreparable harm to regional biodiversity by forcing extinctions of native species, but they have been slower to appreciate problems that can arise from transplanting and mixing well-differentiated genetic stocks within a species. Indeed, fish and game management agencies often sponsor active transplantations of organisms from one geographic region to another for purposes such as bolstering local population sizes, introducing "desirable" genetic traits into an area, or increasing local genetic heterozygosity. Unfortunately, undesirable consequences may also stem from such transplantations, including the possibility of disease or parasite spread, the irretrievable loss of the rich historical genetic records of populations, and the inevitable erosion of overall genetic diversity within a species (much of which was generated and maintained through historical geographic separations). Some transplantation programs may be justifiable, for example, when reintroducing a native species to a former range from which it has been extirpated by human activities. However, a developing perspective in conservation biology is that the burden of proof in any proposed transplantation program normally should rest on advocates of this strategy, rather than on those who would question the desirability of transplantations on the grounds cited above.

DESIGN REGIONAL RESERVES. Comparative molecular analyses of regional biotas have demonstrated that specific geographic areas (such as isolated refugia of the Late Tertiary or Quaternary) have been evolutionary wellsprings for phylogeographic diversity within and among closely related taxa. In many cases, these phylogeographic sources of molecular biodiversity were also recognized in traditional biogeographic appraisals of species' distributions (a conventional basis for defining biogeographic provinces). These types of findings can motivate attempts to ensure the safety and integrity of regional biodiversity hotspots. Such conservation efforts might include implementing strict guidelines to discourage transplantations between phylogeographic provinces. Ideally, they might even entail designation of federally protected reserves, perhaps analogous to National Parks (except that these "phylogeographic parks" would be focused on preserving *biodiversity*, rather than on preserving special geological features of the landscape). Although no such regional perspective on phylogeographic diversity can hope to capture the idiosyncratic population structures and genetic subdivisions of each and every species, it can provide useful broad guidelines for management strategies, particularly as natural environments come under increased pressure and decisions on conservation prioritization become inevitable.

Issues At and Beyond the Species Level

Speciation and conservation biology

Just as issues of close kinship grade into those of population structure and intraspecific phylogeny, so too do the latter grade into issues regarding the speciation process and associated taxonomic judgments. In the field of conservation biology, these taxonomic issues often come into especially sharp focus in discussions of "endangered species" (Box 9.5).

RECOGNITION OF ENDANGERED SPECIES.   Most taxonomic assignments in use today (including formal designations of species and subspecies) were first proposed early in the twentieth century or before, often based on limited phenotypic information and preliminary assessments of geographic variation. How adequately do these traditional taxonomies summarize true genetic biodiversity? For most groups, this question remains to be answered through continued systematic reappraisals, for which various molecular markers are ideally suited. The problems and challenges involved are far more important than a mere concern with nomenclature might at first imply. Taxonomic assignments inevitably shape our most fundamental perceptions about how the biological world is organized. Ineluctably, formal names summarize the biotic units that are perceived and, therefore, discussed. In a conservation context, these perceived entities provide the pool of candidate taxa from which are chosen particular populations toward which special management efforts may be directed (Soltis and Gitzendanner 1998).

A "dusky seaside sparrow" by any other name is just as melanistic in its plumage, but without a taxonomic moniker this local population, formerly endemic to Brevard County, Florida, would probably not have been recognized as a biological unit worthy of special conservation attention. The dusky seaside sparrow was described in the late 1800s as a species

BOX 9.5 Legal Protection for Species Under the ESA
and CITES

In 1966, passage of the Endangered Species Preservation Act initiated federal
efforts in the United States to protect rare species from extinction. Three years
later, attempts to remedy this Act's perceived deficiencies (e.g., lack of habitat
protection) resulted in the Endangered Species Conservation Act of 1969. Again,
shortcomings in the new act were recognized, and in 1973, Congress enacted a
more comprehensive Endangered Species Act (henceforth ESA), which today
remains the country's strongest legal strategy for protection of rare species. The
ESA's stated intent was to "provide a means whereby the ecosystems upon
which endangered species and threatened species depend may be conserved,
[and] to provide a program for the conservation of such … species."

To qualify for ESA protection, a taxon must appear on an official list that is
prepared and updated under the auspices of the U.S. Fish and Wildlife Service
and the National Marine Fisheries Service. Listings may be made for a species, a
subspecies, or a "distinct population segment," the latter sometimes being interpreted as "a group of organisms that represents a segment of biological diversity that shares an evolutionary lineage and contains the potential for a unique
evolutionary future" (NRC 1995). A species is deemed to be endangered if it is
at risk of extinction throughout all or a significant portion of its range, and to be
threatened if it is likely to become endangered in the foreseeable future.
Utilitarian criteria for listing are that the plant or animal in question occur in
numbers or habitats sufficiently depleted to critically threaten its survival. For
example, grizzly bears in the lower 48 states have been listed as threatened,
whereas the much larger Alaska population has not. As of 2003, the lists of
threatened and endangered species included more than 1,250 taxa within U.S.
boundaries and another 500 or more taxa elsewhere. Interestingly, plant species
somewhat outnumber animal species in these lists.

(*Ammodramus nigrescens*) distinct from other seaside sparrows (*A. maritimus*), which were common along the North American Atlantic and Gulf
coasts. Although the dusky was later demoted to formal subspecific status
(*A. m. nigrescens*), its nomenclatural legacy prompted continued conservation focus on this federally "endangered species" when, during the 1960s,
the Brevard County population declined severely due to deterioration of its
salt marsh habitat. In 1987, the last known dusky died in captivity after last-
ditch efforts to save the population through captive breeding failed.

Following the extinction of the dusky seaside sparrow, a molecular
study of nearly the entire seaside sparrow complex (which includes nine
conventionally recognized subspecies) produced a surprise (Avise and
Nelson 1989; Nelson et al. 2000). In terms of mtDNA sequence, the dusky
proved to have been essentially indistinguishable from other Atlantic coast

The ESA has been criticized on several grounds (Geist 1992; Rohlf 1991), including its focus on preserving particular species (rather than ecosystems or biodiversity per se), its frequent emphasis on "charismatic megafauna" (such as many birds, large mammals, and showy plants, perhaps to the neglect of less conspicuous rare taxa), and the considerable time lag often involved in the formal listing process. On the other hand, many of these concerns reflect implementation problems (insufficient funding, lack of political will, incomplete scientific knowledge, etc.) rather than faults in the legislation itself (O'Connell 1992). In the future, the ESA's species-based approach might well be supplemented with legislation designed explicitly to fulfill the broader goals of ecosystem maintenance and biodiversity protection. Nonetheless, at least as an interim measure, the ESA has been a revolutionary, far-sighted, and relatively effective piece of governmental legislation.

CITES (the Convention on International Trade in Endangered Species), also adopted in 1973, has been signed by more than 125 countries. This treaty is another enlightened legal document that focuses on species of conservation concern. Its goal is to "regulate the complex wildlife trade by controlling species-specific trade levels on the basis of biological criteria." Prohibited under the CITES treaty are all commercial and most non-commercial trade in several hundred species that are listed in its oft-updated "Appendix I." Restricted but not entirely disallowed under CITES is commercial trade in many additional threatened species listed in its "Appendix II." "Appendix III" is a list of optional species that countries might wish to protect because these taxa could soon become endangered by trade. As of 2003, about 25,000 plant species and 5,000 animal species were covered by CITES provisions, most of them appearing in Appendix II.

Like ESA, the CITES treaty was a visionary and helpful legal approach to conservation. Its main weaknesses involve implementation difficulties in the treaty's enforcement mechanisms.

birds, but highly distinct in genealogy from birds nesting along the Gulf of Mexico coastline (Figure 9.7). This genetic split probably registers an ancient (Pleistocene) population separation, as further evidenced by a striking similarity between the phylogeographic pattern of the seaside sparrow and those of several other (non-avian) estuarine taxa in the southeastern United States (see Chapter 6). Thus, the traditional taxonomy for the seaside sparrow complex (upon which endangered species designations and management efforts were based) apparently had failed to capture the true phylogenetic (at least matrilineal) partitions within this taxonomic group. They had given special emphasis to a presumed biotic partition that was genealogically shallow or nonexistent, and they had failed to recognize a deep phylogeographic subdivision between most Atlantic and Gulf coast populations.

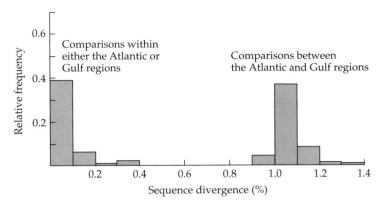

Figure 9.7   Frequency histogram (mismatch distribution) of genetic distances among seaside sparrows. Shown are estimates of mtDNA sequence divergence between pairs of birds from various locales along the Atlantic and Gulf of Mexico coastlines of the southeastern United States.

These molecular findings do not excuse or justify any poor land management practices (chronicled by Walters 1992) that may have led to the dusky seaside sparrow's extinction, nor should publication of the molecular results be interpreted as heartlessness over this population's loss. The extinction of any natural population is regrettable, particularly in this era of rapid deterioration of natural habitats and biodiversity. However, the field of conservation biology finds itself in a difficult situation, with hard choices to be made about which taxa can be saved and how best to allocate limited resources among competing conservation demands. Especially in such circumstances, management decisions should be based on the best available scientific information.

Another example of a misleading taxonomy for an endangered species involved the colonial pocket gopher, *Geomys colonus*, endemic to Camden County, Georgia. This taxon was first described in 1898 on the basis of rather cursory descriptions of pelage and cranial characteristics. The population remained essentially unnoticed and unstudied until the 1960s, when gophers in Camden County were "rediscovered." The population referable to *G. colonus* then consisted of fewer than 100 individuals, and it was listed as an endangered species by the State of Georgia. A molecular genetic survey was subsequently conducted of allozymes, chromosome karyotypes, and mtDNA RFLPs. None of these genetic assays detected a consistent genetic distinction between *G. colonus* and geographically adjacent populations of its presumed sister congener, *G. pinetis* (Laerm et al. 1982). Results were not attributable to a lack of sensitivity in the techniques employed because each molecular method revealed dramatic genetic differences among a broader geographic array of *G. pinetis* populations (particularly those in eastern versus western portions of the species' range; see Figure 6.8). The conclusion in this case was clear: Extant gophers under the name of "*G. colonus*" did not warrant recognition as a distinct species. Either the description of *G. colonus* in 1898 was inappropriate, or an originally valid (i.e., genetically distinct) species had gone extinct early in the twentieth century and was later replaced by *G. pinetis* immigrants into Camden County.

Of course, in principle as well as in practice, no study can prove the null hypothesis that genetic differences between putative taxa are absent. One example of a "valid" species that proved to be nearly indistinguishable in genetic composition from a close relative is the endangered pallid sturgeon (*Scaphirhynchus albus*). In assays of 37 monomorphic and polymorphic allozyme loci, *S. albus* could not be genetically differentiated from its more common congener, the shovelnose sturgeon (*S. platorynchus*; Phelps and Allendorf 1983). However, pronounced morphological differences between these taxa and their sympatric distribution implied that pallid and shovelnose sturgeons nonetheless qualify as a good biological species (the possibility of phenotypic plasticity was deemed unlikely; Phelps and Allendorf 1983). Subsequent assays of microsatellite loci did eventually confirm that pallid and shovelnose sturgeons are genetically distinguishable in sympatry (Tranah et al. 2001), but the broader message is that molecular data (especially when preliminary or based on only one class of markers) should not be interpreted in isolation in making taxonomic or conservation judgments, but instead should be integrated with other available lines of evidence.

Molecular reappraisals of taxonomically suspect species may, of course, also bolster the rationale for special conservation efforts. One case in point involves the nearly extinct silvery minnow (*Hybognathus amarus*), which is endemic to the Rio Grande in the southwestern United States. This species has had a troubled taxonomic history, with some researchers viewing it as a distinct species and other placing it in synonymy with *H. nuchalis* or *H. placitus*. However, based on a survey of 22 allozyme loci, Cook et al. (1992) observed several fixed allelic differences between these taxa, as well as overall levels of genetic distance ($D > 0.10$) somewhat greater than those normally distinguishing conspecific populations in other fish groups. The authors concluded that there was little justification for considering *H. amarus* conspecific with the other species with which it previously had been synonymized.

Another case in point involves the highly endangered Kemp's ridley turtle (*Lepidochelys kempi*), which nests almost exclusively at a single locale in the western Gulf of Mexico and has been the subject of the largest international preservation effort for any marine turtle. The Kemp's ridley has a close relative, the olive ridley (*L. olivacea*), that is among the most abundant of marine turtles and has a nearly global distribution in warm seas. The close morphological similarity between *L. kempi* and *L. olivacea*, and the geographic distributions of these supposed sister taxa (which make little sense under modern conditions of climate and geography; Carr 1967)

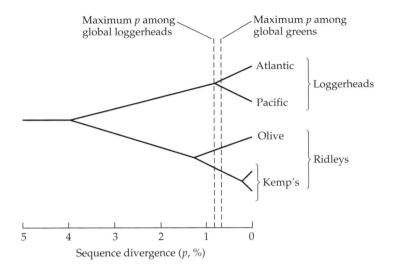

Maximum *p* among
global loggerheads

Maximum *p* among
global greens

Atlantic

Pacific

$\Big\}$ Loggerheads

Olive

$\Big\}$ Ridleys

Kemp's

Sequence divergence (*p*, %)

Figure 9.8   Phylogeny for ridley and loggerhead marine turtles estimated from mtDNA data. Assayed olive ridleys from the Atlantic and Pacific oceans proved indistinguishable and, hence, are plotted here as a single unit. Also shown are the *maximum* levels of mtDNA genetic distance observed among conspecific loggerhead turtles and conspecific green turtles from around the world. (After Bowen et al. 1991.)

raised serious doubts about the evolutionary distinctiveness of *L. kempi*. Nonetheless, reappraisals of this taxonomic assemblage based on mtDNA assays indicated that *L. kempi* is more distinct from *L. olivacea* than assayed Atlantic versus Pacific populations of *L. olivacea* are from one another (Bowen et al. 1991, 1998). Furthermore, the levels of molecular differentiation between the Kemp's and olive ridleys surpassed (slightly) those observed among any conspecific populations of the globally distributed green or loggerhead turtles (Figure 9.8). These findings leave little doubt that *L. kempi* warrants special taxonomic recognition.

Neglected taxonomies for endangered forms can kill, as exemplified by studies of the tuataras of New Zealand. These impressive reptiles were treated by government and management authorities as belonging to a single species, despite molecular and morphological evidence for three distinct (and taxonomically described) groups (Daugherty et al. 1990). Official neglect of this genetic diversity may unwittingly have consigned one form of tuatara (*Sphenodon punctatus reischeki*) to extinction, whereas another distinctive form (*S. guentheri*) appears to have survived to the present only by good fortune. As stated by Daugherty et al. (1990), good taxonomies "are not irrelevant abstractions, but the essential foundations of conservation practice."

In recent years, molecular genetic appraisals of endangered populations and taxa "at the species boundary" have become almost routine (Goldstein et al. 2000). Table 9.4 summarizes several additional examples in which molecular markers have been informative with regard to characterizing genetic relationships in taxonomically problematic groups of special conservation concern.

MOLECULAR FORENSICS AND LAW ENFORCEMENT.   A common challenge in the enforcement of wildlife protection laws is to identify the biological source of blood, carcasses, meat, feathers, or commercial products from endangered or illegally harvested species. Various molecular markers have tremendous utility for such forensic purposes due to their species-diagnostic power. For example, Cronin et al. (1991c) compiled a list of diagnostic molecular characters (mtDNA digestion profiles in this case) for 22 species of large mammals that are frequent objects of illegal poaching. Similarly, Ross et al. (2003) compiled a list of mtDNA sequences for numerous whale and dolphin species in a Web-based "DNA surveillance" program, which can be used as a standard reference against which to compare mtDNA sequences obtained from unknown cetacean material. Especially when used in conjunction with appropriate statistical assignment tests (e.g., Manel et al. 2002), such molecular markers enable researchers and law enforcement agencies to characterize unknown biological material when obvious morphological characters are unavailable for analysis.

An early application of molecular markers to law enforcement occurred in 1978, when a Japanese trawler fishing in United States coastal waters was suspected of illegally harvesting a rockfish species, *Sebastes alutus*. Tissues confiscated by enforcement officers did indeed prove upon protein electrophoretic examination to have come from that species, thereby contradicting claims in the trawler's log that no such specimens had been taken (Utter 1991). A similar case in Texas involved a suspected illegal sale of flathead catfish (*Pylodictis olivaris*). Protein electrophoretic analyses verified the species identity of frozen fish fillets and led to a fine levied against the seller (Harvey 1990). Molecular findings in wildlife forensics can also exonerate the falsely accused. In another case in Texas, electrophoretic analyses of confiscated fillets revealed that fishermen were innocent of suspected illegal possession of red drum (*Sciaenops ocellatus*) and spotted sea trout (*Cynoscion nebulosus*) (Harvey 1990).

Some forensic applications involve distinguishing conspecific populations. For example, a commercial catch of king crab (*Paralithodes camtschatica*) was claimed to have been harvested in a region of Alaska open to fishing, but it proved upon protein electrophoretic examination to have come from a closed area in the northwestern Bering Sea (Seeb et al. 1990). A more peculiar example of molecular forensics in a geographic context involved a bass-fishing tournament in Texas, in which a winning fisherman was suspected of having smuggled in a huge largemouth bass (*Micropterus salmoides*). Tissue samples from the trophy specimen were examined electrophoretically and shown to have come from a genetically distinct Florida subspecies (Philipp et al. 1981) that apparently had indeed been imported illegally (Harvey 1990).

TABLE 9.4 Examples illustrating the wide variety of taxonomically challenging groups of conservation concern for which diverse molecular approaches have been employed in genealogical appraisals

Threatened taxonomic group	Description
Plants	
Acacia trees	cpDNA sequences were used to assess phylogenetic relationships of rare and common species in Western Australia.
Epipactis orchids	Allozyme and cpDNA markers revealed population structures, breeding systems, and genetic relationships of three species.
Eucalyptus trees	Allozymes helped reveal relationships among three taxonomically difficult species endemic to Tasmania.
Zelkova trees	cpDNA sequences revealed strong genetic differentiation between two relict tree congeners in Sicily.
Fungi	
Lentinula edodes	rRNA sequences revealed four highly distinct genetic lineages that had been masquerading asa single morphotypic species.
Invertebrate animals	
Achatinella land snails	mtDNA sequences revealed several distinct evolutionary units of the endangered Hawaiian tree snail (*A. mustelina*).
Amblema freshwater mussels	Allozymes and mtDNA were used to search for diagnostically distinct groups in these and related endangered forms for which morphological features are sparse.
Cambarus cave crayfish	Allozyme data were used to estimate genetic variation and divergence in a difficult taxonomic complex of rare species in the Ozarks.
Cicindela tiger beetles	Phylogenetic lineages and conservation units in these beach-dwelling insects were evaluated by mtDNA (including that isolated from nineteenth-century dried specimens).
Crassostrea oysters	mtDNA and nDNA markers clarified phylogeography in a rare Portuguese oyster and its relationships to a cryptic congener.
Hemileuca buckmoths	Small populations suspected to represent a distinct species proved indistinguishable from other populations in mtDNA and allozyme markers.
Lithasia aquatic snails	mtDNA sequence data were used to reappraise phylogeny and taxonomy in about a dozen recognized taxa, many of which are imperiled.
Lucetta sponges	rDNA sequences revealed four primary, regionally restricted clades that help characterize World Heritage Areas in the western Pacific.
Vertebrate animals	
Acipenseriforme fishes	Molecular sequence data were used to estimate phylogeny in several endangered species of sturgeon and paddlefish.
Charadrius plovers	Geographically disjunct breeding populations of endangered piping plover were shown to have been in extensive and recent genetic contact, as gauged by allozyme comparisons.
Dipsochelys tortoises	Individuals suspected by morphology to be Seychelles tortoises, otherwise extinct, proved upon molecular inspection not to be genetically divergent from common extant tortoises from Aldabra.
Dusicyon foxes	A taxonomically problematic species (*D. fulvipes*) in Peru was shown to be genetically quite distinct from nearest congeners.

Reference
Byrne et al. 2001
Squirrel et al. 2002
Turner et al. 2000
Fineschi et al. 2002
Hibbett and Donoghue 1996
Holland and Hadfield 2002
Mulvey et al. 1997, 1998
Koppleman and Figg, 1995
Goldstein and DeSalle 2003; Vogler 1994
Huvet et al. 2000
Legge et al. 1996
Minton and Lydeard 2003
Wörheide et al. 2002
Birstein et al. 1997; Krieger et al. 2000
Haig and Oring 1988
Palkovacs et al. 2003; see also Austin and Arnold 2002; Austin et al. 2003
Yahnke et al. 1996

(*continued*)

TABLE 9.4 (*continued*)   Examples illustrating the wide variety of taxonomically challenging groups of conservation concern for which diverse molecular approaches have been employed in genealogical appraisals

Threatened taxonomic group	Description
Vertebrate animals	
Equus horses and zebras	mtDNA sequences were used to clarify phylogenetic relationships among six species and eight subspecies of mostly rare equids.
Eubalaena right whales	mtDNA sequences revealed three distinct maternal evolutionary units in a taxonomic complex usually considered to contain two species.
Gazella gazelles	mtDNA sequence analyses called into question whether captive individuals of the otherwise extinct Saudi gazelle truly belong to this taxon.
Icterus orioles	mtDNA sequences were used to assess the phylogeographic status of several endangered oriole taxa in the Lesser Antilles.
Lynx cats	A critically endangered and probably extinct taxonomic species (*L. pardinus*) from Iberia proved to be highly distinct from other *Lynx* species.
Melanotaenia rainbowfish	A taxonomically problematic species (*M. eachamensis*) was shown to be genetically quite distinct from its nearest congeners.
Panthera tigers	Nuclear and mitochondrial markers were used to sort out genetic relationships among five recognized geographic subspecies.
Petrogale rock wallabies	Multifaceted genetic analyses of populations in this taxonomic complex identified a differentiated remnant form in Victoria.
Polioptila gnatcatchers	mtDNA analyses questioned some of the taxonomic designations upon which conservation plans for these birds were based.
Scaphirhynchus sturgeon	Genetic relationships involving three taxonomically problematic species were examined using mtDNA genotyping.
Sternotherus freshwater turtles	A taxonomically problematic species (*S. depressus*) was shown to be genetically quite distinct from its nearest congeners.
Tympanocryptis "dragons"	mtDNA data were used to assess phylogenetic and taxonomic issues in species and subspecies of Australian agamid reptiles.

A global moratorium on commercial harvest of many cetacean species was established by the International Whaling Commission in 1985–1986, but whaling never completely stopped, and whale meat (known as *kujira* in Japan, *gorae* in Korea) continues to be sold in retail outlets, especially in eastern Asia and Scandinavia. One possibility is that this "whale meat" comes from small cetaceans, such as porpoises and dolphins, that still can be harvested legally. Or perhaps the retail material is not from cetaceans, but instead is compressed fishmeal or another seafood substance that has been mislabeled to appear more exotic or pleasing to consumers. In the early 1990s, Baker and Palumbi (1994, 1996; Baker et al. 1996) began purchasing "whale products" from Asian outlets for mtDNA analysis. By comparing unknown samples against a reference database of cetacean mtDNA sequences, they were able to identify the species and sometimes even the geographic source of each sam-

Reference

Oakenfull et al. 2000

Rosenbaum et al. 2000

Hammond et al. 2001

Lovette et al. 1999

Beltrán et al. 1996

Zhu et al. 1998

Cracraft et al. 1998

Browning et al. 2001

Zink et al. 2000

Campton et al. 2000

Walker et al. 1998c

Scott and Keogh, 2000

ple (Figure 9.9). About half of the retail products proved to have come from whale or dolphin species that could plausibly have been harvested under legal permits, but some of the other retail samples came from endangered humpback and fin whales that had been killed illegally.

Analogous research programs for the molecular forensic identification of wildlife products have since been instituted for several other taxonomic groups that include threatened or endangered species (Table 9.5). In 1989, the U.S. Fish and Wildlife Service opened a wildlife forensics laboratory in Ashford, Oregon. The purpose of this first-of-its-kind "Scotland Yard for animals" is to identify wildlife products such as those confiscated from poachers or illegal traders in wildlife products. This laboratory practices traditional morphological forensics (based on "bones and feathers"), but much of its effort involves identification of unknown samples by protein and

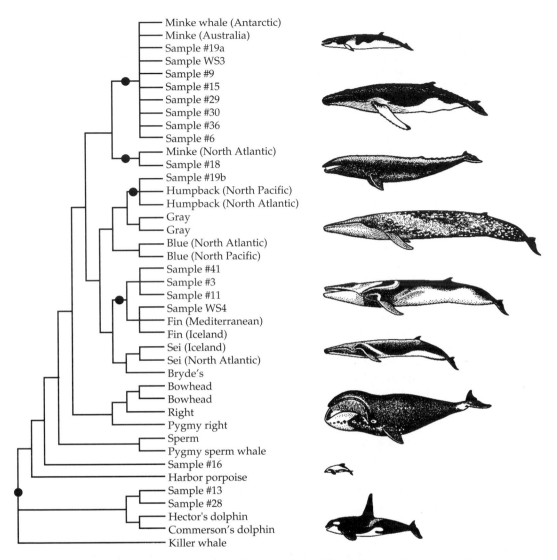

Figure 9.9   Forensic identification of retail "whale meat" products. Shown is a molecular phylogeny (based on mtDNA control region sequences) for representative cetacean species, with black circles indicating > 90% bootstrap support for an indicated clade. Note how this genetic analysis permitted assignment of each retail sample (numbered) to one or another cetacean species or lineage. (After Baker and Palumbi 1994; Frankham et al. 2002.)

DNA evidence. The magnitude of the task is daunting. Unlike the several hundred police crime labs in the United States, which deal with a single species (*Homo sapiens*), the Ashland workers (plus a small cadre of like-minded university researchers) must cope with molecular diagnostics in the entire remainder of the biological world.

Hybridization and introgression

In the context of rare and endangered species, instances of introgressive hybridization, as sometimes documented using molecular markers, have raised both biological and legal issues.

BIOLOGICAL ISSUES.   One evolutionary threat to rare species is genetic swamping through extensive hybridization with related taxa (Levin et al. 1996; Rhymer and Simberloff 1996). An empirical example involves the nearly extinct mahogany tree *Cercocarpus traskiae*, endemic to Santa Catalina Island in Los Angeles County, California. Protein electrophoretic, RAPD, and morphological appraisals of about a dozen remaining adult trees revealed that about one-half of the individuals were products of hybridization between *C. traskiae* and other *Cercocarpus* species that were more abundant on the island (Rieseberg and Gerber 1995; Rieseberg et al. 1989). These genetic discoveries led to two management suggestions intended to lower the probability of further hybridization (Rieseberg et al. 1989): Eliminate individuals of other *Cercocarpus* species from near the remaining pure *C. traskiae* specimens, and transplant seedlings and established cuttings from non-hybrid individuals to more remote areas on Santa Catalina Island.

In another endangered plant—the yellow larkspur (*Delphinium luteum*), whose range is restricted to Bodega Bay, California—questions arose as to whether this localized species might itself be a product of interspecific hybridization. To test the hypothesis that *D. luteum* arose from crosses between two common congeners (*D. decorum* and *D. nudicaule*), Koontz et al. (2001) used allozyme and RAPD techniques. Diagnostic markers for the candidate parental taxa proved not to be additive in *D. luteum*, indicating that this endangered species was probably not of recent hybrid origin.

A probable example of genetic swamping in an animal species involves the declining New Zealand gray duck (*Anas superciliosa*), which has been severely affected by hybridization with introduced mallard ducks (*A. platyrhynchos*) (Rhymer et al. 1994). Near the other end of the continuum, hybridization may be confined to production of first-generation hybrids. An example involves the bull trout (*Salvelinus confluentus*), a federally threatened species that occasionally hybridizes with introduced brook trout (*S. fontinalis*) in the western United States. Analyses using allozyme and mtDNA markers revealed that hybrids beyond the F_1 generation were present, but quite rare (Kanda et al. 2000b; Leary et al. 1993). In such cases, any detrimental effects of hybridization may mostly entail wasted reproductive effort, or perhaps negative social or ecological effects of hybrid animals, rather than genetic introgression per se.

Cutthroat trout (*Oncorhynchus clarki*) native to the western United States and Canada comprise an assemblage of approximately 14 recognized subspecies, many of which are threatened by human habitat alterations and artificial introductions of non-native trout species (Behnke 2002). Most of

TABLE 9.5 Examples of forensic applications of PCR-based DNA assays employed to identify the biological source of the material indicated

Wildlife product	Taxonomic group of concern
Caviar (fish eggs)	Sturgeon (*Acipenser, Huso*)
Body parts	Sharks
Penises	Pinnipeds (seals and allies)
"Turtle meat"	Large freshwater turtles (especially *Macroclemys*)
Scrimshaw	Sperm whales
Feces and hair	Chinese tiger (*Panthera tigris amoyensis*)
Carcasses	Deer
Infants	Chimpanzee (*Pan troglodytes*)

the subspecies have protected legal status, and two already may be extinct. Surveys of allozymes and mtDNA in scores of cutthroat populations have revealed a complex phylogeographic pattern, with some subspecies almost indistinguishable genetically and others as distinct as normal congeneric species (see Allendorf and Leary 1988, and references therein). Molecular markers have also documented extensive hybridization between transplanted cutthroat trout subspecies as well as between native cutthroat and introduced rainbow trout (*O. mykiss*) (Weigel et al. 2002, 2003). For example, introgression from rainbow trout was observed in 7 of 39 assayed populations in Utah (Martin et al. 1985); and in Montana, more than 30 of 80 populations formerly thought to be pure "westslope" cutthroat trout proved upon molecular examination to include products of hybridization with either rainbow trout or the "Yellowstone" cutthroat form. In the Flathead River drainage in Montana (considered one of the last remaining strongholds of native westslope cutthroat trout), only 2 of 19 headwater lakes sampled contained pure populations, and detailed genetic analyses further revealed that the hybridized headwater populations were "leaking" foreign genes into downstream areas (Allendorf and Leary 1988).

Such findings on the introgressed structure of particular cutthroat populations led to two conservation-related concerns (Allendorf and Leary 1988). First, hybrids between genetically differentiated trout often exhibit reduced fitness due to developmental abnormalities (Allendorf and Waples 1996).

Description	Reference
Food samples in retail outlets examined for species of origin; about 25% were mislabeled as to species.	Birstein et al. 1998; DeSalle and Birstein 1996
Identifications made possible for species in worldwide pelagic fisheries.	Shivji et al. 2002
These supposed aphrodisiac samples tested for species of origin; many samples proved not to be from pinnipeds.	Malik et al. 1997
Food samples tested for species of origin; about 25% actually were from alligator.	Roman and Bowen 2000
Molecular methods developed to extract DNA samples from teeth and bones.	Pichler et al. 2001
Unknown fecal samples genotyped and compared against known references to confirm presence of tigers in China.	Wan et al. 2003
DNA fingerprinting used to identify remains of protected cervid species.	Fang and Wan 2002
The geographic origin of orphaned or "refugee" animals was determined.	Goldberg 1997

Second, extensive introgressive hybridization carries the danger of genetic swamping and loss of locally adapted populations. As stated by Allendorf and Leary (1988), "The eventual outcome of widespread introgression and continued introduction of hatchery rainbow trout is the homogenization of western North American trout into a single taxon. Thus, we would exchange all of the diversity within and between many separate lineages, produced by millions of years of evolution … for a single new mongrel species."

In the Canidae (dogs and allies), several studies have used molecular markers to address whether endangered wild species hybridize occasionally among themselves, or perhaps with domestic canines. In the southeastern United States, mtDNA genotypes normally characteristic of domestic dogs (*Canis familiaris*) have been reported in some coyotes (*C. latrans*) (Adams et al. 2003). In northeastern North America, a unidirectional introgression of coyote mtDNA into some gray wolf (*C. lupus*) populations may have occurred following hybridization between gray wolf males and coyote females (Figure 9.10). In eastern Europe, some gray wolves display mtDNA genotypes characteristic of domestic dogs, again probably as a result of hybrid matings involving wolf males (Randi et al. 2000; but see also Vila and Wayne 1999). In Africa, molecular markers indicate that one population of the world's most endangered canid—the Ethiopian wolf (*C. simensis*)—likewise contains genes derived from hybridization with domestic dogs (Gottelli et al. 1994).

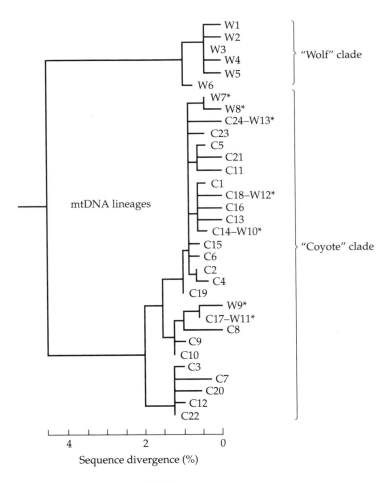

Figure 9.10 Phylogeny for mtDNA genotypes observed in gray wolves and coyotes from North America. Note that whereas several genotypes observed in wolves (W) group separately from those of coyotes (C), others (indicated by asterisks) do not, the exceptions perhaps being attributable to recent introgressive hybridization between these species. Additional data from mtDNA (Wayne et al. 1992) and microsatellites (Roy et al. 1994) were interpreted as further support for this possibility (Wayne 1996). (After Lehman et al. 1991.)

One endangered canid has even been suggested to be a product of past hybridization and introgression. The red wolf (*C. rufus*) formerly ranged throughout the southeastern United States, but declined precipitously after 1900 and became extinct in the wild in the mid-1970s. Molecular analyses of remaining captive animals (as well as museum-preserved skins and blood samples from deceased wild specimens) revealed that extant red wolves are genetically similar to gray wolves, but also contain some alleles possibly derived from hybridization with coyotes (Roy et al. 1996; Wayne 1992;

Wayne and Jenks 1991). Whether this hypothesized introgression occurred in ancient times or recently, following an eastward range expansion of coyotes, is unclear (Nowak and Federoff 1998). Indeed, altogether different interpretations of these data have been advanced (Dowling et al. 1992; Ferrell et al. 1980; Nowak 1992). For example, P. J. Wilson et al. (2000) reported that red wolves are genetically most similar to small-bodied eastern Canadian wolves, with which they might share common ancestry independent of any hybridization with other canids. These uncertainties have produced intense debates over the taxonomy of *C. rufus* (Phillips and Henry 1992) as well as the advisability of management programs that include restoration of a viable wild population in North Carolina (Gittleman and Pimm 1991; Nowak and Federoff 1998).

Based on molecular and other evidence, past or current hybridization between fish species has contributed to the genetic composition of several recognized taxa, including some that are considered threatened or endangered (de Marais et al. 1992; Gerber et al. 2001; Meagher and Dowling 1991; see review in Dowling and Secor 1997). The same can be said of a number of plant species (see review in Rieseberg 1997). Thus, rather than being merely an erosive force that diminishes or swamps preexisting genetic diversity by blurring species' distinctions, introgressive hybridization can sometimes also be viewed as a highly creative evolutionary force (Arnold 1997). Under this view, hybridization can spawn new genotypic diversity, move genetic adaptations between species, and sometimes even generate new species that represent stabilized recombinant lineages.

Another biological context in which hybridization might be a good thing for conservation efforts is exemplified by attempts to save the endangered Florida panther (*Felis concolor coryi*), a subspecies of cougar (also known as puma or mountain lion) from the Florida Everglades. Despite protection from hunting since the late 1960s, the population has continued to decline, due in part to overt genetic defects (including increased frequencies of undescended testicles and defective sperm in males) attendant on intense inbreeding in this small population of only 60–70 animals (O'Brien et al. 1990, 1996). In a theoretical analysis addressing population genetic aspects of the Florida panther's imperiled condition, Hedrick (1995) concluded that a genetic restoration of the population could be achieved by translocating Texas cougars in such a way as to promote about 20% gene flow into the Florida population in an initial generation and about 2%–4% in generations thereafter. In 1995, an introduction program was begun with the release into southwestern Florida of eight Texas cougars, which have since been hybridizing successfully with the native animals (Land and Lacey 2000; Maehr et al. 2002).

In summarizing the consequences of hybridization and introgression for conservation biology, Allendorf et al. (2001) concluded that these phenomena have contributed to the extinction of many species (directly or indirectly), but have also played important creative roles in the evolution of many plant and animal taxa. They also concluded that "any policy that deals with hybrids must be flexible and must recognize that nearly every situation involving hybridization is different enough that general rules are not likely to be effective."

LEGAL ISSUES.   For several years, the "Hybrid Policy" of the Endangered Species Act (Box 9.6) was a flashpoint for legal controversies surrounding the biological topic of hybridization and introgression in conservation programs. This policy, which initially denied formal protection to organisms of hybrid ancestry, served as a basis for challenging several existing endangered species designations and associated management programs. The situation of the red wolf, described above, provides one example. Another involves the gray wolf, which, as mentioned, also appears to have hybridized on occasion with the coyote. When interpreted against the philosophical platform of the original Hybrid Policy, these genetic findings prompted at least one petition to the Interior Department to remove the gray wolf from the list of endangered species in the northern United States. [This petition was then denied by the U.S. Fish and Wildlife Service (Fergus 1991).] Clearly, for several reasons, including those mentioned in Box 9.6, the mere documentation of hybridization involving an endangered population should not be sufficient grounds for removing an endangered or threatened species from the protection rosters.

Species phylogenies and macroevolution

PHYLOGENETIC APPRAISALS.   Phylogenetic considerations at the species level and above can also be of relevance to conservation biology. Such interest may be partly academic. For example, the ancestry and geographic origin of the endangered Hawaiian goose (*Nesochen sandvicensis*) have long been intriguing. Based on mtDNA analyses of living and fossil material, it now appears that the morphologically distinctive Hawaiian goose is allied more closely to the Canada goose (*Branta canadensis*) than to the black brant (*Branta bernicla*) or emperor goose (*Chen canagica*), two other candidate species (Paxinos et al. 2002a,b; Quinn et al. 1991). From this molecular evidence on maternal lineages, it was concluded that the Hawaiian goose's ancestors colonized the islands from North America within at most the last one million years.

North America's black-footed ferret (*Mustela nigripes*) is a highly endangered member of the Mustelidae (weasel and skunk family). An abundant species with a broad distribution a century ago, it was decimated primarily through human eradication of its principal prey base and associate: prairie dogs (*Cynomys* spp.). In 1981, a few remaining specimens of *M. nigripes* were discovered in Wyoming, and molecular analyses (allozymes and microsatellites) showed that this deme had extremely low genetic variability because of severe population bottlenecks (see Table 9.1). The black-footed ferret has more common relatives elsewhere, however, including the steppe polecat (*M. evers-*

BOX 9.6 Hybrid Policies under the Endangered Species Act

In a series of official opinions issued by the Solicitor's Office of the U.S. Department of the Interior beginning in 1977, it was concluded that natural or artificial hybrids between endangered species, subspecies, or populations should not receive protection under the Endangered Species Act (ESA; see Box 9.5). The rationale was that even if these hybrids were themselves to breed, they would not produce purebred offspring of either parental taxon and, hence, would not promote the purposes of the ESA.

These decisions (which became known as the "Hybrid Policy") had serious ramifications. For example, they prompted formal petitions from some land use constituencies to remove certain protected taxa from the endangered species list on the grounds that introgression had compromised the genetic integrity of the listed forms. However, as noted by Grant and Grant (1992), many, if not most, species of plants and animals hybridize at least occasionally in nature, so if hybridizing species fall outside the limits of protection afforded by the ESA, few candidate taxa will ever qualify for protection; and "if rarity increases the chances of interbreeding with a related species, presumably because conspecific mates are scarce, then the species most in need of protection, by virtue of their rarity, are the ones most likely to lose it under current practice, by hybridizing."

O'Brien and Mayr (1991) attacked the Hybrid Policy on additional grounds by claiming that definitional and other operational difficulties had produced "confusion, conflict, and ... certain misinterpretations of the [Endangered Species] Act by well-intentioned government officials." They also concluded that whereas management programs that promote hybridization between distinct species should normally be discouraged, the Hybrid Policy should not be applied to native subspecies or populations because the latter retain a potential to interbreed as part of ongoing ecological and evolutionary processes in nature.

Such criticisms from biologists prompted a withdrawal memorandum from the Solicitor's Office, stating that "the rigid standards set out in those previous opinions [of the Hybrid Policy] should be revisited" and that "the issue of 'hybrids' is more properly a biological issue than a legal one." Consequently, the U.S. Fish and Wildlife Service now recognizes that limited amounts of introgression do not automatically disqualify individuals from "species membership," nor do they necessarily preclude partially introgressed populations from being afforded legal protection under the ESA.

manni) of Siberia. O'Brien et al. (1989) employed allozyme assays to assess the phylogenetic position of *M. nigripes* within Mustelidae (Figure 9.11). These data confirmed that *M. eversmanni* and *M. nigripes* are sister taxa, and also showed that they differ genetically ($D \cong 0.08$) by about as much as do closely related congeners in many other mammalian groups (see Figure 1.2). These two species probably separated about 0.5–2.0 mya (O'Brien et al. 1989).

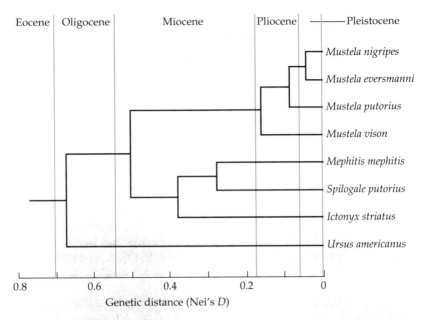

Figure 9.11 Phylogenetic position of the endangered black-footed ferret (*Mustela nigripes*) within the Mustelidae, based on allozyme comparisons. Other species assayed were the Siberian steppe polecat (*M. eversmanni*), European polecat (*M. putorius*), mink (*M. vison*), striped skunk (*Mephitis mephitis*), spotted skunk (*Spilogale putorius*), African striped skunk (*Ictonyx striatus*), and, as an outgroup, the American black bear (*Ursus americanus*, Ursidae). (After O'Brien et al. 1989.)

All seven to eight species of marine turtles are considered either threatened or endangered. Their taxonomy and systematics have been controversial due to phylogenetic uncertainties at levels ranging from population distinctions to deeper evolutionary alliances among species, and some of these uncertainties have had conservation consequences. Examples involving the green and ridley turtles already have been described. Other challenging questions include the following: Is the eastern Pacific black turtle (*Chelonia agassizi*) specifically distinct from the green turtle (*C. mydas*)? Is the spongivorous hawksbill (*Eretmochelys imbricata*) allied phylogenetically to herbivorous green turtles or to carnivorous loggerheads (*Caretta caretta*)? Is the Australian flatback (*Natator depressa*) allied closely to the greens (as its earlier placement within *Chelonia* suggested), or is it a relative of the loggerheads? Molecular markers have provisionally answered such questions (see review in Bowen and Avise 1996). For example, a phylogeny estimated from mtDNA sequences (Figure 9.12) suggested that black turtles fall well within the range of genetic differentiation exhibited among green turtle rookeries worldwide; that the hawksbill is more closely related to the loggerhead complex than to green turtles and hence probably evolved from a carnivorous rather than a

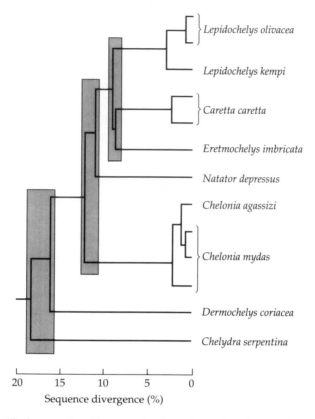

Figure 9.12   Phylogeny for all recognized species of marine turtles (plus the freshwater snapping turtle *C. serpentina* as an outgroup) estimated using sequence data from the mtDNA cytochrome *b* gene. For species represented by more than one sample, the individuals came from different oceanic basins. Exact orders of the nodes within each shaded box are uncertain, differing slightly with alternative methods of data analysis. (After Bowen et al. 1993a.)

herbivorous ancestor; and that the flatback turtle is highly distinct from both the loggerhead and green complexes and is roughly equidistant from both.

Such phylogenetic appraisals of endangered (or other) assemblages are of relevance to conservation biology in the general sense that they provide an understanding of evolutionary relationships among species and higher taxa. Should their ramifications be extended by using phylogenetic position as a criterion for prioritizing taxa with regard to conservation value?

PHYLOGENETICS AND CONSERVATION PRIORITIES.   If all extant species were non-threatened, or if resources available for conservation were unlimited, there would be little need to rank taxa for conservation value. However, all ongoing or contemplated conservation efforts involve establishing priorities. In the broadest sense, any of three underlying rationales justify human efforts to protect biodiversity: aesthetic considerations, the realization that living species provide utilitarian services (Balmford et al. 2002), and an ethical stance that attributes an intrinsic value to life (Crozier 1997; Nixon and Wheeler 1992). In actual management practices, however, several more proximate ranking criteria routinely come to the fore (explicitly or implicitly): rarity, restricted distribution, perceived ecological importance, "charisma" [for example, Clark and May (2002) confirmed that large, attractive, or emotive species attract more attention than small or drab species, thus producing a bias toward "charismatic megabiota" in conservation programs], economics, management feasibility, and phylogenetic distinctiveness.

This last criterion, which can be informed by molecular genetic findings, warrants elaboration. Implicit in the writings of many biologists is the notion that evolutionarily distinct taxa contribute disproportionately to overall biodiversity and thus should be prioritized for conservation efforts. For example, the tuataras (*Sphenodon*) of New Zealand might be deemed to be of exceptional conservation value because they are the sole living members of a very ancient reptilian family (Daugherty et al. 1990). On the other hand, Erwin (1991) suggested the opposite: that such "living fossils" are likely to be evolutionary dead ends, and that members of rapidly speciating clades should be valued more highly by virtue of their greater potential for generating future biodiversity.

May (1990) and Vane-Wright et al. (1991) were among the first to articulate the idea that a taxon's phylogenetic distinctiveness could be quantified and used expressly in priority rankings for conservation. Their proposal was soon refined and elaborated by many workers (Barrowclough 1992; Crozier 1992; Faith 1992, 1993; Williams et al. 1991; see reviews in Crozier 1997; Humphries et al. 1995; Krajewski 1994; May 1994). A key element in such phylogenetic ranking is the concept of "independent evolutionary history" (IEH), relative magnitudes of which are quantitatively assessed as branch lengths in phylograms. In any phylogram (estimated, for example, from molecular data), total IEH is the summed length of all tree branches. When several species within a group are to be rank-ordered for conservation value, the relevant branch lengths are appropriately discounted for branch segments shared with other extant taxa (May 1994). The typical argument is as follows: If conservationists could save only some fraction of living species in a given phylogram, the optimal choice would maximize the sum of independent branch lengths (each counted only once) to be preserved. In practice, this normally means that higher conservation priorities would be given to extant forms that lack close living relatives (such as tuataras), because such forms have had long independent evolutionary histories.

Molecular data are well suited for estimating branch lengths as well as nodal placements in phylograms. Furthermore, to the extent that various

DNA sequences evolve in clocklike fashion (or that non-clocklike behavior is accommodated in the phylogenetic analysis), these branch lengths may also be interpreted as estimates of evolutionary times since shared ancestry. Normally, species to be ranked for conservation value by IEH criteria would belong to a specific taxonomic assemblage, but in principle, the approach could also be used to rank-order species across disparate phylogenetic arrays, such as particular mammals versus particular fishes or particular arthropods.

Figure 9.13 presents empirical phylograms for bears (Ursidae), cats (Felidae), marine turtles (Cheloniidae plus Dermochelyidae), and horseshoe crabs (Limuloidea), all based on DNA sequence comparisons. Suppose that within each of these taxonomic assemblages, three or four candidate species are to be rank-ordered for preservation according to each of five ranking criteria mentioned above: rarity, restricted distribution, ecological significance, charisma, and phylogenetic distinctiveness (using the IEH metric). Assume further that for each taxonomic group, available resources permit only one species to receive conservation attention. Which one should it be?

Several points emerge from Figure 9.13. First, different criteria often rank the same species differently. Among marine turtles, for example, rarity and limited range would demand high conservation priority for Kemp's ridley, whereas the green turtle probably plays a greater ecological role in nature (it is an important marine herbivore), and the magnificent leatherback turtle (*Dermochelys coriacea*) probably has the most charisma and is certainly the most phylogenetically distinctive member of the group. Among felids, the tiger (*Panthera tigris*) probably ranks highest according to rarity and charisma, whereas the serval (*Leptailurus serval*) would get the preservation nod by virtue of having the narrowest range and being phylogenetically most distinctive among the cat species considered.

Second, different ranking criteria do not always associate in the same way. For example, phylogenetic distinctiveness and narrow range jointly support high conservation priority for the giant panda and the serval within their respective clades, but they conflict in the particular marine turtle species they earmark for conservation priority. Third, subjective or otherwise questionable judgments often come into play, as, for example, in ranking brown bears (*Ursus arctos*) versus polar bears (*U. maritimus*), or American horseshoe crabs (*Limulus*) versus Asian horseshoe crabs (*Carcinoscorpius* and *Tachypleus*), according to their ecological significance or charisma.

In principle, prioritization rankings for conservation could also apply across evolutionary groups. Suppose, for example, that again, a total of only four species from Figure 9.13 could be the subject of conservation efforts, but that the choice is no longer constrained to one species from each taxonomic array. Then, the more difficult or subjective of the ranking criteria (ecological significance and charisma) provide little assistance as choic-

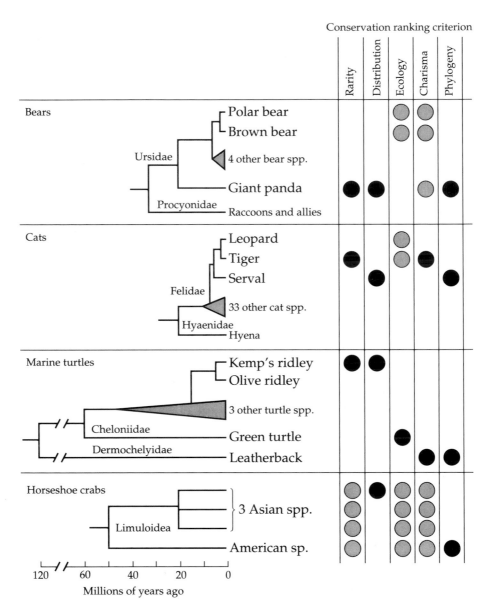

Figure 9.13   How the conservation value of a species can vary according to five conventional ranking criteria. Shown are time-dated phylogenies (estimated from molecular genetic information in conjunction with fossil or other evidence, and depicted on a common temporal scale) for surveyed extant species of bears (O'Brien 1987; O'Brien et al. 1985a), cats (O'Brien et al. 1996), marine turtles (Bowen et al. 1993a; Dutton et al. 1996), and horseshoe crabs (Avise et al. 1994; Lynch 1993). *Within* each taxonomic group, the top-priority species according to each ranking criterion (among those species listed in large type) is indicated by a black circle. Gray circles indicate ties for top ranking. (After Avise 2004d.)

es are forced between preserving, for example, the polar bear, tiger, or leatherback turtle. Even for the objective criteria that are directly comparable across groups and quantifiable (i.e., rarity, restricted distribution, and phylogenetic distinctiveness as measured by IEH), conservationists might well decide not to abide by the final numerical rankings. For example, each of the four living species of horseshoe crabs has a higher IEH score than do any cats or bears, yet I doubt that many people would choose to direct finite conservation resources toward these crabs if this came at the expense of saving tigers or giant pandas.

CONCLUSIONS ABOUT PHYLOGENETIC DISTINCTIVENESS AND CONSERVATION. Notwithstanding the considerable academic attention that has been paid to incorporating phylogenetic distinctiveness as a guide to preservation priorities at supraspecific levels, it appears that this criterion has had little practical effect on conservation efforts at these scales. As illustrated above, phylogenetic distinctiveness in practice often conflicts with various other quantifiable criteria (such as rarity and restricted distribution), as well as with subjective judgments about biological worth, that are not likely to be abandoned as primary bases for conservation prioritization. So, whereas phylogenetic uniqueness may bolster the rationale for particular conservation choices when it agrees with other ranking criteria, it will seldom override those other considerations when they are in conflict. This conclusion holds with even greater force with regard to rank-ordering species across disparate taxonomic groups. Thus, although the phylogenetic distinctiveness of, say, bears and horseshoe crabs can be quantified and objectively compared, more subjective criteria (such as charismatic appeal) will undoubtedly be applied to such "apples and oranges" in most conservation decisions.

There are additional reasons why phylogenetic considerations are unlikely to revolutionize on-the-ground conservation practices at the levels of genera and higher taxa. Based on a quantitative analysis that considered the full phylogenetic panorama of life, Nee and May (1997) concluded that about 80% of life's total independent evolutionary history (IEH) could be preserved even if about 95% of extant species were to go extinct. In more good news and bad news for conservation efforts, their analysis further showed that the fraction of total IEH preserved would not be improved much by intelligent phylogenetic choice, as opposed to random draws, of the species permitted to survive.

Every species alive today traces back through an unbroken chain of ancestry over the eons, irrespective of how many speciation events have intervened along its phylogenetic journey. The mere fact that issues of conservation priority and preservation triage must be raised represents a sad and shameful commentary on how humanity's environmental impacts have endangered Earth's natural genetic heritage.

Conclusion

I want to close this chapter, and the book, by reiterating a sentiment expressed by E. O. Wilson in the quote that opened this chapter. Modern biology has indeed produced a genuinely new way of looking at the world. The molecular perspectives emphasized in this text do not supplant traditional approaches to the study of natural history and evolution, but rather enrich our understanding of life. Therein lies the greatest value of molecular methods in conservation biology, or elsewhere. To the degree that we come to understand and appreciate other organisms, we will increasingly cherish Earth's biological heritage, and our own. As stated by the late Stephen J. Gould (1991), "We cannot win this battle to save species and environments without forging an emotional bond between ourselves and nature as well— for we will not fight to save what we do not love. ... We really must make room for nature in our hearts."

Think back to even a few of the fascinating organisms whose natural histories and evolutionary patterns have been elucidated using molecular markers—honey mushrooms and their giant clones on the floors of northern forests; hybridogenetic live-bearing fishes in the arroyos of northwestern Mexico and the substantial evolutionary ages that some of these unisexual biotypes have achieved; the various sunfish species of the eastern United States and their unsuspected and sometimes devious means of achieving parentage; naked mole-rats in the deserts of Africa, with their eusocial behaviors and tight fabrics of kinship; and female green turtles who, after decades in the open ocean, swim thousands of kilometers to return faithfully to nest at their natal sites. If this book has accomplished nothing else, I hope that it may have engendered an increased awareness, respect, and love of the planet's marvelous genetic diversity.

SUMMARY

1. Many discussions of genetics in conservation biology have centered on the topic of heterozygosity or related measures of the within-population component of genetic variation. Molecular heterozygosity is indeed exceptionally low in many rare or endangered species, presumably because of genetic drift and inbreeding accompanying severe population reductions.

2. Although it is tempting to manage endangered populations for enhancement of genetic variation, in most cases causal links between heterozygosity (as estimated by molecular markers) and fitness have proved difficult to establish. For these and other reasons, some authors have argued that behavioral and demographic issues should take priority over heterozygosity issues in management programs for endangered species. In truth, both classes of concerns are important.

3. At the intraspecific level, molecular markers have found forensic application by allowing researchers to identify and track individuals and determine the gender of particular specimens. Such information, which is often not otherwise evident, can be critical in helping to monitor and manage threatened populations.

4. Molecular approaches can also serve the field of conservation biology by revealing genealogical relationships among populations of rare or endangered species. Such phylogenetic assessments can range from genetic parentage assignments in captive breeding programs, to the identification of management units (MUs) and evolutionarily significant units (ESUs) of particular wild species, to the characterization of major sources of regional phylogeographic diversity around which management guidelines and natural reserves might be established.

5. Particularly in the field of fisheries biology, molecular markers are widely employed to identify and characterize population stocks under exploitation. Contributions of genetic stocks to mixed fisheries can be quantified using molecular markers. Molecular approaches to stock assessment also have stimulated thought about the nature of information provided by inherited markers versus acquired markers (such as physical tags), and about the key distinction between evolutionarily deep versus shallow population genetic structures.

6. Some conservation programs for endangered species have been directed toward taxa whose evolutionary distinctiveness has been questioned by molecular genetic reappraisals. In other cases, endangered "species" that were taxonomically suspect have proved upon molecular reexamination to be highly distinct genetically, thus adding to the rationale for special preservation efforts.

7. By enabling species identification from even small or degraded bits of tissue (such as fish eggs or turtle meat from illegally harvested endangered taxa), molecular markers provide powerful forensic tools for government agencies charged with enforcement of wildlife legislation.

8. Hybridization and genetic introgression, often documented by molecular markers, have raised a variety of biological as well as regulatory issues regarding the status of many rare and endangered species.

9. Molecular appraisals of species-level and higher-level phylograms can be used to quantify the magnitude of any taxon's independent evolutionary history (IEH) or phylogenetic distinctiveness, which often has been promoted as a criterion for conservation prioritization. However, several other important prioritization criteria often conflict with IEH measures and thereby compromise the utility of phylogenetic distinctiveness (especially above the species level) as a pragmatic guide to conservation efforts.

Literature Cited

Aanen, D. K., P. Eggleton, C. Rouland-Lefevre, T. Guldberg-Frøslev, S. Rosendahl and J. J. Boomsma. 2002. The evolution of fungus-growing termites and their mutualistic fungal symbionts. *Proc. Natl. Acad. Sci. USA* 99: 14887–14892.

Abbott, R. J. 1992. Plant invasions, interspecific hybridization and the evolution of new plant taxa. *Trends Ecol. Evol.* 7: 401–405.

Abramoff, P., R. M. Darnell and J. S. Balsano. 1968. Electrophoretic demonstration of the hybrid origin of the gynogenetic teleost *Poecilia formosa*. *Am. Nat.* 102: 555–558.

Achmann, R., K.-G. Heller and J. T. Epplen. 1992. Last-male sperm precedence in the bushcricket *Poecilimon veluchianus* (Orthoptera, Tettigonioidea) demonstrated by DNA fingerprinting. *Mol. Ecol.* 1: 47–54.

Adams, J. R., J. A. Leonard and L. P. Waits. 2003. Widespread occurrence of a domestic dog mitochondrial DNA haplotype in southeastern U.S. coyotes. *Mol. Ecol.* 12: 541–546.

Adams, K. L. and 6 others. 1999. Intracellular gene transfer in action: Dual transcription and multiple silencings of nuclear and mitochondrial *cox2* genes in legumes. *Proc. Natl. Acad. Sci. USA* 96: 13863–13868.

Adams, K. L., D. O. Daley, Y.-L Qui, J. Whelan and J. D. Palmer. 2000. Repeated, recent and diverse transfers of a mitochondrial gene to the nucleus of flowering plants. *Nature* 408: 354–357.

Adams, K. L., H. C. Ong and J. D. Palmer. 2001. Mitochondrial gene transfer in pieces: Fission of the ribosomal protein gene *rpl2* and partial or complete gene transfer to the nucleus. *Mol. Biol. Evol.* 18: 2289–2297.

Adams, K. L., Y.-L. Qiu, M. Stoutemyer and J. D. Palmer. 2002. Punctuated evolution of mitochondrial gene content: High and variable rates of mitochondrial gene loss and transfer to the nucleus during angiosperm evolution. *Proc. Natl. Acad. Sci. USA* 99: 9905–9912.

Adams, W. T., A. R. Griffin and G. F. Moran. 1992. Using paternity analysis to measure effective pollen dispersal in plant populations. *Am. Nat.* 140: 762–780.

Adelman, R., R. L. Saul and B. N. Ames. 1988. Oxidative damage to DNA: Relation to species metabolic rate and life span. *Proc. Natl. Acad. Sci. USA* 85: 2706–2708.

Adkins, R. M. and R. L. Honeycutt. 1991. Molecular phylogeny of the superorder Archonta. *Proc. Natl. Acad. Sci. USA* 88: 10317–10321.

Aguadé, M. and C. H. Langley. 1994. Polymorphism and divergence in regions of low recombination in *Drosophila*. Pp. 67–76 in *Non-Neutral Evolution: Theories and Molecular Data*, B. Golding (ed.). Chapman & Hall, New York.

Aguadé, M., N. Miyashita and C. H. Langley. 1989a. Restriction-map variation at the *Zeste-tko* region in natural populations of *Drosophila melanogaster*. *Mol. Biol. Evol.* 6: 123–130.

Aguadé, M., N. Miyashita and C. H. Langley. 1989b. Reduced variation in the *yellow-achaete-scute* region in natural populations of *Drosophila melanogaster*. *Genetics* 122: 607–615.

Aguinaldo, A. M. A. and 6 others. 1997. Evidence for a clade of nematodes, arthropods, and other moulting animals. *Nature* 387: 489–493.

Aide, T. M. 1986. The influence of wind and animal pollination on variation in outcrossing rates. *Evolution* 40: 434–435.

Ainsworth, E. A., P. J. Tranel, B. G. Drake and S. P. Long. 2003. The clonal structure of *Quercus geminata* revealed by conserved microsatellite loci. *Mol. Ecol.* 12: 527–532.

Akashi, H. 1995. Inferring weak selection from patterns of polymorphism and divergence at "silent" sites in *Drosophila* DNA. *Genetics* 139: 1067–1076.

Alatalo, R. V., L. Gustafsson and A. Lundberg. 1984. High frequency of cuckoldry in pied and collared flycatchers. *Oikos* 42: 41–47.

Albert, V. A., S. E. Williams and M. W. Chase. 1992. Carnivorous plants: Phylogeny and structural evolution. *Science* 257: 1491–1495.

Albertson, R. C., J. T. Streelman and T. D. Kocher. 2003. Directional selection has shaped the oral jaws of Lake Malawi cichlid fishes. *Proc. Natl. Acad. Sci. USA* 100: 5252–5257.

Aldrich, P. R. and J. L. Hamrick. 1998. Reproductive dominance of pasture trees in a fragmented tropical forest mosaic. *Science* 281: 103–105.

Allan, G. J., C. Clark and L. H. Rieseberg. 1997. Distribution of parental DNA markers in *Encelia virginensis* (Asteraceae: Heliantheae), a diploid species of putative hybrid origin. *Pl. Syst. Evol.* 205: 205–221.

Allander, T., S. U. Emerson, R. E. Engle, R. H. Purcell and J. Bukh. 2001. A virus discovery method incorporating DNase treatment and its application to the identification of two bovine paravirus species. *Proc. Natl. Acad. Sci. USA* 98: 11609–11614.

Allard, M. W. and R. L. Honeycutt. 1992. Nucleotide sequence variation in the mitochondrial 12S rDNA gene and the phylogeny of African mole-rats (Rodentia: Bathyergidae). *Mol. Biol. Evol.* 9: 27–40.

Allard, R. W. 1975. The mating system and microevolution. *Genetics* 79: 115–126.

Allard, R. W., G. R. Babbel, M. T. Clegg and A. L. Kahler. 1972. Evidence for coadaptation in *Avena barbata*. *Proc. Natl. Acad. Sci. USA* 69: 3043–3048.

Allen, M. and 7 others. 1998. Mitochondrial DNA sequencing of shed hairs and saliva on robbery caps: Sensitivity and matching probabilities. *J. Forensic Sci.* 43: 453–464.

Allendorf, F. W. 1983. Isolation, gene flow, and genetic differentiation among populations. Pp. 51–65 in *Genetics and Conservation*, C. M. Schonewald-Cox, S. M. Chambers, B. MacBryde and L. Thomas (eds.). Benjamin/Cummings, London.

Allendorf, F. W. 1996. Conservation and genetics of salmonid fishes. Pp. 238–280 in *Conservation Genetics: Case Histories from Nature*, J. C. Avise and J. L. Hamrick (eds.). Chapman & Hall, New York.

Allendorf, F. W. 1998. Conservation and genetics of marine organisms (special issue). *J. Heredity* 89: 377–464.

Allendorf, F. W. and R. F. Leary. 1986. Heterozygosity and fitness in natural populations of animals. Pp. 57–76 in *Conservation Biology: The Science of Scarcity and Diversity*, M. Soulé (ed.). Sinauer Associates, Sunderland, MA.

Allendorf, F. W. and R. F. Leary. 1988. Conservation and distribution of genetic variation in a polytypic species, the cutthroat trout. *Cons. Biol.* 2: 170–184.

Allendorf, F. W. and F. M. Utter. 1979. Population genetics. Pp. 407–454 in *Fish Physiology* Vol. 8, W. S. Hoar, D. J. Randall and J. R. Brett (eds.). Academic Press, New York.

Allendorf, F. W. and R. S. Waples. 1996. Conservation and genetics of salmonid fishes. Pp. 238–280 in *Conservation Genetics: Case Histories from Nature*, J. C. Avise and J. L. Hamrick (eds.). Chapman & Hall, New York.

Allendorf, F. W., R. F. Leary, P. Spruell and J. K. Wenburg. 2001. The problems of hybrids: Setting conservation guidelines. *Trends Ecol. Evol.* 16: 613–622.

Altschul, S. F. and 6 others. 1997. Gapped BLAST and PSI-BLAST: A new generation of protein database search programs. *Nucl. Acids Res.* 25: 3389–3402.

Alvarez, A. J., M. Khanna, G. A. Roranzo and G. Stotsky. 1998. Amplification of DNA bound on clay minerals. *Mol. Ecol.* 7: 775–778.

Alves, P. C., N. Ferrand, F. Suchentrunk and D. J. Harris. 2003. Ancient introgression of *Lepus timidus* mtDNA into *L. granatensis* and *L. europaeus* in the Iberian Peninsula. *Mol. Phylogen. Evol.* 27: 70–80.

Amos, B., J. Barrett and G. A. Dover. 1991a. Breeding system and social structure in the Faroese pilot whale as revealed by DNA fingerprinting. Pp. 255–268 in *Genetic Ecology of Whales and Dolphins*, A. R. Hoelzel (ed.). Black Bear Press, Cambridge, England.

Amos, B., J. Barrett and G. A. Dover. 1991b. Breeding behavior of pilot whales revealed by DNA fingerprinting. *Heredity* 67: 49–55.

Amos, B., C. Schlötterer and D. Tautz. 1993. Social structure of pilot whales revealed by analytical DNA profiling. *Science* 260: 670–672.

Andersen, N. M., L. Cheng, J. Damgaard and F. A. H. Sperling. 2000. Mitochondrial DNA sequence variation and phylogeography of oceanic insects (Hemiptera: Gerridae: *Halobates* spp.). *Marine Biol.* 136: 421–430.

Anderson, E. 1949. *Introgressive Hybridization*. John Wiley and Sons, New York.

Anderson, E. and G. L. Stebbins Jr. 1954. Hybridization as an evolutionary stimulus. *Evolution* 8: 378–388.

Anderson, J. F. and 7 others. 1999. Isolation of West Nile virus from moquitoes, crows, and a Cooper's hawk in Connecticut. *Science* 286: 2331–2337.

Andersson, J. O., A. M. Sjögren, L. A. M. Davis, T. M. Embley and A. J. Roger. 2003. Phylogenetic analyses of diplomonad genes reveal frequent lateral gene transfers affecting eukaryotes. *Curr. Biol.* 13: 94–104.

Andersson, M. 1984. The evolution of eusociality. *Annu. Rev. Ecol. Syst.* 15: 165–189.

Andersson, M. 1999. Hybridization and skua phylogeny. *Proc. R. Soc. Lond. B* 266: 1579–1585.

Andrewartha, H. G. and L. C. Birch. 1954. *The Distribution and Abundance of Animals*. University of Chicago Press, Chicago, IL.

Andrews, P. 1987. Aspects of hominoid phylogeny. Pp. 23–52 in *Molecules and Morphology in Evolution: Conflict or Compromise?* C. Patterson (ed.). Cambridge University Press, Cambridge, England.

Angers, B. and L. Bernatchez. 1997. Complex evolution of a salmonid microsatellite locus and its consequences in inferring allelic divergence from size information. *Mol. Biol. Evol.* 14: 230–238.

Angert, E. R., D. E. Northup, A. L. Reysenbach, A. S. Peek, B. M. Goebel and N. R. Pace. 1998. Molecular phylogenetic analysis of a bacterial community in Sulphur River, Parker Cave, Kentucky. *Am. Miner.* 83: 1583–1592.

Antonovics, J. and A. D. Bradshaw. 1970. Evolution in closely adjacent plant populations. VIII. Clinal patterns at a mine boundary. *Heredity* 25: 349–362.

Antunes, A., A. R. Templeton, R. Guyomard and P. Alexandrino. 2002. The role of nuclear genes in intraspecific evolutionary inference: Genealogy of the *transferrin* gene in the brown trout. *Mol. Biol. Evol.* 19: 1272–1287.

Aoyama, J., M. Nishida and K. Tsukamoto. 2001. Molecular phylogeny and evolution of the freshwater eel, genus *Anguilla*. *Mol. Phylogen. Evol.* 20: 450–459.

Apfelbaum, L. and O. A. Reig. 1989. Allozyme genetic distances and evolutionary relationships in species of akodontine rodents (Cricetidae: Sigmodontinae). *Biol. J. Linn. Soc.* 38: 257–280.

Appels, R. and J. Dvorak. 1982. The wheat ribosomal DNA spacer region: Its structure and variation in populations and among species. *Theoret. Appl. Genet.* 63: 337–348.

Appels, R. and R. L. Honeycutt. 1986. rDNA: Evolution over a billion years. Pp. 81–135 in *DNA Systematics*, Vol. II, S. K. Dutta (ed.). CRC Press, Boca Raton, FL.

Apuya, N. R., B. L. Frazier, P. Keim, E. J. Roth and K. G. Lark. 1988. Restriction fragment length polymorphisms as genetic markers in soybean, *Glycine max* (L.) Merrill. *Theoret. Appl. Genet.* 75: 889–901.

Aquadro, C. F. 1992. Why is the genome variable? Insights from *Drosophila*. *Trends Genet.* 8: 355–362.

Aquadro, C. F. and J. C. Avise. 1981. Genetic divergence between rodent species assessed by using two-dimensional electrophoresis. *Proc. Natl. Acad. Sci. USA* 78: 3784–3788.

Aquadro, C. F. and J. C. Avise. 1982a. Evolutionary genetics of birds. VI. A reexamination of protein divergence using varied electrophoretic conditions. *Evolution* 36: 1003–1019.

Aquadro, C. F. and J. C. Avise. 1982b. An assessment of "hidden" heterogeneity within electromorphs at three enzyme loci in deer mice. *Genetics* 102: 269–284.

Aquadro, C. F. and D. J. Begun. 1993. Evidence for and implications of genetic hitchhiking in the *Drosophila* genome. Pp. 159–178 in *Mechanisms of Molecular Evolution*, N. Takahata and A. Clark (eds.). Sinauer Associates, Sunderland, MA.

Aquadro, C. F. and B. D. Greenberg. 1983. Human mitochondrial DNA variation and evolution: Analysis of nucleotide sequences from seven individuals. *Genetics* 103: 287–312.

Aquadro, C. F., S. F. Desse, M. M. Bland, C. H. Langley and C. C. Laurie-Ahlberg. 1986. Molecular population genetics of the alcohol dehydrogenase gene region of *Drosophila melanogaster*. *Genetics* 114: 1165–1190.

Aquadro, C. F., A. L. Weaver, S. W. Schaeffer and W. W. Anderson. 1991. Molecular evolution of inversions in *Drosophila pseudoobscura*: The amylase gene region. *Proc. Natl. Acad. Sci. USA* 88: 305–309.

Aquadro, C. F., D. J. Begun and E. C. Kindahl. 1994. Selection, recombination, and DNA polymorphism in *Drosophila*. Pp. 46–56 in *Non-neutral Evolution. Theories*

and Molecular Data, B. Golding (ed.). Chapman & Hall, New York.

Aragon, S., A. P. Møller, J. J. Soler and M. Soler. 1999. Molecular phylogeny of cuckoos supports a polyphyletic origin of brood parasitism. *J. Evol. Biol.* 12: 495–506.

Arbogast, B. S., S. V. Edwards, J. Wakeley, P. Beerli and J. B. Slowinski. 2002. Estimating divergence times from molecular data on phylogenetic and population genetic timescales. *Annu. Rev. Ecol. Syst.* 33: 707–740.

Arctander, P. 1999. Mitochondrial recombination? *Science* 284: 2090–2091.

Ardern, S. L. and D. M. Lambert. 1997. Is the black robin in genetic peril? *Mol. Ecol.* 6: 21–28.

Arévalo, E., J. E. Strassmann and D. C. Queller. 1998. Conflicts of interest in social insects: Male reproduction in two species of *Polistes*. *Evolution* 52: 797–805.

Aris-Brosou, S. and L. Excoffier. 1996. The impact of population expansion and mutation rate heterogeneity on DNA sequence polymorphism. *Mol. Biol. Evol.* 13: 494–504.

Arkhipova, I. and M. Meselson. 2000. Transposable elements in sexual and ancient asexual taxa. *Proc. Natl. Acad. Sci. USA* 97: 14473–14477.

Armour, J. A. L. and 9 others. 1996. Minisatellite diversity supports a recent African origin for modern humans. *Nature Genet.* 13: 154–160.

Árnason, Ú., R. Spilliaert, Á. Pálsdóttir and A. Árnason. 1991. Molecular identification of hybrids between the two largest whale species, the blue whale (*Balaenoptera musculus*) and the fin whale (*B. physalus*). *Hereditas* 115: 183–189.

Árnason, E., S. Pálsson and A. Arason. 1992. Gene flow and lack of population differentiation in Atlantic cod, *Gadus morhua* L., from Iceland, and comparison of cod from Norway and Newfoundland. *J. Fish Biol.* 40: 751–770.

Arndt, A. and M. J. Smith. 1998a. Genetic diversity and population structure in two species of sea cucumber: Differing patterns according to mode of development. *Mol. Ecol.* 7: 1053–1064.

Arndt, A. and M. J. Smith. 1998b. Mitochondrial gene rearrangement in the sea cucumber genus *Cucumaria*. *Mol. Biol. Evol.* 15: 1009–1016.

Arnheim, N. 1983. Concerted evolution of multigene families. Pp. 38–61 in *Evolution of Genes and Proteins*, M. Nei (ed.). Sinauer Associates, Sunderland, MA.

Arnheim, N., T. White and W. E. Rainey. 1990. Application of PCR: Organismal and population biology. *BioScience* 40: 174–182.

Arnold, J. 1993. Cytonuclear disequilibria in hybrid zones. *Annu. Rev. Ecol. Syst.* 24: 521–554.

Arnold, K. E., K. J. Orr and R. Griffiths. 2003. Primary sex ratios in birds: Problems with molecular sex identification of undeveloped eggs. *Mol. Ecol.* 12: 3451–3458.

Arnold, M. L. 1992. Natural hybridization as an evolutionary process. *Annu. Rev. Ecol. Syst.* 23: 237–261.

Arnold, M. L. 1997. *Natural Hybridization and Evolution*. Oxford University Press, New York.

Arnold, M. L. 2000. Anderson's paradigm: Louisiana irises and the study of evolutionary phenomena. *Mol. Ecol.* 9: 1687–1698.

Arnold, M. L. and B. D. Bennett. 1993. Natural hybridization in Louisiana irises: Genetic variation and ecological determinants. Pp. 115–139 in *Hybrid Zones and the Evolutionary Process*, R. G. Harrison (ed.). Oxford University Press, Oxford, England.

Arnold, M. L. and S. A. Hodges. 1995. Are natural hybrids fit or unfit relative to their parents? *Trends Ecol. Evol.* 10: 67–71.

Arnold, M. L., D. D. Shaw and N. Contreras. 1987. Ribosomal RNA-encoding DNA introgression across a narrow hybrid zone between two subspecies of grasshopper. *Proc. Natl. Acad. Sci. USA* 84: 3946–3950.

Arnold, M. L., N. Contreras and D. D. Shaw. 1988. Biased gene conversion and asymmetrical introgression between subspecies. *Chromosoma* 96: 368–371.

Arnold, M. L., B. D. Bennett and E. A. Zimmer. 1990a. Natural hybridization between *Iris fulva* and *I. hexagona*: Pattern of ribosomal DNA variation. *Evolution* 44: 1512–1521.

Arnold, M. L., J. L. Hamrick and B. D. Bennett. 1990b. Allozyme variation in Louisiana irises: A test for introgression and hybrid speciation. *Heredity* 65: 297–306.

Arnold, M. L., C. M. Buckner and J. J. Robinson. 1991. Pollen-mediated introgression and hybrid speciation in Louisiana irises. *Proc. Natl. Acad. Sci. USA* 88: 1398–1402.

Arnold, M. L., J. J. Robinson, C. M. Buckner and B. D. Bennett. 1992. Pollen dispersal and interspecific gene flow in Louisiana irises. *Heredity* 68: 399–404.

Arnold, M. L., M. R. Bulger, J. M. Burke, A. L. Hempel and J. H. Williams. 1999. Natural hybridization: How low can you go and still be important? *Ecology* 80: 371–381.

Arntzen, J. W. and G. P. Wallis. 1991. Restricted gene flow in a moving hybrid zone of the newts *Triturus cristatus* and *T. marmoratus* in western France. *Evolution* 45: 805–826.

Ashburner, M., R. Camfield, B. Clarke, D. Thatcher and R. Woodruff. 1979. A genetic analysis of the locus coding for alcohol dehydrogenase, and its adjacent chromosome region, in *Drosophila melanogaster*. Pp. 95–106 in *Eukaryotic Gene Regulation*, R. Axel, T. Maniatis and C. F. Fox (eds.). Academic Press, New York.

Ashley, M. and C. Wills. 1987. Analysis of mitochondrial DNA polymorphisms among channel island deer mice. *Evolution* 41: 854–863.

Ashley, M. and C. Wills. 1989. Mitochondrial-DNA and allozyme divergence patterns are correlated among island deer mice. *Evolution* 43: 646–650.

Ashley, M. V., D. J. Melnick and D. Western. 1990. Conservation genetics of the black rhinoceros (*Diceros bicornis*), I: Evidence from the mitochondrial DNA of three populations. *Cons. Biol.* 4: 71–77.

Ashton, P. A. and R. J. Abbott. 1992. Multiple origins and genetic diversity in the newly arisen allopolyploid species, *Senecio cambrensis* Rosser (Compositae). *Heredity* 68: 25–32.

Asmussen, M. A. and C. J. Basten. 1994. Sampling theory for cytonuclear disequilibria. *Genetics* 138: 1351–1363.

Asmussen, M. A. and M. E. Orive. 2000. The effects of pollen and seed migration on nuclear-dicytoplasmic systems. I. Nonrandom associations and equilibrium structure with both maternal and paternal cytoplasmic inheritance. *Genetics* 155: 813–831.

Asmussen, M. A. and A. Schnabel. 1991. Comparative effects of pollen and seed migration on the cytonuclear structure of plant populations. I. Maternal cytoplasmic inheritance. *Genetics* 128: 639–654.

Asmussen, M. A., J. Arnold and J. C. Avise. 1987. Definition and properties of disequilibrium statistics for associations between nuclear and cytoplasmic genotypes. *Genetics* 115: 755–768.

Asmussen, M. A., J. Arnold and J. C. Avise. 1989. The effects of assortative mating and migration on cytonuclear associations in hybrid zones. *Genetics* 122: 923–934.

Atchison, B. A., P. R. Whitfield and W. Bottomley. 1976. Comparison of chloroplast DNAs by specific fragmentation with *Eco*RI endonuclease. *Mol. Gen. Genet.* 148: 263–269.

Atchley, W. R. and W. M. Fitch. 1991. Gene trees and the origins of inbred strains of mice. *Science* 254: 554–558.

Atkins, M. S., A. P. Teske and O. R. Anderson. 2000. A survey of flagellate diversity at four deep-sea hydrothermal vents in the Eastern Pacific Ocean using structural and molecular approaches. *J. Eukaryot. Microbiol.* 47: 400–411.

Attardi, G. 1985. Animal mitochondrial DNA: An extreme example of genetic economy. *Int. Rev. Cytol.* 93: 93–145.

Aubert, J. and M. Solignac. 1990. Experimental evidence for mitochondrial DNA introgression between *Drosophila* species. *Evolution* 44: 1272–1282.

Austin, D. J. and E. N. Arnold. 2002. Ancient mitochondrial DNA and morphology elucidate an extinct island radiation of Indian Ocean giant tortoises (*Cylindraspis*). *Proc. R. Soc. Lond. B* 268: 2515–2523.

Austin, D. J., E. N. Arnold and R. Bour. 2003. Was there a second adaptive radiation of giant tortoises in the Indian Ocean? Using mitochondrial DNA to investigate speciation and biogeography of *Aldabrachelys* (Reptilia, Testudinidae). *Mol. Ecol.* 12: 1415–1424.

Austin, J. J., A. B. Smith and R. H. Thomas. 1997. Palaeontology in a molecular world: The search for authentic ancient DNA. *Trends Ecol. Evol.* 12: 303–306.

Autumn, K., M. J. Ryan and D. B. Wake. 2002. Integrating historical and mechanistic biology enhances the study of adaptation. *Q. Rev. Biol.* 77: 383–408.

Avery, O. T., C. M. MacLeod and M. McCarty. 1944. Studies on the chemical nature of the substance inducing transformation of Pneumococcal types. I. Induction of transformation by a DNA fraction isolated from *Pneumococcus* type III. *J. Exp. Med.* 79: 137–158.

Avigad, S., B. E. Cohen, S. Bauer, G. Schwartz, M. Frydman, S. L. C. Woo, Y. Niny and Y. Shiloh. 1990. A single origin of phenylketonuria in Yemenite Jews. *Nature* 344: 168–170.

Avise, J. C. 1974. Systematic value of electrophoretic data. *Syst. Zool.* 23: 465–481.

Avise, J. C. 1976. Genetic differentiation during speciation. Pp. 106–122 in *Molecular Evolution*, F. J. Ayala (ed.). Sinauer Associates, Sunderland, MA.

Avise, J. C. 1977a. Is evolution gradual or rectangular? Evidence from living fishes. *Proc. Natl. Acad. Sci. USA* 74: 5083–5087.

Avise, J. C. 1977b. Genic heterozygosity and rate of speciation. *Paleobiology* 3: 422–432.

Avise, J. C. 1983. Protein variation and phylogenetic reconstruction. Pp. 103–130 in *Protein Variation: Adaptive and Taxonomic Significance*, G. Oxford and D. Rollinson (eds.). Systematics Association, British Museum Natural History, London.

Avise, J. C. 1986. Mitochondrial DNA and the evolutionary genetics of higher animals. *Phil. Trans. R. Soc. Lond. B* 312: 325–342.

Avise, J. C. 1987. Identification and interpretation of mitochondrial DNA stocks in marine species. Pp. 105–136 in *Proc. Stock Identification Workshop*, H. Kumpf and E. L. Nakamura (eds.). Publ. National Oceanographic and Atmospheric Administration, Panama City, FL.

Avise, J. C. 1989a. Gene trees and organismal histories: A phylogenetic approach to population biology. *Evolution* 43: 1192–1208.

Avise, J. C. 1989b. Nature's family archives. *Nat. Hist.* 3: 24–27.

Avise, J. C. 1989c. A role for molecular genetics in the recognition and conservation of endangered species. *Trends Ecol. Evol.* 4: 279–281.

Avise, J. C. 1991a. Ten unorthodox perspectives on evolution prompted by comparative population genetic findings on mitochondrial DNA. *Annu. Rev. Genet.* 25: 45–69.

Avise, J. C. 1991b. Matriarchal liberation. *Nature* 352: 192.

Avise, J. C. 1992. Molecular population structure and the biogeographic history of a regional fauna: A case history with lessons for conservation biology. *Oikos* 63: 62–76.

Avise, J. C. 1994. *Molecular Markers, Natural History and Evolution*. Chapman & Hall, New York.

Avise, J. C. 1995. Mitochondrial DNA polymorphism and a connection between genetics and demography of relevance to conservation. *Cons. Biol.* 9: 686–690.

Avise, J. C. 1996. Toward a regional conservation genetics perspective: Phylogeography of faunas in the southeastern United States. Pp. 431–470 in *Conservation Genetics: Case Histories from Nature*, J. C. Avise and J. L. Hamrick (eds.). Chapman & Hall, New York.

Avise, J. C. 1998a. *The Genetic Gods: Evolution and Belief in Human Affairs*. Harvard University Press, Cambridge, MA.

Avise, J. C. 1998b. The history and purview of phylogeography: A personal reflection. *Mol. Ecol.* 7: 371–379.

Avise, J. C. 2000a. *Phylogeography: The History and Formation of Species*. Harvard University Press, Cambridge, MA.

Avise, J. C. 2000b. Cladists in wonderland. *Evolution* 54: 1828–1832.

Avise, J. C. 2001a. *Captivating Life: A Naturalist in the Age of Genetics*. Smithsonian Institution Press, Washington, DC.

Avise, J. C. 2001b. Evolving genomic metaphors: A new look at the language of DNA. *Science* 294: 86–87.

Avise, J. C. 2001c. Cytonuclear genetic signatures of hybridization phenomena: Rationale, utility, and empirical examples from fishes and other aquatic vertebrates. *Rev. Fish Biol. Fisheries* 10: 253–263.

Avise, J. C. (ed.). 2001d. DNA-based profiling of mating systems and reproductive behaviors in poikilothermic vertebrates (special issue). *J. Heredity* 92: 99–211.

Avise, J. C. 2002. *Genetics in the Wild*. Smithsonian Institution Press, Washington, DC.

Avise, J. C. 2003a. The best and the worst of times for evolutionary biology. *BioScience* 53: 247–255.

Avise, J. C. 2003b. Catadromous eels of the North Atlantic: A review of molecular genetic findings relevant to natural history, population structure, speciation, and phylogeny. Pp. 31–48 in *Eel Biology*, K. Aida, K. Tsukammoto and K. Yamauchi (eds.). Springer-Verlag, Tokyo.

Avise, J. C. 2004a. Twenty-five key evolutionary insights from the phylogeographic revolution in population genetics. Pp. 1–18 in *Phylogeography of Southern European Refugia*, S. Weiss and N. Ferrand (eds.). Kluwer, Dordrecht, The Netherlands, *in press*.

Avise, J. C. 2004b. The history and purview of "conservation genetics. " In *Conservation Genetics in the Age of Genomics*, G. Amato, O. Ryder, H. Rosenbaum and R. DeSalle (eds.). Columbia University Press, New York, *in press*.

Avise, J. C. 2004c. *The Hope, Hype, and Reality of Genetic Engineering: Remarkable Stories from Agriculture, Industry, Medicine, and the Environment*. Oxford University Press, New York.

Avise, J. C. 2004d. Conservation phylogenetics. In *Phylogeny and Conservation*, A. Purvis, T. Brooks and J. Gittleman (eds.). Cambridge University Press, Cambridge, England, *in press*.

Avise, J. C. and C. F. Aquadro. 1982. A comparative summary of genetic distances in the vertebrates. *Evol. Biol.* 15: 151–185.

Avise, J. C. and C. F. Aquadro. 1987. Malate dehydrogenase isozymes provide a phylogenetic marker for the Piciformes (woodpeckers and allies). *Auk* 104: 324–328.

Avise, J. C. and F. J. Ayala. 1975. Genetic change and rates of cladogenesis. *Genetics* 81: 757–773.

Avise, J. C. and F. J. Ayala. 1976. Genetic differentiation in speciose versus depauperate phylads: Evidence from the California minnows. *Evolution* 30: 46–58.

Avise, J. C. and R. M. Ball Jr. 1990. Principles of genealogical concordance in species concepts and biological taxonomy. *Oxford Surv. Evol. Biol.* 7: 45–67.

Avise, J. C. and R. M. Ball Jr. 1991. Mitochondrial DNA and avian microevolution. *Acta XX Congressus Internationalis Ornithologici* 1: 514–524.

Avise, J. C. and B. W. Bowen. 1994. Investigating sea turtle migration using DNA markers. *Current Biol.* 4: 882–886.

Avise, J. C. and J. Felley. 1979. Population structure of freshwater fishes I. Genetic variation of bluegill (*Lepomis macrochirus*) populations in man-made reservoirs. *Evolution* 33: 15–26.

Avise, J. C. and J. L. Hamrick (eds.). 1996. *Conservation Genetics: Case Histories from Nature.* Chapman & Hall, New York.

Avise, J. C. and G. C. Johns. 1999. Proposal for a standardized temporal scheme of biological classification for extant species. *Proc. Natl. Acad. Sci. USA* 96: 7358–7363.

Avise, J. C. and G. B. Kitto. 1973. Phosphoglucose isomerase gene duplication in the bony fishes: An evolutionary history. *Biochem. Genet.* 8: 113–132.

Avise, J. C. and R. A. Lansman. 1983. Polymorphism of mitochondrial DNA in populations of higher animals. Pp. 147–164 in *Evolution of Genes and Proteins*, M. Nei and R. K. Koehn (eds.). Sinauer Associates, Sunderland, MA.

Avise, J. C. and J. E. Neigel. 1984. Population biology aspects of histocompatibility polymorphisms in marine invertebrates. Pp. 223–234 in *Genetics: New Frontiers*, V. L. Copra, B. C. Joshi, R. P. Sharma and H. C. Banoal (eds.). Proc. XV Int. Congr. Genet., Vol. 4, Oxford Publ. Co., New Delhi.

Avise, J. C. and W. S. Nelson. 1989. Molecular genetic relationships of the extinct Dusky Seaside Sparrow. *Science* 243: 646–648.

Avise, J. C. and N. C. Saunders. 1984. Hybridization and introgression among species of sunfish (*Lepomis*): Analysis by mitochondrial DNA and allozyme markers. *Genetics* 108: 237–255.

Avise, J. C. and R. K. Selander. 1972. Evolutionary genetics of cave-dwelling fishes of the genus *Astyanax*. *Evolution* 26: 1–19.

Avise, J. C. and D. Y. Shapiro. 1986. Evaluating kinship of newly settled juveniles within social groups of the coral reef fish *Anthias squamipinnis*. *Evolution* 40: 1051–1059.

Avise, J. C. and M. H. Smith. 1974. Biochemical genetics of sunfish. I. Geographic variation and subspecific intergradation in the bluegill, *Lepomis macrochirus*. *Evolution* 28: 42–56.

Avise, J. C. and M. H. Smith. 1977. Gene frequency comparisons between sunfish (Centrarchidae) populations at various stages of evolutionary divergence. *Syst. Zool.* 26: 319–325.

Avise, J. C. and R. C. Vrijenhoek. 1987. Mode of inheritance and variation of mitochondrial DNA in hybridogenetic fishes of the genus *Poeciliopsis*. *Mol. Biol. Evol.* 4: 514–525.

Avise, J. C. and D. Walker. 1998. Pleistocene phylogeographic effects on avian populations and the speciation process. *Proc. R. Soc. Lond. B* 265: 457–463.

Avise, J. C. and D. Walker. 1999. Species realities and numbers in sexual vertebrates: Perspectives from an asexually transmitted genome. *Proc. Natl. Acad. Sci. USA* 96: 992–995.

Avise, J. C. and K. Wollenberg. 1997. Phylogenetics and the origin of species. *Proc. Natl. Acad. Sci. USA* 94: 7748–7755.

Avise, J. C. and R. M. Zink. 1988. Molecular genetic divergence between avian sibling species: King and clapper rails, long-billed and short-billed dowitchers, boat-tailed and great-tailed grackles, and tufted and black-crested titmice. *Auk* 105: 516–528.

Avise, J. C., M. H. Smith, R. K. Selander, T. E. Lawlor and P. R. Ramsey. 1974. Biochemical polymorphism and systematics in the genus *Peromyscus*. V. Insular and mainland species of the subgenus *Haplomylomys*. *Syst. Zool.* 23: 226–238.

Avise, J. C., J. J. Smith and F. J. Ayala. 1975. Adaptive differentiation with little genic change between two native California minnows. *Evolution* 29: 411–426.

Avise, J. C., R. A. Lansman and R. O. Shade. 1979a. The use of restriction endonucleases to measure mitochondrial DNA sequence relatedness in natural populations. I. Population structure and evolution in the genus *Peromyscus*. *Genetics* 92: 279–295.

Avise, J. C., C. Giblin-Davidson, J. Laerm, J. C. Patton and R. A. Lansman. 1979b. Mitochondrial DNA clones and matriarchal phylogeny within and among geographic populations of the pocket gopher, *Geomys pinetis*. *Proc. Natl. Acad. Sci. USA* 76: 6694–6698.

Avise, J. C., M. H. Smith and R. K. Selander. 1979c. Biochemical polymorphism and systematics in the genus *Peromyscus* VII. Geographic differentiation in members of the *truei* and *maniculatus* species groups. *J. Mammal.* 60: 177–192.

Avise, J. C., J. C. Patton and C. F. Aquadro. 1980. Evolutionary genetics of birds II. Conservative protein evolution in North American sparrows and relatives. *Syst. Zool.* 29: 323–334.

Avise, J. C., J. F. Shapira, S. W. Daniel, C. F. Aquadro and R. A. Lansman. 1983. Mitochondrial DNA differentiation during the speciation process in *Peromyscus*. *Mol. Biol. Evol.* 1: 38–56.

Avise, J. C., J. E. Neigel and J. Arnold. 1984a. Demographic influences on mitochondrial DNA lineage survivorship in animal populations. *J. Mol. Evol.* 20: 99–105.

Avise, J. C., E. Bermingham, L. G. Kessler and N. C. Saunders. 1984b. Characterization of mitochondrial DNA variability in a hybrid swarm between subspecies of bluegill sunfish (*Lepomis macrochirus*). *Evolution* 38: 931–941.

Avise, J. C., G. S. Helfman, N. C. Saunders and L. S. Hales. 1986. Mitochondrial DNA differentiation in North Atlantic eels: Population genetic consequences of an unusual life history pattern. *Proc. Natl. Acad. Sci. USA* 83: 4350–4354.

Avise, J. C., J. Arnold, R. M. Ball, E. Bermingham, T. Lamb, J. E. Neigel, C. A. Reeb and N. C. Saunders. 1987a. Intraspecific phylogeography: The mitochondrial DNA bridge between population genetics and systematics. *Annu. Rev. Ecol. Syst.* 18: 489–522.

Avise, J. C., C. A. Reeb and N. C. Saunders. 1987b. Geographic population structure and species differences in mitochondrial DNA of mouthbrooding marine catfishes (Ariidae) and demersal spawning toadfishes (Batrachoididae). *Evolution* 41: 991–1002.

Avise, J. C., R. M. Ball and J. Arnold. 1988. Current versus historical population sizes in vertebrate species with high gene flow: A comparison based on mitochondrial

DNA lineages and inbreeding theory for neutral mutations. *Mol. Biol. Evol.* 5: 331–344.

Avise, J. C., C. D. Ankney and W. S. Nelson. 1990a. Mitochondrial gene trees and the evolutionary relationship of mallard and black ducks. *Evolution* 44: 1109–1119.

Avise, J. C., W. S. Nelson, J. Arnold, R. K. Koehn, G. C. Williams and V. Thorsteinsson. 1990b. The evolutionary genetic status of Icelandic eels. *Evolution* 44: 1254–1262.

Avise, J. C., J. C. Trexler, J. Travis and W. S. Nelson. 1991. *Poecilia mexicana* is the recent female parent of the unisexual fish *P. formosa*. *Evolution* 45: 1530–1533.

Avise, J. C., B. W. Bowen, T. Lamb, A. B. Meylan and E. Bermingham. 1992a. Mitochondrial DNA evolution at a turtle's pace: Evidence for low genetic variability and reduced microevolutionary rate in the Testudines. *Mol. Biol. Evol.* 9: 457–473.

Avise, J. C., R. T. Alisauskas, W. S. Nelson and C. D. Ankney. 1992b. Matriarchal population genetic structure in an avian species with female natal philopatry. *Evolution* 46: 1084–1096.

Avise, J. C., J. M. Quattro and R. C. Vrijenhoek. 1992c. Molecular clones within organismal clones: Mitochondrial DNA phylogenies and the evolutionary histories of unisexual vertebrates. *Evol. Biol.* 26: 225–246.

Avise, J. C., W. S. Nelson and H. Sugita. 1994. A speciational history of "living fossils": Molecular evolutionary history in horseshoe crabs. *Evolution* 48: 1986–2001.

Avise, J. C., P. C. Pierce, M. J. Van Den Avyle, M. H. Smith, W. S. Nelson and M. A. Asmussen. 1997. Cytonuclear introgressive swamping and species turnover of bass after an introduction. *J. Heredity* 88: 14–20.

Avise, J. . C., D. Walker and G. C. Johns. 1998. Speciation durations and Pleistocene effects on vertebrate phylogeography. *Proc. Roy Soc. Lond. B* 265: 1707–1712.

Avise, J. C., W. S. Nelson, B. W. Bowen and D. Walker. 2000. Phylogeography of colonially nesting seabirds, with special reference to global matrilineal patterns in the sooty tern (*Sterna fuscata*). *Mol. Ecol.* 9: 1783–1792.

Avise, J. C. and 10 others. 2002. Genetic mating systems and reproductive natural histories of fishes: Lessons for ecology and evolution. *Annu. Rev. Genet.* 36: 19–45.

Avise, J. C., A. J. Power and D. Walker. 2004. Genetic sex determination, gender identification and pseudohermaphroditism in the knobbed whelk, *Busycon carica* (Mollusca; Melongenidae). *Proc. R. Soc. Lond. B* 271: 641–646.

Awadalla, P., A. Eyre-Walker and J. Maynard Smith. 1999. Linkage disequilibrium and recombination in hominid mitochondrial DNA. *Science* 286: 2524–2525.

Ayala, F. J. 1975. Genetic differentiation during the speciation process. *Evol. Biol.* 8: 1–78.

Ayala, F. J. (ed.) 1976a. *Molecular Evolution*. Sinauer Associates, Sunderland, MA.

Ayala, F. J. 1976b. Molecular genetics and evolution. Pp. 1–20 in *Molecular Evolution*, F. J. Ayala (ed.). Sinauer Associates, Sunderland, MA.

Ayala, F. J. 1982a. The genetic structure of species. Pp. 60–82 in *Perspectives on Evolution*, R. Milkman (ed.). Sinauer Associates, Sunderland, MA.

Ayala, F. J. 1982b. Genetic variation in natural populations: Problem of electrophoretically cryptic alleles. *Proc. Natl. Acad. Sci. USA* 79: 550–554.

Ayala, F. J. 1982c. Of clocks and clades, or a story of old told by genes of now. Pp. 257–301 in *Biochemical Aspects of Evolutionary Biology*, M. H. Nitecki (ed.). University of Chicago Press, Chicago, IL.

Ayala, F. J. 1986. On the virtues and pitfalls of the molecular evolutionary clock. *J. Heredity* 77: 226–235.

Ayala, F. J. 1995. The myth of Eve: Molecular biology and human origins. *Science* 270: 1930–1936.

Ayala, F. J. 1996. HLA sequence polymorphism and the origin of humans. *Science* 274: 1554.

Ayala, F. J. and C. A. Campbell. 1974. Frequency-dependent selection. *Annu. Rev. Ecol. Syst.* 5: 115–138.

Ayala, F. J. and J. R. Powell. 1972a. Enzyme variability in the *Drosophila willistoni* group. VI. Levels of polymorphism and the physiological function of enzymes. *Biochem. Genet.* 7: 331–345.

Ayala, F. J. and J. R. Powell. 1972b. Allozymes as diagnostic characters of sibling species of *Drosophila*. *Proc. Natl. Acad. Sci. USA* 69: 1094–1096.

Ayala, F. J., C. A. Mourão, S. Pérez-Salas, R. Richmond and T. Dobzhansky. 1970. Enzyme variability in the *Drosophila willistoni* group, I. Genetic differentiation among sibling species. *Proc. Natl. Acad. Sci. USA* 67: 225–232.

Ayala, F. J., M. L. Tracey, L. G. Barr, J. F. McDonald and S. Pérez-Salas. 1974. Genetic variation in natural populations of five *Drosophila* species and the hypothesis of the selective neutrality of protein polymorphisms. *Genetics* 77: 343–384.

Ayala, F. J., J. W. Valentine, T. E. Delaca and G. S. Zumwalt. 1975a. Genetic variability of the antarctic brachiopod *Liothyrella notorcadensis* and its bearing on mass extinction hypotheses. *J. Paleontol.* 49: 1–9.

Ayala, F. J., M. L. Tracey, D. Hedgecock and R. C. Richmond. 1975b. Genetic differentiation during the speciation process in *Drosophila*. *Evolution* 28: 576–592.

Ayala, F. José, A. Rzhetsky and F. J. Ayala. 1998. Origin of the metazoan phyla: Molecular clocks confirm paleontological estimates. *Proc. Natl. Acad. Sci. USA* 95: 606–611.

Ayre, D. J. 1982. Inter-genotype aggression in the solitary sea anemone *Actinia tenebrosa*. *Mar. Biol.* 68: 199–205.

Ayre, D. J. 1984. The effects of sexual and asexual reproduction on geographic variation in the sea anemone *Actinia tenebrosa*. *Oecologia* 62: 222–229.

Ayre, D. J. and T. P. Hughes. 2000. Genotypic diversity and gene flow in brooding and spawning corals along the Great Barrier Reef, Australia. *Evolution* 54: 1590–1605.

Ayre, D. J. and J. M. Resing. 1986. Sexual and asexual production of planulae in reef corals. *Mar. Biol.* 90: 187–190.

Ayre, D. J. and B. L. Willis. 1988. Population structure in the coral *Pavona cactus*: Clonal genotypes show little phenotypic plasticity. *Mar. Biol.* 99: 495–505.

Ayre, D. J., A. R. Davis, T. Llorens and M. Billingham. 1997. Genetic evidence for contrasting patterns of dispersal in solitary and colonial ascidians. *Mar. Biol.* 130: 51–62.

Ayres, J. M. and T. H. Clutton-Brock. 1992. River boundaries and species range size in Amazonian primates. *Am. Nat.* 140: 531–537.

Baby, M.-C., L. Bernatchez and J. J. Dodson. 1991. Genetic structure and relationships among anadromous and landlocked populations of rainbow smelt, *Osmerus mordax*, Mitchell, as revealed by mtDNA restriction analysis. *J. Fish Biol.* 39A: 61–68.

Bacilieri, R., A. Ducousso, R. J. Petit and A. Kremer. 1996. Mating system and asymmetric hybridization in a mixed stand of European oaks. *Evolution* 50: 900–908.

Backert, S., R. Lurz and T. Börner. 1996. Electron microscopic investigation of mitochondrial DNA from *Chenopodium album* (L.). *Current Genet.* 29: 427–436.

Backert, S., R. Lurz, O. A. Ayarzabal and T. Börner. 1997. High content, size and distribution of single-stranded DNA in the mitochondria of *Chenopodium album* (L.). *Plant Molec. Biol.* 33: 1037–1050.

Baglione, V., D. Canestrari, J. M. Marcos and J. Ekman. 2003. Kin selection in cooperative alliances of carrion crows. *Science* 300: 1947–1949.

Bahlo, M. and R. C. Griffiths. 2000. Inference from gene trees in a subdivided population. *Theor. Pop. Biol.* 57: 763–773.

Bailey, W. J., J. L. Slightom and M. Goodman. 1992. Rejection of the "flying primate" hypothesis by phylogenetic evidence from the ε-globin gene. *Science* 256: 86–89.

Baker, A. J. (ed.). 2000. *Molecular Methods in Ecology.* Blackwell, Oxford, England.

Baker, A. J. and H. D. Marshall. 1997. Mitochondrial control region sequences as tools for understanding evolution. Pp. 51–82 in *Avian Molecular Evolution and Systematics*, D. P. Mindell (ed.). Academic Press, New York.

Baker, A. J., T. Piersma and L. Rosenmeier. 1994. Unraveling the intraspecific phylogeography of knots *Calidris canutus*: A progress report on the search for genetic markers. *J. Ornithol.* 135: 599–608.

Baker, A. J., C. H. Daugherty, R. Colbourne and J. L. McLennan. 1995. Flightless brown kiwis of New Zealand possess extremely subdivided population structure and cryptic species like small mammals. *Proc. Natl. Acad. Sci. USA* 92: 8254–8258.

Baker, C. S. and S. R. Palumbi. 1994. Which whales are hunted? A molecular genetic approach to monitoring whaling. *Science* 265: 1538–1539.

Baker, C. S. and S. R. Palumbi. 1996. Population structure, molecular systematics, and forensic identification of whales and dolphins. Pp. 10–49 in *Conservation Genetics: Case Histories from Nature*, J. C. Avise and J. L. Hamrick (eds.). Chapman & Hall, New York.

Baker, C. S., S. R. Palumbi, R. H. Lambertsen, M. T. Weinrich, J. Calambokidis and S. J. O'Brien. 1990. Influence of seasonal migration on geographic distribution of mitochondrial DNA haplotypes in humpback whales. *Nature* 34: 238–240.

Baker, C. S., R. H. Lambertsen, M. T. Weinrich, J. Calambokidis, G. Early and S. J. O'Brien. 1991. Molecular genetic identification of the sex of humpback whales (*Megaptera novaeangliae*). Pp. 105–111 in *Genetic Ecology of Whales and Dolphins*, A. R. Hoelzel (ed.). Black Bear Press, Cambridge, England.

Baker, C. S., M. MacCarthy, P. J. Smith, A. P. Perry and G. K. Chambers. 1992. DNA fingerprints of orange roughy, *Hoplostethus atlanticus*: A population comparison. *Mar. Biol.* 113: 561–567.

Baker, C. S. and 13 others. 1993. Abundant mitochondrial DNA variation and world-wide population structure in humpback whales. *Proc. Natl. Acad. Sci. USA* 90: 8239–8243.

Baker, C. S. and 10 others. 1994. Hierarchical structure of mitochondrial DNA gene flow among humpback whales *Megaptera novaeangliae*, world-wide. *Mol. Ecol.* 3: 313–327.

Baker, C. S., F. Cipriano and S. R. Palumbi. 1996. Molecular genetic identification of whale and dolphin products from commercial markets in Korea and Japan. *Mol. Ecol.* 5: 671–685.

Baker, C. S. and 9 others. 1998. Population structure of nuclear and mitochondrial DNA variation among humpback whales in the North Pacific. *Mol. Ecol.* 7: 695–707.

Baker, J. D., P. Moran and R. Ladley. 2003. Nuclear DNA identification of migrating bull trout captured at the Puget Sound Energy diversion dam on the White River, Washington State. *Mol. Ecol.* 12: 557–561.

Baker, R. H. 1968. Habitats and distribution. Pp. 98–126 in *Biology of Peromyscus* (Rodentia), J. A. King (ed.). Special Publication American Society Mammalogy No. 2.

Baker, R. J. and M. A. Bellis. 1995. *Human Sperm Competition.* Chapman & Hall, London.

Baker, R. J., S. K. Davis, R. D. Bradley, M. J. Hamilton and R. A. Van Den Bussche. 1989. Ribosomal-DNA, mitochondrial-DNA, chromosomal, and allozymic studies on a contact zone in the pocket gopher, *Geomys*. *Evolution* 43: 63–75.

Balakrishnan, C. N., S. L. Monforn, A. Gaur, L. Singh and M. D. Sorenson. 2003. Phylogeography and conservation genetics of Eld's deer (*Cervus eldi*). *Mol. Ecol.* 12: 1–10.

Balazs, I., M. Baird, M. Clyne and E. Meade. 1989. Human population genetic studies of five hypervariable DNA loci. *Am. J. Human Genet.* 44: 182–190.

Balazs, I. and 6 others. 1992. Human population genetic studies using hypervariable loci. I. Analysis of Assamese, Australian, Cambodian, Caucasian, Chinese and Melanesian populations. *Genetics* 131: 191–198.

Baldauf, S. L. 2003. The deep roots of Eukaryotes. *Science* 300: 1703–1706.

Baldauf, S. L. and J. D. Palmer. 1990. Evolutionary transfer of the chloroplast *tufA* gene to the nucleus. *Nature* 344: 262–265.

Baldauf, S. L. and J. D. Palmer. 1993. Animals and fungi are each other's closest relatives: Congruent evidence from multiple proteins. *Proc. Natl. Acad. Sci. USA* 90: 11558–11562.

Baldauf, S. L., J. R. Manhart and J. D. Palmer. 1990. Different fates of the chloroplast *tufA* gene following its transfer to the nucleus in green algae. *Proc. Natl. Acad. Sci. USA* 87: 5317–5321.

Baldauf, S. L., A. J. Roger, I. Wenk-Siefert and W. F. Doolittle. 2000. A kingdom-level phylogeny of eukaryotes based on combined protein data. *Science* 290: 972–977.

Baldwin, B. G., D. W. Kyhos and J. Dvorak. 1990. Chloroplast DNA evolution and adaptive radiation in the Hawaiian silversword alliance (Madiinae, Asteraceae). *Annals Missouri Bot. Garden* 77: 96–109.

Baldwin, J. D., A. L. Bass, B. W. Bowen and W. H. Clark Jr. 1998. Molecular phylogeny and biogeography of the marine shrimp *Penaeus*. *Mol. Phylogen. Evol.* 10: 399–407.

Ball, R. M. Jr. and J. C. Avise. 1992. Mitochondrial DNA phylogeographic differentiation among avian populations and the evolutionary significance of subspecies. *Auk* 109: 626–636.

Ball, R. M. Jr., S. Freeman, F. C. James, E. Bermingham and J. C. Avise. 1988. Phylogeographic population structure of red-winged blackbirds assessed by mitochondrial DNA. *Proc. Natl. Acad. Sci. USA* 85: 1558–1562.

Ball, R. M., J. E. Neigel and J. C. Avise. 1990. Gene genealogies within the organismal pedigrees of random-mating populations. *Evolution* 44: 360–370.

Ballinger, S. W. and 7 others. 1992. Southeast Asian mitochondrial DNA analysis reveals genetic continuity of ancient Mongoloid migrations. *Genetics* 130: 139–152.

Ballou, J. D. 1983. Calculating inbreeding coefficients from pedigrees. Pp. 509–520 in *Genetics and Conservation*, C. M. Schonewald-Cox, S. M. Chambers, F. MacBryde and L. Thomas (eds.). Benjamin/ Cummings, Menlo Park, CA.

Ballou, J. D., M. Gilpin and T. J. Foose (eds.). 1995. *Population Management for Survival and Recovery.* Columbia University Press, New York.

Balloux, F. and N. Lugon-Moulin. 2002. The estimation of population differentiation with microsatellite markers. *Mol. Ecol.* 11: 155–165.

Balmford, A. and 18 others. 2002. Economic reasons for conserving wild nature. *Science* 297: 950–953.

Balsano, J. S., R. M. Darnell and P. Abramoff. 1972. Electrophoretic evidence of triploidy associated with populations of the gynogenetic teleost *Poecilia formosa*. *Copeia* 1972: 292–297.

Banks, J. A. and C. W. Birky Jr. 1985. Chloroplast DNA diversity is low in a wild plant, *Lupinus texensis*. *Proc. Natl. Acad. Sci. USA* 82: 6950–6954.

Banks, M. A., V. K. Rashbrook, M. J. Calavetta, C. A. Dean and D. Hedgecock. 2000. Analysis of microsatellite DNA resolves genetic structure and diversity of chinook salmon (*Oncorhynchus tshawytscha*) in California's Central Valley. *Can. J. Fish. Aquat. Sci.* 57: 2368–2373.

Banks, S. C., A. Horsup, A. N. Wilton and A. C. Taylor. 2003. Genetic marker investigation of the source and impact of predation on a highly endangered species. *Mol. Ecol.* 12: 1663–1667.

Bannister, J. V. and M. W. Parker. 1985. The presence of a copper/zinc superoxide dismutase in the bacterium *Photobacterium leiognathi*: A likely case of gene transfer

from eukaryotes to prokaryotes. *Proc. Natl. Acad. Sci. USA* 82: 149–152.

Baragona, M. A. and L. A. Haig-Ladewig. 2000. Multiple paternity in the grass shrimp *Palaemonetes pugio*. *Am. Zool.* 40: 935.

Barber, P. H., S. R. Palumbi, M. V. Erdmann and M. K. Moosa. 2000. A marine Wallace's line? *Nature* 406: 692–693.

Barber, P. H., S. R. Palumbi, M. V. Erdmann and M. K. Moosa. 2002. Sharp genetic breaks among populations of *Haptosquilla pulchella* (Stomatopoda) indicate limits to larval transport: Patterns, causes, and consequences. *Mol. Ecol.* 11: 659–674.

Barker, K., G. F. Barrowclough and J. G. Groth. 2002. Phylogenetic hypothesis for passerine brids: Taxonomic and biogeographic implications of an analysis of nuclear DNA sequence data. *Proc. R. Soc. Lond. B* 269: 295–308.

Barki, Y., D. Gateño, D. Graur and B. Rinkevich. 2002. Soft-coral natural chimerism: A window in ontogeny allows the creation of entities comprised of incongruous parts. *Marine Ecol. Progr. Ser.* 231: 91–99.

Barlow, G. W. 1961. Causes and significance of morphological variation in fishes. *Syst. Zool.* 10: 105–117.

Barnabé, C., S. Brisse and M. Tibayrenc. 2000. Population structure and genetic typing of *Trypanosoma cruzi*, the agent of Chagas' disease: A multilocus enzyme electrophoresis approach. *Parasitology* 120: 513–526.

Barnes, B. V. 1966. The clonal growth habit of American aspens. *Ecology* 47: 439–447.

Barnes, I., P. Matheus, B. Shapiro, D. Jensen and A. Cooper. 2002. Dynamics of Pleistocene population extinctions in Beringian brown bears. *Science* 295: 2267–2270.

Barnes, P. T. and C. C. Laurie-Ahlberg. 1986. Genetic variability of flight metabolism in *Drosophila melanogaster*. III. Effects of *GPDH* allozymes and environmental temperature on power output. *Genetics* 112: 267–294.

Barraclough, T. G. and S. Nee. 2001. Phylogenetics and speciation. *Trends Ecol. Evol.* 16: 391–399.

Barraclough, T. G. and V. Savolainen. 2001. Evolutionary rates and species diversity in flowering plants. *Evolution* 55: 677–683.

Barraclough, T. G. and A. P. Vogler. 2000. Detecting the geographical pattern of speciation from species-level phylogenies. *Am. Nat.* 155: 419–434.

Barraclough, T. G., P. H. Harvey and S. Nee. 1995. Sexual selection and taxonomic diversity in passerine birds. *Proc. R. Soc. Lond. B* 259: 211–215.

Barraclough, T. G., J. E. Hogan and A. P. Vogler. 1999. Testing whether ecological factors promote cladogenesis in a group of tiger beetles (Coleoptera: Cicindelidae). *Proc. R. Soc. Lond. B* 266: 1061–1067.

Barrell, B. G., A. T. Bankier and J. Drouin. 1979. A different genetic code in human mitochondria. *Nature* 282: 189–194.

Barrett, S. C. H. 1989. Mating system evolution and speciation in heterostylous plants. Pp. 257–283 in *Speciation and Its Consequences*, D. Otte and J. A. Endler (eds.). Sinauer Associates, Sunderland, MA.

Barrowclough, G. F. 1983. Biochemical studies of microevolutionary processes. Pp. 223–261 in *Perspectives in Ornithology*, A. H. Brush and G. A. Clark Jr. (eds.). Cambridge University Press, New York.

Barrowclough, G. F. 1992. Systematics, biodiversity, and conservation biology. Pp. 121–143 in *Systematics, Ecology, and the Biodiversity Crisis*, N. Eldredge (ed.). Columbia University Press, New York.

Barrowclough, G. F. and R. J. Gutiérrez. 1990. Genetic variation and differentiation in the spotted owl (*Strix occidentalis*). *Auk* 107: 737–744.

Barrowclough, G. F., R. J. Gutiérrez and J. G. Groth. 1999. Phylogeography of the spotted owl (*Strix occidentalis*) populations based on mitochondrial DNA sequences: Gene flow, genetic structure, and a novel biogeographic pattern. *Evolution* 53: 919–931.

Bartlett, S. E. and W. S. Davidson. 1991. Identification of *Thunnus* species by the polymerase chain reaction and direct sequence analysis of their mitochondrial cytochrome *b* genes. *Can. J. Fish. Aquat. Sci.* 48: 309–317.

Barton, N. H. 1980. The hybrid sink effect. *Heredity* 44: 277–278.

Barton, N. H. 1983. Multilocus clines. *Evolution* 37: 454–471.

Barton, N. H. 1989. Founder effect speciation. Pp. 229–256 in *Speciation and Its Consequences*, D. Otte and J. A. Endler (eds.). Sinauer Associates, Sunderland, MA.

Barton, N. H. 1996. Natural selection and random genetic drift as causes of evolution on islands. *Phil. Trans. R. Soc. Lond. B* 351: 785–795.

Barton, N. H. 2001. The role of hybridization in evolution. *Mol. Ecol.* 10: 551–568.

Barton, N. H. and B. Charlesworth. 1984. Genetic revolutions, founder effects, and speciation. *Annu. Rev. Ecol. Syst.* 15: 133–164.

Barton, N. H. and G. M. Hewitt. 1981. Hybrid zones and speciation. Pp. 109–145 in *Evolution and Speciation. Essays in Honor of MJD White*, W. R. Atchley and D. S. Woodruff (eds.). Cambridge University Press, Cambridge, England.

Barton, N. H. and G. M. Hewitt. 1985. Analysis of hybrid zones. *Annu. Rev. Ecol. Syst.* 16: 113–148.

Barton, N. H. and G. M. Hewitt. 1989. Adaptation, speciation and hybrid zones. *Nature* 341: 497–502.

Barton, N. H. and J. S. Jones. 1983. Mitochondrial DNA: New clues about evolution. *Nature* 306: 317–318.

Barton, N. H. and M. Slatkin. 1986. A quasi-equilibrium theory of the distribution of rare alleles in a subdivided population. *Heredity* 56: 409–415.

Bartz, S. H. 1979. Evolution of eusociality in termites. *Proc. Natl. Acad. Sci. USA* 76: 5764–5768.

Basolo, A. L. 1995. Phylogenetic evidence for the role of a pre-existing bias in sexual selection. *Proc. R. Soc. Lond. B* 259: 307–311.

Bass, A. L. and 6 others. 1996. Testing models of female reproductive migratory behavior and population structure in the Caribbean hawksbill turtle, *Eretmochelys imbricata*, with mtDNA sequences. *Mol. Ecol.* 5: 321–328.

Bass, A. L., C. J. Lagueux and B. W. Bowen. 1998. Origin of green turtles, *Chelonia mydas*, at "sleeping rocks" off the northeast coast of Nicaragua. *Copeia* 1998: 1064–1069.

Basten, C. J., B. S. Weir and Z.-B. Zeng. 2002. *QTL Cartographer: A Reference Manual and Tutorial for QTL Mapping*. North Carolina State University, Raleigh, NC.

Bauch, D., K. Darling, J. Simstich, H. A. Bauch, H. Erienkeuser and D. Kroon. 2003. Palaeoceanographic implications of genetic variation in living North Atlantic *Neogloboquadrina pachyderma*. *Nature* 424: 299–302.

Bauert, M. R., M. Kälin, M. Baltisberger and P. J. Edwards. 1998. No genetic variation detected within isolated relict populations of *Saxifraga cernua* in the Alps using RAPD markers. *Mol. Ecol.* 7: 1519–1527.

Baum, B. R. 1992. Combining trees as a way of combining data sets for phylogenetic inference, and the desirability of combining gene trees. *Taxon* 41: 3–10.

Baum, D. 1994. RbcL and seed-plant phylogeny. *Trends Ecol. Evol.* 9: 39–41.

Baum, D. A. and K. L. Shaw. 1995. Genealogical perspectives on the species problem. Pp. 289–303 in *Experimental and Molecular Approaches to Plant Biosystematics*, P. C. Hoch and A. G. Stephenson (eds.). Missouri Bot. Garden Monogr., no. 53.

Baur, B. 1998. Sperm competition in mollusks. Pp. 255–306 in *Sperm Competition and Sexual Selection*, T. R. Birkhead and A. P. Møller (eds.). Academic Press, London.

Bautz, E. K. and F. A. Bautz. 1964. The influence of non-complementary bases on the stability of ordered polynucleotides. *Proc. Natl. Acad. Sci. USA* 52: 1476–1481.

Baverstock, P. R., S. R. Cole, B. J. Richardson and C. H. Watts. 1979. Electrophoresis and cladistics. *Syst. Zool.* 28: 214–219.

Baverstock, P. R., M. Adams and I. Beveridge. 1985. Biochemical differentiation in bile duct cestodes and their marsupial hosts. *Mol. Biol. Evol.* 2: 321–337.

Baverstock, P. R., J. Birrell and M. Krieg. 1987. Albumin immunologic relationships of the Diprodonta. Pp. 229–234 in *Possums and Opossums: Studies in Evolution*, M. Archer (ed.). Surrey Beatty and Sons, Chipping Norton, Australia.

Bayer, R. J. 1989. Patterns of isozyme variation in the *Antennaria rosea* (Asteraceae: Inuleae) polyploid agamic complex. *Syst. Bot.* 14: 389–397.

Beacham, T. D., S. Pollard and K. D. Le. 2000. Microsatellite DNA population structure and stock identification of steelhead trout (*Oncorhynchus mykiss*) in the Nass and Skeena Rivers in northern British Columbia. *Marine Biotech.* 2: 587–600.

Beacham, T. D. and 10 others. 2001. Evaluation and application of microsatellite and major histocompatibility complex variation for stock identification of coho salmon in British Columbia. *Trans. Am. Fish. Soc.* 130: 1116–1149.

Beacham, T. D., B. McIntosh and C. MacConnachie. 2002. Microsatellite identification of individual sockeye salmon in Barkley Sound, British Columbia. *J. Fish Biol.* 61: 1021–1032.

Beacham, T. D. and 9 others. 2003. Evaluation and application of microsatellites for population identification of Fraser River chinook salmon (*Oncorhynchus tschawytscha*). *Fish. Bull.* 101: 243–259.

Beagley, C. T., R. Okimoto and D. R. Wolstenholme. 1998. The mitochondrial genome of the sea anenome *Metridium senile* (Cnidaria): Introns, a paucity of tRNA genes, and a near-standard genetic code. *Genetics* 148: 1091–1108.

Beardsley, P. M., A. Yen and R. G. Olmstead. 2003. AFLP phylogeny of *Mimulus* section *Erythranthe* and the evolution of hummingbird pollination. *Evolution* 57: 1397–1410.

Beckmann, J. S., Y. Kashi, E. M. Hallerman, A. Nave and M. Soller. 1986. Restriction fragment length polymorphism among Israeli Holstein-Friesian dairy bulls. *Anim. Genet.* 17: 25–38.

Beerli, P. and J. Felsenstein. 1999. Maximum likelihood estimation of migration rates and effective populatoin numbers in two populations using a coalescent approach. *Genetics* 152: 763–773.

Beerli, P. and J. Felsenstein. 2001. Maximum likelihood estimation of a migration matrix and effective population sizes in *n* subpopulations using a coalescent approach. *Proc. Natl. Acad. Sci. USA* 98: 4563–4568.

Begun, D. J. and C. F. Aquadro. 1992. Levels of naturally occurring DNA polymorphism correlate with recombination rates in *D. melanogaster*. *Nature* 356: 519–520.

Begun, D. J. and C. F. Aquadro. 1993. African and North American populations of *Drosophila melanogaster* are very different at the DNA level. *Nature* 365: 548–550.

Begun, D. J., P. Whitley, B. L. Todd, H. M. Waldrip-Dail and A. G. Clark. 2000. Molecular population genetics of male accessory gland proteins in *Drosophila*. *Genetics* 156: 1879–1888.

Behnke, R. J. 2002. *Trout and Salmon of North America*. The Free Press, New York.

Belahbib, N., M.-H. Pemonge, A. Ouassou, H. Sbay, A. Kremer and R. J. Petit. 2001. Frequent cytoplasmic exchanges between oak species that are not closely related: *Quercus suber* and *Q. ilex* in Morocco. *Mol. Ecol.* 10: 2003–2012.

Belfiore, N. M., G. G. Hoffman, R. J. Baker and J. A. DeWoody. 2003. The use of nuclear and mitochondrial single nucleotide polymorphisms to identify cryptic species. *Mol. Ecol.* 12: 2011–2017.

Bell, M. and S. A. Foster. 1994. *The Evolutionary Biology of the Threespine Stickleback*. Oxford University Press, New York.

Belov, K., L. Hellman and D. W. Cooper. 2002. Characterisation of echidna IgM provides insights into the time of divergence of extant mammals. *Devel. and Comp. Immunol.* 26: 831–839.

Beltrán, J. F., J. E. Rice and R. L. Honeycutt. 1996. Taxonomy of the Iberian lynx. *Nature* 379: 407–408.

Beninda-Emonds, O. R. P. and J. L. Gittleman. 2000. Are pinnipeds functionally different from fissiped carnivores? The importance of phylogenetic comparative analyses. *Evolution* 54: 1011–1023.

Beninda-Emonds, O. R. P., J. L Gittleman and A. Purvis. 1999. Building large trees by combining phylogenetic information: A complete phylogeny of the extant Carnivora (Mammalia). *Biol. Rev.* 74: 143–175.

Beninda-Emonds, O. R. P., J. L Gittleman and M. A. Steel. 2002. The (Super)Tree of Life: Procedures, problems, and prospects. *Annu. Rev. Ecol. Syst.* 33: 265–289.

Benirschke, K., J. M. Anderson and L. E. Brownhill. 1962. Marrow chimerism in marmosets. *Science* 138: 513–515.

Benjamin, D. C. and 14 others. 1984. The antigenic structure of proteins: A reappraisal. *Annu. Rev. Immunol.* 2: 67–101.

Bennett, P. M. and I. P. F. Owens. 2002. *Evolutionary Ecology of Birds: Life Histories, Mating Systems, and Extinction*. Oxford University Press, Oxford, England.

Bennett, S., L. J. Alexander, R. H. Crozier and A. G. Mackinley. 1988. Are megabats flying primates? Contrary evidence from a mitochondrial DNA sequence. *Aust. J. Biol. Sci.* 41: 327–332.

Bensasson, D., D. X. Zhang, D. L. Hartl and G. M. Hewitt. 2001. Mitochondrial pseudogenes: Evolution's misplaced witnesses. *Trends Ecol. Evol.* 16: 314–321.

Bensch, S., A. J. Helbig, M. Salomon and I. Seibold. 2002. Amplified fragment length polymorphism analysis identifies hybrids between two subspecies of warblers. *Mol. Ecol.* 11: 473–481.

Benton, M. J. and F. J. Ayala. 2003. Dating the tree of life. *Science* 300: 1698–1700.

Berger, E. 1973. Gene-enzyme variation in three sympatric species of *Littorina*. *Biol. Bull.* 145: 83–90.

Bergthorsson, U., K. L. Adams, B. Thomason and J. D. Palmer. 2003. Widespread horizontal transfer of mitochondrial genes in flowering plants. *Nature* 424: 197–201.

Berlocher, S. H. 1981. A comparison of molecular and morphological data, and phenetic and cladistic methods, in the estimation of phylogeny in *Rhagoletis* (Diptera: Tephritidae). Pp. 1–31 in *Applications of Genetics and Cytology in Insect Systematics and Evolution*, W. M. Stock (ed.). University of Idaho Press, Moscow, ID.

Berlocher, S. H. 1998. Can sympatric speciation via host or habitat shift be proven from phylogenetic and biogeographic evidence? Pp. 99–113 in *Endless Forms: Species and Speciation*, D. J. Howard and S. H. Berlocher (eds.). Oxford University Press, New York.

Berlocher, S. H. 2000. Radiation and divergence in the *Rhagoletis pomonella* species group: Inferences from allozymes. *Evolution* 54: 543–557.

Berlocher, S. H. and G. L. Bush. 1982. An electrophoretic analysis of *Rhagoletis* (Diptera: Tephritidae) phylogeny. *Syst. Zool.* 31: 136–155.

Berlocher, S. H. and J. L. Feder. 2002. Sympatric speciation in phytophagous insects: Moving beyond controversy? *Annu. Rev. Entomol.* 47: 773–815.

Bermingham, E. and J. C. Avise. 1986. Molecular zoogeography of freshwater fishes in the southeastern United States. *Genetics* 113: 939–965.

Bermingham, E. and H. A. Lessios. 1993. Rate variation of protein and mitochondrial DNA evolution as revealed by sea urchins separated by the Isthmus of Panama. *Proc. Natl. Acad. Sci. USA* 90: 2734–2738.

Bermingham, E. and A. P. Martin. 1998. Comparative mtDNA phylogeography of neotropical freshwater fish: Testing shared history to infer the evolutionary landscape of lower Central America. *Mol. Ecol.* 7: 499–517.

Bermingham, E. and C. Moritz. 1998. Comparative phylogeography: Concepts and applications. *Mol. Ecol.* 7: 367–369.

Bermingham, E., T. Lamb and J. C. Avise. 1986. Size polymorphism and heteroplasmy in the mitochondrial DNA of lower vertebrates. *J. Heredity* 77: 249–252.

Bermingham, E., S. H. Forbes, K. Friedland and C. Pla. 1991. Discrimination between Atlantic salmon (*Salmo salar*) of North American and European origin using restriction analyses of mitochondrial DNA. *Can. J. Fish. Aquat. Sci.* 48: 884–893.

Bermingham, E., S. Rohwer, S. Freeman and C. Wood. 1992. Vicariance biogeography in the Pleistocene and speciation in North American wood warblers: A test of Mengel's model. *Proc. Natl. Acad. Sci. USA* 89: 6624–6628.

Bermingham, E., G. Seutin and R. E. Ricklefs. 1996. Regional approaches to conservation biology: RFLPs, DNA sequence and Caribbean birds. Pp. 104–124 in *Molecular Genetic Approaches in Conservation*, T. B. Smith and R. K. Wayne (eds.). Oxford University Press, New York.

Bermingham, E., S. S. McCafferty and A. P. Martin. 1997. Fish biogeography and molecular clocks: Perspectives from the Panamanian Isthmus. Pp. 113–128 in *Molecular Systematics of Fishes*, T. D. Kocher and C. A. Stepien (eds.). Academic Press, New York.

Bernardi, G. 2000. Barriers to gene flow in *Embiotoca jacksoni*, a marine fish lacking a pelagic larval stage. *Evolution* 54: 226–237.

Bernardi, G., P. Sordino and D. A. Powers. 1993. Concordant mitochondrial and nuclear DNA phylogenies for populations of the teleost fish *Fundulus heteroclitus. Proc. Natl. Acad. Sci. USA* 90: 9271–9274.

Bernardi, G., L. Findley and Z. Rocha-Olivares. 2003. Vicariance and dispersal across Baja California in disjunct marine fish populations. *Evolution* 57: 1599–1609.

Bernatchez, L. 1995. A role for molecular systematics in defining evolutionarily significant units (ESU) in fishes. *Am. Fish. Soc. Symp.* 17: 114–132.

Bernatchez, L. 1997. Mitochondrial DNA analysis confirms the existence of two glacial races of rainbow smelt *Osmerus mordax* and their reproductive isolation in the St. Lawrence River estuary (Québec, Canada). *Mol. Ecol.* 6: 73–83.

Bernatchez, L. 2001. The evolutionary history of the brown trout (*Salmo trutta*) inferred from phylogeographic, nested clade, and mismatch analyses of mitochondrial DNA variation. *Evolution* 55: 351–379.

Bernatchez, L. and J. J. Dodson. 1990. Allopatric origin of sympatric populations of lake whitefish (*Coregonus clupeaformis*) as revealed by mitochondrial-DNA restriction analysis. *Evolution* 44: 1263–1271.

Bernatchez, L. and J. J. Dodson. 1991. Phylogeographic structure in mitochondrial DNA of the lake whitefish (*Coregonus clupeaformis*) and its relation to Pleistocene glaciations. *Evolution* 45: 1016–1035.

Bernatchez, L. and J. J. Dodson. 1994. Phylogenetic relationships among Palearctic and Nearctic whitefish (*Coregonus* sp.) populations as revealed by mitochondrial DNA variation. *Can. J. Fish. Aquat. Sci.* 51: 240–251.

Bernatchez, L. and A. Osinov. 1995. Genetic diversity of trout (genus *Salmo*) from its most eastern native range based on mitochondrial DNA and nuclear gene variation. *Mol. Ecol.* 4: 285–297.

Bernatchez, L. and C. C. Wilson. 1998. Comparative phylogeography of Nearctic and Palearctic fishes. *Mol. Ecol.* 7: 431–452.

Bernatchez, L., J. A. Vuorinen, R. A. Bodaly and J. J. Dodson. 1996. Genetic evidence for reproductive isolation and multiple origins of sympatric trophic ecotypes of whitefish (*Coregonus*). *Evolution* 50: 624–635.

Bernstein, I. S. 1976. Dominance, aggression and reproduction in primate societies. *J. Theoret. Biol.* 60: 459–472.

Bernstein, S., L. Throckmorton and J. Hubby. 1973. Still more genetic variability in natural populations. *Proc. Natl. Acad. Sci. USA* 70: 3928–3931.

Berry, R. J. 1996. Small mammal differentiation on islands. *Phil. Trans. R. Soc. Lond.* B 351: 753–764.

Bert, T. M. and W. S. Arnold. 1995. An empirical test of predictions of two competing models for the maintenance and fate of hybrid zones: Both models are supported in a hard-clam hybrid zone. *Evolution* 49: 276–289.

Bertorelle, G. and G. Barbujani. 1995. Analysis of DNA diversity by spatial autocorrelation. *Genetics* 140: 811–819.

Bérubé, M. and P. Pasbøll. 1996. Identification of sex in cetaceans by multiplexing with three ZFX and ZFY specific primers. *Mol. Ecol.* 5: 283–287.

Beverley, S. M. and A. C. Wilson. 1985. Ancient origin for Hawaiian Drosophilinae inferred from protein comparisons. *Proc. Natl. Acad. Sci. USA* 82: 4753–4757.

Bickham, J. W., J. C. Patton and T. R. Loughlin. 1996. High variability for control-region sequences in a marine mammal: Implications for conservation and biogeography of Steller sea lions (*Eumetopias jubatus*). *J. Mammal.* 77: 95–108.

Bidochka, M. J. and 6 others. 1997. Cloned DNA probes distinguish endemic and exotic *Entomophaga grylli* fungal pathotype infections in grasshopper life stages. *Mol. Ecol.* 6: 303–308.

Billington, N. and P. D. N. Hebert. 1991. Mitochondrial DNA diversity in fishes and its implications for introductions. *Can. J. Fish. Aquat. Sci.* 48 (suppl. 1): 80–94.

Bilton, D. T., P. M. Mirol, S. Mascherreti, K. Fredga, J. Zima and J. B. Searle. 1998. Mediterranean Europe as an area of endemism for small mammals rather than a source for northwards postglacial colonization. *Proc. R. Soc. Lond.* B 265: 1219–1226.

Birdsall, D. A. and D. Nash. 1973. Occurrence of successful multiple insemination of females in natural populations of deer mice (*Peromyscus maniculatus*). *Evolution* 27: 106–110.

Birkhead, T. R. 2000. *Promiscuity*. Harvard University Press, Cambridge, MA.

Birkhead, T. R. and A. P. Møller. 1992. *Sperm Competition in Birds*. Academic Press, New York.

Birkhead, T. R. and A. P. Møller. 1993a. Sexual selection and the temporal separation of reproductive events: Sperm storage data from reptiles, birds and mammals. *Biol. J. Linn. Soc.* 50: 295–311.

Birkhead, T. R. and A. P. Møller. 1993b. Female control of paternity. *Trends Ecol. Evol.* 8: 100–104.

Birkhead, T. R. and A. P. Møller (eds.). 1998. *Sperm Competition and Sexual Selection*. Academic Press, London.

Birkhead, T. R., T. Burke, R. Zann, F. M. Hunter and A. P. Krupa. 1990. Extra-pair paternity and intraspecific brood parasitism in wild zebra finches *Taeniopygia guttata*, revealed by DNA fingerprinting. *Behav. Ecol. Sociobiol.* 27: 315–324.

Birky, C. W. Jr. 1978. Transmission genetics of mitochondrial and chloroplasts. *Annu. Rev. Genet.* 12: 471–512.

Birky, C. W. Jr. 1988. Evolution and variation in plant chloroplast and mitochondrial genomes. Pp. 23–53 in *Plant Evolutionary Biology*, L. D. Gottlieb and S. K. Jain (eds.). Chapman & Hall, New York.

Birky, C. W. Jr. 1995. Uniparental inheritance of mitochondrial and chloroplast genes: Mechanisms and evolution. *Proc. Natl. Acad. Sci. USA* 92: 11331–11338.

Birky, C. W. Jr., P. Fuerst and T. Maruyama. 1989. Organelle gene diversity under migration, mutation, and drift: Equilibrium expectations, approach to equilibrium, effects of heteroplasmic cells, and comparison to nuclear genes. *Genetics* 121: 613–627.

Birley, A. J. and J. H. Croft. 1986. Mitochondrial DNAs and phylogenetic relationships. Pp. 107–137 in *DNA Systematics*, S. K. Dutta (ed.). CRC, Boca Raton, FL.

Birstein, V. J., R. Hanner and R. DeSalle. 1997. Phylogeny of Acipenseriformes: Cytogenetic and molecular approaches. *Env. Biol. Fish.* 48: 127–156.

Birstein, V. J., P. Doukakis, B. Sorkin and R. DeSalle. 1998. Population aggregation analysis of three caviar-producing species of sturgeons and implications of the species identification of black caviar. *Cons. Biol.* 12: 766–775.

Birt, T. P. and A. J. Baker. 2000. Polymerase chain reaction. Pp. 50–64 in *Molecular Methods in Ecology*, A. J. Baker (ed.). Blackwell, London.

Birt, T. P., J. M. Green and W. S. Davidson. 1991. Mitochondrial DNA variation reveals genetically distinct sympatric populations of anadromous and nonanadromous Atlantic salmon, *Salmo salar*. *Can. J. Fish. Aquat. Sci.* 48: 577–582.

Bishop, C. E., P. Boursot, B. Baron, F. Bonhomme and D. Hatat. 1985. Most classical *Mus musculus domesticus* laboratory mouse strains carry a *Mus musculus musculus* Y chromosome. *Nature* 315: 70–72.

Bishop, J. D. D. and A. D. Sommerfeldt. 1999. Not like *Botryllus*: Indiscriminate post-metamorphic fusion in a compound ascidian. *Proc. R. Soc. Lond. B* 266: 241–248.

Bishop, J. D. D., A. J. Pemberton and L. R. Noble. 2000. Sperm precedence in a novel context: Mating in a sessile marine invertebrate with dispersing sperm. *Proc. R. Soc. Lond. B* 267: 1107–1113.

Black, R. and M. S. Johnson. 1979. Asexual viviparity and population genetics of *Actinia tenebrosa*. *Mar. Biol.* 53: 27–31.

Blackwelder, R. E. and B. A. Shepherd. 1981. *The Diversity of Animal Reproduction*. CRC Press, Boca Raton, FL.

Blair, W. F. 1950. Ecological factors in speciation of *Peromyscus*. *Evolution* 4: 253–275.

Blair, W. F. 1955. Mating call and stage of speciation in the *Microhyla olivacea-M. carolinensis* complex. *Evolution* 9: 469–480.

Blanchard, J. L. and G. W. Schmidt. 1995. Pervasive migration of organellar DNA to the nucleus in plants. *J. Mol. Evol.* 41: 397–406.

Blanchong, J. A., K. T. Scribner and S. R. Winterstein. 2002. Assignment of individuals to populations: Bayesian methods and multi-locus genotypes. *J. Wildl. Mgt.* 66: 321–329.

Blanco, A. and W. H. Zinkham. 1963. Lactate dehydrogenases in human testes. *Science* 139: 601–602.

Blank, C. E., S. L. Cady and N. R. Pace. 2002. Microbial composition of near-boiling silica-depositing thermal springs throughout Yellowstone National Park. *Appl. Environ. Microbiol.* 68: 5123–5135.

Block, B. A. and J. R. Finnerty. 1994. Endothermy in fishes: A phylogenetic analysis of constraints, predispositions, and selection pressures. *Environ. Biol. Fish.* 40: 283–302.

Block, B. A., J. R. Finnerty, A. F. R. Stewart and J. Kidd. 1993. Evolution of endothermy in fish: Mapping physiological traits on a molecular phylogeny. *Science* 260: 210–214.

Blouin, M. S., J. B. Dame, C. A. Tarrant and C. H. Courtney. 1992. Unusual population genetics of a parasitic nematode: mtDNA variation within and among populations. *Evolution* 46: 470–476.

Blouin, M. S., M. Parsons, V. La Caille and S. Lotz. 1996. Use of microsatellite loci to classify individuals by relatedness. *Mol. Ecol.* 5: 393–401.

Blumer, L. S. 1979. Male parental care in the bony fishes. *Q. Rev. Biol.* 54: 149–161.

Blumer, L. S. 1982. A bibliography and categorization of bony fishes exhibiting parental care. *Zool. J. Linn. Soc.* 76: 1–22.

Boag, P. T. 1987. Effects of nestling diet on growth and adult size of zebra finches (*Poephila guttata*). *Auk* 104: 155–166.

Boehnke, M., N. Arnheim, H. Li and F. S. Collins. 1989. Fine-structure mapping of human chromosomes using the polymerase chain reaction on single sperm: Experimental design considerations. *Am. J. Human Genet.* 45: 21–32.

Boerlin, P., F. Boerlin-Petzold, J. Goudet, C. Durassel, J. L. Pagani, J. P. Chave and J. Bille. 1996. Typing *Candida albicans* oral isolates from human immunodeficiency virus-infected patients by multilocus enzyme electrophoresis and DNA fingerprinting. *J. Clin. Microbiol.* 34: 1235–1248.

Bohonak, A. J. 1999. Dispersal, gene flow and population structure. *Q. Rev. Biol.* 74: 21–45.

Boissinot, S. and P. Boursot. 1997. Discordant phylogeographic patterns between the Y chromosome and mitochondrial DNA in the house mouse: Selection on the Y chromosome? *Genetics* 146: 1019–1034.

Bollinger, E. K. and T. A. Gavin. 1991. Patterns of extrapair fertilizations in bobolinks. *Behav. Ecol. Sociobiol.* 29: 1–7.

Bollmer, J. L., M. E. Irwin, J. P. Reider and P. G. Parker. 1999. Multiple paternity in loggerhead turtle clutches. *Copeia* 1999: 475–478.

Bolten, A. B., K. A. Bjorndal, H. R. Martins, T. Dellinger, M. J. Biscoito, S. E. Encalada and B. W. Bowen. 1998. TransAtlantic developmental migrations of loggerhead sea turtles demonstrated by mtDNA sequence analyses. *Ecol. Appl.* 8: 1–7.

Bond, J. E., M. C. Hedin, M. G. Ramirez and B. D. Opell. 2001. Deep molecular divergence in the absence of morphological and ecological change in the Californian coastal dune endemic trapdoor spider *Aptostichus simus. Mol. Ecol.* 10: 899–910.

Bonhomme, F. and R. K. Selander. 1978. The extent of allelic diversity underlying electrophoretic protein variation in the house mouse. Pp. 569–589 in *Origins of Inbred Mice*, H. C. Morse III (ed.). Academic Press, New York.

Bonnell, M. L. and R. K. Selander. 1974. Elephant seals: Genetic variation and near extinction. *Science* 184: 908–909.

Booke, H. E. 1981. The conundrum of the stock concept: Are nature and nurture definable in fishery science? *Can. J. Fish. Aquat. Sci.* 38: 1479–1480.

Boore, J. L. and W. M. Brown. 2000. Mitochondrial genomes of *Galathaelinum, Helobdella*, and *Platynereis*: Sequence and gene arrangement comparisons indicate that Pogonophora is not a phylum and Annelida and Arthropoda are not sister taxa. *Mol. Biol. Evol.* 17: 87–106.

Boore, J. L., T. M. Collins, D. Stanton, L. L. Daehler and W. M. Brown. 1995. Deducing the pattern of arthropod phylogeny from mitochondrial DNA rearrangements. *Nature* 376: 163–167.

Boore, J. L., D. V. Lavrov and W. M. Brown. 1998. Gene translocation links insects and crustaceans. *Nature* 392: 667–668.

Boore, J. L., L. L. Daehler and W. M. Brown. 1999. Complete sequence, gene arrangement, and genetic code of mitochondrial DNA of the Cephalochordate *Branchiostoma floridae. Mol. Biol. Evol.* 16: 410–418.

Bork, P. and R. F. Doolittle. 1992. Proposed acquisition of an animal protein domain by bacteria. *Proc. Natl. Acad. Sci. USA* 89: 8990–8994.

Bos, M., H. Harmens and K. Vrieling. 1986. Gene flow in *Plantago* I. Gene flow and neighborhood size in *P. lanceolata. Heredity* 56: 43–54.

Bossart, J. L. and D. P. Prowell. 1998. Genetic estimates of population structure and gene flow: Limitations, lessons and new directions. *Trends Ecol. Evol.* 13: 202–206.

Bossuyt, F. and M. C. Milinkovitch. 2001. Amphibians as indicators of early tertiary "out-of-India" dispersal of vertebrates. *Science* 292: 93–95.

Botstein, D., R. L. White, M. Skolnick and R. W. Davis. 1980. Construction of a genetic linkage map in man using restriction fragment length polymorphisms. *Am. J. Human Genet.* 32: 314–331.

Bourke, A. F. G. and N. R. Franks. 1995. *Social Evolution in Ants*. Princeton University Press, Princeton, NJ.

Bouzat, J. L., H. H. Cheng, H. A. Lewin, R. L. Westemeier, J. D. Brawn and K. N. Paige. 1998. Genetic evaluation of a demographic bottleneck in the greater prairie chicken. *Cons. Biol.* 12: 836–843.

Bowcock, A. M., A. Ruiz-Linares, J. Tomfohrde, E. Minch, J. R. Kidd and L. L. Cavalli-Sforza. 1994. High resolution of human evolutionary trees with polymorphic microsatellites. *Nature* 368: 455–457.

Bowe, L. M., G. Coat and C. W. dePamphilis. 2000. Phylogeny of seed plants based on all three genomic compartments: Extant gymnosperms are monophyletic and Gnetales' closest relatives are conifers. *Proc. Natl. Acad. Sci. USA* 97: 4092–4097.

Bowen, B. W. 1995. Voyages of the ancient mariners: Tracking marine turtles with genetic markers. *BioScience* 45: 528–534.

Bowen, B. W. 1998. What is wrong with ESUs? The gap between evolutionary theory and conservation principles. *J. Shellfish Res.* 17: 1355–1358.

Bowen, B. W. 1999. Preserving genes, ecosystems, or species? Healing the fractured foundations of conservation policy. *Mol. Ecol.* 8: S5-S10.

Bowen, B. W. and J. C. Avise. 1990. Genetic structure of Atlantic and Gulf of Mexico populations of sea bass, menhaden, and sturgeon: Influence of zoogeographic factors and life history patterns. *Mar. Biol.* 107: 371–381.

Bowen, B. W. and J. C. Avise. 1996. Conservation genetics of marine turtles. Pp. 190–237 in *Conservation Genetics: Case Histories from Nature*, J. C. Avise and J. L. Hamrick (eds.). Chapman & Hall, New York.

Bowen, B. W. and W. S. Grant. 1997. Phylogeography of the sardines (*Sardinops* spp.): Assessing biogeographic models and population histories in temperate upwelling zones. *Evolution* 51: 1601–1610.

Bowen, B. W. and S. A. Karl. 1997. Population genetics, phylogeography, and molecular evolution. Pp. 29–50 in *The Biology of Sea Turtles*, P. L. Lutz and J. A. Musick (eds.). CRC Press, Boca Raton, Florida.

Bowen, B. W., A. B. Meylan and J. C. Avise. 1989. An odyssey of the green sea turtle: Ascension Island revisited. *Proc. Natl. Acad. Sci. USA* 86: 573–576.

Bowen, B. W., A. B. Meylan and J. C. Avise. 1991. Evolutionary distinctiveness of the endangered Kemp's ridley sea turtle. *Nature* 352: 709–711.

Bowen, B. W., A. B. Meylan, J. P. Ross, C. J. Limpus, G. H. Balazs and J. C. Avise. 1992. Global population structure and natural history of the green turtle (*Chelonia mydas*) in terms of matriarchal phylogeny. *Evolution* 46: 865–881.

Bowen, B. W., W. S. Nelson and J. C. Avise. 1993a. A molecular phylogeny for marine turtles: Trait mapping, rate assessment, and conservation relevance. *Proc. Natl. Acad. Sci. USA* 90: 5574–5577.

Bowen, B. W., J. I. Richardson, A. B. Meylan, D. Margaritoulis, R. Hopkins Murphy and J. C. Avise. 1993b. Population structure of loggerhead turtles

(*Caretta caretta*) in the West Atlantic Ocean and Mediterranean Sea. *Cons. Biol.* 7: 834–844.

Bowen, B. W., N. Kamezaki, C. J. Limpus, G. R. Hughes, A. B. Meylan and J. C. Avise. 1994. Global phylogeography of the loggerhead turtle (*Caretta caretta*) as indicated by mitochondrial DNA genotypes. *Evolution* 48: 1820–1828.

Bowen, B. W., F. A. Abreu-Grobois, G. H. Balazs, N. Kamezaki, C. J. Limpus and R. J. Ferl. 1995. Trans-Pacific migrations of the loggerhead turtle (*Caretta caretta*) demonstrated with mitochondrial DNA markers. *Proc. Natl. Acad. Sci. USA* 92: 3731–3734.

Bowen, B. W., A. M. Clark, F. A. Abreu-Grobois, A. Chaves, H. A. Reichart and R. J. Ferl. 1998. Global phylogeography of the ridley sea turtles (*Lepidochelys* spp.) as inferred from mitochondrial DNA sequences. *Genetica* 101: 179–189.

Bowen, B. W., A. L. Bass, L. A. Rocha, W. S. Grant and D. R. Robertson. 2001. Phylogeography of the trumpetfishes (*Aulostomus*): Ring species complex on a global scale. *Evolution* 55: 1029–1039.

Bowers, J. E., B. A. Chapman, J. Rong and A. H. Paterson. 2003. Unravelling angiosperm genome evolution by phylogenetic analysis of chromosomal duplication events. *Nature* 422: 433–438.

Bowers, J. H., R. J. Baker and M. H. Smith. 1973. Chromosomal, electrophoretic, and breeding studies of selected populations of deer mice (*Peromyscus maniculatus*) and black-eared mice (*P. melanotis*). *Evolution* 27: 378–386.

Bowling, A. T. and O. A. Ryder. 1987. Genetic studies of blood markers in Przewalski's horses. *J. Heredity* 78: 75–80.

Bowman, B. H., J. W. Taylor, A. G. Brownlee, J. Lee, S.-D. Lu and T. J. White. 1992. Molecular evolution of the fungi: Relationships of the Basidiomycetes, Ascomycetes, and Chytridiomycetes. *Mol. Biol. Evol.* 9: 285–296.

Boyce, A. J. 1983. Computation of inbreeding and kinship coefficients on extended pedigrees. *J. Heredity* 74: 400–404.

Boyce, A. J. and C. G. N. Mascie-Taylor (eds.). 1996. *Molecular Biology and Human Diversity*. Cambridge University Press, Cambridge, England.

Boyd, W. C. 1954. Tables and nomograms for calculating chances of excluding paternity. *Am. J. Human Genet.* 6: 426–433.

Boyer, J. F. 1974. Clinal and size-dependent variation at the LAP locus in *Mytilus edulis*. *Biol. Bull.* 147: 535–549.

Bradshaw, H. D. Jr., S. M. Wilbert, K. G. Otto and D. W. Schemske. 1995. Genetic mapping of floral traits associated with reproductive isolation in monkeyflowers (*Mimulus*). *Nature* 376: 762–765.

Bradshaw, H. D. Jr., K. G. Otto, B. E. Frewen, J. K. McKay and D. W. Schemske. 1998. Quantitative trait loci affecting differences in floral morphology between two species of monkeyflower (*Mimulus*). *Genetics* 149: 367–382.

Branco, M., M. Monnerot, N. Ferrand and A. R. Templeton. 2002. Postglacial dispersal of the European rabbit (*Oryctolagus cuniculus*) on the Iberian Peninsula reconstruted from nested clade analysis and mismatch analyses of mitochndrial DNA genetic variation. *Evolution* 56: 792–803.

Braude, S. 2000. Dispersal and new colony formation in wild naked mole-rats: Evidence against inbreeding as the system of mating. *Behav. Ecol.* 11: 7–12.

Breder, C. M. Jr. 1936. The reproductive habits of North American sunfishes (the family Centrarchidae). *Zoologica* 21: 1–47.

Bremer, B. and K. Bremer. 1989. Cladistic analysis of bluegreen procaryote interrelationships and chloroplast origin based on 16S rRNA oligonucleotide catalogues. *J. Evol. Biol.* 2: 13–30.

Brewer, G. J. 1970. *An Introduction to Isozyme Techniques*. Academic Press, New York.

Bridge, D., C. W. Cunningham, B. Schierwater, R. DeSalle and L. W. Buss. 1992. Class-level relationships in the phylum Cnidaria: Evidence from mitochondrial genome structure. *Proc. Natl. Acad. Sci. USA* 89: 8750–8753.

Briggs, J. C. 1958. A list of Florida fishes and their distribution. *Bull. Fla. State Mus. Biol. Sci.* 2: 223–318.

Briggs, J. C. 1974. *Marine Zoogeography*. McGraw-Hill, New York.

Brisse, S., C. Barnabé and M. Tibayrenc. 2000. Identification of six *Trypanosoma cruzi* phylogenetic lineages by random amplified polymorphic DNA and multilocus enzyme electrophoresis. *Int. J. Parasitol.* 30: 35–44.

Britten, H. B. 1996. Meta-analysis of the association between multilocus heterozygosity and fitness. *Evolution* 50: 2158–2164.

Britten, R. J. 1986. Rates of DNA sequence evolution differ between taxonomic groups. *Science* 231: 1393–1398.

Britten, R. J. and E. H. Davidson. 1969. Gene regulation for higher cells: A theory. *Science* 165: 349–357.

Britten, R. J. and E. H. Davidson. 1971. Repetitive and non-repetitive DNA sequences and a speculation on the origins of evolutionary novelty. *Q. Rev. Biol.* 46: 111–138.

Britten, R. J. and D. E. Kohne. 1968. Repeated sequences in DNA. *Science* 161: 529–540.

Britten, R. J., D. E. Graham and B. R. Neufeld. 1974. Analysis of repeating DNA sequences by reassociation. *Methods Enzymol.* 29: 363–418.

Brock, M. K. and B. N. White. 1991. Multifragment alleles in DNA fingerprints of the parrot, *Amazona ventralis*. *J. Heredity* 82: 209–212.

Brockman, H. J., C. Nguyen and W. Potts. 2000. Paternity in horseshoe crabs when spawning in multiple-male groups. *Anim. Behav.* 60: 837–849.

Broderick, D., C. Moritz, J. D. Miller, M. Guinea, R. J. Prince and C. J. Limpus. 1994. Genetic studies of the hawksbill turtle: Evidence for multiple stocks and mixed feeding grounds in Australian waters. *Pacific Cons. Biol.* 1: 123–131.

Brodkorb, P. 1971. Origin and evolution of birds. Pp. 20–55 in *Avian Biology*, Volume I, D. S. Farmer and J. R. King (eds.). Academic Press, New York.

Bromham, L. and M. Cardillo. 2003. Testing the link between the latitudinal gradient in species richness and rates of molecular evolution. *J. Evol. Biol.* 16: 200–207.

Bromham, L., A. Rambaut, R. Fortey, A. Cooper and D. Penny. 1998. Testing the Cambrian explosion hypothesis by using a molecular dating technique. *Proc. Natl. Acad. Sci. USA* 95: 12386–12389.

Bromham, L., M. Woolfit, M. S. Y. Lee and A. Rambaut. 2002. Testing the relationship between morphological and molecular rates of change along phylogenies. *Evolution* 56: 1921–1930.

Brookes, A. J. 1999. The essence of SNPs. *Gene* 234: 177–186.

Brookfield, J. F. Y. 1992. DNA fingerprinting in clonal organisms. *Mol. Ecol.* 1: 21–26.

Brooks, D. R. and D. A. McLennan. 1991. *Phylogeny, Ecology, and Behavior.* University of Chicago Press, Chicago, IL.

Brooks, D. R. and D. A. McLennan. 2002. *The Nature of Diversity: An Evolutionary Voyage of Discovery.* University of Chicago Press, Chicago, IL.

Brosius, J. 1999. Genomes were forged by massive bombardments with retroelements and retrosequences. *Genetica* 107: 209–238.

Brower, A. V. Z. 1994. Rapid morphological radiation and convergence among races of the butterfly *Heliconius erato* inferred from patterns of mitochondrial DNA evolution. *Proc. Natl. Acad. Sci. USA* 91: 6491–6495.

Brower, A. V. Z. and T. M. Boyce. 1991. Mitochondrial DNA variation in monarch butterflies. *Evolution* 45: 1281–1286.

Brown, A. H. D. 1989. Genetic characterization of plant mating systems. Pp. 145–162 in *Plant Population Genetics, Breeding and Genetic Resources*, A. H. D. Brown, M. T. Clegg, A. L. Kahler and B. S. Weir (eds.). Sinauer Associates, Sunderland, MA.

Brown, A. H. D. and R. W. Allard. 1970. Estimation of the mating system in open-pollinated maize populations using isozyme polymorphisms. *Genetics* 66: 133–145.

Brown, D. D., P. C. Wensink and E. Jordan. 1972. A comparison of the ribosomal DNAs of *Xenopus laevis* and *Xenopus mulleri*: Evolution of tandem genes. *J. Mol. Biol.* 63: 57–73.

Brown, G. G. and M. V. Simpson. 1982. Novel features of animal mtDNA evolution as shown by sequences of two rat cytochrome oxidase subunit II genes. *Proc. Natl. Acad. Sci. USA* 79: 3246–3250.

Brown, J. L. 1987. *Helping and Communal Breeding in Birds.* Princeton University Press, Princeton, NJ.

Brown, J. M, W. G. Abrahamson and P. A. Way. 1996. Mitochondrial DNA phylogeography of host races of the goldenrod ball gallmaker, *Eurosta solidaginis* (Diptera: Tephritidae). *Evolution* 50: 777–786.

Brown, J. M., J. H. Leebens-Mack, J. N. Thompson, O. Pellmyr and R. G. Harrison. 1997. Phylogeography and host association in a pollinating seed parasite *Greya politella* (Lepidoptera: Prodoxidae). *Mol. Ecol.* 6: 215–224.

Brown, J. R. 2003. Ancient horizontal gene transfer. *Nature Rev. Genet.* 4: 121–132.

Brown, S. M. and B. A. Houlden. 2000. Conservation genetics of the black rhinoceros (*Diceros bicornis*). *Cons. Genet.* 1: 365–370.

Brown, T. 1992. Ancient DNA reference list. *Ancient DNA News.* 1(1): 36–38.

Brown, W. M. 1980. Polymorphism in mitochondrial DNA of humans as revealed by restriction endonuclease analysis. *Proc. Natl. Acad. Sci. USA* 77: 3605–3609.

Brown, W. M. 1983. Evolution of animal mitochondrial DNA. Pp. 62–88 in *Evolution of Genes and Proteins*, M. Nei and R. K. Koehn (eds.). Sinauer Associates, Sunderland, MA.

Brown, W. M. 1985. The mitochondrial genome of animals. Pp. 95–130 in *Evolutionary Genetics*, R. J. MacIntyre (ed.). Plenum Press, New York.

Brown, W. M. and J. Vinograd. 1974. Restriction endonuclease cleavage maps of animal mitochondrial DNAs. *Proc. Natl. Acad. Sci. USA* 71: 4617–4621.

Brown, W. M. and J. Wright. 1975. Mitochondrial DNA and the origin of parthenogenesis in whiptail lizards (genus *Cnemidophorus*). *Science* 203: 1247–1249.

Brown, W. M. and J. Wright. 1979. Mitochondrial DNA analyses and the origin and relative age of parthenogenetic lizards (genus *Cnemidophorus*). *Science* 203: 1247–1249.

Brown, W. M., M. George Jr. and A. C. Wilson. 1979. Rapid evolution of animal mitochondrial DNA. *Proc. Natl. Acad. Sci. USA* 76: 1967–1971.

Brown, W. M., E. M. Prager, A. Wang and A. C. Wilson. 1982. Mitochondrial DNA sequences of primates: Tempo and mode of evolution. *J. Mol. Evol.* 18: 225–239.

Brown-Gladden, J. G., M. M. Ferguson and J. W. Clayton. 1997. Matriarchal genetic population structure of North American beluga whales *Delhinapterus leucas* (Cetacea: Monodontidae). *Mol. Ecol.* 6: 1033–1046.

Browning, T. L., D. A. Taggart, C. Rummery, R. L. Close and M. D. B. Eldridge. 2001. Multifaceted genetic analysis of the "critically endangered" brush-tailed rock-wallaby *Petrogale penicillata* in Victoria, Australia: Implications for management. *Cons. Genet.* 2: 145–156.

Bruce, E. J. and F. J. Ayala. 1979. Phylogenetic relationships between man and the apes: Electrophoretic evidence. *Evolution* 33: 1040–1056.

Bruford, M. W., O. Hanotte, J. F. Y. Brookfield and T. Burke. 1992. Single-locus and multilocus DNA fingerprinting. Pp. 225–269 in *Molecular Genetic Analysis of Populations*. A Practical Approach, A. R. Hoelzel (ed.). IRL Press, Oxford.

Brumfield, R. T., R. W. Jernigan, D. B. McDonald and M. J. Braun. 2001. Evolutionary implications of divergent clines in an avian (*Manacus*: Aves) hybrid zone. *Evolution* 55: 2070–2087.

Brunk, C. F. and E. C. Olson (eds.). 1990. DNA-DNA hybridization and evolution (special issue). *J. Mol. Evol.* 30: 191–311.

Brunner, I., S. Brodbeck, U. Büchler and C. Sperisen. 2001. Molecular identication of fine roots of trees from the Alps: Reliable and fast DNA extraction and PCR-RFLP analyses of plastid DNA. *Mol. Ecol.* 10: 2079–2987.

Brunner, P. C., M. R. Douglas, A. Osinov, C. C. Wilson and L. Bernatchez. 2001. Holarctic phylogeography of arctic charr (*Salvelinus alpinus* L.) inferred from mitochondrial DNA sequences. *Evolution* 55: 573–586.

Bruns, T. D. and J. D. Palmer. 1989. Evolution of mushroom mitochondrial DNA: *Suillus* and related genera. *J. Mol. Evol.* 28: 349–362.

Bruns, T. D., R. Fogel, T. J. White and J. D. Palmer. 1989. Accelerated evolution of a false-truffle from a mushroom ancestor. *Nature* 339: 140–142.

Bruns, T. D., T. J. White and J. W. Taylor. 1991. Fungal molecular systematics. *Annu. Rev. Ecol. Syst.* 22: 525–564.

Brunsfeld, S. J., D. E. Soltis and P. S. Soltis. 1992. Evolutionary patterns and processes in *Salix* sect. *Longifoliae*: Evidence from chloroplast DNA. *Syst. Bot.* 17: 239–256.

Buchan, J. C., S. C. Alberts, J. B. Silk and J. Altmann. 2003. True paternal care in a multi-male primate society. *Nature* 425: 179–181.

Bucklin, A. and T. D. Kocher. 1996. Source regions for recruitment of *Calanus finmarchicus* to Georges Bank: Evidence from molecular population genetic analysis of mtDNA. *Deep-Sea Res.* 43: 1665–1681.

Bucklin, A., D. Hedgecock and C. Hand. 1984. Genetic evidence of self-fertilization in the sea anemone *Epiactis prolifera*. *Mar. Biol.* 84: 175–182.

Bucklin, A., C. Caudill and M. Guarnieri. 1998. Population genetics and phylogeny of marine planktonic copepods. Pp. 303–317 in *Molecular Approaches to the Study of the Ocean*, K. Cooksey (ed.). Chapman & Hall, London.

Bucklin, A., M. Guarnieri, R. S. Hill, A. M. Bentley and S. Kaartvedt. 1999. Taxonomic and systematic assessment of planktonic copepods using mitochondrial COI sequence variation and competitive, species-specific PCR. *Hydrobiology* 401: 239–254.

Budar, F., P. Touzet and R. De Paepe. 2003. The nucleo-mitochondrial conflict in cytoplasmic male sterilities revisited. *Genetica* 117: 3–16.

Budowle, B., A. M. Giusti, J. S. Waye, F. S. Baechtel, R. M. Fourney, D. E. Adams, L. A. Presley, H. A. Deadman and K. L. Monson. 1991. Fixed-bin analysis for statistical evaluation of continuous distributions of allelic data from VNTR loci, for use in forensic comparisons. *Am. J. Human Genet.* 48: 841–855.

Bull, J. J. 1980. Sex determination in reptiles. *Q. Rev. Biol.* 55: 3–21.

Bunce, M. and 6 others. 2003. Extreme reversed sexual size dimorphism in the extinct New Zealand moa *Dinornis*. *Nature* 425: 172–175.

Burda, H., R. L. Honeycutt, S. Begall, O. Locker-Grütjen and A. Scharff. 2000. Are naked and common mole-rats eusocial and if so, why? *Behav. Ecol. Sociobiol.* 47: 293–303.

Burgoyne, P. S. 1986. Mammalian X and Y crossover. *Nature* 319: 258–259.

Burke, J. M. and M. L. Arnold. 2001. Genetics and the fitness of hybrids. *Annu. Rev. Genet.* 35: 31–52.

Burke, J. M., S. E. Carney and M. L. Arnold. 1998. Hybrid fitness in the Louisiana irises: Analysis of parental and F_1 performance. *Evolution* 52: 37–43.

Burke, J. M., M. R. Bulger, R. A. Wesselingh and M. L. Arnold. 2000. Frequency and spatial patterning of clonal reproduction in Louisiana Iris hybrid populations. *Evolution* 54: 137–144.

Burke, T. 1989. DNA fingerprinting and other methods for the study of mating success. *Trends Ecol. Evol.* 4: 139–144.

Burke, T. (ed.). 1994. Conservation genetics (special issue). *Mol. Ecol.* 3: 277–435.

Burke, T. and M. W. Bruford. 1987. DNA fingerprinting in birds. *Nature* 327: 149–152.

Burke, T., N. B. Davies, M. W. Bruford and B. J. Hatchwell. 1989. Parental care and mating behavior of polyandrous dunnocks *Prunella modularis* related to paternity by DNA fingerprinting. *Nature* 338: 249–251.

Burke, T., O. Hanotte, M. W. Bruford and E. Cairns. 1991. Multilocus and single locus minisatellite analysis in population biological studies. Pp. 154–168 in *DNA Fingerprinting Approaches and Applications*, T. Burke, G. Dolf, A. J. Jeffreys and R. Wolff (eds.). Birkhauser Verlag, Basel, Switzerland.

Burland, T. M., N. C. Bennett, J. U. M. Jarvis and C. G. Faulkes. 2002. Eusociality in African mole-rats: New insights from patterns of genetic relatedness in the Damaraland mole-rat (*Cryptomys damarensis*). *Proc. R. Soc. Lond. B* 269: 1025–1030.

Burley, N. T. and P. G. Parker. 1998. Emerging themes and questions in the study of avian reproductive tactics. *Ornithol. Monogr.* 49: 1–20.

Burley, N. T., P. G. Parker and K. Lundy. 1996. Sexual selection and extrapair fertilization in a socially monogamous passerine, the zebra finch (*Taeniopygia guttata*). *Behav. Ecol.* 7: 218–226.

Burnell, K. L. and S. B. Hedges. 1990. Relationships of West Indian *Anolis* (Sauria: Iguanidae): An approach using slow-evolving protein loci. *Caribbean J. Sci.* 26: 7–30.

Buroker, N. E. 1983. Population genetics of the American oyster *Crassostrea virginica* along the Atlantic coast and the Gulf of Mexico. *Mar. Biol.* 75: 99–112.

Burton, R. S. 1983. Protein polymorphisms and genetic differentiation of marine invertebrate populations. *Mar. Biol. Letters* 4: 193–206.

Burton, R. S. 1985. Mating system of the intertidal copepod *Tigriopus californicus*. *Mar. Biol.* 86: 247–252.

Burton, R. S. 1986. Evolutionary consequences of restricted gene flow among natural populations of the copepod, *Tigriopus californicus*. *Bull. Mar. Sci.* 39: 526–535.

Burton, R. S. 1998. Intraspecific phylogeography across the Point Conception biogeographic boundary. *Evolution* 52: 734–745.

Burton, R. S. and M. W. Feldman. 1981. Population genetics of *Tigriopus californicus*. II. Differentiation among neighboring populations. *Evolution* 35: 1192–1205.

Burton, R. S. and M. W. Feldman. 1983. Physiological effects of an allozyme polymorphism: Glutamate-pyruvate transaminase and response to hyperosmotic stress in the copepod *Tigriopus californicus*. *Biochem. Genet.* 21: 238–251.

Burton, R. S. and B.-N. Lee. 1994. Nuclear and mitochondrial gene genealogies and allozyme polymorphism across a major phylogeographic break in the copepod *Tigriopus californicus*. *Proc. Natl. Acad. Sci. USA* 91: 5197–5201.

Burton, R. S. and S. G. Swisher. 1984. Population structure in the intertidal copepod *Tigriopus californicus* as revealed by field manipulation of allele frequencies. *Oecologia* 65: 108–111.

Burton, R. S., M. W. Feldman and J. W. Curtsinger. 1979. Population genetics of *Tigriopus californicus* (Copepoda: Harpacticoida). I. Population structure along the central California coast. *Mar. Ecol. Progr. Ser.* 1: 29–39.

Busack, C. A. and G. A. E. Gall. 1981. Introgressive hybridization in populations of Paiute cutthroat trout (*Salmo clarki seleniris*). *Can. J. Fish. Aquat. Sci.* 38: 939–951.

Bush, G. L. 1969. Sympatric host race formation and speciation in frugivorous flies of the genus *Rhagoletis* (Diptera, Tephritidae). *Evolution* 23: 237–251.

Bush, G. L. 1975. Modes of animal speciation. *Annu. Rev. Ecol. Syst.* 6: 339–364.

Bushman, F. 2002. *Lateral DNA Transfer: Mechanisms and Consequences.* Cold Spring Harbor Lab Press, Cold Spring Harbor, NY.

Busquet, J., S. H. Strauss, A. H. Doerksen and R. A. Price. 1992. Extensive variation in evolutionary rate of *rbcL* gene sequences among seed plants. *Proc. Natl. Acad. Sci. USA* 89: 7844–7848.

Buss, L. W. 1982. Somatic cell parasitism and the evolution of somatic tissue compatibility. *Proc. Natl. Acad. Sci. USA* 79: 5337–5341.

Buss, L. W. 1983. Evolution, development, and the units of selection. *Proc. Natl. Acad. Sci. USA* 80: 1387–1391.

Buss, L. W. 1985. Uniqueness of the individual revisited. Pp. 467–505 in *Population Biology and Evolution of Clonal Organisms*, J. B. C. Jackson, L. W. Buss and R. E. Cook (eds.). Yale University Press, New Haven, CT.

Bustamante, C. D., R. Nielsen, S. A. Sawyers, K. M. Olsen, M. D. Purugganan and D. L. Hartl. 2002. The cost of inbreeding in Arabidopsis. *Nature* 416: 531–534.

Buth, D. G. 1979. Duplicate gene expression in tetraploid fishes of the tribe Moxostomatini (Cypriniformes, Catostomidae). *Comp. Biochem. Physiol.* 63B: 7–12.

Buth, D. G. 1982. Glucosephosphate-isomerase expression in a tetraploid fish, *Moxostoma lachneri* (Cypriniformes, Catostomidae): Evidence for "retetraploidization"? *Genetica* 57: 171–175.

Buth, D. G. 1983. Duplicate isozyme loci in fishes: Origins, distribution, phyletic consequences, and locus nomenclature. *Isozymes* X: 381–400.

Buth, D. G. 1984. The application of electrophoretic data in systematic studies. *Annu. Rev. Ecol. Syst.* 15: 501–522.

Buth, D. G., R. W. Murphy, M. M. Miyamoto and C. S. Lieb. 1985. Creatine kinases of amphibians and reptiles: Evolutionary and systematic aspects of gene expression. *Copeia* 1985: 279–284.

Butler, J. M. 2001. *Forensic DNA Typing.* Academic Press, New York.

Butlin, R. 1989. Reinforcement of premating isolation. Pp. 158–179 in *Speciation and Its Consequences*, D. Otte and J. A. Endler (eds.). Sinauer Associates, Sunderland, MA.

Butlin, R. 2000. Virgin rotifers. *Trends Ecol. Evol.* 15: 389–390.

Butlin, R. 2002. The costs and benefits of sex: New insights from old asexual lineages. *Nature Rev. Genet.* 3: 311–317.

Byers, D. L. and D. M. Waller. 1999. Do plant populations purge their genetic load? Effects of population size and mating history on inbreeding depression. *Annu. Rev. Ecol. Syst.* 30: 479–513.

Byrne, M., G. Tischler, B. Macdonald D. J. Coates and J. McComb. 2001. Phylogenetic relationships between two rare acacias and their common, widespread relatives in south-western Australia. *Cons. Genet.* 2: 157–166.

Caccone, A. and J. R. Powell. 1987. Molecular evolutionary divergence among North American cave crickets. II. DNA-DNA hybridization. *Evolution* 41: 1215–1238.

Caccone, A. and J. R. Powell. 1989. DNA divergence among hominoids. *Evolution* 43: 925–942.

Caccone, A. and V. Sbordoni. 1987. Molecular evolutionary divergence among North American cave crickets. I. Allozyme variation. *Evolution* 41: 1198–1214.

Caccone, A., G. D. Amato and J. R. Powell. 1987. Intraspecific DNA divergence in *Drosophila*: A study on parthenogenetic *D. mercatorum*. *Mol. Biol. Evol.* 4: 343–350.

Caccone, A., G. D. Amato and J. R. Powell. 1988a. Rates and patterns of scnDNA and mtDNA divergence within the *Drosophila melanogaster* subgroup. *Genetics* 118: 671–683.

Caccone, A., R. DeSalle and J. R. Powell. 1988b. Calibration of the change in thermal stability of DNA duplexes and degree of base-pair mismatch. *J. Mol. Evol.* 27: 212–216.

Caccone, A., J. P. Gibbs, V. Ketmaier, E. Suatoni and J. R. Powell. 1999. Origin and evolutionary relationships of giant Galápagos tortoises. *Proc. Natl. Acad. Sci. USA* 96: 13223–13228.

Caccone, A. and 6 others. 2002. Phylogeography and history of giant Galápagos tortoises. *Evolution* 56: 2052–2066.

Caetano-Anollés, G. and P. M. Gresshoff (eds.). 1997. *DNA Markers: Protocols, Applications, and Overviews.* Wiley, New York.

Caine, E. A. 1986. Carapace epibionts of nesting loggerhead sea turtles: Atlantic coast of U.S.A. *J. Exp. Marine Biol. Ecol.* 95: 15–26.

Calahan, C. M. and C. Gliddon. 1985. Genetic neighborhood sizes in *Primula vulgaris*. *Heredity* 54: 65–70.

Calero, C., O. Ibáñez, M. Mayol and J. A. Rosselló. 1999. Random amplified polymorphic DNA (RAPD) markers detect a single pheontype in *Lysimachia minoricensis* J. J. Rodr. (Primulaceae), a wild extinct plant. *Mol. Ecol.* 8: 2133–2136.

Calie, P. J. and K. W. Hughes. 1987. The consensus land plant chloroplast gene order is present, with two alterations, in the moss *Physcomitrella patens*. *Mol. Gen. Genet.* 208: 335–341.

Calsbeek, R., J. N. Thompson and J. E. Richardson. 2003. Patterns of molecular evolution and diversification in a biodiversity hotspot: The California Floristic Province. *Mol. Ecol.* 12: 1021–1029.

Calvi, B. R., T. J. Hong, S. D. Findley and W. M. Gelbart. 1991. Evidence for a common evolutionary origin of inverted repeat transposons in *Drosophila* and plants: hobo, Activator, and Tam3. *Cell* 66: 465–471.

Cameron, C. B., J. R. Garey and B. J. Swalla. 2000. Evolution of the chordate body plan: New insights from phylogenetic analyses of deuterostome phyla. *Proc. Natl. Acad. Sci. USA* 97: 4469–4474.

Cameron, K. M., K. J. Wurdack and R. W. Jobson. 2002. Molecular evidence for the common origin of snap-traps among carnivorous plants. *Am. J. Bot.* 89: 1503–1509.

Camin, J. H. and R. R. Sokal. 1965. A method for deducing branching sequences in phylogeny. *Evolution* 19: 311–326.

Cammarano, P., P. Palm, R. Creti, E. Ceccarelli, A. M. Sanangelantoni and O. Tiboni. 1992. Early evolutionary relationships among known life forms inferred from elongation factor EF-2/EF-G sequences: Phylogenetic coherence and structure of the Archael domain. *J. Mol. Evol.* 34: 396–405.

Campbell, D., P. Duchesne and L. Bernatchez. 2003. AFLP utility for population assignment studies: Analytical investigation and empirical comparison with microsatellites. *Mol. Ecol.* 12: 1979–1991.

Campton, D. E. 1987. Natural hybridization and introgression in fishes. Pp. 161–192 in *Population Genetics and Fisheries Management*, N. Ryman and F. Utter (eds.). University of Washington Press, Seattle.

Campton, D. E. and F. M. Utter. 1987. Genetic structure of anadromous cutthroat trout (*Salmo clarki clarki*) populations in the Puget Sound area: Evidence for restricted gene flow. *Can. J. Fish. Aquat. Sci.* 44: 573–582.

Campton, D. E., A. L. Bass, F. A. Chapman and B. W. Bowen. 2000. Genetic distinction of pallid, shovelnose, and Alabama sturgeon: Emerging species and the US Endangered Species Act. *Cons. Genet.* 1: 17–32.

Cann, R. L., W. M. Brown and A. C. Wilson. 1984. Polymorphic sites and the mechanism of evolution in human mitochondrial DNA. *Genetics* 106: 479–499.

Cann, R. L., M. Stoneking and A. C. Wilson. 1987. Mitochondrial DNA and human evolution. *Nature* 325: 31–36.

Cannings, C. and E. A. Thompson. 1981. *Genealogical and Genetic Structure*. Cambridge University Press, Cambridge, England.

Cano, R. J., H. N. Poinar, D. W. Roubik and G. O. Poinar Jr. 1992. Enzymatic amplification and nucleotide sequencing of portions of the 18s rRNA gene of the bee *Proplebeia dominicana* (Apidae: Hymenoptera) isolated from 25–40 million year old Dominican amber. *Med. Sci. Res.* 20: 619–622.

Cano, R. J., H. N. Poinar, N. J. Pieniazek, A. Acra and G. O. Poinar Jr. 1993. Amplification and sequencing of DNA from a 120–135-million-year-old weevil. *Nature* 363: 536–538.

Cantatore, P. and C. Saccone. 1987. Organization, structure, and evolution of mammalian mitochondrial genes. *Int. Rev. Cytol.* 108: 149–208.

Cantatore, P., M. N. Gadaleta, M. Roberti, C. Saccone and A. C. Wilson. 1987. Duplication and remoulding of tRNA genes during the evolutionary rearrangement of mitochondrial genomes. *Nature* 329: 853–855.

Caramelli, D. and 10 others. 2003. Evidence for a genetic discontinuity between Neandertals and 24,000-year-old anatomically modern Europeans. *Proc. Natl. Acad. Sci. USA* 100: 6593–6597.

Carlson, T. A. and B. K. Chelm. 1986. Apparent eukaryotic origin of glutamine synthetase II from the bacterium *Bradyrhizobium japonicum*. *Nature* 322: 568–570.

Carney, S. E., S. A. Hodges and M. L. Arnold. 1996. Effects of differential pollen-tube growth on hybridization in the Louisiana irises. *Evolution* 50: 1871–1878.

Carpenter, J. M., J. E. Strassmann, S. Turillazzi, C. R. Hughes, C. R. Solis and R. Cervo. 1993. Phylogenetic relationships among paper wasp social parasites and their hosts (Hymenoptera, Vespidae, Polistinae). *Cladistics* 9: 129–146.

Carr, A. 1967. *So Excellent a Fishe: A Natural History of Sea Turtles*. Scribner, New York.

Carr, A. and P. J. Coleman. 1974. Seafloor spreading theory and the odyssey of the green turtle from Brazil to Ascension Island, Central Atlantic. *Nature* 249: 128–130.

Carr, D. E. and M. R. Dudash. 2003. Recent approaches into the genetic basis of inbreeding depression in plants. *Phil. Trans. R. Soc. Lond. B* 358: 1071–1084.

Carr, S. M. and O. M. Griffith. 1987. Rapid isolation of animal mitochondrial DNA in a small fixed-angle rotor at ultrahigh speed. *Biochem. Genet.* 25: 385–390.

Carr, S. M., S. W. Ballinger, J. N. Derr, L. H. Blankenship and J. W. Bickham. 1986. Mitochondrial DNA analysis of hybridization between sympatric white-tailed deer and mule deer in west Texas. *Proc. Natl. Acad. Sci. USA* 83: 9576–9580.

Carroll, S. B., J. K. Grenier and S. D. Weatherbee. 2001. *From DNA to Diversity: Molecular Genetics and the Evolution of Animal Design*. Blackwell, Malden, MA.

Carson, H. L. 1968. The population flush and its genetic consequences. Pp. 123–137 in *Population Biology and Evolution*, R. C. Lewontin (ed.). Syracuse University Press, Syracuse, NY.

Carson, H. L. 1976. Inference of the time of origin of some *Drosophila* species. *Nature* 259: 395–396.

Carson, H. L. 1990. Increased genetic variance after a population bottleneck. *Trends Ecol. Evol.* 5: 228–230.

Carson, H. L. 1992. The Galapagos that were. *Nature* 355: 202–203.

Carson, H. L. and K. Y. Kaneshiro. 1976. *Drosophila* of Hawaii: Systematics and ecological genetics. *Annu. Rev. Ecol. Syst.* 7: 311–345.

Carson, H. L. and A. R. Templeton. 1984. Genetic revolutions in relation to speciation phenomena: The founding of new populations. *Annu. Rev. Ecol. Syst.* 15: 97–131.

Carter, R. E. 2000. DNA fingerprinting using minisatellite probes. Pp. 113–135 in *Molecular Methods in Ecology*, A. J. Baker (ed.). Blackwell, London.

Carvalho, G. R. (ed.). 1998. *Advances in Molecular Ecology*. IOS Press, Amsterdam.

Carvalho, G. R., N. Maclean, S. D. Wratten, R. E. Carter and J. P. Thurston. 1991. Differentiation of aphid clones using DNA fingerprints from individual aphids. *Proc. R. Soc. Lond. B* 243: 109–114.

Cary, S. G. 1996. PCR-based method for single egg and embryo identification in marine organisms. *BioTechniques* 21: 998–999.

Casanova, M. and 8 others. 1985. A human Y-linked DNA polymorphism and its potential for estimating genetic and evolutionary distance. *Science* 230: 1403–1406.

Castle, P. H. J. 1984. Notacanthiformes and Anguilliformes: Development. Pp. 62–93 in *Ontogeny and Systematics of Fishes*, Spec. Publ. No. 1, Am. Soc. Ichthyologists and Herpetologists. Allen Press, Lawrence, KS.

Castresana, J., G. Feldmaier-Fuchs and S. Pääbo. 1998. Codon reassignment and amino acid composition in hemichordate mitochondria. *Proc. Natl. Acad. Sci. USA* 95: 3703–3707.

Cathey, J. C., J. W. Bickham and J. C. Patton. 1998. Introgressive hybridization and nonconcordant evolutionary history of maternal and paternal lineages in North American deer. *Evolution* 52: 1224–1229.

Catzeflis, F. M., F. H. Sheldon, J. E. Ahlquist and C. G. Sibley. 1987. DNA-DNA hybridization evidence of the rapid rate of muroid rodent DNA evolution. *Mol. Biol. Evol.* 4: 242–253.

Caugant, D. A., B. R. Levin and R. K. Selander. 1981. Genetic diversity and temporal variation in the *E. coli* population of a human host. *Genetics* 98: 467–490.

Caugant, D. A. and 7 others. 1986. Intercontinental spread of a genetically distinctive complex of clones of *Neisseria meningitidis* causing epidemic disease. *Proc. Natl. Acad. Sci. USA* 83: 4927–4931.

Cavalier-Smith, T. 1975. The origin of nuclei and of eukaryotic cells. *Nature* 256: 463–468.

Cavalier-Smith, T. 1983. A 6-kingdom classification and unified phylogeny. Pp. 1027–1034 in *Endocytobiology II: Intracellular Space as an Oligogenetic Ecosystem*, W. Schwemmler and H. E. Schenk (eds.). De Gruyter, Berlin.

Cavalier-Smith, T. (ed.). 1985. *The Evolution of Genome Size*. Wiley, New York.

Cavalier-Smith, T. 2003. Genomic reduction and evolution of novel genetic membranes and protein-targeting machinery in eukaryote-eukaryote chimaeras (meta-algae). *Phil. Trans. R. Soc. Lond. B* 358: 109–134.

Cavalier-Smith, T. and E. E.-Y. Chao. 2003. Phylogeny of Choanozoa, Apusozoa, and other Protozoa and early eukaryote megaevolution. *J. Mol. Evol.* 56: 540–563.

Cavalli-Sforza, L. 1966. Population structure and human evolution. *Proc. R. Soc. Lond. B* 164: 362–379.

Cavalli-Sforza, L. 1997. Genes, peoples, and languages. *Proc. Natl. Acad. Sci. USA* 94: 7719–7724.

Cavalli-Sforza, L. L. 2000. *Genes, People, and Languages*. North Point Press, New York.

Cavener, D. R. and M. T. Clegg. 1981. Evidence for biochemical and physiological differences between enzyme genotypes in *Drosophila melanogaster*. *Proc. Natl. Acad. Sci. USA* 78: 4444–4447.

Cedergren, R., M. W. Gray, Y. Abel and D. Sankoff. 1988. The evolutionary relationships among known life forms. *J. Mol. Evol.* 28: 98–112.

Cercueil, A., E. Bellemain and S. Manel. 2002. PARENTE: Computer program for parentage analysis. *J. Heredity* 93: 458–459.

Chakraborty, R. 1981. The distribution of the number of heterozygous loci in an individual in natural populations. *Genetics* 98: 461–466.

Chakraborty, R. and K. K. Kidd. 1991. The utility of DNA typing in forensic work. *Science* 254: 1735–1739.

Chakraborty, R., D. N. Stivers, Y. Su and B. Budowle. 1999. The utility of short tandem repeat loci beyond human identification: Implications for development of new DNA typing systems. *Electrophoresis* 20: 1682–1696.

Chakravarti, A. 1998. It's raining SNPs, hallelujah? *Nature Genet.* 19: 216–217.

Chamberlain, C. P., S. Bensch, X. Feng, S. Akesson and T. Andersson. 2000. Stable isotopes examined across a migratory divide in Scandanavian willow warblers (*Phylloscopus trochilus trochilus* and *Phylloscopus trochilus acredula*) reflect their African winter quarters. *Proc. R. Soc. Lond. B* 267: 43–48.

Champion, A. B., E. M. Prager, D. Wachter and A. C. Wilson. 1974. Microcomplement fixation. Pp. 397–416 in *Biochemical and Immunological Taxonomy of Animals*, C. A. Wright (ed.). Academic Press, New York.

Chaplin, J. A. and P. D. N. Hebert. 1997. *Cyprinotus incongruens* (Ostracoda): An ancient asexual? *Mol. Ecol.* 6: 155–168.

Chapman, R. W. and D. A. Powers. 1984. A method for rapid isolation of mtDNA from fishes. *Maryland Sea Grant Technical Rep.* UM-SG-TS-84–05 (11 pp.).

Chapman, R. W., J. C. Stephens, R. A. Lansman and J. C. Avise. 1982. Models of mitochondrial DNA transmission genetics and evolution in higher eucaryotes. *Genet. Res.* 40: 41–57.

Charleston, M. A. 1998. Jungles: A new solution to the host/parasite phylogeny reconciliation problem. *Math. Biosci.* 149: 191–223.

Charlesworth, B., R. Lande and M. Slatkin. 1982. A neo-Darwinian commentary on macroevolution. *Evolution* 36: 474–498.

Charlesworth, B., J. A. Coyne and N. H. Barton. 1987. The relative rates of evolution of sex chromosomes and autosomes. *Am. Nat.* 130: 113–146.

Charlesworth, B., M. T. Morgan and D. Charlesworth. 1993. The effects of deleterious mutations on neutral molecular variation. *Genetics* 134: 1289–1303.

Charlesworth, D. 1995. Multi-allelic self-incompatibility polymorphisms in plants. *Bioessays* 17: 31–38.

Charlesworth, D. and B. Charlesworth. 1987. Inbreeding depression and its evolutionary consequences. *Annu. Rev. Ecol. Syst.* 18: 237–268.

Chase, M. R., C. Moller, R. Kessli and K. S. Bawa. 1996. Distant gene flow in tropical trees. *Nature* 383: 398–399.

Chase, M. W. and 41others. 1993. Phylogenetics of seed plants: An analysis of nucleotide sequences from the plastid gene *rbc*L. *Annals Missouri Bot. Gardens* 80: 528–580.

Chat, J., L. Chalak and R. J. Petit. 1999. Strict paternal inheritance of chloroplast DNA and maternal inheritance of mitochondrial DNA in interspecific crosses of kiwifruit. *Theoret. Appl. Genet.* 99: 314–322.

Chaw, S.-M., C. L. Parkinson, Y. Cheng, T. M. Vincent and J. D. Palmer. 2000. Seed plant phylogeny inferred from all three plant genomes: Monophyly of extant gymnosperms and origin of Gnetales from conifers. *Proc. Natl. Acad. Sci. USA* 97: 4086–4091.

Cheliak, W. M. and J. A. Patel. 1984. Electrophoretic identification of clones in trembling aspen. *Can. J. Forest Res.* 14: 740–743.

Chen, F.-C., E. J. Vallender, H. Wang, C.-S. Tzeng and W.-H. Li. 2001. Genomic divergence between human and chimpanzee estimated from large-scale alignments of genomic sequences. *J. Heredity* 92: 481–489.

Chen, W.-J., C. Bonillo and G. Lecointre. 2002. Repeatability of clades as a criterion of reliability: A case study for molecular phylogeny of Acanthomorpha (Teleostei) with larger number of taxa. *Mol. Phylogen. Evol.* 26: 262–288.

Cheng, S., C. Fockler, W. M. Barnes and R. Higuchi. 1994. Effective amplification of long targets from cloned inserts of human genomic DNA. *Proc. Natl. Acad. Sci. USA* 91: 5695–5699.

Cherry, L. M., S. M. Case and A. C. Wilson. 1978. Frog perspective on the morphological divergence between humans and chimpanzees. *Science* 200: 209–211.

Chesser, R. K. 1983. Genetic variability within and among populations of the black-tailed prairie dog. *Evolution* 37: 320–331.

Chesser, R. K., M. W. Smith and M. H. Smith. 1984. Biochemical genetics of mosquitofish populations. I. Incidence and significance of multiple insemination. *Genetica* 64: 77–81.

Childers, W. F. 1967. Hybridization of four species of sunfishes (Centrarchidae). *Bull. Ill. Nat. Hist. Surv.* 29: 159–214.

Childs, M. R., A. A. Echelle and T. E. Dowling. 1996. Development of the hybrid swarm between Pecos pupfish (Cyprinodontidae: *Cyprinodon pecoensis*) and sheepshead minnow (*Cyprinodon variegatus*): A perspective from allozymes and mtDNA. *Evolution* 50: 2014–2022.

Cho, Y. and J. D. Palmer. 1999. Multiple acquisitions via horizontal transfer of a group I intron in the mitochondrial *cox1* gene during evolution in the Araceae family. *Mol. Biol. Evol.* 16: 1155–1165.

Cho, Y., Y.-L. Qiu, P. Kuhlman and J. D. Palmer. 1998. Explosive invasion of plant mitochondria by a group I intron. *Proc. Natl. Acad. Sci. USA* 95: 14244–14249.

Choudhary, M., J. E. Strassmann, D. C. Queller, S. Turillazzi and R. Cervo. 1994. Social parasites in polistine wasps are monophyletic: Implications for sympatric speciation. *Proc. R. Soc. Lond. B* 257: 31–35.

Christie, D. M. and 6 others. 1992. Drowned islands downstream from the Galapagos hotspot imply extended speciation times. *Nature* 355: 246–248.

Cimino, M. C. 1972. Egg production, polyploidization and evolution in a diploid all-female fish of the genus *Poeciliopsis*. *Evolution* 26: 294–306.

Ciofi, C. and 6 others. 1998. Genotyping with microsatellite markers. Pp. 195–205 in *Molecular Tools for Screening Biodiversity*, A. Karp, P. G. Isaac and D. S. Ingram (eds.). Chapman & Hall, London.

Civetta, A., H. M. Waldrip-Dail and A. G. Clark. 2002. An introgression approach to mapping differences in mating success and sperm competitive ability in *Drosophila simulans* and *D. sechellia*. *Genet. Res.* 79: 65–74.

Clark, A. G. 1988. Deterministic theory of heteroplasmy. *Evolution* 42: 621–626.

Clark, A. G. 1990. Inference of haplotypes from PCR-amplified samples of diploid populations. *Mol. Biol. Evol.* 7: 111–122.

Clark, A. G. 1993. Evolutionary inferences from molecular characterization of self-incompatibility alleles. Pp. 79–108 in *Mechanisms of Molecular Evolution*, N. Takahata and A. G. clark (eds.). Sinauer Associates, Sunderland, MA.

Clark, A. G. 1994. Natural selection with nuclear and cytoplasmic transmission. I. A deterministic model. *Genetics* 107: 679–701.

Clark, A. G. and T.-H. Kao. 1991. Excess nonsynonymous substitution at shared polymorphic sites among self-incompatibility alleles of Solanaceae. *Proc. Natl. Acad. Sci. USA* 88: 9823–9827.

Clark, A. G. and R. M. May. 2002. Taxonomic bias in conservation research. *Science* 297: 191–192.

Clark, M. A., N. A. Moran, P. Baumann and J. J. Wernegreen. 2000. Cospeciation between bacterial endosymbionts (*Buchnera*) and a recent radiation of aphids (*Uroleucon*) and pitfalls of testing for phylogenetic congruence. *Evolution* 54: 517–525.

Clarke, B. 1975. The contribution of ecological genetics to evolutionary theory: Detecting the direct effects of natural selection on particular polymorphic loci. *Genetics* 79s: 101–113.

Clarke, K. E., T. E. Rinderer, P. Franck, J. G. Quezada-Euán and B. P. Oldroyd. 2002. The Africanization of honeybees (*Apis mellifera* L.) of the Yucatan: A study of a massive hybridization event across time. *Evolution* 56: 1462–1474.

Clausen, J., D. D. Keck and W. M. Hiesey. 1940. Experimental studies on the nature of species. I. Effect of varied environments on western North American plants. *Carnegie Inst. Wash.* Publ. No. 520.

Clayton, T. M., J. P. Whittaker and C. N. Maguire. 1995. Identification of bodies from the scene of a mass disaster using DNA amplification of short tandem repeat (STR) loci. *Forensic Sci. Int.* 76: 7–15.

Clegg, M. T. 1980. Measuring plant mating systems. *BioScience* 30: 814–818.

Clegg, M. T. and R. W. Allard. 1972. Patterns of genetic differentiation in the slender wild oat species *Avena barbata*. *Proc. Natl. Acad. Sci. USA* 69: 1820–1824.

Clegg, M. T. and R. W. Allard. 1973. Viability versus fecundity selection in the slender wild oat, *Avena barbata* L. *Science* 181: 667–668.

Clegg, M. T. and G. Zurawski. 1992. Chloroplast DNA and the study of plant phylogeny: Present status and future prospects. Pp. 1–13 in *Molecular Systematics of Plants*, P. S. Soltis, D. E. Soltis and J. J. Doyle (eds.). Chapman & Hall, New York.

Clegg, M. T., R. W. Allard and A. L. Kahler. 1972. Is the gene the unit of selection? Evidence from two experimental plant populations. *Proc. Natl. Acad. Sci. USA* 69: 2474–2478.

Clegg, M. T., K. Ritland and G. Zurawski. 1986. Processes of chloroplast DNA evolution. Pp. 275–294 in *Evolutionary Processes and Theory*, S. Karlin and E. Nevo (eds.). Academic Press, New York.

Clegg, S. M., J. F. Kelly, M. Kimura and T. B. Smith. 2003. Combining genetic markers and stable isotopes to reveal population connectivity and migration patterns in a neotropical migrant, Wilson's warbler (*Wilsonia pusilla*). *Mol. Ecol.* 12: 819–830.

Clutton-Brock, T. H. 1991. *The Evolution of Parental Care.* Princeton University Press, Princeton, NJ.

Cockerham, C. C. and B. S. Weir. 1993. Estimation of gene flow from *F*-statistics. *Evolution* 47: 855–863.

Cocks, G. T. and A. C. Wilson. 1972. Enzyme evolution in the Enterobacteriaceae. *J. Bacteriol.* 110: 793–802.

Coffin, J. and 10 others. 1986. Human immunodeficiency viruses. *Science* 232: 697.

Coffin, J., S. Hughes and H. Varmus (eds.). 1997. *Retroviruses.* Cold Spring Harbor Laboratory Press, Cold Spring Harbor, NY.

Coffroth, M. A., H. R. Lasker. 1998. Population structure of a clonal gorgonian coral: The interplay between clonal reproduction and disturbance. *Evolution* 52: 379–393.

Coffroth, M. A. and J. M. Mulawka. 1995. Identification of marine invertebrate larvae by means of PCR-RAPD species-specific markers. *Limnol. Oceanogr.* 40: 181–189.

Coffroth, M. A., H. R. Lasker, M. E. Diamond, J. A. Bruenn and E. Bermingham. 1992. DNA fingerprints of a gorgonian coral: A method for detecting clonal structure in a vegetative species. *Mar. Biol.* 114: 317–325.

Cohen, J. 1994. "Long PCR" leaps into larger DNA sequences. *Science* 263: 1564–1565.

Colborn, J., R. E. Crabtree, J. B. Shaklee, E. Pfeiler and B. W. Bowen. 2001. The evolutionary enigma of bonefishes (*Albula* spp.): Cryptic species and ancient separations in a globally distributed shorefish. *Evolution* 55: 807–820.

Colbourne, J. K., B. D. Neff, J. M. Wright and M. R. Gross. 1996. DNA fingerprinting of bluegill sunfish (*Lepomis macrochirus*) using (GT)$_n$ microsatellietes and its potential for assessment of mating success. *Can. J. Fish. Aquat. Sci.* 53: 342–349.

Colbourne, J. K., T. J. Crease, L. J. Weider and P. D. N. Hebert. 1998. Phylogenetics and evolution of a circumarctic species complex (Cladocera: *Daphnia pulex*). *Biol. J. Linnean Soc.* 65: 347–365.

Cole, B. J. 1983. Multiple mating and the evolution of social behavior in the Hymenoptera. *Behav. Ecol. Sociobiol.* 12: 191–201.

Cole, B. J. and D. C. Wiernasz. 1999. The selective advantage of low relatedness. *Science* 285: 891–893.

Coleman, A. W. 2000. The significance of a coincidence between evolutionary landmarks found in mating affinity and a DNA sequence. *Protist* 15: 1–9.

Coleman, A. W. 2001. Biogeography and speciation in the *Pandorina/Volvulina* (Chlorophyta) superclade. *J. Phycol.* 37: 836–851.

Coleman, A. W. 2002. Microbial eukaryotic dispersal. *Science* 297: 337.

Colgan, D. J. 1999. Phylogenetic studies of marsupials based on phosphoglycerate kinase DNA sequences. *Mol. Phylogen. Evol.* 11: 13–26.

Colgan, D. J., C.-G. Zhang and J. R. Paxton. 2000. Phylogenetic investigations of the Stephanoberyciformes and Beryciformes, particularly whalefishes (Euteleostei: Cetomimidae), based on partial 12S rDNA and 16S rDNA sequences. *Mol. Phylogen. Evol.* 17: 15–25.

Collier, G. E. and R. J. MacIntyre. 1977. Microcomplement fixation studies on the evolution of α-glycerophosphate dehydrogenase within the genus *Drosophila*. *Proc. Natl. Acad. Sci. USA* 74: 684–688.

Collura, R. V. and C. B. Stewart. 1995. Insertions and duplications of mtDNA in the nuclear genomes of Old World monkeys and hominids. *Nature* 378: 485–489.

Collura, R. V., M. R. Auerbach and C. B. Stewart. 1996. A quick, direct method that can differentiate expressed mitochondrial genes from their nuclear pseudogenes. *Curr. Biol.* 6: 1337–1339.

Coltman, D. W. and J. Slate. 2003. Microsatellite measures of inbreeding: A meta-analysis. *Evolution* 57: 971–983.

Coltman, D. W., W. D. Bowen and J. M. Wright. 1998a. Birth weight and neonatal survival of harbour seal pups are positively correlated with genetic variation measured by microsatellites. *Proc. R. Soc. Lond. B* 265: 803–809.

Coltman, D. W., W. D. Bowen and J. M. Wright. 1998b. Male mating success in an aquatically mating pinniped, the harbour seal (*Phoca vitulina*), assessed by microsatellite DNA markers. *Mol. Ecol.* 7: 627–638.

Comes, H. P. and J. W. Kadereit. 1998. The effects of Quaternary climatic changes on plant distribution and evolution. *Trends Plant Sci.* 3: 433–438.

Comstock, K. E. and 6 others. 2002. Patterns of molecular genetic variation among African elephant populations. *Mol. Ecol.* 11: 2489–2498.

Constable, J. J., C. Packer, D. A. Collins and A. E. Pusey. 1995. Nuclear DNA from primate dung. *Nature* 373: 393.

Cook, J. A., K. R. Bestgen, D. L. Propst and T. L. Yates. 1992. Allozymic divergence and systematics of the Rio Grande silvery minnow, *Hybognathus amarus* (Teleostei: Cyprinidae). *Copeia* 1992: 36–44.

Cook, R. E. 1980. Reproduction by duplication. *Nat. Hist.* 89: 88–93.

Cook, R. E. 1983. Clonal plant populations. *Am. Sci.* 71: 244–253.

Cook, R. E. 1985. Growth and development in clonal plant populations. Pp. 259–296 in *Population Biology and Evolution of Clonal Organisms*, J. B. C. Jackson, L. W. Buss and R. E. Cook (eds.). Yale University Press, New Haven, CT.

Cooke, F., C. D. MacInnes and J. P. Prevett. 1975. Gene flow between breeding populations of lesser snow geese. *Auk* 92: 493–510.

Cooke, F., D. T. Parkin and R. F. Rockwell. 1988. Evidence of former allopatry of the two color phases of lesser snow geese (*Chen caerulescens caerulescens*). *Auk* 105: 467–479.

Cooke, P. H. and J. G. Oakeshott. 1989. Amino acid polymorphisms for esterase-6 in *Drosophila melanogaster*. *Proc. Natl. Acad. Sci. USA* 86: 1426–1430.

Cooper, A. 1997. Studies of avian ancient DNA: From Jurassic Park to modern island extinctions. Pp. 345–373 in *Avian Molecular Evolution and Systematics*, D. P. Mindell (ed.). Academic Press, New York.

Cooper, A. and D. Penny. 1997. Mass survival of birds across the Cretaceous-Tertiary boundary: Molecular evidence. *Science* 275: 1109–1113.

Cooper, A. and C. Poinar. 2000. Ancient DNA: Do it right or not at all. *Science* 289: 1139.

Cooper, A., C. Mourer-Chauvire, G. K. Chambers, A. von Haeseler, A. C. Wilson and S. Pääbo. 1992. Independent origins of New Zealand moas and kiwis. *Proc. Natl. Acad. Sci. USA* 89: 8741–8744.

Cooper, A. and 6 others. 1996. Ancient DNA and island endemics. *Nature* 381: 484.

Cooper, A., C. Lalueza-Fox, S. Anderson, A. Rambaut, J. Austin and R. Ward. 2001. Complete mitochondrial genome sequences of two extinct moas clarify ratite evolution. *Nature* 409: 704–707.

Cooper, C. G., P. L. Miller and P. W. H. Holland. 1996. Molecular genetic analysis of sperm competition in the damselfly *Ischnura elegans* (Vander Linden). *Proc. R. Soc. Lond. B* 263: 1343–1349.

Cooper, S. J. B. and G. M. Hewitt. 1993. Nuclear DNA sequence divergence between parapatric subspecies of the grasshopper *Chorthippus parallelus*. *Insect Mol. Biol.* 2: 185–194.

Cooper, S. J. B, K. M. Ibrahim and G. M. Hewitt. 1995. Postglacial expansion and genome subdivision in the European grasshoper *Chorthippus parallelus*. *Mol. Ecol.* 4: 49–60.

Costa, J. T. and K. G. Ross. 1993. Seasonal decline in intracolony genetic relatedness in eastern tent caterpillars: Implications for social evolution. *Behav. Ecol. Sociobiol.* 32: 47–54.

Costa, J. T. and K. G. Ross. 2003. Fitness effects of group merging in a social insect. *Proc. R. Soc. Lond. B* 270: 1697–1702.

Cotton, J. A. and R. D. M. Page. 2002. Going nuclear: Gene family evolution and vertebrate phylogeny reconciled. *Proc. R. Soc. Lond. B* 269: 1555–1561.

Coulson, T. N. and 7 others. 1998. Microstellites reveal heterosis in red deer. *Proc. R. Soc. Lond. B* 265: 489–495.

Cowen, R. K., K. M. M. Lwiza, S. Sponaugle, C. B. Paris and D. B. Olson. 2000. Connectivity of marine populations: Open or closed? *Science* 287: 857–859.

Coyne, J. A. 1982. Gel electrophoresis and cryptic protein variation. *Isozymes* V: 1–32.

Coyne, J. A. 1992. Genetics and speciation. *Nature* 355: 511–515.

Coyne, J. A. 1996. Genetics of differences in pheromonal hydrocarbons between *Drosophila melanogaster* and *D. simulans*. *Genetics* 143: 353–364.

Coyne, J. A. and A. J. Berry. 1994. Effects of the fourth chromosome on the sterility of hybrids between *Drosophila simulans* and its relatives. *J. Heredity* 85: 224–227.

Coyne, J. A. and B. Charlesworth. 1997. Genetics of a pheromonal difference affecting sexual isolation between *Drosophila mauritiana* and *D. seychellia*. *Genetics* 143: 1689–1698.

Coyne, J. A. and H. A. Orr. 1989a. Two rules of speciation. Pp. 180–207 in *Speciation and Its Consequences*, D. Otte and J. A. Endler (eds.). Sinauer Associates, Sunderland, MA.

Coyne, J. A. and H. A. Orr. 1989b. Patterns of speciation in *Drosophila*. *Evolution* 43: 362–381.

Coyne, J. A. and H. A. Orr. 1997. "Patterns of speciation in *Drosophila*" revisited. *Evolution* 51: 295–303.

Coyne, J. A. and H. A. Orr. 1998. The evolutionary genetics of speciation. *Phil. Trans. R. Soc. Lond. B* 353: 287–305.

Coyne, J. A. and T. D. Price. 2000. Little evidence for sympatric speciation in island birds. *Evolution* 54: 2166–2171.

Crabtree, C. B. and D. G. Buth. 1987. Biochemical systematics of the catostomid genus *Catostomus*: Assessment of *C. clarki*, *C. plebeius* and *C. discobolus* including the Zuni sucker, *C. d. yarrowi*. *Copeia* 1987: 843–854.

Cracraft, J. 1983. Species concepts and speciation analysis. Pp. 159–187 in *Current Ornithology*, R. F. Johnston (ed.). Plenum Press, New York.

Cracraft, J. 1986. Origin and evolution of continental biotas: Speciation and historical congruence within the Australian avifauna. *Evolution* 40: 977–996.

Cracraft, J. 1992. Book Review: Phylogeny and Classification of Birds. *Mol. Biol. Evol.* 9: 182–186.

Cracraft, J. and D. P. Mindell. 1989. The early history of modern birds: A comparison of molecular and morphological evidence. Pp. 389–403 in *The Hierarchy of Life*, B. Fernholm, K. Bremer and H. Jornvall (eds.). Elsevier, Amsterdam.

Cracraft, J. and R. O. Prum. 1988. Patterns and processes of diversification: Speciation and historical congruence in some neotropical birds. *Evolution* 42: 603–620.

Cracraft, J., J. Feinstein, J. Vaughn and K. Helm-Bychowski. 1998. Sorting out tigers (*Panthera tigris*): Mitochondrial sequences, nuclear inserts, systematics, and conservation genetics. *Anim. Cons.* 1: 139–150.

Craddock, C., W. R. Hoeh, R. A. Lutz and R. C. Vrijenhoek. 1995. Extensive gene flow among mytilid (*Bathymodiolus thermophilus*) populations from hydrothermal vents of the eastern Pacific. *Marine Biol.* 124: 137–146.

Craig, R. and R. H. Crozier. 1979. Relatedness in the polygynous ant *Myrmecia pilosula*. *Evolution* 33: 335–341.

Crandall, K. A. and A. R. Templeton. 1999. Statistical methods for detecting recombination. Pp. 153–176 in *The Evolution of HIV*, K. A. Crandall (ed.). John Hopkins University Press, Baltimore, MD.

Crandall, K. A., O. R. P. Beninda-Emonds, G. M. Mace and R. K. Wayne. 2000. Considering evolutionary processes in conservation biology. *Trends Ecol. Evol.* 15: 290–295.

Crawford, A. M. and 7 others. 1993. How reliable are sheep pedigrees? *Proc. New Zealand Soc. Anim. Prod.* 53: 363–366.

Crawford, D. J. and R. Ornduff. 1989. Enzyme electrophoresis and evolutionary relationships among three species of *Lasthenia* (Asteraceae: Heliantheae). *Am. J. Bot.* 76: 289–296.

Crease, T. J., D. J. Stanton and P. D. N. Hebert. 1989. Polyphyletic origins of asexuality in *Daphnia pulex*. II. Mitochondrial-DNA variation. *Evolution* 43: 1016–1026.

Crim, J. L., L. D. Spotila, M. O'Connor, R. Reina, C. J. Williams and F. V. Paladino. 2002. The leatherback turtle, *Dermochelys coriacea*, exhibits both polyandry and polygyny. *Mol. Ecol.* 11: 2097–2106.

Crisp, D. J. 1976. Settlement responses in marine organisms. Pp. 83–124 in *Adaptation to the Marine Environment*, R. C. Newell (ed.). Butterworths, London.

Crisp, D. J. 1978. Genetic consequences of different reproductive strategies in marine invertebrates. Pp. 257–273 in *Marine Organisms: Genetics, Ecology, and Evolution*, B. Battaglia and J. A. Beardmore (eds.). Plenum Press, New York.

Cristescu, M. E., P. D. N. Hebert and T. M. Onciu. 2003. Phylogeography of Ponto-Caspian crustaceans: A benthic-planktonic comparison. *Mol. Ecol.* 12: 985–996.

Crnokrak, P. and C. H. Barrett. 2002. Purging the genetic load: A review of the experimental evidence. *Evolution* 56: 2347–2358.

Crnokrak, P. and D. A. Roff. 1999. Inbreeding depression in the wild. *Heredity* 83: 260–270.

Crochet, P.-A. 2000. Genetic structure of avian populations: Allozymes revisited. *Mol. Ecol.* 9: 1463–1469.

Cronin, M. A. 1992. Intraspecific variation in mitochondrial DNA of North American cervids. *J. Mammal.* 73: 70–82.

Cronin, M. A., M. E. Nelson and D. F. Pac. 1991a. Spatial heterogeneity of mitochondrial DNA and allozymes among populations of white-tailed deer and mule deer. *J. Heredity* 82: 118–127.

Cronin, M. A., S. C. Amstrup, G. W. Garner and E. R. Vyse. 1991b. Interspecific and intraspecific mitochondrial DNA variation in North American bears (*Ursus*). *Can. J. Zool.* 69: 2985–2992.

Cronin, M. A., D. A. Palmisciano, E. R. Vyse and D. G. Cameron. 1991c. Mitochondrial DNA in wildlife forensic science: Species identification of tissues. *Wildl. Soc. Bull.* 19: 94–105.

Cronin, M. A., J. B. Grand, D. Esler, D. V. Derksen and K. T. Scribner. 1996. Breeding populations of northern pintails have similar mitochondrial DNA. *Can. J. Zool.* 74: 992–999.

Cronn, R., M. Cedroni, T. Haselkorn, C. Grover and J. F. Wendel. 2002. PCR-mediated recombination in amplification products derived from polyploid cotton. *Theoret. Appl. Genet.* 104: 482–489.

Crosetti, D., W. S. Nelson and J. C. Avise. 1993. Pronounced genetic structure of mitochondrial DNA among populations of the circum-globally distributed grey mullet (*Mugil cephalus* Linnaeus). *J. Fish Biol.* 44: 47–58.

Crosland, M. W. J. 1988. Inability to discriminate between related and unrelated larvae in the ant *Rhytidoponera confusa* (Hymenoptera: Formicidae). *Ann. Entomol. Soc. Am.* 81: 844–850.

Crow, J. F. 1954. Breeding structure of populations. II. Effective population number. Pp. 543–556 in *Statistics and Mathematics in Biology*, T. A. Bancroft, J. W. Gowen and J. L. Lush (eds.). Iowa State College Press, Ames, IA.

Crozier, R. H. 1992. Genetic diversity and the agony of choice. *Biol. Cons.* 61: 11–15.

Crozier, R. H. 1997. Preserving the information content of species: Genetic diversity, phylogeny, and conservation worth. *Annu. Rev. Ecol. Syst.* 28: 243–268.

Crozier, R. H. and P. Luykx. 1985. The evolution of termite eusociality is unlikely to have been based on a male-haploid analogy. *Am. Nat.* 126: 867–869.

Crozier, R. H. and R. E. Page. 1985. On being the right size: Male contributions and multiple mating in social Hymenoptera. *Behav. Ecol. Sociobiol.* 18: 105–115.

Crozier, R. H. and P. Pamilo. 1996. *Evolution of Social Insect Colonies*. Oxford University Press, Oxford, England.

Crozier, R. H., P. Pamilo and Y. C. Crozier. 1984. Relatedness and microgeographic genetic variation in *Rhytidoponera mayri*, an Australian arid zone ant. *Behav. Ecol. Sociobiol.* 15: 143–150.

Crozier, R. H., B. H. Smith and Y. C. Crozier. 1987. Relatedness and population structure of the primitively eusocial bee *Lasioglossum zephyrum* (Hymenoptera: Halictidae) in Kansas. *Evolution* 41: 902–910.

Crozier, R. H., Y. C. Crozier and A. G. Mackinley. 1989. The CO-I and CO-II region of honeybee mitochondrial DNA: Evidence for variation in insect mitochondrial evolutionary rates. *Mol. Biol. Evol.* 6: 399–411.

Cubo, J. and W. Arthur. 2001. Patterns of correlated character evolution in flightless birds: A phylogenetic approach. *Evol. Ecol.* 14: 693–702.

Cuellar, O. 1974. On the origin of parthenogenesis in vertebrates: The cytogenetic factors. *Am. Nat.* 108: 625–648.

Cuellar, O. 1977. Animal parthenogenesis. *Science* 197: 837–843.

Cuellar, O. 1984. Histocompatibility in Hawaiian and Polynesian populations of the parthenogenetic gecko *Lepidodactylus lugubris*. *Evolution* 38: 176–185.

Cunningham, C. W. and T. M. Collins. 1998. Beyond area relationships: Extinction and recolonization in molecular marine biogeography. Pp. 297–321 in *Molecular Approaches to Ecology and Evolution*, R. DeSalle and B. Schierwater (eds.). Birkhäuser, Basel, Switzerland.

Cunningham, C. W., N. W. Blackstone and L. W. Buss. 1992. Evolution of king crabs from hermit crab ancestors. *Nature* 355: 539–542.

Cunningham, M. and C. Moritz. 1998. Genetic effects of forest fragmentation on a rainforest restricted lizard

(Scincidae: *Gnypetoscincus queenslandiae*). *Biol. Cons.* 83: 19–30.

Curach, N. and P. Sunnucks. 1999. Molecular anatomy of an onychophoran: Compartmentalized sperm storage and heterogeneous paternity. *Mol. Ecol.* 8: 1375–1385.

Curie-Cohen, M., D. Yoshihara, L. Luttrell, K. Benforado, J. W. MacCluer and W. H. Stone. 1983. The effects of dominance on mating behavior and paternity in a captive troop of rhesus monkeys (*Macaca mulatta*). *Am. J. Primatol.* 5: 127–138.

Curole, J. P. and T. D. Kocher. 1999. Mitogenomics: Digging deeper with complete mitochondrial genomes. *Trends Ecol. Evol.* 14: 394–398.

Currie, C. R., B. Wong, A. E. Stuart, T. R. Schultz, S. A. Rehner, U. G. Mueller, G.-H. Sung, J. W. Spatafora and N. A. Straus. 2003. Ancient tripartite coevolution in the Attine ant-microbe symbiosis. *Science* 299: 386–388.

Curtis, A. S. G., J. Kerr and N. Knowlton. 1982. Graft rejection in sponges. Genetic structure of accepting and rejecting populations. *Transplantation* 30: 127–133.

Curtis, S. E. and M. T. Clegg. 1984. Molecular evolution of chloroplast DNA sequences. *Mol. Biol. Evol.* 1: 291–301.

Curtis, T. P., W. T. Sloan and J. W. Scannell. 2002. Estimating prokaryotic diversity and its limits. *Proc. Natl. Acad. Sci. USA* 99: 10494–10499.

Cutler, M. G., S. E. Bartlett, S. E. Hartley and W. S. Davidson. 1991. A polymorphism in the ribosomal RNA genes distinguishes Atlantic Salmon (*Salmo salar*) from North America and Europe. *Can. J. Fish. Aquat. Sci.* 48: 1655–1661.

Daemen, E., T. Cross, F. Ollevier and F. A. M. Volckaert. 2001. Analysis of the genetic structure of European eel (*Anguilla anguilla*) using microsatellite and mtDNA markers. *Mar. Biol.* 139: 755–764.

Dallas, J. F. and 8 others. 2003. Similar estimates of population genetic composition and sex ratio derived from carcasses and faeces of Eurasian otter *Lutra lutra*. *Mol. Ecol.* 12: 275–282.

Dallimer, M., C. Blackburn, P. J. Jones and J. M. Pemberton. 2002. Genetic evidence for male biased dispersal in the red-bellied quelea *Quelea quelea*. *Mol. Ecol.* 11: 529–533.

Dancik, B. P. and F. C. Yeh. 1983. Allozyme variability and evolution of lodgepole pine (*Pinus contorta* var. *latifolia*) and jack pine (*Pinus banksiana*) in Alberta. *Can. J. Genet. Cytol.* 25: 57–64.

Danforth, B. N., L. Conway and S. Ji. 2003. Phylogeny of eusocial *Lasioglossum* reveals multiple losses of eusociality within a primitively eusocial clade of bees (Hymenoptera: Halictidae). *Syst. Biol.* 52: 23–26.

Daniels, S. B., K. R. Peterson, L. D. Strausbaugh, M. G. Kidwell and A. Chovnick. 1990. Evidence for horizontal transmission of the P transposable element between *Drosophila* species. *Genetics* 124: 339–355.

Danley, P. D. and T. D. Kocher. 2001. Speciation in rapidly diverging systems: Lessons from Lake Malawi. *Mol. Ecol.* 10: 1075–1086.

Danzmann, R. G. 1997. PROBMAX: A computer program for assigning unknown parentage in pedigree analysis from known genotypic pools of parents and progeny. *J. Heredity* 88: 333.

Darling, K. F., C. M. Wade, I. A. Stewart, D. Kroon, R. Dingle and A. J. Leigh Brown. 2000. Molecular evidence for genetic mixing of Arctic and Antarctic subpolar populations of planktonic foraminifers. *Nature* 405: 43–47.

Darlington, C. D. 1939. *The Evolution of Genetic Systems*. Cambridge, University Press, Cambridge, England.

Darlington, P. J. Jr. 1957. *Zoogeography: The Geological Distributions of Animals*. John Wiley and Sons, New York.

Darlington, P. J. Jr. 1965. *Biogeography of the Southern End of the World*. Harvard University Press, Cambridge, MA.

Darwin, C. 1859. *On the Origin of Species by Means of Natural Selection, of the Preservation of Favored Races in the Struggle for Life*. John Murray, London.

Darwin, C. 1875. *Insectivorous Plants*. John Murray, London.

Darwin, C. 1877. *The Different Forms of Flowers on Plants of the Same Species*. John Murray, London.

da Silva, M. N. F. and J. L. Patton. 1998. Molecular phylogeography and the evolution and conservation of Amazonian mammals. *Mol. Ecol.* 7: 475–486.

Dasmahapatra, K. K. and 6 others. 2002. Inferences from a rapidly moving hybrid zone. *Evolution* 56: 741–753.

Daubin, V., N. A. Moran and H. Ochman. 2003. Phylogenetics and the cohesion of bacterial genomes. *Science* 301: 829–832.

Daugherty, C. H., A. Cree, J. M. Hay and M. B. Thompson. 1990. Neglected taxonomy and continuing extinctions of tuatara (*Sphenodon*). *Nature* 347: 177–179.

Davidson, W. S., T. P. Birt and J. M. Green. 1989. A review of genetic variation in Atlantic salmon, *Salmo salar* L., and its importance for stock identification, enhancement programmes and *Aquaculture*. *J. Fish Biol.* 34: 547–560.

Davies, N., A. Aiello, J. Mallet, A. Pomiankowski and R. E. Silberglied. 1997. Speciation in two neotropical butterflies: Extending Haldane's rule. *Proc. R. Soc. Lond. B* 264: 845–851.

Davies, N., F. X. Villablanca and G. K. Roderick. 1999. Determining the source of individuals: Multilocus genotyping in nonequilibrium population genetics. *Trends Ecol. Evol.* 14: 17–21.

Davies, N. B. 1992. *Dunnock Behaviour and Social Evolution*. Oxford University Press, Oxford, England.

Davis, L. M., T. C. Glenn, R. M. Elsey, H. C. Dessauer and R. H. Sawyer. 2001. Multiple paternity and mating patterns in the American alligator, *Alligator mississippiensis*. *Mol. Ecol.* 10: 1011–1024.

Davis, R. F., C. F. Herreid II and H. L. Short. 1962. Mexican free-tailed bats in Texas. *Ecol. Monogr.* 32: 311–346.

Davis, S. K., J. E. Strassmann, C. Hughes, L. S. Pletscher and A. R. Templeton. 1990. Population structure and kinship in *Polistes* (Hymenoptera, Vespidae): An analysis using ribosomal DNA and protein electrophoresis. *Evolution* 44: 1242–1253.

Dawid, I. B. and A. W. Blackler 1972. Maternal and cytoplasmic inheritance of mitochondrial DNA in *Xenopus*. *Dev. Biol.* 29: 152–161.

Dawkins, R. 1989. *The Selfish Gene* (2nd Edition). Oxford University Press, Oxford, England.

Dawley, R. M. 1992. Clonal hybrids of the common laboratory fish *Fundulus heteroclitus*. *Proc. Natl. Acad. Sci. USA* 89: 2485–2488.

Dawley, R. M. and J. P Bogart (eds.). 1989. *Evolution and Ecology of Unisexual Vertebrates*. New York State Museum, Albany, NY.

Dawson, S. C. and N. R. Pace. 2002. Novel kingdom-level eukaryotic diversity in anoxic environments. *Proc. Natl. Acad. Sci. USA* 99: 8324–8329.

Day, L. 1991. Redefining the tree of life. *Mosaic* 22(4): 47–57.

Day, T. 2000. Sexual selection and the evolution of costly female preferences: Spatial effects. *Evolution* 54: 715–730.

Dayhoff, M. O. 1972. *Atlas of Protein Sequence and Structure*, Vol. 5. National Biomedical Research Foundation, Silver Springs, MD.

Dayhoff, M. O. and R. V. Eck. 1968. *Atlas of Protein Sequence and Structure 1967–68*. National Biomedical Research Foundation, Silver Springs, MD.

DeBry, R. W. and N. A. Slade. 1985. Cladistic analysis of restriction endonuclease cleavage maps within a maximum-likelihood framework. *Syst. Zool.* 34: 21–34.

Decker, M. D., P. G. Parker, D. J. Minchella and K. N. Rabenold. 1993. Monogamy in black vultures: Genetic evidence from DNA fingerprinting. *Behav. Ecol.* 4: 29–35.

Degnan, S. M. 1993. The perils of single gene trees: Mitochondrial versus single-copy nuclear DNA variation in white-eyes (Aves: Zosteropidae). *Mol. Ecol.* 2: 219–225.

Deguilloux, M.-F., M.-H. Pemonge and R. J. Petit. 2002. Novel perspectives in wood certification and forensics: Dry wood as a source of DNA. *Proc. R. Soc. Lond. B* 269: 1039–1046.

Deguilloux, M.-F., M.-H. Pemonge, L. Bertel, A. Kremer and R. J. Petit. 2003. Checking the geographical origin of oak wood: Molecular and statistical tools. *Mol. Ecol.* 12: 1629–1636.

Delmotte, F. and 6 others. 2003. Phylogenetic evidence for hybrid origins of asexual lineages in an aphid species. *Evolution* 57: 1291–1303.

DeLong, E. F., R. B. Frankel and D. A. Bazylinski. 1993. Multiple evolutionary origins of magnetotaxis in bacteria. *Science* 259: 803–805.

Delph, LF. and K. Havens. 1998. Pollen competition in flowering plants. Pp. 147–174 in *Sperm Competition and Sexual Selection*, T. R. Birkhead and A. P. Møller (eds.). Academic Press, London.

Delsuc, F. and 7 others. 2002. Molecular phylogeny of living Xenarthrans and the impact of character and taxon sampling on the placental tree rooting. *Mol. Biol. Evol.* 19: 1656–1671.

Delwiche, C. F. 1999. Tracing the thread of plastic divesity through the tapestry of life. *Am. Nat.* 154: S164-S177.

DeMarais, B. D., T. E. Dowling, M. E. Douglas, M. L. Minckley and P. C. Marsh. 1992. Origin of *Gila seminuda* (Teleostei: Cyprinidae) through introgressive hybridization: Implications for evolution and conservation. *Proc. Natl. Acad. Sci. USA* 89: 2747–2751.

Denaro, M. and 6 others. 1981. Ethnic variation in *Hpa*I endonuclease cleavage patterns of human mitochondrial DNA. *Proc. Natl. Acad. Sci. USA* 78: 5768–5772.

Densmore, L. D. III, C. C. Moritz, J. W. Wright and W. M. Brown. 1989a. Mitochondrial-DNA analyses and the origin and relative age of parthenogenetic lizards (genus *Cnemidophorus*). IV. Nine *sexlineatus*-group unisexuals. *Evolution* 43: 969–983.

Densmore, L. D. III, J. W. Wright and W. M Brown. 1989b. Mitochondrial-DNA analyses and the origin and relative age of parthenogenetic lizards (genus *Cnemidophorus*). II. *C. neomexicanus* and the *C. tesselatus* complex. *Evolution* 43: 943–957.

dePamphilis, C. W. and J. D. Palmer. 1989. Evolution and function of plastid DNA: A review with special reference to non-photosynthetic plants. Pp. 182–202 in *Physiology, Biochemistry, and Genetics of Nongreen Plastids*, C. D. Boyer, J. C. Shannon and R. C. Hardison (eds.). American Society of Plant Physiology, Rockville, MD.

dePamphilis, C. W. and R. Wyatt. 1989. Hybridization and introgression in buckeyes (*Aesculus*: Hippocastanaceae): A review of the evidence and a hypothesis to explain long-distance gene flow. *Syst. Bot.* 14: 593–611.

dePamphilis, C. W. and R. Wyatt. 1990. Electrophoretic confirmation of interspecific hybridization in *Aesculus* (Hippocastanaceae) and the genetic structure of a broad hybrid zone. *Evolution* 44: 1295–1317.

de Queiroz, K. and M. J. Donoghue. 1988. Phylogenetic systematics and the species problem. *Cladistics* 4: 317–338.

de Queiroz, K. and J. Gauthier. 1992. Phylogenetic taxonomy. *Annu. Rev. Ecol. Syst.* 23: 449–480.

de Queiroz, K. and J. Gauthier. 1994. Toward a phylogenetic system of biological nomenclature. *Trends Ecol. Evol.* 9: 27–31.

de Rosa, R. and 7 others. 1999. *Hox* genes in brachiopods and priapulids and protostome evolution. *Nature* 399: 772–776.

Desai, S. M., V. S. Kalyanaraman, J. M. Casey, A. Srinivasan, P. R. Andersen and S. G. Devare. 1986. Molecular cloning and primary nucleotide sequence analysis of a distinct human immunodeficiency virus isolate reveal significant divergence in its genomic sequences. *Proc. Natl. Acad. Sci. USA* 83: 8380–8384.

DeSalle, R. 1992a. The origin and possible time of divergence of the Hawaiian Drosophilidae: Evidence from DNA sequences. *Mol. Biol. Evol.* 9: 905–916.

DeSalle, R. 1992b. The phylogenetic relationships of flies in the family Drosophilidae deduced from mtDNA sequences. *Mol. Phylogen. Evol.* 1: 31–40.

DeSalle, R. and V. J. Birstein. 1996. PCR identification of black caviar. *Nature* 381: 197–198.

DeSalle, R. and L. V. Giddings. 1986. Discordance of nuclear and mitochondrial DNA phylogenies in

Hawaiian *Drosophila. Proc. Natl. Acad. Sci. USA* 83: 6902–6906.

DeSalle, R. and A. R. Templeton. 1988. Founder effects and the rate of mitochondrial DNA evolution in Hawaiian *Drosophila. Evolution* 42: 1076–1084.

DeSalle, R., L. V. Giddings and A. R. Templeton. 1986a. Mitochondrial DNA variability in natural populations of Hawaiian *Drosophila*. I. Methods and levels of variability in *D. silvestris* and *D. heteroneura* populations. *Heredity* 56: 75–86.

DeSalle, R., L. V. Giddings and K. Y. Kaneshiro. 1986b. Mitochondrial DNA variability in natural populations of Hawaiian *Drosophila*. II. Genetic and phylogenetic relationships of natural populations of *D. silvestris* and *D. heteroneura. Heredity* 56: 87–92.

DeSalle, R., T. Freedman, E. M. Prager and A. C. Wilson. 1987. Tempo and mode of sequence evolution in mitochondrial DNA of Hawaiian *Drosophila. J. Mol. Evol.* 26: 157–164.

DeSalle, R., J. Gatesy, W. Wheeler and D. Grimaldi. 1992. DNA sequences from a fossil termite in Oligo-Miocene amber and their phylogenetic implications. *Science* 257: 1933–1936.

Desjardins, P. and R. Morais. 1991. Nucleotide sequence and evolution of coding and noncoding regions of a quail mitochondrial genome. *J. Mol. Evol.* 32: 153–161.

de Souza, M. P. and 6 others. 2001. Identification and characterization of bacteria in a selenium-contaminated hypersaline evaporation pond. *Appl. Environ. Microbiol.* 67: 3785–3794.

Desplanque, B. and 6 others. 2000. The linkage disequilibrium between chloroplast DNA and mitochondrial DNA haplotypes in *Beta vulgaris ssp. maritima* (L.): The usefulness of both genomes for population genetic studies. *Mol. Ecol.* 9: 141–154.

Dessauer, H. C. 2000. Hybridization among western whiptail lizards (*Cnemidophorus tigris*) in southwestern New Mexico: Population genetics, morphology, and ecology in three contact zones. *Bull. Am. Mus. Nat. Hist.* 246: 1–148.

Dessauer, H. C. and C. J. Cole. 1986. Clonal inheritance in parthenogenetic whiptail lizards: Biochemical evidence. *J. Heredity* 77: 8–12.

Dessauer, H. C. and C. J. Cole. 1989. Diversity between and within nominal forms of unisexual teiid lizards. Pp. 49–71 in *Evolution and Ecology of Unisexual Vertebrates*, R. M. Dayley and J. P. Bogart (eds.). New York State Museum, Albany, NY.

Devlin, B. and N. C. Ellstrand. 1990. The development and application of a refined method for estimating gene flow from angiosperm paternity analysis. *Evolution* 44: 248–259.

Devlin, B., N. Risch and K. Roeder. 1991. Estimation of allele frequencies for VNTR loci. *Am. J. Human Genet.* 48: 662–676.

Devlin, B., N. Risch and K. Roeder. 1992. Forensic inference from DNA fingerprints. *J. Am. Stat. Assn.* 87: 337–350.

Devlin, B., C. A. Biagi, T. Y. Yesaki, D. E. Smailus and J. C. Byatt. 2001. Growth of domesticated transgenic fish: A growth-hormone transgene boosts the size of wild but not domesticated trout. *Nature* 409: 781–782.

deVries, H. 1910. *The Mutation Theory*. Translated by J. B. Farmer and A. D. Darbishire. Open Court, Chicago, IL.

DeWinter, A. J. 1992. The genetic basis and evolution of acoustic mate recognition signals in a *Ribautodelphax planthopper* (Homoptera, Delphacidae) I. The female call. *J. Evol. Biol.* 5: 249–265.

DeWoody, J. A. and J. C. Avise. 2000. Microsatellite variation in marine, freshwater and anadromous fishes compared with other animals. *J. Fish Biol.* 56: 461–473.

DeWoody, J. A. and J. C. Avise. 2001. Genetic perspectives on the natural history of fish mating systems. *J. Heredity* 92: 167–172.

DeWoody, J. A., D. E. Fletcher, S. D. Wilkins, W. S. Nelson and J. C. Avise. 1998. Molecular genetic dissection of spawning, parentage, and reproductive tactics in a population of redbreast sunfish, *Lepomis auritus. Evolution* 52: 1802–1810.

DeWoody, J. A., Y. D. DeWoody, A. C. Fiumera and J. C. Avise. 2000a. On the number of reproductives contributing to a half-sib progeny array. *Genet. Res. Camb.* 75: 95–105.

DeWoody, J. A., D. Fletcher, M. Mackiewicz, S. D. Wilkins and J. C. Avise. 2000b. The genetic mating system of spotted sunfish (*Lepomis punctatus*): Mate numbers and the influence of male reproductive parasites. *Mol. Ecol.* 9: 2119–2128.

DeWoody, J. A., D. Fletcher, S. D. Wilkins and J. C. Avise. 2000c. Parentage and nest guarding in the tessellated darter (*Etheostoma olmstedi*) assayed by microsatellite markers (Perciformes: Percidae). *Copeia* 2000: 740–747.

DeWoody, J. A., D. Walker and J. C. Avise. 2000d. Genetic parentage in large half-sib clutches: Theoretical estimates and empirical appraisals. *Genetics* 154: 1907–1912.

DeWoody, J. A., D. E. Fletcher, S. D. Wilkins, W. S. Nelson and J. C. Avise. 2000e. Genetic monogamy and biparental care in an externally fertilizing fish, the largemouth bass (*Micropterus salmoides*). *Proc. R. Soc. Lond. B* 267: 2431–2437.

DeWoody, J. A., D. E. Fletcher, S. D. Wilkins and J. C. Avise. 2001. Genetic documentation of filial cannibalism in nature. *Proc. Natl. Acad. Sci. USA* 98: 5090–5092.

Dewsbury, D. A. 1982. Dominance rank, copulatory behavior, and differential reproduction. *Q. Rev. Biol.* 57: 135–159.

Diamond, J. 1992. The mysterious origin of AIDS. *Nat. Hist.* 101(9): 24–29.

Diamond, J. M. and J. I. Rotter. 1987. Observing the founder effect in human evolution. *Nature* 329: 105–106.

Dickinson, J., J. Haydock, W. Koenig, M. Stanback and F. Pitelka. 1995. Genetic monogamy in single-male groups of acron woodpeckers, *Melanerpes formicivorus. Mol. Ecol.* 4: 765–769.

Dickson, G. W., J. C. Patton, J. R. Holsinger and J. C. Avise. 1979. Genetic variation in cave-dwelling and deep-sea organisms, with emphasis on *Crangonyx antennatus* (Crustacea: Amphipoda) in Virginia. *Brimleyana* 2: 119–130.

Dieckmann, U. and M. Doebeli. 1999. On the origin of species by sympatric speciation. *Nature* 400: 354–357.

Dieckmann, U., H. Metz, M. Doebeli and D. Taut (eds.). 2001. *Adaptive Speciation*. Cambridge University Press, Cambridge, England.

Diehl, S. R. and G. L. Bush. 1989. The role of habitat preference in adaptation and speciation. Pp. 345–365 in *Speciation and Its Consequences*, D. Otte and J. A. Endler (eds.). Sinauer Associates, Sunderland, MA.

Dillon, R. T. Jr. 1988. Evolution from transplants between genetically distinct populations of freshwater snails. *Genetica* 76: 111–119.

DiMichele, L., J. A. DiMichele and D. A. Powers. 1986. Developmental and physiological consequences of genetic variation at enzyme synthesizing loci in *Fundulus heteroclitus*. *Am. Zool.* 26: 201–210.

DiMichele, L., K. Paynter and D. A. Powers. 1991. Lactate dehydrogenase-B allozymes directly effect development of *Fundulus heteroclitus*. *Science* 253: 898–900.

Dimmick, W. W., M. J. Ghedotti, M. J. Grose, A. M. Maglia, D. J. Meinhardt and D. S. Pennock. 1999. The importance of systematic biology in defining units of conservation. *Cons. Biol.* 13: 653–660.

Dinerstein, E. and G. F. McCracken. 1990. Endangered greater one-horned rhinoceros carry high levels of genetic variation. *Cons. Biol.* 4: 417–422.

DiRienzo, A. and A. C. Wilson. 1991. Branching pattern in the evolutionary tree for human mitochondrial DNA. *Proc. Natl. Acad. Sci. USA* 88: 1597–1601.

Ditchfield, A. D. 2000. The comparative phylogeography of Neotropical mammals: Patterns of intraspecific mitochondrial DNA variation among bats contrasted to nonvolant small mammals. *Mol. Ecol.* 9: 1307–1318.

Dixon, A., D. Ross, S. L. C. O'Malley and T. Burke. 1994. Paternal investment inversely related to degree of extra-pair paternity in the reed bunting. *Nature* 371: 698–700.

Dizon, A. E., C. Lockyer, W. F. Perrin, D. P. Demaster and J. Sisson. 1992. Rethinking the stock concept: A phylogeographic approach. *Cons. Biol.* 6: 24–36.

Dobson, F. S., R. K. Chesser, J. L. Hoogland, D. W. Sugg and D. W. Foltz. 1997. Do black-tailed prairie dogs minimize inbreeding? *Evolution* 51: 970–978.

Dobson, F. S., R. K. Chesser, J. L. Hoogland, D. W. Sugg and D. W. Foltz. 1998. Breeding groups and gene dynamics in a socially structured population of prairie dogs. *J. Mammal.* 79: 671–680.

Dobzhansky, T. 1937. *Genetics and the Origin of Species*. Columbia University Press, New York.

Dobzhansky, T. 1940. Speciation as a stage in evolutionary divergence. *Am. Nat.* 74: 312–321.

Dobzhansky, T. 1951. *Genetics and the Origin of Species* (3rd Edition). Columbia University Press, New York.

Dobzhansky, T. 1955. A review of some fundamental concepts and problems of population genetics. Cold Spring Harbor Symp. *Quant. Biol.* 20: 1–15.

Dobzhansky, T. 1970. *Genetics of the Evolutionary Process*. Columbia University Press, New York.

Dobzhansky, T. 1973. Nothing in biology makes sense except in the light of evolution. *Am. Biol. Teacher* 35: 125–129.

Dobzhansky, T. 1974. Genetic analysis of hybrid sterility within the species *Drosophila pseudoobscura*. *Hereditas* 77: 81–88.

Dobzhansky, T. 1976. Organismic and molecular aspects of species formation. Pp. 95–105 in *Molecular Evolution*, F. J. Ayala (ed.). Sinauer Associates, Sunderland, MA.

Dodd, B. E. 1985. DNA fingerprinting in matters of family and crime. *Nature* 318: 506–507.

Dodds, K. G., M. L. Tate, J. C. McEwen and A. M. Crawford. 1996. Exclusion probabilities for pedigree testing farm animals. *Theoret. Appl. Genet.* 92: 966–975.

Doebeli, M. 1996. A quantitative genetic competition model for sympatric speciation. *J. Evol. Biol.* 9: 893–909.

Doebley, J. F. 1989. Molecular evidence for a missing wild relative of maize and the introgression of its chloroplast genome into *Zea perennis*. *Evolution* 43: 1555–1558.

Doherty, P. J., S. Planes and P. Mather. 1995. Gene flow and larval duration in seven species of fish from the Great Barrier Reef. *Ecology* 76: 2373–2391.

Dole, J. A. and M. Sun. 1992. Field and genetic survey of the endangered Butte County meadowfoam: *Limnanthes floccosa* subsp. *californica* (Limnanthaceae). *Cons. Biol.* 6: 549–558.

Dong, J. and 6 others. 1992. Paternal chloroplast DNA inheritance in *Pinus contorta* and *Pinus banksiana*: Independence of parental species or cross direction. *J. Heredity* 83: 419–422.

Donnelly, P. and S. Tavaré (eds.). 1997. *Progress in Population Genetics and Human Evolution*. Springer-Verlag, New York.

Donoghue, M. J. 1985. A critique of the biological species concept and recommendations for a phylogenetic alternative. *Bryologist* 88: 172–181.

Doolittle, R. F. 1987. The evolution of the vertebrate plasma proteins. *Biol. Bull.* 172: 269–283.

Doolittle, R. F., D.-F. Feng, M. S. Johnson and M. A. McClure. 1989. Origins and evolutionary relationships of retroviruses. *Q. Rev. Biol.* 64: 1–30.

Doolittle, R. F., D.-F. Feng, K. L. Anderson and M. R. Alberro. 1990. A naturally occurring horizontal gene transfer from a eukaryote to a prokaryote. *J. Mol. Evol.* 31: 383–388.

Doolittle, R. F., D.-F. Feng, S. Tsang, G. Cho and E. Little. 1996. Determining divergence times of the major kingdoms of living organisms with a protein clock. *Science* 271: 470–477.

Doolittle, W. F. 1998. You are what you eat: A gene transfer ratchet could account for bacterial genes in eukaryotic nuclear genomes. *Trends Genet.* 14: 307–311.

Doolittle, W. F. 1999. Phylogenetic classification and the universal tree. *Science* 284: 2124–2128.

Doolittle, W. F., Y. Boucher, C. L. Nesbø, C. J. Douady, J. O. Andersson and A. J. Roger. 2003. How big is the iceberg of which organellar genes in nuclear genomes are but the tip? *Phil. Trans. R. Soc. Lond. B* 358: 39–58.

Doran, G. H., D. N. Dickel, W. E. Ballinger Jr., O. F. Agee, P. J. Laipis and W. W. Hauswirth. 1986. Anatomical, cellular and molecular analysis of 8,000-yr-old human brain tissue from the Windover archaeological site. *Nature* 323: 803–806.

Dorit, R. L., H. Akashi and W. Gilbert. 1995. Absence of polymorphism at the ZFY locus on human chromosome 22. *Nature* 402: 489–495.

Doty, P., J. Marmur, J. Eigner and C. Schildkraut. 1960. Strand separation and specific recombination in deoxyribonucleic acids: Physical chemical studies. *Proc. Natl. Acad. Sci. USA* 46: 461–476.

Douady, C. J., M. Dosay, M. S. Shivji and M. J. Stanhope. 2003. Molecular phylogenetic evidence refuting the hypothesis of Batoidea (rays and skates) as derived sharks. *Mol. Phylogen. Evol.* 26: 215–221.

Double, M. and P. Olsen. 1997. Simplified polymerase chain reaction (PCR)-based sexing assists conservation of an endangered owl, Norfolk Island boobook *Ninox novaeseelandiae undulata. Bird Cons. Int.* 7: 283–286.

Double, M. C. and A. Cockburn. 2000. Pre-dawn infidelity: Females control extra-pair mating in superb fairywrens. *Proc. R. Soc. Lond. B* 267: 465–470.

Double, M. C., A. Cockburn, S. C. Barry and P. E. Smouse. 1997. Exclusion probabilities for single-locus paternity analysis when related males compete for matings. *Mol. Ecol.* 6: 1155–1166.

Douglas, M. E. and J. C. Avise. 1982. Speciation rates and morphological divergence in fishes: Tests of gradual versus rectangular modes of evolutionary change. *Evolution* 36: 224–232.

Dover, G. A. 1982. Molecular drive: A cohesive mode of species evolution. *Nature* 299: 111–117.

Dow, B. D. and M. V. Ashley. 1998. High levels of gene flow in bur oak revealed by paternity analysis using microsatellites. *J. Heredity* 89: 62–70.

Dowling, T. E. and B. D. DeMarais. 1993. Evolutionary significance of introgressive hybridization in cyprinid fishes. *Nature* 362: 444–446.

Dowling, T. E. and W. R. Hoeh. 1991. The extent of introgression outside the contact zone between *Notropis cornutus* and *Notropis chrysocephalus* (Teleostei: Cyprinidae). *Evolution* 45: 944–956.

Dowling, T. E. and C. L. Secor. 1997. The role of hybridization and introgression in the diversification of animals. *Annu. Rev. Ecol. Syst.* 28: 593–619.

Dowling, T. E., G. R. Smith and W. M. Brown. 1989. Reproductive isolation and introgression between *Notropis cornutus* and *Notropis chrysocephalus* (family Cyprinidae): Comparison of morphology, allozymes, and mitochondrial DNA. *Evolution* 43: 620–634.

Dowling, T. E., W. L. Minckley, M. E. Douglas, P. C. Marsh and B. D. Demarais. 1992. Response to Wayne, Nowak, and Phillips and Henry: Use of molecular characters in conservation biology. *Cons. Biol.* 6: 600–603.

Dowling, T. E., W. L. Minckley and P. C. Marsh. 1996a. Mitochondrial DNA diversity within and among populations of razorback sucker, *Xyrauchen texanus*, as determined by restriction endonuclease analysis. *Copeia* 1996: 542–550.

Dowling, T. E., W. L. Minckley, P. C. Marsh and E. S. Goldstein. 1996b. Mitochondrial DNA variability in the endangered razorback sucker (*Xyrauchen texanus*): Analysis of hatchery stocks and implications for captive propagation. *Cons. Biol.* 10: 120–127.

Dowling, T. E., C. Moritz, J. D. Palmer and L. H. Rieseberg. 1996c. Nucleic acids III: Analysis of fragments and restriction sites. Pp. 249–320 in *Molecular Systematics* (2nd Edition), D. M. Hillis, C. Moritz and B. K. Mable (eds.). Sinauer Associates, Sunderland, MA.

Dowling, T. E., R. E. Broughton and B. D. DeMarais. 1997. Significant role for historical effects in the evolution of reproductive isolation: Evidence from patterns of introgression between the cyprinid fishes, *Luxilus cornutus* and *Luxilus chrysocephalus. Evolution* 51: 1574–1583.

Downie, S. R. and J. D. Palmer. 1992. Use of chloroplast DNA rearrangements in reconstructing plant phylogeny. Pp. 14–35 in *Molecular Systematics of Plants*, P. S. Soltis, D. E. Soltis and J. J. Doyle (eds.). Chapman & Hall, New York.

Downie, S. R., R. G. Olmstead, G. Zurawski, D. E. Soltis, P. S. Soltis, J. C. Watson and J. D. Palmer. 1991. Six independent losses of the chloroplast DNA *rpl*2 intron in dicotyledons: Molecular and phylogenetic implications. *Evolution* 45: 1245–1259.

Doyle, J. J. 1992. Gene trees and species trees: Molecular systematics as one-character taxonomy. *Syst. Bot.* 17: 144–163.

Doyle, J. J., J. L. Doyle, A. H. D. Brown and J. P. Grace. 1990. Multiple origins of polyploids in the *Glycine tabacina* complex inferred from chloroplast DNA polymorphism. *Proc. Natl. Acad. Sci. USA* 87: 714–717.

Doyle, J. J., M. Lavin and A. Bruneau. 1992. Contributions of molecular data to papilionoid legume systematics. Pp. 223–251 in *Molecular Systematics of Plants*, P. S. Soltis, D. E. Soltis and J. J. Doyle (eds.). Chapman & Hall, New York.

DuBose, R. F., D. E. Dykhuisen and D. L. Hartl. 1988. Genetic exchange among natural isolates of bacteria: Recombination within the *phoA* gene of *Escherichia coli. Proc. Natl. Acad. Sci. USA* 85: 7036–7040.

Duchesne, P., M.-H. Godbout and L. Bernatchez. 2002. PAPA (Package for the Analysis of Parental Allocation): A computer program for simulated and real parental allocation. *Mol. Ecol. Notes* 2: 191–193.

Duellman, W. E. and D. M. Hillis. 1987. Marsupial frogs (Anura: Hylidae: *Gastrotheca*) of the Ecuadorian Andes: Resolution of taxonomic problems and phylogenetic relationships. *Herpetologica* 43: 135–167.

Duffy, J. E. 1996. Eusociality in a coral-reef shrimp. *Nature* 381: 512–514.

Duffy, J. E., C. L. Morrison and R. Ríos. 2000. Multiple origins of eusociality among sponge-dwelling shrimps (*Synalpheus*). *Evolution* 54: 503–516.

Duffy, J. E., C. L. Morrison and K. S. MacDonald. 2002. Colony defense and behavioral differentiation in the eusocial shrimp *Synalpheus regalis. Behav. Ecol. Sociobiol.* 51: 488–495.

Duggins, G. F. Jr., A. A. Karlin, T. A. Mousseau and K. G. Relyea. 1995. Analysis of a hybrid zone in *Fundulis majalis* in a northeastern Florida ecotone. *Heredity* 74: 117–128.

Dumolin-Lapegue, S., A. Kremer and R. J. Petit. 1999a. Are chloroplast and mitochondrial DNA variation species independent in oaks? *Evolution* 53: 1406–1413.

Dumolin-Lapegue, S., M.-H. Pemonge, L. Geilly, P. Taberlet and R. J. Petit. 1999b. Amplification of oak DNA from ancient and modern wood. *Mol. Ecol.* 8: 2137–2140.

Dunning, A. M., P. Talmud and S. E. Humphries. 1988. Errors in the polymerase chain reaction. *Nucleic Acids Res.* 16: 10393.

Dutton, P. H., S. K. Davis, T. Guerra and D. Owens. 1996. Molecular phylogeny for marine turtles based on sequences of the ND4-leucine tRNA and control regions of mitochondrial DNA. *Mol. Phylogen. Evol.* 5: 511–521.

Dutton, P. H., B. W. Bowen, D. W. Owens, A. Barragan and S. K. Davis. 1999. Global phylogeography of the leatherback turtle (*Dermochelys coriacea*). *J. Zool. Lond.* 248: 397–409.

Duvall, M. R. and others. 1993. Phylogenetic hypotheses for the monocotyledons constructed from *rbc*L sequence data. *Annals Missouri Bot. Garden* 80: 607–619.

Duvall, S. W., I. S. Bernstein and T. P. Gordon. 1976. Paternity and status in a rhesus monkey group. *J. Reprod. Fert.* 47: 25–31.

Duvernell, D. D. and N. Aspinwall. 1995. Introgression of *Luxilus cornutus* mtDNA into allopatric populations of *Luxilus chrysocephalus* (Teleostei: Cyprinidae) in Missouri and Arkansas. *Mol. Ecol.* 4: 173–181.

Dvornyk, V., O. Vinogradova and E. Nevo. 2003. Origin and evolution of circadian clock genes in prokaryotes. *Proc. Natl. Acad. Sci. USA* 100: 2495–2500.

Eanes, W. F. 1994. Patterns of polymorphism and between species divergence in the enzymes of central metabolism. Pp. 18–28 in *Non-Neutral Evolution*, B. Golding (ed.). Chapman & Hall, New York.

Eanes, W. F. and R. K. Koehn. 1978. Relationship between subunit size and number of rare electrophoretic alleles in human enzymes. *Biochem. Genet.* 16: 971–985.

Eanes, W. F., M. Kirchner and J. Yoon. 1993. Evidence for adaptive evolution of the *G6pd* gene in the *Drosophila melanogaster* and *Drosophila simulans* lineages. *Proc. Natl. Acad. Sci. USA* 90: 7475–7479.

East, M. I., T. Burke, K. Wilhelm, C. Greig and H. Hofer. 2002. Sexual conflicts in spotted hyenas: Male and female mating tactics and their reproductive outcome with respect to age, social status and tenure. *Proc. R. Soc. Lond. B* 270: 1247–1254.

Easteal, S. 1991. The relative rate of DNA evolution in primates. *Mol. Biol. Evol.* 8: 115–127.

Eaves, A. A. and A. R. Palmer. 2003. Widespread cloning in echinoderm larvae. *Nature* 425: 146.

Eberhard, W. G. 1980. Evolutionary consequences of intracellular organelle competition. *Q. Rev. Biol.* 55: 231–249.

Eberhard, W. G. 1998. Female roles in sperm competition. Pp. 91–116 in *Sperm Competition and Sexual Selection*, T. R. Birkhead and A. P. Møller (eds.). Academic Press, London.

Echelle, A. A. and P. J. Connor. 1989. Rapid, geographically extensive genetic introgression after secondary contact between two pupfish species (Cyprinodon, Cyprinodontidae). *Evolution* 43: 717–727.

Echelle, A. A. and A. F. Echelle. 1984. Evolutionary genetics of a "species flock: " atherinid fishes on the Mesa Central of Mexico. Pp. 93–110 in *Evolution of Fish Species Flocks*, A. A. Echelle and I. Kornfield (eds.). University of Maine Press, Orono, ME.

Echelle, A. A. and I. Kornfield (eds.). 1984. *Evolution of Fish Species Flocks*. University of Maine Press, Orono, ME.

Echelle, A. A. and D. T. Mosier. 1981. All-female fish: A cryptic species of *Menidia* (Atherinidae). *Science* 212: 1411–1413.

Echelle, A. A., A. F. Echelle and D. R. Edds. 1987. Population structure of four pupfish species (Cyprinodontidae: *Cyprinodon*) from the Chihuahuan desert region of New Mexico and Texas: Allozymic variation. *Copeia* 1987: 668–681.

Echelle, A. A., T. E. Dowling, C. C. Moritz and W. M. Brown. 1989. Mitochondrial-DNA diversity and the origin of the *Menidia clarkhubbsi* complex of unisexual fishes (Atherinidae). *Evolution* 43: 984–993.

Echelle, A. A., C. W. Hoagstrom, A. F. Echelle and J. E. Brooks. 1997. Expanded occurrence of genetically introgressed pupfish (Cyprinodontidae: *Cyprinodon pecosensis* X *variegatus*) in New Mexico. *Southwest. Nat.* 42: 336–339.

Echelle, A. F., A. A. Echelle and D. R. Edds. 1989. Conservation genetics of a spring-dwelling desert fish, the Pecos gambusia (*Gambusia nobilis*, Poeciliidae). *Cons. Biol.* 3: 159–169.

Edman, J. C., J. A. Kovacs, H. Masur, D. V. Santi, H. J. Elwood and M. L. Sogin. 1988. Ribosomal RNA sequence shows *Pneumocystis carinii* to be a member of the fungi. *Nature* 334: 519–522.

Edmands, S. 1999. Heterosis and outbreeding depression in interpopulation crosses spanning a wide range of divergence. *Evolution* 53: 1757–1768.

Edmands, S., P. E. Moberg and R. S. Burton. 1996. Allozyme and mitochondrial DNA evidence of population subdivision in the purple sea urchin *Strongylocentrotus purpuratus*. *Marine Biol.* 126: 443–450.

Edwards, K. J. 1998. Ramdomly amplified polymorphic DNAs (RAPDs). Pp. 171–205 in *Molecular Tools for Screening Biodiversity*, A. Karp, P. G. Isaac and D. S. Ingram (eds.). Chapman & Hall, London.

Edwards, S. V. and P. Beerli. 2000. Perspective: Gene divergence, population divergence, and the variance in coalescence time in phylogeographic studies. *Evolution* 54: 1839–1854.

Edwards, S. V., J. Nusser and J. Gasper. 2000. Characterization and evolution of major histocompatibility complex (MHC) genes in non-model organisms, with examples from birds. Pp. 168–207 in *Molecular Methods in Ecology*, A. J. Baker (ed.). Blackwell, London.

Edwardson, J. R. 1970. Cytoplasmic male sterility. *Bot. Rev.* 36: 341–420.

Eggert, L. S., C. A. Rasner and D. S. Woodruff. 2002. The evolution and phylogeography of the African elephant inferred from mitochondrial DNA sequence and nuclear microsatellite markers. *Proc. R. Soc. Lond. B* 269: 1993–2006.

Ehrlich, P. R. 1975. The population biology of coral reef fishes. *Annu. Rev. Ecol. Syst.* 6: 211–248.

Ehrlich, P. R. 1992. Population biology of checkerspot butterflies and the preservation of global biodiversity. *Oikos* 63: 6–12.

Ehrlich, P. R. and A. H. Ehrlich. 1991. *Healing the Planet.* Addison-Wesley, Reading, MA.

Ehrlich, P. R. and P. H. Raven. 1969. Differentiation of populations. *Science* 165: 1228–1232.

Ehrlich, P. R. and E. O. Wilson. 1991. Biodiversity studies: Science and policy. *Science* 253: 758–761.

Eisen, J. A. 2000. Horizontal gene transfer among microbial genomes: New insights from complete genome analysis. *Curr. Opin. Genet. Develop.* 10: 606–611.

Eizirik,, E. and 7 others. 1998. Phylogeographic patterns and evolution of the mitochondrial DNA control region in two neotropical cats (Mammalia, Felidae). *J. Mol. Evol.* 47: 613–624.

Eldredge, N. and J. Cracraft. 1980. *Phylogenetic Patterns and the Evolutionary Process.* Columbia University Press, New York.

Eldredge, N. and S. J. Gould. 1972. Punctuated equilibria: An alternative to phyletic gradualism. Pp. 82–115 in *Models in Paleobiology,* T. J. M. Schopf (ed.). Freeman, Cooper and Co., San Francisco, CA.

Ellegren, H. 1991. Fingerprinting birds' DNA with a synthetic polynucleotide probe $(TG)_n$. *Auk* 108: 956–958.

Ellegren, H. 1996. First gene on the avian W chromosome (CHD) provides a tag for universal sexing of non-ratite birds. *Proc. R. Soc. Lond. B* 263: 1635–1644.

Ellegren, H. and B. C. Sheldon. 1997. New tools for sex identification and the study of sex allocation in birds. *Trends Ecol. Evol.* 12: 255–259.

Ellegren, H., G. Hartman, M. Johansson and L. Andersson. 1993. Major histocompatibility complex monomorphism and low levels of DNA fingerprinting variability in a reintroduced and rapidly expanding population of beavers. *Proc. Natl. Acad. Sci. USA* 90: 8150–8153.

Ellis, N., A. Taylor, B. O. Bengtsson, J. Kidd, J. Rogers and P. Goodfellow. 1990. Population structure of the human pseudoautosomal boundary. *Nature* 34: 663–665.

Ellis, S. and U. S. Seal. 1995. Tools of the trade to aid decision-making for species survival. *Biodiversity and Cons.* 4: 553–572.

Ellstrand, N. C. 1984. Multiple paternity within the fruits of the wild radish, *Raphanus sativus. Am. Nat.* 123: 819–828.

Ellstrand, N. C. 2003. *Dangerous Liaisons? When Cultivated Plants Mate with their Wild Relatives.* Johns Hopkins University Press, Baltimore, MD.

Ellstrand, N. C. and D. L. Marshall. 1985. Interpopulation gene flow by pollen in wild radish, *Raphanus sativus. Am. Nat.* 126: 606–616.

Ellstrand, N. C. and M. L. Roose. 1987. Patterns of genotypic diversity in clonal plant species. *Am. J. Bot.* 74: 123–131.

Ellstrand, N. C., R. Whitkus and L. H. Rieseberg. 1996. Distribution of spontaneous plant hybrids. *Proc. Natl. Acad. Sci. USA* 93: 5090–5093.

Elson, J. L., R. M. Andrews, P. F. Chinnery, R. N. Lightowlers, D. M. Turnbull and N. Howell. 2001. Analysis of European mtDNAs for recombination. *Am. J. Human Genet.* 68: 145–153.

Emerson, B. C. 2002. Evolution on oceanic islands: Molecular phylogenetic approaches to understanding pattern and process. *Mol. Ecol.* 11: 951–966.

Emery, A. M., I. J. Wilson, S. Craig, P. R. Boyle and L. R. Noble. 2001. Assignment of paternity groups without access to parental genotypes: Multiple mating and developmental plasticity in squid. *Mol. Ecol.* 10: 1265–1278.

Emms, S. K. and M. L. Arnold. 1997. The effect of habitat on parental and hybrid fitness: Transplant experiments with Louisiana irises. *Evolution* 51: 1112–1119.

Emms, S. K., S. A. Hodges and M. L. Arnold. 1996. Pollen-tube competition, siring success, and consistent asymmetric hybridization in Louisiana irises. *Evolution* 50: 2201–2206.

Enard, W. and 12 others. 2002a. Intra- and interspecific variation in primate gene expression patterns. *Science* 296: 340–343.

Enard, W. and 7 others. 2002b. Molecular evolution of *FOXP2,* a gene involved in speech and language. *Nature* 418: 869–872.

Encalada, S. E., P. N. Lahanas, K. A. Bjorndal, A. B. Bolten, M. M. Miyamoto and B. W. Bowen. 1996. Phylogeography and population structure of the Atlantic and Mediterranean green turtle *Chelonia mydas*: A mitochondrial DNA control region sequence assessment. *Mol. Ecol.* 5: 473–483.

Encalada, S. E. and 7 others. 1998. Population structure of loggerhead turtle (*Caretta caretta*) nesting colonies in the Atlantic and Mediterranean as inferred from mitochondrial DNA control region sequences. *Marine Biol.* 130: 567–575.

Endler, J. A. 1977. *Geographic Variation, Speciation, and Clines.* Princeton University Press, Princeton, NJ.

Enghoff, H. 1995. Historical biogeography of the Holarctic: Area relationships, ancestral areas, and dispersal of non-marine animals. *Cladistics* 11: 223–263.

Ennis, P. D., J. Zemmour, R. D. Salter and P. Parham. 1990. Rapid cloning of HLA-A,B cDNA by using the polymerase chain reaction: Frequency and nature of errors produced in amplification. *Proc. Natl. Acad. Sci. USA* 87: 2833–2837.

Ennos, R. A. and M. T. Clegg. 1982. Effect of population substructuring on estimates of outcrossing rate in plant populations. *Heredity* 48: 283–292.

Ennos, R. A. and K. W. Swales. 1987. Estimation of the mating system in a fungal pathogen *Crumenulopsis sororia* (Karst.) Groves using isozyme markers. *Heredity* 59: 423–430.

Epperson, B. K. 1993. Recent advances in correlation studies of spatial patterns of genetic variation. *Evol. Biol.* 27: 95–155.

Epperson, B. K. 2003. *Geographical Genetics*. Princeton University Press, Princeton, NJ.

Epperson, B. K. and M. T. Clegg. 1987. First-pollination primacy and pollen selection in the morning glory, *Ipomoea purpurea*. *Heredity* 58: 5–14.

Epstein, M. P., W. L. Duren and M. Boehnke. 2000. Improved inference of relationships for pairs of individuals. *Am. J. Human Genet.* 67: 1219–1231.

Ereshefsky, M. 2001. *The Poverty of the Linnaean Hierarchy*. Cambridge University Press, Cambridge, England.

Ericson, P. G. P., U. S. Johansson and T. J. Parsons. 2000. Major divisions of oscines revealed by insertions in the nuclear gene c-*myc*: A novel gene in avian phylogenetics. *Auk* 117: 1077–1086.

Ericson, P. G. P. and 6 others. 2002. A Gondwanan origin of passerine birds supported by DNA sequences of the endemic New Zealand wrens. *Proc. R. Soc. Lond. B* 269: 235–241.

Erlich, H. A. and N. Arnheim. 1992. Genetic analysis using the polymerase chain reaction. *Annu. Rev. Genet.* 26: 479–506.

Erlich, H. A., D. Gelfand and J. J. Sninsky. 1991. Recent advances in the polymerase chain reaction. *Science* 252: 1643–1651.

Ernest, H. B., M. C. T. Penedo, B. P. May, M. Syvanen and W. M. Boyce. 2000. Molecular tracking of mountain lions in the Yosemite Valley region of California: Genetic analysis using microsatellites and faecal DNA. *Mol. Ecol.* 9: 433–441.

Erpenbeck, D., J. A. J. Breeuwer, H. C. van der Velde and R. W. M. van Soest. 2002. Unravelling host and symbiont phylogenies of halichondrid sponges (Demospongiae, Porifera) using a mitochondrial marker. *Marine Biol.* 141: 377–386.

Erwin, D., J. W. Valentine and D. Jablonski. 1997. The origin of animal body plans. *Am. Sci.* 85: 126–137.

Erwin, T. L. 1991. An evolutionary basis for conservation strategies. *Science* 256: 193–197.

Escalante, A. A., A. A. Lal and F. J. Ayala. 1998. Genetic polymorphism and natural selection in the malaria parasite *Plasmodium falciparum*. *Genetics* 149: 189–202.

Estoup, A. and B. Angers. 1998. Microsatellites and minisatellites for molecular ecology: Theoretical and empirical considerations. Pp. 55–86 in *Advances in Molecular Ecology*, G. Carvalho (ed.). IOS Press, Amsterdam.

Estoup, A., L. Garnery, M. Solignac and J.-M. Cornuet. 1995. Microsatellite variation in honey bee (*Apis mellifera* L.) populations: Hierarchical genetic structure and test of the infinite allele and stepwise mutation models. *Genetics* 140: 679–695.

Estoup, A., P. Jarne and J.-M. Cornuet. 2002. Homoplasy and mutation model at microsatellite loci and their consequences for population genetics analysis. *Mol. Ecol.* 11: 1591–1604.

Evans, B. J., J. Supriatna and D. J. Melnick. 2001. Hybridization and population genetics of two macaque species in Sulawesi, Indonesia. *Evolution* 55: 1686–1702.

Evans, B. J., J. Supriatna, N. Andayani, M. I. Setiadi, D. C. Cannatella and D. J. Melnick. 2003. Monkeys and toads define areas of endemism on Sulawesi. *Evolution* 57: 1436–1433.

Evans, B. S., R. W. G. White and R. D. Ward. 1998. Genetic identification of asteroid larvae from Tasmania, Australia, by PCR-RFLP. *Mol. Ecol.* 7: 1077–1082.

Evarts, S. and C. J. Williams. 1987. Multiple paternity in a wild population of mallards. *Auk* 104: 597–602.

Excoffier, L. 1990. Evolution of human mitochondrial DNA: Evidence for departure from a pure neutral model of populations at equilibrium. *J. Mol. Evol.* 30: 125–139.

Eyre-Walker, A. 1997. Differentiating between selection and mutation bias. *Genetics* 147: 1983–1987.

Eyre-Walker, A., N. H. Smith and J. Maynard Smith. 1999. How clonal are human mitochondria? *Proc. R. Soc. Lond. B* 266: 477–483.

Faaborg, J. and 7 others. 1995. Confirmation of cooperative polyandry in the Galapagos hawk (*Buteo galapagoensis*). *Behav. Ecol. Sociobiol.* 36: 83–90.

Fabiani, A., A. R. Hoelzel, F. Galimberti and M. M. C. Muelbert. 2003. Long-range paternal gene flow in the southern elephant seal. *Science* 299: 676.

Faith, D. P. 1992. Conservation evaluation and phylogenetic diversity. *Biol. Cons.* 61: 1–10.

Faith, D. P. 1993. Systematics and conservation: On predicting the feature diversity of subsets of taxa. *Cladistics* 8: 361–373.

Faith, D. P. 1994. Genetic diversity and taxonomic priorities for conservation. *Biol. Cons.* 68: 69–74.

Falk, D. A. and K. E. Holsinger (eds.). 1991. *Genetics and Conservation of Rare Plants*. Oxford University Press, New York.

Fang, S. and Q. Wan. 2002. A genetic fingerprinting test for identifying carcasses of protected deer species in China. *Biol. Cons.* 103: 371–372.

Farias, I. P., G. Ortí, I. Sampaio, H. Schneider and A. Meyer. 1999. Mitochondrial DNA phylogeny of the family Cichlidae: Monophyly and high genetic divergence of the neotropical assemblage. *J. Mol. Evol.* 48: 703–711.

Farrell, B. D. 1998. "Inordinate fondness" explained: Why are there so many beetles? *Science* 281: 555–559.

Farrell, L. E., J. Roman and M. E. Sunquist. 2000. Dietary separation of sympatric carnivores identified by molecular analysis of scat. *Mol. Ecol.* 9: 1583–1590.

Farris, J. S. 1970. Methods for computing Wagner trees. *Syst. Zool.* 34: 21–34.

Farris, J. S. 1971. The hypothesis of nonspecificity and taxonomic congruence. *Annu. Rev. Ecol. Syst.* 2: 227–302.

Farris, J. S. 1972. Estimating phylogenetic trees from distance matrices. *Am. Nat.* 106: 645–668.

Farris, J. S. 1973. On comparing the shapes of taxomomic trees. *Syst. Zool.* 22: 50–54.

Farris, J. S. 1977. Phylogenetic analysis under Dollo's Law. *Syst. Zool.* 26: 77–88.

Faulkes, C. G. and 6 others. 1997. Micro- and macrogeographical genetic structure of colonies of naked mole-rats *Heterocephalus glaber*. *Mol. Ecol.* 6: 615–628.

Feder, J. L., C. A. Chilcote and G. L. Bush. 1988. Genetic differentiation between sympatric host races of the apple maggot fly *Rhagoletis pomonella*. *Nature* 336: 61–64.

Feder, J. L., C. A. Chilcote and G. L. Bush. 1990a. The geographic pattern of genetic differentiation between host associated populations of *Rhagoletis pomonella* (Diptera: Tephritidae) in the eastern United States and Canada. *Evolution* 44: 570–594.

Feder, J. L., C. A. Chilcote and G. L. Bush. 1990b. Regional, local and microgeographic allele frequency variation between apple and hawthorne populations of *Rhagoletis pomonella* in western Michigan. *Evolution* 44: 595–608.

Feder, J. L., J. B. Roethele, B. Wlazlo and S. H. Berlocher. 1997. Selective maintenance of allozyme differences among sympatric host races of the apple maggot fly. *Proc. Natl. Acad. Sci. USA* 94: 11417–11421.

Feder, J. L. and 9 others. 2003. Allopatric genetic origins for sympatric host-plant shifts and race formation in *Rhagoletis*. *Proc. Natl. Acad. Sci. USA* 100: 10314–10319.

Feduccia, A. 1995. Explosive evolution in Tertiary birds and mammals. *Science* 267: 637–638.

Feduccia, A. 1996. *The Origin and Evolution of Birds*. Yale University Press, New Haven, CT.

Fel-Clair, F. and 7 others. 1996. Genomic incompatibilities in the hybrid zone between house mice in Denmark: Evidence from steep and non-coincident chromosomal clines for Robertsonian fusions. *Genet. Res.* 67: 123–134.

Fel-Clair, F., J. Catalan, T. Lenormand and J. Britton-Davidian. 1998. Centromeric incompatibilities in the hybrid zone between house mouse subspecies from Denmark: Evidence from patterns of NOR activity. *Evolution* 52: 592–603.

Feldman, R. A., L. A. Freed and R. L. Cann. 1995. A PCR test for avian malaria in Hawaiian birds. *Mol. Ecol.* 4: 663–673.

Fell, J. W., A. Statzell-Tallman, M. J. Lutz and C. P. Kurtzman. 1992. Partial rRNA sequences in marine yeasts: A model for identification of marine eukaryotes. *Mol. Mar. Biol. Biotech.* 1: 175–186.

Felsenstein, J. 1974. The evolutionary advantage of recombination. *Genetics* 78: 737–756.

Felsenstein, J. 1978a. The number of evolutionary trees. *Syst. Zool.* 27: 27–33.

Felsenstein, J. 1978b. Cases in which parsimony and compatibility methods will be positively misleading. *Syst. Zool.* 27: 401–410.

Felsenstein, J. 1981a. Evolutionary trees from DNA sequences: A maximum likelihood approach. *J. Mol. Evol.* 17: 368–376.

Felsenstein, J. 1981b. Skepticism towards Santa Rosalia, or why are there so few kinds of animals. *Evolution* 35: 124–138.

Felsenstein, J. 1983. Parsimony in systematics: Biological and statistical issues. *Annu. Rev. Ecol. Syst.* 14: 313–333.

Felsenstein, J. 1985a. Confidence limits on phylogenies: An approach using the bootstrap. *Evolution* 39: 783–791.

Felsenstein, J. 1985b. Phylogenies and the comparative method. *Am. Nat.* 125: 1–15.

Felsenstein, J. 1992. Estimating effective population size from samples of sequences: Inefficiency of pairwise and segregating sites as compared to phylogenetic estimates. *Genet. Res. Camb.* 59: 139–147.

Fenster, C. B. 1991. Gene flow in *Chamaecrista fasciculata* (Leguminosae). I. Gene dispersal. *Evolution* 45: 398–409.

Fenton, B. and 7 others. 1995. Species identification of *Cecidophyopsis* mites (Arari: Eriophyidae) from different *Ribes* species and countries using molecular genetics. *Mol. Ecol.* 4: 385–387.

Fergus, C. 1991. The Florida panther verges on extinction. *Science* 251: 1178–1180.

Ferguson, A. 1989. Genetic differences among brown trout, *Salmo trutta*, stocks and their importance for the conservation and management of the species. *Freshwat. Biol.* 21: 35–46.

Ferguson, M. M. 1992. Enzyme heterozygosity and growth in rainbow trout: Genetic and physiological explanations. *Heredity* 68: 115–122.

Ferguson, M. M. and L. R. Drahushchak. 1990. Disease resistance and enzyme heterozygosity in rainbow trout. *Heredity* 64: 413–418.

Fernando, P. and D. J. Melnick. 2001. Molecular sexing eutherian mammals. *Mol. Ecol. Notes* 1: 350–353.

Fernando, P., T. N. C. Vidya, C. Rajapakse, A. Dangolla and D. J. Melnick. 2003a. Reliable noninvasive genotyping: Fantasy or reality? *J. Heredity* 94: 115–123.

Fernando, P., and 9 others. 2003b. DNA analysis indicates that Asian elephants are native to Borneo and are therefore a high priority for conservation. *PloS Biol.* 1: 110–115.

Ferraris, J. D. and S. R. Palumbi (eds.). 1996. *Molecular Zoology*. Wiley-Liss, New York.

Ferrell, R. E., D. C. Morizot, J. Horn and C. J. Carley. 1980. Biochemical markers in a species endangered by introgression: The Red Wolf. *Biochem. Genet.* 18: 39–49.

Ferris, S. D. and W. J. Berg. 1986. The utility of mitochondrial DNA in fish genetics and fishery management. Pp. 277–299 in *Population Genetics and Fishery Management*, N. Ryman and F. Utter (eds.). University of Washington Press, Seattle, WA.

Ferris, S. D. and G. S. Whitt. 1977. Loss of duplicate gene expression after polyploidization. *Nature* 265: 258–260.

Ferris, S. D. and G. S. Whitt. 1978. Phylogeny of tetraploid catostomid fishes based on the loss of duplicate gene expression. *Syst. Zool.* 27: 189–206.

Ferris, S. D. and G. S. Whitt. 1979. Evolution of the differential regulation of duplicate genes after polyploidization. *J. Mol. Evol.* 12: 267–317.

Ferris, S. D., R. D. Sage, C.-M. Huang, J. T. Nielsen, U. Ritte and A. C. Wilson. 1983a. Flow of mitochondrial DNA across a species boundary. *Proc. Natl. Acad. Sci. USA* 80: 2290–2294.

Ferris, S. D., R. D. Sage, E. M. Prager, U. Ritte and A. C. Wilson. 1983b. Mitochondrial DNA evolution in mice. *Genetics* 105: 681–721.

Fiala, K. L. and R. R. Sokal. 1985. Factors determining the accuracy of cladogram estimation: Evaluation using computer simulation. *Evolution* 39: 609–622.

Field, K. G. and 7 others. 1988. Molecular phylogeny of the animal kingdom. *Science* 239: 748–753.

Figueroa, F., E. Gunther and J. Klein. 1988. MHC polymorphisms pre-dating speciation. *Nature* 335: 265–271.

Fillatre, E. K., P. Etherton and D. D. Heath. 2003. Bimodal run distribution in a northern population of sockeye salmon (*Oncorhynchus nerka*): Life history and genetic analysis on a temporal scale. *Mol. Ecol.* 12: 1793–1805.

Fineschi, S. and 8 others. 2002. Molecular markers reveal a strong genetic differentiation between two European relic tree species: *Selkova abelicea* (Lam.) Boissier and *Z. sicula* Di Pasquale, Garfi and Quézel (Ulmaceae). *Cons. Genet.* 3: 145–153.

Finlay, B. J. 2002. Global dispersal of free-living microbial eukaryotic species. *Science* 296: 1061–1063.

Finnegan, D. J. 1983. Retroviruses and transposable elements: Which came first? *Nature* 302: 105–106.

Finnegan, D. J. 1989. Eukaryotic transposable elements and genome evolution. *Trends Genet.* 5: 103–107.

Finnerty, V. and F. H. Collins. 1988. Ribosomal DNA probes for identification of member species of the *Anopheles gambiae* complex. *Fla. Entomol.* 71: 288–294.

Firestone, K. B., M. S. Elphinstone, W. B. Sherwin and B. A. Houlden. 1999. Phylogeographical population structure of tiger quolls *Dasyurus maculatus* (Dasyuridae: Marsupialia), an endangered carnivorous marsupial. *Mol. Ecol.* 8: 1613–1625.

Fisher, R. A. 1930. *The Genetical Theory of Natural Selection.* Clarendon Press, Oxford.

Fisher, S. E., J. B. Shaklee, S. D. Ferris and G. S. Whitt. 1980. Evolution of five multilocus isozyme systems in the chordates. *Genetica* 52: 73–85.

Fisher, S. E. and G. S. Whitt. 1978. Evolution of isozyme loci and their differential tissue expression. Creatine kinase as a model system. *J. Mol. Evol.* 12: 25–55.

Fitch, W. M. 1970. Distinguishing homologous from analogous proteins. *Syst. Zool.* 19: 99–113.

Fitch, W. M. 1971. Toward defining the course of evolution: Minimal change for a specific tree topology. *Syst. Zool.* 20: 406–416.

Fitch, W. M. 1976. Molecular evolutionary clocks. Pp. 160–178 in *Molecular Evolution*, F. J. Ayala (ed.). Sinauer Associates, Sunderland, MA.

Fitch, W. M. and E. Margoliash. 1967. Construction of phylogenetic trees. *Science* 155: 279–284.

Fitze, P. S., M. Kölliker and H. Richner. 2003. Effects of common origin and common environment on nestling plumage coloration in the great tit (*Parus major*). *Evolution* 57: 144–150.

Fitzgerald, J. R. and J. M. Musser. 2001. Evolutionary genomics of pathogenic bacteria. *Trends Microbiol.* 9: 547–553.

Fitzgerald, J. R., D. E. Sturdevant, S. M. Mackie, S. R. Gill and J. M. Musser. 2001. Evolutionary genomics of *Staphylococcus aureus*: Insights into the origin of methicillin-resistant strains and the toxic shock syndrome epidemic. *Proc. Natl. Acad. Sci. USA* 98: 8821–8826.

Fitzsimmons, N. N. 1998. Single paternity of clutches and sperm storge in the promiscuous green turtle (*Chelonia mydas*). *Mol. Ecol.* 7: 575–584.

Fiumera, A. C., Y. D. DeWoody, J. A. DeWoody, M. A. Asmussen and J. C. Avise. 2001. Accuracy and precision of methods to estimate the number of parents contributing to a half-sib progeny array. *J. Heredity* 92: 120–126.

Fiumera, A. C., J. A. DeWoody, M. A. Asmussen and J. C. Avise. 2002a. Estimating the proportion of offspring attributable to candidate adults. *Evol. Ecol.* 16: 549–565.

Fiumera, A. C., B. A. Porter, G. D. Grossman and J. C. Avise. 2002b. Intensive genetic assessment of the mating system and reproductive success in a semi-closed population of the mottled sculpin, *Cottus bairdi. Mol. Ecol.* 11: 2367–2377.

Flavell, A. J. 1992. *Ty1-copia* group retrotransposons and the evolution of retroelements in the eukaryotes. *Genetica* 86: 203–214.

Flavell, R. B. 1986. Repetitive DNA and chromosome evolution in plants. *Phil. Trans. R. Soc. Lond. B* 13: 335–340.

Fleischer, R. C. 1996. Application of molecular methods to the assessment of genetic mating systems in vertebrates. Pp. 133–161 in *Molecular Zoology: Advances, Strategies, and Protocols*, J. D. Ferraris and S. R. Palumbi (eds.). Wiley-Liss, New York.

Fleischer, R. C., E. A. Perry, K. Muralidharan, E. E. Stevens and C. M. Wemmer. 2001. Phylogeography of the Asian elephant (*Elephas maximus*) based on mitochondrial DNA. *Evolution* 55: 1882–1892.

Fletcher, D. J. C. and C. D. Michener (eds.). 1987. *Kin recognition in animals.* John Wiley and Sons, New York.

Floyd, R., E. Abebe, A. Papert and M. Blaxter. 2002. Molecular barcodes for soil nematode identification. *Mol. Ecol.* 11: 839–850.

Flynn, J. J., M. A. Nedbal, J. W. Dragoo and R. L. Honeycutt. 2000. Whence the red panda? *Mol. Phylogen. Evol.* 17: 190–199.

Foerster, K., K. Delhey, A. Johnsen, J. T. Lifjeld and B. Kempanaers. 2003. Females increase offspring heterozygosity and fitness through extra-pair matings. *Nature* 425: 714–717.

Foltz, D. W. 1981. Genetic evidence for long-term monogamy in a small rodent, *Peromyscus polionotus. Am. Nat.* 117: 665–675.

Foltz, D. W. and J. L. Hoogland. 1981. Analysis of the mating system in the black-tailed prairie dog (*Cynomys ludovicianus*) by likelihood of paternity. *J. Mammal.* 62: 706–712.

Foltz, D. W. and J. L. Hoogland. 1983. Genetic evidence of outbreeding in the black-tailed prairie dog (*Cynomys ludovicianus*). *Evolution* 37: 273–281.

Foltz, D. W. and P. L. Schwagmeyer. 1989. Sperm competition in the thirteen-lined ground squirrel: Differential fertilization success under field conditions. *Am. Nat.* 133: 257–265.

Foltz, D. W., H. Ochman, J. S. Jones, S. M. Evangelisti and R. K. Selander. 1982. Genetic population structure and breeding systems in arionid slugs (Mollusca: Pulmonata). *Biol. J. Linn. Soc.* 17: 225–241.

Foltz, D. W., H. Ochman and R. K. Selander. 1984. Genetic diversity and breeding systems in terrestrial slugs of the families Limacidae and Arionidae. *Malacologia* 25: 593–605.

Foote, C. J., C. C. Wood and R. E. Withler. 1989. Biochemical genetic composition of sockeye salmon and kokanee, the anadromous and nonanadromous forms of *Oncorhynchus nerka*. *Can. J. Fish. Aquat. Sci.* 46: 149–158.

Forbes, A. T., N. T. Demetriades, J. A. H. Benzie and E. Ballment. 1999. Allozyme frequencies indicate little geographic variation among stocks of giant tiger prawn *Penaeus monodon* in the south-west Indian Ocean. *S. Afr. J. Marine Sci.* 21: 271–277.

Forbes, S. H. and F. W. Allendorf. 1991. Associations between mitochondrial and nuclear genotypes in cutthroat trout hybrid swarms. *Evolution* 45: 1332–1349.

Ford, E. B. 1964. *Ecological Genetics*. Metheun, London.

Ford, H. 1985. Life history strategies in two coexisting agamospecies of dandelion. *Biol. J. Linn. Soc.* 25: 169–186.

Ford, H. and A. J. Richards. 1985. Isozyme variation within and between *Taraxacum* agamospecies in a single locality. *Heredity* 55: 289–291.

Ford, M. J. 2002. Applications of selective neutrality tests to molecular ecology. *Mol. Ecol.* 11: 1245–1262.

Ford, V. S. and L. D. Gottlieb. 1999. Molecular characterization of *PgiC* in a tetraploid plant and its diploid relatives. *Evolution* 53: 1060–1067.

Ford, V. S. and L. D. Gottlieb. 2002. Single mutations silence *PgiC2* genes in two very recent allotetraploid species of *Clarkia*. *Evolution* 56: 699–707.

Ford, V. S., B. R. Thomas and L. D. Gottlieb. 1995. The same duplication accounts for the *PgiC* genes in *Clarkia xantiana* and *C. lewisi* (Onagraceae). *Syst. Bot.* 20: 147–160.

Foreman, L, A. Smith and I. Evett. 1997. Bayesian analysis of DNA profiling data in forensic identification applications. *J. R. Stat. Soc. A* 160: 429–469.

Forey, P. L., C. J. Humphries and R. I. Vane-Wright (eds.). 1994. *Systematics and Conservation Evaluation*. Clarendon Press, Oxford, England.

Forsthoefel, N. R., H. J. Bohnert and S. E. Smith. 1992. Discordant inheritance of mitochondrial and plastid DNA in diverse alfalfa genotypes. *J. Heredity* 83: 342–345.

Forterre, P., C. Bouthier de la Tour, H. Philippe and M. Duguet. 2000. Reverse gyrase from hyperthermophiles: Probable transfer of a thermoadaptation trait from Archaea to Bacteria. *Trends Genet.* 16: 152–154.

Fortunato, A., J. E. Strassmann, L. Santorelli and D. C. Queller. 2003. Co-occurrence in nature of different clones of the social amoeba, *Dictyostelium discoideum*. *Mol. Ecol.* 12: 1031–1038.

Fos, M., M. A. Dominguez, A. Latorre and A. Moya. 1990. Mitochondrial DNA evolution in experimental populations of *Drosophila subobscura*. *Proc. Natl. Acad. Sci. USA* 87: 4198–4201.

Foster, E. A. and 7 others. 1998. Jefferson fathered slave's last child. *Nature* 396: 27–28.

Foster, K. R., A. Fortunato, J. E. Strassmann and D. C. Queller. 2002. The costs and benefits of being a chimera. *Proc. R. Soc. Lond. B* 269: 2357–2362.

Fox, G. E., L. J. Magrum, W. E. Balch, R. S. Wolfe and C. R. Woese. 1977. Classification of methanogenic bacteria by 16S ribosomal RNA characterization. *Proc. Natl. Acad. Sci. USA* 74: 4537–4541.

Fox, G. E. and 18 others. 1980. The phylogeny of prokaryotes. *Science* 209: 457–463.

Fox, T. D. 1983. Mitochondrial genes in the nucleus. *Nature* 301: 371–372.

Frankel, O. H. 1980. Evolutionary change in small populations. Pp. 135–150 in *Conservation Biology: An Evolutionary-Ecological Perspective*, M. E. Soulé and B. A. Wilcox (eds.). Sinauer Associates, Sunderland, MA.

Frankel, O. H. and J. G. Hawkes (eds.). 1975. *Crop Genetic Resources for Today and Tomorrow*. Cambridge University Press, Cambridge, England.

Frankel, O. H. and M. Soulé. 1981. *Conservation and Evolution*. Cambridge University Press, Cambridge, England.

Frankham, R. 1995. Conservation genetics. *Annu. Rev. Genet.* 29: 305–327.

Frankham, R., J. D. Ballou and D. A. Briscoe. 2002. *An Introduction to Conservation Genetics*. Cambridge University Press, Cambridge, England.

Fraser, D. J. and L. Bernatchez. 2001. Adaptive evolutionary conservation: Towards a unified concept for defining conservation units. *Mol. Ecol.* 10: 2741–2752.

Freeland, J. R. and P. T. Boag. 1999. The mitochondrial and nuclear genetic heterogeneity of the phenotypically diverse Darwin's ground finches. *Evolution* 33: 1553–1563.

Freeman, D. C., J. H. Graham, D. W. Bryd, E. D. McArthur and W. A. Turner. 1995. Narrow hybrid zone between two subspecies of big sagebrush, *Artemesia tridentata* (Asteraceae). III. Developmental instability. *Am. J. Bot.* 82: 1144–1152.

Freeman, D. C., H. Wang, S. Sanderson and E. D. McArthur. 1999. Characterization of a narrow hybrid zone between two subspecies of big sagebrush, *Artemesia tridentata* (Asteraceae). VII. Community and demographic analyses. *Evol. Ecol. Res.* 1: 487–502.

Frey, J. E. and B. Frey. 1995. Molecular identification of six species of scale insects (*Quadraspidiotus* sp.) by RAPD-PCR: Assessing the field specificity of pheromone traps. *Mol. Ecol.* 4: 777–780.

Friedman, S. T. and W. T. Adams. 1985. Estimation of gene flow into two seed orchards of loblolly pine (*Pinus taeda*). *Theoret. Appl. Genet.* 69: 609–615.

Fries, R., A. Eggen and G. Stranzinger. 1990. The bovine genome contains polymorphic microsatellites. *Genomics* 8: 403–406.

Friesen, V. 2000. Introns. Pp. 274–294 in *Molecular Methods in Ecology*, A. J. Baker (ed.). Blackwell, London.

Fritsch, P. and L. H. Rieseberg. 1992. High outcrossing rates maintain male and hermaphroditic individuals in populations of the flowering plant *Datisca glomerata*. *Nature* 359: 633–636.

Frost, D. R. and D. M. Hillis. 1990. Species in concept and practice: Herpetological applications. *Herpetologica* 46: 87–104.

Frumhoff, P. C. and H. K. Reeve. 1994. Using phylogenies to test hypotheses of adaptation: A critique of some current proposals. *Evolution* 48: 172–180.

Fryer, G. 1997. Biological implications of a suggested Late Pleistocene desiccation of Lake Victoria. *Hydrobiologia* 354: 177–182.

Fryer, G. 2001. On the age and origin of the species flock of haplochromine cichlid fishes of Lake Victoria. *Proc. R. Soc. Lond. B* 268: 1147–1152.

Fryer, G. and T. D. Iles. 1972. *The Cichlid Fishes of the Great Lakes of Africa.* TFH, Neptune City, NJ.

Fu, Y.-X. 1994a. A phylogenetic estimator of effecive population size or mutation rate. *Genetics* 136: 685–692.

Fu, Y.-X. 1994b. Estimating effective population size or mutation rate using the frequencies of mutations of various classes in a sample of DNA sequences. *Genetics* 138: 1375–1386.

Fu, Y.-X. 1997. Statistical tests of neutrality of mutations against population growth, hitchhiking and background selection. *Genetics* 147: 915–925.

Fuhrman, J. A., K. McCallum and A. A. Davis. 1992. Novel major archaebacterial group from marine plankton. *Nature* 356: 148–149.

Fullerton, S. M., R. M. Harding, A. J. Boyce and J. B. Clegg. 1994. Molecular and population genetic analysis of allelic sequence diversity at the human β-globin locus. *Proc. Natl. Acad. Sci. USA* 91: 1805–1809.

Funduyga, R. E., T. J. Lott and J. Arnold. 2002. Population structure of *Candida albicans*, a member of the human flora, as determined by microsatellite loci. *Infect., Genet., and Evol.* 2: 57–68.

Funk, D. J. and K. E. Omland. 2003. Species-level paraphyly and polyphyly: Frequency, causes, and consequencess, with insights from animal mitochondrial DNA. *Annu. Rev. Ecol. Evol. Syst.* 34: 397–423.

Funk, D. J., D. J. Futuyma, G. Ortí and A. Meyer. 1995. A history of host associations and evolutionary diversification for *Ophraella* (Coleoptera: Chrysomelidae): New evidence from mitochondrial DNA. *Evolution* 49: 1008–1017.

Funk, D. J., L. Helbling, J. J. Wernegreen and N. A. Moran. 2000. Intraspecific phylogenetic congruence among multiple symbiont genomes. *Proc. R. Soc. Lond. B* 267: 2517–2521.

Funk, D. J., K. E. Filchak and J. L. Feder. 2002. Herbivorous insects: Model systems for the comparative study of speciation ecology. *Genetica* 116: 251–267.

Furnier, G. R., P. Knowles, M. A. Clyde and B. P. Dancik. 1987. Effects of avian seed dispersal on the genetic structure of whitebark pine populations. *Evolution* 41: 607–612.

Furnier, G. R., M. P. Cummings and M. T. Clegg. 1990. Evolution of the avacados as revealed by DNA restriction fragment variation. *J. Heredity* 81: 183–188.

Futuyma, D. J. and G. C. Mayer. 1980. Non-allopatric speciation in animals. *Syst. Zool.* 29: 254–271.

Gaffney, P. M. and B. McGee. 1992. Multiple paternity in *Crepidula fornicata* (Linnaeus). *Veliger* 35: 12–15.

Gagneux, P., D. S. Wooduff and C. Boesch. 1997. Furtive mating in female chimpanzees. *Nature* 387: 358–359.

Gaillard, J.-M., D. Allainé, D. Pontier, . G. Yoccoz and D. E. L. Promislow. 1994. Sensecence in natural populations of mammals: A reanalysis. *Evolution* 48: 509–516.

Galeotti, P., A. Pilastro, G. Tavecchia, A. Bonetti and L. Congiu. 1997. Genetic similarity in long-eared owl communal winter roosts: A DNA fingerprinting study. *Mol. Ecol.* 6: 429–435.

Galindo, B. E., V. D. Vacquier and W. J. Swanson. 2003. Positive selection in the egg receptor for abalone sperm lysin. *Proc. Natl. Acad. Sci. USA* 100: 4639–4643.

Gallez, G. P. and L. D. Gottlieb. 1982. Genetic evidence for the hybrid origin of the diploid plant *Stephanomeria diegensis. Evolution* 36: 1158–1167.

Gallo, R. C. 1987. The AIDS virus. *Sci. Am.* 256(1): 46–56.

Galloway, L. F., J. R. Etterson and J. L. Hamrick. 2003. Outcrossing rate and inbreeding depression in the herbaceous autotetrapolid, *Campanula americana. Heredity* 90: 308–315.

Gamow, G. 1954. Possible relation between deoxyribonucleic acid and protein structures. *Nature* 173: 318.

Ganders, F. R. 1989. Adaptive radiation in Hawaiian *Bidens.* Pp. 99–112 in *Genetics, Speciation and the Founder Principle,* L. V. Giddings, K. Y. Kaneshiro and W. W. Anderson (eds.). Oxford University Press, New York.

Gandolfi, A., I. R. Sanders, V. Rossi and P. Menozzi. 2003. Evidence of recombination in putative ancient asexuals. *Mol. Biol. Evol.* 20: 754–761.

Gantt, J. S., S. L. Baldauf, P. J. Calie, N. F. Weeden and J. D. Palmer. 1991. Transfer of *rpl22* to the nucleus greatly preceded its loss from the chloroplast and involved the gain of an intron. *The EMBO J.* 10: 3073–3078.

Gao, F. and 11 others. 1999. Origin of HIV-1 in the chimpanzee *Pan troglodytes troglodytes. Nature* 397: 436–440.

Garant, D., J. J. Dodson and L. Bernatchez. 2001. A genetic evaluation of mating system and determinants of individual reproductive success in Atlantic salmon (*Salmo salar*). *J. Heredity* 92: 137–145.

Garber, R. A. and J. W. Morris. 1983. General equations for the average power of exclusion for genetic systems of *n* codominant alleles in one-parent and in no-parent cases of disputed parentage. Pp. 277–280 in *Inclusion Probabilities in Parentage Testing,* R. H. Walker (ed.). American Assn. Blood Banks, Arlington, VA.

Garcia-Rodriguez, A. I. and 9 others. 1998. Phylogeography of the West Indian manatee (*Trichechus manatus*): How many populations and how many taxa? *Mol. Ecol.* 7: 1137–1149.

Garcia-Vallvé, A. Romeu and J. Palau. 2000. Horizontal gene transfer of glycosyl hydrolases of the rumen fungi. *Mol. Biol. Evol.* 17: 352–361.

Garcia-Vasquez, E., P. Moran, J. L Martinez, J. Perez, B. de Gaudemar and E. Beall. 2001. Alternative mating strategies in Atlantic salmon and brown trout. *J. Heredity* 92: 146–149.

Garland, T., P. H. Harvey and A. R. Ives. 1992. Procedures for the analysis of comparative data using phylogenetically independent contrasts. *Syst. Biol.* 41: 8–32.

Garner, K. J. and O. A. Ryder. 1996. Mitochondrial DNA diversity in gorillas. *Mol. Phylogen. Evol.* 6: 39–48.

Garner, T. W. J. and 6 others. 2002. Geographic variation of multiple paternity in the common garter snake (*Thamnophis sirtalis*). *Copeia* 2002: 15–23.

Garrigan, D. and P. W. Hedrick. 2003. Detecting adaptive molecular polymorphism: Lessons from the MHC. *Evolution* 57: 1707–1722.

Garrigan, D., P. C. Marsh and T. E. Dowling. 2002. Long-term effective population size of three endangered Colorado River fishes. *Anim. Cons.* 5: 95–102.

Garten, C. T. Jr. 1977. Relationships between exploratory behaviour and genic heterozygosity in the oldfield mouse. *Anim. Behav.* 25: 328–332.

Garton, D. W., R. K. Koehn and T. M. Scott. 1984. Multiple-locus heterozygosity and the physiological energetics of growth in the coot clam, *Mulinia lateralis*, from a natural population. *Genetics* 108: 445–455.

Gartside, D. F., J. S. Rogers and H. C. Dessauer. 1977. Speciation with little genic and morphological differentiation in the ribbon snakes *Thamnophis proximus* and *T. sauritus* (Colubridae). *Copeia* 1977: 697–705.

Garza, J. and N. Freimer. 1996. Homoplasy for size at microsatellite loci in humans and chimpanzees. *Genome Res.* 6: 211–217.

Garza, J. C, J. Dallas, D. Duryadi, S. Gerasimov, H. Croset and P. Boursot. 1997. Social structure of the mound-building mouse *Mus spicilegus* revealed by genetic analysis with microsatellites. *Mol. Ecol.* 6: 1009–1017.

Gastony, G. J. 1986. Electrophoretic evidence for the origin of fern species by unreduced spores. *Am. J. Bot.* 73: 1563–1569.

Gatesy, J. 1997. More DNA support for a Cetacea/Hippopotamidae clade: The blood-clotting protein gene γ-fibrinogen. *Mol. Biol. Evol.* 14: 537–543.

Gaudet, J., J. Julien, J. F. Lafay and Y. Brygoo. 1989. Phylogeny of some *Fusarium* species, as determined by large-subunit rRNA sequence comparisons. *Mol. Biol. Evol.* 6: 227–242.

Gaunt, M. W. and 11 others. 2003. Mechanism of genetic exchange in American trypanosomes. *Nature* 421: 936–939.

Gaut, B. S., S. V. Muse, W. D. Clark and M. T. Clegg. 1992. Relative rates of nucleotide substitution at the *rbcL* locus of monocotyledonous plants. *J. Mol. Evol.* 35: 292–303.

Gavrilets, S. 2000. Rapid evolution of reproductive barriers driven by sexual conflict. *Nature* 403: 886–889.

Geeta, R. 2003. The origin and maintenance of nuclear endosperms: Viewing development through a phylogenetic lens. *Proc. R. Soc. Lond. B* 270: 29–35.

Geist, V. 1992. Endangered species and the law. *Nature* 357: 274–276.

Geller, J. B., E. D. Walton, E. D. Grosholz and G. M. Ruiz. 1997. Cryptic invasions of the crab *Caricinus* detected by molecular phylogeography. *Mol. Ecol.* 6: 901–906.

Gellissen, G., J. Y. Bradfield, B. N. White and G. R. Wyatt. 1983. Mitochondrial DNA sequences in the nuclear genome of a locust. *Nature* 301: 631–634.

Gelter, H. P. and H. Tegelström. 1992. High frequency of extra-pair paternity in Swedish pied flycatchers revealed by allozyme electrophoresis and DNA fingerprinting. *Behav. Ecol. Sociobiol.* 31: 1–7.

Gelter, H. P., H. Tegelström and L. Gustafsson. 1992. Evidence from hatching success and DNA fingerprinting for the fertility of hybrid pied X collared flycatchers *Ficedula hypoleuca* X *albicollis*. *Ibis* 134: 62–68.

Genereux, D. P. and J. M. Logsdon Jr. 2003. Much ado about bacteria-to-vertebrate lateral gene transfer. *Trends Genet.* 19: 191–195.

George, M. Jr. and O. A. Ryder. 1986. Mitochondrial DNA evolution in the genus *Equus*. *Mol. Biol. Evol.* 3: 535–546.

Georges, M., P. Cochaux, A. S. Lequarre, M. W. Young and G. Vassart. 1987. DNA fingerprinting in man using a mouse probe related to part of the *Drosophila* "Per" gene. *Nucleic Acids Res.* 15: 7193.

Gerber, A. S., C. A. Tibbets and T. E. Dowling. 2001. The role of introgressive hybridization in the evolution of the *Gila robusta* complex (Teleostei: Cyprinidae). *Evolution* 55: 2028–2039.

Gerbi, S. A. 1985. Evolution of ribosomal RNA. Pp. 419–518 in *Molecular Evolutionary Genetics*, R. J. MacIntyre (ed.). Plenum Press, New York.

Gerhart, J. and M. Kirschner. 1997. *Cells, Embryos, and Evolution*. Blackwell, Malden, MA.

Gerloff, U. and 6 others. 1995. Amplification of hypervariable simple sequence repeats (microsatellites) from excremental DNA of wild living bonobos (*Pan paniscus*). *Mol. Ecol.* 4: 515–518.

Gertsch, P., P. Pamilo and S.-L. Varvio. 1995. Microsatellites reveal high genetic diversity within colonies of *Camponotus* ants. *Mol. Ecol.* 4: 257–260.

Geyer, C. J., E. A. Thompson and O. A. Ryder. 1989. Gene survival in the Asian wild horse (*Equus przewalskii*): II. Gene survival in the whole population, in subgroups, and through history. *Zoo Biol.* 8: 313–329.

Geyer, L. B. and S. R. Palumbi. 2003. Reproductive character diaplacement and the genetics of gamete recognition in tropical sea urchins. *Evolution* 57: 1049–1060.

Gibbons, A. 1991. Looking for the father of us all. *Science* 251: 378–380.

Gibbs, H. L. and P. J. Weatherhead. 2001. Insights into population ecology and sexual selection in snakes through the application of DNA-based genetic markers. *J. Heredity* 92: 173–179.

Gibbs, H. L., P. J. Weatherhead, P. T. Boag, B. N. White, L. M. Tabak and D. J. Hoysak. 1990. Realized reproductive success of polygynous red-winged blackbirds revealed by DNA markers. *Science* 250: 1394–1397.

Gibbs, H. L., P. T. Boag, B. N. White, P. J. Weatherhead and L. M. Tabak. 1991. Detection of a hypervariable DNA locus in birds by hybridization with a mouse MHC probe. *Mol. Biol. Evol.* 8: 433–446.

Gibbs, H. L., M. D. Sorenson, K. Marchetti, M. L. Brooke, N. B. Davies and H. Nakamura. 2000a. Genetic evidence for female host-specific races of the common cuckoo. *Nature* 407: 183–186.

Gibbs, H. L., R. J. G. Dawson and K. A. Hobson. 2000b. Limited differentiation in microsatellite DNA variatoin among northern populations of the yellow warbler: Evidence for male-biased gene flow? *Mol. Ecol.* 9: 2137–2147.

Gibson, A. R. and J. B. Falls. 1975. Evidence for multiple insemination in the common garter snake, *Thamnophis sirtalis*. *Can. J. Zool.* 53: 1362–1368.

Giddings, L. V., K. Y. Kaneshiro and W. W. Anderson (eds.). 1989. *Genetics, Speciation and the Founder Principle.* Oxford University Press, New York.

Gilbert, D. A., N. Lehman, S. J. O'Brien and R. K. Wayne. 1990. Genetic fingerprinting reflects population differentiation in the California Channel Island fox. *Nature* 344: 764–767.

Gilbert, D. A., C. Packer, A. E. Pusey, J. C. Stephens and S. J. O'Brien. 1991. Analytical DNA fingerprinting in lions: Parentage, genetic diversity, and kinship. *J. Heredity* 82: 378–386.

Gilbert, D. G. and R. C. Richmond. 1982. Studies of esterase 6 in *Drosophila melanogaster* XII. Evidence for temperature selection of *Est 6* and *Adh* alleles. *Genetica* 58: 109–119.

Gilbert, W. 1978. Why genes in pieces? *Nature* 271: 501.

Giles, E. and S. H. Ambrose. 1986. Are we all out of Africa? *Nature* 322: 21–22.

Giles, R. E., H. Blanc, H. M. Cann and D. C. Wallace. 1980. Maternal inheritance of human mitochondrial DNA. *Proc. Natl. Acad. Sci. USA* 77: 6715–6719.

Gill, F. B. 1990. *Ornithology.* W. H. Freeman and Co., New York.

Gill, P., A. J. Jeffreys and D. J. Werrett. 1985. Forensic application of DNA "fingerprints." *Nature* 318: 577–579.

Gill, P. and 8 others 1994. Identification of the remains of the Romanov family by DNA analysis. *Nature Genet.* 6: 130–135.

Gilles, P. N., D. J. Wu, C. B. Foster, P. J. Billon and S. J. Sambrook. 1999. Single nucleotide polymorphic discrimination by an electronic dot blot assay on semiconductor microchips. *Nat. Biotech.* 17: 365–370.

Gillespie, J. H. 1986. Variability of evolutionary rates of DNA. *Genetics* 113: 1077–1091.

Gillespie, J. H. 1987. Molecular evolution and the neutral allele theory. *Oxford Surv. Evol. Biol.* 4: 10–37.

Gillespie, J. H. 1988. More on the overdispersed molecular clock. *Genetics* 118: 385–386.

Gillespie, J. H. 1991. *The Causes of Molecular Evolution.* Oxford University Press, New York.

Gillespie, J. H. 2001. Is the population size of a species relevant to its evolution? *Evolution* 55: 2161–2169.

Gillespie, J. H. and C. H. Langley. 1974. A general model to account for enzyme variation in natural populations. *Genetics* 76: 837–884.

Gilley, D. C. 2003. Absence of nepotism in the harassment of duelling queens by honeybee workers. *Proc. R. Soc. Lond. B* 270: 1045–1049.

Gillham, N. W. 1978. *Organelle Heredity.* Raven Press, New York.

Gilpin, M. E. and M. E. Soulé. 1986. Minimum viable populations: Processes of species extinction. Pp. 19–34 in *Conservation Biology: The Science of Scarcity and Diversity*, M. E. Soulé (ed.). Sinauer Associates, Sunderland, MA.

Gilpin, M. E. and C. Wills. 1991. MHC and captive breeding: A rebuttal. *Cons. Biol.* 5: 554–555.

Giordano, R., J. J. Jackson and H. M. Robertson. 1997. The role of *Wolbachia* bacteria in reproductive incompatibilities and hybrid zones of *Diabrotica* beetles and *Gryllus* crickets. *Proc. Natl. Acad. Sci. USA* 94: 11439–11444.

Giorgi, P. P. 1992. Sex and the male stick insect. *Nature* 357: 444–445.

Giovannoni, S. J., S. Turner, G. J. Olsen, S. Barns, D. J. Lane and N. R. Pace. 1988. Evolutionary relationships among cyanobacteria and green chloroplasts. *J. Bacteriol.* 170: 3584–3592.

Giovannoni, S. J., T. B. Britschgi, C. L. Moyer and K. G. Field. 1990. Genetic diversity in Sargasso Sea bacterioplankton. *Nature* 345: 60–63.

Giribet, G. 2002. Current advances in the phylogenetic reconstructoin of metazoan evolution. A new paradigm for the Cambrian explosion? *Mol. Phylogen. Evol.* 24: 345–357.

Girman, D. J. and 9 others. 1993. Molecular genetic and morphological analyses of the African wild dog (*Lycaon pictus*). *J. Heredity* 84: 450–459.

Gittleman, J. L. and S. L. Pimm. 1991. Crying wolf in North America. *Nature* 351: 524–525.

Gittleman, J. L., C. G. Anderson, M. Kot and H.-K. Luh. 1996a. Comparative tests of evolutionary lability and rates using molecular phylogenies. Pp. 289–307 in *New Uses for New Phylogenies*, P. H. Harvey, A. J. Leigh Brown, J. Maynard Smith and S. Nee (eds.). Oxford University Press, Oxford, England.

Gittleman, J. L., C. G. Anderson, M. Kot and H.-K. Luh. 1996b. Phylogenetic lability and rates of evolution: A comparison of behavioral, morphological and life-history traits. Pp. 166–205 in *Phylogenies and the Comparative Method in Animal Behavior*, E. P. Martins (ed). Oxford University Press, New York.

Givnish, T. J. and K. J. Sytsma (eds.). 1997. *Molecular Evolution and Adaptive Radiation.* Cambridge University Press, Cambridge, England.

Glaubitz, J. C., O. E. Rhodes Jr. and J. A. DeWoody. 2003. Prospects for inferring pairwise relationships with single nucleotide polymorphisms. *Mol. Ecol.* 12: 1039–1047.

Goddard, K. A., R. M. Dawley and T. E. Dowling. 1989. Origin and genetic relationships of diploid, triploid, and diploid-triploid mosaic biotypes in the *Phoxinus eos-neogaeus* unisexual complex. Pp. 268–280 in *Evolution and Ecology of Unisexual Vertebrates*, R. Dawley and J. Bogart (eds.). New York State Museum, Albany, NY.

Godt, M. J., J. Walker and J. L. Hamrick. 1997. Genetic diversity in the endangered lily *Harperocallis flava* and a close relative, *Tofieldia recemosa*. *Cons. Biol.* 11: 361–366.

Gogarten, J. P. 1995. The early evolution of cellular life. *Trends Ecol. Evol.* 10: 147–151.

Gogarten, J. P. 2003. Gene transfer: Gene swapping craze reaches eukaryotes. *Current Biol.* 13: 53–54.

Gogarten, J. P. and 12 others. 1989. Evolution of the vacuolar H^+-ATPase: Implications for the origin of eukaryotes. *Proc. Natl. Acad. Sci. USA* 86: 6661–6665.

Gogarten, J. P., W. F. Doolittle and J. G. Lawrence. 2002. Prokaryotic evolution in light of gene transfer. *Mol. Biol. Evol.* 19: 2226–2238.

Gokool, S., C. F. Curtis and D. F. Smith. 1993. Analysis of mosquito bloodmeals by DNA profiling. *Medical Vet. Entomol.* 7: 208–215.

Gold, J. R. and L. R. Richardson. 1998. Mitochondrial DNA diversification and population structure in fishes from the Gulf of Mexico and western Atlantic. *J. Heredity* 89: 404–414.

Goldberg, T. L. 1997. Inferring the geographic origins of "refugee" chimpanzees in Uganda from mitochondrial DNA sequences. *Cons. Biol.* 11: 1441–1446.

Goldberg, T. L. and M. Ruvolo. 1997. Molecular phylogenetics and historical biogeography of east African chimpanzees. *Biol. J. Linn. Soc.* 61: 301–324.

Golding, B. (ed.). 1994. *Non-Neutral Evolution: Theories and Molecular Data.* Chapman & Hall, New York.

Golding, G. B. and R. Gupta. 1995. Protein-based phylogenies support a chimeric origin of the eukaryotic genome. *Mol. Biol. Evol.* 12: 1–6.

Goldman, D., P. R. Giri and S. J. O'Brien. 1989. Molecular genetic-distance estimates among the Ursidae as indicated by one- and two-dimensional protein electrophoresis. *Evolution* 43: 282–295.

Goldschmidt, R. 1940. *The Material Basis of Evolution.* Yale University Press, New Haven, CT.

Goldstein, D. B. and C. Schlötterer (eds.). 1999. *Microsatellites: Evolution and Applications.* Oxford University Press, Oxford, England.

Goldstein, D. B., A. R. Linares, L. L. Cavalli-Sforza and M. W. Feldman. 1995a. An evaluation of genetic distances for use with microsatellite loci. *Genetics* 139: 463–471.

Goldstein, D. B., A. R. Linares, L. L. Cavalli-Sforza and M. W. Feldman. 1995b. Genetic absolute dating based on microsatellites and the origin of modern humans. *Proc. Natl. Acad. Sci. USA* 92: 6723–6727.

Goldstein, P. Z. and R. DeSalle. 2003. Calibrating phylogenetic species formation in a threatened insect using DNA from historical specimens. *Mol. Ecol.* 12: 1993–1998.

Goldstein, P. Z, R. DeSalle, G. Amato and A. P. Vogler. 2000. Conservation genetics at the species boundary. *Cons. Biol.* 14: 120–131.

Golenberg, E. M. 1989. Migration patterns and the development of multilocus associations in a selfing annual, *Triticum dicoccoides. Evolution* 43: 595–606.

Golenberg, E. M. and 6 others. 1990. Chloroplast DNA sequence from a Miocene *Magnolia* species. *Nature* 344: 656–658.

Gomendio, M., A. H. Harcourt and E. R. S. Roldán. 1998. Sperm competition in mammals. Pp. 667–756 in *Sperm Competition and Sexual Selection*, T. R. Birkhead and A. P. Møller (eds.). Academic Press, London.

Gómez, A., M. Serra, G. R. Carvalho and D. H. Lunt. 2002. Speciation in ancient cryptic species complexes: Evidence from the molecular phylogeny of *Brachionus plicatilis* (Rotifera). *Evolution* 56: 1431–1444.

González-Villaseñor, L. I. and D. A. Powers. 1990. Mitochondrial-DNA restriction-site polymorphisms in the teleost *Fundulus heteroclitus* support secondary intergradation. *Evolution* 44: 27–37.

Gooch, J. L. 1975. Mechanisms of evolution and population genetics. Pp. 349–409 in *Marine Ecology*, Vol. 2, Part 1, O. Kinne (ed.). John Wiley and Sons, London.

Good, D. A. and J. W. Wright. 1984. Allozymes and the hybrid origin of the parthenogenetic lizard *Cnemidophorus exsanguis. Experientia* 40: 1012–1014.

Good, J. M., J. R. Demboski, D. W. Nagorsen and J. Sullivan. 2003. Phylogeography and introgressive hybridization: Chipmunks (genus *Tamias*) in the northern Rocky Mountains. *Evolution* 57: 1900–1916.

Goodisman, M. A. D. and M. A. Asmussen. 1997. Cytonuclear theory for haplodiploid species and X-linked genes. I. Hardy-Weinberg dynamics and continent-island, hybrid zone models. *Genetics* 147: 321–338.

Goodisman, M. A. D., K. G. Ross and M. A. Asmussen. 2000. A formal assessment of gene flow and selection in the fire ant *Solenopsis invicta. Evolution* 54: 606–616.

Goodman, M. 1962. Immunochemistry of the primates and primate evolution. *Ann. N. Y. Acad. Sci.* 102: 219–234.

Goodman, M. 1963. Serological analysis of the systematics of recent hominoids. *Human Biol.* 35: 377–424.

Goodman, M., J. Barnabas, G. Matsuda and G. W. Moor. 1971. Molecular evolution in the descent of man. *Nature* 233: 604–613.

Goodman, M., J. Czelusniak and J. E. Beeber. 1985. Phylogeny of primates and other eutherian orders: A cladistic analysis using amino acid and nucleotide sequence data. *Cladistics* 1: 171–185.

Goodman, M. and 7 others. 1990. Primate evolution at the DNA level and a classification of hominoids. *J. Mol. Evol.* 30: 260–266.

Goodnight, K. F. and D. C. Queller. 1999. Computer software for performing likelihood tests of pedigree relationships using genetic markers. *Mol. Ecol.* 8: 1231–1234.

Goossens, B., L. P. Watts and P. Taberlet. 1998. Plucked hair samples as a source of DNA: Reliability of dinucleotide microsatellite genotyping. *Mol. Ecol.* 7: 1237–1241.

Gore, P. L., B. M. Potts, P. W. Volker and J. Megalos. 1990. Unilateral cross-incompatibility in *Eucalyptus*: The case of hybridization between *E. globulus* and *E. nitens. Aust. J. Bot.* 38: 383–394.

Gorr, T., T. Kleinschmidt and H. Fricke. 1991. Close tetrapod relationship of the coelacanth *Latimeria* indicated by haemoglobin sequences. *Nature* 351: 394–397.

Gottelli, D. and 7 others. 1994. Molecular genetics of the most endangered canid: The Ethiopian wolf *Canis simensis. Mol. Ecol.* 3: 301–312.

Gottlieb, L. D. 1973a. Enzyme differentiation and phylogeny in *Clarkia franciscana, C. rubicunda* and *C. amoena. Evolution* 27: 205–214.

Gottlieb, L. D. 1973b. Genetic differentiation, sympatric speciation and the origin of a diploid species of *Stephanomeria. Am. J. Bot.* 60: 545–553.

Gottlieb, L. D. 1974. Genetic confirmation of the origin of *Clarkia lingulata. Evolution* 28: 244–250.

Gottlieb, L. D. 1977. Electrophoretic evidence and plant systematics. *Annals Missouri Bot. Gardens* 64: 161–180.

Gottlieb, L. D. 1981. Electrophoretic evidence and plant populations. *Progr. Phytochem.* 7: 2–46.

Gottlieb, L. D. 1988. Towards molecular genetics in *Clarkia*: Gene duplications and molecular characterization of *PGI* genes. *Annals Missouri Bot. Gardens* 75: 1169–1179.

Gottlieb, L. D. and V. S. Ford. 1996. Phylogenetic relationships among the sections of *Clarkia* (Onagraceae) inferred from the nucleotide sequences of *PgiC*. *Syst. Bot.* 21: 45–62.

Gottlieb, L. D. and V. S. Ford. 1997. A recently silenced, duplicate *PgiC* locus in *Clarkia*. *Mol. Biol. Evol.* 14: 125–132.

Gottlieb, L. D. and G. Pilz. 1976. Genetic similarity between *Gaura longiflora* and its obligately outcrossing derivative *G. demareei*. *Syst. Bot.* 1: 181–187.

Gould, S. J. 1977. *Ontogeny and Phylogeny*. Belknap Press, Cambridge, MA.

Gould, S. J. 1980. Is a new and general theory of evolution emerging? *Paleobiology* 6: 119–130.

Gould, S. J. 1985. A clock of evolution. *Nat. Hist.* 94(4): 12–25.

Gould, S. J. 1991. Unenchanted evening. *Nat. Hist.* 100(5): 7–14.

Gould, S. J. 1992. We are all monkeys' uncles. *Nat. Hist.* 101(6): 14–21.

Gould, S. J. and N. Eldredge. 1977. Punctuated equilibria: The tempo and mode of evolution reconsidered. *Paleobiology* 3: 115–151.

Gould, S. J. and R. C. Lewontin. 1979. The spandrels of San Marco and the Panglossian paradigm: A critique of the adaptationist programme. *Proc. R. Soc. Lond. B* 205: 581–598.

Gouy, M. and W.-H. Li. 1989. Phylogenetic analysis based on rRNA sequences supports the archaebacterial rather than the eocyte tree. *Nature* 339: 145–147.

Gowaty, P. A. 1996. Field studies of parental care in birds: New data focus questions on variation among females. Pp. 476–531 in *Advances in the Study of Behavior*, C. T. Snowdon and J. S. Rosenblatt (eds.). Academic Press, New York.

Gowaty, P. A. and W. C. Bridges. 1991a. Nestbox availability affects extra-pair fertilizations and conspecific nest parasitism in eastern bluebirds, *Sialia sialis*. *Anim. Behav.* 41: 661–675.

Gowaty, P. A. and W. C. Bridges. 1991b. Behavioral, demographic, and environmental correlates of extra-pair fertilizations in eastern bluebirds, *Sialia sialis*. *Behav. Ecol.* 2: 339–350.

Gowaty, P. A. and A. A. Karlin. 1984. Multiple maternity and paternity in single broods of apparently monogamous eastern bluebirds (*Sialia sialis*). *Behav. Ecol. Sociobiol.* 15: 91–95.

Grachev, M. A. and 8 others. 1992. Comparative study of two protein-coding regions of mitochondrial DNA from three endemic sculpins (Cottoidei) of Lake Baikal. *J. Mol. Evol.* 34: 85–90.

Graham, J., G. R. Squire, B. Marshall and R. E. Harrison. 1997. Spatially dependent genetic diversity within and between colonies of wild raspberry (*Rubus idaeus*) detected using RAPD markers. *Mol. Ecol.* 6: 1001–1008.

Graham, J. H., D. C. Freeman and E. D. McArthur. 1995. Narrow hybrid zone between two subspecies of big sagebrush (*Artemisia tridentata*: Asteraceae). II. Selection gradients and hybrid fitness. *Am. J. Bot.* 82: 709–716.

Grant, B. R. and P. R. Grant. 2003. What Darwin's finches can teach us about the evolutionary origin and regulation of biodiversity. *BioScience* 53: 965–975.

Grant, P. R. and B. R. Grant. 1992. Hybridization of bird species. *Science* 256: 193–197.

Grant, V. 1963. *The Origin of Adaptations*. Columbia University Press, New York.

Grant, V. 1981. *Plant Speciation* (2nd Edition). Columbia University Press, New York.

Grant, W. S. 1987. Genetic divergence between congeneric Atlantic and Pacific Ocean fishes. Pp. 225–246 in *Population Genetics and Fisheries Management*, N. Ryman and F. Utter (eds.). University of Washington Press, Seattle.

Gräser, Y., M. Volovsek, J. Arrington, G. Schönian, W. Presber, T. G. Mitchell and R. Vilgalys. 1996. Molecular markers reveal that population structure of the human pathogen *Candida albicans* exhibits both clonality and recombination. *Proc. Natl. Acad. Sci. USA* 93: 12473–12477.

Grassle, J. P. and J. F. Grassle. 1976. Sibling species in the marine pollution indicator *Capitella* (Polychaeta). *Science* 192: 567–569.

Graur, D. and D. G. Higgins. 1994. Molecular evidence for the inclusion of Cetaceans within the order Artiodactyla. *Mol. Biol. Evol.* 11: 357–364.

Graur, D. and W.-H. Li. 2000. *Fundamentals of Molecular Evolution* (2nd Edition). Sinauer Associates, Sunderland, MA.

Graves, J. E. 1996. Conservation genetics of fishes in the pelagic marine realm. Pp. 335–366 in *Conservation Genetics: Case Histories from Nature*, J. C. Avise and J. L. Hamrick (eds.). Chapman & Hall, New York.

Graves, J. E. 1998. Molecular insights into the population structures of cosmopolitan marine fishes. *J. Heredity* 89: 427–437.

Graves, J. E., M. A. Simovich and K. M. Schaefer. 1988. Electrophoretic identification of early juvenile yellowfin tuna, *Thunnus albacares*. *Fish. Bull.* 86: 835–838.

Graves, J. E., M. J. Curtis, P. A. Oeth and R. S. Waples. 1989. Biochemical genetics of Southern California basses of the genus *Paralabrax*: Specific identification of fresh and ethanol-preserved individual eggs and early larvae. *Fish. Bull.* 88: 59–66.

Gray, E. M. 1998. Intraspecific variation in extra-pair behavior of red-winged blackbirds (*Agelaius phoeniceus*). *Ornithol. Monogr.* 49: 61–80.

Gray, M. W. 1989. Origin and evolution of mitochondrial DNA. *Annu. Rev. Cell Biol.* 5: 25–50.

Gray, M. W. 1992. The endosymbiont hypothesis revisited. *Int. Rev. Cytol.* 141: 233–357.

Gray, M. W., R. Cedergren, Y. Abel and D. Sankoff. 1989. On the evolutionary origin of the plant mitochondrion and its genome. *Proc. Natl. Acad. Sci. USA* 86: 2267–2271.

Greenberg, B. D., J. E. Newbold and A. Sugino. 1983. Intraspecific nucleotide sequence variability surrounding the origin of replication in human mitochondrial DNA. *Gene* 21: 33–49.

Greenlaw, J. S. 1993. Behavioral and morphological diversification in Sharp-tailed Sparrows (*Ammodramus caudacutus*) of the Atlantic coast. *Auk* 110: 286–303.

Greenwood, P. H. 1980. Towards a phyletic classification of the "genus" *Haplochromis* (Pisces Cichlidae) and related taxa. Part 2. The species from Lake Victoria, Nabugabo, Edward, George and Kivu. *Bull. Brit. Mus. Nat. Hist. (Zool.)* 39: 1–101.

Greenwood, P. H. 1981. *The Haplochromine Fishes of the East African Lakes.* Cornell University Press, Ithaca, N. Y.

Greenwood, P. J. 1980. Mating systems, philopatry and dispersal in birds and mammals. *Anim. Behav.* 28: 1140–1162.

Greenwood, P. J. and P. H. Harvey. 1982. The natal and breeding dispersal of birds. *Annu. Rev. Ecol. Syst.* 13: 1–21.

Grewal, R. P. and 10 others. 1999. French Machado-Joseph disease patients do not exhibit gametic segregation distortion: A sperm typing analysis. *Human Mol. Genet.* 8: 1779–1784.

Griffin, A. R., I. P. Burgess and L. Wolf. 1988. Patterns of natural and manipulated hybridization in the genus *Eucalyptus* L'Herit.- a review. *Aust. J. Bot.* 36: 41–66.

Griffith, S. C. 2000. High fidelity on islands: A comparative study of extra-pair paternity in passerine birds. *Behav. Ecol.* 11: 265–273.

Griffith, S. C., I. P. F. Owens and K. A. Thuman. 2002. Extra pair paternity in birds: A review of interspecific variation and adaptive function. *Mol. Ecol.* 11: 2195–2212.

Griffiths, R. 2000. Sex identification using DNA markers. Pp. 295–321 in *Molecular Methods in Ecology*, A. J. Baker (ed.). Blackwell, London.

Griffiths, R. and P. Holland. 1990. A novel avian W chromosome DNA repeat sequence in the lesser black-backed gull (*Larus fuscus*). *Chromosoma* 99: 243–250.

Griffiths, R. and S. Tavaré. 1997. Computational methods for the coalescent. Pp. 165–182 in *Progress in Population Genetics and Human Evolution*, P. Donnelly and S. Tavaré (eds.). Springer-Verlag, New York.

Griffiths, R. and B. Tiwari. 1995. Sex of the last Spix's Macaw. *Nature* 375: 454.

Griffiths, R., S. Daan and C. Dijkstra. 1996. Sex identification in birds using two CHD genes. *Proc. R. Soc. Lond. B* 263: 1249–1254.

Griffiths, R., M. C. Double, K. Orr and R. J. G. Dawson. 1998. A DNA test to sex most birds. *Mol. Ecol.* 7: 1071–1075.

Grivet, D. and R. J. Petit. 2002. Phylogeography of the common ivy (*Hedera* sp.) in Europe: Genetic differentiation through space and time. *Mol. Ecol.* 11: 1351–1362.

Groman, J. D. and O. Pellmyr. 2000. Rapid evolution and specialization following host colonization in a yucca moth. *J. Evol. Biol.* 13: 223–236.

Gromko, M. H., D. G. Gilbert and R. C. Richmond. 1984. Sperm transfer and use in the multiple mating system of *Drosophila.* Pp. 371–426 in *Sperm Competition and the Evolution of Animal Mating Systems*, R. L. Smith (ed.). Academic Press, New York.

Groombridge, B. (ed.). 1992. *Global Biodiversity: Status of the Earth's Living Resources.* Chapman & Hall, New York.

Grosberg, R. K. 1988. The evolution of allorecognition specificity in clonal invertebrates. *Q. Rev. Biol.* 63: 377–412.

Grosberg, R. K. 1991. Sperm-mediated gene flow and the genetic structure of a population of the colonial ascidian *Botryllus schlosseri. Evolution* 45: 130–142.

Grosberg, R. K. and M. W. Hart. 2000. Mate selection and the evolution of highly polymorphic self/nonself recognition genes. *Science* 289: 2111–2114.

Grosberg, R. K. and J. F. Quinn. 1986. The genetic control and consequences of kin recognition by the larvae of a colonial marine invertebrate. *Nature* 322: 456–459.

Grosberg, R. K. and R. R. Strathmann. 1998. One cell, two cell, red cell, blue cell: The persistence of a unicellular state in multicellular life histories. *Trends Ecol. Evol.* 13: 112–116.

Grosberg, R. K., D. R. Levitan and B. B. Cameron. 1996. Evolutionary genetics of allorecognition in the colonial hydroid *Hydractinia symbiolongicarpus. Evolution* 50: 2221–2240.

Grosberg, R. K., M. W. Hart and D. R. Levitan. 1997. Is allorecognition specificity in *Hydractinia symbiolongicarpus* controlled by a single gene? *Genetics* 145: 857–860.

Gross, M. R. 1979. Cuckoldry in sunfishes (*Lepomis*: Centrarchidae). *Can. J. Zool.* 57: 1507–1509.

Gross, M. R. 1996. Alternative reproductive strategies and tactics: Diversity within sexes. *Trends Ecol. Evol.* 11: 92–98.

Gross, M. R. and E. L. Charnov. 1980. Alternative male life histories in bluegill sunfish. *Proc. Natl. Acad. Sci. USA* 77: 6937–6940.

Grudzien, T. A. and B. J. Turner. 1984. Genic identity and geographic differentiation of trophically dichotomous *Ilyodon* (Teleostei: Goodeidae). *Copeia* 1984: 102–107.

Grunwald, L. and S. J. Adler (eds.). 1999. *Letters of the Century: America 1900–1999.* The Dial Press, New York.

Guillemette, J. G. and P. N. Lewis. 1983. Detection of sub-nanogram quantities of DNA and RNA on native and denaturing polyacrylamide and agarose gels by silver staining. *Electrophoresis* 4: 92–94.

Guinand, B., A. Topchy, K. S. Page, M. K. Burnham-Curtis, W. F. Punch and K. T. Scribner. 2002. Comparisons of likelihood and machine learning methods of individual classification. *J. Heredity* 93: 260–269.

Gupta, R. S. 1998. Protein phylogenies and signature sequences: A reappraisal of evolutionary relationships among archaebacteria, eubacteria, and eukaryotes. *Micro. Mol. Biol. Rev.* 62: 1435–1491.

Guries, R. P. and F. T. Ledig. 1982. Genetic diversity and population structure in pitch pine (*Pinus rigida* Mill.). *Evolution* 36: 387–402.

Gyllensten, U. B. 1985. The genetic structure of fish: Differences in the intraspecific distribution of biochemical genetic variation between marine, anadromous, and freshwater species. *J. Fish Biol.* 26: 691–699.

Gyllensten, U. B. and A. C. Wilson. 1987a. Mitochondrial DNA of salmonids. Pp. 301–317 in *Population Genetics and Fisheries Management*, N. Ryman and F. Utter (eds.). University of Washington Press, Seattle.

Gyllensten, U. B. and A. C. Wilson. 1987b. Interspecific mitochondrial DNA transfer and the colonization of Scandinavia by mice. *Genet. Res. Camb.* 49: 25–29.

Gyllensten, U. B., D. Wharton and A. C. Wilson. 1985a. Maternal inheritance of mitochondrial DNA during backcrossing of two species of mice. *J. Heredity* 76: 321–324.

Gyllensten, U. B., R. F. Leary, F. W. Allendorf and A. C. Wilson. 1985b. Introgression between two cutthroat trout subspecies with substantial karyotypic, nuclear and mitochondrial genomic divergence. *Genetics* 111: 905–915.

Gyllensten, U. B., S. Jakobsson and H. Temrin. 1990. No evidence for illegitimate young in monogamous and polygynous warblers. *Nature* 343: 168–170.

Gyllensten, U. B., D. Wharton, A. Josefsson and A. C. Wilson. 1991. Paternal inheritance of mitochondrial DNA in mice. *Nature* 352: 255–257.

Haag, E. S. and J. R. True. 2001. From mutants to mechanisms? Assessing the candidate gene paradigm in evolutionary biology. *Evolution* 55: 1077–1084.

Hachtel, W. 1980. Maternal inheritance of chloroplast DNA in some *Oenothera* species. *J. Heredity* 71: 191–194.

Haddrath, O. and A. J. Baker. 2001. Complete mitochondrial DNA genome sequences of extinct birds: Ratite phylogenetics and the vicariance biogeography hypothesis. *Proc. R. Soc. Lond. B* 268: 939–945.

Hadly, E. A., M. K. Kohn, J. A. Leonard and R. K. Wayne. 1998. A genetic record of population isolation in pocket gophers during Holocene climatic change. *Proc. Natl. Acad. Sci. USA* 95: 6893–6896.

Hadrys, H., M. Balick and B. Schierwater. 1992. Applications of random amplified polymorphic DNA (RAPD) in molecular ecology. *Mol. Ecol.* 1: 55–63.

Haffer, J. 1969. Speciation in Amazonian forest birds. *Science* 165: 131–137.

Hafner, M. S., J. W. Demastes, D. J. Hafner, T. A. Spradling, P. D. Sudman and S. A. Nadler. 1998. Age and movement of a hybrid zone: Implications for dispersal distance in pocket gophers and their chewing lice. *Evolution* 52: 278–282.

Hafner, M. S., J. C. Hafner, J. L. Patton and M. F. Smith. 1987. Macrogeographic patterns of genetic differentiation in the pocket gopher (*Thomomys umbrinus*). *Syst. Zool.* 36: 18–34.

Hagelberg, E., I. C. Gray and A. J. Jeffreys. 1991. Identification of the skeletal remains of a murder victim by DNA analysis. *Nature* 352: 427–429.

Hahn, B. H., G. M. Shaw, K. M. De Cock and P. M. Sharp. 2000. AIDS as a zoonosis: Science and public health implications. *Science* 287: 607–614.

Haig, D. 1999. What is a marmoset? *Am. J. Primatol.* 49: 285–296.

Haig, S. M. and J. C. Avise. 1996. Avian conservation genetics. Pp. 160–189 in *Conservation Genetics: Case Histories from Nature*, J. C. Avise and J. L. Hamrick (eds.). Chapman & Hall, New York.

Haig, S. M. and L. W. Oring. 1988. Genetic differentiation of piping plovers across North America. *Auk* 105: 260–267.

Haig, S. M., J. D. Ballou and S. R. Derrickson. 1990. Management options for preserving genetic diversity: Reintroduction of Guam rails to the wild. *Cons. Biol.* 4: 290–300.

Haig, S. M., J. R. Walters and J. H. Plissner. 1994. Genetic evidence for monogamy in the cooperatively breeding red-cockaded woodpecker. *Behav. Ecol. Sociobiol.* 34: 295–303.

Haig, S. M., C. L. Gratto-Trevor, T. D. Mullins and M. A. Colwell. 1997. Population differentiation of western hemisphere shorebirds throughout the annual cycle. *Mol. Ecol.* 6: 413–427.

Haig, S. M., R. S. Wagner, E. D. Forsman and T. D. Mullins. 2001. Geographic variation and genetic structure in spotted owls. *Cons. Genet.* 2: 25–40.

Haldane, J. B. S. 1922. Sex ratio and unisexual sterility of hybrid animals. *J. Genet.* 12: 101–109.

Haldane, J. B. S. 1932. *The Causes of Evolution.* Longmans and Green, London.

Hale, L. R. and R. S. Singh. 1986. Extensive variation and heteroplasmy in size of mitochondrial DNA among geographic populations of *Drosophila melanogaster*. *Proc. Natl. Acad. Sci. USA* 83: 8813–8817.

Hale, L. R. and R. S. Singh. 1987. Mitochondrial DNA variation and genetic structure in populations of *Drosophila melanogaster*. *Mol. Biol. Evol.* 4: 622–637.

Hale, L. R. and R. S. Singh. 1991. A comprehensive study of genic variation in natural populations of *Drosophila melanogaster*. IV. Mitochondrial DNA variation and the role of history *vs.* selection in the genetic structure of geographic populations. *Genetics* 129: 102–117.

Hall, B. G. 2004. *Phylogenetic Trees Made Easy: A How-To Manual* (2nd Edition). Sinauer Associates, Sunderland, MA.

Hall, H. G. 1990. Parental analysis of introgressive hybridization between African and European honeybees using nuclear DNA RFLP's. *Genetics* 125: 611–621.

Hall, H. G. and K. Muralidharan. 1989. Evidence from mitochondrial DNA that African honey bees spread as continuous maternal lineages. *Nature* 339: 211–213.

Hall, H. G. and D. R. Smith. 1991. Distinguishing African and European honeybee matrilines using amplified mitochondrial DNA. *Proc. Natl. Acad. Sci. USA* 88: 4548–4552.

Hall, J. P. and D. J. Harvey. 2002. The phylogeography of Amazonia revisited: New evidence from riodinid butterflies. *Evolution* 56: 1489–1497.

Halliday, T. 1998. Sperm competition in amphibians. Pp. 503–578 in *Sperm Competition and Sexual Selection*, T. R. Birkhead and A. P. Møller (eds.). Academic Press, London.

Halpern, B. S. 2003. The impact of marine reserves: Do reserves work and does reserve size matter? *Ecol. Appl.* 13: S117-S137.

Halpern, B. S. and R. R. Warner. 2003. Matching marine reserve design to reserve objectives. *Proc. R. Soc. Lond. B* 270: 1871–1878.

Hamada, H., M. G. Petrino, T. Kakunaga, M. Seidman and B. D. Stollar. 1984. Characterization of genomic poly (dT-dG) poly (dC-dA) sequences: Structure, organization, and conformation. *Mol. Cell. Biol.* 4: 2610–2621.

Hamada, M., N. Takasaki, J. D. Reist, A. L. DeCicco, A. Goto and N. Okada. 1998. Detection of the ongoing sorting of ancestrally polymorphic SINEs toward fixation or loss in populations of two species of charr during speciation. *Genetics* 150: 301–311.

Hamblin, M. T. and C. F. Aquadro. 1999. DNA sequence variation and the recombinational landscape in *Drosophila pseudoobscura*: A study of the second chromosome. *Genetics* 153: 859–869.

Hamby, R. K. and E. A. Zimmer. 1992. Ribosomal RNA as a phylogenetic tool in plant systematics. Pp. 50–91 in *Molecular Systematics of Plants*, P. S. Soltis, D. E. Soltis and J. J. Doyle (eds.). Chapman & Hall, New York.

Hames, B. D. and S. J. Higgins (eds.). 1985. *Nucleic Acid Hybridization: A Practical Approach*. IRL Press, Oxford, England.

Hamilton, W. D. 1964. The genetical evolution of social behavior. *J. Theoret. Biol.* 7: 1–52.

Hamilton, W. D. 1990. Mate choice near and far. *Am. Zool.* 30: 341–352.

Hammer, M. F. 1995. A recent common ancestry for human Y chromosomes. *Nature* 378: 376–378.

Hammond, R. L., W. Macesero, B. Flores, O. B. Mohammed, T. Wacher and M. W. Bruford. 2001. Phylogenetic reanalysis of the Saudi gazelle and its implications for conservation. *Cons. Biol.* 15: 1123–1133.

Hamrick, J. L. and R. W. Allard. 1972. Microgeographical variation in allozyme frequencies in *Avena barbata*. *Proc. Natl. Acad. Sci. USA* 69: 2100–2104.

Hamrick, J. L. and M. J. W. Godt. 1989. Allozyme diversity in plants. Pp. 43–63 in *Plant Population Genetics, Breeding and Genetic Resources*, A. H. D. Brown, M. T. Clegg, A. L. Kahler and B. S. Weir (eds.). Sinauer Associates, Sunderland, MA.

Hamrick, J. L. and M. J. W. Godt. 1996. Conservation genetics of endemic plant species. Pp. 281–304 in *Conservation Genetics: Case Histories from Nature*, J. C. Avise and J. L. Hamrick (eds.). Chapman & Hall, New York.

Hamrick, J. L. and M. D. Loveless. 1989. The genetic structure of tropical tree populations: Associations with reproductive biology. Pp. 129–146 in *Plant Evolutionary Ecology*, J. H. Bock and Y. B. Linhart (eds.). Westview Press, Boulder, CO.

Hamrick, J. L. and D. A. Murawski. 1990. The breeding structure of tropical tree populations. *Plant Species Biol.* 5: 157–165.

Hamrick, J. L., Y. B. Linhart and J. B. Mitton. 1979. Relationships between life history characteristics and

electrophoretically detectable genetic variation in plants. *Annu. Rev. Ecol. Syst.* 10: 173–200.

Hamrick, J. L., H. M. Blanton and K. J. Hamrick. 1989. Genetic structure of geographically marginal populations of ponderosa pine. *Am. J. Bot.* 76: 1559–1568.

Hamrick, J. L., M. J. W. Godt and S. L. Sherman-Broyles. 1992. Factors influencing levels of genetic diversity in woody plant species. *New Forests* 6: 95–124.

Hänfling, B., B. Hellemans, F. A. M. Volkaert and G. R. Carvalho. 2002. Late glacial history of the cold-adapted freshwater fish *Cottus gobio*, revealed by microsatellites. *Mol. Ecol.* 11: 1717–1729.

Hanken, J. and P. W. Sherman. 1981. Multiple paternity in Belding's ground squirrel litters. *Science* 212: 351–353.

Hänni, C., V. Laudet, D. Stehelin and P. Taberlet. 1994. Tracking the origins of the cave bear (*Ursus spelaeus*) by mitochondrial DNA sequencing. *Proc. Natl. Acad. Sci. USA* 91: 12336–12340.

Hannonen, M. and L. Sundström. 2003. Worker nepotism among polygynous ants. *Nature* 421: 910.

Hanotte, O., T. Burke, J. A. L. Armour and A. J. Jeffreys. 1991. Cloning, characterization and evolution of Indian peafowl *Pavo cristatus* minisatellite loci. Pp. 193–216 in *DNA Fingerprinting Approaches and Applications*, T. Burke, G. Dolf, A. J. Jeffreys and R. Wolff (eds.). Birkhauser Verlag, Basel, Switzerland.

Hanotte, O., M. W. Bruford and T. Burke. 1992a. Multilocus DNA fingerprints in gallinaceous birds: General approach and problems. *Heredity* 68: 481–494.

Hanotte, O., E. Cairns, T. Robson, M. C. Double and T. Burke. 1992b. Cross-species hybridization of a single-locus minisatellite probe in passerine birds. *Mol. Ecol.* 1: 127–130.

Hanski, I. A. and M. E. Gilpin (eds.). 1997. *Metapopulation Biology: Ecology, Genetics, and Evolution*. Academic Press, New York.

Hanson, M. R. 1991. Plant mitochondrial mutations and male sterility. *Annu. Rev. Genet.* 25: 461–486.

Hanson, M. R. and O. Folkerts. 1992. Structure and function of the higher plant mitochondrial genome. *Int. Rev. Cytol.* 141: 129–172.

Hansson, B. and L. Westerberg. 2002. On the correlation between heterozygosity and fitness in natural populations. *Mol. Ecol.* 11: 2467–2474.

Hansson, B., B. Bensch, D. Hasselquist and M. Åkesson. 2001. Microsatellite diversity predicts recruitment in sibling great reed warblers. *Proc. R. Soc. Lond. B* 268: 1287–1291.

Harding, R. M., S. M. Fullerton, R. C. Griffiths, J. Bond, M. J. Cox, J. A. Schneider, D. S. Moulin and J. B. Clegg. 1997. Archaic African and Asian lineages in the genetic ancestry of modern humans. *Am. J. Human Genet.* 60: 772–789.

Hardy, J., A. Singleton and K. Gwinn-Hardy. 2003. Ethnic differences and disease phenotypes. *Science* 300: 739–740.

Hardy, O. J. 2003. Estimation of pairwise relatedness between individuals and characterization of isolation-by-distance processes using dominant genetic markers. *Mol. Ecol.* 12: 1577–1588.

Hare, M. P. 2001. Prospects for nuclear gene phylogeography. *Trends Ecol. Evol.* 16: 700–706.

Hare, M. P. and J. C. Avise. 1996. Molecular genetic analysis of a stepped multilocus cline in the American oyster (*Crassostrea virginica*). *Evolution* 50: 2305–2315.

Hare, M. P. and J. C. Avise. 1998. Population structure in the American oyster as inferred by nuclear gene genealogies. *Mol. Biol. Evol.* 15: 119–128.

Hare, M. P and G. F. Shields. 1992. Mitochondrial-DNA variation in the polytypic Alaskan song sparrow. *Auk* 109: 126–132.

Hare, M. P., S. A. Karl and J. C. Avise. 1996. Anonymous nuclear DNA markers in the American oyster and their implications for the heterozygote deficiency phenomenon in marine bivalves. *Mol. Biol. Evol.* 13: 334–345.

Hare, P. E. (ed.). 1980. *The Biogeochemistry of Amino Acids*. John Wiley and Sons, New York.

Härlid, A., A. Janke and U. Arnason. 1998. The complete mitochondrial genome of *Rhea americana* and early avian divergences. *J. Mol. Evol.* 46: 669–679.

Harper, J. L. 1977. *The Population Biology of Plants*. Academic Press, New York.

Harper, J. L. 1985. Modules, branches, and the capture of resources. Pp. 1–33 in *Population Biology and Evolution of Clonal Organisms*, J. B. C. Jackson, L. W. Buss and R. E. Cook (eds.). Yale University Press, New Haven, CT.

Harrington, R. W. Jr. 1961. Oviparous hermaphroditic fish with internal self-fertilization. *Science* 134: 1749–1750.

Harrington, R. W. Jr. and K. D. Kallman. 1968. The homozygosity of clones of the self-fertilizing hermaphroditic fish *Rivulus marmoratus* Poey (Cyprinodontidae, Atheriniformes). *Am. Nat.* 102: 337–343.

Harris, H. 1966. Enzyme polymorphisms in man. *Proc. R. Soc. Lond. B* 164: 298–310.

Harris, H. and D. A. Hopkinson. 1976. *Handbook of Enzyme Electrophoresis in Human Genetics*. North-Holland, Amsterdam.

Harris, S. A. and R. Ingram. 1991. Chloroplast DNA and biosystematics: The effects of intraspecific diversity and plastid transmission. *Taxon* 40: 393–412.

Harrison, R. G. 1979. Speciation in North American field crickets: Evidence from electrophoretic comparisons. *Evolution* 33: 1009–1023.

Harrison, R. G. 1989. Animal mitochondrial DNA as a genetic marker in population and evolutionary biology. *Trends Ecol. Evol.* 4: 6–11.

Harrison, R. G. 1990. Hybrid zones: Windows on evolutionary process. *Oxford Surv. Evol. Biol.* 7: 69–128.

Harrison, R. G. 1991. Molecular changes at speciation. *Annu. Rev. Ecol. Syst.* 22: 281–308.

Harrison, R. G. (ed.). 1993. *Hybrid Zones and the Evolutionary Process*. Oxford University Press, Oxford, England.

Harrison, R. G. 1998. Linking evolutionary pattern and process. Pp. 19–31 in *Endless Forms: Species and Speciation*, D. J. Howard and S. H. Berlocher (eds.). Oxford University Press, New York.

Harrison, R. G. and S. M. Bogdanowicz. 1997. Patterns of variation and linkage disequilibrium in field crickets across a narrow hybrid zone. *Evolution* 51: 493–505.

Harrison, R. G. and D. M. Rand. 1989. Mosaic hybrid zones and the nature of species boundaries. Pp. 111–133 in *Speciation and Its Consequences*, D. Otte and J. A. Endler (eds.). Sinauer Associates, Sunderland, MA.

Harrison, R. G., D. M. Rand and W. C. Wheeler. 1985. Mitochondrial DNA size variation within individual crickets. *Science* 228: 1446–1448.

Harrison, R. G., D. M. Rand and W. C. Wheeler. 1987. Mitochondrial DNA variation in field crickets across a narrow hybrid zone. *Mol. Biol. Evol.* 4: 144–158.

Harry, J. L. and D. A. Briscoe. 1988. Multiple paternity in the loggerhead turtle (*Caretta caretta*). *J. Heredity* 79: 96–99.

Hart, M. W. and R. K. Grosberg. 1999. Kin interactions in a colonial hydrozoan (*Hydractinia symbiolongicarpus*): Population structure on a mobile landscape. *Evolution* 53: 793–805.

Hartl, D. L. and A. G. Clark. 1989. *Principles of Population Genetics* (2nd Edition). Sinauer Associates, Sunderland, MA.

Hartl, D. L. and D. E. Dykhuizen. 1984. The population genetics of *Escherichia coli*. *Annu. Rev. Genet.* 18: 31–68.

Hartley, I. R., M. Shepherd, T. Robson and T. Burke. 1993. Reproductive success of polygynous male corn buntings (*Miliaria calandra*) as confirmed by DNA fingerprinting. *Behav. Ecol.* 4: 310–317.

Hartman, H. and A. Fedorov. 2002. The origin of the eukaryotic cell: A genomic investigation. *Proc. Natl. Acad. Sci. USA* 99: 1420–1425.

Harvey, P. H. and M. D. Pagel. 1991. *The Comparative Method in Evolutionary Biology*. Oxford University Press, Oxford, England.

Harvey, P. H. and A. Purvis. 1991. Comparative methods for explaining adaptations. *Nature* 351: 619–624.

Harvey, P. H. and H. Steers. 1999. One use of phylogenies for conservation biologists: Inferring population history from gene sequences. Pp. 101–120 in *Genetics and the Extinction of Species*, L. F. Landweber and A. P. Dobson (eds.). Princeton University Press, Princeton, NJ.

Harvey, P. H., R. M. May and S. Nee. 1994. Phylogenies without fossils. *Evolution* 48: 523–529.

Harvey, P. H., A. F. Read and S. Nee. 1995. Why ecologists need to be phylogenetically challenged. *J. Ecol.* 83: 535–536.

Harvey, P. H., A. J. Leigh Brown, J. Maynard Smith and S. Nee (eds.). 1996. *New Uses for New Phylogenies*. Oxford University Press, Oxford, England.

Harvey, W. D. 1990. Electrophoretic techniques in forensics and law enforcement. Pp. 313–321 in *Electrophoretic and Isoelectric Focusing Techniques in Fisheries Management*, D. H. Whitmore (ed.). CRC Press, Boca Raton, FL.

Hasegawa, M. 1990. Phylogeny and molecular evolution in primates. *Japan J. Genet.* 65: 243–266.

Hasegawa, M. and W. M. Fitch. 1996. Dating the cenancestor of organisms. *Science* 274: 1750.

Hasegawa, M. and S. Horai. 1991. Time of the deepest root for polymorphism in human mitochondrial DNA. *J. Mol. Evol.* 32: 37–42.

Hasegawa, M. and H. Kishino. 1989. Heterogeneity of tempo and mode of mitochondrial DNA evolution among mammalian orders. *Japan J. Genet.* 64: 243–258.

Hasegawa, M., J. Adachi and M. C. Milinkovitch. 1997. Novel phylogeny of whales supported by total molecular evidence. *J. Mol. Evol.* 44: S117-S120.

Hashimoto, T., E. Otaka, J. Adachi, K. Mizuta and M. Hasegawa. 1993. The giant panda is closer to a bear, judged by α- and β-hemoglobin sequences. *J. Mol. Evol.* 36: 282–289.

Hasslequist, D., S. Bensch and T. von Schantz. 1996. Correlation between male song repertoire, extra-pair paternity and offspring survival in the great reed warbler. *Nature* 381: 229–232.

Hasson, W. and W. F. Eanes. 1996. Contrasting histories of three gene regions associated with *IN(3L) Payne* of *Drosophila melanogaster. Genetics* 144: 1565–1575.

Hauser, L., A. R. Beaumont, G. T. H. Marshall and R. J. Wyatt. 1991. Effects of sea trout stocking on the population genetics of landlocked brown trout, *Salmo trutta* L., in the Conwy River system, North Wales, U. K. *J. Fish Biol.* 39A: 109–116.

Havey, M. J., J. D. McCreight, B. Rhodes and G. Taurick. 1998. Differential transmission of the *Cucumis* organellar genomes. *Theoret. Appl. Genet.* 97: 122–128.

Hawthorne, D. J. and S. Via. 2001. Genetic linkage of ecological specialization and reproductive isolation in pea aphids. *Nature* 412: 904–907.

Hayden, M. R., H. C. Hopkins, M. MacRae and P. H. Brighton. 1980. The origin of Huntington's chorea in the Afrikaner population of South Africa. *S. African Med. J.* 58: 197–200.

Haydock, J., P. G. Parker and K. N. Rabenold. 1996. Extra-pair paternity uncommon in the cooperatively breeding bicolored wren. *Behav. Ecol. Sociobiol.* 38: 1–16.

Heard, S. B. 1996. Patterns in phylogenetic tree balance with variable and evolving speciation rates. *Evolution* 50: 2141–2148.

Heath, D. J., J. R. Radford, B. J. Riddoch and D. Childs. 1990. Multiple mating in a natural population of the isopod *Sphaeroma rugicauda*; evidence from distorted ratios in offspring. *Heredity* 64: 81–85.

Hebert, P. D. N. 1974a. Enzyme variability in natural populations of *Daphnia magna* III. Genotypic frequencies in intermittent populations. *Genetics* 77: 335–341.

Hebert, P. D. N. 1974b. Ecological differences between genotypes in a natural population of *Daphnia magna. Heredity* 33: 327–337.

Hebert, P. D. N. 1974c. Enzyme variability in natural populations of *Daphnia magna* II. Genotypic frequencies in permanent populations. *Genetics* 77: 323–334.

Hebert, P. D. N. and T. J. Crease. 1980. Clonal existence in *Daphnia pulex* (Leydig): Another planktonic paradox. *Science* 207: 1363–1365.

Hebert, P. D. N. and R. D. Ward. 1972. Inheritance during parthenogenesis in *Daphnia magna. Genetics* 71: 639–642.

Hebert, P. D. N. and R. D. Ward. 1976. Enzyme variability in natural populations of *Daphnia magna. Heredity* 36: 331–341.

Hebert, P. D. N., M. J. Beaton, S. S. Schwartz and D. J. Stanton. 1989. Polyphyletic origin of asexuality in *Daphnia pulex*. I. Breeding-system variation and levels of clonal diversity. *Evolution* 43: 1004–1015.

Hebert, P. D. N., A. Cywinska, S. L. Ball and J. R. deWaard. 2003. Biological identifications through DNA barcodes. *Proc. R. Soc. Lond. B* 270: 313–321.

Heckman, D. S., D. M. Geiser, B. R. Eidell, R. L. Stauffer, N. L. Kardos and S. B. Hedges. 2001. Molecular evidence for the early colonization of land by fungi and plants. *Science* 293: 1129–1133.

Hedgecock, D. 1979. Biochemical genetic variation and evidence of speciation in *Chthamalus* barnacles of the tropical eastern Pacific Ocean. *Mar. Biol.* 54: 207–214.

Hedgecock, D. 1986. Is gene flow from pelagic larval dispersal important in the adaptation and evolution of marine invertebrates? *Bull. Mar. Sci.* 39: 550–564.

Hedgecock, D. 1994a. Does variance in reproductive success limit effective population sizes of marine organisms? Pp. 122–134 in *Genetics and Evolution of Aquatic Organisms*, A. R. Beaumont (ed.). Chapman & Hall, London.

Hedgecock, D. 1994b. Temporal and spatial genetic structure of marine animal populations in the California current. *CalCOFI Reports* 35: 73–81.

Hedgecock, D. and F. Sly. 1990. Genetic drift and effective population sizes of hatchery-propagated stocks of the Pacific oyster, *Crassostrea gigas. Aquaculture* 88: 21–38.

Hedgecock, D., V. Chow and R. S. Waples. 1992. Effective population numbers of shellfish broodstocks estimated from temporal variance in allelic frequencies. *Aquaculture* 108: 215–232.

Hedges, R. W. 1972. The pattern of evolutionary change in bacteria. *Heredity* 28: 39–48.

Hedges, S. B. 1989. An island radiation: Allozyme evolution in Jamaican frogs of the genus *Eleutherodactylus* (Leptodactylidae). *Caribbean J. Sci.* 25: 123–147.

Hedges, S. B. 1992. The number of replications needed for accurate estimation of the bootstrap *P* value in phylogenetic studies. *Mol. Biol. Evol.* 9: 366–369.

Hedges, S. B. 1996. Historical biogeography of West Indian vertebrates. *Annu. Rev. Ecol. Syst.* 27: 163–196.

Hedges, S. B. 2001. Afrotheria: Plate tectonics meets genomics. *Proc. Nat. Acad. Sci. USA* 98: 1–2.

Hedges, S. B. and K. L. Burnell. 1990. The Jamaican radiation of *Anolis* (Sauria: Iguanidae): An analysis of relationships and biogeography using sequential electrophoresis. *Caribbean J. Sci.* 26: 31–44.

Hedges, S. B. and S. Kumar. 2002. Vertebrate genomes compared. *Science* 297: 1283–1285.

Hedges, S. B. and L. L. Poling. 1999. A molecular phylogeny for reptiles. *Science* 283: 998–1001.

Hedges, S. B. and C. G. Sibley. 1994. Molecules vs morphology in avian evolution: The case of the "pelecaniform" birds. *Proc. Natl. Acad. Sci. USA* 91: 9861–9865.

Hedges, S. B., K. D. Moberg and L. R. Maxson. 1990. Tetrapod phylogeny inferred from 18S and 28S ribosomal RNA sequences and a review of the evidence for amniote relationships. *Mol. Biol. Evol.* 7: 607–633.

Hedges, S. B., R. L. Bezy and L. R. Maxson. 1991. Phylogenetic relationships and biogeography of Xantusiid lizards, inferred from mitochondrial DNA sequences. *Mol. Biol. Evol.* 8: 767–780.

Hedges, S. B., J. P. Bogart and L. R. Maxson. 1992a. Ancestry of unisexual salamanders. *Nature* 356: 708–710.

Hedges, S. B., C. A. Hass and L. R. Maxson. 1992b. Caribbean biogeography: Molecular evidence for dispersal in West Indian terrestrial vertebrates. *Proc. Natl. Acad. Sci. USA* 89: 1909–1913.

Hedges, S. B., S. Kumar and K. Tamura. 1992c. Human origins and analysis of mitochondrial DNA sequences. *Science* 255: 737–739.

Hedges, S. B., C. A. Hass and L. R. Maxson. 1993. Relations of fish and tetrapods. *Nature* 363: 501–502.

Hedges, S. B., P. H. Parker, C. G. Sibley and S. Kumar. 1996. Continental breakup and the ordinal diversification of birds and mammals. *Nature* 381: 226–229.

Hedges, S. B., H. Chen, S. Kumar, D. Y-C. Wang, A. S. Thompson and H. Watanabe. 2001. A genomic timescale for the origin of eukaryotes. *BMC Evol. Biol.* 2001: 1–4.

Hedrick, P. W. 1986. Genetic polymorphisms in heterogeneous environments: A decade later. *Annu. Rev. Ecol. Syst.* 17: 535–566.

Hedrick, P. W. 1992. Shooting the RAPDs. *Nature* 355: 679–680.

Hedrick, P. W. 1995. Gene flow and genetic restoration: The Florida panther as a case study. *Cons. Biol.* 9: 996–1007.

Hedrick, P. W. 1999. Highly variable loci and their interpretation in evolution and conservation. *Evolution* 53: 313–318.

Hedrick, P. W. and S. T. Kalinowski. 2000. Inbreeding depression in conservation biology. *Annu. Rev. Ecol. Syst.* 31: 139–162.

Hedrick, P. W. and P. S. Miller. 1992. Conservation genetics: Techniques and fundamentals. *Ecol. Applic.* 2: 30–46.

Hedrick, P. W., P. F. Brussard, F. W. Allendorf, J. A. Beardmore and S. Orzack. 1986. Protein variation, fitness, and captive propagation. *Zoo Biol.* 5: 91–99.

Hedrick, P. W., T. S. Whittam and P. Parham. 1991. Heterozygosity at individual amino acid sites: Extremely high levels for *HLA-A* and *-B* genes. *Proc. Natl. Acad. Sci. USA* 88: 5897–5901.

Hedrick, P. W., T. E. Dowling, W. L. Minckley, C. A. Tibbets, B. D. DeMarais and P. C. Marsh. 2000. Establishing a captive broodstock for the endangered bonytail chub (*Gila elegans*). *J. Heredity* 91: 35–39.

Hedrick, P. W., R. Fredrickson and H. Ellegren. 2001. Evaluation of d^2, a microsatellite measure of inbreeding and outbreeding, in wolves with a known pedigree. *Evolution* 55: 1254–1260.

Heg, D. and R. van Treuren. 1998. Female-female cooperation in polygynous oystercatchers. *Nature* 391: 687–691.

Heinemann, J. A. 1991. Genetics of gene transfer between species. *Trends Genet.* 7: 181–185.

Heinemann, J. A. and G. F. Sprague Jr. 1989. Bacterial conjugative plasmids mobilize DNA transfer between bacteria and yeast. *Nature* 340: 205–209.

Helbig, A. J., M. Salomon, S. Bensch and I. Seibold. 2001. Male-biased gene flow across an avian hybrid zone: Evidence from mitochondrial and microsatellite DNA. *J. Evol. Biol.* 14: 277–287.

Hellberg, M. E. 1996. Dependence of gene flow on geographic distance in two solitary corals with different larval dispersal capabilities. *Evolution* 50: 1167–1175.

Hellmann, I., S. Zöllner, W. Enard, I. Ebersberger, B. Nickel and S. Pääbo. 2003. Selection on human genes as revealed from comparisons to chimpanzee cDNA. *Genome Res.* 13: 831–837.

Hendy, M. D. and D. Penny. 1989. A framework for the quantitative study of evolutionary trees. *Syst. Zool.* 38: 297–309.

Hennig, W. 1966. *Phylogenetic Systematics*. University of Illinois Press, Urbana.

Henshaw, M. T., J. E. Strassmann and D. C. Queller. 2001. Swarm-founding in the polistine wasps: The importance of finding many microsatellite loci in studies of adaptation. *Mol. Ecol.* 10: 185–191.

Henson, S. A. and R. R. Warner. 1997. Male and female alternative reproductive behaviors in fishes: A new approach using intersexual dynamics. *Annu. Rev. Ecol. Syst.* 28: 571–592.

Hepper, P. G. (ed.). 1991. *Kin Recognition*. Cambridge University Press, New York.

Herbers, J. M. 1986. Nest site limitation and facultative polygyny in the ant *Leptothorax longispinosus*. *Behav. Ecol. Sociobiol.* 19: 115–122.

Herbots, H. M. 1997. The structured coalescent. Pp. 231–235 in *Progress in Human Population Genetics and Human Evolution*, P. Donnelly and S. Tavaré (eds.). Springer-Verlag, New York.

Hermanutz, L. A., D. J. Innes and I. M. Weis. 1989. Clonal structure of arctic dwarf birch (*Betula glandulosa*) at its northern limit. *Am. J. Bot.* 76: 755–761.

Hewitt, G. M. 1988. Hybrid zones: Natural laboratories for evolutionary studies. *Trends Ecol. Evol.* 3: 158–166.

Hewitt, G. M. 1989. The subdivision of species by hybrid zones. Pp. 85–110 in *Speciation and Its Consequences*, D. Otte and J. A. Endler (eds.). Sinauer Associates, Sunderland, MA.

Hewitt, G. M. 1996. Some genetic consequences of ice ages, and their role in divergence and speciation. *Biol. J. Linn. Soc.* 58: 247–276.

Hewitt, G. M. 1999. Post-glacial recolonization of European biota. *Biol. J. Linn. Soc.* 68: 87–112.

Hewitt, G. M. 2000. The genetic legacy of the Quaternary ice ages. *Nature* 405: 907–913.

Hewitt, G. M. 2001. Speciation, hybrid zones and phylogeography—or seeing genes in space and time. *Mol. Ecol.* 10: 537–549.

Hey, J. 1994. Bridging phylogenetics and population genetics with gene tree models. Pp. 435–449 in *Molecular Ecology and Evolution: Approaches and Applications*, B. Schierwater, B. Streit, G. P. Wagner and R. DeSalle (eds.). Birkhaüser Verlag, Basel, Switzerland.

Hey, J. 2001. *Genes, Categories and Species*. Oxford University Press, New York.

Hey, J. and R. M. Kliman. 1993. Population genetics and phylogenetics of DNA sequence variation at multiple loci within the *Drosophila melanogaster* species complex. *Mol. Biol. Evol.* 10: 804–822.

Hey, J. and C. A. Machado. 2003. The study of structured populations: New hope for a difficult and divided science. *Nature Rev. Genet.* 4: 535–543.

Hey, J., R. S. Waples, M. L. Arnold, R. K. Butlin and R. G. Harrison. 2003. Understanding and confronting species uncertainty in biology and conservation. *Trends Ecol. Evol.* 18: 597–603.

Hey, J., Y.-J. Won, A. Sivasundar, R. Nielsen and J. A. Markert. 2004. Using nuclear haplotypes with microsatellites to study gene flow between recently separated populations. *Mol. Ecol., in press*.

Heyward, A. J. and J. A. Stoddart. 1985. Genetic structure of two species of *Montipora* on a patch reef: Conflicting results from electrophoresis and histocompatibility. *Mar. Biol.* 85: 117–121.

Heywood, J. S. 1991. Spatial analysis of genetic variation in plant populations. *Annu. Rev. Ecol. Syst.* 22: 335–355.

Hibbett, D. S. and M. Binder. 2002. Evolution of complex fruiting-body morphologies in homobasidiomycetes. *Proc. R. Soc. Lond. B* 269: 1963–1969.

Hibbett, D. S. and M. J. Donoghue. 1996. Implications of phylogenetic studies for conservation of genetic diversity in shiitake mushrooms. *Cons. Biol.* 10: 1321–1327.

Hickey, D. A. 1982. Selfish DNA: A sexually-transmitted nuclear parasite. *Genetics* 101: 519–531.

Hickey, D. A. and M. D. McLean. 1980. Selection for ethanol tolerance and Adh allozymes in natural populations of *Drosophila melanogaster*. *Genet. Res.* 36: 11–15.

Hickey, R. J., M. A. Vincent and S. I. Guttman. 1991. Genetic variation in running buffalo clover (*Trifolium stoloniferum*, Fabaceae). *Cons. Biol.* 5: 309–316.

Hickman, C. S. and J. H. Lipps. 1985. Geologic youth of Galapagos Islands confirmed by marine stratigraphy and paleontology. *Science* 227: 1578–1580.

Hidaka, M., K. Yurugi, S. Sunagawa and R. A. Kinzie. 1997. Contact reactions between young colonies of the coral *Pocillopora damicornis*. *Coral Reefs* 16: 13–20.

Hift, R. J. and 9 others. 1997. Variegate porphyria in South Africa, 1688–1996: New developments in an old disease. *S. Afr. Med. J.* 87: 722–731.

Higashi, M., G. Takimoto and N. Yamamura. 1999. Sympatric speciation by sexual selection. *Nature* 402: 523–526.

Higgs, D. R. and 12 others. 1986. Analysis of the human α-globin gene cluster reveals a highly informative genetic locus. *Proc. Natl. Acad. Sci. USA* 83: 5165–5169.

Highton, R. 1984. A new species of woodland salamander of the *Plethodon glutinosus* group from the southern Appalachian Mountains. *Brimleyana* 9: 1–20.

Highton, R., G. C. Maha and L. R. Maxson. 1989. Biochemical evolution in the slimy salamanders of the *Plethodon glutinosus* complex in the eastern United States. *Illinois Biol. Monogr.* 57: 1–153.

Higuchi, R., B. Bowman, M. Freiberger, O. A. Ryder and A. C. Wilson. 1984. DNA sequence from the quagga, an extinct member of the horse family. *Nature* 312: 282–284.

Higuchi, R. G., L. A. Wrischnik, E. Oakes, M. George, B. Tong and A. C. Wilson. 1987. Mitochondrial DNA of the extinct quagga: Relatedness and extent of post-mortem change. *J. Mol. Evol.* 25: 283–287.

Hilbish, T. J. and R. K. Koehn. 1985. The physiological basis of natural selection at the *LAP* locus. *Evolution* 39: 1302–1317.

Hilbish, T. J., L. E. Deaton and R. K. Koehn. 1982. Effect of an allozyme polymorphism on regulation of cell volume. *Nature* 298: 688–689.

Hildemann, W. H., R. L. Raison, G. Cheung, C. J. Hull, L. Akaka and J. Okamoto. 1977. Immunological specificity and memory in a scleractinian coral. *Nature* 270: 219–223.

Hill, W. G. 1987. DNA fingerprints applied to animal and bird populations. *Nature* 327: 98–99.

Hillis, D. M. 1987. Molecular versus morphological approaches to systematics. *Annu. Rev. Ecol. Syst.* 18: 23–42.

Hillis, D. M. 1989. Genetic consequences of partial self-fertilization on populations of *Liguus fasciatus* (Mollusca: Pulmonata: Bulimulidae). *Am. Malacolog. Bull.* 7: 7–12.

Hillis, D. M. and J. J. Bull. 1991. Of genes and genomes. *Science* 254: 528–558.

Hillis, D. M. and M. T. Dixon. 1989. Vertebrate phylogeny: Evidence from 28S ribosomal DNA sequences. Pp. 355–367 in *The Hierarchy of Life*, B. Fernholm, K. Bremer and H. Jornvall (eds.). Elsevier, Amsterdam.

Hillis, D. M. and M. T. Dixon. 1991. Ribosomal DNA: Molecular evolution and phylogenetic inference. *Q. Rev. Biol.* 66: 411–453.

Hillis, D. M. and C. Moritz (eds.). 1990. *Molecular Systematics*. Sinauer Associates, Sunderland, MA.

Hillis, D. M., A. Larson, S. K. Davis and E. A. Zimmer. 1990. Nucleic acids III: Sequencing. Pp. 318–370 in *Molecular Systematics*, D. M. Hillis and C. Moritz (eds.). Sinauer Associates, Sunderland, MA.

Hillis, D. M., M. T. Dixon and A. L. Jones. 1991. Minimal genetic variation in a morphologically diverse species (Florida tree snail, *Liguus fasciatus*). *J. Heredity* 82: 282–286.

Hillis, D. M., J. J. Bull, M. E. White, M. R. Badgett and I. J. Molineux. 1992. Experimental phylogenetics: Generation of a known phylogeny. *Science* 255: 589–592.

Hillis, D. M., C. Moritz and B. K. Mable (eds.). 1996. *Molecular Systematics* (2nd Edition). Sinauer Associates, Sunderland, MA.

Hindar, K., B. Jonsson, N. Ryman and G. Ståhl. 1991. Genetic relationships among landlocked, resident, and anadromous brown trout, *Salmo trutta* L. *Heredity* 66: 83–91.

Hirt, R. P., J. M. Logsdon Jr., B. Healy, M. W. Dorey, W. F. Doolittle and T. M. Embley. 1999. Microsporidia are relatd to fungi: Evidence from the largest subunit of RNA polymerase II and other proteins. *Proc. Natl. Acad. Sci. USA* 96: 580–585.

Hobbs, R. J. 1992. The role of corridors in conservation: Solution or bandwagon? *Trends Ecol. Evol.* 7: 389–392.

Hobson, K. A. 1999. Tracing origins and migration of wildlife using stable isotopes: A review. *Oecologia* 120: 314–326.

Hodges, S. A. and M. A. Arnold. 1994. Columbines: A geographically widespread species flock. *Proc. Natl. Acad. Sci. USA* 91: 5129–5232.

Hodges, S. A. and M. A. Arnold. 1995. Spurring plant diversification: Are floral nectar spurs a key innovation? *Proc. R. Soc. Lond. B* 262: 343–348.

Hodges, S. A., J. M. Burke and M. L. Arnold. 1996. Natural formation of *Iris* hybrids: Experimental evidence on the establishment of hybrid zones. *Evolution* 50: 2504–2509.

Hoeh, W. R., K. H. Blakley and W. M. Brown. 1991. Heteroplasmy suggests limited biparental inheritance of *Mytilus* mitochondrial DNA. *Science* 251: 1488–1490.

Hoeh, W. R., D. T. Stewart, C. Saavedra, W. B. Sutherland and E. Zouros. 1997. Phylogenetic evidence for role-reversals of gender-associated mitochondrial DNA in *Mytilus* (Bivalvia; Mytilidae). *Mol. Biol. Evol.* 14: 959–967.

Hoeh, W. R., D. T. Stewart and S. I. Guttman. 2002. High fidelity of mitochondrial transmission under the doubly uniparental mode of inheritance in freshwater mussels (Bivalvia: Unionoidea). *Evolution* 56: 2252–2261.

Hoelzel, A. R. (ed.). 1991a. *Genetic Ecology of Whales and Dolphins*. Black Bear Press, Cambridge, England.

Hoelzel, A. R. 1991b. Analysis of regional mitochondrial DNA variation in the killer whale; implications for cetacean conservation. Pp. 225–233 in *Genetic Ecology of Whales and Dolphins*, A. R. Hoelzel (ed.). Black Bear Press, Cambridge, England.

Hoelzel, A. R. (ed.). 1992. *Molecular Genetic Analysis of Populations: A Practical Approach*. IRL Press, Oxford, England.

Hoelzel, A. R. 1994. Genetics and ecology of whales and dolphins. *Annu. Rev. Ecol. Syst.* 25: 377–399.

Hoelzel, A. R. 1998. Genetic structure of cetacean populations in sympatry, parapatry, and mixed assemblages: Implications for conservation policy. *J. Heredity* 89: 451–458.

Hoelzel, A. R. 1999. Impact of population bottlenecks on genetic variation and the importance of life-history; a case study of the northern elephant seal. *Biol. J. Linn. Soc.* 68: 23–39.

Hoelzel, A. R. and G. A. Dover. 1991a. *Molecular Genetic Ecology*. Oxford University Press, Oxford, England.

Hoelzel, A. R. and G. A. Dover. 1991b. Genetic differentiation between sympatric killer whale populations. *Heredity* 66: 191–196.

Hoelzel, A. R., J. Halley, C. Campagna, T. Arnbom, B. Le Boeuf, S. J. O'Brien, K. Ralls and G. A. Dover. 1993. Elephant seal genetic variation and the use of simulation models to investigate historical population bottlenecks. *J. Heredity* 84: 443–449.

Hoelzel, A. R., C. W. Potter and P. B. Best. 1998a. Genetic differentiation between parapatric "nearshore" and "offshore" populations of the bottlenose dolphin. *Proc. R. Soc. Lond. B* 265: 1177–1183.

Hoelzel, A. R., M. Dahlheim and S. J. Stern. 1998b. Low genetic variation among killer whales (*Orcinus orca*) in the eastern North Pacific and genetic differentiation between foraging specialists. *J. Heredity* 89: 121–128.

Hoelzel, A. R., B. J. Le Boeuf, J. Reiter and C. Campagna. 1999. Alpha-male paternity in elephant seals. *Behav. Ecol. Sociobiol.* 46: 298–306.

Hoelzel, A. R., C. Campagna and T. Arnbom. 2001. Genetic and morphometric differentiation between island and mainland southern elephant seal populations. *Proc. R. Soc. Lond. B* 268: 325–332.

Hoelzel, A. R., A. Natoli, M. E. Dalheim, C. Olavarria, R. W. Baird and N. A. Black. 2002a. Low worldwide genetic diversity in the killer whale (*Orcinus orca*): Implications for demographic history. *Proc. R. Soc. Lond. B* 269: 1467–1473.

Hoelzel, A. R., R. C. Fleischer, C. Campagna, B. J. Le Boeuf and G. Alvord. 2002b. Impact of a population bottleneck on symmetry and genetic diversity in the northern elephant seal. *J. Evol. Biol.* 15: 567–575.

Hoelzer, G. A. 1997. Inferring phylogenies from mtDNA variation: Mitochondrial-gene trees versus nuclear-gene trees revisited. *Evolution* 51: 622–626.

Hoffman, R. J. 1983. The mating system of the terrestrial slug *Deroceras laeve*. *Evolution* 37: 423–425.

Hoffman, R. J. 1986. Variation in contributions of asexual reproduction to the genetic structure of populations of the sea anemone *Metridium senile*. *Evolution* 40: 357–365.

Hoffmann, R. J., J. L. Boore and W. M. Brown. 1992. A novel mitochondrial genome organization for the blue mussel, *Mytilus edulis*. *Genetics* 131: 397–412.

Hofreiter, M., H. N. Poinar, W. G. Spaulding, K. Bauer, P. S. Martin, G. Possnert and S. Pääbo. 2000. A molecular analysis of ground sloth diet through the last glaciation. *Mol. Ecol.* 9: 1975–1984.

Hofreiter, M., D. Serre, H. N. Poinar, M. Kuch and S. Pääbo. 2001. Ancient DNA. *Nature Rev. Genet.* 2: 353–359.

Hofreiter, M. and 12 others. 2002. Ancient DNA analyses reveal high mitoochondrial DNA sequence diversity and parallel morphological evolution of Late Pleistocene cave bears. *Mol. Biol. Evol.* 19: 1244–1250.

Holder, M. T., M. V. Erdmann, T. P. Wilcox, R. L. Caldwell and D. M. Hillis. 1999. Two living coelacanth species? *Proc. Natl. Acad. Sci. USA* 96: 12616–12620.

Holland, B. S. and M. G. Hadfield. 2002. Islands within an island: Phylogeography and conservation genetics of the endangered Hawaiian tree snail *Achatinella mustelina*. *Mol. Ecol.* 11: 365–375.

Holland, M. M. and T. J. Parsons. 1999. Validation and use of mitochondrial DNA sequence analysis for forensic casework: A review. *Forensic Sci. Rev.* 11: 21–50.

Hölldobler, B. and E. O. Wilson. 1990. *The Ants*. Belknap Press, Cambridge, MA.

Hollingsworth, P. M., R. M. Bateman and R. J. Gornall (eds.). 1999. *Molecular Systematics and Plant Evolution*. Taylor & Francis, New York.

Hollocher, H. 1996. Island hopping in *Drosophila*: Patterns and processes. *Phil. Trans. R. Soc. Lond. B* 351: 735–743.

Hollocher, H. and C.-I. Wu. 1996. The genetics of reproductive isolation in the *Drosophlia simulans* clade: X vs. autosomal effects and male versus female effects. *Genetics* 143: 1243–1255.

Holmes, E. C., L. Q. Zhang, P. Simmonds, C. A. Ludlam and A. J. L. Brown. 1992. Convergent and divergent sequence evolution in the surface envelope glycoprotein of human immunodeficiency virus type 1 within a single infected patient. *Proc. Natl. Acad. Sci. USA* 89: 4835–4839.

Holmes, W. G. and P. W. Sherman. 1982. The ontogeny of kin recognition in two species of ground squirrels. *Am. Zool.* 22: 491–517.

Holmquist, G. 1989. Evolution of chromosome bands: Molecular ecology of noncoding DNA. *J. Mol. Evol.* 28: 469–486.

Holt, R. A. and 122 others. 2002. The genome sequence of the malaria mosquito *Anopheles gambiae*. *Science* 298: 129–149.

Honeycutt, R. L. 1992. Naked mole-rats. *Am. Sci.* 80: 43–53.

Hoogland, J. L. 1982. Prairie dogs avoid extreme inbreeding. *Science* 215: 1639–1641.

Hoogland, J. L. and D. W. Foltz. 1982. Variance in male and female reproductive success in a harem-polygynous mammal, the black-tailed prairie dog (Sciuridae: *Cynomys ludovicianus*). *Behav. Ecol. Sociobiol.* 11: 155–163.

Hooper, R. E. and M. T. Siva-Jothy. 1996. Last male sperm precedence in a damselfly demonstrated by RAPD profiling. *Mol. Ecol.* 5: 449–452.

Horai, S., Y. Satta, K. Hayasaka, R. Kondo, T. Inoue, T. Ishida, S. Hayashi and N. Takahata. 1992. Man's place in Hominoidea revealed by mitochondrial DNA genealogy. *J. Mol. Evol.* 35: 32–43.

Hori, H. and S. Osawa. 1987. Origin and evolution of organisms as deduced from 5S ribosomal RNA sequences. *Mol. Biol. Evol.* 4: 445–472.

Hori, H., B.-L. Lim and S. Osawa. 1985. Evolution of green plants as deduced from 5S rRNA sequences. *Proc. Natl. Acad. Sci. USA* 82: 820–823.

Horn, G. T., B. Richards and K. W. Klinger. 1989. Amplification of a highly polymorphic VNTR segment by the polymerase chain reaction. *Nucleic Acids Res.* 17: 2140.

Horowitz, J. J. and G. S. Whitt. 1972. Evolution of a nervous system specific lactate dehydrogenase isozyme in fish. *J. Exp. Zool.* 180: 13–31.

Höss, M., M. Kohn, S. Pääbo, F. Knauer and W. Schröder. 1992. Excrement analysis by PCR. *Nature* 359: 199.

Höss, M., A. Dilling, A. Currant and S. Pääbo. 1996. Molecular phylogeny of the extinct ground sloth *Mylodon darwinii. Proc. Natl. Acad. Sci. USA* 93: 181–185.

Houck, M. A., J. B. Clark, K. R. Peterson and M. G. Kidwell. 1991. Possible horizontal transfer of *Drosophila* genes by the mite *Proctolaelaps regalis. Science* 253: 1125–1129.

Houde, P., A. Cooper, E. Leslie, A. E. Strand and G. A. Montaño. 1997. Phylogeny and evolution of 12S rDNA in Gruiformes (Aves). Pp. 121–158 in *Avian Molecular Evolution and Systematics*, D. Mindell (ed.). Academic Press, San Diego, CA.

Houlden, B. A., L. Woodworth and K. Humphreys. 1997. Captive breeding, paternity determination, and genetic variation in chimpanzees (*Pan troglodytes*) in the Australasian region. *Primates* 38: 241–347.

Howard, D. J. 1998. Unanswered questions and future direction in the study of speciation. Pp. 439–448 in *Endless Forms: Species and Speciation*, D. J. Howard and S. H. Berlocher (eds.). Oxford University Press, New York.

Howard, D. J. and S. H. Berlocher (eds.). 1998. *Endess Forms: Species and Speciation.* Oxford University Press, New York.

Howard, D. J., M. Reece, P. G. Gregory, J. Chu and M. L. Cain. 1998. The evolution of barriers to fertilization between closely related organisms. Pp. 279–288 in *Endless Forms: Species and Speciation*, D. J. Howard and S. H. Berlocher (eds.). Oxford University Press, New York.

Howarth, B. 1974. Sperm storage as a function of the female reproductive tract. Pp. 237–270 in *The Oviduct and Its Functions*, A. D. Johnson and C. E. Foley (eds). Academic Press, New York.

Howe, C. J., T. J. Beanland, A. W. D. Larkum and P. J. Lockhart. 1992. Plastid origins. *Trends Ecol. Evol.* 7: 378–383.

Hu, Y.-P., R. A. Lutz and R. C. Vrijenhoek. 1992. Electrophoretic identification and genetic analysis of bivalve larvae. *Mar. Biol.* 113: 227–230.

Huang, C. Y., M. A. Ayliffe and J. N. Timmis. 2003. Direct measurement of the transfer rate of chloroplast DNA into the nucleus. *Nature* 422: 72–76.

Huang, S., Y. C. Chiang, B. A. Schaal, C. H. Chou and T. Y. Chiang. 2001. Organelle DNA phylogeography of *Cycas taitungensis*, a relict species in Taiwan. *Mol. Ecol.* 10: 2669–2681.

Huang, W., Y.-X. Fu, B. H.-J. Chang, X. Gu, L. B. Jorde and W.-H. Li. 1998. Sequence variation in ZFX introns in human populations. *Mol. Biol. Evol.* 15: 138–142.

Hubbs, C. L. 1955. Hybridization between fish species in nature. *Syst. Zool.* 4: 1–20.

Hubbs, C. L. and L. C. Hubbs. 1933. The increased growth, predominant maleness, and apparent infertility of hybrid sunfishes. *Pap. Mich. Acad. Sci. Arts Lett.* 17: 613–641.

Hubby, J. L. and L. H. Throckmorton. 1968. Protein differences in *Drosophila*. IV. A study of sibling species. *Am. Nat.* 102: 193–205.

Hudson, R. R. 1983. Testing the constant-rate neutral allele model with protein sequence data. *Evolution* 37: 203–217.

Hudson, R. R. 1990. Gene genealogies and the coalescent process. *Oxford Surv. Evol. Biol.* 7: 1–44.

Hudson, R. R. 1998. Island models and the coalescent process. *Mol. Ecol.* 7: 413–418.

Hudson, R. R. and J. A. Coyne. 2002. Mathematical consequences of the genealogical species concept. *Evolution* 56: 1557–1565.

Hudson, R. R., M. Kreitman and M. Aguadé. 1987. A test of neutral molecular evolution based on nucleotide data. *Genetics* 116: 153–159.

Hudson, R. R., A. G. Sáez and F. J. Ayala. 1997. DNA variation at the SOD locus of *Drosophila melanogaster*: An unfolding story of natural selection. *Proc. Natl. Acad. Sci. USA* 94: 7725–7729.

Huelsenbeck, J. P. 2000. *McBayes: Bayesian Inferences of Phylogeny* (software). University of Rochester, Rochester, NY.

Hufford, K. M. and S. J. Mazer. 2003. Plant ecotypes: Genetic differentiation in the age of ecological restoration. *Trends Ecol. Evol.* 18: 147–155.

Hughes, A. L. 1991. MHC polymorphism and the design of captive breeding programs. *Cons. Biol.* 5: 249–251.

Hughes, A. L. 1992. Avian species described on the basis of DNA only. *Trends Ecol. Evol.* 7: 2–3.

Hughes, A. L. and M. Nei. 1988. Pattern of nucleotide substitution at major histocompatibility complex class I loci reveals overdominant selection. *Nature* 335: 167–170.

Hughes, A. L. and M. Nei. 1989. Nucleotide substitution at major histocompatibility complex class II loci: Evidence for overdominant selection. *Proc. Natl. Acad. Sci. USA* 86: 958–962.

Hughes, J. and A. J. Richards. 1988. The genetic structure of populations of sexual and asexual *Taraxacum* (dandelions). *Heredity* 60: 161–171.

Hughes, J. and A. J. Richards. 1989. Isozymes, and the status of *Taraxacum* (Asteraceae) agamospecies. *Bot. J. Linn. Soc.* 99: 365–376.

Hughes, M. B. and J. C. Lucchesi. 1977. Genetic rescue of a lethal "null" activity allele of 6-phosphogluconate dehydrogenase in *Drosophila melanogaster*. *Science* 196: 1114–1115.

Hull, C. and A. Johnson. 1999. Identification of a mating type-like locus in the asexual pathogenic yeast *Candida albicans*. *Science* 285: 1271–1275.

Hull, C., R. M. Raisner and A. D. Johnson. 2000. Evidence for mating of the "asexual" yeast *Candida albicans* in a mammalian host. *Science* 289: 307–310.

Hull, D. L. 1988. *Science as a Process*. University of Chicago Press, Chicago, IL.

Hull, D. L. 1997. The ideal species concept—and why we can't get it. Pp. 357–380 in *Species: The Units of Biodiversity*, M. F. Claridge, H. A. Dawah and M. R. Wilson (eds.). Chapman & Hall, New York.

Humphries, C. J. and L. Parenti. 1986. *Cladistics-Biogeography*. Oxford Monographs on Biogeography, Clarendon Press, Oxford, England.

Humphries, C. J., P. H. Williams and R. I. Vane-Wright. 1995. Measuring biodiversity value for conservation. *Annu. Rev. Ecol. Syst.* 26: 93–111.

Humphries, J. M. 1984. Genetics of speciation in pupfishes from Laguna Chichancanab, Mexico. Pp. 129–139 in *Evolution of Fish Species Flocks*, A. A. Echelle and I. Kornfield (eds.). University of Maine Press, Orono.

Hunt, A. and D. J. Ayre. 1989. Population structure in the sexually reproducing sea anemone *Oulactis muscosa*. *Mar. Biol.* 102: 537–544.

Hunt, W. G. and R. K. Selander. 1973. Biochemical genetics of hybridisation in European house mice. *Heredity* 31: 11–33.

Hunter, C. L. 1985. Assessment of clonal diversity and population structure of *Porites compressa* (Cnideria, Scleractinia). *Proc. 5th Int. Coral Reef Symp.* 6: 69–74.

Hunter, R. L. and C. L. Markert. 1957. Histochemical demonstration of enzymes separated by zone electrophoresis in starch gels. *Science* 125: 1294–1295.

Hurles, M. E. and M. A. Jobling. 2001. Haploid chromosomes in molecular ecology: Lessons from the human Y. *Mol. Ecol.* 10: 1599–1613.

Hurst, L. and H. Ellegren. 1998. Sex biases in the mutation rate. *Trends Genet.* 14: 446–452.

Hutchison, C. A. III, J. E. Newbold, S. S. Potter and M. H. Edgell. 1974. Maternal inheritance of mammalian mitochondrial DNA. *Nature* 251: 536–538.

Hutchinson, W. F., G. R. Carvalho and S. I. Rogers. 1999. A non-destructive technique for the recovery of DNA from dried fish otoliths for subsequent molecular genetic analysis. *Mol. Ecol.* 8: 893–894.

Hutton, J. R. and J. G. Wetmur. 1973. Effect of chemical modification on the rate of renaturation of deoxyribonucleic acid. Deaminated and glyoxalated deoxyribonucleic acid. *Biochemistry* 12: 558–563.

Huvet, A., S. Lapegue, A. Mogoulas and P. Boudry. 2000. Mitochondrial and nuclear DNA phylogeography of *Crassostrea angulata*, the Portuguese oyster endangered in Europe. *Cons. Genet.* 1: 251–262.

Huynen, L., C. D. Millar and D. M. Lambert. 2002. A DNA test to sex ratite birds. *Mol. Ecol.* 11: 851–856.

Huynen, L., C. D. Millar, R. P. Scofield and D. M. Lambert. 2003. Nuclear DNA sequences detect species limits in ancient moa. *Nature* 425: 175–178.

Hyman, B. C., J. L. Beck and K. C. Weiss. 1988. Sequence amplification and gene arrangement in parasitic nematode mitochondrial DNA. *Genetics* 120: 707–712.

Ihssen, P. E., H. E. Booke, J. M. Casselman, J. M. McGlade, N. R. Payne and F. M. Utter. 1981. Stock identification: Materials and methods. *Can. J. Fish. Aquat. Sci.* 38: 1838–1855.

Imsirodou, A., H. Hardy, N. Maudling, G. Amoutzias and J.-M. Zaldivar Comenges. 2003. Web database of molecular genetic data from fish stocks. *J. Heredity* 94: 265–270.

Ingman, M., H. Kaessmann, S. Pääbo and U. Gyllensten. 2000. Mitochondrial genome variation and the origin of modern humans. *Nature* 408: 708–713.

Innes, D. J. 1987. Genetic structure of asexually reproducing *Enteromorpha linza* (Ulvales: Chlorophyta) in Long Island Sound. *Mar. Biol.* 94: 459–467.

Innes, D. J. and C. Yarish. 1984. Genetic evidence for the occurrence of asexual reproduction in populations of *Enteromorpha linza* (L.) J. Ag. (Chlorophyta, Ulvales) from Long Island Sound. *Phycologia* 23: 311–320.

Innis, M. A., K. B. Myambo, D. H. Gelfand and M. A. D. Brow. 1988. DNA sequencing with *Thermus aquaticus* DNA polymerase and direct sequencing of polymrease chain reaction-amplified DNA. *Proc. Natl. Acad. Sci. USA* 85: 9436–9440.

Innis, M. A., D. H. Gelfand, J. J. Sninsky and T. J. White. 1990. *PCR Protocols: A Guide to Methods and Applications.* Academic Press, New York.

Ioerger, T. R., A. G. Clark, T.-H. Kao. 1990. Polymorphism at the self-incompatibility locus in Solanaceae predates speciation. *Proc. Natl. Acad. Sci. USA* 87: 9732–9735.

Irwin, D. E. 2002. Phylogeographic breaks without geographic barriers to gene flow. *Evolution* 56: 2383–2394.

Irwin, D. E., S. Bensch and T. D. Price. 2001a. Speciation in a ring. *Nature* 409: 333–337.

Irwin, D. E., J. H. Irwin and T. D. Price. 2001b. Ring species as bridges between microevolution and speciation. *Genetica* 112: 223–243.

Irwin, D. M., T. D. Kocher and A. C. Wilson. 1991. Evolution of the cytochrome *b* gene of mammals. *J. Mol. Evol.* 32: 128–144.

Irwin, R. E. 1996. The phylogenetic content of avian courtship display and song evolution. Pp. 234–252 in *Phylogenies and the Comparative Method in Animal Behavior*, E. P. Martins (ed.). Oxford University Press, New York.

Istock, C. A., K. E. Duncan, N. Ferguson and X. Zhou. 1992. Sexuality in a natural population of bacteria: *Bacillus subtilis* challenges the clonal paradigm. *Mol. Ecol.* 1: 95–103.

Iwabe, N., K.-I. Kuma, M. Hasegawa, S. Osawa and T. Miyata. 1989. Evolutionary relationship of archaebacteria, eubacteria, and eukaryotes inferred from phylogenetic trees of duplicated genes. *Proc. Natl. Acad. Sci. USA* 86: 9355–9359.

Jaarola, M. and H. Tegelström. 1995. Colonization history of north European field voles (*Microtus agrestis*) revealed by mitochondrial DNA. *Mol. Ecol.* 4: 299–310.

Jaarola, M. and H. Tegelström. 1996. Mitochondrial DNA variation in the field vole (*Microtus agrestis*): Regional population structure and colonization history. *Evolution* 50: 2073–2085.

Jablonski, D. 1986. Larval ecology and macroevolution in marine invertebrates. *Bull. Mar. Sci.* 39: 565–587.

Jackman, T., J. B. Losos, A. Larson and K. de Queiroz. 1997. Phylogenetic studies of convergent adaptive radiations in Caribbean *Anolis* lizards. Pp. 535–557 in *Molecular Evolution and Adaptive Radiation*, T. J. Givnish and K. Systma (eds.). Cambridge University Press, Cambridge, England.

Jackman, T. R. and D. B. Wake. 1994. Evolutionary and historical analysis of protein variation in the blotched forms of salamanders of the *Ensatina* complex (Amphibia: Plethodontidae). *Evolution* 48: 876–897.

Jackson, H. D. 1999. Chloroplast DNA evidence for reticulate evolution in *Eucalyptus* (Myrtaceae). *Mol. Ecol.* 8: 739–751.

Jackson, J. B. C. 1985. Distribution and ecology of clonal and aclonal benthic invertebrates. Pp. 297–355 in *Population Biology and Evolution of Clonal Organisms*, J. B. C. Jackson, L. W. Buss and R. E. Cook (eds.). Yale University Press, New Haven, CT.

Jackson, J. B. C. 1986. Modes of dispersal of clonal benthic invertebrates: Consequences for species' distributions and genetic structure of local populations. *Bull. Mar. Sci.* 39: 588–606.

Jackson, J. B. C., L. W. Buss and R. E. Cook (eds.). 1985. *Population Biology and Evolution of Clonal Organisms.* Yale University Press, New Haven, CT.

Jackson, R. B., C. R. Linder, M. Lynch, M. Purugganan, S. Somerville and S. S. Thayer. 2002. Linking molecular insight and ecological research. *Trends Ecol. Evol.* 17: 409–414.

Jacobson, D. J. and T. R. Gordon. 1990. Variability of mitochondrial DNA as an indicator of relationships between populations of *Fusarium oxysporum f. sp. melonis*. *Mycol. Res.* 94: 734–744.

Jaenike, J. and R. K. Selander. 1979. Evolution and ecology of parthenogenesis in earthworms. *Am. Zool.* 19: 729–737.

Jaenike, J., E. D. Parker Jr. and R. K. Selander. 1980. Clonal niche structure in the parthenogenetic earthworm *Octolasion tyrtaeum*. *Am. Nat.* 116: 196–205.

Jain, R., M. C. Rivera, J. E. Moore and J. A. Lake. 2002. Horizontal gene transfer in microbial genome evolution. *Theoret. Pop. Biol.* 61: 489–495.

Jain, R., M. C. Rivera, J. E. Moore and J. A. Lake. 2003. Horizontal gene transfer accelerates genome innovation and evolution. *Mol. Biol. Evol.* 20: 1598–1602.

Jain, S. K. and R. W. Allard. 1966. The effects of linkage, epistasis and inbreeding on population changes under selection. *Genetics* 53: 633–659.

James, F. C. 1983. Environmental component of morphological differentiation in birds. *Science* 221: 184–186.

Jamieson, A. 1965. The genetics of transferrins in cattle. *Heredity* 20: 419–441.

Jamieson, A. and S. C. S. Taylor. 1997. Comparisons of three probability formulae for parentage exclusion. *Anim. Genet.* 28: 397–400.

Janczewski, D. N., N. Yuhki, D. A. Gilbert, G. T. Jefferson and S. J. O'Brien. 1992. Molecular phylogenetic inference from saber-toothed cat fossils of Rancho La Brea. *Proc. Natl. Acad. Sci. USA* 89: 9769–9773.

Janke, A., G. Feldmaier-Fuchs, W. K. Thomas, A. von Haeseler and S. Pääbo. 1994. The marsupial mitochondrial genome and the evolution of placental mammals. *Genetics* 137: 243–256.

Janke, A., O. Magnell, G. Wieczorek, M. Westerman and U. Arnason. 2002. Phylogenetic analysis of 18S rRNA and the mitochondrial genomes of the wombat, *Vombatus ursinus*, and the spiny anteater, *Tachyglossus aculeatus*: Increased support for the Marsupionta hypothesis. *J. Mol. Evol.* 54: 71–80.

Jansen, R. K. and J. D. Palmer. 1987. A chloroplast DNA inversion marks an ancient evolutionary split in the sunflower family (Asteraceae). *Proc. Natl. Acad. Sci. USA* 84: 5818–5822.

Jansen, R. K. and J. D. Palmer. 1988. Phylogenetic implications of chloroplast DNA restriction site variation in the Mutisieae (Asteraceae). *Am. J. Bot.* 75: 753–766.

Jansen, R. K. and 6 others. 1992. Chloroplast DNA variation in the Asteraceae: Phylogenetic and evolutionary implications. Pp. 252–279 in *Molecular Systematics of Plants*, P. S. Soltis, D. E. Soltis and J. J. Doyle (eds.). Chapman & Hall, New York.

Janson, K. 1987. Allozyme and shell variation in two marine snails (*Littorina*, Prosobranchia) with different dispersal abilities. *Biol. J. Linn. Soc.* 30: 245–256.

Janzen, D. H. 1979. How to be a fig. *Annu. Rev. Ecol. Syst.* 10: 13–51.

Jarman, A. P. and R. A. Wells. 1989. Hypervariable minisatellites: Recombinators or innocent bystanders? *Trends Genet.* 5: 367–371.

Jarman, A. P., R. D. Nichols, D. J. Weatherall, J. B. Clegg and D. R. Higgs. 1986. Molecular characterization of a hypervariable region downstream of the human α-globin gene cluster. *The EMBO J.* 5: 1857–1863.

Jarman, S. N. and N. G. Elliott. 2000. DNA evidence for morphological and cryptic Cenozoic speciation in the Anaspididae, 'living fossils' from the Triassic. *J. Evol. Biol.* 13: 624–633.

Jarman, S. N, N. J. Gales, M. Tierney, P. C. Gill and N. G. Elliott. 2002. A DNA-based method for identification of krill species and its application to analyzing the diet of marine vertebrate predators. *Mol. Ecol.* 11: 2679–2690.

Jarne, P. and B. Delay. 1991. Population genetics of freshwater snails. *Trends Ecol. Evol.* 6: 383–386.

Jarne, P., B. Delay, C. Bellec, G. Roizes and G. Cuny. 1990. DNA fingerprinting in schistosome-vector snails. *Biochem. Genet.* 28: 577–583.

Jarne, P., B. Delay, C. Bellec, G. Roizes and G. Cuny. 1992. Analysis of mating systems in the schistosome-vector hermaphrodite snail *Bulinus globosus* by DNA fingerprinting. *Heredity* 68: 141–146.

Jarvis, J. U. M. 1981. Eusociality in a mammal: Cooperative breeding in naked mole-rat colonies. *Science* 212: 571–573.

Jeffreys, A. J. 1987. Highly variable minisatellites and DNA fingerprints. *Biochem. Soc. Trans.* 15: 309–317.

Jeffreys, A. J. and D. B. Morton. 1987. DNA fingerprints of dogs and cats. *Anim. Genet.* 18: 1–15.

Jeffreys, A. J., V. Wilson and S. L. Thein. 1985a. Hypervariable "minisatellite" regions in human DNA. *Nature* 314: 67–73.

Jeffreys, A. J., V. Wilson and S. L. Thein. 1985b. Individual-specific "fingerprints" of human DNA. *Nature* 316: 76–79.

Jeffreys, A. J., J. F. Y. Brookfield and R. Semenoff. 1985c. Positive identification of an immigration test-case using human DNA fingerprints. *Nature* 317: 818–819.

Jeffreys, A. J., V. Wilson, R. Kelly, B. A. Taylor and G. Bulfield. 1987. Mouse DNA "fingerprints": Analysis of chromosome localization and germ-line stability of hypervariable loci in recombinant inbred strains. *Nucleic Acids Res.* 15: 2823–2836.

Jeffreys, A. J., V. Wilson, R. Neumann and J. Keyte. 1988a. Amplification of human minisatellites by the polymerase chain reaction: Towards DNA fingerprinting of single cells. *Nucleic Acids Res.* 16: 10953–10971.

Jeffreys, A. J., N. J. Royle, V. Wilson and Z. Wong. 1988b. Spontaneous mutation rates to new length alleles at tandem-repetitive hypervariable loci in human DNA. *Nature* 332: 278–281.

Jeffreys, A. J., R. Neumann and V. Wilson. 1990. Repeat unit sequence variation in minisatellites: A novel source of DNA polymorphism for studying variation and mutation by single molecule analysis. *Cell* 60: 473–485.

Jeffreys, A. J., A. MacLeod, K. Tamaki, D. L. Neil and D. G. Monckton. 1991. Minisatellite repeat coding as a digital approach to DNA typing. *Nature* 354: 204–209.

Jenkins, G. M., A. Rambaut, O. G. Pybus and E. C. Holmes. 2002. Rates of molecular evolution in RNA viruses: A quantitative phylogenetic analysis. *J. Mol. Evol.* 54: 152–161.

Jermann, T. M., J. G. Opitz, J. Stackhouse and S. A. Benner. 1995. Reconstructing the evolutoinary history of the artiodactyl ribonuclease superfamily. *Nature* 374: 57–59.

Jermiin, L. S., V. Loeschcke, V. Simonsen and V. Mahler. 1991. Electrophoretic and morphometric analyses of two sibling species pairs in *Trachyphloeus* (Coleoptera: Curculionidae). *Entomol. Scandinavica* 22: 159–170.

Jin, L. and M. Nei. 1991. Relative efficiencies of the maximum-parsimony and distance-matrix methods of phylogeny construction for restriction data. *Mol. Biol. Evol.* 8: 356–365.

Jockusch, E. L. 1997. Geographic variation and phenotypic plasticity of number of trunk vertebrae in slender salamanders, *Batrachoseps* (Caudata: Plethodontidae). *Evolution* 51: 1966–1982.

John, U., R. A. Fensome and L. K. Medlin. 2003. The appliction of a molecular clock based on molecular sequences and the fossil record to explain biogeographic distributions within the *Alexandrium tamarense* "species complex" (Dinophyceae). *Mol. Biol. Evol.* 20: 1015–1027.

Johns, G. C. and J. C. Avise. 1998a. A comparative summary of genetic distances in the vertebrates from the mitochondrial cytochrome *b* gene. *Mol. Biol. Evol.* 15: 1481–1490.

Johns, G. C. and J. C. Avise. 1998b. Tests for ancient species flocks based on molecular phylogenetic appraisals of *Sebastes* rockfishes and other marine fishes. *Evolution* 52: 1135–1146.

Johnson, F. M., C. G. Kanapi, R. H. Richardson, M. R. Wheeler and W. S. Stone. 1966. An analysis of polymorphisms among isozyme loci in dark and light *Drosophila ananassae* strains from American and Western Samoa. *Proc. Natl. Acad. Sci. USA* 56: 119–125.

Johnson, G. 1976a. Hidden alleles at the α-glycerophosphate dehydrogenase locus in *Colias* butterflies. *Genetics* 83: 149–167.

Johnson, G. 1976b. Genetic polymorphism and enzyme function. Pp. 46–59 in *Molecular Evolution*, F. J. Ayala (ed.). Sinauer Associates, Sunderland, MA.

Johnson, G. 1977. Evaluation of the charge state model of electrophoretic mobility: Comparison of the gel sieving behavior of alleles at the esterase-5 locus of *Drosophila pseudoobscura*. *Genetics* 87: 139–157.

Johnson, K. P., R. J. Adams, R. D. M. Page and D. H. Clayton. 2003. When do parasites fail to speciate in response to host speciation? *Syst. Biol.* 52: 37–47.

Johnson, M. S. and R. Black. 1982. Chaotic genetic patchiness in an intertidal limpet, *Siphonaria* sp. *Mar. Biol.* 70: 157–164.

Johnson, M. S. and T. J. Threlfall. 1987. Fissiparity and population genetics of *Coscinasterias calamaria*. *Mar. Biol.* 93: 517–525.

Johnson, M. S., D. C. Wallace, S. D. Ferris, M. C. Rattazzi and L. L. Cavalli-Sforza. 1983. Radiation of human mitochondrial DNA types analyzed by restriction endonuclease cleavage patterns. *J. Mol. Evol.* 19: 255–271.

Johnson, N. D. and C. Cicero. 2002. The role of ecologic diversification in sibling speciation of *Empidonax* flycatchers (Tyrannidae): Multigene evidence from mtDNA. *Mol. Ecol.* 11: 2065–2081.

Johnson, N. D. and C. Cicero. 2004. Speciation times of North American birds. *Evolution, in press*.

Johnson, N. D., J. V. Remsen Jr. and C. Cicero. 1999. Resolution of the debate over species concepts in ornithology: A new comprehensive biologic species concept. Pp. 1470–1482 in *Proc. 22nd Inter. Ornithol. Congr.*, N. J. Adams and R. H. Slotow (eds.). Birdlife, Johannesburg, South Africa.

Johnson, S. G. and E. Bragg. 1999. Age and polyphyletic origins of hybrid and spontaneous parthenogenetic *Campeloma* (Gastropoda: Viviparidae) from the southeastern United States. *Evolution* 53: 1769–1781.

Johnson, T. and 7 others. 1996. Late Pleistocene desiccation of Lake Victoria and rapid evolution of cichlid fishes. *Science* 273: 1091–1093.

Johnston, J. A., R. A. Wesselingh, A. C. Bouck, L. A. Donovan and M. L. Arnold. 2001. Intimately linked or hardly speaking? The relationship between genotype and environmental gradients in a Louisiana iris hybrid population. *Mol. Ecol.* 10: 673–681.

Jones, A. G. 2001. GERUD1. 0: A computer program for the reconstruction of parental genotypes from progeny arrays using multilocus data. *Mol. Ecol. Notes* 1: 215–218.

Jones, A. G. and W. R. Ardren. 2003. Methods of parentage analysis in natural populations. *Mol. Ecol.* 12: 2511–2523.

Jones, A. G. and J. C. Avise. 1997a. Microsatellite analysis of maternity and the mating system in the Gulf pipefish *Syngnathus scovelli*, a species with male pregnancy and sex-role reversal. *Mol. Ecol.* 6: 203–213.

Jones, A. G. and J. C. Avise. 1997b. Polygynandry in the dusky pipefish *Syngnathus floridae* revealed by microsatellite DNA markers. *Evolution* 51: 1611–1622.

Jones, A. G. and J. C. Avise. 2001. Mating systems and sexual selection in male-pregnant pipefishes and seahorses: Insights from microsatellite-based studies of maternity. *J. Heredity* 92: 150–158.

Jones, A. G., S. Östlund-Nilsson and J. C. Avise. 1998a. A microsatellite assessment of sneaked fertilizations and egg thievery in the fifteenspine stickleback. *Evolution* 52: 848–858.

Jones, A. G., C. Kvarnemo, G. I. Moore, L. W. Simmons and J. C. Avise. 1998b. Microsatellite evidence for monogamy and sex-biased recombination in the Western Australian seahorse *Hippocampus angustus*. *Mol. Ecol.* 7: 1497–1505.

Jones, A. G., G. Rosenqvist, A. Berglund and J. C. Avise. 1999a. Clustered microsatellite mutations in the pipefish *Syngnathus typhle*. *Genetics* 152: 1057–1063.

Jones, A. G., G. Rosenqvist, A. Berglund and J. C. Avise. 1999b. The genetic mating system of a sex-role-reversed pipefish (*Syngnathus typhle*): A molecular inquiry. *Behav. Ecol. Sociobiol.* 46: 357–365.

Jones, A. G., G. Rosenqvist, A. Berglund, S. J. Arnold and J. C. Avise. 2000. The Bateman gradient and the cause of sexual selection in a sex-role-reversed pipefish. *Proc. R. Soc. Lond. B* 267: 677–680.

Jones, A. G., D. Walker, C. Kvarnemo, K. Lindström and J. C. Avise. 2001a. Surprising similarity of sneaking rates and genetic mating patterns in two populations of the sand goby experiencing disparate sexual selection regimes. *Mol. Ecol.* 10: 461–469.

Jones, A. G., D. Walker, C. Kvarnemo, K. Lindström and J. C. Avise. 2001b. How cuckoldry can decrease the opportunity for sexual selection: Data and theory from a genetic parentage analysis of the sand goby, *Pomatoschistus minutus*. *Proc. Natl. Acad. Sci. USA* 98: 9151–9156.

Jones, A. G., D. Walker and J. C. Avise. 2001c. Genetic evidence for extreme polyandry and extraordinary sex-role reversal in a pipefish. *Proc. R. Soc. Lond. B* 268: 2531–2535.

Jones, A. G., G. I. Moore, C. Kvarnemo, D. Walker and J. C. Avise. 2003. Sympatric speciation as a consequence of male pregnancy in seahorses. *Proc. Natl. Acad. Sci. USA* 100: 6598–6603.

Jones, C. S., H. Tegelström, D. S. Latchman and R. J. Berry. 1988. An improved rapid method for mitochondrial DNA isolation suitable for use in the study of closely related populations. *Biochem. Genet.* 26: 83–88.

Jones, C. S., B. Okamura and L. R. Noble. 1994. Parent and larval RAPD fingerprints reveal outcrossing in freshwater bryozoans. *Mol. Ecol.* 3: 193–199.

Jones, G. P., M. J. Milicich, M. J. Emslie and C. Lunow. 1999. Self-recruitment in a coral reef fish population. *Nature* 402: 802–804.

Jones, J. S., S. H. Bryant, R. C. Lewontin, J. A. Moore and T. Prout. 1981. Gene flow and the geographical distribution of a molecular polymorphism in *Drosophila pseudoobscura*. *Genetics* 98: 157–178.

Jones, K. L. and 6 others. 2002. Refining the whooping crane studbook by incorporating microsatellite DNA and leg-banding analyses. *Cons. Biol.* 16: 789–799.

Jordan, I. K. and J. F. McDonald. 2002. A biologically active family of human endogenous retroviruses evolved from an ancient inactive lineage. *Genome Letters* 2002(1): 1–5.

Jordan, I. K., L. V. Matyunina and J. F. McDonald. 1999. Evidence for the recent horizontal transfer of long terminal repeat retrotransposon. *Proc. Natl. Acad. Sci. USA* 96: 12621–12625.

Jordan, W. C. and A. F. Youngson. 1992. The use of genetic marking to assess the reproductive success of mature male Atlantic salmon parr (*Salmo salar* L.) under natural spawning conditions. *J. Fish Biol.* 41: 613–618.

Joseph, L. and C. Moritz. 1994. Mitochondrial DNA phylogeography of birds in eastern Australian rainforests: First fragments. *Aust. J. Zool.* 42: 385–403.

Joseph, L., C. Moritz and A. Hugall. 1995. Molecular support for vicariance as a source of diversity in rainforest. *Proc. R. Soc. Lond. B.* 261: 177–182.

Jorgensen, R. A. and P. D. Cluster. 1988. Modes and tempos in the evolution of nuclear ribosomal DNA: New characters for evolutionary studies and new markers for genetic and population studies. *Annals Missouri Bot. Gardens* 75: 1238–1247.

Judson, O. P. and B. B. Normark. 1996. Ancient asexual scandals. *Trends Ecol. Evol.* 11: 41–46.

Jukes, T. H. and C. R. Cantor. 1969. Evolution of protein molecules. Pp. 21–132 in *Mammalian Protein Metabolism*, H. N. Munro (ed.). Academic Press, New York.

Kalinowski, S. T. 2002. Evolutionary and statistical properties of three genetic distances. *Mol. Ecol.* 11: 1263–1273.

Kallman, K. D. and R. W. Harrington Jr. 1964. Evidence for the existence of homozygous clones in the self-fertilizing hermaphroditic teleost *Rivulus marmoratus* (Poey). *Biol. Bull.* 126: 101–114.

Kaluthinal, R. and R. S. Singh. 1998. Cytological characterization of premeiotic versus postmeiotic defects producing hybrid male sterility among sibling species of the *Drosophila melanogaster* complex. *Evolution* 52: 1067–1079.

Kambysellis, M. P. and E. M. Craddock. 1997. Ecological and reproductive shifts in the diversification of the endemic Hawaiian *Drosophila*. Pp. 475–509 in *Molecular Evolution and Adaptive Radiations*, T. J. Givnish and K. J. Sytsma (eds.). Cambridge University Press, Cambridge, England.

Kanda, N., R. F. Leary, P. Spruell and F. W. Allendorf. 2002a. Molecular genetic markers identifying hybridization between the Colorado River greenback-cutthroat trout complex and Yellowstone cutthroat trout or rainbow trout. *Trans. Am. Fish. Soc.* 131: 312–319.

Kanda, N., R. F. Leary and F. W. Allendorf. 2002b. Evidence of introgressive hybridization between bull trout and brook trout. *Trans. Am. Fish. Soc.* 131: 772–782.

Kann, L. M. and K. Wishner. 1996. Genetic population structure of the copepod *Calanus finmarchicus* in the Gulf of Maine: Allozyme and amplified mitochondrial DNA variation. *Marine Biol.* 125: 65–75.

Kapitonov, V. V. and J. Jurka. 2003. The esterase and PHD domains in CR1-like non-LTR retrotransposons. *Mol. Biol. Evol.* 20: 38–46.

Kärkkäinen, K., V. Korski and O. Savolainen. 1996. Geographical variation in the inbreeding depression of Scots pine. *Evolution* 50: 111–119.

Karl, S. A. and J. C. Avise. 1992. Balancing selection at allozyme loci in oysters: Implications from nuclear RFLPs. *Science* 256: 100–102.

Karl, S. A. and J. C. Avise. 1993. PCR-based assays of Mendelian polymorphisms from anonymous single-copy nuclear DNA: Techniques and applications for population genetics. *Mol. Biol. Evol.* 10: 342–361.

Karl, S. A., B. W. Bowen and J. C. Avise. 1992. Global population genetic structure and male-mediated gene flow in the green turtle (*Chelonia mydas*): RFLP analyses of anonymous nuclear loci. *Genetics* 131: 163–173.

Karp, A. and K. J. Edwards. 1997. DNA markers: A global overview. Pp. 1–13 in *DNA Markers: Protocols, Applications, and Overviews*, G. Caetano-Anollés and P. M. Gresshoff (eds.). Wiley, New York.

Karp, A., P. G. Isaac and D. S. Ingram. 1998. *Molecular Tools for Screening Biodiversity*. Chapman & Hall, London.

Kassler, T. W. and 8 others. 2002. Molecular and morphological analyses of the black basses (*Mircopterus*): Implications for taxonomy and conservation. *Am. Fish. Soc. Symp.* 31: 291–322.

Katz, L. A. 1999. The tangled web: Gene genealogies and the origin of eukaryotes. *Am. Nat.* 154: S138-S145.

Katz, L. A. and M. L. Sogin (eds.). 1999. Special Symposium: Evolutionary relationships among eukaryotes. *Am. Nat.* 154: S93-S188.

Kawamura, S., H. Tanabe, Y. Watanabe, K. Kurosaki, N. Saitou and S. Ueda. 1991. Evolutionary rate of immunoglobulin alpha noncoding region is greater in hominoids than in Old World monkeys. *Mol. Biol. Evol.* 8: 743–752.

Kawecki, T. J. 1996. Sympatric speciation driven by beneficial mutations. *Proc. R. Soc. Lond. B* 263: 1515–1520.

Kawecki, T. J. 1997. Sympatric speciation by habitat specialization driven by deleterious mutations. *Evolution* 51: 1751–1763.

Kayser, M., S. Brauer and M. Stoneking. 2003. A genome scan to detect candidate regions influenced by local natural selection in human populations. *Mol. Biol. Evol.* 20: 893–900.

Ke, Y. and 22 others. 2001. African origin of modern humans in East Asia: A tale of 12,000 Y chromosomes. *Science* 292: 1151–1153.

Keane, B., W. P. J. Dittus and D. J. Melnick. 1997. Paternity assessment in wild groups of toque macaques *Macaca sinica* at Polonnaruwa, Sir Lanka using molecular markers. *Mol. Ecol.* 6: 267–282.

Keeling, P. J., M. A. Luker and J. D. Palmer. 2000. Evidence from beta-tubulin that microsporidia evolved from within the fungi. *Mol. Biol. Evol.* 17: 23–31.

Kelly, J. F. and D. M. Finch. 1998. Tracking migrant songbirds with stable isotopes. *Trends Ecol. Evol.* 13: 48–49.

Kempenaers, B., G. R. Verheyen, M. Van den Broeck, T. Burke, C. Van Broeckhoven and A. A. Dhondt. 1992. Extra-pair paternity results from female preference for high-quality males in the blue tit. *Nature* 357: 494–496.

Kempenaers, B., B. Congdon, P. Boag and R. J. Robertson. 1999. Extrapair paternity and egg hatchability in tree swallows: Evidence for the genetic compatibility hypothesis? *Behav. Ecol.* 10: 304–311.

Kemperman, J. A. and B. V. Barnes. 1976. Clone size in American aspens. *Can. J. Bot.* 54: 2603–2607.

Kendal, W. S. 2003. An exponential dispersion model for the distribution of human single nulceotide polymorphisms. *Mol. Biol. Evol.* 20: 579–590.

Kennedy, P. K., M. L. Kennedy and M. H. Smith. 1985. Microgeographic genetic organization of populations of largemouth bass and two other species in a reservoir. *Copeia* 1985: 118–125.

Kennedy, P. K., M. L. Kennedy, E. G. Zimmerman, R. K. Chesser and M. H. Smith. 1986. Biochemical genetics of mosquitofish. V. Perturbation effects on genetic organization of populations. *Copeia* 1986: 937–945.

Kerth, G., F. Mayer and E. Petit. 2002. Extreme sex-biased dispersal in the communally breeding, nonmigratory Bechstein's bat (*Myotis bechsteinii*). *Mol. Ecol.* 11: 1491–1498.

Kessler, C. 1987. Class II restriction endonucleases. Pp. 225–279 in *Cytogenetics*, G. Obe and A. Basler (eds.). Springer-Verlag, Berlin.

Ketterson, E. D., and 6 others. 1998. The relative impact of extra-pair fertilizations on variation in male and female reproductive success in dark-eyed juncos (*Junco hyemalis*). *Ornithol. Monogr.* 49: 81–102.

Key, K. H. L. 1968. The concept of stasipatric speciation. *Syst. Zool.* 17: 14–22.

Kichler, K., M. T. Holder, S. K. Davis, R. Márquez-M and D. W. Owens. 1999. Detection of multiple paternity in the Kemp's ridley sea turtle with limited sampling. *Mol. Ecol.* 8: 819–830.

Kidwell, M. G. 1992. Horizontal transfer of *P* elements and other short inverted repeat transposons. *Genetica* 86: 275–286.

Kidwell, M. G. 1993. Lateral transfer in natural populations of eukaryotes. *Annu. Rev. Genet.* 27: 235–256.

Kidwell, S. M. and S. M. Holland. 2002. The quality of the fossil record: Implications for evolutionary analyses. *Annu. Rev. Ecol. Syst.* 33: 561–588.

Kim, K.-J., R. K. Jansen, R. S. Wallace, H. J. Michaels and J. D. Palmer. 1992. Phylogenetic implications of *rbcL* sequence variation in the Asteraceae. *Annals Missouri Bot. Gardens* 79: 428–445.

Kim, S.-C. and L. H. Rieseberg. 1999. Genetic architecture of species differences in annual sunflowers: Implications for adaptive trait introgression. *Genetics* 153: 965–977.

Kim, S.-C. and L. H. Rieseberg. 2001. The contribution of epistasis to species differences in annual sunflowers. *Mol. Ecol.* 10: 683–690.

Kim, S. S. and S. K. Narang. 1990. Restriction site polymorphism of mtDNA for differentiating *Anopheles quadrimaculatus* (Say) sibling species. *Korean J. Appl. Entomol.* 29: 132–135.

Kimura, M. 1968a. Evolutionary rate at the molecular level. *Nature* 217: 624–626.

Kimura, M. 1968b. Genetic variability maintained in a finite population due to mutational production of neutral and nearly neutral isoalleles. *Genet. Res.* 11: 247–269.

Kimura, M. 1980. A simple method for estimating evolutionary rate of base substitutions through comparative studies of nucleotide sequences. *J. Mol. Evol.* 16: 111–120.

Kimura, M. 1983. *The Neutral Theory of Molecular Evolution*. Cambridge University Press, Cambridge, England.

Kimura, M. 1990. The present status of the neutral theory. Pp. 1–16 in *Population Biology of Genes and Molecules*, N. Takahata and J. F. Crow (eds.). Baifukan, Tokyo.

Kimura, M. 1991. Recent developments of the neutral theory viewed from the Wrightian tradition of theoretical population genetics. *Proc. Natl. Acad. Sci. USA* 88: 5969–5973.

Kimura, M. and T. Ohta. 1971. *Theoretical Aspects of Population Genetics*. Princeton University Press, Princeton, NJ.

Kimwele, C. N. and J. A. Graves. 2003. A molecular genetic analysis of the communal nesting of the ostrich (*Struthio camelus*). *Mol. Ecol.* 12: 229–236.

King, J. L. and T. H. Jukes. 1969. Non-Darwinian evolution: Random fixation of selectively neutral mutations. *Science* 164: 788–798.

King, L. M. 1993. Origins of genotypic variation in North American dandelions inferred from ribosomal DNA and chloroplast DNA restriction enzyme analysis. *Evolution* 47: 136–151.

King, M.-C. and A. C. Wilson. 1975. Evolution at two levels in humans and chimpanzees. *Science* 188: 107–116.

King, P. S. 1987. Macro- and microgeographic structure of a spatially subdivided beetle species in nature. *Evolution* 41: 401–416.

King, R. C. and W. D. Stansfield. 1990. *A Dictionary of Genetics*. Oxford University Press, New York.

King, T. L, M. S. Eackles, B. Gjetvaj and W. R. Hoeh. 1999. Intraspecific phylogeography of *Lasmigona subviridis* (Bivalia: Unionidae): Conservation implications of range discontinuity. *Mol. Ecol.* 8: S65-S78.

King, T. M., M. Williams and D. M. Lambert. 2000. Dams, ducks and DNA: Identifying the effects of a hydroelectric scheme on New Zealand's endangered blue duck. *Cons. Genet.* 1: 103–113.

Kirby, L. T. 1990. *DNA Fingerprinting*. Stockton Press, New York.

Kirkpatrick, K. J. and H. D. Wilson. 1988. Interspecific gene flow in Cucurbita: *C. texana* vs. *C. pepo*. *Am. J. Bot.* 75: 519–527.

Kirkpatrick, M. and R. K. Selander. 1979. Genetics of speciation in lake whitefishes in the Allegash basin. *Evolution* 33: 478–485.

Kirkpartick, M. and M. Slatkin. 1993. Searching for evolutionary patterns in the shape of a phylogenetic tree. *Evolution* 47: 1171–1181.

Kirsch, J. A. W. 1977. The comparative serology of Marsupialia, and a classification of marsupials. *Aust. J. Zool. Suppl. Ser.* 52: 1–152.

Kirsch, J. A. W., M. S. Springer, C. Krajewski, M. Archer, K. Aplin and A. W. Dickerman. 1990. DNA/DNA hybridization studies of the carnivorous marsupials. I: The intergeneric relationships of bandicoots (Marsupialia: Perameloidea). *J. Mol. Evol.* 30: 434–448.

Kishino, H. and M. Hasegawa. 1989. Evaluation of the maximum likelihood estimate of the evolutionary tree topologies from DNA sequence data, and the branching order in Hominoidea. *J. Mol. Evol.* 29: 170–179.

Kishino, H., T. Miyata and M. Hasegawa. 1990. Maximum likelihood inference of protein phylogeny and the origin of chloroplasts. *J. Mol. Evol.* 31: 151–160.

Kittayapong, P., W. Jamnongluk, A. Thipaksorn, J. R. Milne and C. Sindhusake. 2003. *Wolbachia* infection complexity among insects in the tropical rice-field community. *Mol. Ecol.* 12: 1049–1060.

Kivisild, T. and 13 others. 2000. Questioning evidence for recombination in human mitochondrial DNA. *Science* 288: 1931.

Klein, C. A., O. Schmidt-Kittler, J. A. Schardt, K. Pantel, M. R. Speicher and G. Riethmüller. 1999. Comparative genomic hybridizatoin, loss of heterozygosity, and DNA sequence analysis of single cells. *Proc. Natl. Acad. Sci. USA* 96: 4494–4499.

Klein, J. 1986. *Natural History of the Major Histocompatibility Complex*. Wiley-Interscience, New York.

Klein, J., Y. Satta, N. Takahata and C. O'hUigin. 1993. Trans-specific MHC polymorphism and the origin of species in primates. *J. Med. Primatol.* 22: 57–64.

Klein, J, A. Sato, S. Nagl and C. O'hUigin. 1998. Transspecies polymorphism. *Annu. Rev. Ecol. Syst.* 29: 1–21.

Klein, N. K. and W. M. Brown. 1995. Intraspecific molecular phylogeny in the yellow warbler (*Dendroica petechia*), and implications for avian biogeography in the West Indies. *Evolution* 48: 1914–1932.

Klein, N. K. and R. B. Payne. 1998. Evolutionary associations of brood parasitic finches (*Vidua*) and their host races: Analyses of mitochondrial DNA restriction sites. *Evolution* 52: 566–582.

Klicka, J. and R. M. Zink. 1997. The importance of recent Ice Ages in speciation: A failed paradigm. *Science* 277: 1666–1669.

Klicka, J. and R. M. Zink. 1999. Pleistocene effects on North American songbird evolution. *Proc. R. Soc. Lond. B* 266: 695–670.

Klotz, M. G. and P. C. Loewen. 2003. The molecular evolution of catalatic hydroperoxidases: Evidence for multiple lateral transfer of genes between Prokaryota and from bacteria into Eukaryota. *Mol. Biol. Evol.* 20: 1098–1112.

Kluge, A. G. and J. S. Farris. 1969. Quantitative phyletics and the evolution of anurans. *Syst. Zool.* 18: 1–32.

Knight, R. D., S. J. Freeland and L. F. Landweber. 2001a. Rewiring the keyboard: Evolvability of the genetic code. *Nature Rev. Genet.* 2: 49–58.

Knight, R. D., L. F. Landweber and M. Yarus. 2001b. How mitochondria redefine the code. *J. Mol. Evol.* 53: 299–313.

Knobloch, I. W. 1972. Intergeneric hybridization in flowering plants. *Taxon* 21: 97–103.

Knoll, A. H. 1992. The early evolution of eukaryotes: A geological perspective. *Science* 256: 622–627.

Knoll, A. H. 1999. A new molecular window on early life. *Science* 285: 1025–1026.

Knoll, A. H. and S. B. Carroll. 1999. Early animal evolution: Emerging views from comparative biology and geology. *Science* 284: 2129–2137.

Knowles, L. L. 2001. Did the Pleistocene glaciations promote divergence? Tests of explicit refugial models in montane grasshoppers. *Mol. Ecol.* 10: 691–701.

Knowles, L. L. and W. P. Maddison. 2002. Statistical phylogeography. *Mol. Ecol.* 11: 2623–2635.

Knowlton, N. 1993. Sibling species in the sea. *Annu. Rev. Ecol. Syst.* 24: 189–216.

Knowlton, N. and S. R. Greenwell. 1984. Male sperm competition avoidance mechanisms: The influence of female interests. Pp. 61–84 in *Sperm Competition and the Evolution of Animal Mating Systems*, R. L. Smith (ed.). Academic Press, New York.

Knowlton, N. and B. D. Keller. 1986. Larvae which fall short of their potential: Highly localized recruitment in an alpheid shrimp with extended larval development. *Bull. Mar. Sci.* 39: 213–223.

Knowlton, N., E. Weil, L. A. Weight and H. M. Guzman. 1992. Sibling species in *Montastraea annularis*, coral bleaching, and the coral climate record. *Science* 255: 330–333.

Knowlton, N., L. A. Weight, L. A. Solórzano, D. K. Mills and E. Bermingham. 1993. Divergence of proteins, mitochondrial DNA, and reproductive compatibility across the Isthmus of Panama. *Science* 260: 1629–1632.

Kocher, T. D. and C. A. Stepien (eds.). 1997. *Molecular Systematics of Fishes*. Academic Press, New York.

Kocher, T. D. and 6 others. 1989. Dynamics of mitochondrial DNA evolution in animals: Amplification and sequencing with conserved primers. *Proc. Natl. Acad. Sci. USA* 86: 6196–6200.

Kocher, T. D., J. A. Conroy, K. R. McKaye and J. R. Stauffer. 1993. Similar morphologies of cichlid fish in Lakes Tanganyika and Malawi are due to convergence. *Mol. Phylogen. Evol.* 2: 158–165.

Kochert, G. 1989. *Introduction to RFLP Mapping and Plant Breeding Applications*. Special Publication Rockefeller Foundation, New York.

Koehn, R. K. 1978. Physiology and biochemistry of enzyme variation: The interface of ecology and population genetics. Pp. 51–72 in *Ecological Genetics: The Interface*, P. Brussard (ed.). Springer-Verlag, New York.

Koehn, R. K. and W. F. Eanes. 1976. An analysis of allelic diversity in natural populations of *Drosophila*: The correlation of rare alleles with heterozygosity. Pp. 377–390 in *Population Genetics and Ecology*, S. Karlin and E. Nevo (eds.). Academic Press, New York.

Koehn, R. K. and W. F. Eanes. 1978. Molecular structure and protein variation within and among populations. *Evol. Biol.* 11: 39–100.

Koehn, R. K. and T. J. Hilbish. 1987. The adaptive importance of genetic variation. *Am. Sci.* 75: 134–141.

Koehn, R. K. and F. W. Immerman. 1981. Biochemical studies of animopeptidase in *Mytilus edulis*. I. Dependence of enzyme activity on season, tissue and genotype. *Biochem. Genet.* 19: 1115–1142.

Koehn, R. K. and G. C. Williams. 1978. Genetic differentiation without isolation in the American eel, *Anguilla rostrata*. II. Temporal stability of geographic variation. *Evolution* 32: 624–637.

Koehn, R. K., R. Milkman and J. B. Mitton. 1976. Population genetics of marine pelecypods. IV. Selection, migration and genetic differentiation in the blue mussel, *Mytilus edulis*. *Evolution* 30: 2–32.

Koehn, R. K., A. J. Zera and J. G. Hall. 1983. Enzyme polymorphism and natural selection. Pp. 115–136 in *Evolution of Genes and Proteins*, M. Nei and R. K. Koehn (eds.). Sinauer Associates, Sunderland, MA.

Koehn, R. K., W. J. Diehl and T. M. Scott. 1988. The differential contribution by individual enzymes of glycolysis and protein catabolism to the relationship between heterozygosity and growth rate in the coot clam, *Mulinia lateralis*. *Genetics* 118: 121–130.

Kohn, M., F. Knauer, A. Stoffella, W. Schröder and S. Pääbo. 1995. Conservation genetics of the European brown bear: A study using excremental PCR of nuclear and mitochondrial sequences. *Mol. Ecol.* 4: 95–103.

Kohn, M. H. and R. K. Wayne. 1997. Facts from feces revisited. *Trends Ecol. Evol.* 12: 223–227.

Kohn, M. H., E. C. York, D. A. Kamradt, G. Haught, R. M. Sauvajot and R. K. Wayne. 1999. Estimating population size by genotyping faeces. *Proc. R. Soc. Lond. B* 266: 657–663.

Kohne, D. E. 1970. Evolution of higher organism DNA. *Q. Rev. Biophys.* 33: 327–375.

Kojima, K., J. Gillespie and Y. N. Tobari. 1970. A profile of *Drosophila* species' enzymes assayed by electrophoresis. I. Number of alleles, heterozygosities, and linkage disequilibrium in glucose-metabolizing systems and some other enzymes. *Biochem. Genet.* 4: 627–637.

Kokko, H., E. Ranta, G. Ruxton and P. Lundberg. 2002. Sexually transmitted disease and the evolution of mating systems. *Evolution* 56: 1091–1100.

Kokko, H., R. Brooks, M. D. Jennions and J. Morley. 2003. The evolution of mate choice and mating biases. *Proc. R. Soc. Lond. B* 270: 653–664.

Komdeur, J., S. Daan, J. Tinbergen and S. Mateman. 1997. Extreme adaptive modification in sex ratio of the Seychelles warbler's eggs. *Nature* 385: 522–525.

Kondo, R., Y. Satta, E. T. Matsuura, H. Ishiwa, N. Takahata and S. I. Chigusa. 1990. Incomplete maternal transmission of mitochondrial DNA in *Drosophila*. *Genetics* 126: 657–663.

Kondo, T., Y. Tsumura, T. Kawahara and M. Okamura. 1998. Inheritance of chloroplast and mitochondrial DNA in interspecific hybrids of *Chamaecyparis* spp. *Breeding Sci.* 78: 177–179.

Kondrashov, A. S. 1986. Multilocus models of sympatric speciation. III. Computer simulations. *Theor. Pop. Biol.* 29: 1–15.

Kondrashov, A. S. and F. A. Kondrashov. 1999. Interactions among quantitative traits in the course of sympatric speciation. *Nature* 400: 351–354.

Kondrashov, A. S. and M. V. Mina. 1986. Sympatric speciation: When is it possible? *Biol. J. Linn. Soc.* 27: 201–223.

Konkle, B. R. and D. P. Philipp. 1992. Asymmetric hybridization between two species of sunfishes (*Lepomis*: Centrarchidae). *Mol. Ecol.* 1: 215–222.

Koonin, E. V., K. S. Makarova and L. Aravind. 2001. Horizontal gene transfer in prokaryotes: Quantification and classification. *Annu. Rev. Microbiol.* 55: 709–742.

Koontz, J. A., P. S. Soltis and S. J. Brunsfeld. 2001. Genetic diversity and tests of the hybrid origin of the endangered yellow larkspur. *Cons. Biol.* 15: 1608–1618.

Koop, B. F., M. Goodman, P. Zu, J. L. Chan and J. L. Slighton. 1986. Primate η-globin DNA sequences and man's place among the great apes. *Nature* 319: 234–238.

Koop, B. F., D. A. Tagle, M. Goodman and J. L. Slightom. 1989. A molecular view of primate phylogeny and important systematic and evolutionary questions. *Mol. Biol. Evol.* 6: 580–612.

Koopman, K. F. 1950. Natural selection for reproductive isolation between *Drosophila pseudoobscura* and *Drosophila persimilis*. *Evolution* 4: 135–148.

Koppelman, J. B. and D. E. Figg. 1995. Genetic estimates of variability and relatedness for conservation of an Ozark cave crayfish species complex. *Cons. Biol.* 9: 1288–1294.

Korber, B. and 8 others. 2000. Timing the ancestor of the HIV-1 pandemic strains. *Science* 288: 1789–1796.

Kornegay, J. R., T. D. Kocher, L. A. Williams and A. C. Wilson. 1993. Pathways of lysozyme evolution inferred from the sequences of cytochrome *b* in birds. *J. Mol. Evol.* 37: 367–379.

Kornfield, I. L. and K. E. Carpenter. 1984. Cyprinids of Lake Lanao, Philippines: Taxonomic validity, evolutionary rates and speciation scenarios. Pp. 69–84 in *Evolution of Fish Species Flocks*, A. A. Echelle and I. Kornfield (eds.). University of Maine Press, Orono, ME.

Kornfield, I. L. and R. K. Koehn. 1975. Genetic variation and speciation in New World cichlids. *Evolution* 29: 427–437.

Kornfield, I. L., D. C. Smith, P. S. Gagnon and J. N. Taylor. 1982. The cichlid fish of Cuatro Ciénegas, Mexico: Direct evidence of conspecificity among distinct trophic morphs. *Evolution* 36: 658–664.

Korpelainen, H. 2002. A genetic method to resolve gender complements investigations on sex ratios in *Rumex acetosa*. *Mol. Ecol.* 11: 2151–2156.

Koski, L. B., R. A. Morton and G. B. Golding. 2001. Codon bias and base composition are poor indicators of horizontally transferred genes. *Mol. Biol. Evol.* 18: 404–412.

Kotlík, P. and P. Berrebi. 2001. Phylogeography of the barbel (*Barbus barbus*) assessed by mitochondrial DNA variation. *Mol. Ecol.* 10: 2177–2186.

Kotlík, P., N. G. Bogutskaya and F. G. Ekmekci. 2004. Circum Black Sea phylogeography of *Barbus* freshwater fishes: Divergence in the Pontic glacial refugium. *Mol. Ecol.* 13: 87–95.

Koufopanou, V., A. Burt and J. W. Taylor. 1997. Concordance of gene genealogies reveals reproductive isolation in the pathogenic fungus *Coccidioides immitis*. *Proc. Natl. Acad. Sci. USA* 94: 5478–5482.

Kozol, A. J., J. F. A. Traniello and S. M. Williams. 1994. Genetic variation in the endangered burying beetle *Nicrophorus americanus* (Coleoptera: Silphidae). *Ann. Ent. Soc. Am.* 87: 928–935.

Krajewski, C. 1994. Phylogenetic measures of biodiversity: A comparison and critique. *Biol. Cons.* 69: 33–39.

Krajewski, C., L. Buckley and M. Westerman. 1997. DNA phylogeny of the marsupial wolf resolved. *Proc. R. Soc. Lond. B* 264: 911–917.

Krane, D. E., R. W. Allen, S. A. Sawyer, D. A. Petrov and D. L. Hartl. 1992. Genetic differences at four DNA typing loci in Finnish, Italian, and mixed Caucasian populations. *Proc. Natl. Acad. Sci. USA* 89: 10583–10587.

Kraus, F. and M. M. Miyamoto. 1990. Mitochondrial genotype of a unisexual salamander of hybrid origin is unrelated to either of its nuclear haplotypes. *Proc. Natl. Acad. Sci. USA* 87: 2235–2238.

Kreiswirth, B. and 7 others. 1993. Evidence for a clonal origin of methicillin resistance in *Staphylococcus aureus*. *Science* 259: 227–230.

Kreitman, M. 1987. Molecular population genetics. *Oxford Surv. Evol. Biol.* 4: 38–60.

Kreitman, M. 1991. Detecting selection at the level of DNA. Pp. 204–221 in *Evolution at the Molecular Level*, R. K. Selander, A. G. Clark and T. S. Whittam (eds.). Sinauer Associates, Sunderland, MA.

Kreitman, M. and R. R. Hudson. 1991. Inferring the evolutionary histories of the *Adh* and *Adh-dup* loci in *Drosophila melanogaster* from patterns of polymorphism and divergence. *Genetics* 127: 565–582.

Kretzmann, M. B. and 6 others. 1997. Low genetic variability in the Hawaiian monk seal. *Cons. Biol.* 11: 482–490.

Krieber, M. and M. R. Rose. 1986. Molecular aspects of the species barrier. *Annu. Rev. Ecol. Syst.* 17: 465–485.

Krieger, J. and P. A. Fuerst. 2002. Evidence for a slowed rate of molecular evolution in the order Acipenseriformes. *Mol. Biol. Evol.* 19: 891–897.

Krieger, J., P. A. Fuerst and T. M. Cavender. 2000. Phylogenetic relationships of the North American sturgeons (order Acipenseriformes) based on mitochondrial DNA sequences. *Mol. Phylogen. Evol.* 16: 64–72.

Krings, M., A. Stone, R. W. Schmitz, H. Krainitzki, M. Stoneking and S. Pääbo. 1997. Neanderthal DNA sequences and the origin of modern humans. *Cell* 90: 19–30.

Krings, M. and 9 others. 2000. A view of Neanderthal genetic diversity. *Nature Genetics* 26: 144–146.

Kroon, A. M. and C. Saccone (eds.). 1980. *The Organization and Expression of the Mitochondrial Genome*. Elsevier, New York.

Krueger, C. C. and B. May. 1987. Stock identification of naturalized brown trout in Lake Superior tributaries: Differentiation based on allozyme data. *Trans. Am. Fish. Soc.* 116: 785–794.

Krzywinski, J., R. C. Wilkerson and N. J. Besansky. 2001. Toward understanding anophelinae (Diptera, Culicidae) phylogeny: Insights from nuclear single-copy genes and the weight of evidence. *Syst. Biol.* 50: 540–556.

Kubo, N., K. Harada, A. Hirai and K. Kadowaki. 1999. A single nuclear transcript encoding mitochondrial RPS14 and SDHB of rice is processed by alternative splicing: Common use of the same mitochondrial targeting signal for different proteins. *Proc. Natl. Acad. Sci. USA* 96: 9207–9211.

Kubo, T. and Y. Isawa. 1995. Inferring the rates of branching and extinction from molecular phylogenies. *Evolution* 49: 694–704.

Kuch, M., N. Rohland, J. L. Betancourt, C. Latorre, S. Steppan and H. N. Poinar. 2002. Molecular analysis of a 11,700-year-old rodent midden from the Atacama Desert, Chile. *Mol. Ecol.* 11: 913–924.

Kuhner, M. K., J. Yamato and J. Felsenstein. 2000. Maximum likelihood estimation of recombination rates from population data. *Genetics* 156: 1393–1401.

Kuhnlein, U., D. Zadworny, Y. Dawe, R. W. Fairfull and J. S. Gavora. 1990. Assessment of inbreeding by DNA fingerprinting: Development of a calibration curve using defined strains of chickens. *Genetics* 125: 161–165.

Kumar, S. and S. B. Hedges. 1998. A molecular timescale for verebrate evolution. *Nature* 392: 917–920.

Kumar, S. and S. Subramanian. 2002. Mutation rates in mammalian genomes. *Proc. Natl. Acad. Sci. USA* 99: 803–808.

Kumar, S., K. Tamura, I. Jakobsen and M. Nei. 2000. *MEGA: Molecular Evolutionary Genetic Analysis*. Arizona State University, Tempe, AZ.

Kumazaki, T., H. Hori and S. Osawa. 1983. Phylogeny of protozoa deduced from 5S rRNA sequences. *J. Mol. Evol.* 19: 411–419.

Kumazama, Y. and M. Nishida. 1999. Complete mitochondrial DNA sequences of the green turtle and blue-tailed mole skink: Statistical evidence for Archosaurian affinity of turtles. *Mol. Biol. Evol.* 16: 784–792.

Kusakabe, T., K. W. Makabe and N. Satoh. 1992. Tunicate muscle actin genes: Structure and organization as a gene cluster. *J. Mol. Biol.* 227: 955–960.

Kuzoff, R. K., D. E. Soltis, L. Hufford and P. S. Soltis. 1999. Phylogenetic relationships within *Lithophragma* (Saxifragaceae): Hybridization, allopolyploidy, and ovary diversification. *Syst. Bot.* 24: 598–615.

Kwiatowski, J. and F. J. Ayala. 1999. Phylogeny of *Drosophila* and related genera: Conflict between molecular and anatomical analyses. *Mol. Phylogen. Evol.* 13: 319–328.

Kwiatowski, J., D. Skarecky, S. Hernandez, D. Pham, F. Quijas and F. J. Ayala. 1991. High fidelity of the polymerase chain reaction. *Mol. Biol. Evol.* 8: 884–887.

Lack, D. 1968. *Ecological Adaptations for Breeding in Birds*. Chapman & Hall, London.

Lacson, J. M. 1992. Minimal genetic variation among samples of six species of coral reef fishes collected at La Parguera, Puerto Rico, and Discovery Bay, Jamaica. *Mar. Biol.* 112: 327–331.

Lacy, R. C. 1980. The evolution of eusociality in termites: A haplodiploid analogy? *Am. Nat.* 116: 449–451.

Lacy, R. C. 1989. Analysis of founder representation in pedigrees: Founder equivalents and founder genome equivalents. *Zoo Biol.* 8: 111–123.

Lacy, R. C. 1992. The effects of inbreeding on isolated populations: Are minimum viable population sizes predictable? Pp. 277–296 in *Conservation Biology*, P. L. Fiedler and S. K. Jain (eds.). Chapman & Hall, New York.

Lacy, R. C., G. Alaks and A. Walsh. 1996. Hierarchical analysis of inbreeding depression in *Peromyscus polionotus. Evolution* 50: 2187–2200.

Ladoukakis, E. D. and E. Zouros. 2001. Direct evidence for homologous recombination in mussel (*Mytilus galloprovincialis*) mitochondrial DNA. *Mol. Biol. Evol.* 18: 1168–1175.

Laerm, J., J. C. Avise, J. C. Patton and R. A. Lansman. 1982. Genetic determination of the status of an endangered species of pocket gopher in Georgia. *J. Wildlife Mgt.* 46: 513–518.

Laguerre, G., L. Rigottier-Gois and P. Lemanceau. 1994. Fluorescent *Pseudomonas* species categorized by using polymerase chain reaction (PCR)/restriction fragment analysis of 16S rDNA. *Mol. Ecol.* 3: 479–487.

Lahanas, P. N. and 6 others. 1998. Genetic composition of a green turtle (*Chelonia mydas*) feeding ground population: Evidence for multiple origins. *Mar. Biol.* 130: 345–352.

Laidlaw, H. H. Jr. and R. E. Page Jr. 1984. Polyandry in honey bees (*Apis mellifera* L.): Sperm utilization and intracolony genetic relationships. *Genetics* 108: 985–997.

Laikre, L. and N. Ryman. 1991. Inbreeding depression in a captive wolf (*Canis lupus*) population. *Cons. Biol.* 5: 33–40.

Laikre, L. and 7 others. 2002. Spatial and temporal population structure of sea trout at the Island of Gotland, Sweden, delineated from mitochondrial DNA. *J. Fish Biol.* 60: 49–71.

Laipis, P. J., M. J. Van de Walle and W. W. Hauswirth. 1988. Unequal partitioning of bovine mitochondrial genotypes among siblings. *Proc. Natl. Acad. Sci. USA* 85: 8107–8110.

Laird, C. D., B. L. McConaughy and B. J. McCarthy. 1969. Rate of fixation of nucleotide substitutions in evolution. *Nature* 224: 149–154.

LaJeunesse, T. C. 2001. Investigating the biodiversity, ecology, and phylogeny of endosymbiotic dinoflagellates in the genus *Symbiodinium* using the internal transcribed spacer region: In search of a "species" level marker. *J. Phycol.* 37: 866–880.

Lake, J. A. 1991. Tracing origins with molecular sequences: Metazoan and eukaryotic beginnings. *Trends Biochem. Sci.* 16: 46–50.

Lalueza-Fox, C., J. Bertranpetit, J. A. Alcover, N. Shailer and E. Hagelberg. 2000. Mitochondrial DNA from *Myotragus balearicus*, an extinct bovid from the Balearic Islands. *J. Exp. Zool.* 288: 56–62.

Lamb, T. and J. C. Avise. 1986. Directional introgression of mitochondrial DNA in a hybrid population of tree frogs: The influence of mating behavior. *Proc. Natl. Acad. Sci. USA* 83: 2526–2530.

Lamb, T. and J. C. Avise. 1987. Morphological variability in genetically defined categories of anuran hybrids. *Evolution* 41: 157–165.

Lamb, T. and J. C. Avise. 1992. Molecular and population genetic aspects of mitochondrial DNA variability in the diamondback terrapin, *Malaclemys terrapin. J. Heredity* 83: 262–269.

Lamb, T., J. C. Avise and J. W. Gibbons. 1989. Phylogeographic patterns in mitochondrial DNA of the desert tortoise (*Xerobates agassizi*), and evolutionary relationships among the North American gopher tortoises. *Evolution* 43: 76–87.

Lamb, T., T. R. Jones and J. C. Avise. 1992. Phylogeographic histories of representative herpetofauna of the desert southwest: Mitochondrial DNA variation in the chuckwalla (*Sauromalus obesus*) and desert iguana (*Dipsosaurus dorsalis*). *J. Evol. Biol.* 5: 465–480.

Lambert, D. M., P. A. Ritchie, C. D. Millar, B. Holland, A. J. Drummond and C. Baroni. 2002. Rates of evolution in ancient DNA from Adélie penguins. *Science* 295: 2270–2273.

Lanctot, R. B., K. T. Scribner, B. Kempenaers and P. J. Weatherhead. 1997. Lekking without a paradox in the buff-breasted sandpiper. *Am. Nat.* 149: 1051–1070.

Lanctot, R. B., P. J. Weatherhead, B. Kempenaers and K. T. Scribner. 1998. Male traits, mating tactics and reproductive success in the buff-breasted sandpiper, *Tryngites subruficollis. Anim. Behav.* 56: 419–432.

Land, E. D. and R. C. Lacey. 2000. Introgression level achieved through Florida panther genetic restoration. *Endangered Species Update* 17: 100–105.

Lande, R. 1981. The minimum number of genes contributing to quantitative variation between and within populations. *Genetics* 99: 541–553.

Lande, R. 1988. Genetics and demography in biological conservation. *Science* 241: 1455–1460.

Lande, R. 1995. Mutation and conservation. *Cons. Biol.* 9: 782–791.

Lande, R. 1999. Extinction risks from anthropogenic, ecological, and genetic factors. Pp. 1–22 in *Genetics and Extinction of Species*, L. F. Landweber and A. P. Dobson (eds.). Princeton University Press, Princeton, NJ.

Lande, R. and D. W. Schemske. 1985. The evolution of self-fertilization and inbreeding depression in plants. I. Genetic models. *Evolution* 39: 24–40.

Lander, E. S. 1989. DNA fingerprinting on trial. *Nature* 339: 501–505.

Lander, E. S. 1991. Research on DNA typing catching up with courtroom application. *Am. J. Human Genet.* 48: 819–823.

Lander, E. S. and D. Botstein. 1989. Mapping Mendelian factors underlying quantitative traits using RFLP linkage maps. *Genetics* 121: 185–199.

Lander, E. S. and B. Budowle. 1994. DNA fingerprinting dispute laid to rest. *Nature* 371: 735–738.

Lander, E. S. and N. J. Schork. 1994. Genetic dissection of complex traits. *Science* 265: 2037–2047.

Lander, E. S. and 243 others. 2001. Initial sequencing and analysis of the human genome. *Nature* 409: 860–921.

Landweber, L. F. and A. P. Dobson (eds.). 1999. *Genetics and the Extinction of Species*. Princeton University Press, Princeton, NJ.

Lang, D. F., M. W. Gray and G. Burger. 1999. Mitochondrial genome evolution and the origin of eukaryotes. *Annu. Rev. Genet.* 33: 351–397.

Langley, C. H. and C. F. Aquadro. 1987. Restriction-map variation in natural populations of *Drosophila melanogaster*: *White*-locus region. *Mol. Biol. Evol.* 4: 651–663.

Langley, C. H. and W. M. Fitch. 1974. An examination of the constancy of the rate of molecular evolution. *J. Mol. Evol.* 3: 161–177.

Langley, C. H., A. E. Shrimpton, T. Yamazaki, N. Miyashita, Y. Matsuo and C. F. Aquadro. 1988. Naturally-occurring variation in the restriction map of the *Amy* region of *Drosophila melanogaster*. *Genetics* 119: 619–629.

Lank, D. B., P. Mineau, R. F. Rockwell and F. Cooke. 1989. Intraspecific nest parasitism and extra-pair copulation in lesser snow geese. *Anim. Behav.* 37: 74–89.

Lansman, R. A., R. O. Shade, J. F. Shapira and J. C. Avise. 1981. The use of restriction endonucleases to measure mitochondrial DNA sequence relatedness in natural populations. III. Techniques and potential applications. *J. Mol. Evol.* 17: 214–226.

Lansman, R. A., J. C. Avise, C. F. Aquadro, J. F. Shapira and S. W. Daniel. 1983. Extensive genetic variation in mitochondrial DNA's among geographic populations of the deer mouse, *Peromyscus maniculatus*. *Evolution* 37: 1–16.

Lanyon, S. M. 1992. Interspecific brood parasitism in blackbirds (Icterinae): A phylogenetic perspective. *Science* 255: 77–79.

Lanyon, S. M. and J. G. Hall. 1994. Re-examination of barbet monophyly using mitochondrial-DNA sequence data. *Auk* 111: 389–397.

Lanyon, S. M. and K. E. Omland. 1999. A molecular phylogeny of the blackbirds (Icteridae): Five lineages revealed by cytochrome-*b* sequence data. *Auk* 116: 629–639.

Lanyon, S. M. and R. M. Zink. 1987. Genetic variability in piciform birds: Monophyly and generic and familial relationships. *Auk* 104: 724–732.

Larsen, A. H., J. Sigurjónsson, N. Øien, G. Vikingsson and P. Palsbøll. 1996. Population genetic analysis of nuclear and mitochondrial loci in skin biopsies collected from central and northeastern North Atlantic humpback whales (*Megaptera novaeangliae*): Population identity and migratory destinations. *Proc. R. Soc. Lond. B* 263: 1611–1618.

Larson, A. 1984. Neontological inferences of evolutionary pattern and process in the salamander family Plethodontidae. *Evol. Biol.* 17: 119–127.

Larson, A. 1989. The relationship between speciation and morphological evolution. Pp. 579–598 in *Speciation and Its Consequences*, D. Otte and J. A. Endler (eds.). Sinauer Associates, Sunderland, MA.

Larson, A., D. B. Wake and K. P. Yanev. 1984. Measuring gene flow among populations having high levels of genetic fragmentation. *Genetics* 106: 293–308.

Lasker, G. W. 1985. *Surnames and Genetic Structure*. Cambridge University Press, New York.

Latorre, A., A. Moya and F. J. Ayala. 1986. Evolution of mitochondrial DNA in *Drosophila subobscura*. *Proc. Natl. Acad. Sci. USA* 83: 8649–8653.

Latta, R. G. and J. B. Mitton. 1997. A comparison of population differentiation across four classes of gene marker in limber pine (*Pinus flexilis* James). *Genetics* 146: 1153–1163.

Latta, R. G. and J. B. Mitton. 1999. Historical separation and present gene flow though a zone of secondary contact in ponderosa pine. *Evolution* 53: 769–776.

Lauder, G. V. and S. M. Reilly. 1996. The mechanistic bases of behavioral evolution: A multivariate analysis of musculoskeletal function. Pp. 104–137 in *Phylogenies and the Comparative Method in Animal Behavior*, E. P. Martins (ed.). Oxford University Press, New York.

Laughlin, T. F., B. A. Lubinski, E. H. Park, D. S. Taylor and B. J. Turner. 1995. Clonal stability and mutation in the self-fertilizing hermaphroditic fish, *Rivulus marmoratus*. *J. Heredity* 86: 399–402.

Launey, S. and D. Hedgecock. 2001. High genetic load in the Pacific oyster *Crassostrea gigas*. *Genetics* 159: 255–265.

Laurent, L. and 7 others. 1993. Genetic studies of relationships between Mediterranean and Atlantic populations of loggerhead turtle *Caretta caretta* (Cheloniidae). Implications for conservation. *Compt. Rend. Acad. Sci. Paris* 316: 1233–1239.

Laurent, L. and 17 others. 1998. Molecular resolution of marine turtle stock composition in fishery bycatch: A case study in the Mediterranean. *Mol. Ecol.* 7: 1529–1542.

Lavery, S., C. Moritz and D. R. Fielder. 1996. Genetic patterns suggest exponential growth in a declining species. *Mol. Biol. Evol.* 13: 1106–1113.

Lavin, M., J. J. Doyle and J. D. Palmer. 1990. Evolutionary significance of the loss of chloroplast-DNA inverted repeat in the Leguminosae subfamily Papilionoideae. *Evolution* 44: 390–402.

Lavrov, D. V., J. L. Boore and W. M. Brown. 2002. Complete mtDNA sequences from two millipedes suggest a new model for mitochondial gene rearrangements: Duplication and nonrandom loss. *Mol. Biol. Evol.* 19: 163–169.

Lawlor, D. A., F. E. Ward, P. D. Ennis, A. P. Jackson and P. Parham. 1988. HLA-A and B polymorphisms predate the divergence of humans and chimpanzees. *Nature* 335: 268–271.

Lawrence, J. G. 2002. Gene transfer in bacteria: Speciation without species? *Theoret. Pop. Biol.* 61: 449–460.

Lawrence, J. G. and D. L. Hartl. 1992. Inference of horizontal genetic transfer from molecular data: An approach using the bootstrap. *Genetics* 131: 753–760.

Lawrence, J. G. and H. Ochman. 1998. Molecular archaeology of the *Escherichia coli* genome. *Proc. Natl. Acad. Sci. USA* 95: 9413–9417.

Layton, C. R. and F. R. Ganders. 1984. The genetic consequences of contrasting breeding systems in *Plectritis* (Valerianaceae). *Evolution* 38: 1308–1325.

Leakey, R. and R. Lewin. 1995. *The Sixth Extinction: Biodiversity and its Survival*. Doubleday, New York.

Learn, G. L. and B. A. Schaal. 1987. Population subdivision for rDNA repeat variants in *Clematis fremontii*. *Evolution* 41: 433–437.

Leary, R. F., F. W. Allendorf, S. R. Phelps and K. L. Knudsen. 1984. Introgression between west-slope cutthroat and rainbow trout in Clark Fork River drainage, Montana. *Proc. Mont. Acad. Sci.* 43: 1–18.

Leary, R. F., F. W. Allendorf, K. L. Knudsen and G. H. Thorgaard. 1985. Heterozygosity and developmental stability in gynogenetic diploid and triploid rainbow trout. *Heredity* 54: 219–225.

Leary, R. F., F. W. Allendorf and S. H. Forbes. 1993. Conservation genetics of bull trout in the Columbia and Klamath River drainages. *Cons. Biol.* 7: 856–865.

Ledig, F. T. and M. T. Conkle. 1983. Gene diversity and genetic structure in a narrow endemic, Torrey pine (*Pinus torreyana* Parry ex Carr). *Evolution* 37: 79–85.

Ledig, F. T., R. P. Guries and B. A. Bonefield. 1983. The relation of growth to heterozygosity in pitch pine. *Evolution* 37: 1227–1238.

Lee, K., J. Feinstein and J. Cracraft. 1997. The phylogeny of ratite birds: Resolving conflicts between molecular and morphological data sets. Pp. 173–195 in *Avian Molecular Evolution and Systematics*, D. P. Mindell (ed.). Academic Press, New York.

Lee, W.-J. and T. D. Kocher. 1995. Complete sequence of a sea lamprey (*Petromyzon marinus*) mitochondrial genome: Early establishment of the vertebrate genome organization. *Genetics* 139: 873–887.

Lee, Y.-H., T. Ota and V. D. Vacquier. 1995. Positive selection is a general phenomenon in the evolution of abalone sperm lysin. *Mol. Biol. Evol.* 12: 231–238.

Lee, Y. M., D. J. Friedman and F. J. Ayala. 1985. Superoxide dismutase: An evolutionary puzzle. *Proc. Natl. Acad. Sci. USA* 82: 824–828.

Legge, J. T., R. Roush, R. DeSalle, A. P. Vogler and B. May. 1996. Genetic criteria for establishing evolutionarily significant units in Cryan's buckmoth. *Cons. Biol.* 10: 85–98.

Lehman, N. and R. K. Wayne. 1991. Analysis of coyote mitochondrial DNA genotype frequencies: Estimation of the effective number of alleles. *Genetics* 128: 405–416.

Lehman, N., A. Eisenhawer, K. Hansen, D. L. Mech, R. O. Peterson, J. P. Gogan and R. K. Wayne. 1991. Introgression of coyote mitochondrial DNA into sympatric North American gray wolf populations. *Evolution* 45: 104–119.

Lehmann, T., W. A. Hawley, H. Grebert and F. H. Collins. 1998. The effective population size of *Anopheles gambiae* in Kenya: Implications for population structure. *Mol. Biol. Evol.* 15: 264–276.

Lehmann, T. and 8 others. 2003. Population structure of *Anopheles gambiae* in Africa. *J. Heredity* 94: 133–147.

Leigh Brown, A. J. and C. H. Langley. 1979. Re-evaluation of level of genic heterozygosity in natural populations of *Drosophila melanogaster* by two-dimensional electrophoresis. *Proc. Natl. Acad. Sci. USA* 76: 2381–2384.

Leinaas, H. P. 1983. A haplodiploid analogy in the evolution of termite eusociality? Reply to Lacy. *Am. Nat.* 121: 302–304.

Lemen, C. A. and P. W. Freeman. 1981. A test of macroevolutionary problems with neontological data. *Paleobiology* 7: 311–315.

Lemen, C. A. and P. W. Freeman. 1989. Testing macroevolutionary hypotheses with cladistic analysis: Evidence against rectangular evolution. *Evolution* 43: 1538–1554.

Lemey, P., O. G. Pybus, B. Wang, N. K. Saksena, M. Salemi and A.-M. Vandamme. 2003. Tracing the origin and history of the HIV-2 epidemic. *Proc. Natl. Acad. Sci. USA* 100: 6588–6592.

Lens, L., P. Galbusera, R. Brooks, E. Waiyaki and T. Schenck. 1998. Highly skewed sex ratios in the critically endangered Taita thrush as revealed by CHD genes. *Biodiv. Cons.* 7: 869–873.

Lenski, R. E. 1995. Molecules are more that markers: New directions in molecular microbial ecology. *Mol. Ecol.* 4: 643–651.

Leonard, J. A., R. K. Wayne and A. Cooper. 2000. Population genetics of Ice Age brown bears. *Proc. Natl. Acad. Sci. USA* 97: 1651–1654.

Leone, C. A. 1964. *Taxonomic Biochemistry and Serology.* Ronald Press, New York.

Lerat, E., V. Daubin and N. A. Moran. 2003. From gene trees to organismal phylogeny in prokaryotes: The case of the γ-proteobacteria. *PLoS Biol.* 1: 101–109.

Lerner, I. M. 1954. *Genetic Homeostasis.* Oliver and Boyd, Edinburgh, Scotland.

Lesica, P., R. F. Leary, F. W. Allendorf and D. E. Bilderback. 1988. Lack of genic diversity within and among populations of an endangered plant, *Howellia aquatilis. Cons. Biol.* 2: 275–282.

Lessa, E. P. 1990. Multidimensional analysis of geographic genetic structure. *Syst. Zool.* 39: 242–252.

Lessa, E. P. 1993. Analysis of DNA sequence variation at the population level by polymerase chain reaction and denaturing gradient gel electrophoresis. *Meth. Enzymol.* 224: 419–428.

Lessa, E. P. and G. Applebaum. 1993. Screening techniques for detecting allelic variation in DNA sequences. *Mol. Ecol.* 2: 119–129.

Lessa, E. P., J. A. Cook and J. L. Patton. 2003. Genetic footprints of demographic expansion in North America, but not Amazonia, during the Late Quaternary. *Proc. Natl. Acad. Sci. USA* 100: 10331–10334.

Lessells, C. M. and A. C. Mateman. 1998. Sexing birds using random amplified polymorphic DNA (RAPD) markers. *Mol. Ecol.* 7: 187–195.

Lessios, H. A. 1979. Use of Panamanian sea urchins to test the molecular clock. *Nature* 280: 599–601.

Lessios, H. A. 1981. Divergence in allopatry: Molecular and morphological differentiation between sea urchins separated by the Isthmus of Panama. *Evolution* 35: 618–634.

Lessios, H. A. 1998. The first stage of speciation as seen in organisms separated by the Isthmus of Panama. Pp. 186–201 in *Endless Forms: Species and Speciation.* D. J. Howard and S. H. Berlocher (eds.). Oxford University Press, New York.

Lessios, H. A., B. D. Kessing and D. R. Robinson. 1998. Massive gene flow across the world's most potent marine biogeographic barrier. *Proc. R. Soc. Lond. B* 265: 583–588.

Letcher, B. H. and T. L. King. 1999. Targeted stock identification using multilocus genotype 'familyprinting.' *Fisheries Res.* 43: 99–111.

Lettink, M., I. G. Jamieson, C. D. Millar and D. M. Lambert. 2002. Mating system and genetic variation in the endangered New Zealand takahe. *Cons. Genet.* 3: 427–434.

Leunissen, J. A. M. and W. W. de Jong. 1986. Copper/zinc superoxide dismutase: How likely is gene transfer from ponyfish to *Photobacterium leiognathi*? *J. Mol. Evol.* 23: 250–258.

Levene, H. 1953. Genetic equilibrium when more than one ecological niche is available. *Am. Nat.* 87: 331–333.

Levin, B. R. 1981. Periodic selection, infectious gene exchange and the genetic structure of *E. coli* populations. *Genetics* 99: 1–23.

Levin, D. A. 1979. The nature of plant species. *Science* 204: 381–384.

Levin, D. A. and H. W. Kerster. 1971. Neighborhood structure in plants under diverse reproductive methods. *Am. Nat.* 105: 345–354.

Levin, D. A. and H. W. Kerster. 1974. Gene flow in seed plants. *Evol. Biol.* 7: 139–220.

Levin, D. A., J. Francisco-Ortega and R. K. Jansen. 1996. Hybridization and the extinction of rare plant species. *Cons. Biol.* 10: 10–16.

Levin, L. A. 1990. A review of methods for labeling and tracking marine invertebrate larvae. *Ophelia* 32: 115–144.

Levine, H. 1953. Genetic equilibrium when more than one ecological niche is available. *Am. Nat.* 87: 331–333.

Levins, R. 1968. *Evolution in Changing Environments.* Princeton University Press, Princeton, NJ.

Levy, H. 1996. *And the Blood Cried Out.* Basic Books, New York.

Lewin, B. 1999. *Genes VII.* Oxford University Press, New York.

Lewin, R. 1999. *Human Evolution.* Blackwell, Malden, MA.

Lewis, R. E. Jr. and J. M. Cruse. 1992. DNA typing in human parentage using multilocus and single-locus probes. Pp. 3–17 in *Paternity in Primates: Genetic Tests and Theories*, R. D. Martin, A. F. Dixson and E. J. Wickings (eds.). Karger, Basel, Switzerland.

Lewis, W. 1980. *Polyploidy: Biological Relevance.* Plenum Press, New York.

Lewontin, R. C. 1972. The apportionment of human diversity. *Evol. Biol.* 6: 381–398.

Lewontin, R. C. 1974. *The Genetic Basis of Evolutionary Change.* Columbia University Press, New York.

Lewontin, R. C. 1985. Population genetics. *Annu. Rev. Genet.* 19: 81–102.

Lewontin, R. C. 1988. On measures of gametic disequilibrium. *Genetics* 120: 849–852.

Lewontin, R. C. 1991. Twenty-five years ago in GENETICS: Electrophoresis in the development of evolutionary genetics: Milestone or millstone? *Genetics* 128: 657–662.

Lewontin, R. C. and L. C. Birch. 1966. Hybridization as a source of variation for adaptation to new environments. *Evolution* 10: 315–336.

Lewontin, R. C. and D. L. Hartl. 1991. Population genetics in forensic DNA typing. *Science* 254: 1745–1750.

Lewontin, R. C. and J. L. Hubby. 1966. A molecular approach to the study of genic heterozygosity in natural populations. II. Amount of variation and degree of heterozygosity in natural populations of *Drosophila psuedoobscura*. *Genetics* 54: 595–609.

Lewontin, R. C. and J. Krakauer. 1973. Distribution of gene frequency as a test of the theory of the selective neutrality of polymorphisms. *Genetics* 74: 175–195.

Lewontin, R. C. and J. Krakauer. 1975. Testing the heterogeneity of F values. *Genetics* 80: 397–398.

Li, G. and D. Hedgecock. 1998. Genetic heterogeneity, detected by PCR-SSCP, among samples of larval Pacific oysters (*Crassostrea gigas*) supports the hypothesis of large variance in reproductive success. *Can. J. Fish. Aquat. Sci.* 55: 1025–1033.

Li, H., U. B. Gyllensten, X. Cui, R. K. Saiki, H. A. Erlich and N. Arnheim. 1988. Amplification and analysis of DNA sequences in single human sperm and diploid cells. *Nature* 335: 414–417.

Li, S. K. and D. H. Owings. 1978. Sexual selection in the three-spined stickleback. II. Nest raiding during the courtship phase. *Behaviour* 64: 298–304.

Li, W.-H. 1978. Maintenance of genetic variability under the joint effect of mutation, selection, and random genetic drift. *Genetics* 90: 349–382.

Li, W.-H. 1997. *Molecular Evolution.* Sinauer Associates, Sunderland, MA.

Li, W.-H. and D. Graur. 1991. *Fundamentals of Molecular Evolution.* Sinauer Associates, Sunderland, MA.

Li, W.-H. and L. A. Sadler. 1991. Low nucleotide diversity in man. *Genetics* 129: 513–523.

Li, W.-H. and M. Tanimura. 1987. The molecular clock runs more slowly in man than in apes and monkeys. *Nature* 326: 93–96.

Li, W.-H., M. Tanimura and P. M. Sharp. 1987. An evaluation of the molecular clock hypothesis using mammalian DNA sequences. *J. Mol. Evol.* 25: 330–342.

Li, W.-H., M. Tanimura and P. M. Sharp. 1988. Rates and dates of divergence between AIDS virus nucleotide sequences. *Mol. Biol. Evol.* 5: 313–330.

Li, Y. C., A. B. Korol, T. Fahima, A. Beiles and E. Nevo. 2002. Microsatellites: Genomic distribution, putative functions and mutational mechanisms: A review. *Mol. Ecol.* 11: 2453–2465.

Li, Y. Y., C. Hengstenberg and B. Maisch. 1995. Whole mitochondrial genome amplification reveals basal level multiple deletions in mtDNA of patients with dilated cardiomyopathy. *Biochem. Biophys. Res. Comm.* 210: 211–218.

Lidholm, J., A. E. Szmidt, J. E. Hallgren and P. Gustafsson. 1988. The chloroplast genomes of conifers lack one of the rRNA-encoding inverted repeats. *Mol. Gen. Genet.* 212: 6–10.

Liebherr, J. K. 1988. Gene flow in ground beetles (Coleoptera: Carabidae) of differing habitat preference and flight-wing development. *Evolution* 42: 129–137.

Lien, S., N. E. Cockett, H. Klungland, N. Arnheim, M. Georges and L. Gomez-Raya. 1999. High-resolution genetic map of the sheep callipyge region: Linkage heterogeneity among rams detected by sperm typing. *Anim. Genet.* 30: 42–46.

Ligon, J. D. 1999. *The Evolution of Avian Breeding Systems.* Oxford University Press, Oxford, England.

Lindahl, T. 1993a. Instability and decay of the primary structure of DNA. *Nature* 362: 709–715.

Lindahl, T. 1993b. Antediluvian DNA. *Nature* 365: 700.

Linder, C. R., L. A. Moore and R. B. Jackson. 2000. A universal molecular method for identifying underground plant parts to species. *Mol. Ecol.* 9: 1549–1559.

Linn, C. Jr., J. L. Feder, S. Nojima, H. R. Dambroski, S. H. Berlocher and W. Roelofs. 2003. Fruit odor discrimination and sympatric host formation in *Rhagoletis. Proc. Natl. Acad. Sci. USA* 100: 11490–11493.

Linn, S. and W. Arber, 1968. Host specificity of DNA produced by *Escherichia coli.* X. In vitro restriction of phage fd replicative form. *Proc. Natl. Acad. Sci. USA* 59: 1300–1306.

Linnaeus, C. 1759. *Systema Naturae.* Reprinted 1964 by Wheldon and Wesley, Ltd., New York.

Lintas, C., J. Hirano and S. Archer. 1998. Genetic variation of the European eel (*Anguilla anguilla*). *Mol. Mar. Biol. Biotechnol.* 7: 263–269.

Liou, L. W. and T. D. Price. 1994. Speciation by reinforcement of premating isolation. *Evolution* 48: 1451–1459.

Liston, A., L. H. Rieseberg and T. S. Elias. 1989. Genetic similarity is high between intercontinental disjunct species of *Senecio* (Asteraceae). *Am. J. Bot.* 76: 383–388.

Liston, A., L. H. Rieseberg and T. S. Elias. 1990. Functional androdioecy in the flowering plant *Datisca glomerata. Nature* 343: 641–642.

Litt, M. and J. A. Luty. 1989. A hypervariable microsatellite revealed by in vitro amplification of a dinucleotide repeat within the cardiac muscle actin gene. *Am. J. Human Genet.* 44: 387–401.

Little, T. J. and P. D. N. Hebert. 1996. Ancient asexuals: Scandal or artifact. *Trends Ecol. Evol.* 11: 296.

Liu, H.-P., J. B. Mitton and S.-K. Wu. 1996. Paternal mitochondrial DNA differentiation far exceeds maternal DNA and allozyme differentiation in the freshwater mussel *Anodonda grandis grandis. Evolution* 50: 952–957.

Liu, J., J. M. Mercer, L. F. Stam, G. C. Gibson, Z.-B. Zeng and C. C. Laurie. 1996. Genetic analysis of a morphological shape difference in the male genitalia of *Drosophila simulans* and *D. mauritiana. Genetics* 142: 1129–1145.

Liu, L. L., D. W. Foltz and W. B. Stickle. 1991. Genetic population structure of the southern oyster drill *Stramonita* (= *Thais*) *haemostoma. Mar. Biol.* 111: 71–79.

Lobo, J. A., M. A. Del Lama and M. A. Mestriner. 1989. Population differentiation and racial admixture in the Africanized honeybee (*Apis mellifera* L.). *Evolution* 43: 794–802.

Locke, D. P. and 6 others. 2003. Large-scale variation among human and great ape genomes determined by array comparative genomic hybridization. *Genome Res.* 13: 347–357.

Lockhart, P. J., T. J. Beanland, C. J. Howe and A. W. D. Larkum. 1992. Sequence of *Prochloron didemni atpBE* and the inference of chloroplast origins. *Proc. Natl. Acad. Sci. USA* 89: 2742–2746.

Loeschcke, V., J. Tomiuk and S. K. Jain (eds.). 1994. *Conservation Genetics.* Birkhäuser Verlag, Basel, Switzerland.

Long, A. D., R. F. Lyman, C. H. Langley and T. F. C. Mackay. 1998. Two sites in the *Delta* gene region contribute to naturally occurring variation in bristle number in *Drosophila melanogaster. Genetics* 149: 999–1017.

Long, E. O. and I. B. Dawid. 1980. Repeated sequences in eukaryotes. *Annu. Rev. Biochem.* 49: 727–764.

Longmire, J. L., P. M. Kraemer, N. C. Brown, L. C. Hardekopf and L. L. Deaven. 1990. A new multi-locus DNA fingerprinting probe: pV47–2. *Nucleic Acids Res.* 18: 1658.

Longmire, J. L., G. F. Gee, C. L. Hardekopf and G. A. Mark. 1992. Establishing paternity in whooping cranes (*Grus americana*) by DNA analysis. *Auk* 109: 522–529.

Lopez, J. V., R. Kersanach, S. A. Rehner and N. Knowlton. 1999. Molecular determination of species boundaries in corals: Genetic analysis of the *Mostastrea annularis* complex using Amplified Fragment Length Polymorphisms and a microsatellite marker. *Biol. Bull.* 196: 80–93.

Lopez, T. J., E. D. Hauselman, L. R. Maxson and J. W. Wright. 1992. Preliminary analysis of phylogenetic relationships among Galapagos Island lizards of the genus *Tropidurus. Amphibia-Reptilia* 13: 327–339.

Loreille, O., L. Orlando, M. Patou-Mathis, M. Philippe, P. Taberlet and C. Hänni. 2001. Ancient DNA analysis reveals divergence of the cave bear, *Ursus spelaeus,* and brown bear, *Ursus arctos,* lineages. *Curr. Biol.* 11: 200–203.

Losos, J. B. 1996. Phylogenetic perspectives on community ecology. *Ecology* 77: 1344–1345.

Losos, J. B. and F. R. Adler. 1995. Stumped by trees? A generalized null model for patterns of organismal diversity. *Am. Nat.* 145: 329–342.

Losos, J. B., T. R. Jackman, A. Larson, K. deQueiroz and L. R. Schettino. 1998. Contingency and determinism in replicated adaptive radiations of island lizards. *Science* 279: 2115–2118.

Lotka, A. J. 1931. Population analysis: The extinction of families. I. *J. Washington Acad. Sci.* 21: 377–380.

Loughry, W. J., P. A. Prodöhl, C. M. McDonough and J. C. Avise. 1998a. Polyembryony in armadillos. *Am. Sci.* 86: 274–279.

Loughry, W. J., P. A. Prodöhl, C. M. McDonough, W. S. Nelson and J. C. Avise. 1998b. Correlates of reproductive success in a population of nine-banded armadillos. *Can. J. Zool.* 76: 1815–1821.

Loveless, M. D. and J. L. Hamrick. 1984. Ecological determinants of genetic structure in plant populations. *Annu. Rev. Ecol. Syst.* 15: 65–95.

Loveless, M. D., J. L. Hamrick and R. B. Foster. 1998. Population structure and mating system in *Tachigali versicolor,* a monocarpic Neotropical tree. *Heredity* 81: 134–143.

Lovette, I. J. and E. Bermingham. 1999. Explosive speciation in the New World *Dendroica* warblers. *Proc. R. Soc. Lond. B* 266: 1629–1636.

Lovette, I. J. and E. Bermingham. 2000. c-*mos* variation in songbirds: Molecular evolution, phylogenetic implications, and comparisons with mitochondrial differentiation. *Mol. Biol. Evol.* 17: 1569–1577.

Lovette, I. J., E. Bermingham, G. Seutin and R. E. Ricklefs. 1998. Evolutionary differentiation in three endemic West Indian warblers. *Auk* 115: 890–903.

Lovette, I. J., E. Bermingham and R. E. Ricklefs. 1999. Mitochondrial DNA phylogeography and the conservation of endangered Lesser Antillean *Icterus* orioles. *Cons. Biol.* 13: 1088–1096.

Lowenstein, J. M., V. M. Sarich and B. J. Richardson. 1981. Albumin systematics of the extinct mammoth and Tasmanian wolf. *Nature* 291: 409–411.

Loy, A., M. Corti and S. Cataudella. 1999. Variation in gill-raker number during growth of the sea bass, *Dicentrarchus labrax* (Perciformes: Moronidae), reared at different salinities. *Env. Biol. Fishes* 55: 391–398.

Lu, G., D. J. Basley and L. Bernatchez. 2001. Contrasting patterns of mitochondrial DNA and microsatellite introgressive hybridization between lineages of lake whitefish (*Coregonus clupeaformis*): Relevance for speciation. *Mol. Ecol.* 10: 965–985.

Lubchenko, J., S. R. Palumbi, S. D. Gaines and S. Andelman. 2003. Plugging a hole in the ocean: The emerging science of marine reserves. *Ecol. Appl.* 13: S3-S7.

Lubick, N. 2003. New count of old whales adds up to big debate. *Science* 301: 451.

Lubinski, B. A., W. P. Davis, D. S. Taylor and B. J. Turner. 1995. Outcrossing in a natural population of a self-fertilizing fish. *J. Heredity* 86: 469–473.

Luikart, G. and P. R. England. 1999. Statistical analysis of microsatellite data. *Trends Ecol. Evol.* 14: 253–256.

Luikart, G., W. B. Sherwin, B. M. Steele and F. W. Allendorf. 1998. Usefulness of molecular markers for detecting population bottlenecks via monitoring genetic change. *Mol. Ecol.* 7: 963–974.

Lundy, K. J., P. G. Parker and A. Zahavi. 1998. Reproduction by subordinates in cooperatively breeding Arabian babblers is uncommon but predictable. *Behav. Ecol. Sociobiol.* 43: 173–180.

Lunt, D. H. and B. C. Hyman. 1997. Animal mitochondrial DNA recombination. *Nature* 387: 247.

Lunt, D. H., L. E. Whipple and B. C. Hyman. 1998. Mitochondrial DNA variable number tandem repeats (VNTRs): Utility and problems in molecular ecology. *Mol. Ecol.* 7: 1441–1455.

Luo, Z. W., C.-I. Wu and M. J. Kearsey. 2002. Precision and high-resolution mapping of quantitative trait loci by use of recurrent selection, backcross or intercross schemes. *Genetics* 161: 915–929.

Lutzoni, F. and M. Pagel. 1997. Accelerated evolution as a consequence of transitions to mutualism. *Proc. Natl. Acad. Sci. USA* 94: 11422–11427.

Lydeard, C. and K. J. Roe. 1997. The phylogenetic utility of the mitochondrial cytochrome *b* gene for inferring relationships among Actinopterygian fishes. Pp. 285–303 in *Molecular Systematics of Fishes*, T. D. Kocher and C. A. Stepien (eds.). Academic Press, New York.

Lynch, M. 1988. Estimation of relatedness by DNA fingerprinting. *Mol. Biol. Evol.* 5: 584–589.

Lynch, M. 1993. A method for calibrating molecular clocks and its application to animal mitochondrial DNA. *Genetics* 135: 1197–1208.

Lynch, M. 1999. The age and relationships of the major animal phyla. *Evolution* 53: 319–325.

Lynch, M. and T. J. Crease. 1990. The analysis of population survey data on DNA sequence variation. *Mol. Biol. Evol.* 7: 377–394.

Lynch, M. and R. Lande. 1998. The critical effective size for a genetically secure population. *Anim. Cons.* 1: 70–72.

Lynch, M. and K. Ritland. 1999. Estimation of pairwise relatedness with molecular markers. *Genetics* 152: 1753–1766.

Lynch, M. and J. B. Walsh. 1998. *Genetics and Analysis of Quantitative Traits.* Sinauer Associates, Sunderland, MA.

Lyrholm, T., O. Leimar, B. Johanneson and U. Gyllensten. 1999. Sex-biased dispersal in sperm whales: Contrasting mitochondrial and nuclear genetic structure of global populations. *Proc. R. Soc. Lond. B* 266: 347–354.

Macdonald, S. J. and D. B. Goldstein. 1999. A quantitative genetic analysis of male sexual traits distinguishing the sibling species *Drosophila simulans* and *D. seychellia*. *Genetics* 153: 1683–1699.

Mace, G. M., J. L. Gittleman and A. Purvis. 2003. Preserving the tree of life. *Science* 300: 1707–1709.

Macey, J. R., A. Larson, N. B. Ananjeva, Z. Fang and T. J. Papenfuss. 1997. Two novel gene orders and the role of light-strand replication in rearrangement of the vertebrate mitochondrial genome. *Mol. Biol. Evol.* 14: 91–104.

Macey, J. R., J. A. Schulte II, A. Larson and T. J. Papenfuss. 1998. Tandem duplication via light-strand synthesis may provide a precursor for mitochondrial genome rearrangement. *Mol. Biol. Evol.* 15: 71–75.

Machado, C. A. and F. J. Ayala. 2001. Nucleotide sequences provide evidence of genetic exchange among distantly related lineages of *Trypanosoma cruzi*. *Proc. Natl. Acad. Sci. USA* 98: 7396–7401.

Machado, C. A. and J. Hey. 2003. The causes of phylogenetic conflict in a classic *Drosophila* species group. *Proc. R. Soc. Lond. B* 270: 1193–1202.

Machado, C. A., E. Jousselin, F. Kjellberg, S. G. Compton and E. A. Herre. 2001. Phylogenetic relationships, historical biogeography and character evolution of fig-pollinating wasps. *Proc. R. Soc. Lond. B* 268: 685–694.

Machado, C. A., R. M. Kliman, J. A. Markert and J. Hey. 2002. Inferring the history of speciation from multilocus DNA sequence data: The case of *Drosophila pseudoobscura* and close relatives. *Mol. Biol. Evol.* 19: 472–488.

MacIntyre, R. J. 1976. Evolution and ecological value of duplicate genes. *Annu. Rev. Ecol. Syst.* 7: 421–468.

MacIntyre, R. J., M. R. Dean and G. Batt. 1978. Evolution of acid phosphatase-1 in the genus *Drosophila*. Immunological studies. *J. Mol. Evol.* 12: 121–142.

Mack, P. D., N. K. Priest and D. E. L. Promislow. 2003. Female age and sperm competition: Last-male precedence declines as female age increases. *Proc. R. Soc. Lond. B* 270: 159–165.

Mackay, T. F. C. 2001. Quantitative trait loci in *Drosophila*. *Nature Rev. Genet.* 2: 11–20.

Mackiewicz, M., D. E. Fletcher, S. D. Wilkins, J. A. DeWoody and J. C. Avise. 2002. A genetic assessment of parentage in a natural population of dollar sunfish (*Lepomis marginatus*) based on microsatellite markers. *Mol. Ecol.* 11: 1877–1883.

MacNeil, D. and C. Strobeck. 1987. Evolutionary relationships among colonies of Columbian ground squirrels as shown by mitochondrial DNA. *Evolution* 41: 873–881.

MacRae, A. F. and W. W. Anderson. 1988. Evidence for non-neutrality of mitochondrial DNA haplotypes in *Drosophila pseudoobscura*. *Genetics* 120: 485–494.

Maddison, D. R. 1991. African origins of human mitochondrial DNA reexamined. *Syst. Zool.* 40: 355–363.

Maddison, D. R. 1994. Phylogenetic methods for inferring the evolutionary history and processes of change in discretely valued characters. *Annu. Rev. Entomol.* 39: 267–292.

Maddison, D. R. 1995. Calculating the probability distribution of ancestral states reconstructed by parsimony on phylogenetic trees. *Syst. Biol.* 44: 474–481.

Maddison, W. P. 1995. Phylogenetic histories within and among species. Pp. 273–287 in *Experimental and Molecular Approaches to Plant Biosystematics*, P. C. Hoch and A. G. Stephenson (eds.). Monogr. Syst. Missouri Bot. Gardens 53, St. Louis, MO.

Maddison, W. P. 1997. Gene trees in species trees. *Syst. Biol.* 46: 523–536.

Maddox, G. D., R. E. Cook, P. H. Wimberger and S. Gardescu. 1989. Clone structure in four *Solidago altissima* (Asteraceae) populations: Rhizome connections within genotypes. *Am. J. Bot.* 76: 318–326.

Madsen, O. and 9 others. 2001. Parallel adaptive radiations in two major clades of placental mammals. *Nature* 409: 610–614.

Maeda, N., C.-I. Wu, J. Bliska and J. Reneke. 1988. Molecular evolution of intergenic DNA in higher primates: Pattern of DNA changes, molecular clock, and evolution of repetitive sequences. *Mol. Biol. Evol.* 5: 1–20.

Maehr, D. S., E. D. Land, D. B. Shindle, O. L. Bass and T. S. Hoctor. 2002. Florida panther dispersal and conservation. *Biol. Cons.* 106: 187–197.

Maes, G. E. and F. A. M. Volckaert. 2002. Clinal genetic variation and isolation by distance in the European eel *Anguilla anguilla* (L.). *Biol. J. Linn. Soc.* 77: 509–521.

Magee, B. B. and P. T. Magee. 2000. Induction of mating in *Candida albicans* by construction of *MTLa* and *MTLα* strains. *Science* 289: 310–313.

Maggioncalda, A. N. and R. M. Sapolsky. 2002. Disturbing behaviors of the orangutan. *Sci. Am.* 286(6): 60–65.

Maha, G. C., R. Highton and L. R. Maxson. 1989. Biochemical evolution in the slimy salamanders of the *Plethodon glutinosus* complex in the eastern United States. *Illinois Biol. Monogr.* 57: 81–150.

Maki, M. and S. Horie. 1999. Random amplified polymorphic DNA (RAPD) markers reveal less genetic variation in the endangered plant *Cerastium fischerianum* var. *molle* than in the widespread conspecific *C. fischerianum* var. *fischerianum* (Caryophyllceae). *Mol. Ecol.* 8: 145–150.

Maldonado, M. 1998. Do chimeric sponges have improved chances of survival? *Marine Ecol. Progr. Ser.* 164: 301–306.

Malik, S., P. J. Wilson, R. J. Smith, D. M. Lavigne and B. N. White. 1997. Pinniped penises in trade: A molecular-genetic investigation. *Cons. Biol.* 11: 1365–1374.

Mallatt, J. and J. Sullivan. 1998. 28S and 18S rDNA sequences support the monophyly of lampreys and hagfishes. *Mol. Biol. Evol.* 15: 1706–1718.

Mallatt, J. and C. J. Winchell. 2002. Testing the new animal phylogeny: First use of combined large-subunit and small-subunit rRNA gene sequences to classify the Protostomes. *Mol. Biol. Evol.* 19: 289–301.

Mallet, J. 1995. A species definition for the Modern Synthesis. *Trends Ecol. Evol.* 10: 294–299.

Mandel, M. J., C. L. Ross and R. G. Harrison. 2001. Do *Wolbachia* infections play a role in unidirectional incompatibilities in a field cricket hybrid zone? *Mol. Ecol.* 10: 703–709.

Manel, S., P. Berthier and G. Luikart. 2002. Detecting wildlife poaching: Identifying the origin of individuals with Bayesian assignment tests and multilocus genotypes. *Cons. Biol.* 16: 650–659.

Manhart, J. R. and J. D. Palmer. 1990. The gain of two chloroplast tRNA introns marks the green algal ancestors of land plants. *Nature* 345: 268–270.

Maniatis, T. and 7 others. 1978. The isolation of structural genes from libraries of eucaryotic DNA. *Cell* 15: 687–701.

Mank, J. E. and J. C. Avise. 2003. Microsatellite variation and differentiation in North Atlantic eels. *J. Heredity* 94: 310–314.

Mannarelli, B. M. and C. P. Kurtzman. 1998. Rapid identification of *Candida albicans* and other human pathogenic yeasts by using short oligonucleotides in a PCR. *J. Clin. Microbiol.* 36: 1634–1641.

Mantovani, B. and V. Scali. 1992. Hybridogenesis and androgenesis in the stick-insect *Bacillus rossius-grandii benazzii* (Insecta, Phasmatodea). *Evolution* 46: 783–796.

Mantovani, B., V. Scali and F. Tinti. 1991. Allozyme analysis and phyletic relationships of two new stick-insects from north-west Sicily: *Bacillus grandii benazzii* and *B. rossius-grandii benazzii* (Insecta Phasmatodea). *J. Evol. Biol.* 4: 279–290.

Mantovani, B., M. Passamonti and V. Scali. 2001. The mitochondrial cytochrome oxidase II gene in *Bacillus* stick insects: Ancestry of hybrids, androgenesis, and phylogenetic relationships. *Mol. Phylogen. Evol.* 19: 157–163.

Marchant, A. D. 1988. Apparent introgression of mitochondrial DNA across a narrow hybrid zone in the *Caledia captiva* species-complex. *Heredity* 60: 39–46.

Marchetti, K., H. Nakamura and H. L. Gibbs. 1998. Host-race formation in the common cuckoo. *Science* 282: 471–472.

Marchinko, K. B. 2003. Dramatic phenotypic plasticity in barnacle legs (*Balanus glandula* Darwin): Magnitude, age dependence, and speed of response. *Evolution* 57: 1281–1290.

Margoliash, E. 1963. Primary structure and evolution of cytochrome *c*. *Proc. Natl. Acad. Sci. USA* 50: 672–679.

Margulis, L. 1970. *Origin of Eukaryotic Cells*. Yale University Press, New Haven, CT.

Margulis, L. 1981. *Symbiosis in Cell Evolution: Life and Its Environment in the Early Earth*. W. H. Freeman and Co., San Francisco.

Margulis, L. 1995. *Symbiosis in Cell Evolution: Microbial Communities in the Archaen and Proterozoic Eons* (2nd Edition). W. H. Freeman and Co., San Francisco.

Margulis, L. 1996. Archael-eubacterial mergers in the origin of Eukarya: Phylogenetic classification of life. *Proc. Natl. Acad. Sci. USA* 93: 1071–1076.

Margulis, L. and D. Sagan. 2002. *Acquiring Genomes: A Theory of the Origin of Species*. Basic Books, New York.

Marjoram, P. and P. Donnelly. 1997. Human demography and the time since mitochondrial Eve. Pp. 107–131 in *Progress in Population Genetics and Human Evolution*, P. Donnelly and S. Tavaré (eds.). Springer-Verlag, New York.

Markert, C. L. and I. Faulhaber. 1965. Lactate dehydrogenase isozyme patterns of fish. *J. Exp. Zool.* 159: 319–332.

Markert, C. L., J. B. Shaklee and G. S. Whitt. 1975. Evolution of a gene. *Science* 189: 105–114.

Marko, P. B. 2002. Fossil calibration of molecular clocks and the divergence times of geminate species pairs separated by the Isthmus of Panama. *Mol. Biol. Evol.* 19: 2005–2021.

Mark Welch, D. B. M. and M. S. Meselson. 2000. Evidence for the evolution of bdelloid rotifers without sexual reproduction or genetic exchange. *Science* 288: 1211–1214.

Mark Welch, D. B. M. and M. S. Meselson. 2001. Rates of nucleotide substitution in sexual and anciently asexual rotifers. *Proc. Natl. Acad. Sci. USA* 98: 6720–6724.

Marra, P. P., K. A. Hobson and R. T. Holmes. 1998. Linking winter and summer events in a migratory bird by using stable-carbon isotopes. *Science* 282: 1884–1886.

Marshall, D. E. and N. C. Ellstrand. 1985. Proximal causes of multiple paternity in wild radish, *Raphanus sativus*. *Am. Nat.* 126: 596–605.

Marshall, D. E. and N. C. Ellstrand. 1986. Sexual selection in *Raphanus sativus*: Experimental data on nonrandom fertilization, maternal choice, and consequences of multiple paternity. *Am. Nat.* 127: 446–461.

Marshall, T. C., J. Slate, L. E. B. Kruuk and J. M. Pemberton. 1998. Statistical confidence for likelihood-based paternity inference in natural populations. *Mol. Ecol.* 7: 639–655.

Marshall, T. C., P. Sunnocks, J. A. Spalton, A. Greth and J. M. Pemberton. 1999. Use of genetic data for conservation management: The case of the Arabian oryx. *Anim. Cons.* 2: 269–278.

Martens, K., G. Rossetti and D. J. Horne. 2003. How ancient are ancient asexuals? *Proc. R. Soc. Lond. B* 270: 723–729.

Martin, A. P. and S. R. Palumbi. 1993. Body size, metabolic rate, generation time, and the molecular clock. *Proc. Natl. Acad. Sci. USA* 90: 4087–4091.

Martin, A. P., G. J. P. Naylor and S. R. Palumbi. 1992. Rates of mitochondrial DNA evolution in sharks are slow compared with mammals. *Nature* 357: 153–155.

Martin, B., J. Nienhuis, G. King and A. Schaefer. 1989. Restriction fragment length polymorphisms associated with water use efficiency in tomato. *Science* 243: 1725–1728.

Martin, E. D. 1975. *Breeding Endangered Species in Captivity*. Academic Press, London.

Martin, J. P. and I. Fridovich. 1981. Evidence for a natural gene transfer from the ponyfish to its bioluminescent bacterial symbiont *Photobacter leiognathi*. The close relationship between bacteriocuprein and the copper-zinc superoxide dismutase of teleost fishes. *J. Biol. Chem.* 256: 6080–6089.

Martin, L. J. and M. B. Cruzan. 1999. Patterns of hybridization in the *Piriqueta caroliniana* complex in central Florida: Evidence for an expanding hybrid zone. *Evolution* 53: 1037–1049.

Martin, M. A., D. K. Shiozawa, E. J. Loudenslager and J. N. Jensen. 1985. Electrophoretic study of cutthroat trout populations in Utah. *Great Basin Nat.* 45: 677–687.

Martin, O. Y. and D. J. Hosken. 2003. The evolution of reproductive isolation through sexual conflict. *Nature* 423: 979–982.

Martin, R. D., A. F. Dixson and E. J. Wickings (eds.). 1992. *Paternity in Primates: Genetic Tests and Theories*. Karger, Basel, Switzerland.

Martin, W., C. C. Somerville and S. Loiseaux-de Goer. 1992. Molecular phylogenies of plastid origins and algal evolution. *J. Mol. Evol.* 35: 385–404.

Martinez, J., L. J. Dugaiczyk and R. Zielinski. 2001. Human genetic disorders, a phylogenetic perspective. *J. Mol. Biol.* 308: 587–596.

Martins, E. P. 1995. Phylogenies and comparative data: A microevolutionary perspective. *Phil. Trans. R. Soc. Lond. B* 349: 85–91.

Martins, E. P. (ed.). 1996. *Phylogenies and the Comparative Method in Animal Behavior*. Oxford University Press, New York.

Martins, E. P. and T. F. Hansen. 1996. The statistical analysis of interspecific data: A review and evaluation of phylogenetic comparative methods. Pp. 22–75 in *Phylogenies and the Comparative Method in Animal Behavior*, E. P. Martins (ed.). Oxford University Press, New York.

Martins, E. P. and T. F. Hansen. 1997. Phylogenies and the comparative method: A general approach to incorporating phylogenetic information into the analysis of interspecific data. *Am. Nat.* 149: 646–667.

Martinsen, G. D., T. G. Whittam, R. L. Turek and P. Keim. 2001. Hybrid populations selectively filter gene introgression between species. *Evolution* 55: 1325–1335.

Martyniuk, J. and J. Jaenike. 1982. Multiple mating and sperm usage patterns in natural populations of

Prolinyphia marginata (Aranae: Linyphiidae). *Ann. Entomol. Soc. Am.* 75: 516–518.

Maruyama, T. and M. Kimura. 1980. Genetic variability and effective population size when local extinction and recolonization of subpopulations are frequent. *Proc. Natl. Acad. Sci. USA* 77: 6710–6714.

Marx, P. A., P. G. Alcabes and E. Drucker. 2001. Serial human passage of simian immunodeficiency virus by unsterile injections and the emergence of epidemic human immunodeficiency virus in Africa. *Proc. R. Soc. Lond. B* 356: 911–920.

Massey, L. K. and J. L. Hamrick. 1999. Breeding structure of a *Yucca filamentosa* (Agavaceae) population. *Evolution* 53: 1293–1298.

Mathews, L. M., C. D. Schubart, J. E. Neigel and D. L. Felder. 2002. Genetic, ecological, and behavioural divergence between two sibling snapping shrimp species (Crustacea: Decapoda: *Alpheus*). *Mol. Ecol.* 11: 1427–1437.

Matson, R. H. 1989. Distribution of the testis-specific LDH-X among avian taxa with comments on the evolution of the LDH gene family. *Syst. Zool.* 38: 106–115.

Matsuhashi, T., R. Masuda, T. Mano, K. Murata and A. Aiurzaniin. 2001. Phylogenetic relationships among worldwide populations of the brown bear *Ursus arctos*. *Zool. Sci.* 18: 1137–1143.

Matthes, M. C., A. Daly and K. J. Edwards. 1998. Amplified fragment length polymorphism (AFLP). Pp. 183–190 in *Molecular Tools for Screening Biodiversity*, A. Karp, P. G. Isaac and D. S. Ingram (eds.). Chapman & Hall, London.

Mau, B., M. Newton and B. Larget. 1999. Bayesian phylogenetic inference via Markov chain Monte Carlo methods. *Biometrics* 55: 1–12.

Mauck, R. A., T. A. Waite and P. G. Parker. 1995. Monogamy in Leach's storm-petrel: DNA-fingerprinting evidence. *Auk* 112: 473–482.

Maxam, A. M. and W. Gilbert. 1977. A new method for sequencing DNA. *Proc. Natl. Acad. Sci. USA* 74: 560–564.

Maxam, A. M. and W. Gilbert. 1980. Sequencing end-labeled DNA with base-specific chemical cleavages. *Meth. Enzymol.* 65: 499–559.

Maxson, L. R. and R. D. Maxson. 1986. Micro-complement fixation: A quantitative estimator of protein evolution. *Mol. Biol. Evol.* 3: 375–388.

Maxson, L. R. and R. D. Maxson. 1990. Proteins II: Immunological techniques. Pp. 127–155 in *Molecular Systematics*, D. M. Hillis and C. Moritz (eds.). Sinauer Associates, Sunderland, MA.

Maxson, L. R. and J. D. Roberts. 1984. Albumin and Australian frogs: Molecular data a challenge to speciation model. *Science* 225: 957–958.

Maxson, L. R. and A. C. Wilson. 1975. Albumin evolution and organismal evolution in tree frogs (Hylidae). *Syst. Zool.* 24: 1–15.

Maxson, L. R., E. Pepper and R. D. Maxson. 1977. Immunological resolution of a diploid-tetraploid species complex of tree frogs. *Science* 197: 1012–1013.

May, R. M. 1990. Taxonomy as destiny. *Nature* 347: 129–130.

May, R. M. 1994. Conceptual aspects of the quantification of the extent of biological diversity. *Phil. Trans. R. Soc. Lond. B* 345: 13–20.

Mayden, R. L. 1986. Speciose and depauperate phylads and tests of punctuated and gradual evolution: Fact or artifact? *Syst. Zool.* 35: 147–152.

Mayden, R. L. 1997. A hierarchy of species concepts: The denouement in the saga of the species problem. Pp. 381–424 in *Species: The Units of Biodiversity*, M. F. Claridge, H. A. Dawah and M. R. Wilson (eds.). Chapman & Hall, New York.

Maynard Smith, J. 1966. Sympatric speciation. *Am. Nat.* 100: 637–650.

Maynard Smith, J. 1978. *The Evolution of Sex*. Cambridge University Press, Cambridge, England.

Maynard Smith, J. 1986. Contemplating life without sex. *Nature* 324: 300–301.

Maynard Smith, J. 1990. The Y of human relationships. *Nature* 344: 591–592.

Maynard Smith, J. 1992. Age and the unisexual lineage. *Nature* 356: 661–662.

Maynard Smith, J. and N. H. Smith. 1998. Detecting recombination from gene trees. *Mol. Biol. Evol.* 15: 590–599.

Maynard Smith, J. and E. Szathmáry. 1995. *The Major Transitions in Evolution*. W. H. Freeman and Co., New York.

Maynard Smith, J., N. H. Smith, M. O'Rourke and B. G. Spratt. 1993. How clonal are bacteria? *Proc. Natl. Acad. Sci. USA* 90: 4384–4388.

Mayr, E. 1954. Change of genetic environment and evolution. Pp. 157–180 in *Evolution as a Process*, J. Huxley, A. C. Hardy and E. B. Ford (eds.). Allen and Unwin, London.

Mayr, E. 1963. *Animal Species and Evolution*. Harvard University Press, Cambridge, MA.

Mayr, E. 1990. A natural system of organisms. *Nature* 348: 491.

Mayr, E. 1998. Two empires or three? *Proc. Natl. Acad. Sci. USA* 95: 9720–9723.

Mayr, E. and P. A. Ashlock. 1991. *Principles of Systematic Zoology*. McGraw-Hill, New York.

Mazodier, P. and J. Davies. 1991. Gene transfer between distantly related bacteria. *Annu. Rev. Genet.* 25: 147–171.

McCarthy, E. M. and J. F. McDonald. 2003. LTR_STRUC: A novel search and identification program for LTR retrotransposons. *Bioinformatics* 19: 362–367.

McCauley, D. E. 1995. The use of chloroplast DNA polymorphism in studies of gene flow in plants. *Trends Ecol. Evol.* 10: 198–202.

McCauley, D. E. 1998. The genetic structure of a gynodioecious plant: Nuclear and cytoplasmic genes. *Evolution* 52: 255–260.

McCauley, D. E. and R. O'Donnell. 1984. The effect of multiple mating on genetic relatedness in larval aggregations of the imported willow leaf beetle (*Plagiodera versicolora*, Coleoptera: Chrysomelidae). *Behav. Ecol. Sociobiol.* 15: 287–291.

McCauley, D. E., J. E. Stevens, P. A. Peroni and J. A. Raveill. 1996. The spatial distribution of chloroplast DNA and allozyme polymorphisms within a population of *Silene alba* (Caryophyllaceae). *Am. J. Bot.* 83: 727–731.

McClenaghan, L. R. Jr. and T. J. O'Shea. 1988. Genetic variability in the Florida manatee (*Trichechus manatus*). *J. Mammal.* 69: 481–488.

McClenaghan, L. R. Jr., M. H. Smith and M. W. Smith. 1985. Biochemical genetics of mosquitofish. IV. Changes in allele frequencies through time and space. *Evolution* 39: 451–459.

McClenaghan, L. R. Jr., J. Berger and H. D. Truesdale. 1990. Founding lineages and genic variability in plains bison (*Bison bison*) from Badlands National Park, South Dakota. *Cons. Biol.* 4: 285–289.

McConnell, T. J., W. S. Talbot, R. A. McIndoe and E. K. Wakeland. 1988. The origin of MHC class II gene polymorphism within the genus *Mus. Nature* 332: 651–654.

McCorquodale, D. B. 1988. Relatedness among nestmates in a primitively social wasp, *Cerceris antipodes* (Hymenoptera: Sphecidae). *Behav. Ecol. Sociobiol.* 23: 401–406.

McCoy, E. D. and K. L. Heck Jr. 1976. Biogeography of corals, sea grasses, and mangroves: An alternative to the center of origin concept. *Syst. Zool.* 25: 201–210.

McCoy, E. E., A. G. Jones and J. C. Avise. 2001. The genetic mating system and tests for cuckoldry in a pipefish species in which males fertilize eggs and brood offspring externally. *Mol. Ecol.* 10: 1793–1800.

McCracken, G. F. 1984. Communal nursing in Mexican free-tailed bat maternal colonies. *Science* 223: 1090–1091.

McCracken, G. F. and J. W. Bradbury. 1977. Paternity and genetic heterogeneity in the polygynous bat, *Phyllostomus hastatus. Science* 198: 303–306.

McCracken, G. F. and J. W. Bradbury. 1981. Social organization and kinship in the polygynous bat *Phyllostomus hastatus. Behav. Ecol. Sociobiol.* 8: 11–34.

McCracken, K. G., W. P. Johnson and F. H. Sheldon. 2001. Molecular population genetics, phylogeography, and conservation biology of the mottled duck (*Anas fulvigula*). *Cons. Genet.* 2: 87–102.

McCune, A. R. 1997. How fast is speciation? Molecular, geological, and phylogenetic evidence from adaptive radiations of fishes. Pp. 585–610 in *Molecular Evolution and Adaptive Radiation*, T. J. Givnish and K. J. Sytsma (eds.). Cambridge University Press, Cambridge, England.

McCune, A. R. and N. R. Lovejoy. 1998. The relative rate of sympatric and allopatric speciation in fishes: Tests using DNA sequence divergence between sister species and among clades. Pp. 172–185 in *Endless Forms: Species and Speciation*, D. J. Howard and S. H. Berlocher (eds.). Oxford University Press, New York.

McDermott, J. M., B. A. McDonald, R. W. Allard and R. K. Webster. 1989. Genetic variability for pathogenicity, isozyme, ribosomal DNA and colony color variants in populations of *Rhynchosporium secalis. Genetics* 122: 561–565.

McDonald, D. B., W. K. Potts, J. W. Fitzpatrick and G. E. Woolfenden. 1999. Contrasting genetic structures in sister species of North American scrub-jays. *Proc. R. Soc. Lond. B* 266: 1117–1125.

McDonald, J. and M. Kreitman. 1991. Adaptive protein evolution at the *adh* locus in *Drosophila. Nature* 351: 652–654.

McDonald, J. F. 1983. The molecular basis of adaptation: A critical review of relevant ideas and observations. *Annu. Rev. Ecol. Syst.* 14: 77–102.

McDonald, J. F. 1989. The potential evolutionary significance of retroviral-like transposable elements in peripheral populations. Pp. 190–205 in *Evolutionary Biology of Transient Unstable Populations*, A. Fontdevila (ed.). Springer-Verlag, New York.

McDonald, J. F. 1990. Macroevolution and retroviral elements. *BioScience* 40: 183–191.

McDonald, J. F. 1998. Transposable elements, gene silencing, and macroevolution. *Trends Ecol. Evol.* 13: 94–95.

McDonald, J. F. and F. J. Ayala. 1974. Genetic response to environmental heterogeneity. *Nature* 250: 572–574.

McDowell, R. and S. Prakash. 1976. Allelic heterogeneity within allozymes separated by electrophoresis in *Drosophila pseudoobscura. Proc. Natl. Acad. Sci. USA* 73: 4150–4153.

McFadden, C. S. 1997. Contributions of sexual and asexual reproduction to population structure in the clonal soft coral, *Alcyonium rudyi. Evolution* 51: 112–126.

McFadden, C. S., R. K. Grosberg, B. B. Cameron, D. P. Karlton and D. Secord. 1997. Genetic relationships within and between clonal and solitary forms of the sea anemone *Anthopleura elegantissima* revisited: Evidence for the existence of two species. *Mar. Biol.* 128: 127–139.

McGovern, T. M. and M. E. Hellberg. 2003. Cryptic species, cryptic endosymbionts, and geographical variation in chemical defenses in the bryozoan *Bugula neritina. Mol. Ecol.* 12: 1207–1215.

McGraw, E. A., J. Li, R. K. Selander and T. S. Whittam. 1999. Molecular evolution and mosaic structure of α, β, and γ intimins of pathogenic *Escherichia coli. Mol. Biol. Evol.* 16: 12–22.

McGuigan, K., K. McDonald, K. Parris and C. Moritz. 1998. Mitochondrial DNA diversity and historical biogeography of a wet forest-restricted frog (*Litoria pearsoniana*) from mid-east Australia. *Mol. Ecol.* 7: 175–186.

McGuire, G., F. Wright and M. J. Prentice. 1997. A graphical method for detecting recombination in phylogenetic data sets. *Mol. Biol. Evol.* 14: 1125–1131.

McKinney, F., K. M. Cheng and D. J. Bruggers. 1984. Sperm competition in apparently monogamous birds. Pp. 523–545 in *Sperm Competition and the Evolution of Animal Mating Systems*, R. L. Smith (ed.). Academic Press, New York.

McKinnon, G. E., R. E. Vaillancourt, P. A. Tilvard and B. M. Potts. 2001. Maternal inheritance of the chloroplast genome in *Eucalyptus globulus* and interspecific hybrids. *Genome* 44: 831–835.

McKitrick, M. C. 1990. Genetic evidence for multiple parentage in eastern kingbirds (*Tyrannus tyrannus*). *Behav. Ecol. Sociobiol.* 26: 149–155.

McKitrick, M. C. and R. M. Zink. 1988. Species concepts in ornithology. *Condor* 90: 1–14.

McKusick, V. A. (ed.). 1998. *Mendelian Inheritance in Man* (12th Edition). Johns Hopkins University Press, Baltimore, MD.

McLean, I. G., S. D. Kayes, J. O. Murie, L. S. Davis and D. M. Lambert. 2000. Genetic monogamy mirrors social monogamy in the Fiordland crested penguin. *New Zealand J. Zool.* 27: 311–316.

McLean, J. E., P. Bentzen and T. P. Quinn. 2003. Differential reproductive success of sympatric, naturally spawning hatchery and wild steelhead trout (*Oncorhynchus mykiss*) through the adult stage. *Can. J. Fish. Aquat. Sci.* 60: 433–440.

McMillan, W. O. and S. R. Palumbi. 1995. Concordant evolutionary patterns among Indo-west Pacific butterflyfishes. *Proc. R. Soc. Lond. B* 260: 229–236.

McMillan, W. O., R. A. Raff and S. R. Palumbi. 1992. Population genetic consequences of developmental evolution in sea urchins (genus *Heliocidaris*). *Evolution* 46: 1299–1312.

McNeilly, T. and J. Antonovics. 1968. Evolution in closely adjacent plant populations. IV. Barriers to gene flow. *Heredity* 23: 205–218.

McPeek, M. A. 1995. Testing hypotheses about evolutionary change on single branches of a phylogeny using evolutionary contrasts. *Am. Nat.* 145: 686–703.

McPheron, B. A., D. C. Smith and S. H. Berlocher. 1988. Genetic differences between host races of *Rhagoletis pomonella*. *Nature* 336: 64–67.

Meagher, S. and T. E. Dowling. 1991. Hybridization between the cyprinid fishes *Luxilus albeolus*, *L. cornutus* and *L. cerasinus* with comments on the proposed hybrid origin of *L. albeolus*. *Copeia* 1991: 979–991.

Meagher, T. R. 1986. Analysis of paternity within a natural population of *Chamaelirium luteum*. I. Identification of most-likely male parents. *Am. Nat.* 128: 199–215.

Meagher, T. R. 1991. Analysis of paternity within a natural population of *Chamaelirium luteum*. II. Patterns of male reproductive success. *Am. Nat.* 137: 738–752.

Meagher, T. R. and E. Thompson. 1987. Analysis of parentage for naturally established seedlings of *Chamaelirium luteum* (Liliaceae). *Ecology* 68: 803–812.

Medina, M., A. G. Collins, J. D. Silberman and M. L. Sogin. 2001. Evaluating hypotheses of basal animal phylogeny using complete sequences of large and small subunit rRNA. *Proc. Natl. Acad. Sci. USA* 17: 9707–9712.

Meek, S. B., R. J. Robertson and P. T. Boag. 1994. Extrapair paternity and intraspecific brood parasitism in eastern bluebirds revealed by DNA fingerprinting. *Auk* 111: 739–744.

Meffe, G. K. and C. R. Carroll. 1997. *Principles of Conservation Biology*. Sinauer Associates, Sunderland, MA.

Meffe, G. K. and R. C. Vrijenhoek. 1988. Conservation genetics in the management of desert fishes. *Cons. Biol.* 2: 157–169.

Melnick, D. J. and G. A. Hoelzer. 1992. Differences in male and female macaque dispersal lead to contrasting distributions of nuclear and mitochondrial DNA variation. *Int. J. Primatol.* 13: 379–393.

Melson, K. E. 1990. Legal and ethical considerations. Pp. 189–215 in *DNA Fingerprinting*, L. T. Kirby (ed.). Stockton Press, New York.

Mendelson, T. C. 2003. Sexual isolation evolves faster than hybrid inviability in a diverse and sexually dimorphic genus of fish (Percidae: *Etheostoma*). *Evolution* 57: 317–327.

Meng, A., R. E. Carter and D. T. Parkin. 1990. The variability of DNA fingerprints in three species of swan. *Heredity* 64: 73–80.

Mengel, R. N. 1964. The probable history of species formation in some northern wood warblers (Parulidae). *Living Bird* 3: 9–43.

Menotti-Raymond, M. and S. J. O'Brien. 1993. Dating the genetic bottleneck of the African cheetah. *Proc. Natl. Acad. Sci. USA* 90: 3172–3176.

Mercer, J. M. and V. L. Roth. 2003. The effects of Cenozoic global change on squirrel phylogeny. *Science* 299: 1568–1572.

Merenlender, A. M., D. S. Woodruff, O. A. Ryder, R. Kock and J. Váhala. 1989. Allozyme variation and differentiation in African and Indian rhinoceroses. *J. Heredity* 80: 377–382.

Merilä, J., M Björkland and A. J. Baker. 1997. Historical demography and present day population structure of the greenfinch, *Carduelis chloris*: An analysis of mtDNA control-region sequences. *Evolution* 51: 946–956.

Merriweather, A. and D. Kaestle. 1999. Mitochondrial recombination? *Science* 285: 835.

Merriwether, D. A. and 7 others. 1991. The structure of human mitochondrial DNA variation. *J. Mol. Evol.* 33: 543–555.

Mes, T. H. M. and 6 others. 2002. Detection of genetically divergent clone mates in apomictic dandelions. *Mol. Ecol.* 11: 253–265.

Meselson, M. and R. Yuan. 1968. DNA restriction enzyme from *E. coli*. *Nature* 217: 1110–1114.

Messier, W. and C.-B. Smith. 1997. Episodic adaptive evolution of primate lysozymes. *Nature* 385: 151–154.

Metcalf, A. E., L. Nunney and B. C. Hyman. 2001. Geographic patterns of genetic differentiation within the restricted range of the endangered Stephens' kangaroo rat *Dipodomys stephensi*. *Evolution* 55: 1233–1244.

Metcalf, R. A. and G. S. Whitt. 1977. Intra-nest relatedness in the social wasp *Polistes metricus*. A genetic analysis. *Behav. Ecol. Sociobiol.* 2: 339–351.

Metz, E. C. and S. R. Palumbi. 1996. Positive selection and sequence rearrangements generate extensive polymorphism in the gamete recognition protein bindin. *Mol. Biol. Evol.* 13: 397–406.

Metzker, M. L., D. P. Mindell, X.-M. Liu, R. G. Ptak, R. A. Gibbs and D. M. Hillis 2002. Molecular evidence of HIV-1 transmission in a criminal case. *Proc. Natl. Acad. Sci. USA* 99: 14292–14297.

Metzlaff, M., T. Borner and R. Hagemann. 1981. Variations of chloroplast DNAs in the genus *Pelargonium* and their biparental inheritance. *Theoret. Appl. Genet.* 60: 37–41.

Meyer, A. 1987. Phenotypic plasticity and heterochrony in *Cichlasoma managuense* (Pisces, Cichlidae) and their implications for speciation in cichlid fishes. *Evolution* 41: 1357–1369.

Meyer, A. 1994a. DNA technology and phylogeny of fish: Molecular phylogenetic studies of fishes. Pp. 219–249 in *Evolution and Genetics of Aquatic Organisms*, A. R. Beaumont (ed.), Chapman & Hall, New York.

Meyer, A. 1994b. Shortcomings of the cytochrome *b* gene as a molecular marker. *Trends Ecol. Evol.* 9: 278–280.

Meyer, A. and S. I. Dolven. 1992. Molecules, fossils, and the origin of tetrapods. *J. Mol. Evol.* 35: 102–113.

Meyer, A. and A. C. Wilson. 1990. Origin of tetrapods inferred from their mitochondrial DNA affiliation to lungfish. *J. Mol. Evol.* 31: 359–364.

Meyer, A. and R. Zardoya. 2003. Recent advances in the (molecular) phylogeny of vertebrates. *Annu. Rev. Ecol. Evol. Syst.* 34: 311–338.

Meyer, A., T. D. Kocher, P. Basasibwaki and A. C. Wilson. 1990. Monophyletic origin of Lake Victoria cichlid fishes suggested by mitochondrial DNA sequences. *Nature* 347: 550–553.

Meyer, A., J. M. Morrissey and M. Schartl. 1994. Recurrent origin of a sexually selected trait in *Xiphophorus* fishes inferred from a molecular phylogeny. *Nature* 368: 539–542.

Meylan, A. B., B. W. Bowen and J. C. Avise. 1990. A genetic test of the natal homing versus social facilitation models for green turtle migration. *Science* 248: 724–727.

Michener, G. R. 1983. Kin identification, matriarchies, and the evolution of sociality in ground-dwelling sciurids. Pp. 528–572 in *Advances in the Study of Mammalian Behavior*, J. F. Eisenberg and D. G. Kleiman (eds.). American Society Mammalogy, Special Publ. No. 7.

Michiels, N. K. 1998. Mating conflicts and sperm competition in simultaneous hermaphrodites. Pp. 219–254 in *Sperm Competition and Sexual Selection*, T. R. Birkhead and A. P. Møller (eds.). Academic Press, London.

Michod, R. E. 1999. *Darwinian Dynamics: Evolutionary Transitions in Fitness and Individuality*. Princeton University Press, Princeton, NJ.

Michod, R. E. and W. W. Anderson. 1979. Measures of genetic relationship and the concept of inclusive fitness. *Am. Nat.* 114: 637–647.

Mickevich, M. F. 1978. Taxonomic congruence. *Syst. Zool.* 27: 143–158.

Miles, S. J. 1978. Enzyme variations in the *Anopheles gambiae* group of species (Diptera, Culicidae). *Bull. Entomol. Res.* 68: 85–96.

Milinkovitch, M. C. 1992. DNA-DNA hybridization support ungulate ancestry of Cetacea. *J. Evol. Biol.* 5: 149–160.

Milinkovitch, M. C. 1995. Molecular phylogeny of cetaceans prompts revision of morphological transformations. *Trends Ecol. Evol.* 10: 328–334.

Milinkovitch, M. C. and J. G. M. Thewissen. 1997. Even-toed fingerprints on whale ancestry. *Nature* 388: 622–623.

Milinkovitch, M. C., G. Ortí and A. Meyer. 1993. Revised phylogeny of whales suggested by mitochondrial ribosomal DNA sequences. *Nature* 361: 346–348.

Milinkovitch, M. C., A. Meyer and J. R. Powell. 1994a. Phylogeny of all major groups of cetaceans based on DNA sequences from three mitochondrial genes. *Mol. Biol. Evol.* 11: 939–948.

Milinkovitch, M. C., G. Ortí and A. Meyer. 1994b. Novel phylogeny of whales revisited but not revised. *Mol. Biol. Evol.* 12: 518–520.

Milinkovitch, M. C., R. G. LeDuc, J. Adachi, F. Farnir, M. Georges and M. Hasegawa. 1996. Effects of character weighting and species sampling on phylogeny reconstruction: A case study based on DNA sequence data in Cetaceans. *Genetics* 144: 1817–1833.

Milkman, R. 1973. Electrophoretic variation in *Escherichia coli* from natural sources. *Science* 182: 1024–1026.

Milkman, R. 1975. Allozyme variation in *E. coli* of diverse natural origins. *Isozymes* IV: 273–285.

Milkman, R. 1976. Further studies on thermostability variation within electrophoretic mobility classes of enzymes. *Biochem. Genet.* 14: 383–387.

Milkman, R. and M. M. Bridges. 1990. Molecular evolution of the *E. coli* chromosome. III. Clonal frames. *Genetics* 126: 505–517.

Milkman, R. and A. Stoltzfus. 1988. Molecular evolution of the *E. coli* chromosome. II. Clonal segments. *Genetics* 120: 359–366.

Milkman, R. and R. R. Zeitler. 1974. Concurrent multiple paternity in natural and laboratory populations of *Drosophila melanogaster*. *Genetics* 78: 1191–1193.

Millar, C. D., D. M. Lambert, A. R. Bellamy, P. M. Stapleton and E. C. Young. 1992. Sex-specific restriction fragments and sex ratios revealed by DNA fingerprinting in the brown skua. *J. Heredity* 83: 350–355.

Millar, C. D., D. M. Lambert, S. Anderson and J. L Halverson. 1996. Molecular sexing of the communally breeding pukeko: An important ecological tool. *Mol. Ecol.* 5: 289–293.

Millar, C. D., C. E. M. Reed, J. L. Halverson and D. M. Lambert. 1997. Captive management and molecular sexing of endangered avian species: An application to the black stilt *Himantopus novaezelandiae* and hybrids. *Biol. Cons.* 82: 81–86.

Millar, C. I. 1983. A steep cline in *Pinus muricata*. *Evolution* 37: 311–319.

Millar, R. B. 1987. Maximum likelihood estimation of mixed stock fishery composition. *Can. J. Fish. Aquat. Sci.* 44: 583–590.

Millen, R. S. and 12 others. 2001. Many parallel losses of *infA* from chloroplast DNA during angiosperm evolution with multiple independent transfers to the nucleus. *The Plant Cell* 13: 645–658.

Miller, B. R., M. B. Crabtree and H. M. Savage. 1996. Phylogeny of fourteen *Culex* mosquito species, including the *Culex pipiens* complex, inferred from the internal transcribed spacers of ribosomal DNA. *Insect Mol. Biol.* 5: 93–107.

Miller, P. S. and P. W. Hedrick. 1991. MHC polymorphism and the design of captive breeding programs: Simple solutions are not the answer. *Cons. Biol.* 5: 556–558.

Miller, S., R. W. Pearcy and E. Berger. 1975. Polymorphism at the alpha-glycerophosphate dehydrogenase locus in *Drosophila melanogaster*. I. Properties of adult allozymes. *Biochem. Genet.* 13: 175–188.

Milligan, B. G., J. N. Hampton and J. D. Palmer. 1989. Dispersed repeats and structural reorganization in subclover chloroplast DNA. *Mol. Biol. Evol.* 6: 355–368.

Minckley, W. L. and 6 others. 2003. A conservation plan for native fishes of the lower Colorado River. *BioScience* 53: 219–234.

Mindell, D. P. (ed.). 1997. *Avian Molecular Evolution and Systematics*. Academic Press, New York.

Mindell, D. P. and R. L. Honeycutt. 1990. Ribosomal RNA in vertebrates: Evolution and phylogenetic applications. *Annu. Rev. Ecol. Syst.* 21: 541–566.

Mindell, D. P. and J. W. Sites Jr. 1987. Tissue expression patterns of avian isozymes: A preliminary study of phylogenetic applications. *Syst. Zool.* 36: 137–152.

Mindell, D. P. and C. E. Thacker. 1996. Rates of molecular evolution: Phylogenetic issues and applications. *Annu. Rev. Ecol. Syst.* 27: 279–303.

Mindell, D. P., J. W. Sites Jr. and D. Graur. 1990. Mode of allozyme evolution: Increased genetic distance associated with speciation events. *J. Evol. Biol.* 3: 125–131.

Mindell, D. P., C. W. Dick and R. J. Baker. 1991. Phylogenetic relationships among megabats, microbats, and primates. *Proc. Natl. Acad. Sci. USA* 88: 10322–10326.

Mindell, D. P. and 6 others. 1997. Phylogenetic relationships among and within select avian orders based on mitochondrial DNA. Pp. 213–247 in *Avian Molecular Evolution and Systematics*, D. P. Mindell (ed.). Academic Press, San Diego, CA.

Mindell, D. P., M. D. Sorenson, D. E. Dimcheff. 1998. Multiple independent origins of mitochondrial gene order in birds. *Proc. Natl. Acad. Sci. USA* 95: 10693–10697.

Mindell, D. P., M. D. Sorenson, D. E. Dimcheff, M. Hasegawa, J. C. Ast and T. Yuri. 1999. Interordinal relationships of birds and other reptiles based on whole mitochondrial genomes. *Syst. Biol.* 48: 138–152.

Minton, R. L. and C. Lydeard. 2003. Phylogeny, taxonomy, genetics and global heritage ranks of an imperilled, freshwater snail genus *Lithasia* (Pleuroceridae). *Mol. Ecol.* 12: 75–87.

Mishler, B. D. and M. J. Donoghue. 1982. Species concepts: A case for pluralism. *Syst. Zool.* 31: 491–503.

Mitchell, S. E., S. K. Narang, A. F. Cockburn, J. A. Seawright and M. Goldenthal. 1993. Mitochondrial and ribosomal DNA variation among members of the *Anopheles quadrimaculatus* (Diptera: Culicidae) species complex. *Genome* 35: 939–950.

Mitra, S., H. Landel and S. Pruett-Jones. 1996. Species richness covaries with mating system in birds. *Auk* 113: 544–551.

Mitton, J. B. 1993. Theory and data pertinent to the relationship between heterozygosity and fitness. Pp. 17–41 in *The Natural History of Inbreeding and Outbreeding*, N. Thornhill (ed.). University of Chicago Press, Chicago, IL.

Mitton, J. B. 1994. Molecular approaches to population biology. *Annu. Rev. Ecol. Syst.* 25: 45–69.

Mitton, J. B. 1997. *Selection in Natural Populations*. Oxford University Press, New York.

Mitton, J. B. and M. C. Grant. 1984. Associations among protein heterozygosity, growth rate, and developmental homeostasis. *Annu. Rev. Ecol. Syst.* 15: 479–499.

Mitton, J. B. and R. K. Koehn. 1975. Genetic organization and adaptive response of allozymes to ecological variables in *Fundulus heteroclitus*. *Genetics* 79: 97–111.

Mitton, J. B. and R. K. Koehn. 1985. Shell shape variation in the blue mussel, *Mytilus edulis* L., and its association with enzyme heterozygosity. *J. Exp. Mar. Biol. Ecol.* 90: 73–90.

Mitton, J. B. and W. M. Lewis Jr. 1989. Relationships between genetic variability and life-history features of bony fishes. *Evolution* 43: 1712–1723.

Mitton, J. B. and B. A. Pierce. 1980. The distribution of individual heterozygosity in natural populations. *Genetics* 95: 1043–1054.

Miya, M. and 11 others. 2003. Major patterns of higher teleostean phylogenies: A new perspective based on 100 complete mitochondrial DNA sequences. *Mol. Phylogen. Evol.* 26: 121–138.

Miyaki, C. M., S. R. Matioli, T. Burke and A. Qajntal. 1998. Parrot evolution and paleographical events: Mitochondrial DNA evidence. *Mol. Biol. Evol.* 15: 544–551.

Miyamoto, M. M. and J. Cracraft. 1991. *Phylogenetic Analysis of DNA Sequences*. Oxford University Press, New York.

Miyamoto, M. M. and M. Goodman. 1986. Biomolecular systematics of Eutherian mammals: Phylogenetic patterns and classification. *Syst. Zool.* 35: 230–240.

Miyamoto, M. M. and M. Goodman. 1990. DNA systematics and evolution of primates. *Annu. Rev. Ecol. Syst.* 21: 197–220.

Mizrokhi, L. J. and A. M. Mazo. 1990. Evidence for horizontal transmission of the mobile element *jockey* between distant *Drosophila* species. *Proc. Natl. Acad. Sci. USA* 87: 9216–9220.

Mock, D. W. 1983. On the study of avian mating systems. Pp. 55–84 in *Perspectives in Ornithology*, A. H. Brush and G. A. Clark (eds.). Cambridge University Press, London.

Molbo, D., C. A. Machado, J. G. Stevenster, L. Keller and E. A. Herre. 2003. Cryptic species of fig-pollinating wasps: Implications for the evolution of the fig-wasp mutualism, sex allocation, and precision of adaptation. *Proc. Natl. Acad. Sci. USA* 100: 5867–5872.

Møller, A. P. 1992. Female swallow preference for symmetrical male sexual ornaments. *Nature* 357: 238–240.

Møller, A. P. 1997. Immune defence, extra-pair paternity, and sexual selection in birds. *Proc. R. Soc. Lond. B* 264: 561–566.

Møller, A. P. 1998. Sperm competition and sexual selection. Pp. 91–116 in *Sperm Competition and Sexual Selection*, T. R. Birkhead and A. P. Møller (eds.). Academic Press, London.

Møller, A. P. 2000. Male parental care, female reproductive success, and extrapair paternity. *Behav. Ecol.* 11: 161–168.

Møller, A. P. and R. V. Alatalo. 1999. Good-genes effects in sexual selection. *Proc. R. Soc. Lond. B* 266: 85–91.

Møller, A. P. and T. R. Birkhead. 1993a. Cuckoldry and sociality: A comparative study of birds. *Am. Nat.* 142: 118–140.

Møller, A. P. and T. R. Birkhead. 1993b. Certainty of paternity covaries with paternal care in birds. *Behav. Ecol. Sociobiol.* 33: 261–268.

Møller, A. P. and T. R. Birkhead. 1994. The evolution of plumage brightness in birds is related to extrapair paternity. *Evolution* 48: 1089–1100.

Møller, A. P. and J. V. Briskie. 1995. Extra-pair paternity, sperm competition and the evolution of testis size in birds. *Behav. Ecol. Sociobiol.* 36: 357–365.

Møller, A. P. and J. J. Cuervo. 1998. Speciation and feather ornamentation in birds. *Evolution* 52: 859–869.

Møller, A. P. and J. J. Cuervo. 2000. The evolution of paternity and paternal care in birds. *Behav. Ecol.* 11: 472–485.

Møller, A. P. and P. Ninni. 1998. Sperm competition and sexual selection: A meta-analysis of paternity studies of birds. *Behav. Ecol. Sociobiol.* 43: 345–358.

Møller, A. P. and H. Tegelström. 1997. Extra-pair paternity and tail ornamentation in the barn swallow *Hirundo rustica*. *Behav. Ecol. Sociobiol.* 41: 353–360.

Møller, A. P., A. Barbosa, J. J. Cuervo, F. De Lope, S. Merino and N. Saino. 1998. Sexual selection and tail streamers in the barn swallow. *Proc. R. Soc. Lond. B* 265: 409–414.

Montagna, W. 1942. The Sharp-tailed sparrows of the Atlantic coast. *Wilson Bull.* 54: 107–120.

Monteiro, W., J. M. G. Almeida. Jr. and B. S. Dias. 1984. Sperm sharing in *Biomphalaria* snails: A new behavioural strategy. *Nature* 308: 727–729.

Montgomery, M. E., J. D. Ballou, R. K. Nurthen, P. R. England, D. A. Briscoe and R. Frankham. 1997. Minimizing kinship in captive breeding programs. *Zoo Biol.* 16: 377–389.

Moody, M. D. 1989. DNA analysis in forensic science. *BioScience* 39: 31–36.

Mooers, A. Ø. and S. J. Heard. 1997. Evolutionary process from phylogenetic tree shape. *Q. Rev. Biol.* 72: 31–54.

Moore, W. S. 1995. Inferring phylogenies from mtDNA variation: Mitochondrial-gene trees versus nuclear-gene trees. *Evolution* 49: 718–726.

Moore, W. S. 1997. Mitochondrial-gene trees versus nuclear-gene trees, a reply to Hoelzer. *Evolution* 51: 627–629.

Moore, W. S. and D. B. Buchanan. 1985. Stability of the northern flicker hybrid zone in historical times: Implications for adaptive speciation theory. *Evolution* 39: 135–151.

Moore, W. S. and V. R. DeFilippis. 1997. The window of taxonomic resolution for phylogenies based on mitochondrial cytochrome *b*. Pp. 83–119 in *Avian Molecular Evolution and Systematics*, D. P. Mindell (ed.). Academic Press, New York.

Mopper, S., J. B. Mitton, T. G. Whitham, N. S. Cobb and K. M. Christensen. 1991. Genetic differentiation and heterozygosity in pinyon pine associated with resistance to herbivory and environmental stress. *Evolution* 45: 989–999.

Morales, J. C., J. C. Patton and J. W. Bickham. 1993. Partial endonuclease digestion mapping of restriction sites using PCR-amplified DNA. *PCR Methods and Appl.* 2: 228–233.

Moran, N. A. 2001. The coevolution of bacterial endosymbionts and phloem-feeding insects. *Ann. Missouri Bot. Gard.* 88: 35–44.

Moran, N. A., M. A. Munson, P. Baumann and H. Ishikawa. 1993. A molecular clock in endosymbiotic bacteria is calibrated using the insect hosts. *Proc. R. Soc. Lond. B* 253: 167–171.

Moran, N. A., M. A. Kaplan, M. Gelsey, T. Murphy and E. Scholes. 1998. Phylogenetics and evolution of the aphid genus *Uroleucon* based on mitochondrial and nuclear DNA sequences. *Syst. Entomol.* 24: 85–93.

Moran, P. and I. Kornfield. 1993. Retention of an ancestral polymorphism in the Mbuna species flock (Teleostei: Cichlidae) of Lake Malawi. *Mol. Biol. Evol.* 10: 1015–1029.

Moran, P., A. M. Pendas, E. Garcia-Vazquez and J. Izquierdo. 1991. Failure of a stocking policy, of hatchery reared brown trout, *Salmo trutta* L., in Asturias, Spain, detected using *LDH-5** as a genetic marker. *J. Fish Biol.* 39A: 117–121.

Moran, P., A. M. Pendas, E. Beall and E. Garcia-Vasquez. 1996. Genetic assessment of the reproductive success of Atlantic salmon precocious parr by means of VNTR loci. *Heredity* 77: 655–660.

Morden, C. W., C. F. Delwiche, M. Kuhsel and J. D. Palmer. 1992. Gene phylogenies and the endosymbiotic origin of plastids. *BioSystems* 28: 75–90.

Morgan, J. A. T. and 6 others. 2002. A phylogeny of planorbid snails, with implications for the evolution of *Schistosoma* parasites. *Mol. Phylogen. Evol.* 25: 477–488.

Morgan, R. P. 1975. Distinguishing larval white perch and striped bass by electrophoresis. *Chesapeake Sci.* 16: 68–70.

Morgan, T. H. 1919. *The Physical Basis of Heredity*. Lippincott, Philadelphia, PA.

Morgante, M., A. Pfeiffer, I. Jurman, G. Paglia and A. M. Oliviera. 1998. Isolation of microsatellite markers in plants. Pp. 288–296 in *Molecular Tools for Screening Biodiversity*, A. Karp, P. G. Isaac and D. S. Ingram (eds.). Chapman & Hall, London.

Morin, P. A. and D. S. Woodruff. 1996. Noninvasive genotyping for vertebrate conservation. Pp. 298–313 in *Molecular Genetic Approaches in Conservation*, T. B. Smith and R. K. Wayne (eds.). Oxford University Press, New York.

Morin, P. A., J. J. Moore and D. S. Woodruff. 1992. Identification of chimpanzee subspecies with DNA from hair and allele-specific probes. *Proc. R. Soc. Lond. B* 249: 293–297.

Morin, P. A., J. J. Moore, R. Chakraborty, L. Jin, J. Goodall and D. S. Woodruff. 1994a. Kin selection, social-structure, gene flow, and the evolution of chimpanzees. *Science* 265: 1193–1201.

Morin, P. A., J. Wallis, J. J. Moore and D. S. Woodruff. 1994b. Paternity exclusion in a community of wild chimpanzees using hypervariable simple sequence repeats. *Mol. Ecol.* 3: 469–478.

Morin, P. A., K. E. Chambers, C. Boesch and L. Vigilant. 2001. Quantitative polymerase chain reaction analysis of DNA from noninvasive samples for accurate microsatellite genotyping of wild chimpanzees (*Pan troglodytes verus*). *Mol. Ecol.* 10: 1835–1844.

Morita, T. and 10 others. 1992. Evolution of the mouse *t* haplotype: Recent and worldwide introgression to *Mus musculus*. *Proc. Natl. Acad. Sci. USA* 89: 6851–6855.

Moritz, C. 1991. The origin and evolution of partheno-genesis in *Heteronotia binoei* (Gekkonidae): Evidence for recent and localized origins of widespread clones. *Genetics* 129: 211–219.

Moritz, C. 1994. Applications of mitochondrial DNA analysis in conservation: A critical review. *Mol. Ecol.* 3: 401–411.

Moritz, C. 2002. Strategies to protect biological diversity and the evolutionary processes that sustain it. *Syst. Biol.* 51: 238–254.

Moritz, C. and W. M. Brown. 1986. Tandem duplication of D-loop and ribosomal RNA sequences in lizard mitochondrial DNA. *Science* 233: 1425–1427.

Moritz, C. and W. M. Brown. 1987. Tandem duplications in animal mitochondrial DNAs: Variation in incidence and gene content among lizards. *Proc. Natl. Acad. Sci. USA* 84: 7183–7187.

Moritz, C. and D. P. Faith. 1998. Comparative phylogeog-raphy and the identification of genetically divergent areas for conservation. *Mol. Ecol.* 7: 419–429.

Moritz, C., T. E. Dowling and W. M. Brown. 1987. Evolution of animal mitochondrial DNA: Relevance for population biology and systematics. *Annu. Rev. Ecol. Syst.* 18: 269–292.

Moritz, C. C., J. W. Wright and W. M. Brown. 1989. Mitochondrial-DNA analyses and the origin and rela-tive age of parthenogenetic lizards (genus *Cnemidophorus*). III. *C. velox* and *C. exsanguis*. *Evolution* 43: 958–968.

Moritz, C. C., C. J. Schneider and D. B. Wake. 1992a. Evolutionary relationships within the *Ensatina eschscholtzii* complex confirm the ring species interpre-tation. *Syst. Biol.* 41: 273–291.

Moritz, C. C., J. W. Wright and W. M. Brown. 1992b. Mitochondrial-DNA analyses and the origin and rela-tive age of parthenogenetic lizards (genus *Cnemidophorus*). Phylogenetic constraints on hybrid origins. *Evolution* 46: 184–192.

Moritz, C. C., J. L. Patton, C. J. Schneider and T. B. Smith. 2000. Diversification of rainforest faunas: An integrated molecular approach. *Annu. Rev. Ecol. Syst.* 31: 533–563.

Moriyama, E. N. and J. R. Powell. 1996. Intraspecific nuclear DNA variation in *Drosophila*. *Mol. Biol. Evol.* 13: 261–277.

Morris, D. C., M. P. Schwarz, S. J. B. Cooper and L. A. Mound. 2002. Phylogenetics of Australian *Acacia* thrips: The evolution of behaviour and ecology. *Mol. Phylogen. Evol.* 25: 278–292.

Morton, E. S., L. Forman and M. Braun. 1990. Extrapair fertilizations and the evolution of colonial breeding in purple martins. *Auk* 107: 275–283.

Morton, N. E. 1992. Genetic structure of forensic popula-tions. *Proc. Natl. Acad. Sci. USA* 89: 2556–2560.

Mountain, J. L. and 6 others. 2002. SNPSTRs: Empirically derived, rapidly typed, autosomal haplotypes for inference of population history and mutational processes. *Genome Res.* 12: 1766–1772.

Mourier, T., A. J. Hansen, E. Willerslev and P. Arctander. 2001. The Human Genome Project reveals a continu-ous transfer of large mitochondrial fragments to the nucleus. *Mol. Biol. Evol.* 18: 1833–1837.

Moya, A., A. Galiana and F. J. Ayala. 1995. Founder-effect speciation theory: Failure of experimental corrobora-tion. *Proc. Natl. Acad. Sci. USA* 92: 3983–3986.

Mueller, U. G. and L. L. Wolfenbarger. 1999. AFLP geno-typing and fingerprinting. *Trends Ecol. Evol.* 14: 389–394.

Mueller, U. G., S. A. Rehner and T. R. Schultz. 1998. The evolution of agriculture in ants. *Science* 281: 2034–2038.

Muir, W. M. and R. D. Howard. 2001. Fitness components and ecological risk of transgenic release: A model using Japanese medaka (*Orizias latipes*). *Am. Nat.* 158: 1–16.

Mulder, R. A., P. O. Dunn, A. Cocburn, K. A. Lazenby-Cohen and M. J. Howell. 1994. Helpers liberate female fairy wrens from constraints on extra-pair mate choice. *Proc. R. Soc. Lond. B* 255: 223–229.

Muller, H. J. 1950. Our load of mutations. *Am. J. Human Genet.* 2: 111–176.

Muller, H. J. 1964. The relevance of mutation to muta-tional advance. *Mutat. Res.* 1: 2–9.

Müller-Starck, G. 1998. Isozymes. Pp. 75–81 in *Molecular Tools for Screening Biodiversity*, A. Karp, P. G. Isaac and D. S. Ingram (eds.). Chapman & Hall, London.

Mullis, K. B. 1990. The unusual origin of the polymerase chain reaction. *Sci. Am.* 262(4): 56–65.

Mullis, K. and F. Faloona. 1987. Specific synthesis of DNA in vitro via a polymerase catalyzed chain reaction. *Methods Enzymol.* 155: 335–350.

Mullis, K., F. Faloona, S. Scharf, R. Saiki, G. Horn and H. Erlich. 1986. Specific enzymatic amplification of DNA in vitro: The polymerase chain reaction. *Cold Spring Harb. Symp. Quant. Biol.* 51: 263–273.

Mulvey, J. and R. C. Vrijenhoek. 1981. Multiple paternity in the hermaphroditic snail, *Biomphalaria obstructa*. *J. Heredity* 72: 308–312.

Mulvey, M. and 6 others. 1997. Conservation genetics of North American freshwater mussels *Amblema* and *Megalonaias*. *Cons. Genet.* 11: 868–878.

Mulvey, M., H.-P. Liu and K. L. Kandl. 1998. Application of molecular genetic markers to conservation of fresh-water mussels. *J. Shellfish Res.* 17: 1395–1405.

Mundy, N. I., C. S. Winchell and D. S. Woodruff. 1997. Genetic differences between the endangered San Clemente Island loggerhead shrike *Lanius ludovicianus mearnsi* and two neighbouring subspecies demonstrat-ed by mtDNA control region and cytochrome *b* sequence variation. *Mol. Ecol.* 6: 29–37.

Munstermann, L. E. 1988. Biochemical systematics of nine nearctic *Aedes* mosquitoes (subgenus *Ochlerotatus*, *annulipes* group B). Pp. 135–147 in *Biosystematics of Haematophagous Insects*, M. W. Service (ed.). Clarendon, Oxford, England.

Munstermann, L. E. 1995. Mosquito systematics: Current status, new trends, associated complications. *J. Vector Ecol.* 20: 129–138.

Murata, S., N. Takasaki, M. Saitoh, H. Tachida and N. Okada. 1996. Details of retropositional genome dynamics that provide a rationale for a generic division: The distinct branching of all the Pacific salmon and trout (*Oncorhynchus*) from the Atlantic salmon and trout (*Salmo*). *Genetics* 142: 915–926.

Murawski, D. A. and J. L. Hamrick. 1990. Local genetic and clonal structure in the tropical terrestrial bromeliad, *Aechmea magdalenae*. *Am. J. Bot.* 77: 1201–1208.

Murphy, R. W. 1988. The problematic phylogenetic analysis of interlocus heteropolymer isozyme characters: A case study from sea snakes and cobras. *Can. J. Zool.* 66: 2628–2633.

Murphy, R. W., J. W. Sites Jr., D. G. Buth and C. H. Haufler. 1996. Proteins: Isozyme electrophoresis. Pp. 51–120 in *Molecular Systematics* (2nd Edition), D. M. Hillis, C. Moritz and B. K. Maple (eds.). Sinauer Associates, Sunderland, MA.

Murphy, W. J. and G. E. Collier. 1997. A molecular phylogeny for Aplocheiloid fishes (Atherinomorpha, Cyprinodontiformes): The role of vicariance and the origins of annualism. *Mol. Biol. Evol.* 14: 790–799.

Murphy, W. J., E. Eizirik, W. E. Johnson, Y. P. Zhang, O. A. Ryder and S. J. O'Brien. 2001. Molecular phylogenetics and the origins of placental mammals. *Nature* 409: 614–618.

Murray, J., O. C. Stine and M. S. Johnson. 1991. The evolution of mitochondrial DNA in *Partula*. *Heredity* 66: 93–104.

Musser, J. M., D. M. Granoff, P. E. Pattison and R. K. Selander. 1985. A population genetic framework for the study of invasive diseases caused by serotype b strains of *Haemophilus influenzae*. *Proc. Natl. Acad. Sci. USA* 82: 5078–5082.

Musser, J. M., S. J. Barenkamp, D. M. Granoff and R. K. Selander. 1986. Genetic relationships of serologically nontypable and serotype b strains of *Haemophilus influenzae*. *Infect. Immunol.* 52: 183–191.

Musser, J. M., D. A. Bemis, H. Ishikawa and R. K. Selander. 1987. Clonal diversity and host distribution in *Bordetella bronchiseptica*. *J. Bacteriol.* 169: 2793–2803.

Myers, R. M., T. Maniatis and L. S. Lerman. 1986. Detection and localization of single base changes by denaturing gradient gel electrophoresis. *Meth. Enzymol.* 155: 501–527.

Nachman, M. W., H. E. Hoekstra and S. L. D'Agostino. 2003. The genetic basis of adaptive melanism in pocket mice. *Proc. Natl. Acad. Sci. USA* 100: 5268–5273.

Nagl, S., H. Tichy, W. E. Mayer, N. Takezaki, N. Takahata and J. Klein. 2000. The origin and age of haplochromine fishes in Lake Victoria, East Africa. *Proc. R. Soc. Lond. B* 267: 1049–1061.

Nairn, C. J. and R. J. Ferl. 1988. The complete nucleotide sequence of the small-subunit ribosomal RNA coding region for the cycad *Zamia pumila*: Phylogenetic implications. *J. Mol. Evol.* 27: 133–141.

Nakamura, Y. and 10 others. 1987. Variable number of tandem repeat (VNTR) markers for human gene mapping. *Science* 235: 1616–1622.

Narang, S. K., P. E. Kaiser and J. A. Seawright. 1989a. Dichotomous electrophoretic key for identification of sibling species A, B, and C of the *Anopheles quadrimaculatus* complex (Diptera: Culicidae). *J. Med. Entomol.* 26: 94–99.

Narang, S. K., P. E. Kaiser and J. A. Seawright. 1989b. Identification of species D, a new member of the *Anopheles quadrimaculatus* species complex: A biochemical key. *J. Am. Mosquito Control Assn.* 5: 317–324.

Narang, S. K., S. R. Toniolo, J. A. Seawright and P. E. Kaiser. 1989c. Genetic differentiation among sibling species A, B, and C of the *Anopheles quadrimaculatus* complex (Diptera: Culicidae). *Ann. Entomol. Soc. Am.* 82: 508–515.

Nason, J. D. and J. L. Hamrick. 1997. Reproductive and genetic consequences of forest fragmentation: Two case studies of Neotropical canopy trees. *J. Heredity* 88: 264–266.

Nasser, J. M., J. L. Hamrick and T. H. Fleming. 2001. Genetic variation and population structure of the mixed-mating cactus, *Melocactus curvispinus* (Cactaceae). *Heredity* 87: 69–79.

National Research Council. 1992. *DNA Technology in Forensic Science*. National Academy Press, Washington, DC.

National Research Council. 1995. *Science and the Endangered Species Act*. National Academy Press, Washington, DC.

National Research Council. 1996. *The Evaluation of Forensic DNA Evidence*. National Academy Press, Washington, DC.

National Research Council. 1999. *Sustaining Marine Fisheries*. National Academy Press, Washington, DC.

National Research Council. 2000a. *Marine Protected Areas: Tools for Sustaining Ocean Ecosystems*. National Academy Press, Washington, DC.

National Research Council. 2000b. *Clean Coastal Waters: Understanding and Reducing the Effects of Nutrient Pollution*. National Academy Press, Washington, DC.

Nauta, M. and F. Weising. 1996. Constraints on allelic size at microsatellite loci: Implications for genetic differentiation. *Genetics* 143: 1021–1032.

Navarro, A. and N. H. Barton. 2003a. Accumulating postzygotic isolation genes in parapatry: A new twist on chromosomal speciation. *Evolution* 57: 447–459.

Navarro, A. and N. H. Barton. 2003b. Chromosomal speciation and molecular divergence: Accelerated evolution in rearranged chromosomes. *Science* 300: 321–324.

Navarro, A., E. Betrán, A. Barbadilla and A. Ruiz. 1997. Recombination and gene flux caused by gene conversion and crossing over in inversion heterokaryotypes. *Genetics* 146: 695–709.

Navarro, A., A. Barbadilla and A. Ruiz. 2000. Effect of inversion polymorphism on the neutral nucleotide variability of linked chromosomal regions in *Drosophila*. *Genetics* 155: 685–698.

Navidi, W. and N. Arnheim. 1999. Combining data from polymerase chain reaction DNA typing experiments:

Applications to sperm typing data. *J. Am. Statis. Assn.* 94: 726–733.

Naylor, G. V. P and W. M. Brown. 1998. Amphioxus mitochondrial DNA, chordate phylogeny, and the limits of inference based on comparisons of sequences. *Syst. Biol.* 47: 61–76.

Nedbal, M. A. and D. P. Philipp. 1994. Differentiation of mitochondrial DNA in largemouth bass. *Trans. Am. Fish. Soc.* 123: 460–468.

Nee, S. 2001. Inferring speciation rates from phylogenies. *Evolution* 55: 661–668.

Nee, S and R. M. May. 1997. Extinction and the loss of evolutionary history. *Science* 278: 692–694.

Nee, S., R. M. May and P. H. Harvey. 1994a. The reconstructed evolutionary process. *Phil. Trans. R. Soc. Lond. B* 344: 305–311.

Nee, S., R. M. May and P. H. Harvey. 1994b. Extinction rates can be estimated from molecular phylogenies. *Phil. Trans. R. Soc. Lond. B* 344: 77–82.

Nee, S., E. C. Holmes, A. Rambaut and P. H. Harvey. 1996a. Inferring population history from molecular phylogenies. Pp. 66–80 in *New Uses for New Phylogenies*, P. H. Harvey, A. J. Leigh Brown, J. Maynard Smith and S. Nee (eds.). Oxford University Press, New York.

Nee, S., A. F. Read and P. H. Harvey. 1996b. Why phylogenies are necessary for comparative analysis. Pp. 399–411 in *Phylogenies and the Comparative Method in Animal Behavior*, E. P. Martins (ed.). Oxford University Press, New York.

Neff, B. D. 2001. Genetic paternity analysis and breeding success in bluegill sunfish (*Lepomis macrochirus*). *J. Heredity* 92: 111–119.

Neff, B. D. and M. R. Gross. 2001. Dynamic adjustment of parental care in response to perceived paternity. *Proc. R. Soc. Lond. B* 268: 1559–1565.

Neff, B. D. and T. E. Pitcher. 2002. Assessing the statistical power of genetic analyses to detect multiple mating in fishes. *J. Fish Biol.* 61: 739–750.

Neff, B. D., J. Repka and M. R. Gross. 2000a. Parentage analysis with incomplete sampling of candidate parents and offspring. *Mol. Ecol.* 9: 515–528.

Neff, B. D., J. Repka and M. R. Gross. 2000b. Statistical confidence in parentage analysis with incomplete samping: How many loci and offspring are needed? *Mol. Ecol.* 9: 529–539.

Neff, B. D., T. E. Pitcher and J. Repka. 2002. A Bayseian model for assessing the freqeuncy of multiple mating in nature. *J. Heredity* 93: 406–414.

Nei, M. 1972. Genetic distance between populations. *Am. Nat.* 106: 283–292.

Nei, M. 1973. Analysis of gene diversity in subdivided populations. *Proc. Natl. Acad. Sci. USA* 70: 3321–3323.

Nei, M. 1975. *Molecular Population Genetics and Evolution*. North-Holland, Amsterdam.

Nei, M. 1977. *F*-statistics and analysis of gene diversity in subdivided populations. *Ann. Human Genet. London* 41: 225–233.

Nei, M. 1978. Estimation of average heterozygosity and genetic distance from a small number of individuals. *Genetics* 23: 341–369.

Nei, M. 1983. Genetic polymorphism and the role of mutation in evolution. Pp. 165–190 in *Evolution of Genes and Proteins*, M. Nei and R. K. Koehn (eds.). Sinauer Associates, Sunderland, MA.

Nei, M. 1985. Human evolution at the molecular level. Pp. 41–64 in *Population Genetics and Molecular Evolution*, T. Ohta and K. Aoki (eds.). Japan Science Society Press and Springer-Verlag, Tokyo.

Nei, M. 1987. *Molecular Evolutionary Genetics*. Columbia University Press, New York.

Nei, M. and D. Graur. 1984. Extent of protein polymorphism and the neutral mutation theory. *Evol. Biol.* 17: 73–118.

Nei, M. and A. L. Hughes. 1991. Polymorphism and evolution of the major histocompatibility complex loci in mammals. Pp. 222–247 in *Evolution at the Molecular Level*, R. K. Selander, A. G. Clark and T. S. Whittam (eds.). Sinauer Associates, Sunderland, MA.

Nei, M. and R. K. Koehn (eds.). 1983. *Evolution of Genes and Proteins*. Sinauer Associates, Sunderland, MA.

Nei, M. and S. Kumar. 2000. *Molecular Evolution and Phylogenetics*. Oxford University Press, Oxford, England.

Nei, M. and W.-H. Li. 1979. Mathematical model for studying genetic variation in terms of restriction endonucleases. *Proc. Natl. Acad. Sci. USA* 76: 5269–5273.

Nei, M. and G. Livshits. 1990. Evolutionary relationships of Europeans, Asians, and Africans at the molecular level. Pp. 251–265 in *Population Biology of Genes and Molecules*, N. Takahata and J. F. Crow (eds.). Baifukan, Tokyo.

Nei, M. and T. Maruyama. 1975. Lewontin-Krakauer test for neutral genes. *Genetics* 80: 395.

Nei, M. and A. K. Roychoudhury. 1982. Genetic relationship and evolution of human races. *Evol. Biol.* 14: 1–59.

Nei, M. and F. Tajima. 1981. DNA polymorphism detectable by restriction endonucleases. *Genetics* 97: 145–163.

Nei, M. and F. Tajima. 1985. Evolutionary change of restriction cleavage sites and phylogenetic inference for man and apes. *Mol. Biol. Evol.* 2: 189–205.

Nei, M. and N. Takahata. 1993. Effective population size, genetic diversity, and coalescence time in subdivided populations. *J. Mol. Evol.* 37: 240–244.

Nei, M. and N. Takezaki. 1996. The root of the phylogenetic tree of human populations. *Mol. Biol. Evol.* 13: 170–177.

Nei, M., T. Maruyama and R. Chakraborty. 1975. The bottleneck effect and genetic variability in populations. *Evolution* 29: 1–10.

Nei, M., J. C. Stephens and N. Saitou. 1985. Methods for computing the standard errors of branching points in an evolutionary tree and their application to molecular data from humans and apes. *Mol. Biol. Evol.* 2: 66–85.

Nei, M., F. Tajima and Y. Tateno. 1983. Accuracy of estimated phylogenetic trees from molecular data. II. Gene frequency data. *J. Mol. Evol.* 19: 153–170.

Nei, M., P. Xu and G. Glazko. 2001. Estimation of divergence times from multiprotein sequences for a few mammalian species and several distantly related organisms. *Proc. Natl. Acad. Sci. USA* 98: 2497–2502.

Neigel, J. E. 1996. Estimation of effective population size and migration parameters from genetic data. Pp. 329–346 in *Molecular Genetic Approaches in Conservation*, T. B. Smith and R. K. Wayne (eds.). Oxford University Press, New York.

Neigel, J. E. 1997. A comparison of alternative strategies for estimating gene flow from genetic markers. *Annu. Rev. Ecol. Syst.* 28: 105–128.

Neigel, J. E. 2002. Is *FST* obsolete? *Cons. Genet.* 3: 167–173.

Neigel, J. E. and J. C. Avise. 1983a. Clonal diversity and population structure in a reef-building coral, *Acropora cervicornis*: Self-recognition analysis and demographic interpretation. *Evolution* 37: 437–453.

Neigel, J. E. and J. C. Avise. 1983b. Histocompatibility bioassays of population structure in marine sponges. *J. Heredity* 74: 134–140.

Neigel, J. E. and J. C. Avise. 1985. The precision of histocompatibility response in clonal recognition in tropical marine sponges. *Evolution* 39: 724–732.

Neigel, J. E. and J. C. Avise. 1986. Phylogenetic relationships of mitochondrial DNA under various demographic models of speciation. Pp. 515–534 in *Evolutionary Processes and Theory*, E. Nevo and S. Karlin (eds.). Academic Press, New York.

Neigel, J. E. and J. C. Avise. 1993. Application of a random-walk model to geographic distributions of animal mitochondrial DNA variation. *Genetics* 135: 1209–1220.

Neigel, J. E., R. M. Ball Jr. and J. C. Avise. 1991. Estimation of single generation migration distances from geographic variation in animal mitochondrial DNA. *Evolution* 45: 423–432.

Nelson, G. and N. I. Platnick. 1981. *Systematics and Biogeography: Cladistics and Vicariance*. Columbia University Press, New York.

Nelson, G. and D. E. Rosen (eds.). 1981. *Vicariance Biogeography, A Critique*. Columbia University Press, New York.

Nelson, K. and D. Hedgecock. 1977. Electrophoretic evidence of multiple paternity in the lobster *Homarus americanus* (Milne-Edwards). *Am. Nat.* 111: 361–365.

Nelson, K., R. J. Baker and R. L. Honeycutt. 1987. Mitochondrial DNA and protein differentiation between hybridizing cytotypes of the white-footed mouse, *Peromyscus leucopus*. *Evolution* 41: 864–872.

Nelson, K., T. S. Whittam and R. K. Selander. 1991. Nucleotide polymorphism and evolution in the glyceraldehyde-3-phosphate dehydrogenase gene (*gapA*) in natural populations of *Salmonella* and *Escherichia coli*. *Proc. Natl. Acad. Sci. USA* 88: 6667–6671.

Nelson, K. E. and 28 others. 1999. Evidence for lateral gene transfer between Archaea and Bacteria from genome sequences of *Thermotoga maritima*. *Nature* 399: 323–329.

Nelson, W. S., P. A. Pródohl and J. C. Avise. 1996. Development and application of long-PCR for the assay of full-length animal mitochondrial DNA. *Mol. Ecol.* 5: 807–810.

Nelson, W. S., T. Dean and J. C. Avise. 2000. Matrilineal history of the endangered Cape Sable seaside sparrow inferred from mitochondrial DNA polymorphism. *Mol. Ecol.* 9: 809–813.

Nevo, E. 1978. Genetic variation in natural populations: Patterns and theory. *Theoret. Pop. Biol.* 13: 121–177.

Nevo, E. and C. R. Shaw. 1972. Genetic variation in a subterranean mammal, *Spalax ehrenbergi*. *Biochem. Genet.* 7: 235–241.

Nevo, E., Y. J. Kim, C. R. Shaw and C. S. Thaeler. 1974. Genetic variation, selection, and speciation in *Thomomys talpoides* pocket gophers. *Evolution* 28: 1–23.

Newton, A. C., C. E. Caten and R. Johnson. 1985. Variation for isozymes and double-stranded RNA among isolates of *Puccinia striiformis* and two other cereal rusts. *Plant Pathol.* 34: 235–247.

Newton, C. R. and 7 others. 1989. Analysis of any point mutation in DNA. The amplification refractory mutation system (ARMS). *Nucl. Acids Res.* 17: 2503–2516.

Newton, I. 2003. *The Speciation and Biogeography of Birds*. Academic Press, Amsterdam.

Nichols, R. 2001. Gene trees and species trees are not the same. *Trends Ecol. Evol.* 16: 358–364.

Nichols, R. A. and D. J. Balding. 1991. Effects of population structure on DNA fingerprint analysis in forensic science. *Heredity* 66: 297–302.

Nielsen, E. E., M. M. Hansen and V. Loeschcke. 1997. Analysis of microsatellite DNA from old scale samples of Atlantic salmon *Salmo salar*: A comparison of genetic composition over 60 years. *Mol. Ecol.* 6: 487–492.

Nielsen, J. L., M. C. Fountain and J. M. Wright. 1997. Biogeographic analysis of Pacific trout (*Oncorhynchus mykiss*) in California and Mexico based on mitochondrial DNA and nuclear microsatellites. Pp. 53–69 in *Molecular Systematics of Fishes*, T. D. Kocher and C. A. Stepien (eds.). Academic Press, San Diego, California.

Nikaido, M., A. P. Rooney and N. Okada. 1999. Phylogenetic relationships among cetartiodactyls based on insertions of short and long interspersed elements: Hippopotamuses are the closest extant relatives of whales. *Proc. Natl. Acad. Sci. USA* 96: 10261–10266.

Nikaido, M., H. Nishihara, Y. Hukumoto and N. Okada. 2003. Ancient SINEs from African endemic mammals. *Mol. Biol. Evol.* 20: 522–527.

Nixon, K. C. and Q. D. Wheeler. 1992. Measures of phylogenetic diversity. Pp. 216–234 in *Extinction and Phylogeny*, M. J. Novacek (ed.). Columbia University Press, New York.

Noor, M. A. F. 1999. Reinforcement and other consequences of sympatry. *Heredity* 83: 503–508.

Noor, M. A. F., K. L. Grams, L. A. Bertucci and J. Reiland. 2001. Chromosomal inversions and the reproductive isolation of species. *Proc. Natl. Acad. Sci. USA* 98: 12084–12088.

Normark, B. B., A. R. McCune and R. G. Harrison. 1991. Phylogenetic relationships of neopterygian fishes, inferred from mitochondrial DNA sequences. *Mol. Biol. Evol.* 8: 819–834.

Noro, M., R. Masuda, I. A. Dubrovo, M. C. Yoshida and M. Kato. 1998. Molecular phylogenetic inference of the woolly mammoth *Mammuthus primigenius*, based on complete sequences of mitochondrial cytochrome *b* and 12S ribosomal RNA genes. *J. Mol. Evol.* 46: 314–326.

Novacek, M. J. 1992. Mammalian phylogeny: Shaking the tree. *Nature* 356: 121–125.

Nowak, R. M. 1992. The red wolf is not a hybrid. *Cons. Biol.* 6: 593–595.

Nowak, R. M. and N. E. Federoff. 1998. Validity of the red wolf: Response to Roy et al. *Cons. Biol.* 12: 722–725.

Nozaki, H. and 9 others. 2003. The phylogenetic position of red algae revealed by multiple nuclear genes from mitochondria-containing eukaryotes and an alternative hypothesis on the origin of plastids. *J. Mol. Evol.* 56: 485–497.

Nugent, J. M. and J. D. Palmer. 1991. RNA-mediated transfer of the gene *coxII* from the mitochondrion to the nucleus during flowering plant evolution. *Cell* 66: 473–481.

Nurminsky, D. I., M. V. Nurminskaya, D. De Aguiar and D. L. Hartl. 1998. Selective sweep of a newly evolved sperm-specific gene in *Drosophila. Nature* 396: 572–575.

Nuttall, G. H. F. 1904. *Blood Immunity and Blood Relationship.* Cambridge University Press, Cambridge, England.

Nybom, H. and B. A. Schaal. 1990. DNA "fingerprints" reveal genotypic distributions in natural populations of blackberries and raspberries (*Rubus*, Rosaceae). *Am. J. Bot.* 77: 883–888.

Nybom, H., J. Ramser, D. Kaemmer, G. Kahl and K. Weising. 1992. Oligonucleotide DNA fingerprinting detects a multiallelic locus in box elder (*Acer negundo*). *Mol. Ecol.* 1: 65–67.

Nyffeler, R. 1999. A new ordinal classification of the flowering plants. *Trends Ecol. Evol.* 14: 168–170.

Oakenfull, E. A., H. N. Lim and O. A. Ryder. 2000. A survey of equid mitochondrial DNA: Implications for the evolution, genetic diversity and conservation of *Equus. Cons. Genet.* 1: 341–355.

Oakeshott, J. G., J. B. Gibson, P. R. Anderson, W. R. Knibb, D. G. Anderson and G. K. Chambers. 1982. Alcohol dehydrogenase and glycerol-3-phosphate dehydrogenase clines in *Drosophila melanogaster* on different continents. *Evolution* 36: 86–96.

Oakley, T. H. and C. W. Cunningham. 2002. Molecular phylogenetic evidence for the independent evolutionary origin of an arthropod compound eye. *Proc. Natl. Acad. Sci. USA* 99: 1426–1430.

Obermiller, L. E. and E. Pfeiler. 2003. Phylogenetic relationships of elopomorph fishes inferred from mitochondrial ribosomal DNA sequences. *Mol. Phylogen. Evol.* 26: 202–214.

O'Brien, S. J. 1987. The ancestry of the giant panda. *Sci. Am.* 257(5): 102–107.

O'Brien, S. J. and J. F. Evermann. 1988. Interactive influence of infectious disease and genetic diversity in natural populations. *Trends Ecol. Evol.* 3: 254–259.

O'Brien, S. J. and R. J. MacIntyre. 1972. The alpha-glycerophosphate cycle in *Drosophila melanogaster*. I. Biochemical and developmental aspects. *Biochem. Genet.* 7: 141–161.

O'Brien, S. J. and E. Mayr. 1991. Bureaucratic mischief: Recognizing endangered species and subspecies. *Science* 251: 1187–1188.

O'Brien, S. J., D. E. Wildt, D. Goldman, C. R. Merril and M. Bush. 1983. The cheetah is depauperate in genetic variation. *Science* 221: 459–462.

O'Brien, S. J., W. G. Nash, D. E. Wildt, M. E. Bush and R. E. Benveniste. 1985a. A molecular solution to the riddle of the giant panda's phylogeny. *Nature* 317: 140–144.

O'Brien, S. J. and 9 others. 1985b. Genetic basis for species vulnerability in the cheetah. *Science* 227: 1428–1434.

O'Brien, S. J. and 6 others. 1987. East African cheetahs: Evidence for two population bottlenecks? *Proc. Natl. Acad. Sci. USA* 84: 508–511.

O'Brien, S. J., J. S. Martenson, M. A. Eichelberger, E. T. Thorne and F. Wright. 1989. Genetic variation and molecular systematics of the black-footed ferret. Pp. 21–33 in *Conservation Biology and the Black-Footed Ferret*, U. S. Seal, E. T. Thorne, M. A. Bogan and S. H. Anderson (eds.). Yale University Press, New Haven, CT.

O'Brien, S. J. and 9 others. 1990. Genetic introgression within the Florida panther *Felis concolor coryi. Natl. Geog. Res.* 6: 485–494.

O'Brien, S. J. and 10 others. 1996. Conservation genetics in the Felidae. Pp. 50–74 in *Conservation Genetics: Case Histories from Nature*, J. C. Avise and J. L. Hamrick (eds.). Chapman & Hall, New York.

O'Brien, S. J. and 9 others. 1999. The promise of comparative genomics in mammals. *Science* 286: 458–481.

Ochman, H. and R. K. Selander. 1984. Evidence for clonal population structure in *Escherichia coli. Proc. Natl. Acad. Sci. USA* 81: 198–201.

Ochman, H. and A. C. Wilson. 1987. Evolution in bacteria: Evidence for a universal substitution rate in cellular genomes. *J. Mol. Evol.* 26: 74–86.

Ochman, H., J. G. Lawrence and E. A. Groisman. 2000. Lateral gene transfer and the nature of bacterial innovation. *Nature* 405: 299–304.

O'Connell, M. 1992. Response to: "Six biological reasons why the endangered species act doesn't work and what to do about it. " *Cons. Biol.* 6: 140–143.

O'Connor, D. and R. Shine. 2003. Lizards in 'nuclear families': A novel reptilian social system in *Egernia saxatilis* (Scincidae). *Mol. Ecol.* 12: 743–752.

O'Corry-Crowe, G. M., R. S. Suydam, A. Rosenberg, K. J. Frost and A. E. Dizon. 1997. Phylogeography, population structure and dispersal patterns of the beluga whale *Delphinapterus leucas* in the western Nearctic revealed by mitochondrial DNA. *Mol. Ecol.* 6: 955–970.

Oddou-Muratorio, R. J. Petit, B. Le Guerroue, D. Guesneet and B. Demesure. 2001. Pollen- versus seed-mediated gene flow in a scattered forest tree species. *Evolution* 55: 1123–1135.

Odgers, W. A., M. J. Healy and J. G. Oakeshott. 1995. Nucleotide polymorphism in the 5′ promoter region of esterase 6 in *Drosophila melanogaster* and its relationship to enzyme activity variation. *Genetics* 141: 215–222.

O'Farrell, P. H. 1975. High resolution two-dimensional electrophoresis of proteins. *J. Biol. Chem.* 250: 4007–4021.

O'Foighil, D. and M. J. Smith. 1996. Phylogeography of an asexual marine clam complex, *Lasaea*, in the northeastern Pacific based on cytochrome oxidase III sequence variation. *Mol. Phylogen. Evol.* 6: 134–142.

Ohlsson, O. T., H. G. Smith, L. Raberg and D. Hasselquist. 2002. Pheasant sexual ornaments reflect nutritional conditions during early growth. *Proc. R. Soc. Lond. B* 269: 21–27.

Ohno, S. 1967. *Sex Chromosomes and Sex-Linked Genes*. Springer-Verlag, New York.

Ohno, S. 1970. *Evolution by Gene Duplication*. Springer-Verlag, New York.

Ohta, T. 1980. *Evolution and Variation of Multigene Families*. Springer-Verlag, Berlin.

Ohta, T. 1984. Some models of gene conversion for treating the evolution of multigene families. *Genetics* 106: 517–528.

Ohta, T. 1992a. The nearly neutral theory of molecular evolution. *Annu. Rev. Ecol. Syst.* 23: 263–286.

Ohta, T. 1992b. The meaning of natural selection revisited at the molecular level. *Trends Ecol. Evol.* 7: 311–312.

Ohta, T. 1993. An examination of the generation-time effect on molecular evolution. *Proc. Natl. Acad. Sci. USA* 90: 10676–10680.

Ohta, T. 2000. Evolution of gene families. *Gene* 259: 45–52.

Ohta, T. and G. A. Dover. 1983. Population genetics of multi-gene families that are dispersed into two or more chromosomes. *Proc. Natl. Acad. Sci. USA* 80: 4079–4083.

Ohta, T. and M. Kimura. 1971. On the constancy of the evolutionary rate of cistrons. *J. Mol. Evol.* 1: 18–25.

Ohta, T. and M. Kimura. 1973. A model of mutation appropriate to estimate the number of electrophoretically detectable alleles in a finite population. *Genet. Res.* 22: 201–204.

Ohta, T. and H. Tachida. 1990. Theoretical study of near neutrality. I. Heterozygosity and rate of mutant substitution. *Genetics* 126: 219–229.

O'hUigin, C. and W.-H. Li. 1992. The molecular clock ticks regularly in muroid rodents and hamsters. *J. Mol. Evol.* 35: 377–384.

Ohyama, K. and 12 others. 1986. Chloroplast gene organization deduced from complete sequence of liverwort *Marchantia polymorpha* chloroplast DNA. *Nature* 322: 572–574.

Oleksiak, M. F., G. A. Churchill and D. L. Crawford. 2002. Variation in gene expression within and among natural populations. *Nature Genetics* 32: 261–266.

Oliveira, R. P., N. E. Broude, A. M. Macedo, C. R. Cantor, C. L. Smith and S. D. J. Pena. 1998. Probing the genetic population structure of *Trypanosoma cruzi* with polymorphic microsatellites. *Proc. Natl. Acad. Sci. USA* 95: 3776–3780.

Olivera, B. M. 2002. *Conus* venom peptides: Reflections from the biology of clades and species. *Annu. Rev. Ecol. Syst.* 33: 25–47.

Olmstead, R. G. 1995. Species concepts and plesiomorphic species. *Syst. Biol.* 20: 623–630.

Olmstead, R. G., R. K. Jansen, H. J. Michaels, S. R. Downie and J. D. Palmer. 1990. Chloroplast DNA and phylogenetic studies in the Asteridae. Pp. 119–134 in *Biological Approaches and Evolutionary Trends in Plants*, S. Kawano (ed.). Academic Press, San Diego, CA.

Olmstead, R. G., H. J. Michaels, K. M. Scott and J. D. Palmer. 1992. Monophyly of the Asteridae and identification of their major lineages inferred from DNA sequences of *rbcL*. *Annals Missouri Bot. Gardens* 79: 249–265.

Olsen, G. J. and C. R. Woese. 1997. Archael genomics: An overview. *Cell* 89: 991–994.

Olson, M. S. and D. E. McCauley. 2002. Mitochondrial DNA diversity, population structure, and gender association in the gynodioecious plant *Silene vulgaris*. *Evolution* 56: 253–262.

Olson, R. R., J. A. Runstadler and T. D. Kocher. 1991. Whose larvae? *Nature* 351: 357–358.

Olson, S. 2002. Seeking the signs of selection. *Science* 298: 1324–1325.

Olsson, M. and T. Madsen. 2001. Promiscuity in sand lizards (*Lacerta agilis*) and adder snakes (*Vipera berus*): Causes and consequences. *J. Heredity* 92: 190–197.

Olsson, M. and R. Shine. 1997. Advantages of multiple matings to females: A test of the infertility hypothesis using lizards. *Evolution* 51: 1684–1688.

Omland, K. E. 1997. Correlated rates of molecular and morphological evolution. *Evolution* 51: 1381–1393.

Omland, K. E. and S. M. Lanyon. 2000. Reconstructing plumage evolution in orioles (*Icterus*): Repeated convergence and reversal in patterns. *Evolution* 54: 2119–2133.

Omland, K. E., S. M. Lanyon and S. J. Fritz. 1999. A molecular phylogeny of the New World orioles (Icterus): The importance of dense taxon sampling. *Mol. Phylogen. Evol.* 12: 224–239.

O'Neill, S. L., A. A. Hoffmann and J. H. Werren (eds.). 1997 *Influential passengers: Inherited Microorganisms and Arthropod Reproduction*. Oxford University Press, Oxford, England.

Oppliger, A., Y. Naciri-Graven, G. Ribi and D. J. Hosken. 2003. Sperm length influences fertilization success during sperm competition in the snail *Viviparus ater*. *Mol. Ecol.* 12: 485–492.

O'Riain, M. J., J. U. M. Jarvis and C. G. Faukes. 1996. A dispersive morph in the naked mole-rat. *Nature* 380: 619–621.

Oring, L. W., R. C. Fleischer, J. M. Reed and K. E. Marsden. 1992. Cuckoldry through stored sperm in the sequentially polyandrous spotted sandpiper. *Nature* 359: 631–633.

Orita, M., Y. Suzuki, T. Sekiya and K. Hayashi. 1989. Rapid and sensitive detection of point mutations and DNA polymorphisms using the polymerase chain reaction. *Genomics* 5: 874–879.

Orive, M. E. and M. A. Asmussen. 2000. The effects of pollen and seed migration on nuclear-dicytoplasmic systems. II. A new method for estimating plant gene flow from joint nuclear-cytoplasmic data. *Genetics* 155: 833–854.

Orlando, L. and 6 others. 2002. Ancient DNA and the population genetics of cave bears (*Ursus spelaeus*) through space and time. *Mol. Biol. Evol.* 19: 1920–1933.

Orr, H. A. 1987. Genetics of male and female sterility in hybrids of *Drosophila pseudoobscura* and *D. persimilis*. *Genetics* 116: 555–563.

Orr, H. A. 1992. Mapping and characterization of a "speciation gene" in *Drosophila*. *Genet. Res. Camb.* 59: 73–80.

Orr, H. A. 1997. Haldane's rule. *Annu. Rev. Ecol. Syst.* 28: 195–218.

Orr, H. A. 2001. The genetics of species differences. *Trends Ecol. Evol.* 16: 343–350.

Orr, H. A. 2002. The population genetics of adaptation: The adaptation of DNA sequences. *Evolution* 56: 1317–1330.

Orr, H. A. and S. Irving. 2001. Complex epistasis and the genetic basis of hybrid sterility in the *Drosophila pseudoobscura* Bogota-USA hybridization. *Genetics* 158: 1089–1100.

Orr, H. A. and L. H. Orr. 1996. Waiting for speciation: The effect of population subdivision on the time to speciation. *Evolution* 50: 1742–1749.

Orr, M. R. and T. B. Smith. 1998. Ecology and speciation. *Trends Ecol. Evol.* 13: 502–506.

Ortí, G. and A. Meyer. 1997. The radiation of characiform fishes and the limits of resolution of mitochondrial ribosomal DNA sequences. *Syst. Biol.* 46: 75–100.

Ortí, G., M. P. Hare and J. C. Avise. 1997a. Detection and isolation of nuclear haplotypes by PCR-SSCP. *Mol. Ecol.* 6: 575–580.

Ortí, G., D. E. Pearse and J. C. Avise. 1997b. Phylogenetic assessment of length variation at a microsatellite locus. *Proc. Natl. Acad. Sci. USA* 94: 10745–10749.

O'Ryan, C., J. R. B. Flamand and E. H. Harley. 1994. Mitochondrial DNA variation in black rhinoceros (*Diceros bicornis*): Conservation management implications. *Cons. Biol.* 8: 495–500.

Osborne, M. J., J. A. Norman, L. Christidis and N. D. Murray. 2000. Genetic distinctness of isolated populations of an endangered marsupial, the mountain pygmy-possum, *Burramys parvus*. *Mol. Ecol.* 9: 609–613.

Osentoski, M. F. and T. Lamb. 1995. Intraspecific phylogeography of the gopher tortoise, *Gopherus polyphemus*: RFLP analysis of amplified mtDNA segments. *Mol. Ecol.* 4: 709–718.

Oste, C. 1988. Polymerase chain reaction. *BioTechniques* 6: 162–167.

Ou, C.-Y. and 15 others. 1992. Molecular epidemiology of HIV transmission in a dental practice. *Science* 256: 1165–1171.

Ovchinnikov, I. V., A. Göthenstein, G. P. Romanova, V. M. Kharitinov, K. Lidén and W. Goodwin. 2000. Molecular analysis of Neanderthal DNA from the northern Caucasus. *Nature* 404: 490.

Ovenden, J. R. 1990. Mitochondrial DNA and marine stock assessment: A review. *Aust. J. Mar. Freshwater Res.* 41: 835–853.

Ovenden, J. R. and R. W. G. White. 1990. Mitochondrial and allozyme genetics of incipient speciation in a land-locked population of *Galaxias truttaceus* (Pisces: Galaxiidae). *Genetics* 124: 701–716.

Ovenden, J. R., D. J. Brasher and R. W. G. White. 1992. Mitochondrial DNA analyses of the red rock lobster *Jasus edwardsii* supports an apparent absence of population subdivision throughout Australasia. *Mar. Biol.* 112: 319–326.

Overath, R. D. and M. A. Asmussen. 2000a. The cytonuclear effects of facultative apomixis. I. Disequilibrium dynamics in diploid populations. *Theor. Pop. Biol.* 58: 107–121.

Overath, R. D. and M. A. Asmussen. 2000b. The cytonuclear effects of facultative apomixis. II. Definitions and dynamics of disequilibria in tetraploid populations. *Theor. Pop. Biol.* 58: 123–142.

Owens, I. P. F. and P. M. Bennett. 1995. Ancient ecological diversification explains life-history variation among living birds. *Proc. R. Soc. Lond. B* 261: 227–232.

Owens, K. N., M. Harvey-Blankenship and M.-C. King. 2002. Genomic sequencing in the service of human rights. *Int. J. Epidemiology* 31: 53–58.

Owens, P. F., A. Dixon, T. Burke and D. B. A. Thompson. 1995. Strategic paternity assurance in the sex-role reversed Eurasian dotterel (*Charadrius morinellus*): Behavioral and genetic evidence. *Behav. Ecol.* 6: 14–26.

Ozawa, T., S. Hayashi and V. M. Mikhelson. 1997. Phylogenetic position of mammoth and Steller's sea cow within Tethytheria demonstrated by mitochondrial DNA sequences. *J. Mol. Evol.* 44: 406–413.

Pääbo, S. 1985. Molecular cloning of ancient mummy DNA. *Nature* 314: 644–645.

Pääbo, S. 1989. Ancient DNA: Extraction, characterization, molecular cloning and enzymatic amplification. *Proc. Natl. Acad. Sci. USA* 86: 1939–1943.

Pääbo, S. 2003. The mosaic that is our genome. *Nature* 421: 409–412.

Pääbo, S. and A. C. Wilson. 1988. Polymerase chain reaction reveals cloning artefacts. *Nature* 334: 387–388.

Pääbo, S., R. G. Higuchi and A. C. Wilson. 1989. Ancient DNA and the polymerase chain reaction. *J. Biol. Chem.* 264: 9709–9712.

Pääbo, S., W. K. Thomas, K. M. Whitfield, Y. Kumazawa and A. C. Wilson. 1991. Rearrangements of mitochondrial transfer RNA genes in marsupials. *J. Mol. Evol.* 33: 426–430.

Pace, N. R. 1997. A molecular view of microbial diversity and the biosphere. *Science* 276: 734–740.

Packer, C., D. A. Gilbert, A. E. Pussey and S. J. O'Brien. 1991. A molecular genetic analysis of kinship and cooperation in African lions. *Nature* 351: 562–564.

Packer, L. 1991. The evolution of social behavior and nest architecture in sweat bees of the subgenus *Evylaeus* (Hymenoptera: Halictidae): A phylogenetic approach. *Behav. Ecol. Sociobiol.* 29: 153–160.

Paetkau, D. 1999. Using genetics to identify intraspecific conservation units: A critique of current methods. *Cons. Biol.* 13: 1507–1509.

Paetkau, D. and C. Strobeck. 1994. Microsatellite analysis of genetic variation in black bear populations. *Mol. Ecol.* 3: 489–495.

Paetkau, D., W. Calvert, J. Stirling and C. Strobeck. 1995. Microsatellite analysis of population structure in Canadian polar bears. *Mol. Ecol.* 4: 347–354.

Paetkau, D., G. F. Shields and C. Strobeck. 1998. Gene flow between insular, coastal and interior populations of brown bears in Alaska. *Mol. Ecol.* 7: 1283–1292.

Page, D. C. and 6 others. 1982. Single copy sequence hybridizes to polymorphic and homologous loci on human X and Y chromosomes. *Proc. Natl. Acad. Sci. USA* 79: 5352–5356.

Page, D. C., M. E. Harper, J. Love and D. Botstein. 1984. Occurrence of a transposition from the X-chromosome long arm to the Y-chromosome short arm during human evolution. *Nature* 311: 119–123.

Page, D. C., A. de la Chapelle and J. Weissenbach. 1985. Chromosome Y-specific DNA in related human XX males. *Nature* 315: 224–226.

Page, R. D. M. 1993. Parasites, phylogeny and cospeciation. *Int. J. Parasitol.* 23: 499–506.

Page, R. D. M. 1994. Parallel phylogenies: Reconstructing the history of host-parasite assemblages. *Cladistics* 10: 155–173.

Page, R. D. M. 1998. GENE TREE: Comparing gene and species phylogenies using reconciled trees. *Bioinformatics* 14: 819–820.

Page, R. D. M. 2000. Extracting species trees from complex gene trees: Reconciled trees and vertebrate phylogeny. *Mol. Phylogen. . Evol.* 14: 89–106.

Page, R. D. M. (ed.). 2003. *Tangled Trees: Phylogeny, Cospeciation and Coevolution.* University of Chicago Press, Chicago, IL.

Page, R. D. M. and M. A. Charleston. 1997. From gene to organismal phylogeny: Reconciled trees and the gene tree species tree problem. *Mol. Phylogen. Evol.* 7: 231–240.

Page, R. D. M. and M. A. Charleston. 1998. Trees within trees: Phylogeny and historical associations. *Trends Ecol. Evol.* 13: 356–359.

Page, R. D. M. and E. C. Holmes. 1998. *Molecular Evolution: A Phylogenetic Approach.* Blackwell, Oxford, England.

Page, R. D. M., P. L. M. Lee, S. A. Becher, R. Griffiths and D. H. Clayton. 1998. A different tempo of mitochondrial DNA evolution in birds and their parasitic lice. *Mol. Phylogen. Evol.* 9: 276–293.

Pagel, M. D. 1992. A method for the analysis of comparative data. *J. Theoret. Biol.* 156: 431–442.

Pagel, M. D. 1994. Detecting correlated evolution on phylogenies: A general method for the comparative analysis of discrete characters. *Proc. R. Soc. Lond. B* 255: 37–45.

Paige, K. N., W. C. Capman and P. Jennetten. 1991. Mitochondrial inheritance patterns across a cottonwood hybrid zone: Cytonuclear disequilibria and hybrid zone dynamics. *Evolution* 45: 1360–1369.

Palkovacs, E. P., M. Marschner, C. Ciofi, J. Gerlach and A. Caccone. 2003. Are the native giant tortoises from the Seychelles really extinct? A genetic perspective based on mtDNA and microsatellite data. *Mol. Ecol.* 12: 1403–1413.

Palmer, A. R. and C. Strobeck. 1986. Fluctuating asymmetry: Measurement, analysis, patterns. *Annu. Rev. Ecol. Syst.* 17: 391–421.

Palmer, J. D. 1985. Evolution of chloroplast and mitochondrial DNA in plants and algae. Pp 131–240 in *Molecular Evolutionary Genetics,* R. J. MacIntyre (ed.). Plenum Press, New York.

Palmer, J. D. 1987. Chloroplast DNA evolution and biosystematic uses of chloroplast DNA variation. *Am. Nat.* 130: S6-S29.

Palmer, J. D. 1990. Contrasting modes and tempos of genome evolution in land plant organelles. *Trends. Genet.* 6: 115–120.

Palmer, J. D. 1992. Mitochondrial DNA in plant systematics: Applications and limitations. Pp. 36–49 in *Molecular Systematics of Plants,* P. S. Soltis, D. E. Soltis and J. J. Doyle (eds.). Chapman & Hall, New York.

Palmer, J. D. 2003. The symbiotic birth and spread of plastids: How many times and whodunit? *J. Phycol.* 39: 4–11.

Palmer, J. D. and L. A. Herbon. 1986. Tripartite mitochondrial genomes of *Brassica* and *Raphanus*: Reversal of repeat configurations by inversion. *Nucleic Acids Res.* 14: 9755–9765.

Palmer, J. D. and L. A. Herbon. 1988. Plant mitochondrial DNA evolves rapidly in structure, but slowly in sequence. *J. Mol. Evol.* 28: 87–97.

Palmer, J. D. and C. R. Shields. 1984. Tripartite structure of the *Brassica campestris* mitochondrial genome. *Nature* 307: 437–440.

Palmer, J. D. and W. F. Thompson. 1981. Rearrangements in the chloroplast genomes of mung bean and pea. *Proc. Natl. Acad. Sci. USA* 78: 5533–5537.

Palmer, J. D. and D. Zamir. 1982. Chloroplast DNA evolution and phylogenetic relationships in *Lycopersicon. Proc. Natl. Acad. Sci. USA* 79: 5006–5010.

Palmer, J. D., C. R. Shields, D. B. Cohen and T. J. Orten. 1983. Chloroplast DNA evolution and the origin of amphidiploid *Brassica* species. *Theoret. Appl. Genet.* 65: 181–189.

Palmer, J. D., R. A. Jorgenson and W. F. Thompson. 1985. Chloroplast DNA variation and evolution in *Pisum*: Patterns of change and phylogenetic analysis. *Genetics* 109: 195–213.

Palmer, J. D., R. K. Jansen, H. J. Michaels, M. W. Chase and J. R. Manhart. 1988a. Chloroplast DNA variation and plant phylogeny. *Annals Missouri Bot. Gardens* 75: 1180–1206.

Palmer, J. D., B. Osorio and W. F. Thompson. 1988b. Evolutionary significance of inversions in legume chloroplast DNAs. *Curr. Genet.* 14: 65–74.

Palmer, K. S., D. C. Rostal, J. S. Grumbles and M. Mulvey. 1998. Long-term sperm storage in the desert tortoise (*Gopherus agassizii*). *Copeia* 1998: 702–705.

Palomeres, F., J. A. Godoy, A. Piraz, S. J. O'Brien and W. E. Johnson. 2002. Faecal genetic analysis to determine the presence and distribution of elusive carnivores: Design and feasibility for the Iberian lynx. *Mol. Ecol.* 11: 2171–2182.

Palsbøll, P. J. 1999. Genetic tagging: Contemporary molecular ecology. *Biol. J. Linn. Soc.* 68: 3–22.

Palsbøll, P. J. and 8 others. 1995. Distribution of mtDNA haplotypes in North Atlantic humpback whales: The influence of behaviour on population structure. *Mar. Ecol. Progr. Ser.* 116: 1–10.

Palsbøll, P. J., M. P. Heide-Jørgensen and R. Dietz. 1997a. Population structure and seasonal movements of narwhals, *Monodon monoceros*, determined from mtDNA analysis. *Heredity* 78: 284–292.

Palsbøll, P. J. and 18 others. 1997b. Genetic tagging of humpback whales. *Nature* 388: 676–679.

Palumbi, S. R. 1994. Reproductive isolation, genetic divergence, and speciation in the sea. *Annu. Rev. Ecol. Syst.* 25: 547–572.

Palumbi, S. R. 1995. Using genetics as an indirect estimator of larval dispersal. Pp. 369–387 in *Marine Invertebrate Larvae*, L. McEdward (ed.). CRC Press, Boca Raton, Florida.

Palumbi, S. R. 1996a. Nucleic acids II: The polymerase chain reaction. Pp. 205–247 in *Molecular Systematics* (2nd Edition), D. M. Hillis, C. Moritz and B. K. Mable (eds.). Sinauer Associates, Sunderland, MA.

Palumbi, S. R. 1996b. Macrospatial genetic structure and speciation in marine taxa with high dispersal abilities. Pp. 101–117 in *Molecular Zoology*, J. D. Ferraris and S. R. Palumbi (eds.). Wiley-Liss, New York.

Palumbi, S. R. 1998. Species formation and the evolution of gamete recognition loci. Pp. 271–278 in *Endless Forms: Species and Speciation*, D. J. Howard and S. H. Berlocher (eds.). Oxford University Press, New York.

Palumbi, S. R. 1999. All males are not created equal: Fertility differences depend on bindin genotype in tropical sea urchins. *Proc. Natl. Acad. Sci. USA* 96: 12632–12637.

Palumbi, S. R. 2001. The ecology of marine protected areas. Pp. 509–530 in *Marine Ecology: The New Synthesis*, M. Bertness, S. Gaines and M. Hixon (eds.). Sinauer Associates, Sunderland, MA.

Palumbi, S. R. 2003. Population genetics, demographic connectivity, and the design of marine reserves. *Ecol. Appl.* 13: S146-S158.

Palumbi, S. R. and C. S. Baker. 1994. Contrasting population structure from nuclear intron sequences and mtDNA of humpback whales. *Mol. Biol. Evol.* 11: 426–435.

Palumbi, S. R. and J. Benzie. 1991. Large mitochondrial DNA differences between morphologically similar Penaeid shrimp. *Mol. Mar. Biol. Biotech.* 1: 27–34.

Palumbi, S. R. and A. C. Wilson. 1990. Mitochondrial DNA diversity in the sea urchins *Strongylocentrotus purpuratus* and *S. droebachiensis*. *Evolution* 44: 403–415.

Palumbi, S. R., A. Martin, S. Romano, W. O. McMillan, L. Stice and G. Grabowsky. 1991. *The Simple Fool's Guide to PCR*, Version 2. University of Hawaii Zoology Department, Honolulu, HI.

Palumbi, S. R., G. Grabowsky, T. Duda, L. Geyer and N. Tachino. 1997. Speciation and population genetic structure in tropical Pacific sea urchins. *Evolution* 51: 1506–1517.

Palumbi, S. R., F. Cipriano and M. P. Hare. 2001. Predicting nuclear gene coalescence from mitochondrial data: The three-times rule. *Evolution* 55: 859–868.

Palva, T. K. and E. T. Palva. 1985. Rapid isolation of animal mitochondrial DNA by alkaline extraction. *FEBS Letters* 192: 267–270.

Pamilo, P. 1981. Genetic organization of *Formica sanguinea* populations. *Behav. Ecol. Sociobiol.* 9: 45–50.

Pamilo, P. 1982. Genetic population structure in polygynous *Formica* ants. *Heredity* 48: 95–106.

Pamilo, P. 1984a. Genotypic correlation and regression in social groups: Multiple alleles, multiple loci and subdivided populations. *Genetics* 107: 307–320.

Pamilo, P. 1984b. Genetic relatedness and evolution of insect sociality. *Behav. Ecol. Sociobiol.* 15: 241–248.

Pamilo, P. and R. H. Crozier. 1982. Measuring genetic relatedness in natural populations: Methodology. *Theoret. Pop. Biol.* 21: 171–193.

Pamilo, P. and M. Nei. 1988. Relationships between gene trees and species trees. *Mol. Biol. Evol.* 5: 568–583.

Pamilo, P. and S.-L. Varvio-Aho. 1979. Genetic structure of nests in the ant *Formica sanguinea*. *Behav. Ecol. Sociobiol.* 6: 91–98.

Pancer, Z., H. Gershon and B. Rinkevich. 1995. Coexistence and possible parasitism of somatic and germ cell lines in chimeras of the colonial urochordate *Botryllus schlosseri*. *Biol. Bull.* 189: 106–112.

Panhuis, T. M., R. Butlin, M. Zuk and T. Tregenza. 2001. Sexual selection and speciation. *Trends Ecol. Evol.* 16: 364–371.

Pannell, J. R. 2003. Coalescence in a metapopulation with recurrent local extinction and recolonization. *Evolution* 57: 949–961.

Parham, P. and T. Ohta. 1996. Population biology of antigen presentation by MHC Class I molecules. *Science* 272: 67–72.

Parker, G. A. 1970. Sperm competition and its evolutionary consequences in the insects. *Biol. Rev.* 45: 525–567.

Parker, G. A. 1984. Sperm competition and the evolution of male mating strategies. Pp. 1–60 in *Sperm Competition and the Evolution of Animal Mating Systems*, R. L. Smith (ed.). Academic Press, New York.

Parker, K. C. and J. L. Hamrick. 1992. Genetic diversity and clonal structure in a columnar cactus, *Lophocereus schottii*. *Am. J. Bot.* 79: 86–96.

Parker, M. A. 2002. Conflicting phylogeographic patterns in rRNA and *nifD* indicate regionally restricted gene transfer in *Bradyrhizobium*. *Microbiology* 148: 2557–2565.

Parker, M. A. and J. M. Spoerke. 1998. Geographic structure of lineage associations in a plant-bacterial mutualism. *J. Evol. Biol.* 11: 549–562.

Parker, P. G., T. A. Waite and T. Peare. 1996. Paternity studies in animal populations. Pp. 413–423 in *Molecular Genetic Approaches in Animal Conservation*, T. B. Smith and R. K. Wayne (eds.). Oxford University Press, New York.

Parker, P. G., A. A. Snow, M. D. Schug, G. C. Booton and P. A. Fuerst. 1998. What molecules can tell us about populations: Choosing and using a molecular marker. *Ecology* 79: 261–382.

Parkinson, C. L., K. L. Adams and J. D. Palmer. 1999. Multigene analyses identify the three earliest lineages of extant flowering plants. *Current Biol.* 9: 1485–1488.

Pascual, M., J. Balanya, A. Latorre and L. Serra. 1997. Diagnosis of sibling species of *Drosophila* involved in the colonization of North America by *D. subobscura*. *Mol. Ecol.* 6: 293–296.

Pasdar, M., D. P. Philipp and G. S. Whitt. 1984. Linkage relationships of nine enzyme loci in sunfishes (*Lepomis*; Centrarchidae). *Genetics* 107: 435–446.

Pastene, L. A., K. Numachi and K. Tsukamoto. 1991. Examination of reproductive success of transplanted stocks in an amphidromous fish, *Plecoglossus altivelis* (Temmink et Schlegel) using mitochondrial DNA and isozyme markers. *J. Fish Biol.* 39A: 93–100.

Pastorini, J., U. Thalmann and R. D. Martin. 2003. A molecular approach to comparative phylogeography of extant malagasy lemurs. *Proc. Natl. Acad. Sci. USA* 100: 5879–5884.

Patenaude, N. J., J. S. Quinn, P. Beland, M. Kingsley and B. N. White. 1994. Genetic variation of the St. Lawrence beluga whale population assessed by DNA fingerprinting. *Mol. Ecol.* 3: 375–381.

Paterson, A. H. (ed.). 1998. *Molecular Dissection of Complex Traits*. CRC Press, Boca Raton, FL.

Paterson, A. H., E. Lander, J. Hewitt, S. Peterson, S. Lincoln and S. Tanksley. 1988. Resolution of quantitative traits into Mendelian factors by using a complete linkage map of restriction length polymorphisms. *Nature* 335: 721–726.

Paterson, A. M., G. P. Wallis, L. J. Wallis and R. D. Gray. 2000. Seabird and louse coevolution: Complex histories revealed by 12S rRNA sequences and reconciliation analysis. *Syst. Biol.* 49: 383–399.

Paterson, H. E. H. 1985. The recognition concept of species. Pp. 21–29 in *Species and Speciation*, E. S. Vrba (ed.). Transvaal Museum Monograph No. 4, Pretoria, South Africa.

Paton, T., O. Haddrath and A. J. Baker. 2002. Complete mitochondrial DNA genome sequences show that modern birds are not descended from transitional shorebirds. *Proc. R. Soc. Lond. B* 269: 839–846.

Patterson, C. (ed.) 1987. *Molecules and Morphology in Evolution: Conflict or Compromise?* Cambridge University Press, Cambridge, England.

Patton, J. C. and J. C. Avise. 1983. An empirical evaluation of qualitative Hennigian analyses of protein electrophoretic data. *J. Mol. Evol.* 19: 244–254.

Patton, J. L. and M. S. Hafner. 1983. Biosystematics of the native rodents of the Galapagos archipelago. Pp. 539–568 in *Patterns of Evolution in Galapagos Organisms*, R. I. Bowman, M. Berson and A. E. Leviton (eds.). American Association for the Advancement of Science, San Francisco, CA.

Patton, J. L. and M. F. Smith. 1981. Molecular evolution in *Thomomys*: Phyletic systematics, paraphyly, and rates of evolution. *J. Mammal.* 62: 493–500.

Patton, J. L. and M. F. Smith. 1989. Population structure and the genetic and morphological divergence among pocket gopher species (genus *Thomomys*). Pp. 284–304 in *Speciation and Its Consequences*, D. Otte and J. A. Endler (eds.). Sinauer Associates, Sunderland, MA.

Patton, J. L., M. N. F. da Silva and J. R. Malcolm. 1994a. Hierarchical genetic structure and gene flow in three sympatric species of Amazonian rodents. *Mol. Ecol.* 5: 229–238.

Patton, J. L., M. N. F. da Silva and J. R. Malcolm. 1994b. Gene genealogy and differentiation among arboreal spiny rats (Rodentia: Echimyidae) of the Amazon basin: A test of the riverine barrier hypothesis. *Evolution* 48: 1314–1323.

Patton, J. L, M. N. F. da Silva, M. C. Lara and M. A. Mustrangi. 1997. Diversity, differentiation, and the historical biogeography of non-volant small mammals of the neotropical forests. Pp. 455–465 in *Tropical Forest Remnants: Ecology, Management, and Conservation of Fragmented Communities*, W. F. Laurance and R. O. Bierregaard Jr. (eds.). University of Chicago Press, Chicago, IL.

Paxinos, E. E., C. McIntosh, K. Ralls and R. Fleischer. 1997. A noninvasive method for distinguishing among canid species: Amplification and enzyme restriction of DNA from dung. *Mol. Ecol.* 6: 483–486.

Paxinos, E. E., H. F. James, S. L. Olson, J. D. Ballou, L. A. Leonard and R. C. Fleischer. 2002a. Prehistoric decline of genetic diversity in the nene. *Science* 296: 1827.

Paxinos, E. E., H. F. James, S. L. Olson, M. D. Sorenson, J. Jackson and R. C. Fleischer. 2002b. MtDNA from fossils reveals a radiation of Hawaiian geese recently derived from the Canada goose (*Branta canadensis*). *Proc. Natl. Acad. Sci. USA* 99: 1399–1404.

Paxton, R. J., M. Ayasse, J. Field and A. Soro. 2002. Complex sociogenetic organization and reproductive skew in a primitively eusocial sweat bee, *Lasioglossum malachurum*, as revealed by microsatellites. *Mol. Ecol.* 11: 2405–2416.

Pearce, J. M., R. L. Fields and K. T. Scribner. 1997. Nest materials as a source of genetic data for avian ecological studies. *J. Field Ornithol.* 68: 471–481.

Pearse, D. E. and J. C. Avise. 2001. Turtle matings systems: Behavior, sperm storage, and genetic paternity. *J. Heredity* 92: 206–211.

Pearse, D. E., F. J. Janzen and J. C. Avise. 2001a. Genetic markers substantiate long-term storage and utilization of sperm by female painted turtles. *Heredity* 86: 378–384.

Pearse, D. E., C. M. Eckerman, F. J. Janzen and J. C. Avise. 2001b. A genetic analogue of 'mark-recapture' methods for estimating population size: An approach based on molecular parentage assessments. *Mol. Ecol.* 10: 2711–2718.

Pearse, D. E., F. J. Janzen and J. C. Avise. 2002. Multiple paternity, sperm storage, and reproductive success of female and male painted turtles (*Chrysemys picta*) in nature. *Behav. Ecol. Sociobiol.* 51: 164–171.

Pearson, B. 1983. Intra-colonial relatedness amongst workers in a population of nests of the polygynous ant, *Myrmica rubra* Latreille. *Behav. Ecol. Sociobiol.* 12: 1–4.

Pearson, C. V. M., A. D. Rogers and M. Sheader. 2002. The genetic structure of the rare lagoonal sea anemone, *Nematostella vectensis* Stephenson (Cnidaria; Anthozoa) in the United Kingdom based on RAPD analysis. *Mol. Ecol.* 11: 2285–2293.

Pedersen, K., J. Arlinger, L. Hallbeck and C. Pettersson. 1996. Diversity and distribution of subterranean bacteria in groundwater at Oklo in Gabon, Africa, as determined by 16S rRNA gene sequencing. *Mol. Ecol.* 5: 427–436.

Peichel, C. L., K. S. Nereng, K. A. Ohgi, B. L. E. Cole, P. F. Colosimo, C. A. Buerkle, D. Schluter and D. M. Kingsley. 2001. The genetic architecture of divergence between threespine stickleback species. *Nature* 414: 901–905.

Pella, J. J. and G. B. Milner. 1987. Use of genetic marks in stock composition analysis. Pp. 247–276 in *Population Genetics and Fishery Management*, N. Ryman and F. Utter (eds.). University of Washington Press, Seattle, WA.

Pemberton, J. M., D. W. Coltman, T. N. Coulson and J. Slate. 1999. Using microsatellites to measure the fitness consequences of inbreeding and outbreeding. Pp. 151–164 in *Microsatellites: Evolution and Application*, D. B. Goldstein and C. Schlötterer (eds.). Oxford University Press, Oxford, England.

Pennisi, E. 2001. Preparing the ground for a modern 'Tree of Life.' *Science* 293: 1979–1980.

Pereira, S. L. 2000. Mitochondrial genome organization and vertebrate phylogenetics. *Genet. Mol. Biol.* 23: 745–752.

Peres, C. A., J. L. Patton and M. N. F. da Silva. 1996. Riverine barriers and gene flow in Amazonian saddle-backed tamarins. *Folia Primatoligica* 67: 113–124.

Perez, D. E., C.-I. Wu, N. A. Johnson and M.-L. Wu. 1993. Genetics of reproductive isolation in the *Drosophila simulans* clade: DNA marker-assisted mapping and characterization of a hybrid-male sterility gene. *Genetics* 133: 261–275.

Pérez, T., J. Albornoz and A. Domínguez. 1998. An evaluation of RAPD fragment reproducibility and nature. *Mol. Ecol.* 7: 1347–1357.

Pergams, O. R. W., W. M. Barnes and D. Nyberg. 2003. Rapid change in mouse mitochondrial DNA. *Nature* 423: 397.

Perkins, H. D. and A. J. Howells. 1992. Genomic sequences with homology to the *P* element of *Drosophila melanogaster* occur in the blowfly *Lucilia cuprina*. *Proc. Natl. Acad. Sci. USA* 89: 10753–10757.

Perkins, S. L., S. M. Osgood and J. J. Schall. 1998. Use of PCR for detection of subpatent infections of lizard malaria: Implications for epizootiology. *Mol. Ecol.* 7: 1587–1590.

Perlman, S. J., G. S. Spicer, D. D. Shoemaker and J. Jaenike. 2003. Associations between mycophagous *Drosophila* and their *Howardula* nematode parasites: A worldwide phylogenetic shuffle. *Mol. Ecol.* 12: 237–249.

Pernin, P., A. Ataya and M. L. Cariou. 1992. Genetic structure of natural populations of the free-living amoeba, *Naegleria lovaniensis*. Evidence for sexual reproduction. *Heredity* 68: 173–181.

Perry, W. L., J. L. Feder, G. Dwyer and D. M. Lodge. 2001. Hybrid zone dynamics and species replacement between *Orconectes* crayfishes in a northern Wisconsin lake. *Evolution* 55: 1153–1166.

Peterson, K. J. and D. J. Eernisse. 2001. Animal phylogeny and the ancestry of bilaterians: Inferences from morphology and 18S rDNA gene sequences. *Evol. and Develop.* 3: 170–205.

Petit, R. and G. G. Verdramin. 2004. Plant phylogeography based on organelle genes: An introduction. In *Phylogeography of Southern European Refugia*, S. Weiss and N. Ferrand (eds.). Kluwer, Dordrecht, The Netherlands, *in press*.

Petit, R., R. Bialozyt, S. Brewer, R. Cheddadi and B. Comps. 2001. From spatial patterns of genetic diversity to postglacial migration processes in forest trees. Pp. 295–318 in *Integrating Ecology and Evolution in a Spatial Context*, J. Silvertown and J. Antonovics (eds.). Blackwell, Oxford, England.

Petit, R. J. and 29 others. 2002. Chloroplast DNA variation in European white oaks: Phylogeography and patterns of diversity based on data from over 2600 populations. *Forest Ecol. and Mgt.* 156: 5–26.

Petit, R. J. and 27 others. 2003a. Identification of refugia and post-glacial colonisation routes of European white oaks based on chloroplast DNA and fossil pollen evidence. *Forest Ecol. and Mgt.* 156: 49–74.

Petit, R. J. and 16 others. 2003b. Glacial refugia: Hotspots but not melting pots of genetic diversity. *Science* 300: 1563–1565.

Petren, K., B. R. Grant and P. R. Grant. 1999. A phylogeny of Darwin's finches based on microsatellite DNA length variation. *Proc. R. Soc. Lond. B* 266: 321–329.

Petri, B., S. Pääbo, A. von Haeseler and D. Tautz. 1997. Paternity assessment and population subdivision in a natural population of the larger mouse-eared bat *Myotis myotis*. *Mol. Ecol.* 6: 235–242.

Petrie, M. and B. Kempenaers. 1998. Extra-pair paternity in birds: Explaining variation between species and populations. *Trends Ecol. Evol.* 13: 52–58.

Petrie, M. and A. P. Møller. 1991. Laying eggs in others' nests: Intraspecific brood parasitism in birds. *Trends Ecol. Evol.* 6: 315–320.

Petrie, M., C. Doums and A. P. Møller. 1998. The degree of extra-pair paternity increases with genetic variability. *Proc. Natl. Acad. Sci. USA* 95: 9390–9395.

Petter, S. C., D. B. Miles and M. M. White. 1990. Genetic evidence of mixed reproductive strategy in a monogamous bird. *The Condor* 92: 702–708.

Pettigrew, J. D. 1986. Flying primates? Megabats have the advanced pathway from eye to midbrain. *Science* 231: 1304–1306.

Pettigrew, J. D. 1991. Wings or brain? Convergent evolution in the origins of bats. *Syst. Zool.* 40: 199–216.

Pfrender, M. E., K. Spitze, J. Hicks, K. Morgan, L. Latta and M. Lynch. 2000. Lack of concordance between genetic diversity estimates at the molecular and quantitative-trait levels. *Cons. Genet.* 1: 263–269.

Phelps, S. R. and F. W. Allendorf. 1983. Genetic identity of pallid and shovelnose sturgeon (*Scaphirhynchus albus* and *S. platorynchus*). *Copeia* 1983: 696–700.

Philipp, D. P. and M. R. Gross. 1994. Genetic evidence for cuckoldry in bluegill *Lepomis macrochirus*. *Mol. Ecol.* 3: 563–569.

Philipp, D. P., W. F. Childers and G. S. Whitt. 1981. Management implications for different genetic stocks of largemouth bass (*Micropterus salmoides*) in the United States. *Can. J. Fish. Aquat. Sci.* 38: 1715–1723.

Philipp, D. P., H. R. Parker and G. S. Whitt. 1983a. Evolution of gene regulation: Isozymic analysis of patterns of gene expression during hybrid fish development. *Isozymes* X: 193–237.

Philipp, D. P., W. F. Childers and G. S. Whitt. 1983b. A biochemical genetic evaluation of the northern and Florida subspecies of largemouth bass. *Trans. Am. Fish. Soc.* 112: 1–20.

Phillips, M. K. and V. G. Henry. 1992. Comments on red wolf taxonomy. *Cons. Biol.* 6: 596–599.

Phillips, R. L. and I. K. Vasil. 2001. *DNA-based Markers in Plants.* Kluwer, Boston, MA.

Piano, F., E. M. Craddock and M. P. Kambysellis. 1997. Phylogeny of the island populations of the Hawaiian *Drosophila grimshawi* complex: Evidence from combined data. *Mol. Phylogen. Evol.* 7: 173–184.

Pichler, F. B., S. M. Dawson, E. Slooten and C. S. Baker. 1998. Geographic isolation of Hector's dolphin populations as described by mitochondrial DNA sequences. *Cons. Biol.* 12: 676–682.

Pichler, F. B., M. L. Dalebout and C. S. Baker. 2001. Nondestructive DNA extraction from sperm whale teeth and scrimshaw. *Mol. Ecol. Notes* 1: 106–109.

Pierce, B. A. and J. B. Mitton. 1982. Allozyme heterozygosity and growth in the tiger salamander, *Ambystoma tigrinum. J. Heredity* 73: 250–253.

Piertney, S. B., A. D. C. MacColl, P. J. Bacon, P. A. Racey, X. Lambin and J. F. Dallas. 2000. Matrilineal genetic structure and female-mediated gene flow in red grouse (*Lagopus lagopus scoticus*): An analysis using mitochondrial DNA. *Evolution* 54: 279–289.

Pigeon, D., A. Chouinard and L. Bernatchez. 1997. Multiple modes of speciation involved in the parallel evolution of sympatric morphotypes of lake whitefish (*Coregonus clupeaformis*, Salmonidae). *Evolution* 51: 196–205.

Pilbeam, D. 1984. The descent of hominoids and hominids. *Sci. Am.* 250 (3): 84–96.

Pinjon, E., D. Sullivan, I. Salkin, D. Shanley and D. Coleman. 1998. Simple, inexpensive, reliable methods for differentiation of *Candida dublinensis* from *Candida albicans. J. Clin. Microbiol.* 36: 2093–2095.

Place, A. R. and D. A. Powers. 1979. Genetic variation and relative catalytic efficiencies: Lactate dehydrogenase B allozymes of *Fundulus heteroclitus. Proc. Natl. Acad. Sci. USA* 76: 2354–2358.

Place, A. R. and D. A. Powers. 1984. The LDH-B allozymes of *Fundulus heteroclitus*: II. Kinetic analyses. *J. Biol. Chem.* 259: 1309–1318.

Planes, S. and P. J. Doherty. 1997. Genetic and color interactions at a contact zone of *Acanthochromis polyacanthus*: A marine fish lacking pelagic larvae. *Evolution* 51: 1232–1243.

Plante, Y., P. T. Boag and B. N. White. 1989. Microgeographic variation in mitochondrial DNA of meadow voles (*Microtus pennsylvanicus*) in relation to population density. *Evolution* 43: 1522–1537.

Pleasants, J. M. and J. F. Wendel. 1989. Genetic diversity in a clonal narrow endemic, *Erythronium propullans*, and its widespread progenitor, *Erythronium albidum. Am. J. Bot.* 76: 1136–1151.

Pochon, X., J. Pawlowski, L. Zaninetti and R. Rowan. 2001. High genetic diversity and relative specificity among *Symbiodinium*-like endosymbiotic dinoflagellates in soritid foraminiferans. *Marine Biol.* 139: 1069–1078.

Podar, M., S. H. D. Haddock, M. L. Sogin and G. R. Harbison. 2001. A molecular phylogenetic framework for the phylum Ctenophora using 18S rRNA genes. *Mol. Phylogen. Evol.* 21: 218–230.

Poinar, H. N. 2002. The genetic secrets some fossils hold. *Accounts Chem. Res.* 35: 676–684.

Poinar, H. N., M. Höss, J. L. Bada and S. Pääbo. 1996. Amino acid racemization and the preservation of ancient DNA. *Science* 272: 864–866.

Poinar, H. N. and 8 others. 1998. Molecular coproscopy: Dung and diet of the extinct ground sloth *Nothrotheriops shastensis. Science* 281: 402–406.

Poinar, H. N., M. Kuch and S. Pääbo. 2001a. Molecular analyses of oral polio vaccine samples. *Science* 292: 743–744.

Poinar, H. N. and 9 others. 2001b. A molecular analysis of dietary diversity for three archaic Native Americans. *Proc. Natl. Acad. Sci. USA* 98: 4317–4322.

Polans, N. O. 1983. Enzyme polymorphisms in Galapagos finches. Pp. 219–236 in *Patterns of Evolution in Galapagos Organisms*, R. I. Bowman, M. Berson and A. E. Leviton (eds.). American Association for the Advancement of *Science*, San Francisco, CA.

Pollock, K. H., J. D. Nichols, C. Brownie and J. E. Hines. 1990. Statistical inference for capture-recapture experiments. *Wildl. Monogr.* 107: 1–97.

Pongratz, N. and N. K. Michiels. 2003. High multiple paternity and low last-male sperm precedence in a hermaphroditic planarian flatworm: Consequences for reciprocity patterns. *Mol. Ecol.* 12: 1425–1433.

Popadic, A. and W. W. Anderson. 1994. The history of a genetic system. *Proc. Natl. Acad. Sci. USA* 91: 6819–6823.

Popadic, A. and W. W. Anderson. 1995. Evidence for gene conversion in the amylase multigene family of *Drosophila pseudoobscura. Mol. Biol. Evol.* 12: 564–572.

Popadic, A., D. Popadic and W. W. Anderson. 1995. Interchromosomal exchange of genetic information between gene arrangements on the third chromosome of *Drosophila pseudoobscura. Mol. Biol. Evol.* 12: 938–943.

Pope, L. C., A. Sharp and C. Moritz. 1996. Population structure of the yellow-footed rock-wallaby *Petrogale xanthopus* (Gray, 1854) inferred from mtDNA sequences and microsatellite loci. *Mol. Ecol.* 5: 629–640.

Pope, L. C., A. Estoup and C. Moritz. 2000. Phylogeography and population structure of an ecotonal marsupial, *Bettongia tropica*, determined using mtDNA and microsatellites. *Mol. Ecol.* 9: 2041–2053.

Pope, T. R. 1990. The reproductive consequences of male cooperation in the red howler monkey: Paternity exclusion in multi-male and single-male troops using genetic markers. *Behav. Ecol. Sociobiol.* 27: 439–446.

Pope, T. R. 1996. Socioecology, population fragmentation, and patterns of genetic loss in endangered primates. Pp. 119–159 in *Conservation Genetics: Case Histories from*

Nature, J. C. Avise and J. L. Hamrick (eds.). Chapman & Hall, New York.

Popper, K. R. 1968. *The Logic of Scientific Discovery*. Harper Torchbooks, New York.

Porteous, L. A., R. J. Seidler and L. S. Watrud. 1997. An improved method for purifying DNA from soil for polymerase chain reaction amplification and molecular ecology applications. *Mol. Ecol.* 6: 787–791.

Porter, B. A., A. C. Fiumera and J. C. Avise. 2002. Egg mimicry and allopaternal care: Two mate-attracting tactics by which nesting striped darter (*Etheostoma virgatum*) males enhance reproductive success. *Behav. Ecol. Sociobiol.* 51: 350–359.

Porterfield, J. C., L. M. Page and T. J. Near. 1999. Phylogenetic relationships among fantail darters (Percidae: *Etheostoma: Catonotus*): Total evidence analysis of morphological and molecular data. *Copeia* 1999: 551–564.

Posada, D. 2002. Evaluation of methods for detecting recombination from DNA sequences: Empirical data. *Mol. Biol. Evol.* 19: 708–717.

Potts, B. M. and J. B. Reid. 1985. Analysis of a hybrid swarm between *Eucalyptus risdonii* and *E. amygdalina*. *Aust. J. Bot.* 33: 543–562.

Potvin, C. and L. Bernatchez. 2001. Lacustrine spatial distribution of landlocked Atlantic salmon populations assessed across generations by multilocus individual assignment and mixed-stock analysis. *Mol. Ecol.* 10: 2375–2388.

Powell, J. R. 1971. Genetic polymorphisms in varied environments. *Science* 174: 1035–1036.

Powell, J. R. 1975. Protein variation in natural populations of animals. *Evol. Biol.* 3: 79–119.

Powell, J. R. 1983. Interspecific cytoplasmic gene flow in the absence of nuclear gene flow: Evidence from *Drosophila*. *Proc. Natl. Acad. Sci. USA* 80: 492–495.

Powell, J. R. 1991. Monophyly/paraphyly/polyphyly and gene/species trees: An example from *Drosophila*. *Mol. Biol. Evol.* 8: 892–896.

Powell, J. R. and C. E. Taylor. 1979. Genetic variation in ecologically diverse environments. *Am. Sci.* 67: 590–596.

Powell, J. R. and M. C. Zuninga. 1983. A simplified procedure for studying mtDNA polymorphisms. *Biochem. Genet.* 21: 1051–1055.

Powell, J. R., W. J. Tabachnick and J. Arnold. 1980. Genetics and the origin of a vector population: *Aedes aegypti*, a case study. *Science* 208: 1385–1387.

Powell, J. R., A. Caccone, G. D. Amato and C. Yoon. 1986. Rates of nucleotide substitutions in *Drosophila* mitochondrial DNA and nuclear DNA are similar. *Proc. Natl. Acad. Sci. USA* 83: 9090–9093.

Powers, D. A. 1991. Evolutionary genetics of fish. *Adv. Genet.* 29: 120–228.

Powers, D. A. and 6 others. 1986. Genetic variation in *Fundulus heteroclitus*: Geographic distribution. *Am. Zoo.* 26: 131–144.

Powers, D. A., T. Lauerman, D. Crawford and L. DiMichele. 1991a. Genetic mechanisms for adapting to a changing environment. *Annu. Rev. Genet.* 25: 629–659.

Powers, D. A., T. Lauerman, D. Crawford, M. Smith, I. González-Villaseñor and L. DiMichele. 1991b. The evolutionary significance of genetic variation at enzyme synthesizing loci in the teleost *Fundulus heteroclitus*. *J. Fish Biol.* 39A: 169–184.

Prager, E. M. and A. C. Wilson. 1975. Slow evolutionary loss of the potential for interspecific hybridization in birds: A manifestation of slow regulatory evolution. *Proc. Natl. Acad. Sci. USA* 72: 200–204.

Prager, E. M. and A. C. Wilson. 1976. Congruency of phylogenies derived from different proteins. *J. Mol. Evol.* 9: 45–57.

Prager, E. M. and A. C. Wilson. 1978. Construction of phylogenetic trees for proteins and nucleic acids: Empirical evaluation of alternative matrix methods. *J. Mol. Evol.* 11: 129–142.

Prager, E. M. and A. C. Wilson. 1993. Information content of immunological distances. *Methods Enzymol.* 224: 140–152.

Prager, E. M., A. H. Brush, R. A. Nolan, M. Nakanishi and A. C. Wilson. 1974. Slow evolution of transferrin and albumin in birds according to micro-complement fixation analysis. *J. Mol. Evol.* 3: 243–262.

Prager, E. M., A. C. Wilson, D. T. Osuga and R. E. Feeney. 1976. Evolution of flightless land birds on southern continents: Transferrin comparison shows monophyletic origin of ratites. *J. Mol. Evol.* 8: 283–294.

Prager, E. M., A. C. Wilson, J. M. Lowenstein and V. M. Sarich. 1980. Mammoth albumin. *Science* 209: 287–289.

Prager, E. M. and 8 others. 1993. Mitochondrial DNA sequence diversity and the colonization of Scandinavia by house mice from East Holstein. *Biol. J. Linn. Soc.* 50: 85–122.

Prakash, S. 1977. Allelic variants at the xanthine dehydrogenase locus affecting enzyme activity in *Drosophila pseudoobscura*. *Genetics* 87: 159–168.

Presgraves, D. C. 1997. Patterns of postzygotic isolation in Lepidoptera. *Evolution* 56: 1168–1183.

Presgraves, D. C., L. Balagopalan, S. M. Abmayr and H. A. Orr. 2003. Adaptive evolution drives divergence of a hybrid inviability gene between two species of *Drosophila*. *Nature* 423: 715–719.

Preziosi, R. F. and D. J. Fairbairn. 1992. Genetic population structure and levels of gene flow in the stream dwelling waterstrider, *Aquarius* (=*Gerris*) *remigis* (Hemiptera: Gerridae). *Evolution* 46: 430–444.

Price, D. K., G. E. Collier and C. F. Thompson. 1989. Multiple parentage in broods of house wrens: Genetic evidence. *J. Heredity* 80: 1–5.

Price, J. P. and D. A. Clague. 2002. How old is the Hawaiian biota? Geology and phylogeny suggest recent divergence. *Proc. R. Soc. Lond. B* 269: 2429–2435.

Price, M. V. and N. M. Wasser. 1979. Pollen dispersal and optimal outcrossing in *Delphinium nelsoni*. *Nature* 277: 294–297.

Primmer, C. R., H. Ellegren, N. Saino and A. P. Møller. 1996. Directional evolution in germline microsatellite mutations. *Nature Genet.* 13: 391–393.

Pring, D. R. and D. M. Lonsdale. 1985. Molecular biology of higher plant mitochondrial DNA. *Int. Rev. Cytol.* 97: 1–46.

Prinsloo, P. and T. J. Robinson. 1992. Geographic mitochondrial DNA variation in the rock hyrax, *Procavia capensis*. *Mol. Biol. Evol.* 9: 447–456.

Pritchard, J. K., M. Stephens and P. Donnelly. 2000. Inference of population structure using multilocus genotypic data. *Genetics* 155: 945–959.

Prodöhl, P. A., J. B. Taggart and A. Ferguson. 1992. Genetic variability within and among sympatric brown trout (*Salmo trutta*) populations: Multi-locus DNA fingerprint analysis. *Hereditas* 117: 45–50.

Prodöhl, P. A., W. J. Loughry, C. M. McDonough, W. S. Nelson and J. C. Avise. 1996. Molecular documentation of polyembryony and the micro-spatial dispersion of clonal sibships in the nine-banded armadillo, *Dasypus novemcinctus*. *Proc. R. Soc. Lond. B* 263: 1643–1649.

Prodöhl, P. A., W. J. Loughry, C. M. McDonough, W. S. Nelson and J. C. Avise. 1998. Genetic maternity and paternity in a local population of armadillos assessed by microsatellite DNA markers and field data. *Am. Nat.* 151: 7–19.

Promislow, D. E. L., I. K. Jordan and J. F. McDonald. 1999. Genomic demography: A life-history analysis of transposable element evolution. *Proc. R. Soc. Lond. B* 266: 1555–1560.

Provine, W. B. 1986. *Sewell Wright and Evolutionary Biology*. University of Chicago Press, Chicago, IL.

Provine, W. B. 1989. Founder effects and genetic revolutions in microevolution and speciation: An historical perspective. Pp. 43–76 in *Genetics, Speciation and the Founder Principle*, L. V. Giddings, K. Y. Kaneshiro and W. W. Anderson (eds.). Oxford University Press, New York.

Pühler, G. and 7 others. 1989. Archaebacterial DNA-dependent RNA polymerases testify to the evolution of the eukaryotic nuclear genome. *Proc. Natl. Acad. Sci. USA* 86: 4569–4573.

Pujol, C. and 8 others. 1993. The yeast *Candida albicans* has a clonal mode of reproduction in a population of infected human immunodeficiency virus-positive patients. *Proc. Natl. Acad. Sci. USA* 90: 9456–9459.

Pullium, H. R. 1988. Sources, sinks, and population regulation. *Am. Nat.* 132: 652–661.

Puorto, G., M. da Graca Salomao, R. D. G. Theakston, R. S. Thorpe, D. A. Warrell and W. Wüster. 2001. Combining mitochondrial DNA sequences and morphological data to infer species boundaries: Phylogeography of lanceheaded pitvipers in the Brazilian Atlantic forest, and the status of *Bothrops pradoi* (Squamata: Serpentes: Viperidae). *J. Evol. Biol.* 14: 527–538.

Purugganan, M. and G. Gibson (organizers). 2003. Genes in ecology (special issue). *Mol. Ecol.* 12: 1109–1337.

Purugganan, M. D. and J. I. Suddith. 1998. Molecular population genetics of the *Arabidopsis* CAULIFLOWER regulatory gene: Nonneutral evolution and naturally occurring variation in floral homeotic evolution. *Proc. Natl. Acad. Sci. USA* 95: 8130–8134.

Purugganan, M. D. and S. R. Wessler. 1995. Transposon signatures: Species-specific molecular markers that utilize a class of multiple-copy nuclear DNA. *Mol. Ecol.* 4: 265–269.

Purvis, A. 1995. A modification to Baum and Ragan's method for combining phylogenetic trees. *Syst. Biol.* 44: 251–255.

Purvis, A. and A. Rambaut. 1995. Comparative Analyses by Independent Contrasts (CAIC): An Apple Macintosh application for analysing comparative data. *Computer Appl. Biosci.* 11: 247–251.

Qian, J., S.-W. Kwon and M. A. Parker. 2003. rRNA and *nifD* phylogeny of *Bradyrhizobium* from sites across the Pacific Basin. *FEMS Microbiol. Letters* 219: 159–165.

Quattro, J. M. and R. C. Vrijenhoek. 1989. Fitness differences among remnant populations of the endangered Sonoran topminnow. *Science* 245: 976–978.

Quattro, J. M., J. C. Avise and R. C. Vrijenhoek. 1991. Molecular evidence for multiple origins of hybridogenetic fish clones (Poeciliidae: *Poeciliopsis*). *Genetics* 127: 391–398.

Quattro, J. M., J. C. Avise and R. C. Vrijenhoek. 1992a. An ancient clonal lineage in the fish genus *Poeciliopsis* (Atheriniformes: Poeciliidae). *Proc. Natl. Acad. Sci. USA* 89: 348–352.

Quattro, J. M., J. C. Avise and R. C. Vrijenhoek. 1992b. Mode of origin and sources of genotypic diversity in triploid fish clones (*Poeciliopsis*: Poeciliidae). *Genetics* 130: 621–628.

Quattro, J. M., H. A. Woods and D. A. Powers. 1993. Sequence analysis of teleost retina-specific lactate dehydrogenase C: Evolutionary implications for the vertebrate lactate dehydrogenase gene family. *Proc. Natl. Acad. Sci. USA* 90: 242–246.

Queller, D. C. 1989. Inclusive fitness in a nutshell. *Oxford Surv. Evol. Biol.* 6: 73–109.

Queller, D. C. 1996. The measurement and meaning of inclusive fitness. *Anim. Behav.* 51: 229–232.

Queller, D. C. 2000. Relatedness and the fraternal major transitions. *Proc. R. Soc. Lond. B* 355: 1647–1655.

Queller, D. C. and K. F. Goodnight. 1989. Estimating relatedness using genetic markers. *Evolution* 43: 258–275.

Queller, D. C. and J. E. Strassmann. 1998. Kin selection and social insects. *BioScience* 48: 165–175.

Queller, D. C. and J. E. Strassmann. 2002. The many selves of social insects. *Science* 296: 311–313.

Queller, D. C., J. E. Strassmann and C. R. Hughes. 1988. Genetic relatedness in colonies of tropical wasps with multiple queens. *Science* 242: 1155–1157.

Queller, D. C., J. E. Strassmann and C. R. Hughes. 1993. Microsatellites and kinship. *Trends Ecol. Evol.* 8: 285–288.

Queller, D. C., F. Zacchi, R. Cervo, S. Turillazzi, M. T. Henshaw, L. A. Santorelli and J. E. Strassmann. 2000. Unrelated helpers at the nest. *Nature* 405: 784–787.

Queller, D. C., E. Ponte, S. Bozzaro and J. E. Strassmann. 2003. Single-gene greenbeard effects in the social amoeba *Dictyostelium discoideum*. *Science* 299: 105–106.

Queney, G., N. Ferrand, S. Weiss, F. Mougel and M. Monnerot. 2001. Stationary distributions of microsatellite loci between divergent population groups of the European rabbit (*Oryctolagus cuniculus*). *Mol. Biol. Evol.* 18: 2169–2178.

Quesada, H., C. M. Benyon and D. O. F. Skibinski. 1995. A mitochondrial DNA discontinuity in the mussel *Mytilus galloprovincialis* Lmk: Pleistocene vicariance biogeography and secondary introgradation. *Mol. Biol. Evol.* 12: 521–524.

Quesada, H., C. Gallagher, D. A. G. Skibinski and D. O. F. Skibinski. 1998. Patterns of polymorphism and gene flow of gender-associated mitochondrial DNA lineages in European mussel populations. *Mol. Ecol.* 7: 1041–1051.

Questiau, S. 1999. How can sexual selection promote population divergence? *Ethl. Ecol. Evol.* 11: 313–324.

Questiau, S., M.-C. Eybert, A. R. Gaginskaya, L. Gielly and P. Taberlet. 1998. Recent divergence between two morphologically differentiated subspecies of bluethroat (Aves: Muscicapidae: *Luscinia svecica*) inferred from mitochondrial DNA sequence variation. *Mol. Ecol.* 7: 239–245.

Qui, Y.-L., Y. Cho, J. C. Cox and J. D. Palmer. 1998. The gain of three mitochondrial introns identifies liverworts as the earliest land plants. *Nature* 394: 671–674.

Qui, Y.-L. and 9 others. 1999. The earliest angiosperms: Evidence from mitochondrial, plastid, and nuclear genomes. *Nature* 402: 404–407.

Quinn, J. S., G. E. Woolfenden, J. W. Fitzpatrick and B. N. White. 1999. Multi-locus DNA fingerprinting supports genetic monogamy in Florida scrub-jays. *Behav. Ecol. Sociobiol.* 45: 1–10.

Quinn, T. P., C. C. Wood, L. Margolis, B. E. Riddell and K. D. Hyatt. 1987. Homing in wild sockeye salmon (*Oncorhynchus nerka*) populations as inferred from differences in parasite prevalance and allozyme allele frequencies. *Can. J. Fish. Aquat. Sci.* 44: 1963–1968.

Quinn, T. W. 1988. DNA sequence variation in the Lesser Snow Goose, *Anser caerulescens caerulescens*. Ph. D. Dissertation, Queen's University, Kingston, Ontario, Canada.

Quinn, T. W. 1992. The genetic legacy of mother goose: Phylogeographic patterns of lesser snow goose *Chen caerulescens caerulescens* maternal lineages. *Mol. Ecol.* 1: 105–117.

Quinn, T. W. and D. P. Mindell. 1996. Mitochondrial gene order adjacent to the control region in crododile, turtle, and tuatara. *Mol. Phylogen. Evol.* 5: 344–351.

Quinn, T. W. and B. N. White. 1987. Identification of restriction-fragment-length polymorphisms in genomic DNA of the lesser snow goose (*Anser caerulescens caerulescens*). *Mol. Biol. Evol.* 4: 126–143.

Quinn, T. W., J. S. Quinn, F. Cooke and B. N. White. 1987. DNA marker analysis detects multiple maternity and paternity in single broods of the lesser snow goose. *Nature* 326: 392–394.

Quinn, T. W., J. C. Davies, F. Cooke and B. N. White. 1989. Genetic analysis of offspring of a female-female pair in the lesser snow goose (*Chen c. caerulescens*). *Auk* 106: 177–184.

Quinn, T. W., F. Cooke and B. N. White. 1990. Molecular sexing of geese using a cloned Z chromosomal sequence with homology to the W chromosome. *Auk* 107: 199–202.

Quinn, T. W., G. F. Shields and A. C. Wilson. 1991. Affinities of the Hawaiian goose based on two types of mitochondrial DNA data. *Auk* 108: 585–593.

Rabenold, P. P., K. N. Rabenold, W. H. Piper, J. Haydock and S. W. Zack. 1990. Shared paternity revealed by genetic analysis in cooperatively breeding tropical wrens. *Nature* 348: 538–540.

Rabenold, P. P., W. H. Piper, M. D. Decker and D. J. Minchella. 1991. Polymorphic minisatellite amplified on avian W chromosome. *Genome* 34: 489–493.

Racine, R. R. and C. H. Langley. 1980. Genetic heterozygosity in a natural population of *Mus musculus* assessed using two-dimensional electrophoresis. *Nature* 283: 855–857.

Ragan, M. A. 1992. Phylogenetic inference based on matrix representation of trees. *Mol. Phylogen. Evol.* 1: 53–58.

Ragan, M. A. 2001. On surrogate methods for detecting lateral gene transfer. *FEMS Microbiol. Letters* 201: 187–191.

Ralin, D. B. 1976. Behavioral and genetic differentiation in a diploid-tetraploid cryptic species complex of treefrogs. *Herpetol. Rev.* 7: 97–98.

Ralls, K., K. Brugger and J. Ballou. 1979. Inbreeding and juvenile mortality in small populations of ungulates. *Science* 206: 1101–1103.

Ralls, K., J. D. Ballou and A. Templeton. 1988. Estimates of lethal equivalents and the cost of inbreeding in mammals. *Cons. Biol.* 2: 185–193.

Ramshaw, J. A. M., J. A. Coyne and R. C. Lewontin. 1979. The sensitivity of gel electrophoresis as a detector of genetic variation. *Genetics* 93: 1019–1037.

Rand, A. L. 1948. Glaciation, an isolating factor in speciation. *Evolution* 2: 314–321.

Rand, D. M. 1994. Thermal habit, metabolic rate and the evolution of mitochondrial DNA. *Trends Ecol. Evol.* 9: 125–131.

Rand, D. M. 2001. The units of selection on mitochondrial DNA. *Annu. Rev. Ecol. Syst.* 32: 415–448.

Rand, D. M. and R. G. Harrison. 1986. Mitochondrial DNA transmission genetics in crickets. *Genetics* 114: 955–970.

Rand, D. M. and R. G. Harrison. 1989. Ecological genetics of a mosaic hybrid zone: Mitochondrial, nuclear, and reproductive differentiation of crickets by soil type. *Evolution* 43: 432–449.

Randi, E. 2000. Mitochondrial DNA. Pp. 136–167 in *Molecular Methods in Ecology*, A. J. Baker (ed.). Blackwell, London.

Randi, E., V. Lucchini, M. F. Christensen, N. Mucci, S. M. Funk, G. Dolf and V. Loeschcke. 2000. Mitochondrial DNA variability in Italian and East European wolves: Detecting the consequences of small population size and hybridization. *Cons. Biol.* 14: 464–473.

Rannala, B. and G. Bertorelle. 2001. Using linked markers to infer the age of a mutation. *Human Mut.* 18: 87–100.

Rannala, B. and J. L. Mountain. 1997. Detecting immigration by using multilocus genotypes. *Proc. Natl. Acad. Sci. USA* 94: 9197–9201.

Rannala, B. and Z. H. Yang. 1996. Probability distribution of molecular evolutionary trees: A new method of phylogenetic inference. *J. Mol. Evol.* 43: 304–311.

Rapacz, J. and 6 others. 1991. Identification of the ancestral haplotype for apolipoprotein B suggests an African origin of *Homo sapiens sapiens* and traces their subsequent migration to Europe and the Pacific. *Proc. Natl. Acad. Sci. USA* 88: 1403–1406.

Rasheed, B. K. A., E. C. Whisenant, R. Fernandez, H. Ostrer and Y. M. Bhatnagar. 1991. A Y-chromosomal DNA fragment is conserved in human and chimpanzee. *Mol. Biol. Evol.* 8: 416–432.

Rassman, K. 1997. Evolutionary age of the Galápagos iguanas predates the age of the present Galápagos Islands. *Mol. Phylogen. Evol.* 7: 158–172.

Raubeson, L. A. and R. K. Jansen. 1992a. Chloroplast DNA evidence on the ancient evolutionary split in vascular land plants. *Science* 255: 1697–1699.

Raubeson, L. A. and R. K. Jansen. 1992b. A rare chloroplast-DNA structural mutation is shared by all conifers. *Biochem. Syst. Ecol.* 20: 17–24.

Raven, P. H. 1979. A survey of reproductive biology in Onagraceae. *N. Zeal. J. Bot.* 17: 575–593.

Rawson, P. D. and R. S. Burton. 2002. Functional coadaptation between cytochrome *c* and cytochrome *c* oxidase within allopatric populations of a marine copepod. *Proc. Natl. Acad. Sci. USA* 99: 12955–12958.

Rawson, P. D. and T. J. Hilbish. 1998. Asymmetric introgression of mitochondrial DNA among European populations of blue mussels (*Mytilus* spp.). *Evolution* 52: 100–108.

Raymond, J., O. Zhaxybayeva, J. P. Gogarten, S. Y. Gerdes and R. E. Blankenship. 2002. Whole-genome analysis of photosynthetic prokaryotes. *Science* 298: 1616–1620.

Raymond, M., A. Callaghan, P. Fort and N. Pasteur. 1991. Worldwide migration of amplified resistance genes in mosquitoes. *Nature* 350: 151–153.

Recipon, H., R. Perasso, A. Adoutte and F. Quetier. 1992. ATP synthase subunit c/III/9 gene sequences as a tool for interkingdom and metaphytes molecular phylogenies. *J. Mol. Evol.* 34: 292–303.

Redenbach, Z. and E. B. Taylor. 2002. Evidence for historical introgression along a contact zone between two species of char (Pisces: Salmonidae) in northwestern North America. *Evolution* 56: 1021–1035.

Reeb, C. A. and J. C. Avise. 1990. A genetic discontinuity in a continuously distributed species: Mitochondrial DNA in the American oyster, *Crassostrea virginica*. *Genetics* 124: 397–406.

Reed, J. Z., D. J. Tollit, P. M. Thompson and W. Amos. 1997. Molecular scatology: The use of molecular genetic analysis to assign species, sex and individual identity to seal faeces. *Mol. Ecol.* 6: 225–234.

Reed, K. M. and J. W. Sites Jr. 1995. Female fecundity in a hybrid zone between two chromosomal races of the *Sceloporus grammicus* complex (Sauria, Phrynosomatidae). *Evolution* 49: 61–69.

Reed, K. M., I. F. Greenbaum and J. W. Sites Jr. 1995a. Cytogenetic analysis of chromosomal intermediates from a hybrid zone between two chromosomal races of the *Sceloporus grammicus* complex (Sauria, Phyrnosomatidae). *Evolution* 49: 37–47.

Reed, K. M., I. F. Greenbaum and J. W. Sites Jr. 1995b. Dynamics of a novel chromosomal polymorphism within a hybrid zone between two chromosomal races of the *Sceloporus grammicus* complex (Suaria, Phrynosomatidae). *Evolution* 49: 48–60.

Reeder, T. W. and R. R. Montanucci. 2001. Phylogenetic analysis of the horned lizards (Phrynosomatidae: *Phrynosoma*): Evidence from mitochondrial DNA and morphology. *Copeia* 2001: 309–323.

Reeve, H. K., D. F. Westneat, W. A. Noon, P. W. Sherman and C. F. Aquadro. 1990. DNA "fingerprinting" reveals high levels of inbreeding in colonies of the eusocial naked mole-rat. *Proc. Natl. Acad. Sci. USA* 87: 2496–2500.

Reeve, H. K., D. F. Westneat and D. C. Queller. 1992. Estimating average within-group relatedness from DNA fingerprints. *Mol. Ecol.* 1: 223–232.

Regnery, R. L., C. L. Spruill and B. D. Plikaytis. 1991. Genotypic identification of rickettsiae and estimation of intraspecies sequence divergence for portions of two rickettsial genes. *J. Bacteriol.* 173: 1576–1589.

Reich, D. E. and D. B. Goldstein. 1998. Genetic evidence for a Paleolithic human population expansion in Africa. *Proc. Natl. Acad. Sci. USA* 95: 8119–8123.

Reid, S. D., C. J. Herbelin, A. C. Bumbaugh, R. K. Selander and T. S. Whittam. 2000. Parallel evolution of virulence in pathogenic *E. coli*. *Nature* 406: 64–67.

Resing, J. M. and D. J. Ayre. 1985. The usefulness of the tissue grafting bioassay as an indicator of clonal identity in scleractinian corals. *Proc. Fifth Int. Coral Reef Congr.* 6: 75–81.

Rest, J. S. and D. P. Mindell. 2003. Retroids in Archaea: Phylogeny and lateral origins. *Mol. Biol. Evol.* 20: 1134–1142.

Revollo, S. and 7 others. 1998. *Trypanosoma cruzi*: Impact of clonal evolution of the parasite on its biological and medical properties. *Exp. Parasitol.* 89: 30–39.

Reznick, D. N., M. Mateos and M. J. Springer. 2002. Independent origins and rapid evolution of the placenta in the fish genus *Poeciliopsis*. *Science* 298: 1018–1020.

Rhodes, O. E. Jr., R. K. Chesser and M. H. Smith (eds.). 1996. *Population Dynamics in Ecological Space and Time*. University of Chicago Press, Chicago, IL.

Rhymer, J. M. and D. Simberloff. 1996. Extinction by hybridization and introgression. *Annu. Rev. Ecol. Syst.* 27: 83–109.

Rhymer, J. M., M. J. Williams and M. J. Braun. 1994. Mitochondrial analysis of gene flow between New Zealand mallards (*Anas platyrhynchos*) and grey ducks (*A. superciliosa*). *Auk* 111: 970–978.

Ribble, D. O. 1991. The monogamous mating system of *Peromyscus californicus* as revealed by DNA fingerprinting. *Behav. Ecol. Sociobiol.* 29: 161–166.

Ribeiro, S. and B. Golding. 1998. The mosaic nature of the eukaryotic nucleus. *Mol. Biol. Evol.* 15: 779–788.

Rice, W. R. 1998. Intergenomic conflict, interlocus antagonistic coevolution, and the evolution of reproductive isolation. Pp. 261–270 in *Endless Forms: Species and Speciation*, D. J. Howard and S. H. Berlocher (eds.). Oxford University Press, New York.

Richards, R. I. and 10 others. 1992. Evidence of founder chromosomes in fragile X syndrome. *Nature Genetics* 1: 257–260.

Richardson, D. S. and T. Burke. 1999. Extra-pair paternity in relation to male age in Bullock's orioles. *Mol. Ecol.* 8: 2115–2126.

Richie, M. and R. Butlin (eds.). 2001. Phylogeography, hybridization and speciation (special issue). *Mol. Ecol.* 10: 536–806.

Richie, M. G. and S. D. F. Phillips. 1998. The genetics of sexual isolation. Pp. 291–308 in *Endless Forms: Species and Speciation*, D. J. Howard and S. H. Berlocher (eds.). Oxford University Press, New York.

Richman, A. D. and J. R. Kohn. 1996. Learning from rejection: The evolutionary biology of single-locus incompatibility. *Trends Ecol. Evol.* 11: 497–502.

Richman, A. D. and T. Price. 1992. Evolution of ecological differences in the Old World leaf warblers. *Nature* 355: 817–821.

Richmond, R. C., D. G. Gilbert, K. B. Sheehan, M. H. Gromko and F. M. Butterworth. 1980. Esterase 6 and reproduction in *Drosophila melanogaster*. *Science* 207: 1483–1485.

Rick, C. M. and P. G. Smith. 1953. Novel variation in tomato species hybrids. *Am. Nat.* 87: 359–373.

Rick, C. M., E. Kesicki, J. F. Forbes and M. Holle. 1976. Genetic and biosystematic studies of two new sibling species of *Lycopersicon* from interandean Peru. *Theoret. Appl. Genet.* 47: 55–68.

Ricklefs, R. E. 1980. Phyletic gradualism vs. punctuated equilibrium: Applicability of neontological data. *Paleobiology* 6: 271–275.

Rico, C., U. Kuhnlein and G. J. Fitzgerrald. 1992. Male reproductive tactics in the threespine stickleback: An evaluation by DNA fingerprinting. *Mol. Ecol.* 1: 79–87.

Riddle, B. R. and R. L. Honeycutt. 1990. Historical biogeography in North American arid regions: An approach using mitochondrial-DNA phylogeny in grasshopper mice (genus *Onychomys*). *Evolution* 44: 1–15.

Riddle, B. R., D. J. Hafner, L. F. Alexander and J. R. Jaeger. 2000. Cryptic vicariance in the historical assembly of a Baja California Peninsular desert biota. *Proc. Natl. Acad. Sci. USA* 97: 14438–14443.

Rieseberg, L. H. 1991. Homoploid reticulate evolution in *Helianthus* (Asteraceae): Evidence from ribosomal genes. *Am. J. Bot.* 78: 1218–1237.

Rieseberg, L. H. 1995. The role of hybridization in evolution: Old wine in new skins. *Am. J. Bot.* 82: 944–953.

Rieseberg, L. H. 1996. Homology among RAPD fragments in interspecific comparisons. *Mol. Ecol.* 5: 99–105.

Rieseberg, L. H. 1997. Hybrid origins of plant species. *Annu. Rev. Ecol. Syst.* 28: 359–389.

Rieseberg, L. H. 2001. Chromosomal rearrangements and speciation. *Trends Ecol. Evol.* 16: 351–358.

Rieseberg, L. H. and L. Brouillet. 1994. Are many plant species paraphyletic? *Taxon* 43: 21–32.

Rieseberg, L. H. and S. J. Brunsfeld. 1992. Molecular evidence and plant introgression. Pp. 151- 176 in *Molecular Systematics of Plants*, P. S. Soltis, D. E. Soltis and J. J. Doyle (eds.). Chapman & Hall, New York.

Rieseberg, L. H. and M. F. Doyle. 1989. Tetrasomic segregation in the naturally occurring autotetraploid *Allium nevii* (Alliaceae). *Hereditas* 111: 31–36.

Rieseberg, L. H. and D. Gerber. 1995. Hybridization in the Catalina Island mountain mahogany (*Cercocarpus traskiae*): RAPD evidence. *Cons. Biol.* 9: 199–203.

Rieseberg, L. H. and J. D. Morefield. 1995. Character expression, phylogenetic reconstruction, and the detection of reticulate evolution. Pp. 333–354 in *Experimental and Molecular Approaches to Plant Biosystematics*, P. C. Hoch and A. G. Stephenson (eds.). Monogr. Syst. Bot. Missouri Bot. Gardens 53.

Rieseberg, L. H. and D. E. Soltis. 1991. Phylogenetic consequences of cytoplasmic gene flow in plants. *Evol. Trends Plants* 5: 65–84.

Rieseberg, L. H. and S. M. Swensen. 1996. Conservation genetics of endangered island plants. Pp. 305–334 in *Conservation Genetics: Case Histories from Nature*, J. C. Avise and J. L. Hamrick (eds.). Chapman & Hall, New York.

Rieseberg, L. H. and J. F. Wendel. 1993. Introgression and its consequences in plants. Pp. 70–109 in *Hybrid Zones and the Evolutionary Process*, R. G. Harrison (ed.). Oxford University Press, Oxford, England.

Rieseberg, L. H., D. E. Soltis and J. D. Palmer. 1988. A molecular reexamination of introgression between *Helianthus annuus* and *H. bolanderi* (Compositae). *Evolution* 42: 227–238.

Rieseberg, L. H., S. Zona, L. Aberbom and T. D. Martin. 1989. Hybridization in the island endemic, Catalina mahogany. *Cons. Biol.* 3: 52–58.

Rieseberg, L. H., R. Carter and S. Zona. 1990a. Molecular tests of the hypothesized hybrid origin of two diploid *Helianthus* species (Asteraceae). *Evolution* 44: 1498–1511.

Rieseberg, L. H., S. Beckstrom-Sternberg and K. Doan. 1990b. *Helianthus annuus* spp. *texanum* has chloroplast DNA and nuclear ribosomal RNA genes of *Helianthus debilis* spp. *cucumerifolius*. *Proc. Natl. Acad. Sci. USA* 87: 593–597.

Rieseberg, L. H., S. M. Beckstrom-Sternberg, A. Liston and D. M. Arias. 1991. Phylogenetic and systematic inferences from chloroplast DNA and isozyme variation in *Helianthus* sect. *Helianthus* (Asteraceae). *Syst. Bot.* 16: 50–76.

Rieseberg, L. H., M. A. Hanson and C. T. Philbrick. 1992. Androdioecy is derived from dioecy in Datiscaceae: Evidence from restriction site mapping of PCR-amplified chloroplast DNA fragments. *Syst. Bot.* 17: 324–326.

Rieseberg, L. H., C. R. Linder and G. J. Seiler. 1995a. Chromosomal and genic barriers to introgression in *Helianthus*. *Genetics* 141: 1163–1171.

Rieseberg, L. H., A. M. Desrochers and S. J. Youn. 1995b. Interspecific pollen competition as a reproductive barrier between sympatric species of *Helianthus* (Asteraceae). *Am. J. Bot.* 82: 515–519.

Rieseberg, L. H., C. Van Fossen and A. M. Desrochers. 1995c. Hybrid speciation accompanied by genomic reorganization in wild sunflowers. *Nature* 375: 313–316.

Rieseberg, L. H., D. M. Arias, M. C. Ungerer, C. R. Linder and B. Sinervo. 1996a. The effects of mating design on introgression between chromosomally divergent sunflower species. *Theoret. Appl. Genet.* 93: 633–644.

Rieseberg, L. H., J. Whitton and C. R. Linder. 1996b. Molecular marker incongruence in plant hybrid zones and phylogenetic trees. *Acta Bot. Neerl.* 45: 243–262.

Rieseberg, L. H., B. Sinervo, C. R. Linder, M. C. Ungerer and D. M. Arias. 1996c. Role of gene interactions in hybrid speciation: Evidence from ancient and experimental hybrids. *Science* 272: 741–744.

Rieseberg, L. H., J. Whitton and K. Gardner. 1999. Hybrid zones and the genetic architecture of a barrier to gene flow between two sunflower species. *Genetics* 152: 713–727.

Rieseberg, L. H. and 9 others. 2003. Major ecological transitions in wild sunflowers facilitated by hybridization. *Science* 301: 1211–1216.

Riley, M. A., M. E. Hallas and R. C. Lewontin. 1989. Distinguishing the forces controlling genetic variation at the *Xdh* locus in *Drosophila pseudoobscura*. *Genetics* 123: 359–369.

Rinderer, T. E., J. A. Stelzer, B. P. Oldroyd, S. M. Buco and W. L. Rubink. 1991. Hybridization between European and Africanized honey bees in the neotropical Yucutan peninsula. *Science* 253: 309–311.

Risch, N. J. and B. Devlin. 1992. On the probability of matching DNA fingerprints. *Science* 255: 717–720.

Rising, J. D. and J. C. Avise. 1993. Application of genealogical concordance principles to the taxonomy and evolutionary history of the Sharp-tailed Sparrow (*Ammodramus caudacutus*). *Auk* 110: 844–856.

Ritchison, G., P. H. Klatt and D. F. Westneat. 1994. Mate guarding and extra-pair paternity in northern cardinals. *Condor* 96: 1055–1063.

Ritland, C. and K. Ritland. 2000. DNA-fragment markers in plants. Pp. 208–234 in *Molecular Methods in Ecology*, A. J. Baker (ed.). Blackwell, London.

Ritland, K. 1996. Estimators for pairwise relatedness and individual inbreeding coefficients. *Genet. Res.* 67: 175–185.

Ritland, K. 2000. Marker-inferred relatedness as a tool for detecting heritability in nature. *Mol. Ecol.* 9: 1195–1204.

Ritland, K and M. T. Clegg. 1987. Evolutionary analyses of plant DNA sequences. *Am. Nat.* 130: S74-S100.

Ritland, K. and S. K. Jain. 1981. A model for the estimation of outcrossing rate and gene frequencies using m independent loci. *Heredity* 47: 35–52.

Rivera, M. C., R. Jain, J. E. Moore and J. A. Lake. 1998. Genomic evidence for two functionally distinct gene classes. *Proc. Natl. Acad. Sci. USA* 95: 6239–6244.

Roberts, J. D. and L. R. Maxson. 1985. Tertiary speciation models in Australian anurans: Molecular data challenge Pleistocene scenario. *Evolution* 39: 325–334.

Roberts, J. R. 1984. Restriction and modification enzymes and their recognition sequences. *Nucleic Acids Res.* 12: r167-r204.

Roberts, L. 1991. Fight erupts over DNA fingerprinting. *Science* 254: 1721–1723.

Robertson, A. 1975. Remarks on the Lewontin-Krakauer test. *Genetics* 80: 396.

Robertson, B. C., C. D. Millar, E. O. Minot, D. V. Merton and D. M. Lambert. 2000. Sexing the critically endangered kakapo *Strigops habroptilus*. *Emu* 100: 336–339.

Robichaux, R. H., E. A. Friar and D. W. Mount. 1997. Molecular genetic consequences of a population bottleneck associated with reintroduction of the Mauna Kea silversword (*Argyroxiphium sandwicense* spp. *sandwicense* [Asteraceae]). *Cons. Biol.* 11: 1140–1146.

Robinson, N. A. 1995. Implications from mitochondrial DNA for management to conserve the eastern barred bandicoot (*Perameles gunnii*). *Cons. Biol.* 9: 114–125.

Robinson, T. J., A. D. Bastos, K. M. Halanych and B. Herzig. 1996. Mitochondrial DNA sequence relationships of the extinct blue antelope *Hippotragus leucophaeus*. *Naturwissenschaften* 83: 178–182.

Roca, A. L., N. Georgiadis, J. Pecon-Slattery and S. J. O'Brien. 2001. Genetic evidence for two species of elephant in Africa. *Science* 293: 1473–1477.

Rodriguez, R. L. and M. D. Greenfield. 2003. Genetic variance and phenotypic plasticity in a component of female mate choice in an ultrasonic moth. *Evolution* 57: 1304–1313.

Rodriguez-Trelles, F., R. Tarrio and F. J. Ayala. 2002. A methodological bias toward overestimation of molecular evolutionary time scales. *Proc. Natl. Acad. Sci. USA* 99: 8112–8115.

Roeder, K., M. Escobar, J. B. Kadane and I. Balazs. 1998. Measuring heterogeneity in forensic databases using hierarchical Bayes models. *Biometrika* 85: 269–287.

Roger, A. J. 1999. Reconstructing early events in eukaryotic evolution. *Am. Nat.* 154: S146–163.

Roger, A. J. and 6 others. 1998. A mitochondrial-like chaperonin 60 gene in *Giardia lamblia*: Evidence that diplomonads once harbored an endosymbiont related to the progenitor of mitochondria. *Proc. Natl. Acad. Sci. USA* 95: 229–234.

Rogers, A. R. and H. Harpending. 1992. Population growth makes waves in the distribution of pairwise genetic differences. *Mol. Biol. Evol.* 9: 552–569.

Rogers, J. S. 1972. Measures of genetic similarity and genetic distance. *Studies in Genetics VII*, University of Texas Publ. 7213: 145–153.

Rogers, J. S. 1994. Central moments and probability distributions of Colless's coefficient of tree imbalance. *Evolution* 48: 2026–2036.

Rogers, S. O., S. Honda and A. J. Bendich. 1986. Variation in the ribosomal RNA genes among individuals of *Vicia faba*. *Plant. Mol. Biol.* 6: 339–345.

Rognon, X. and R. Guyomard. 2003. Large extent of mitochondrial DNA transfer from *Oreochromis aureus* to *O. niloticus* in West Africa. *Mol. Ecol.* 12: 435–445.

Rogstad, S. H., J. C. Patton II and B. A. Schaal. 1988. M13 repeat probe detects minisatellite-like sequences in gymnosperms and angiosperms. *Proc. Natl. Acad. Sci. USA* 85: 9176–9178.

Rogstad, S. H., H. Nybom and B. A. Schaal. 1991. The tetrapod "DNA fingerprinting" M13 repeat probe reveals genetic diversity and clonal growth in quaking aspen (*Populus tremuloides*, Salicaceae). *Plant Syst. Evol.* 175: 115–123.

Rohde, K. 1992. Latitudinal gradients in species-diversity: The search for the primary cause. *Oikos* 65: 514–527.

Rohlf, D. J. 1991. Six biological reasons why the Endangered Species Act doesn't work—and what to do about it. *Cons. Biol.* 5: 273–282.

Rohwer, S., E. Bermingham and C. Wood. 2001. Plumage and mtDNA haplotype variation across a moving hybrid zone. *Evolution* 55: 405–422.

Rollinson, D. 1986. Reproductive strategies of some species of *Bulinus*. *Proc. 8th Malacological Congr. Budapest 1983*: 221–226.

Rollinson, D., R. A. Kane and J. R. L. Lines. 1989. An analysis of fertilization in *Bulinus cernicus* (Gastropoda: Planorbidae). *J. Zool.* 217: 295–310.

Roman, J. and B. W. Bowen. 2000. The mock turtle syndrome: Genetic identification of turtle meat purchased in the south-eastern United States of America. *Anim. Cons.* 3: 61–65.

Roman, J. and S. R. Palumbi. 2003. Whales before whaling in the North Atlantic. *Science* 301: 508–510.

Roman, J., S. Santhuff, P. Moler and B. W. Bowen. 1999. Cryptic evolution and population structure in the alligator snapping turtle (*Macroclemys temminckii*). *Cons. Biol.* 13: 1–9.

Romano, S. L. and S. R. Palumbi. 1996. Evolution of scleractinian corals inferred from molecular systematics. *Science* 271: 640–642.

Ronquist, R. 1997. Dispersal-vicariance analysis: A new approach to the quantification of historical biogeography. *Syst. Biol.* 46: 195–203.

Rooney, A. P., R. L. Honeycutt, S. K. Davis and J. N. Derr. 1999. Evaluating a putative bottleneck in a population of bowhead whales from patterns of microsatellite diversity and genetic disequilibria. *J. Mol. Evol.* 49: 682–690.

Roose, M. L. and L. D. Gottlieb. 1976. Genetic and biochemical consequences of polyploidy in *Tragopogon*. *Evolution* 30: 818–830.

Rose, M. R. and W. F. Doolittle. 1983. Molecular biological mechanisms of speciation. *Science* 220: 157–162.

Rosel, R. W. and T. D. Kocher. 2002. DNA-based identification of larval cod in stomach contents of predatory fishes. *J. Exp. Marine Biol. Ecol.* 267: 75–88.

Rosen, D. E. 1978. Vicariant patterns and historical explanation in biogeography. *Syst. Zool.* 27: 159–188.

Rosenbaum, H. C., M. G. Egan, P. J. Clapham, R. L. Brownell Jr. and R. DeSalle. 1997. An effective method for isolating DNA from historical specimens of baleen. *Mol. Ecol.* 6: 677–681.

Rosenbaum, H. C. and 19 others. 2000. World-wide genetic differentiation of *Eubalaena*: Questioning the number of right whale species. *Mol. Ecol.* 9: 1793–1802.

Rosenberg, N. A. 2003. The shapes of neutral gene genealogies in two species: Probabilities of monophyly, paraphyly, and polyphyly in a coalescent model. *Evolution* 57: 1465–1477.

Rosenblatt, R. H. and R. S. Waples. 1986. A genetic comparison of allopatric populations of shore fish species from the eastern and central Pacific Ocean: Dispersal or vicariance? *Copeia* 1986: 275–284.

Rosenblum, L. L., J. Supriatna and D. J. Melnick. 1997. Phylogeographic analysis of pigtail macaque populations (*Macaca nemestrina*) inferred from mitochondrial DNA. *Am. J. Physical Anthropol.* 104: 35–45.

Rosewich, U. L. and H. C. Kistler. 2000. Role of horizontal gene transfer in the evolution of fungi. *Annu. Rev. Phytopathol.* 38: 325–363.

Ross, H. A. and 8 others. 2003. DNA surveillance: Web-based molecular identification of whales, dolphins, and porpoises. *J. Heredity* 94: 111–114.

Ross, K. G. and J. M. Carpenter. 1991. Population genetic structure, relatedness, and breeding systems. Pp. 451–479 in *The Social Biology of Wasps*, K. G. Ross and R. W. Matthews (eds.). Comstock Publ., Ithaca, NY.

Ross, K. G. and D. J. C. Fletcher. 1985. Comparative study of genetic and social structure in two forms of the fire ant *Solenopsis invicta* (Hymenoptera: Formicidae). *Behav. Ecol. Sociobiol.* 17: 349–356.

Ross, K. G. and R. W. Matthews. 1989a. Population genetic structure and social evolution in the sphecid wasp *Microstigmus comes*. *Am. Nat.* 134: 574–598.

Ross, K. G. and R. W. Matthews. 1989b. New evidence for eusociality in the sphecid wasp *Microstigmus comes*. *Anim. Behav.* 38: 613–619.

Rossetto, M., G. Jezierski, S. D. Hooper and K. W. Dixon. 1999. Conservation genetics and clonality in two critically endangered eucalypts from the highly endemic south-western Australian flora. *Biol. Cons.* 88: 321–331.

Rossiter, S. J., G. Jones, R. D. Ransome and E. M. Barratt. 2001. Outbreeding increases offspring survival in wild greater horseshoe bats (*Rhinolophus ferrumequinum*). *Proc. R. Soc. Lond. B* 268: 1055–1061.

Rotte, C. and W. Martin. 2001. Does endosymbiosis explain the origin of the nucleus? *Nat. Cell Biol.* 3: E173-E174.

Rowan, R. and D. A. Powers. 1991. A molecular genetic classification of zooxanthellae and the evolution of animal-algal symbioses. *Science* 251: 1348–1351.

Rowan, R. and D. A. Powers. 1992. Ribosomal RNA sequences and the diversity of symbiotic dinoflagellates (zooxanthellae). *Proc. Natl. Acad. Sci. USA* 89: 3639–3643.

Roy, B. A. 2001. Patterns of association between crucifers and their flower-mimic pathogens: Host jumps are more common than coevolution or cospeciation. *Evolution* 55: 41–53.

Roy, B. A. and L. H. Rieseberg. 1989. Evidence for apomixis in *Arabis*. *J. Heredity* 80: 506–508.

Roy, M. S., E. Geffen, D. Smith, E. Ostrander and R. K. Wayne. 1994. Patterns of differentiation and hybridization in North American wolf-like canids revealed by analysis of microsatellite loci. *Mol. Biol. Evol.* 11: 553–570.

Roy, M. S., E. Geffen, D. Smith and R. K. Wayne. 1996. Molecular genetics of pre-1940 red wolves. *Cons. Biol.* 10: 1413–1424.

Ruano, G., K. H. Kidd and J. C. Stephens. 1990. Haplotype of multiple polymorphisms resolved by enzymatic amplification of single DNA molecules. *Proc. Natl. Acad. Sci. USA* 87: 6296–6300.

Rubenstein, D. R. and 6 others. 2002. Linking breeding and wintering ranges of a migratory songbird using stable isotopes. *Science* 295: 1062–1065.

Rüber, L., E. Verheyen and A. Meyer. 1999. Replicated evolution of trophic specializations in an endemic cichlid fish lineage from Lake Tanganyika. *Proc. Natl. Acad. Sci. USA* 96: 10230–10235.

Rudloe, A. 1979. *Limulus polyphemus*: A review of the ecologically significant literature. Pp. 27–35 in *Biomedical Applications of the Horseshoe Crab* (Limulidae), E. Cohen (ed.). Alan R. Liss, New York.

Ruedi, M. M. F. Smith and J. L. Patton. 1997. Phylogenetic evidence of mitochondrial DNA introgression among pocket gophers in New Mexico (family Geomyidae). *Mol. Ecol.* 6: 453–462.

Ruggiero, M. V., R. Turk and G. Procaccini. 2002. Genetic identity and homozygosity in north-Adriatic populations of *Posidonia oceanica*: An ancient, post-glacial clone? *Cons. Genet.* 3: 71–74.

Rundle, H. D., A. O. Mooers and M. C. Whitlock. 1998. Single founder-flush events and the evolution of reproductive isolation. *Evolution* 52: 1850–1855.

Rundle, H. D., L. Nagel, J. W. Boughman and D. Schluter. 2000. Natural selection and parallel speciation in sympatric sticklebacks. *Science* 287: 306–308.

Ruter, B, J. L. Hamrick and B. W. Wood. 2000. Outcrossing rates and relatedness estimates in pecan (*Carya illinoinensis*) populations. *J. Heredity* 91: 72–75.

Ruvolo, M., S. Zehr, M. von Dornum, D. Pan, B. Chang and J. Lin. 1993. Mitochondrial COII sequences and modern human origins. *Mol. Biol. Evol.* 10: 1115–1135.

Ruvolo, M., D. Pan, S. Zehr, T. Goldberg, T. R. Disotell and M. von Dornum. 1994. Gene trees and hominoid phylogeny. *Proc. Natl. Acad. Sci. USA* 91: 8900–8904.

Ruzzante, D. E. 1998. A comparison of several measures of genetic distances and population structure with microsatellite data: Bias and sampling variance. *Can. J. Fish. Aquat. Sci.* 55: 1–14.

Ruzzante, D. E., C. T. Taggart, R. W. Doyle and D. Cook. 2001. Stability in the historical pattern of genetic structure of Newfoundland cod (*Gadus morhua*) despite the catastrophic decline in population size from 1964 to 1994. *Cons. Genet.* 2: 257–269.

Ryder, O. A. 1986. Species conservation and the dilemma of subspecies. *Trends Ecol. Evol.* 1: 9–10.

Ryder, O. A., A. McLaren, S. Brenner, Y.-P. Zhang and K. Benirschke. 2000. DNA banks for endangered animal species. *Science* 288: 275–277.

Ryman, N. 1983. Patterns of distribution of biochemical genetic variation in salmonids: Differences between species. *Aquaculture* 33: 1–21.

Ryman, N. and F. Utter. 1987. *Population Genetics and Fishery Management*. University of Washington Press, Seattle.

Ryskov, A. P., A. G. Jincharadze, M. I. Prosnyak, P. L. Ivanov and S. A. Limborska. 1988. M13 phage DNA as a universal marker for DNA fingerprinting of animals, plants and microorganisms. *FEBS Letters* 233: 388–392.

Sætre, G.-P., M. Kral, S. Bures and R. A. Ims. 1999. Dynamics of a clinal hybrid zone and a comparison with island hybrid zones of flycatchers. *J. Zool. London* 247: 53–64.

Sætre, G.-P. and others. 2001. Speciation, introgressive hybridization and nonlinear rate of molecular evolution in flycatchers. *Mol. Ecol.* 10: 737–749.

Sage, R. D. and R. K. Selander. 1975. Trophic radiation through polymorphism in cichlid fishes. *Proc. Natl. Acad. Sci. USA* 72: 4669–4673.

Sage, R. D. and J. O. Wolff. 1986. Pleistocene glaciations, fluctuating ranges, and low genetic variability in a large mammal (*Ovis dalli*). *Evolution* 40: 1092–1095.

Sage, R. D., P. V. Loiselle, P. Basasibwaki and A. C. Wilson. 1984. Molecular versus morphological change among cichlid fishes of Lake Victoria. Pp. 185–201 in *Evolution of Fish Species Flocks*, A. A. Echelle and I. Kornfield (eds.). University of Maine Press, Orono, ME.

Sage, R. D., D. Heyneman, K.-C. Lim and A. C. Wilson. 1986. Wormy mice in a hybrid zone. *Nature* 324: 60–63.

Saghai-Maroof, M. A., K. M. Soliman, R. A. Jorgensen and R. W. Allard. 1984. Ribosomal DNA spacer-length polymorphisms in barley: Mendelian inheritance, chromosomal location, and population dynamics. *Proc. Natl. Acad. Sci. USA* 81: 8014–8018.

Saiki, R. K. and 6 others. 1985. Enzymatic amplification of β-globin genomic sequences and restriction site analysis for diagnosis of sickle cell anemia. *Science* 230: 1350–1354.

Saiki, R. K. and 7 others. 1988. Primer-directed enzymatic amplification of DNA with a thermostable DNA polymerase. *Science* 239: 487–491.

Saino, N., C. R. Primmer, H. Ellegren and A. P. Møller. 1997. An experimental study of paternity and tail ornamentation in the barn swallow (*Hirundo rustica*). *Evolution* 51: 562–570.

Saitou, N. and M. Nei. 1986. The number of nucleotides required to determine the branching order of three species, with special reference to the human-chimpanzee-gorilla divergence. *J. Mol. Evol.* 24: 189–204.

Saitou, N. and M. Nei. 1987. The neighbor-joining method: A new method for reconstructing phylogenetic trees. *Mol. Biol. Evol.* 4: 406–425.

Sakisaka, Y., T. Yahara, I. Miura and E. Kasuya. 2000. Maternal control of sex ratio in *Rana rugosa*: Evidence from DNA sexing. *Mol. Ecol.* 9: 1711–1715.

Salama, M., W. Sandine and S. Giovannoni. 1991. Development and application of oligonucleotide probes for identification of *Lactococcus lactis* subsp. *cremoris*. *Appl. Environ. Microbiol.* 57: 1313–1318.

Saltonstall, K., G. Amato and J. Powell. 1998. Mitochondrial DNA variability in Grauer's gorillas of Kahuzi-Biega National Park. *J. Heredity* 89: 129–135.

Salzberg, S. L., O. White, J. Peterson and J. A. Eisen. 2001. Microbial genes in the human genome: Lateral transfer or gene loss? *Science* 292: 1903–1906.

Salzburger, W., S. Baric and C. Sturmbauer. 2002. Speciation via introgressive hybridization in East African cichlids? *Mol. Ecol.* 11: 619–625.

Sambrook, J., E. F. Fritsch and T. Maniatis. 1989. *Molecular Cloning* (2nd Edition). Cold Spring Harbor Lab Press, Cold Spring Harbor, NY.

Sampsell, B. and S. Sims. 1982. Effect of *adh* genotype and heat stress on alcohol tolerance in *Drosophila melanogaster*. *Nature* 296: 853–855.

Sanderson, M. J. 2002. Estimating absolute rates of molecular evolution and divergence times: A penalized likelihood approach. *Mol. Biol. Evol.* 19: 101–109.

Sanderson, M. J., A. Purvis and C. Henze. 1998. Phylogenetic supertrees: Assembling the trees of life. *Trends Ecol. Evol.* 13: 105–109.

Sanderson, M. J., A. C. Driskell, R. H. Ree, O. Eulenstein and S. Langley. 2003. Obtaining maximal concatenated phylogenetic data sets from large sequence databases. *Mol. Biol. Evol.* 20: 1036–1042.

Sanford, E., M. R. Roth, G. C. Johns, J. P. Wares and G. N. Somero. 2003. Local selection and latitudinal variation in a marine predator-prey interaction. *Science* 300: 1135–1137.

Sanger, F., S. Nicklen and A. R. Coulson. 1977. DNA sequencing with chain-terminating inhibitors. *Proc. Natl. Acad. Sci. USA* 74: 5463–5467.

Sankoff, D., G. Leduc, N. Antoine, B. Paquin, B. F. Lang and R. Cedergren. 1992. Gene order comparisons for phylogenetic inference: Evolution of the mitochondrial genome. *Proc. Natl. Acad. Sci. USA* 89: 6575–6579.

SanMiguel, P. and 10 others. 1996. Nested retrotransposons in the intergenic regions of the maize genome. *Science* 274: 765–768.

Santos, S., D. J. Taylor, R. A. Zinzie III, M. Hidaka, K. Sakai and M. A. Coffroth. 2002. Molecular phylogeny of symbiotic dinoflagellates inferred from partial chloroplast large subunit (23S)-rDNA sequences. *Mol. Phylogen. Evol.* 23: 97–111.

Saperstein, D. A. and J. M. Nickerson. 1991. Restriction fragment length polymorphism analysis using PCR coupled to restriction digests. *BioTechniques* 10: 488–489.

Sarich, V. M. 1973. The giant panda is a bear. *Nature* 245: 218–220.

Sarich, V. M. and A. C. Wilson. 1966. Quantitative immunochemistry and the evolution of primate albumins: Micro-complement fixation. *Science* 154: 1563–1566.

Sarich, V. M. and A. C. Wilson. 1967. Immunological time scale for hominid evolution. *Science* 158: 1200–1203.

Sarich, V. M. and A. C. Wilson. 1973. Generation time and genomic evolution in primates. *Science* 179: 1144–1147.

Sarich, V. M., C. W. Schmid and J. Marks. 1989. DNA hybridization as a guide to phylogenies: A critical analysis. *Cladistics* 5: 3–32.

Sarver, S. K., M. C. Landrum and D. W. Foltz. 1992. Genetics and taxonomy of ribbed mussels (*Geukensia* spp.). *Marine Biol.* 113: 385–390.

Sassaman, C. 1978. Mating systems in porcellionid isopods: Multiple paternity and sperm mixing in *Porcellio scaber* Latr. *Heredity* 41: 385–397.

Sato, A. and 6 others. 1999. Phylogeny of Darwin's finches as revealed by mtDNA sequences. *Proc. Natl. Acad. Sci. USA* 96: 5101–5106.

Sato, A., H. Tichy, C. O'hUigin, P. R. Grant, B. R. Grant and J. Klein. 2001. On the origin of Darwin's finches. *Mol. Biol. Evol.* 18: 299–311.

Sato, A. and 7 others. 2002. Persistence of *Mhc* heterozygosity in homozygous clonal killifish, *Rivulus marmoratus*: Implications for the origin of hermaphroditism. *Genetics* 162: 1791–1803.

Sattler, G. D. and M. J. Braun. 2000. Morphometric variation as an indicator of genetic interactions between black-capped and Carolina chickadees at a contact zone in the Appalachian mountains. *Auk* 117: 427–444.

Saunders, N. C., L. G. Kessler and J. C. Avise. 1986. Genetic variation and geographic differentiation in mitochondrial DNA of the horseshoe crab, *Limulus polyphemus*. *Genetics* 112: 613–627.

Sawyer, S. A. and D. L. Hartl. 1992. Population genetics of polymorphism and divergence. *Genetics* 132: 1161–1176.

Scanlan, B. E., L. R. Maxson and W. E. Duellman. 1980. Albumin evolution in marsupial frogs (Hylidae: *Gastrotheca*). *Evolution* 34: 222–229.

Schaal, B. A. 1980. Measurement of gene flow in *Lupinus texensis*. *Nature* 284: 450–451.

Schaal, B. A. 1985. Genetic variation in plant populations: From demography to DNA. Pp. 321–342 in *Structure and Functioning of Plant Populations*, J. Haeck and J. Woldendorp (eds.). North-Holland, Amsterdam.

Schaal, B. A., W. J. Leverich and J. Nieto-Sotelo. 1987. Ribosomal DNA variation in the native plant *Phlox divaricata*. *Mol. Biol. Evol.* 4: 611–621.

Schaal, B. A., S. L. O'Kane Jr. and S. H. Rogstad. 1991. DNA variation in plant populations. *Trends Ecol. Evol.* 6: 329–332.

Schaal, B. A., D. A. Hayworth, K. M. Olsen, J. T. Rauscher and W. A. Smith. 1998. Phylogeographic studies in plants: Problems and prospects. *Mol. Ecol.* 7: 465–474.

Schaal, B. A., J. F. Gaskin and A. L. Caicedo. 2003. Phylogeography, haplotype trees, and invasive plant species. *J. Heredity* 94: 197–204.

Schaeffer, S. W. and E. L. Miller. 1992. Molecular population genetics of an electrophoretically monomorphic protein in the alcohol dehydrogenase region of *Drosophila pseudoobscura*. *Genetics* 132: 163–178.

Schaeffer, S. W., C. F. Aquadro and W. W. Anderson. 1987. Restriction-map variation in the alcohol dehydrogenase region of *Drosophila pseudoobscura*. *Mol. Biol. Evol.* 4: 254–265.

Schaeffer, S. W., C. F. Aquadro and C. H. Langley. 1988. Restriction-map variation in the *Notch* region of *Drosophila melanogaster*. *Mol. Biol. Evol.* 5: 30–40.

Schaeffer, S. W., C. S. Walthour, D. M. Toleno, A. T. Olek and E. L. Miller. 2001. Protein variation in ADH and ADH-RELATED in *Drosophila pseudoobscura*: Linkage disequilibrium between single nucleotide polymorphisms and protein alleles. *Genetics* 159: 673–687.

Schaeffer, S. W. and 8 others. 2003. Evolutionary genomics of inversions in *Drosophila pseudoobscura*: Evidence for epistasis. *Proc. Natl. Acad. Sci. USA* 100: 8319–8324.

Schaller, G. B. 1972. *The Serengeti Lion*. University of Chicago Press, Chicago, IL.

Scharf, S. J., G. T. Horn and H. A. Erlich. 1986. Direct cloning and sequence analysis of enzymatically amplified genomic sequences. *Science* 233: 1076–1078.

Scheller, G., M. T. Conkle and L. Griswald. 1985. Local differentiation among Mediterranean populations of Aleppo pine in their isozymes. *Silv. Genet.* 35: 11–19.

Scheltema, R. S. 1986. On dispersal and planktonic larvae of benthic invertebrates: An eclectic overview and summary of problems. *Bull. Mar. Sci.* 39: 290–322.

Schemske, D. W. and R. Lande. 1985. The evolution of self-fertilization and inbreeding depression in plants. II. Empirical observations. *Evolution* 39: 41–52.

Schliewen, U. K., D. Tautz and S. Pääbo. 1994. Sympatric speciation suggested by monophyly of crater lake cichlids. *Nature* 368: 629–632.

Schlötterer, C., B. Amos and D. Tautz. 1991. Conservation of polymorphic simple sequence loci in cetacean species. *Nature* 354: 63–65.

Schluter, D. 1996a. Ecological causes of adaptive radiation. *Am. Nat.* 148: S40-S64.

Schluter, D. 1996b. Ecological speciation in postglacial fishes. *Phil Trans. R. Soc. Lond. B* 351: 807–814.

Schluter, D. 2000. *The Ecology of Adaptive Radiation*. Oxford University Press, Oxford, England.

Schluter, D. 2001. Ecology and the origin of species. *Trends Ecol. Evol.* 16: 372–380.

Schluter, D., T. Price, A. Mooers and D. Ludwig. 1997. Likelihood of ancestor states in adaptive evolution. *Evolution* 51: 1699–1711.

Schmidt, T. M., E. F. DeLong and N. R. Pace. 1991. Analysis of a marine picoplankton community by 16S rRNA gene cloning and sequencing. *J. Bacteriol.* 173: 4371–4378.

Schnabel, A. and M. A. Asmussen. 1989. Definition and properties of disequilibria within nuclear-mitochondrial-chloroplast and other nuclear-dicytoplasmic systems. *Genetics* 123: 199–215.

Schneider, C. J., M. Cunningham and C. Moritz. 1998. Comparative phylogeography and the history of endemic vertebrates in the wet tropics rainforests of Australia. *Mol. Ecol.* 7: 487–498.

Schneider-Broussard, R. and J. E. Neigel. 1997. A large-subunit mitochondrial DNA sequence translocated to the nuclear genome of two stone crabs (*Menippe*). *Mol. Biol. Evol.* 14: 156–165.

Schoen, D. J. 1988. Mating system estimation via the one pollen parent model with the progeny array as the unit of observation. *Heredity* 60: 439–444.

Schoen, D. J. and A. H. D. Brown. 1991. Intraspecific variation in population gene diversity and effective population size correlates with the mating system in plants. *Proc. Natl. Acad. Sci. USA* 88: 4494–4497.

Schoen, D. J. and M. T. Clegg. 1984. Estimation of mating system parameters when outcrossing events are correlated. *Proc. Natl. Acad. Sci. USA* 81: 5258–5262.

Schön, I. and K. Martens. 1998. DNA repair in ancient asexuals: A new solution to an old problem? *J. Nat. Hist.* 32: 943–948.

Schön, I. and K. Martens. 2002. Are ancient asexuals less burdened? Selfish DNA, transposons and reproductive mode. *J. Nat. Hist.* 36: 379–390.

Schön, I. and K. Martens. 2003. No slave to sex. *Proc. R. Soc. Lond. B* 270: 827–833.

Schonewald-Cox, C. M., S. M. Chambers, B. MacBryde and W. L. Thomas (eds.). 1983. *Genetics and Conservation*. Benjamin/Cummings, Menlo Park, CA.

Schubart, C. D., R. Diesel and S. B. Hedges. 1998. Rapid evolution to terrestrial life in Jamaican crabs. *Nature* 393: 363–364.

Schubbert, R., D. Renz, B. Schmitz and W. Doerfler. 1997. Foreign (M13) DNA ingested by mice reaches peripheral leukocytes, spleen, and liver via the intestinal wall mucosa and can be covalently linked to mouse DNA. *Proc. Natl. Acad. Sci. USA* 94: 961–966.

Schug, M. D., T. F. C. Mackay and C. F. Aquadro. 1997. Low mutation rates of microsatellite loci in *Drosophila melanogaster*. *Nature Genet.* 15: 99–102.

Schulte, J. A. II, J. Melville and A. Larson. 2003. Molecular phylogenetic evidence for ancient divergence of lizard taxa on either side of Wallace's line. *Proc. R. Soc. Lond. B* 270: 597–603.

Schulte, P. M., M. Gómez-Chiarri and D. A. Powers. 1997. Structural and functional differences in the promoter and 5' flanking region of *Ldh-B* within and between populations of the teleost *Fundulus heteroclitus*. *Genetics* 145: 759–769.

Schultz, R. J. 1969. Hybridization, unisexuality and polyploidy in the teleost *Poeciliopsis* (Poeciliidae) and other vertebrates. *Am. Nat.* 103: 605–619.

Schultz, R. J. 1973. Unisexual fish: Laboratory synthesis of a "species." *Science* 179: 180–181.

Schuster, W. S. and J. B. Mitton. 1991. Relatedness within clusters of a bird-dispersed pine and the potential for kin interactions. *Heredity* 67: 41–48.

Schuster, W. S. and J. B. Mitton. 2000. Paternity and gene dispersal in limber pine (*Pinus flexilis* James). *Heredity* 84: 348–361.

Schwagmeyer, P. L. 1988. Scramble-competition polygyny in an asocial mammal: Male mobility and mating success. *Am. Nat.* 131: 885–892.

Schwartz, F. J. 1981. *World Literature to Fish Hybrids, with an Analysis by Family, Species, and Hybrid: Supplement I.* NOAA Tech. Report NMFS SSRF-750, U.S. Dept. of Commerce.

Schwartz, M. K., D. A. Tallmon and G. Luikart. 1998. Review of DNA-based census and effective population size estimators. *Anim. Cons.* 1: 293–299.

Schwartz, M. P. 1987. Intra-colony relatedness and sociality in the allodapine bee *Exoneura bicolor*. *Behav. Ecol. Sociobiol.* 21: 387–392.

Schwartz, R. M. and M. O. Dayhoff. 1978. Origins of prokaryotes, eukaryotes, mitochondria, and chloroplasts. *Science* 199: 395–403.

Schwarzbach, A. E. and L. H. Rieseberg. 2002. Likely multiple origins of a diploid hybrid sunflower species. *Mol. Ecol.* 11: 1703–1715.

Scott, I. A. W. and J. S. Keogh. 2000. Conservatoin genetics of the endangered grassland earless dragon *Tym-*

panocryptis pinquicolla (Reptilia: Agamidae) in southeastern Australia. *Cons. Genet.* 1: 357–363.

Scribner, K. T. 1991. Heterozygosity as an indicator of fitness and historical population demography. Pp. 77–84 in *Genetics and Wildlife Conservation*, E. Randi (ed.). Supplemento alle Richerche di Biologia della Selvaggina Vol. XVIII, Bologna, Italy.

Scribner, K. T. and J. C. Avise. 1993a. Cytonuclear genetic architecture in mosquitofish populations, and the possible roles of introgressive hybridization. *Mol. Ecol.* 2: 139–149.

Scribner, K. T. and J. C. Avise. 1993b. Demographic and life-history characteristics influence the cytonuclear genetic composition of mosquitofish hybrid populations. Pp. 280–290 in *Genetics and Evolution of Aquatic Organisms*, A. R. Beaumont (ed.). Chapman & Hall, New York.

Scribner, K. T. and J. C. Avise. 1994a. "Population cage" experiments with a vertebrate: The temporal demography and cytonuclear genetics of hybridization in *Gambusia* fishes. *Evolution* 48: 155–171.

Scribner, K. T. and J. C. Avise. 1994b. Cytonuclear genetics of experimental fish hybrid zones inside Biosphere 2. *Proc. Natl. Acad. Sci. USA* 91: 5066–5069.

Scribner, K. T. and T. D. Bowman. 1998. Microsatellites identify depredated waterfowl remains from glaucous gull stomachs. *Mol. Ecol.* 7: 1401–1405.

Scribner, K. T. and J. M. Pearce. 2000. Microsatellites: Evolutionary and methodological background and empirical evaluations at individual, population and phylogenetic levels. Pp. 235–273 in *Molecular Methods in Ecology*, A. J. Baker (ed.). Blackwell, London.

Scribner, K. T. and M. H. Smith. 1990. Genetic variability and antler development. Pp. 457–469 in *Pronghorns, Horns, and Antlers*, G. A. Bubenik and B. Bubenik (eds.). Springer-Verlag, New York.

Scribner, K. T., P. A. Crane, W. J. Spearman and L. W. Webb. 1998. DNA and allozyme markers provide concordant estimates of population differentiation: Analyses of U.S. and Canadian populations of Yukon River fall-run chum salmon (*Oncorhynchus keta*). *Can. J. Fish. Aquat. Sci.* 55: 1748–1758.

Scribner, K. T., K. S. Page and M. L. Barton. 2001a. Hybridization in freshwater fishes: A review of case studies and cytonuclear methods of biological inference. *Rev. Fish Biol. Fisheries* 10: 293–323.

Scribner, K. T., M. R. Petersen, R. L. Fields, S. L. Talbot, J. M. Pearce and R. K. Chesser. 2001b. Sex-biased gene flow in spectacled eiders (Anatidae): Inferences from molecular markers with contrasting modes of inheritance. *Evolution* 55: 2105–2115.

Scriver, C. R., W. S. Sly, B. Childs, A. L. Beaudet, D. Valle, K. W. Kinzler and B. Vogelstein (eds.). 2000. *The Metabolic and Molecular Basis of Inherited Disease* (8th Edition). McGraw-Hill, New York.

Searcy, W. A., K. Yasukawa and S. Lanyon. 1999. Evolution of polygyny in the ancestors of red-winged blackbirds. *Auk* 116: 5–19.

Sears, C. J., B. W. Bowen, R. W. Chapman, S. B. Galloway, S. R. Hopkins-Murphy and C. M. Woodley. 1995. Demographic composition of the juvenile loggerhead sea turtle (*Caretta caretta*) feeding population off Charleston, South Carolina: Evidence from mitochondial DNA markers. *Mar. Biol.* 123: 869–874.

Sebens, K. P. 1984. Agonistic behavior in the intertidal sea anemone *Anthopleura xanthogrammica*. *Biol. Bull.* 166: 457–472.

Seber, G. A. F. 1982. *The Estimation of Animal Abundance and Related Parameters*. Macmillan, New York.

Seddon, J. M., P. R. Baverstock and A. Georges. 1998. The rate of mitochondrial 12S rRNA sequences is similar in freshwater turtles and marsupials. *J. Mol. Evol.* 46: 460–464.

See, D., V. Kanazin, H. Talbert and T. Blake. 2000. Electrophoretic detection of single-nucleotide polymorphisms. *BioTechniques* 28: 710–716.

Seeb, J. E., G. H. Kruse, L. W. Seeb and R. G. Weck. 1990. Genetic structure of red king crab populations in Alaska facilitates enforcement of fishing regulations. *Alaska Sea Grant College Program Report* 90–04: 91–502.

Seehausen, O. 2002. Patterns of fish radiation are compatible with Pleistocene desiccation of Lake Victoria and 14,600 year history for its cichlid species flock. *Proc. R. Soc. Lond. B* 269: 491–497.

Seehausen, O., J. J. M. van Alphen and F. Witte. 1997. Cichlid fish diversity threatened by eutrophication that curbs sexual selection. *Science* 277: 1808–1811.

Seehausen, O. and 8 others. 2003. Nuclear markers reveal unexpected genetic variation and a Congolese-Nilotic origin of the Lake Victoria cichlid species flock. *Proc. R. Soc. Lond. B* 270: 129–137.

Seibold, I. and A. J. Helbig. 1995. Evolutionary history of New and Old World vultures inferred from nucleotide sequences of their mitochondrial cytochrome *b* gene. *Phil. Trans. R. Soc. Lond. B* 350: 163–178.

Seidler, R. J. and J. K. Fredrickson. 1995. Introduction: Molecular microbial ecology (special issue). *Mol. Ecol.* 4: 533–534.

Selander, R. K. 1970. Behavior and genetic variation in natural populations. *Am. Zool.* 10: 53–66.

Selander, R. K. 1975. Stochastic factors in the genetic structure of populations. Pp. 284–332 in *Proceedings of the Eighth International Conference on Numerical Taxonomy*, G. F. Estabrook (ed.). W. H. Freeman and Co., San Francisco.

Selander, R. K. 1976. Genic variation in natural populations. Pp. 21–45 in *Molecular Evolution*, F. J. Ayala (ed.). Sinauer Associates, Sunderland, MA.

Selander, R. K. 1982. Phylogeny. Pp. 32–59 in *Perspectives on Evolution*, R. Milkman (ed.). Sinauer Associates, Sunderland, MA.

Selander, R. K. and R. O. Hudson. 1976. Animal population structure under close inbreeding: The land snail *Rumina* in southern France. *Am. Nat.* 110: 695–718.

Selander, R. K. and D. W. Kaufman. 1973a. Genic variability and strategies of adaptation in animals. *Proc. Natl. Acad. Sci. USA* 70: 1875–1877.

Selander, R. K. and D. W. Kaufman. 1973b. Self-fertilization and genetic population structure in a colonizing land snail. *Proc. Natl. Acad. Sci. USA* 70: 1186–1190.

Selander, R. K. and D. W. Kaufman. 1975a. Genetic population structure and breeding systems. *Isozymes* IV: 27–48.

Selander, R. K. and D. W. Kaufman. 1975b. Genetic structure of populations of the brown snail (*Helix aspersa*). I. Microgeographic variation. *Evolution* 29: 385–401.

Selander, R. K. and B. R. Levin. 1980. Genetic diversity and structure in *Escherichia coli* populations. *Science* 210: 545–547.

Selander, R. K. and J. M. Musser. 1990. The population genetics of bacterial pathogenesis. Pp. 11–36 in *Molecular Basis of Bacterial Pathogenesis*, B. H. Iglewski and V. L. Clark (eds.). Academic Press, Orlando, FL.

Selander, R. K. and T. S.Whittam. 1983. Protein polymorphism and the genetic structure of populations. Pp. 89–114 in *Evolution of Genes and Proteins*, M. Nei and R. K. Koehn (eds.). Sinauer Associates, Sunderland, MA.

Selander, R. K., W. G. Hunt and S. Y. Yang. 1969. Protein polymorphism and genic heterozygosity in two European subspecies of the house mouse. *Evolution* 23: 379–390.

Selander, R. K., M. H. Smith, S. Y. Yang, W. E. Johnson and J. B. Gentry. 1971. Biochemical polymorphism and systematics in the genus *Peromyscus*. II. Genic heterozygosity and genetic similarity among populations of the old-field mouse (*Peromyscus polionotus*). *Studies in Genetics VI*, University of Texas Publ. 7103: 49–90.

Selander, R. K., D. W. Kaufman and R. S. Ralin. 1974. Self-fertilization in the terrestrial snail *Rumina decollata*. *Veliger* 16: 265–270.

Selander, R. K. and 6 others. 1985. Genetic structure of populations of *Legionella pneumophila*. *J. Bacteriol.* 163: 1021–1037.

Selander, R. K., D. A. Caugant and T. S. Whittam. 1987a. Genetic structure and variation in natural populations of *Escherichia coli*. Pp. 1625–1648 in *Escherichia coli and Salmonella typhimurium Cellular and Molecular Biology*, Vol. 2, F. C. Neidhardt et al. (eds.). American Society for Microbiology, Washington, DC.

Selander, R. K., J. M. Musser, D. A. Caugant, M. N. Gilmour and T. S. Whittam. 1987b. Population genetics of pathogenic bacteria. *Microbial Pathogen.* 3: 1–7.

Selander, R. K., P. Beltran and N. H. Smith. 1991a. Evolutionary genetics of *Salmonella*. Pp. 25–57 in *Evolution at the Molecular Level*, R. K. Selander, A. G. Clark and T. S. Whittam (eds.). Sinauer Associates, Sunderland, MA.

Selander, R. K., A. G. Clark and T. S. Whittam (eds.). 1991b. *Evolution at the Molecular Level*. Sinauer Associates, Sunderland, MA.

Sena, L., M. Vallinoto, I. Sampaio, H. Schneider, S. F. Ferran and M. P. C. Schneider. 2002. Mitochondrial COII gene sequences provide new insights into the phylogeny of marmoset species groups (Callitrichidae, Primates). *Folia Primat.* 73: 240–251.

Sene, F. M. and H. L. Carson. 1977. Genetic variation in Hawaiian *Drosophila*. IV. Allozymic similarity between *D. sylvestris* and *D. heteroneura* from the island of Hawaii. *Genetics* 86: 187–198.

Seutin, G., N. K. Klein, R. E. Ricklefs and E. Bermingham. 1994. Historical biogeography of the bananaquit (*Coereba flaveola*) in the Caribbean region: A mitochondrial DNA assessment. *Evolution* 49: 1041–1061.

Shaklee, J. B. 1983. The utilization of isozymes as gene markers in fisheries management and conservation. *Isozymes: Curr. Top. Biol. Med. Res.* 11: 213–247.

Shaklee, J. B. 1984. Genetic variation and population structure in the damselfish, *Stegastes fasciolatus*, throughout the Hawaiian archipelago. *Copeia* 1984: 629–640.

Shaklee, J. B. and P. Bentzen. 1998. Genetic identification of stocks of marine fish and shellfish. *Bull. Marine Sci.* 62: 589–621.

Shaklee, J. B., K. L. Kepes and G. S. Whitt. 1973. Specialized lactate dehydrogenase isozymes: The molecular and genetic basis for the unique eye and liver LDHs of teleost fishes. *J. Exp. Zool.* 185: 217–240.

Shaklee, J. B., C. S. Tamaru and R. S. Waples. 1982. Speciation and evolution of marine fishes studied by the electrophoretic analysis of proteins. *Pacific Sci.* 36: 141–157.

Shaklee, J. B., D. C. Klaybor, S. Young and B. A. White. 1991. Genetic stock structure of odd-year pink salmon, *Oncorhynchus gorbuscha* (Walbaum), from Washington and British Columbia and potential mixed-stock fisheries applications. *J. Fish Biol.* 39A: 21–34.

Shapiro, D. Y. 1983. On the possibility of kin groups in coral reef fishes. Pp. 39–45 in *Ecology of Deep and Shallow Coral Reefs*, M. L. Reaka (ed.). Coral Reef Symp. Ser. Undersea Res. Vol. 1, National Oceanographic and Atmospheric Administration, Washington, DC.

Sharp, P. M. 1997. In search of molecular Darwinism. *Nature* 385: 111–112.

Sharp, P. M., A. T. Lloyd and D. G. Higgins. 1991. Coelacanth's relationships. *Nature* 353: 218–219.

Shatters, R. G. and M. L. Kahn. 1989. Glutamine synthetase II in *Rhizobium*: Reexamination of the proposed horizontal transfer of DNA from eukaryotes to prokaryotes. *J. Mol. Evol.* 29: 422–428.

Shaw, A. J. 2000. Molecular phylogeography and cryptic speciation in the mosses, *Mielichhoferia elongata* and *M. mielichhoferiana* (Bryaceae). *Mol. Ecol.* 9: 595–608.

Shaw, C. R. and R. Prasad. 1970. Starch gel electrophoresis of enzymes: A compilation of recipes. *Biochem. Genet.* 4: 297–320.

Shaw, D. V., A. L. Kahler and R. W. Allard. 1981. A multilocus estimator of the mating system parameters in plant populations. *Proc. Natl. Acad. Sci. USA* 78: 1298–1302.

Shaw, K. L. 1996. Polygenic inheritance of a behavioral phenotype: Interspecific genetics of song in the Hawaiian cricket genus *Laupala*. *Evolution* 50: 256–266.

Shaw, K. L. 2000. Specific genetics of mate recognition: Inheritance of female acoustic preference in Hawaiian crickets. *Evolution* 54: 1303–1312.

Shaw, K. L. 2002. Conflict between nuclear and mitochondrial DNA phylogenies of a recent species radiation: What mtDNA reveals and conceals about modes of speciation in Hawaiian crickets. *Proc. Natl. Acad. Sci. USA* 99: 16122–16127.

Shedlock, A. M. and N. Okada. 2000. SINE insertions: Powerful tools for molecular systematics. *BioEssays* 22: 148–160.

Sheldon, F. H. and A. H. Bledsoe. 1989. Indexes to the reassociation and stability of solution DNA hybrids. *J. Mol. Evol.* 29: 328–343.

Sheppard, W. S., T. E. Rinderer, J. A. Mazzoli, J. A. Stelzer and H. Shimanuki. 1991. Gene flow between African- and European-derived honey bee populations in Argentina. *Nature* 349: 782–784.

Sheridan, M. and R. H. Tamarin. 1986. Kinships in a natural meadow vole population. *Behav. Ecol. Sociobiol.* 19: 207–211.

Sherman, P. W. and M. L. Morton. 1988. Extra-pair fertilizations in mountain white-crowned sparrows. *Behav. Ecol. Sociobiol.* 22: 413–420.

Sherman, P. W., J. U. M. Jarvis and R. D. Alexander (eds.). 1991. *The Biology of the Naked Mole-Rat*. Princeton University Press, Princeton, NJ.

Sherratt, D. J. 1995. *Mobile Genetic Elements*. Oxford University Press, New York.

Sherwin, W. B., N. D. Murray, J. A. M. Graves and P. R. Brown. 1991. Measurement of genetic variation in endangered populations: Bandicoots (Marsupialia: Peramelidae) as an example. *Cons. Biol.* 5: 103–108.

Shields, G. F. and J. R. Gust. 1995. Lack of geographic structure in mitochondrial DNA sequences of Bering Sea walleye pollack, *Theragra chalcogramma*. *Mol. Mar. Biol. Biotech.* 4: 69–82.

Shields, G. F. and A. C. Wilson. 1987a. Calibration of mitochondrial DNA evolution in geese. *J. Mol. Evol.* 24: 212–217.

Shields, G. F. and A. C. Wilson. 1987b. Subspecies of the Canada goose (*Branta canadensis*) have distinct mitochondrial DNAs. *Evolution* 41: 662–666.

Shields, G. F. and 8 others. 2000. Phylogeography of mitochondrial DNA variation in brown bears and polar bears. *Mol. Phylogen. Evol.* 15: 319–326.

Shimamura, M. and 8 others. 1997. Molecular evidence from retroposons that whales form a clade within even-toed ungulates. *Nature* 388: 666–670.

Shine, R. 1999. Why is sex determined by nest temperature in many reptiles? *Trends Ecol. Evol.* 14: 186–189.

Shingleton, A. W. and D. L. Stern. 2002. Molecular phylogenetic evidence for multiple gains or losses of ant mutualism within the aphid genus *Chaitophorus*. *Mol. Phylogen. Evol.* 26: 26–35.

Shinozaki, K. and 22 others. 1986. The complete nucleotide sequence of the tobacco chloroplast genome: Its gene organization and expression. *The EMBO J.* 5: 2043–2049.

Shirihai, H., G. Gargallo and A. J. Helbig. 2001. *Sylvia Warblers: Identification, Taxonomy and Phylogeny of the Genus Sylvia*. A. & C. Black, London.

Shively, C. and D. G. Smith. 1985. Social status and reproductive success of male *Macaca fascicularis*. *Am. J. Primatol.* 9: 129–135.

Shivji, M., S. Clarke, M. Park, L. Natanson, N. Kohler and M. Stanhope. 2002. Genetic identification of pelagic shark body parts for conservation and trade monitoring. *Cons. Biol.* 16: 1036–1047.

Shoemaker, D. D, C. A. Machado, D. Molbo, J. H. Werren, D. M. Windsor and E. A. Herre. 2002. The distribution of *Wolbachia* in fig wasps: Correlations with host phylogeny, ecology and population structure. *Proc. R. Soc. Lond. B* 269: 2257–2267.

Shore, J. S. and M. Triassi. 1998. Paternally baised cpDNA inheritance in *Turnera ulmifolia* (Turneraceae). *Am. J. Bot.* 85: 328–332.

Shows, T. B. 1983. Human genome organization of enzyme loci and metabolic diseases. *Isozymes* X: 323–339.

Shulman, M. J. and E. Bermingham. 1995. Early life histories, ocean currents, and the population genetics of Caribbean reef fishes. *Evolution* 49: 897–910.

Shykoff, J. A. and P. Schmid-Hempel. 1991a. Genetic relatedness and eusociality: Parasite-mediated selection on the genetic composition of groups. *Behav. Ecol. Sociobiol.* 28: 371–376.

Shykoff, J. A. and P. Schmid-Hempel. 1991b. Parasites and the advantage of genetic variability within social insect colonies. *Proc. R. Soc. Lond. B* 243: 55–58.

Sibley, C. G. 1991. Phylogeny and classification of birds from DNA comparisons. *Acta XX Congressus Internationalis Ornithologici* 1: 111–126.

Sibley, C. G. and J. E. Ahlquist. 1981. The phylogeny and relationships of the ratite birds as indicated by DNA-DNA hybridization. Pp. 301–335 in *Evolution Today*, G. G. E. Scudder and J. L. Reveal (eds.). Hunt Institue Botanical Document, Pittsburgh, PA.

Sibley, C. G. and J. E. Ahlquist. 1984. The phylogeny of the hominoid primates, as indicated by DNA-DNA hybridization. *J. Mol. Evol.* 20: 2–15.

Sibley, C. G. and J. E. Ahlquist. 1986. Reconstructing bird phylogeny by comparing DNA's. *Sci. Am.* 254(2): 82–93.

Sibley, C. G. and J. E. Ahlquist. 1987. DNA hybridization evidence of hominoid phylogeny: Results from an expanded data set. *J. Mol. Evol.* 26: 99–121.

#Sibley, C. G. and J. E. Ahlquist. 1990. *Phylogeny and Classification of Birds*. Yale University Press, New Haven, CT.

Sibley, C. G., J. E. Ahlquist and B. L. Monroe Jr. 1988. A classification of the living birds of the world based on DNA-DNA hybridization studies. *Auk* 105: 409–423.

Sibley, C. G., J. A. Comstock and J. E. Ahlquist. 1990. DNA hybridization evidence of hominoid phylogeny: A reanalysis of the data. *J. Mol. Evol.* 30: 202–236.

Sibley, L. D. and J. C. Boothroyd. 1992. Virulent strains of *Toxoplasma gondii* comprise a single clonal lineage. *Nature* 359: 82–85.

Sidell, B. D., R. G. Otto and D. A. Powers. 1978. A biochemical method for distinction of striped bass and white perch larvae. *Copeia* 1978: 340–343.

Siegel-Causey, D. 1997. Phylogeny of the Pelecaniformes: Molecular systematics of a primitive group. Pp. 159–171 in *Avian Molecular Evolution and Systematics*, D. Mindell (ed.). Academic Press, San Diego, CA.

Signer, E. N., C. R. Schmidt and A. J. Jeffreys. 1994. DNA variability and parentage testing in captive Waldripp ibises. *Mol. Ecol.* 3: 291–300.

Signer, E. N., G. Anzenberger and A. J. Jeffreys. 2000. Chimaeric and constitutive DNA fingerprints in the common marmoset (*Callithrix jacchus*). *Primates* 41: 49–61.

Signorovitch, J. and R. Nielsen. 2002. PATRI: Paternity inference using genetic data. *Bioinformatics* 18: 341–342.

Silander, J. A. Jr. 1985. Microevolution in clonal plants. Pp. 107–152 in *Population Biology and Evolution of Clonal Organisms*, J. B. C. Jackson, L. W. Buss and R. E. Cook (eds.). Yale University Press, New Haven, CT.

Silberman, J. D. and P. J. Walsh. 1992. Species identification of spiny lobster phyllosome larvae via ribosomal DNA analysis. *Mol. Mar. Biol. Biotech.* 1: 195–205.

Simberloff, D. 1988. The contribution of population and community biology to conservation science. *Annu. Rev. Ecol. Syst.* 19: 473–511.

Simberloff, D. and J. Cox. 1987. Consequences and costs of conservation corridors. *Cons. Biol.* 1: 63–71.

Simberloff, D., J. A. Farr, J. Cox and D. W. Mehlman. 1992. Movement corridors: Conservation bargains or poor investments. *Cons. Biol.* 6: 493–504.

Simmons, G. M. 1992. Horizontal transfer of *hobo* transposable elements within the *Drosophila melanogaster* species complex: Evidence from DNA sequencing. *Mol. Biol. Evol.* 9: 1050–1060.

Simmons, L. W. and M. T. Siva-Jothy. 1998. Sperm competition in insects: Mechanisms and the potential for selection. Pp. 341–434 in *Sperm Competition and Sexual Selection*, T. R. Birkhead and A. P. Møller (eds.). Academic Press, London.

Simon, C. M. 1979. Evolution of periodical cicadas: Phylogenetic inferences based on allozyme data. *Syst. Zool.* 28: 22–39.

Simon, C. M., F. Frati, A. Beckenbach, B. Crespi, H. Liu and P. Flook. 1994. Evolution, weighting, and phylogenetic utility of mitochondrial gene sequences and a compilation of conserved polymerase chain reaction primers. *Annals Entomol. Soc. Am.* 87: 651–701.

Simon, J.-C. and 6 others. 1999. Reproductive mode and population genetic structure of the cereal aphid *Sitobion avenae* studied using phenotypic and microsatellite markers. *Mol. Ecol.* 8: 531–545.

Simpson, G. G. 1940. Mammals and land bridges. *J. Washington Acad. Sci.* 30: 137–163.

Simpson, G. G. 1944. *Tempo and Mode in Evolution*. Columbia University Press, New York.

Simpson, G. G. 1945. The principles of classification and a classification of mammals. *Bull. Am. Mus. Nat. Hist.* 85: 1–350.

Simpson, G. G. 1951. The species concept. *Evolution* 5: 285–298.

Sinclair, A. H. and 9 others. 1990. A gene from the human sex-determining region encodes a protein with homology to a conserved DNA-binding motif. *Nature* 346: 240–244.

Sinclair, E. A. 2001. Phylgeographic variation in the quokka, *Setonix brachyurus* (Marsupialia: Macropodidae): Implications for conservation. *Anim. Cons.* 4: 325–333.

Sinervo, B. and J. Clobert. 2003. Morphs, dispersal behavior, genetic similarity, and the evolution of cooperation. *Science* 300: 1949–1951.

Sinervo, B. and C. M. Lively. 1996. The rock-paper-scissors game and the evolution of alternative male strategies. *Nature* 380: 240–243.

Singh, R. S. and L. R. Rhomberg. 1987. A comprehensive study of genic variation in natural populations of *Drosophila melanogaster*. I. Estimates of gene flow from rare alleles. *Genetics* 115: 313–322.

Singh, S. M. and E. Zouros. 1978. Genetic variability associated with growth rate in the American oyster (*Crassostrea virginica*). *Evolution* 32: 342–353.

Sites, J. W. Jr. and I. F. Greenbaum. 1983. Chromosome evolution in the iguanid lizard *Sceloporus grammicus*. II. Allozyme variation. *Evolution* 37: 54–65.

Sites, J. W. Jr. and C. Moritz. 1987. Chromosomal evolution and speciation revisited. *Syst. Zool.* 36: 153–174.

Sites, J. W. Jr., R. L. Bezy and P. Thompson. 1986. Nonrandom heteropolymer expression of lactate dehydrogenase isozymes in the lizard family Xantusiidae. *Biochem. Syst. Ecol.* 14: 539–545.

Skaala, Ø. and G. Nævdal. 1989. Genetic differentiation between freshwater resident and anadromous brown trout, *Salmo trutta*, within watercourses. *J. Fish Biol.* 34: 597–605.

Skrochowska, S. 1969. Migrations of the sea-trout (*Salmo trutta* L.), brown trout (*Salmo trutta* m. *fario* L.), and their crosses. *Pol. Arch. Hydrobiol.* 16: 125–192.

Slade, R. W. and C. Moritz. 1998. Phylogeography of *Bufo marinus* from its natural and introduced range. *Proc. R. Soc. Lond. B* 265: 769–777.

Slade, R. W., C. Moritz, A. R. Hoelzel and H. R. Burton. 1998. Molecular population genetics of the southern elephant seal *Mirounga leonina*. *Genetics* 149: 1945–1957.

Slate, J. and J. M. Pemberton. 2002. Comparing molecular measures for detecting inbreeding depression. *J. Evol. Biol.* 15: 20–31.

Slate, J., L. E. B. Kruuk, T. C. Marshall, J. M. Pemberton and T. H. Clutton-Brock. 2000. Inbreeding depression influences lifetime breeding success in a wild population of red deer (*Cervus elaphus*). *Proc. R. Soc. Lond. B* 267: 1657–1662.

Slatkin, M. 1985a. Rare alleles as indicators of gene flow. *Evolution* 39: 53–65.

Slatkin, M. 1985b. Gene flow in natural populations. *Annu. Rev. Ecol. Syst.* 16: 393–430.

Slatkin, M. 1987. Gene flow and the geographic structure of natural populations. *Science* 236: 787–792.

Slatkin, M. 1995. A measure of population subdivision based on microsatellite allele frequencies. *Genetics* 139: 457–462.

Slatkin, M. 1996. In defense of founder-flush theories of speciation. *Am. Nat.* 147: 493–505.

Slatkin, M. and N. H. Barton. 1989. A comparison of three indirect methods for estimating average levels of gene flow. *Evolution* 43: 1349–1368.

Slatkin, M. and R. R. Hudson. 1991. Pairwise comparisons of mitochondrial DNA sequences in stable and exponentially growing populations. *Genetics* 129: 555–562.

Slatkin, M. and W. P. Maddison. 1989. A cladistic measure of gene flow inferred from the phylogenies of alleles. *Genetics* 123: 603–613.

Slattery, J. P. and S. J. O'Brien. 1995. Molecular phylogeny of the red panda (*Ailurus fulgens*). *J. Heredity* 86: 413–422.

Slowinski, J. B. and R. D. M. Page. 1999. How should phylogenies be inferred from sequence data? *Syst. Biol.* 48: 814–825.

Small, M. F. and D. G. Smith. 1982. The relationship between maternal and paternal rank in rhesus macaques (*Macaca mulatta*). *Anim. Behav.* 30: 626–633.

Smit, A. F. 1999. Interspersed repeats and other mementos of transposable elements in mammalian genomes. *Curr. Opin. Genet. Develop.* 9: 657–663.

Smith, A. B. 1992. Echinoderm phylogeny: Morphology and molecules approach accord. *Trends Ecol. Evol.* 7: 224–239.

Smith, D. G. 1981. The association between rank and reproductive success in male rhesus monkeys. *Am. J. Primatol.* 1: 83–90.

Smith, D. G. and S. Smith. 1988. Parental rank and reproductive success of natal rhesus males. *Anim. Behav.* 36: 554–562.

Smith, D. R., O. R. Taylor and W. M. Brown. 1989. Neotropical Africanized honey bees have African mitochondrial DNA. *Nature* 339: 213–215.

Smith, E. F. G., P. Arctander, J. Fjeldsa and O. G. Amir. 1991. A new species of shrike (Laniidae: *Laniarius*) from Somalia, verified by DNA sequence data from the only known individual. *Ibis* 133: 227–235.

Smith, H. G., R. Montgomerie, T. Poldmaa, B. N. White and P. T. Boag. 1991. DNA fingerprinting reveals relation between tail ornaments and cuckoldry in barn swallows, *Hirundo rustica*. *Behav. Ecol.* 2: 90–98.

Smith, J. J. and G. L. Bush. 1997. Phylogeny of the genus *Rhagoletis* (Diptera: Tephritidae) inferred from DNA sequences of mitochondrial cytochrome oxidase II. *Mol. Phylogen. Evol.* 7: 33–43.

Smith, M. F., W. K. Thomas and J. L. Patton. 1992. Mitochondrial DNA-like sequence in the nuclear genome of an akodontine rodent. *Mol. Biol. Evol.* 9: 204–215.

Smith, M. H., C. T. Garten Jr. and P. R. Ramsey. 1975. Genic heterozygosity and population dynamics in small mammals. *Isozymes* IV: 85–102.

Smith, M. H., K. T. Scribner, J. D. Hernandez and M. C. Wooten. 1989. Demographic, spatial, and temporal genetic variation in *Gambusia*. Pp. 235–257 in *Ecology and Evolution of Livebearing Fishes* (Poeciliidae), G. K. Meffe and F. F. Snelson Jr. (eds.). Prentice Hall, Englewood Cliffs, NJ.

Smith, M. J., A. Arndt, S. Gorski and E. Fajber. 1993. The phylogeny of echinoderm classes based on mitochondrial gene arrangements. *J. Mol. Evol.* 36: 545–554.

Smith, M. L., L. C. Duchesne, J. N. Bruhn and J. B. Anderson. 1990. Mitochondrial genetics in a natural population of the plant pathogen *Armillaria*. *Genetics* 126: 575–582.

Smith, M. L., J. N. Bruhn and J. B. Anderson. 1992. The fungus *Armillaria bulbosa* is among the largest and oldest living organisms. *Nature* 356: 428–431.

Smith, M. W. and R. F. Doolittle. 1992. Anomalous phylogeny involving the enzyme glucose-6-phosphate isomerase. *J. Mol. Evol.* 34: 544–545.

Smith, M. W., D.-F. Feng and R. F. Doolittle. 1992. Evolution by acquisition: The case for horizontal gene transfers. *Trends Biochem. Sci.* 17: 489–493.

Smith, N. G. C. and A. Eyre-Walker. 2002. Adaptive protein evolution in *Drosophila*. *Nature* 415: 1022–1024.

Smith, P. J. 1986. Genetic similarity between samples of the orange roughy *Hoplostethus atlanticus* from the Tasman Sea, south-west Pacific Ocean and north-east Atlantic Ocean. *Mar. Biol.* 91: 173–180.

Smith, P. J. and P. G. Benson. 1980. Electrophoretic identification of larval and O-group flounders (*Rhombosolea* spp.) from Wellington Harbour, *N. Z. J. Mar. Freshwater Res.* 14: 401–404.

Smith, P. J. and J. Crossland. 1977. Identification of larvae of snapper, *Chrysophrys auratus* Forster, by electrophoretic separation of tissue enzymes. *N. Z. Mar. Freshwater Res.* 11: 795–798.

Smith, P. J. and Y. Fujio. 1982. Genetic variation in marine teleosts: High variability in habitat specialists and low variability in habitat generalists. *Mar. Biol.* 69: 7–20.

Smith, R. L. (ed.). 1984. *Sperm Competition and the Evolution of Animal Mating Systems*. Academic Press, New York.

Smith, R. L. and K. J. Sytsma. 1990. Evolution of *Populus nigra* L. (sect. *Aigeiros*): Introgressive hybridization and the chloroplast contribution of *Populus alba* L. (sect. *Populus*). *Am. J. Bot.* 77: 1176–1187.

Smith, S. C., R. R. Racine and C. H. Langley. 1980. Lack of genic variation in the abundant proteins of human kidney. *Genetics* 96: 967–974.

Smith, T. B. and R. K. Wayne (eds.). 1996. *Molecular Genetic Approaches in Conservation*. Oxford University Press, New York.

Smith, T. B., K. Holder, D. Girman, K. O'Keefe, B. Larson and Y. Chan. 2000. Comparative avian phylogeography of Cameroon and Equatorial Guinea mountains: Implications for conservation. *Mol. Ecol.* 9: 1505–1516.

Smithies, O. 1955. Zone electrophoresis in starch gels: Group variations in the serum proteins of normal individuals. *Biochem. J.* 61: 629–641.

Smithies, O. and P. A. Powers. 1986. Gene conversions and their relation to homologous chromosome pairing. *Phil. Trans. R. Soc. Lond.* B312: 291–302.

Smouse, P. E. 1986. The fitness consequences of multiple-locus heterozygosity under the inbreeding and multiplicative overdominance and inbreeding depression models. *Evolution* 40: 946–957.

Smouse, P. E. and T. R. Meagher. 1994. Genetic analysis of male reproductive contributions in *Chamaelium luteum* (L.) Gray (Liliaceae). *Genetics* 136: 313–322.

Smouse, P. E., R. S. Waples and J. A. Tworek. 1990. A genetic mixture analysis for use with incomplete source population-data. *Can. J. Fish. Aquat. Sci.* 47: 620–634.

Smyth, C. A. and J. L. Hamrick. 1987. Realized gene flow via pollen in artificial populations of the musk thistle, *Carduus nutans* L. *Evolution* 41: 613–619.

Sneath, P. H. A. and R. R. Sokal. 1973. *Numerical Taxonomy*. W. H. Freeman and Co., San Francisco.

Snell, W. J. 1990. Adhesion and signalling during fertilization in multicellular and unicellular organisms. *Curr. Biol.* 2: 821–832.

Snow, A. A. 1990. Effects of pollen-load size and number of donors on sporophyte fitness in wild radish (*Raphanus raphanistrum*). *Am. Nat.* 136: 742–758.

Snow, A. A. and T. P. Spira. 1991. Differential pollen-tube growth rates and nonrandom fertilization in *Hibiscus moscheutos* (Malvaceae). *Am. J. Bot.* 78: 1419–1426.

Sober, E. 1983. Parsimony in systematics: Philosophical issues. *Annu. Rev. Ecol. Syst.* 14: 335–357.

Sogin, M. L. 1991. Early evolution and the origin of eukaryotes. *Curr. Opinion Genet. Develop.* 1: 457–463.

Sogin, M. L., U. Edman and H. Elwood. 1989. A single kingdom of eukaryotes. Pp. 133–143 in *The Hierarchy of Life*, B. Fernholm, K. Bremer and H. Jornvall (eds.). Elsevier, Amsterdam.

Sokal, R. R. and N. L. Oden. 1978a. Spatial autocorrelation in biology. 1. Methodology. *Biol. J. Linn. Soc.* 10: 199–228.

Sokal, R. R. and N. L. Oden. 1978b. Spatial autocorrelation in biology. 2. Some biological implications and four applications of evolutionary and ecological interest. *Biol. J. Linn. Soc.* 10: 229–249.

Sokal, R. R. and P. H. A. Sneath. 1963. *Principles of Numerical Taxonomy*. W. H. Freeman and Co., San Francisco.

Solé-Cava, A. M. and J. P. Thorpe. 1986. Genetic differentiation between morphotypes of the marine sponge *Suberites ficus* (Demospongiae: Hadromerida). *Mar. Biol.* 93: 247–253.

Solé-Cava, A. M. and J. P. Thorpe. 1989. Biochemical genetics of genetic variation in marine lower invertebrates. *Biochem. Genet.* 27: 303–312.

Solignac, M. and M. Monnerot. 1986. Race formation, speciation, and introgression within *Drosophila simulans*, *D. mauritiana* and *D. sechellia* inferred from mitochondrial DNA analysis. *Evolution* 40: 531–539.

Solignac, M., J. Génermont, M. Monnerot and J.-C. Mounolou. 1984. Genetics of mitochondria in *Drosophila*: mtDNA inheritance in heteroplasmic strains of *D. mauritiana*. *Mol. Gen. Genet.* 197: 183–188.

Solignac, M., J. Génermont, M. Monnerot and J.-C. Mounolou. 1987. *Drosophila* mitochondrial genetics: Evolution of heteroplasmy through germ line cell divisions. *Genetics* 117: 687–696.

Soltis, D. E. and M. A. Gitzendanner. 1999. Molecular systematics and the conservation of rare species. *Cons. Biol.* 13: 471–483.

Soltis, D. E. and R. K. Kuzoff. 1995. Discordance between nuclear and chloroplast phylogenies in the *Heuchera* group (Saxifragaceae). *Evolution* 49: 727–742.

Soltis, D. E., P. S. Soltis and B. D. Ness. 1989. Chloroplast-DNA variation and multiple origins of autopolyploidy in *Heuchera micrantha* (Saxifragaceae). *Evolution* 43: 650–656.

Soltis, D. E., P. S. Soltis, M. T. Clegg and M. Durbin. 1990. *rbcL* sequence divergence and phylogenetic relationships in Saxifragaceae sensu lato. *Proc. Natl. Acad. Sci. USA* 87: 4640–4644.

Soltis, D. E., P. S. Soltis, T. G. Collier and M. L. Edgerton. 1991. Chloroplast DNA variation within and among genera of the *Heuchera* group: Evidence for chloroplast capture and paraphyly. *Am. J. Bot.* 78: 1091–1112.

Soltis, D. E., P. S. Soltis and B. G. Milligan. 1992. Intraspecific chloroplast DNA variation: Systematic and phylogenetic implications. Pp. 117–150 in *Molecular Systematics of Plants*, P. S. Soltis, D. E. Soltis and J. J. Doyle (eds.). Chapman & Hall, New York.

Soltis, D. E., M. A. Gitzendanner, D. D. Strenge and P. S. Soltis. 1997. Chloroplast DNA intraspecific phylogeography of plants from the Pacific Northwest of North America. *Plant Syst. Evol.* 206: 353–373.

Soltis, D. E., P. S. Soltis and J. J. Doyle (eds.). 1998. *Molecular Systematics of Plants II. DNA Sequencing*. Kluwer, Boston, MA.

Soltis, P. S., D. E. Soltis and L. D. Gottlieb. 1987. Phosphoglucomutase gene duplications in *Clarkia* (Onagraceae) and their phylogenetic implications. *Evolution* 41: 667–671.

Soltis, P. S., D. E. Soltis and C. J. Smiley. 1992a. An *rbcL* sequence from a Miocene *Taxodium* (bald cypress). *Proc. Natl. Acad. Sci. USA* 89: 449–451.

Soltis, P. S., D. E. Soltis, T. L. Tucker and F. A. Lang. 1992b. Allozyme variability is absent in the narrow endemic *Bensoniella oregona* (Saxifragaceae). *Cons. Biol.* 6: 131–134.

Soltis, P. S., D. E. Soltis and J. J. Doyle (eds.). 1992c. *Molecular Systematics of Plants*. Chapman & Hall, New York.

Soltis, P. S., D. E. Soltis and M. W. Chase. 1999. Angiosperm phylogeny inferred from multiple genes as a tool for comparative biology. *Nature* 402: 402–404.

Sommerfeldt, A. D. and J. D. D. Bishop. 1999. Random amplified polymorphic DNA (RAPD) analysis reveals extensive natural chimerism in a marine protochordate. *Mol. Ecol.* 8: 885–890.

Sorenson, M. D. and R. C. Fleischer. 1996. Multiple independent transpositions of mitochondrial DNA control region sequences to the nucleus. *Proc. Natl. Acad. Sci. USA* 93: 15239–15243.

Sorenson, M. D., J. C. Ast, D. E. Dimcheff, T. Yuri and D. P. Mindell. 1999a. Primers for a PCR-based approach to mitochondrial genome sequencing in birds and other vertebrates. *Mol. Phylogen. Evol.* 12: 105–114.

Sorenson, M. D. and 6 others. 1999b. Relationships of the extinct moa-nalos, flightless Hawaiian waterfowl, based on ancient DNA. *Proc. R. Soc. Lond. B* 266: 2187–2193.

Sorenson, M. D., E. Oneal, J. García-Moreno and D. P. Mindell. 2003. More taxa, more characters: The hoatzin problem is still unresolved. *Mol. Biol. Evol.* 20: 1484–1499.

Sosef, M. S. M. 1997. Hierarchical models, reticulate evolution and the inevitability of paraphyletic supraspecific taxa. *Taxon* 46: 75–85.

Sota, T. and A. P. Vogler. 2001. Incongruence of mitochondrial and nuclear gene trees in the carabid beetles *Ohomopterus*. *Syst. Biol.* 50: 39–59.

Soucy, S. and J. Travis. 2003. Multiple paternity and population genetic structure in natural populations of the poeciliid fish, *Heterndria formosa*. *J. Evol. Biol.* 16: 1328–1336.

Soulé, M. 1976. Allozyme variation: Its determinants in space and time. Pp. 60–77 in *Molecular Evolution*, F. J. Ayala (ed.). Sinauer Associates, Sunderland, MA.

Soulé, M. 1980. Thresholds for survival: Maintaining fitness and evolutionary potential. Pp. 111–124 in *Conservation Biology: An Evolutionary-Ecological Perspective*, M. E. Soulé and B. A. Wilcox (eds.). Sinauer Associates, Sunderland, MA.

Soulé, M. (ed.). 1986. *Conservation Biology: The Science of Scarcity and Diversity*. Sinauer Associates, Sunderland, MA.

Soulé, M. (ed.). 1987. *Viable Populations for Conservation*. Cambridge University Press, New York.

Soulé, M. and K. Kohm (eds.). 1989. *Research Priorities for Conservation Biology*. Island Press, Washington, DC.

Soulé, M. and B. R. Stewart. 1970. The "niche-variation" hypothesis: A test and alternatives. *Am. Nat.* 104: 85–97.

Soulé, M. and B. A. Wilcox. 1980. *Conservation Biology: An Ecological-Evolutionary Perspective*. Sinauer Associates, Sunderland, MA.

Soumalainen, E., A. Saura and J. Lokki. 1976. Evolution of parthenogenetic insects. *Evol. Biol.* 9: 209–257.

Sourdis, J. and C. Krimbas. 1987. Accuracy of phylogenetic trees estimated from DNA sequence data. *Mol. Biol. Evol.* 4: 159–166.

Southern, E. M. 1975. Detection of specific sequences among DNA fragments separated by gel electrophoresis. *J. Mol. Biol.* 98: 503–517.

Southern, S. O., P. J. Southern and A. E. Dizon. 1988. Molecular characterization of a cloned dolphin mitochondrial genome. *J. Mol. Evol.* 28: 32–42.

Sparrow, A. H., H. J. Price and A. G. Underbrink. 1972. A survey of DNA content per cell and per chromosome of prokaryotic and eukaryotic organisms: Some evolutionary considerations. *Brookhaven Symp. Biol.* 23: 451–494.

Spencer, C. C., C. E. Howell, A. R. Wright and D. E. L. Promislow. 2003. Testing an 'aging gene' in long-lived *Drosophila* strains: Increased longevity depends on sex and genetic background. *Aging Cell* 2: 123–130.

Sperisen, C., U. Büchler, F. Gugerli, G. Matyas, T. Geburek and G. G. Vendramin. 2001. Tandem repeats in plant mitochondrial genomes: Application to the analysis of population differentiation in the conifer Norway spruce. *Mol. Ecol.* 10: 257–263.

Spiess, E. B. and C. M. Wilke. 1984. Still another attempt to achieve assortative mating by disruptive selection in *Drosophila*. *Evolution* 38: 505–515.

Spilliaert, R., G. Vikingsson, U. Arnason, A. Pálsdóttir, J. Sigurjónsson and A. Arnason. 1991. Species hybridization between a female blue whale (*Balaenoptera musculus*) and a male fin whale (*B. physalus*): Molecular and morphological documentation. *J. Heredity* 82: 269–274.

Spirito, F. 1998. The role of chromosomal change in speciation. Pp. 320–329 in *Endless Forms: Species and Speciation*, D. J. Howard and S. H. Berlocher (eds.). Oxford University Press, New York.

Spolsky, C. M. and T. Uzzell. 1984. Natural interspecies transfer of mitochondrial DNA in amphibians. *Proc. Natl. Acad. Sci. USA* 81: 5802–5805.

Spolsky, C. M. and T. Uzzell. 1986. Evolutionary history of the hybridogenetic hybrid from *Rana esculenta* as deduced from mtDNA analyses. *Mol. Biol. Evol.* 3: 44–56.

Spolsky, C. M., C. A. Phillips and T. Uzzell. 1992. Antiquity of clonal salamander lineages revealed by mitochondrial DNA. *Nature* 356: 706–708.

Spong, G., M. Johansson and M. Björkland. 2000. High genetic variation in leopards indicates large and long-term stable effective population size. *Mol. Ecol.* 9: 1773–1782.

Sprague, G. F. Jr. 1991. Genetic exchange between kingdoms. *Curr. Opin. Genet. Dev.* 1: 530–533.

Spratt, B. G. and M. C. J. Maiden. 1999. Bacterial population genetics, evolution and epidemiology. *Phil. Trans. R. Soc. Lond. B* 354: 701–710.

Springer, M. and J. A. W. Kirsch. 1989. Rates of single-copy DNA evolution in phalangeriform marsupials. *Mol. Biol. Evol.* 6: 331–341.

Springer, M. and J. A. W. Kirsch. 1991. DNA hybridization, the compression effect, and the radiation of diprotodontian marsupials. *Syst. Zool.* 40: 131–151.

Springer, M. and C. Krajewski. 1989. DNA hybridization in animal taxonomy: A critique from first principles. *Q. Rev. Biol.* 64: 291–318.

Springer, M., J. A. W. Kirsch, K. Aplin and T. Flannery. 1990. DNA hybridization, cladistics, and the phylogeny of phalangerid marsupials. *J. Mol. Evol.* 30: 298–311.

Springer, M., E. H. Davidson and R. J. Britten. 1992. Calculation of sequence divergence from the thermal stability of DNA duplexes. *J. Mol. Evol.* 34: 379–382.

Springer, M. S., G. C. Cleven, O. Madsen, W. W. de Jong, V. G. Waddell, H. M. Amrine and M. J. Stanhope. 1997. Endemic African mammals shake the phylogenetic tree. *Nature* 388: 61–64.

Springer, M. S. and 6 others. 1998. The origin of the Australasian marsupial fauna and the phylogenetic affinities of the enigmatic monito del monte and marsupial mole. *Proc. R. Soc. Lond. B* 265: 2381–2386.

Springer, M. S., E. C. Teeling, O. Madsen, M. J. Stanhope and W. de Jong. 2001. Integrated fossil and molecular data reconstruct bat echolocation. *Proc. Natl. Acad. Sci. USA* 98: 6241–6246.

Springer, M. S., W. J. Murphy, E. Eizirik and S. J. O'Brien. 2003. Placental mammal diversification and the Cretaceous-Tertiary boundary. *Proc. Natl. Acad. Sci. USA* 100: 1056–1061.

Squirrell, J., P. M. Hollingsworth, R. M. Bateman, M. C. Tebbitt and M. L. Hollingsworth. 2002. Taxonomic complexity and breeding system transitions: Conservation genetics of the *Epipactis leptochila* complex (Orchidaceae). *Mol. Ecol.* 11: 1957–1964.

Squirrell, J. and 6 others. 2003. How much effort is required to isolate nuclear microsatellites from plants? *Mol. Ecol.* 12: 1339–1348.

Stacy, E. A., J. L. Harmick, J. D. Nason, S. P. Hubbell, R. B. Foster and R. Condit. 1996. Pollen dispersal in low-density populations of three Neotropical tree species. *Am. Nat.* 148: 275–298.

Stahl, E. A., G. Dwyer, R. Mauricio, M. Kreitman and J. Bergelson. 1999. Dynamics of disease resistance polymorphism at the *Rpm1* locus of *Arabidopsis*. *Nature* 400: 667–671.

Ståhl, G. 1987. Genetic population structure of Atlantic salmon. Pp. 121–140 in *Population Genetics and Fisheries Management*, N. Ryman and F. Utter (eds.). University of Washington Press, Seattle, WA.

Stallings, R. L., A. F. Ford, D. Nelson, D. C. Torney, C. E. Hildebrand and R. K. Moyzis. 1991. Evolution and distribution of (GT)n repetitive sequences in mammalian genomes. *Genomics* 10: 807–815.

Stangel, P. W., J. A. Rodgers Jr. and A. L. Bryan. 1990. Genetic variation and population structure of the Florida wood stork. *Auk* 107: 614–619.

Stangel, P. W., M. R. Lennartz and M. H. Smith. 1992. Genetic variation and population structure of red-cockaded woodpeckers. *Cons. Biol.* 6: 283–292.

Stanhope, M. J. and 7 others. 1998a. Molecular evidence for multiple origins of Insectivora and for a new order of endemic African insectivore mammals. *Proc. Natl. Acad. Sci. USA* 95: 9967–9972.

Stanhope, M. J., O. Madsen, V. G. Waddell, G. C. Cleven, W. W. de Jong and M. S. Springer. 1998b. Highly congruent molecular support for a diverse superordinal clade of endemic African mammals. *Mol. Phylogen. Evol.* 9: 501–508.

Stanhope, M. J., A. Lupas, M. J. Italia, K. K. Koretke, C. Volker and J. R. Brown. 2001. Phylogenetic analyses do not support horizontal gene transfers from bacteria to vertebrates. *Nature* 411: 940–944.

Stanley, S. M. 1975. A theory of evolution above the species level. *Proc. Natl. Acad. Sci. USA* 72: 646–650.

Stapel, S. O., J. A. M. Leunissen, M. Versteeg, J. Wattel and W. W. deJong. 1984. Ratites as oldest offshoot of avian stem: Evidence from α-crystallin A sequences. *Nature* 311: 257–259.

Stebbins, G. L. 1950. *Variation and Evolution in Plants*. Columbia University Press, New York.

Stebbins, G. L. 1970. Adaptive radiation in angiosperms. I. Pollination mechanisms. *Annu. Rev. Ecol. Syst.* 1: 307–326.

Stebbins, G. L. 1989. Plant speciation and the founder principle. Pp. 113–125 in *Genetics, Speciation and the Founder Principle*, L. V. Giddings, K. Y. Kaneshiro and W. W. Anderson (eds.). Oxford University Press, New York.

Steffens, G. J. and 6 others. 1983. The primary structure of Cu-Zn superoxide dismutase from *Photobacterium leiognathi*: Evidence for a separate evolution of Cu-Zn superoxide dismutase in bacteria. *Hoppe-Seyler's Z. Physiol Chem.* 364: 675–690.

Stephan, W. and C. H. Langley. 1992. Evolutionary consequences of DNA mismatch inhibited repair opportunity. *Genetics* 132: 567–574.

Stephens, J. C. 1985. Statistical methods of DNA sequence analysis: Detection of intragenic recombination or gene conversion. *Mol. Biol. Evol.* 2: 539–556.

Stephens, J. C. and M. Nei. 1985. Phylogenetic analysis of polymorphic DNA sequences at the *Adh* locus in *Drosophila melanogaster* and its sibling species. *J. Mol. Evol.* 22: 289–300.

Stephens, J. C., M. L. Cavanaugh, M. I. Gradie, M. L. Mador and K. K. Kidd. 1990a. Mapping the human genome: Current status. *Science* 250: 237–244.

Stephens, J. C., J. Rogers and G. Ruano. 1990b. Theoretical underpinning of the single-molecule-dilution (SMD) method of direct haplotype resolution. *Am. J. Human Genet.* 46: 1149–1155.

Stephens, J. C., N. J. Smith and P. Donnelly. 2001a. A new statistical method for haplotype reconstruction from population data. *Am. J. Human Genet.* 68: 978–989.

Stephens, J. C. and 27 others. 2001b. Haplotype variation and linkage disequilibrium in 313 human genes. *Science* 293: 489–493.

Stern, B. R. and D. G. Smith. 1984. Sexual behaviour and paternity in three captive groups of rhesus monkeys (*Macaca mulatta*). *Anim. Behav.* 32: 23–32.

Stern, D. B. and J. D. Palmer. 1984. Extensive and widespread homologies between mitochondrial DNA and chloroplast DNA in plants. *Proc. Natl. Acad. Sci. USA* 81: 1946–1950.

Stern, D. L. and W. A. Foster. 1996. The evolution of soldiers in aphids. *Biol. Rev. Camb. Philos. Soc.* 71: 27–79.

Stevens, P. F. 1980. Evolutionary polarity of character states. *Annu. Rev. Ecol. Syst.* 11: 333–358.

Stewart, I. J., T. P. Quinn and P. Bentzen. 2003. Evidence for fine-scale natal homing among island beach spawning sockeye salmon, *Oncorhynchus nerka*. *Environ. Biol. Fish.* 67: 77–85.

Stewart, J. and A. Lister. 2001. Cryptic northern refugia and the origins of the modern biota. *Trends Ecol. Evol.* 16: 608–613.

Stiassny, M. L. and A. Meyer. 1999. Cichlids of the rift lakes. *Sci. Am.* 280(2): 64–69.

Stille, M., B. Stille and P. Douwes. 1991. Polygyny, relatedness and nest founding in the polygynous myrmecine ant *Leptothorax acervorum* (Hymenoptera; Formicidae). *Behav. Ecol. Sociobiol.* 28: 91–96.

Stiller, J. W., D. C. Reel and J. C. Johnson. 2003. A single origin of plastids revisited: Convergent evolution in organellar genome content. *J. Phycol.* 39: 95–105.

Stock, D. W., K. D. Moberg, L. R. Maxson and G. S. Whitt. 1991. A phylogenetic analysis of the 18S ribosomal RNA sequence of the coelacanth *Latimeria chalumnae*. *Env. Biol. Fishes* 32: 99–117.

Stoddart, J. A. 1983a. Asexual production of planulae in the coral *Pocillopora damicornis*. *Mar. Biol.* 76: 279–288.

Stoddart, J. A. 1983b. The accumulation of genetic variation in a parthenogenetic snail. *Evolution* 37: 546–554.

Stoddart, J. A. 1984a. Genetic structure within populations of the coral *Pocillopora damicornis*. *Mar. Biol.* 81: 19–30.

Stoddart, J. A. 1984b. Genetic differentiation amongst populations of the coral *Pocillopora damicornis* off southwestern Australia. *Coral Reefs* 3: 149–156.

Stoddart, J. A., R. C. Babcock and A. J. Heyward. 1988. Self-fertilization and maternal enzymes in the planulae of the coral *Goniastrea favulus. Mar. Biol.* 99: 489–494.

Stoneking, M. 1995. Ancient DNA: How do you know when you have it and what can you do with it? *Am. J. Human Genet.* 57: 1259–1262.

Stoneking, M., K. Bhatia and A. C. Wilson. 1986. Mitochondrial DNA variation in eastern highlanders of Papua New Guinea. Pp. 87–100 in *Genetic Variation and Its Maintenance*, D. F. Roberts and G. F. DeStefano (eds.). Cambridge University Press, Cambridge, England.

Stoneking, M., D. Hedgecock, R. G. Higuchi, L. Vigilant and H. A. Erlich. 1991. Population variation of human mtDNA control region sequences detected by enzymatic amplification and sequence-specific oligonucleotide probes. *Am. J. Human Genet.* 48: 370–382.

Stoner, D. S. and I. L. Weissman. 1996. Somatic and germ cell parasitism in a colonial ascidian: Possible role for a highly polymorphic allorecognition system. *Proc. Natl. Acad. Sci. USA* 93: 15254–15259.

Stoner, D. S., B. Rinkovich and I. L. Weissman. 1999. Heritable germ and somatic cell lineage competitions in chimeric colonial protochordates. *Proc. Natl. Acad. Sci. USA* 96: 9148–9153.

Stouthamer, R., J. A. J. Breeuwer, R. F. Luck and J. H. Werren. 1993. Molecular identification of microorganisms associated with parthenogenesis. *Nature* 361: 66–68.

Straney, D. O. 1981. The stream of heredity: Genetics in the study of phylogeny. Pp. 100–138 in *Mammalian Population Genetics*, M. H. Smith and J. Joule (eds.). University of Georgia Press, Athens, GA.

Strassmann, J. E. 1996. Selective altruism towards closer over more distant relatives in colonies of the primitively eusocial wasp, *Polistes*. Pp. 190–201 in *Natural History and Evolution of Paper Wasps*, M. D. Breed and R. E. Page (eds.). Westview Press, Boulder, CO.

Strassmann, J. E. 2001. The rarity of multiple mating by females in the social Hymenoptera. *Insectes Sociaux* 48: 1–13.

Strassmann, J. E., C. R. Hughes, D. C. Queller, S. Turillazzi, R. Cervo, S. K. Davis and K. F. Goodnight. 1989. Genetic relatedness in primitively eusocial wasps. *Nature* 342: 268–269.

Strassmann, J. E., C. R. Hughes, S. Turillazzi, C. R. Solís and D. C. Queller. 1994. Genetic relatedness and incipient eusociality in stenogastrine wasps. *Anim. Behav.* 48: 813–821.

Strassmann, J. E., C. R. Solís, J. M. Peters and D. C. Queller. 1996. Strategies for finding and using highly polymorphic DNA microsatellite loci for studies of genetic relatedness and pedigrees. Pp. 163–180 in *Molecular Zoology: Advances, Strategies and Protocols*, J. D. Ferraris and S. R. Palumbi (eds.). John Wiley & Sons, New York.

Strassmann, J. E. and 7 others. 1997. Absence of within-colony kin discrimination in behavioural interactions of swarm-founding wasps. *Proc. R. Soc. Lond. B* 264: 1565–1570.

Strassmann, J. E., P. Seppä and D. C. Queller. 2000a. Absence of within-colony kin discrimination: Foundresses of the social wasp, *Polistes carolina*, do not prefer their own larvae. *Naturwissenschaften* 87: 266–269.

Strassmann, J. E., Y. Zhu and D. C. Queller. 2000b. Altruism and social cheating in the social amoeba *Dictyostelium discoideum. Nature* 408: 965–967.

Strauss, S. H., J. D. Palmer, G. T. Howe and A. H. Doerksen. 1988. Chloroplast genomes of two conifers lack a large inverted repeat and are extensively rearranged. *Proc. Natl. Acad. Sci. USA* 85: 3898–3902.

Strauss, S. Y. 1994. Levels of herbivory in host hybrid zones. *Trends Ecol. Evol.* 9: 209–214.

Strecker, U., L. Bernatchez and H. Wilkens. 2003. Genetic divergence between cave and surface populations of *Astyanax* in Mexico (Characidae, Teleostei). *Mol. Ecol.* 12: 699–710.

Streelman, J. T. and P. D. Danley. 2003. The stages of vertebrate evolutionary radiation. *Trends Ecol. Evol.* 18: 126–131.

Streelman, J. T., R. Zardoya, A. Meyer and S. A. Karl. 1998. Multilocus phylogeny of cichlid fishes (Pisces: Perciformes): Evolutionary comparison of microsatellite and single-copy nuclear loci. *Mol. Biol. Evol.* 15: 798–808.

Streelman, J. T., R. C. Alberston and T. D. Kocher. 2003. Genome mapping of the orange blotch color pattern in cichlid fishes. *Mol. Ecol.* 12: 2465–2471.

Strimmer, K. and A. von Haeseler. 1996. Quartet puzzling: A quartet maximum likelihood method for reconstructing tree topologies. *Mol. Biol. Evol.* 13: 964–969.

Stringer, C. B. and P. Andrews. 1988. Genetic and fossil evidence for the origin of modern humans. *Science* 239: 1263–1268.

Studier, J. A. and K. J. Keppler. 1988. A note on the neighbor-joining algorithm of Saitou and Nei. *Mol. Biol. Evol.* 5: 729–731.

Sturmbauer, C. and A. Meyer. 1992. Genetic divergence, speciation and morphological stasis in a lineage of African cichlid fishes. *Nature* 358: 578–581.

Stutchbury, B. J., J. Rhymer and E. S. Morton. 1994. Extrapair paternity in hooded warblers. *Behav. Ecol.* 5: 384–392.

Su, C., D. Evans, R. H. Cole, J. C. Kissinger, J. W. Ajioka and L. D. Sibley. 2003. Recent expansion of *Toxoplasma* through enhanced oral transmission. *Science* 299: 414–416.

Sugden, A. M., B. R. Jasny, E. Culotta and E. Pennisi. 2003. Charting the evolutionary history of life. *Science* 300: 1691.

Sullivan, J., K. E. Holsinger and C. Simon. 1995. Among-site rate variation and phylogenetic analysis of 12S rRNA in sigmodontine rodents. *Mol. Biol. Evol.* 12: 988–1001.

Sun, F., N. Arnheim and M. S. Waterman. 1995. Whole genome amplification of single cells: Mathematical analysis of PEP and tagged PCR. *Nucl. Acids Res.* 23: 3034–3040.

Sundström, L. and J. Boomsma. 2001. Conflicts and alliances in insect families. *Heredity* 86: 515–521.

Sunnucks, P., A. C. C. Wilson, I. B. Beheregaray, K. Zenger, J. French and A. C. Taylor. 2000. SSCP is not so difficult: The application and utility of single-stranded conformation polymorphism in evolutionary biology and molecular ecology. *Mol. Ecol.* 9: 1699–1710.

Suwamura, K., A. W. Davis and C.-I. Wu. 2000. Genetic analysis of speciation by means of introgression into *Drosophila melanogaster*. *Proc. Natl. Acad. Sci. USA* 97: 2652–2655.

Suyama, Y., K. Obayashi and I. Hayashi. 2000. Clonal structure in a dwarf bamboo (*Sasa senanensis*) population inferred from amplified fragment length polymorphism (AFLP) fingerprints. *Mol. Ecol.* 9: 901–906.

Swain, D. P. and C. J. Foote. 1999. Stocks and chameleons: The use of phenotypic variation in stock identification. *Fish. Res.* 43: 113–128.

Swanson, S. M., C. F. Aquadro and V. D. Vacquier. 2001a. Polymorphism in abalone fertilization proteins is consistent with the neutral evolution of the egg's receptor for lysin (VERL) and positive Darwinian selection of sperm lysin. *Mol. Biol. Evol.* 18: 376–383.

Swanson, S. M., Z. H. Zhang, M. F. Wolfner and C. F. Aquadro. 2001b. Positive Darwinian selection drives the evolution of several female reproductive proteins in mammals. *Proc. Natl. Acad. Sci. USA* 98: 2509–2514.

Swanson, S. M., A. G. Clark, H. M. Waldrip-Dail, M. F. Wolfner and C. F. Aquadro. 2001c. Evolutionary EST analysis identifies rapidly evolving reproductive proteins in *Drosophila*. *Proc. Natl. Acad. Sci. USA* 98: 7375–7379.

Swanson, W. J. and V. D. Vacquier. 1998. Concerted evolution in an egg receptor for a rapidly evolving abalone sperm protein. *Science* 281: 710–712.

Swart, M. K. J. and J. W. H. Ferguson. 1997. Conservation implications of genetic differentiation in southern African populations of black rhinoceros (*Diceros bicornis*). *Cons. Biol.* 11: 79–83.

Swatschek, I., D. Ristow and M. Wink. 1994. Mate fidelity and parentage in Cory's shearwater *Calonectris diomedea*: Field studies and DNA fingerprinting. *Mol. Ecol.* 3: 259–262.

Swearer, S. E., J. E. Caselle, D. W. Lea and R. R. Warner. 1999. Larval retention and recruitment in an island population of a coral-reef fish. *Nature* 402: 799–802.

Swensen, S. M., J. N. Luthi and L. H. Rieseberg. 1998. Datiscaceae revisited: Monophyly and the sequence of breeding system evolution. *Syst. Bot.* 23: 157–169.

Swift, C. C., C. R. Gilbert, S. A. Bortone, G. H. Burgess and R. W. Yerger. 1985. Zoogeography of the southeastern United States: Savannah River to Lake Ponchartrain. Pp. 213–265 in *Zoogeography of the North American Freshwater Fishes*, C. H. Hocutt and E. O Wiley (eds.). John Wiley & Sons, New York.

Swofford, D. L. 2000. *PAUP*: Phylogenetic Analysis Using Parsimony and Other Methods* (software). Sinauer Associates, Sunderland, MA.

Swofford, D. L., G. J. Olsen, P. J. Waddell and D. M. Hillis. 1996. Phylogenetic Inference. Pp. 407–514 in *Molecular Systematics* (2nd Edition), D. M. Hillis, C. Moritz and B. K. Mable (eds.). Sinauer Associates, Sunderland, MA.

Symondson, W. O. C. 2002. Molecular identification of prey in predator diets. *Mol. Ecol.* 11: 627–641.

Symula, R., R. Schulte and K. Summers. 2001. Molecular phylogenetic evidence for a mimetic radiation in Peruvian poison frogs supports a Müllerian mimicry hypothesis. *Proc. R. Soc. Lond. B* 268: 2415–2421.

Syren, R. M. and P. Luykx. 1977. Permanent segmental interchange complex in the termite *Incisitermes schwarzi*. *Nature* 266: 167–168.

Sytsma, K. J. and J. F. Smith. 1992. Molecular systematics of Onagraceae: Examples from *Clarkia* and *Fuchsia*. Pp. 295–323 in *Molecular Systematics of Plants*, P. S. Soltis, D. E. Soltis and J. J. Doyle (eds.). Chapman & Hall, New York.

Syvänen, A.-C. 1999. From gels to chips: Minisequencing primer extension for analysis of point mutations and single nucleotide polymorphisms. *Human Mutat.* 13: 1–10.

Syvanen, M. 2002. On the occurrence of horizontal gene transfer among an arbitrarily chosen group of 26 genes. *J. Mol. Evol.* 54: 258–266.

Syvanen, M. and C. I. Kado (eds.). 2002. *Horizontal Gene Transfer*. Academic Press, London.

Szmidt, A. E., T. Alden and J.-E. Hallgren. 1987. Paternal inheritance of chloroplast DNA in *Larix*. *Plant Mol. Biol.* 9: 59–64.

Szymura, J. M., C. Spolsky and T. Uzzell. 1985. Concordant change in mitochondrial and nuclear genes in a hybrid zone between two frog species (genus *Bombina*). *Experientia* 41: 1469–1470.

Tabachnick, W. J. and J. R. Powell. 1978. Genetic structure of the East African domestic population of *Aedes aegypti*. *Nature* 272: 535–537.

Tabachnick, W. J., L. E. Munstermann and J. R. Powell. 1979. Genetic distinctness of sympatric forms of *Aedes aegypti* in east Africa. *Evolution* 33: 287–295.

Taberlet, P. and J. Bouvet. 1991. A single plucked feather as a source of DNA for bird genetic studies. *Auk* 108: 959–960.

Taberlet, P. and J. Bouvet. 1992. Bear conservation genetics. *Nature* 358: 197.

Taberlet, P. and J. Bouvet. 1994. Mitochondrial DNA polymorphism, phylogeography, and conservation genetics of the brown bear *Ursus arctos* in Europe. *Proc. R. Soc. Lond. B* 255: 195–200.

Taberlet, P. and R. Cheddadi. 2002. Quaternary refugia and persistence of biodiversity. *Science* 297: 2009–2010.

Taberlet, P. and L. Fumagalli. 1996. Owl pellets as a source for genetic studies of small mammals. *Mol. Ecol.* 5: 301–305.

Taberlet, P. and G. Luikart. 1999. Non-invasive genetic sampling and individual identification. *Biol. J. Linn. Soc.* 68: 41–55.

Taberlet, P., H. Mattock, C. Dubois-Paganon and J. Bouvet. 1993. Sexing free-ranging brown bears *Ursos arctos* using hairs found in the field. *Mol. Ecol.* 2: 399–403.

Taberlet, P. and 8 others. 1997. Noninvasive genetic tracking of the endangered Pyrenean brown bear population. *Mol. Ecol.* 6: 869–876.

Taberlet, P., L. Fumagalli, A. G. West Saucy and J. F. Cosson. 1998. Comparative phylogeography and postglacial colonization routes in Europe. *Mol. Ecol.* 7: 453–464.

Taborsky, M. 2001. The evolution of bourgeois, parasitic and cooperative reproductive behaviors in fishes. *J. Heredity* 92: 100–109.

Taggart, D. A., W. G. Breed, P. D. Temple-Smith, A. Purvis and G. Shimmin. 1998. Reproduction, mating strategies and sperm competition in marsupials and monotremes. Pp. 623–666 in *Sperm Competition and Sexual Selection*, T. R. Birkhead and A. P. Møller (eds.). Academic Press, London.

Taib, Z. 1997. Branching processes and evolution. Pp. 321–329 in *Progress in Population Genetics and Human Evolution*, P. Donnelly and S. Tavaré (eds.). Springer-Verlag, New York.

Tajima, F. 1983. Evolutionary relationships of DNA sequences in finite populations. *Genetics* 105: 437–460.

Tajima, F. 1989a. Statistical method for testing the neutral mutation hypothesis by DNA polymorphism. *Genetics* 123: 585–595.

Tajima, F. 1989b. The effect of change in population size on DNA polymorphism. *Genetics* 123: 597–601.

Tajima, F. 1993. Simple methods for testing the molecular evolutionary clock hypothesis. *Genetics* 135: 599–607.

Takahashi, K., Y. Terai, M. Nishida and N. Okada. 1998. A novel family of short interspered repetitive elements (SINEs) from cichlids: The patterns of insertion of SINEs at orthologous loci support the proposed monophyly of four major groups of cichlid fishes in Lake Tanganyika. *Mol. Biol. Evol.* 15: 391–407.

Takahata, N. 1985. Population genetics of extranuclear genomes: A model and review. Pp. 195–212 in *Population Genetics and Molecular Evolution*, T. Ohta and K. Aoki (eds.). Springer-Verlag, Berlin.

Takahata, N. 1988. More on the episodic clock. *Genetics* 118: 387–388.

Takahata, N. 1989. Gene genealogy in three related populations: Consistency probability between gene and population trees. *Genetics* 122: 957–966.

Takahata, N. 1993. Allelic genealogy and human evolution. *Mol. Biol. Evol.* 10: 2–22.

Takahata, N. 1995. A genetic perspective on the origin and history of humans. *Annu. Rev. Ecol. Syst.* 26: 343–372.

Takahata, N. and A. G. Clark (eds.). 1993. *Mechanisms of Molecular Evolution*. Sinauer Associates, Sunderland, MA.

Takahata, N. and M. Nei. 1990. Allelic genealogy under overdominant and frequency-dependent selection and polymorphism of major histocompatibility complex loci. *Genetics* 124: 967–978.

Takahata, N. and S. R. Palumbi. 1985. Extranuclear differentiation and gene flow in the finite island model. *Genetics* 109: 441–457.

Takahata, N., Y. Satta and J. Klein. 1992. Polymorphism and balancing selection at major histocompatibility complex loci. *Genetics* 130: 925–938.

Takahata, N., S.-H. Lee and Y. Satta. 2001. Testing multiregionality of modern human origins. *Mol. Biol. Evol.* 18: 172–183.

Talbot, S. L. and G. F. Shields. 1996a. Phylogeography of brown bears (*Ursos arctos*) of Alaska and paraphyly within the Ursidae. *Mol. Phylogen. Evol.* 5: 477–494.

Talbot, S. L. and G. F. Shields. 1996b. A phylogeny for the bears (Ursidae) inferred from complete sequences of three mitochondrial genes. *Mol. Phylogen. Evol.* 5: 567–575.

Tamarin, R. H., M. Sheridan and C. K. Levy. 1983. Determining matrilineal kinship in natural populations of rodents using radionuclides. *Can. J. Zool.* 61: 271–274.

Tamas, I. and 8 others. 2002. 50 million years of genomic stasis in endosymbiotic bacteria. *Science* 296: 2376–2379.

Tanaka, T. and M. Nei. 1989. Positive Darwinian selection observed at the variable region genes of immunoglobulins. *Mol. Biol. Evol.* 6: 447–459.

Tanksley, S. D. 1993. Mapping polygenes. *Annu. Rev. Genet.* 27: 205–233.

Tashian, R. and G. Lasker (eds.). 1996. Molecular anthropology: Toward a new evolutionary paradigm (special issue). *Mol. Phylogen. Evol.* 5: 1–285.

Tateno, Y., M. Nei and F. Tajima. 1982. Accuracy of estimated phylogenetic trees from molecular data. I. Distantly related species. *J. Mol. Evol.* 18: 387–404.

Tauber, C. A. and M. J. Tauber. 1989. Sympatric speciation in insects: Perception and perspective. Pp. 307–344 in *Speciation and Its Consequences*, D. Otte and J. A. Endler (eds.). Sinauer Associates, Sunderland, MA.

Tautz, D. 1989. Hypervariability of simple sequences as a general source for polymorphic DNA markers. *Nucleic Acids Res.* 17: 6463–6471.

Tautz, D., P. Arctander, A. Minelli, R. H. Thomas and A. P. Vogler. 2003. A plea for DNA taxonomy. *Trends Ecol. Evol.* 18: 70–74.

Taylor, A. C., A. Horsup, C. N. Johnson, P. Sunnucks and B. Sherwin. 1997. Relatedness structure detected by microsatellite analysis and attempted pedigree reconstruction in an endangered marsupial, the northern hairy-nosed wombat *Lasiorhinus krefftii*. *Mol. Ecol.* 6: 9–19.

Taylor, E. B. and P. Bentzen. 1993. Evidence for multiple origins and sympatric divergence of trophic ecotypes of smelt (*Osmerus*) in Northeastern North America. *Evolution* 47: 813–832.

Taylor, M. S. and M. E. Hellberg. 2003. Genetic evidence for local retention of pelagic larvae in a Caribbean reef fish. *Science* 299: 107–109.

Teeling, E. C., M. Scally, D. J. Kao, M. L. Romagnoli, M. S. Springer and M. J. Stanhope. 2000. Molecular evidence regarding the origin of echolocation and flight in bats. *Nature* 403: 188–192.

Teeling, E. C., O. Madsen, R. A. Van Den Bussche, W. W. de Jong, M. J. Stanhope and M. S. Springer. 2002. Microbat paraphyly and the convergent evolution of a key innovation in Old World rhinolophoid microbats. *Proc. Natl. Acad. Sci. USA* 99: 1431–1436.

Tegelström, H. 1987a. Genetic variability in mitochondrial DNA in a regional population of the great tit (*Parus major*). *Biochem. Genet.* 25: 95–110.

Tegelström, H. 1987b. Transfer of mitochondrial DNA from the northern red-backed vole (*Clethrionomys rutilus*) to the bank vole (*C. glareolus*). *J. Mol. Evol.* 24: 218–227.

Tegelström, H. and H. P. Gelter. 1990. Haldane's rule and sex biassed gene flow between two hybridizing flycatcher species (*Ficedula albicollis* and *F. hypoleuca*, Aves: Muscicapidae). *Evolution* 44: 2012–2021.

Tegelström, H., P.-I. Wyoni, H. Gelter and M. Jaarola. 1988. Concordant divergence in proteins and mitochondrial DNA between two vole species in the genus *Clethrionomys*. *Biochem. Genet.* 26: 223–237.

Tegelström, H., J. Searle, J. Brookfield and S. Mercer. 1991. Multiple paternity in wild common shrews (*Sorex araneus*) is confirmed by DNA-fingerprinting. *Heredity* 66: 373–379.

Templeton, A. R. 1980a. Modes of speciation and inferences based on genetic distances. *Evolution* 34: 719–729.

Templeton, A. R. 1980b. The theory of speciation via the founder principle. *Genetics* 94: 1011–1038.

Templeton, A. R. 1983. Phylogenetic inference from restriction endonuclease cleavage site maps with particular reference to the humans and apes. *Evolution* 37: 221–244.

Templeton, A. R. 1987. Nonparametric inference from restriction cleavage sites. *Mol. Biol. Evol.* 4: 315–319.

Templeton, A. R. 1989. The meaning of species and speciation: A genetic perspective. Pp. 3–27 in *Speciation and Its Consequences*, D. Otte and J. A. Endler (eds.). Sinauer Associates, Sunderland, MA.

Templeton, A. R. 1992. Human origins and analysis of mitochondrial DNA sequences. *Science* 255: 737.

Templeton, A. R. 1993. The "Eve" hypothesis: A genetic critique and reanalysis. *Am. Anthropol.* 95: 51–72.

Templeton, A. R. 1994. The role of molecular genetics in speciation studies. Pp. 455–477 in *Molecular Approaches to Ecology and Evolution*, B. Schierwater, B. Streit, G. P. Wagner and R. DeSalle (eds.). Birkhäuser Verlag, Basel, Switzerland.

Templeton, A. R. 1996. Gene lineages and human evolution. *Science* 272: 1363.

Templeton, A. R. 1998. Nested clade analysis of phylogeographic data: Testing hypotheses about gene flow and population history. *Mol. Ecol.* 7: 381–397.

Templeton, A. R. and N. J. Georgiadis. 1996. A landscape approach to conservation genetics: Conserving evolutionary processes in the African Bovidae. Pp. 398–430 in *Conservation Genetics: Case Histories from Nature*, J. C. Avise and J. L. Hamrick (eds.). Chapman & Hall, New York.

Templeton, A. R. and B. Read. 1984. Factors eliminating inbreeding depression in a captive herd of Speke's gazelle (*Gazella spekei*). *Zoo Biol.* 3: 177–199.

Tennessen, J. A. and K. R. Zamudio. 2003. Early male reproductive advantage, multiple paternity and sperm storage in an amphibian aggregate breeder. *Mol. Ecol.* 12: 1567–1576.

Theisen, B. F. 1978. Allozyme clines and evidence of strong selection in three loci in *Mytilus edulis* (Bivalvia) from Danish waters. *Ophelia* 17: 135–142.

Thelen, G. C. and F. W. Allendorf. 2001. Heterozygosity-fitness correlations in rainbow trout: Effects of allozyme loci or associative overdominance? *Evolution* 55: 1180–1187.

Thoday, J. M. and J. B. Gibson. 1962. Isolation by disruptive selection. *Nature* 193: 1164–1166.

Thomas, R. H. and J. A. Hunt. 1991. The molecular evolution of the alcohol dehydrogenase locus and the phylogeny of Hawaiian *Drosophila*. *Mol. Biol. Evol.* 8: 687–702.

Thomas, R. H., W. Schaffner, A. C. Wilson and S. Pääbo. 1989. DNA phylogeny of the extinct marsupial wolf. *Nature* 340: 465–467.

Thomas, W. K. and A. T. Beckenbach. 1989. Variation in salmonid mitochondrial DNA: Evolutionary constraints and mechanisms of substitution. *J. Mol. Evol.* 29: 233–245.

Thomas, W. K., S. Pääbo, F. X. Villiblanca and A. C. Wilson. 1990. Spatial and temporal continuity of kangaroo rat populations shown by sequencing mitochondrial DNA from museum specimens. *J. Mol. Evol.* 31: 101–112.

Thomaz, D., E. Beall and T. Burke. 1997. Alternative reproductive tactics in Atlantic salmon: Factors affecting mature parr success. *Proc. R. Soc. Lond. B* 264: 219–226.

Thompson, C. E., E. B. Taylor and J. D. McPhail. 1997. Parallel evolution of lake-stream pairs of threespine sticklebacks (*Gasterosteus*) inferred from mitochondrial DNA variation. *Evolution* 51: 1955–1965.

Thompson, J. D., E. A. Herre, J. L. Hamrick and J. L. Stone. 1991. Genetic mosaics in strangler fig trees: Implications for tropical conservation. *Science* 254: 1214–1216.

Thorne, J., H. Kishino and I. Painter. 1998. Estimating the rate of evolution of the rate of molecular evolution. *Mol. Biol. Evol.* 15: 1647–1657.

Thornhill, N. W. (ed.). 1993. *The Natural History of Inbreeding and Outbreeding*. University of Chicago Press, Chicago, IL.

Thorpe, J. P. 1982. The molecular clock hypothesis: Biochemical evolution, genetic differentiation and systematics. *Annu. Rev. Ecol. Syst.* 13: 139–168.

Thorpe, R. S. 1996. The use of DNA divergence to help determine the correlates of evolution of morphological characters. *Evolution* 50: 524–531.

Thorpe, R. S., D. P. McGregor and A. M. Cummings. 1993. Population evolution of western Canary Island lizards (*Gallotia galloti*): 4-base endonuclease restriction fragment length polymorphisms of mitochondrial DNA. *Biol. J. Linn. Soc.* 49: 219–227.

Thorpe, R. S., D. P. McGregor, A. M. Cumming and W. C. Jordan. 1994. DNA evolution and colonization sequence of island lizards in relation to geological history: mtDNA RFLP, cytochrome *b*, cytochrome oxidase, 12S rRNA sequence, and nuclear RAPD analysis. *Evolution* 48: 230–240.

Thorsness, P. E. and E. R. Weber. 1996. Escape and migration of nucleic acids between chloroplasts, mitochondria, and the nucleus. *Int. Rev. Cytol.* 165: 207–234.

Thorson, G. 1961. Length of pelagic larval life in marine bottom invertebrates as related to larval transport by ocean currents. Pp. 455–474 in *Oceanography*, M. Sears (ed.). American Association for the Advancement of Science, Washington, DC.

Thresher, R. E. and E. B. Brothers. 1985. Reproductive ecology and biogeography of Indo-West Pacific angelfishes (Pisces: Pomacentridae). *Evolution* 39: 878–887.

Throckmorton, L. 1975. The phylogeny, ecology and geography of *Drosophila*. Pp. 421–469 in *Handbook of Genetics*, Vol. 3, R. C. King (ed.). Plenum Press, New York.

Thulin, C.-G., M. Jaarola and H. Tegelström. 1997. The occurrence of mountain hare mitochondrial DNA in wild brown hares. *Mol. Ecol.* 6: 463–467.

Tibayrenc, M. 1995. Population genetics of parasitic protozoa and other microorganisms. *Adv. Parasitol.* 36: 47–115.

Tibayrenc, M. 1998. Genetic epidemiology of parasitic protozoa and other infectious agents: The need for an integrated approach. *Int. J. Parasitol.* 28: 85–104.

Tibayrenc, M. and F. J. Ayala. 1987. Forte correlation entre classification isoenzymatique et variabilite de l'ADN kinetoplastique chez *Trypanosoma cruzi*. *Comp. Rend. Acad. des Sci.* 304: 89–92.

Tibayrenc, M. and F. J. Ayala. 1988. Isozyme variability in *Trypanosoma cruzi*, the agent of Chagas' disease: Genetical, taxonomical, and epidemiological significance. *Evolution* 42: 277–292.

Tibayrenc, M. and F. J. Ayala. 1991. Towards a population genetics of microorganisms: The clonal theory of parasitic protozoa. *Parasitol. Today* 7: 228–232.

Tibayrenc, M. and F. J. Ayala. 2002. The clonal theory of parasitic protozoa: 12 years on. *Trends Parasitol.* 18: 405–410.

Tibayrenc, M., P. Ward, A. Moya and F. J. Ayala. 1986. Natural populations of *Trypanosoma cruzi*, the agent of Chagas' disease, have a complex multiclonal structure. *Proc. Natl. Acad. Sci. USA* 83: 115–119.

Tibayrenc, M., F. Kjellberg and F. J. Ayala. 1990. A clonal theory of parasitic protozoa: The population structures of *Entamoeba, Giardia, Leishmania, Naegleria, Plasmodium, Trichomonas*, and *Trypanosoma* and their medical and taxonomical consequences. *Proc. Natl. Acad. Sci. USA* 87: 2414–2418.

Tibayrenc, M., F. Kjellberg, J. Arnaud, B. Oury, S. F. Breniere, M.-L. Darde and F. J. Ayala. 1991a. Are eukaryotic microorganisms clonal or sexual? A population genetics vantage. *Proc. Natl. Acad. Sci. USA* 88: 5129–5133.

Tibayrenc, M., F. Kjellberg and F. J. Ayala. 1991b. The clonal theory of parasitic protozoa. *BioScience* 41: 767–774.

Tikel, D., D. Blair and H. D. Marsh. 1996. Marine mammal faeces as a source of DNA. *Mol. Ecol.* 5: 456–457.

Tilley, S. G. and J. S. Hausman. 1976. Allozymic variation and occurrence of multiple inseminations in populations of the salamander *Desmognathus ochrophaeus*. *Copeia* 1976: 734–741.

Ting, C.-T., S.-C. Tsaur and C.-I. Wu. 2000. The phylogeny of closely related species as revealed by the genealogy of a speciation gene, *Odysseus*. *Proc. Natl. Acad. Sci. USA* 97: 5313–5316.

Tishkoff, S. A. and 14 others. 1996. Global patterns of linkage disequilibrium at the CD4 locus and modern human origins. *Science* 271: 1380–1387.

Tomiuk, J. and V. Loeschcke. 2003. Comments on : evolutionary and statistical properties of three genetic distances. *Mol. Ecol.* 12: 2275–2277.

Tomiuk, J., B. Guldbrandsten and V. Loeschcke. 1998. Population differentiation through mutation and drift: A comparison of genetic identity measures. *Genetica* 103: 545–558.

Torimaru, T., N. Tomaru, N. Nishimura and S. Yamamoto. 2003. Clonal diversity and genetic differentiation in *Ilex leucoclada* M. patches in an old-growth beech forest. *Mol. Ecol.* 12: 809–818.

Török, J., G. Michl, L. Z. Garamszegi and J. Barna. 2003. Repeated inseminations required for natural fertility in a wild bird population. *Proc. R. Soc. Lond. B* 270: 641–647.

Torroni, A. and 10 others. 1992. Native American mitochondrial DNA analysis indicates that the Amerind and the Nadene populations were founded by two independent migrations. *Genetics* 130: 153–162.

Tosi, A. J., J. C. Morales and D. J. Melnick. 2002. Y-chromosome and mitochondrial markers in *Macaca fascicularis* indicate introgression with Indochinese *M. mulatta* and a biogeographic barrier in the Isthmus of Kra. *Int. J. Primatol.* 23: 161–178.

Tosi, A. J., J. C. Morales and D. J. Melnick. 2003. Paternal, maternal, and biparental molecular markers provide unique windows onto the evolutionary history of macaque monkeys. *Evolution* 57: 1419–1435.

Tovar, J. and 8 others. 2003. Mitochondrial remnant organelles of *Giardia* function in iron-sulphur protein maturation. *Nature* 426: 172–176.

Tracey, M. L., K. Nelson, D. Hedgecock, R. A. Shleser and M. L. Pressick. 1975. Biochemical genetics of lobsters: Genetic variation and the structure of American lobster (*Homerus americanus*) populations. *J. Fish. Res. Bd. Can.* 32: 2091–2101.

Tranah, G. J., H. L. Kincaid, C. C. Krueger, D. E. Campton and B. May. 2001. Reproductive isolation in sympatric populations of pallid and shovelnose sturgeon. *North Am. J. Fish. Mgt.* 21: 367–373.

Travis, J., J. C. Trexler and M. Mulvey. 1990. Multiple paternity and its correlates in female *Poecilia latipinna* (Poeciliidae). *Copeia* 1990: 722–729.

Tregenza, T. and N. Wedell. 2000. Genetic compatibility, mate choice and patterns of parentage: An invited review. *Mol. Ecol.* 9: 1013–1027.

Trewick, S. A. 2000. Mitochondrial DNA sequences support allozyme evidence for cryptic radiation of New Zealand *Peripatoides* (Onychophora). *Mol. Ecol.* 9: 269–282.

Trexler, J. C., J. Travis and A. Dinep. 1997. Variation among populations of the sailfin molly in the rate of concurrent multiple paternity and its implications for mating-system evolution. *Behav. Ecol. Sociobiol.* 40: 297–305.

Triggs, S. J., R. G. Powlesland and C. H. Daugherty. 1989. Genetic variation and conservation of kakapo (*Strigops habroptilus*: Psittaciformes). *Cons. Biol.* 3: 92–96.

Triggs, S. J., M. J. Williams, S. J. Marshall and G. K. Chambers. 1992. Genetic structure of blue duck (*Hymenolaimus malacorhynchos*) populations revealed by DNA fingerprinting. *Auk* 109: 80–89.

Trivers, R. L. 1972. Parental investment and sexual selection. Pp. 136–179 in *Sexual Selection and the Descent of Man, 1871–1971*, B. Campbell (ed.). Aldine Press, Chicago, IL.

Tucker, P. K., R. D. Sage, J. Warner, A. C. Wilson and E. M. Eicher. 1992. Abrupt cline for sex chromosomes in a hybrid zone between two species of mice. *Evolution* 46: 1146–1163.

Tudge, C. 2000. *The Variety of Life*. Oxford University Press, Oxford, England.

Tuomisto, H., K. Ruokolainen, R. Kalliola, A. Linna, W. Danjoy and Z. Rodriguez. 1995. Dissecting Amazonian biodiversity. *Science* 269: 63–66.

Turelli, M. and H. A. Orr. 2000. Dominance, epistasis and the genetics of postzygotic isolation. *Genetics* 154: 1663–1679.

Turelli, M., N. H. Barton and J. A. Coyne. 2001. Theory and speciation. *Trends Ecol. Evol.* 16: 330–343.

Turgeon, J. and P. D. N. Hebert. 1994. Evolutionary interactions between sexual and all-female taxa of *Cyprinotus* (Ostracoda: Cyprididae). *Evolution* 48: 1855–1865.

Turmel, M., C. Otis and C. Lemieux. 2002. The chloroplast and mitochondrial genome sequences of the charophyte *Chaetosphaeridium globosum*: Insights into the timing of the events that restructured organelle DNAs within the green algal lineage that led to land plants. *Proc. Natl. Acad. Sci. USA* 99: 11275–11280.

Turner, B. J. 1974. Genetic divergence of Death Valley pupfish species: Biochemical versus morphological evidence. *Evolution* 28: 281–294.

Turner, B. J. 1982. The evolutionary genetics of a unisexual fish, *Poecilia formosa*. Pp. 265–305 in *Mechanisms of Speciation*, C. Barigozzi (ed.). Alan R. Liss, New York.

Turner, B. J. 1983. Genic variation and differentiation of remnant natural populations of the desert pupfish, *Cyprinodon macularius*. *Evolution* 37: 690–700.

Turner, B. J. and D. J. Grosse. 1980. Trophic differentiation in *Ilyodon*, a genus of stream-dwelling goodeid fishes: Speciation versus ecological polymorphism. *Evolution* 34: 259–270.

Turner, B. J., J. F. Endler Jr., T. F. Laughlin and W. P. Davis. 1990. Genetic variation in clonal vertebrates detected by simple-sequence DNA fingerprinting. *Proc. Natl. Acad. Sci. USA* 87: 5653–5657.

Turner, B. J., J. F. Endler Jr., T. F. Laughlin, W. P. Davis and D. S. Taylor. 1992. Extreme clonal diversity and divergence in populations of a selfing hermaphroditic fish. *Proc. Natl. Acad. Sci. USA* 89: 10643–10647.

Turner, C., R. J. E. Wiltshire, B. M. Potts and R. E. Vaillancourt. 2000. Allozyme variation and conservation of the Tasmanian endemics, *Eucalyptus risdonii, E. tenuiramis* and *E. coccifera*. *Cons. Genet.* 1: 209–216.

Turner, G. F., O. Seehausen, M. E. Knight, C. J. Allender and R. L. Robinson. 2001. How many species of cichlid fishes are there in African lakes? *Mol. Ecol.* 10: 793–806.

Turner, M. E., J. C. Stephens and W. W. Anderson. 1982. Homozygosity and patch structure in plant populations as a result of nearest-neighbor pollination. *Proc. Natl. Acad. Sci. USA* 79: 203–207.

Turner, T. F., J. C. Trexler, D. N. Kuhn and H. W. Robison. 1996. Life-history variation and comparative phylogeography of darters (Pisces: Percidae) from the North American central highlands. *Evolution* 50: 2023–2036.

Twiddy, S. S., E. C. Holmes and A. Rambaut. 2003. Inferring the rate and time-scale of dengue virus evolution. *Mol. Biol. Evol.* 20: 122–129.

Upholt, W. B. 1977. Estimation of DNA sequence divergence from comparison of restriction endonuclease digests. *Nucleic Acids Res.* 4: 1257–1265.

Utami, S. S., B. Goosens, M. W. Bruford, J. R. de Ruiter and J. van Hooff. 2002. Male bimaturism and reproductive success in Sumatran orang-utans. *Behav. Ecol.* 13: 643–652.

Utter, F. 1991. Biochemical genetics and fishery management: An historical perspective. *J. Fish Biol.* 39A: 1–20.

Uzzell, T. and C. Spolsky. 1981. Two data sets: Alternative explanations and interpretations. *Ann. N. Y. Acad. Sci.* 361: 481–499.

Vacquier, V. D. and Y.-H. Lee. 1993. Abalone sperm lysin: An unusual mode of evolution of a gamete recognition protein. *Zygote* 1: 181–196.

Vacquier, V. D., W. J. Swanson and H. Hellberg. 1995. What have we learned about sea urchin sperm bindin? *Dev. Growth Differ.* 37: 1–10.

Valdés, A. M. and D. Piñero. 1992. Phylogenetic estimation of plasmid exchange in bacteria. *Evolution* 46: 641–656.

Valdés, A. M., M. Slatkin and N. B. Freimer. 1993. Allele frequencies at microsatellite loci: The stepwise mutation model revisited. *Genetics* 133: 737–749.

Valentine, J. W. 1976. Genetic strategies of adaptation. Pp. 78–94 in *Molecular Evolution*, F. J. Ayala, (ed.). Sinauer Associates, Sunderland, MA.

Valentine, J. W. and F. J. Ayala. 1974. Genetic variation in *Frieleia halli*, a deep-sea brachiopod. *Deep-Sea Res.* 22: 37–44.

Valsecchi, E., D. Glockner-Farrari, M. Ferrari and W. Amos. 1998. Molecular analysis of the efficiency of sloughed skin sampling in whale population genetics. *Mol. Ecol.* 7: 1419–1422.

VandeBerg, J. L., M. J. Aivaliotis, L. E. Williams and C. R. Abee. 1990. Biochemical genetic markers of squirrel monkeys and their use for pedigree validation. *Biochem. Genet.* 28: 41–56.

Van de Casteele, T., P. Galbusera and E. Matthysen. 2001. A comparison of microsatellite-based pairwise relatedness estimators. *Mol. Ecol.* 10: 1539–1549.

van Delden, W. 1982. The alcohol dehydrogenase polymorphism in *Drosophila melanogaster*: Selection at an enzyme locus. *Evol. Biol.* 15: 187–211.

Van Den Bussche, R. A., R. J. Baker, J. P. Huelsenbeck and D. M. Hillis. 1998. Base compositional bias and phylogenetic analyses: A test of the "flying DNA" hypothesis. *Mol. Phylogen. Evol.* 13: 408–416.

Van den Eynde, H. and 6 others. 1988. The 5S RNA sequences of a red algal rhodoplast and a gymnosperm chloroplast. Implications for the evolution of plastids and cyanobacteria. *J. Mol. Evol.* 27: 126–132.

Van de Peer, Y., J.-M. Neefs and R. De Wachter. 1990. Small ribosomal subunit RNA sequences, evolutionary relationships among different life forms, and mitochondrial origins. *J. Mol. Evol.* 30: 463–473.

Van Dijk, M. A. M., O. Madsen, F. Catzeflis, M. J. Stanhope, W. W. de Jong and M. Pagel. 2001. Protein sequence signatures support the African clade of mammals. *Proc. Natl. Acad. Sci. USA* 98: 188–193.

Vane-Wright, R. I., C. J. Humphries and P. H. Williams. 1991. What to protect: Systematics and the agony of choice. *Biol. Cons.* 55: 235–254.

Vanlerberghe, F., B. Dod, P. Boursot, M. Bellis and F. Bonhomme. 1986. Absence of Y-chromosome introgression across the hybrid zone between *Mus musculus domesticus* and *Mus musculus musculus*. *Genet. Res. Camb.* 48: 191–197.

Vanlerberghe, F., P. Boursot, J. T. Nielsen and F. Bonhomme. 1988. A steep cline for mitochondrial DNA in Danish mice. *Genet. Res. Camb.* 52: 185–193.

van Oppen, M. J. H., C. Rico, G. F. Turner and G. M. Hewitt. 2000. Extensive homoplasy, nonstepwise mutations, and shared ancestral polymorphisms at a complex microsatellite locus in Lake Malawi cichlids. *Mol. Biol. Evol.* 17: 489–498.

van Oppen, M. J. H., B. L. Willis and D. J. Miller. 1999. Atypically low rate of cytochrome *b* evolution in the scleractinian coral genus *Acropora*. *Proc. R. Soc. Lond. B* 266: 179–183.

van Pijlen, I. A., B. Amos and G. A. Dover. 1991. Multilocus DNA fingerprinting applied to population studies of the Minke whale *Balaenoptera acutorostrata*. Pp. 245–254 in *Genetic Ecology of Whales and Dolphins*, A. R. Hoelzel (ed.). Black Bear Press, Cambridge, England.

van Tuinen, M. and S. B. Hedges. 2001. Calibration of avian molecular clocks. *Mol. Biol. Evol.* 18: 206–213.

van Tuinen, M., C. G. Sibley and S. B. Hedges. 1998. Phylogeny and biogeography of ratite birds inferred from DNA sequences of the mitochondrial ribosomal genes. *Mol. Biol. Evol.* 15: 370–376.

van Tuinen, M., C. G. Sibley and S. B. Hedges. 2000. The early history of modern birds inferred from DNA sequences of nuclear and mitochondrial ribosomal genes. *Mol. Biol. Evol.* 17: 451–457.

Van Valen, L. 1962. A study of fluctuating asymmetry. *Evolution* 16: 125–142.

Van Wagner, C. E. and A. J. Baker. 1990. Association between mitochondrial DNA and morphological evolution in Canada geese. *J. Mol. Evol.* 31: 373–382.

Varvio, S.-L., R. Chakraborty and M. Nei. 1986. Genetic variation in subdivided populations and conservation genetics. *Heredity* 57: 189–198.

Vassart, G., M. Georges, R. Monsieur, H. Brocas, A. S. Lequarre and D. Christophe. 1987. A sequence in M13 phage detects hypervariable minisatellites in human and animal DNA. *Science* 235: 683–684.

Vawter, L. and W. M. Brown. 1986. Nuclear and mitochondrial DNA comparisons reveal extreme rate variation in the molecular clock. *Science* 234: 194–196.

Vawter, A. T., R. Rosenblatt and G. C. Gorman. 1980. Genetic divergence among fishes of the eastern Pacific and the Caribbean: Support for the molecular clock. *Evolution* 34: 705–711.

Vázquez, D. P. and J. L. Gittleman. 1998. Biodiversity and conservation: Does phylogeny matter? *Current Biol.* 8: R379-R381.

Vedel, F., F. Quétier and M. Bayen. 1976. Specific cleavage of chloroplast DNA from higher plants by *Eco*RI restriction nuclease. *Nature* 263: 440–442.

Vege, S. and S. F. McCracken. 2001. Microsatellite genotypes of big brown bats (*Eptesicus fuscus*: Vespertilionidae, Chiroptera) obtained from their feces. *Acta Chiropterologica* 3: 237–244.

Venter, J. C. and 273 others. 2001. The sequence of the human genome. *Science* 291: 1304–1351.

Verheyen, E., W. Salzburger, J. Snoeks and A. Meyer. 2003. Origin of the superflock of cichlid fishes from Lake Victoria, East Africa. *Science* 300: 325–329.

Verspoor, E. and J. Hammar. 1991. Introgressive hybridization in fishes: The biochemical evidence. *J. Fish Biol.* 39A: 309–334.

Via, S. 1999. Reproductive isolation between sympatric races of pea aphids. I. Gene flow restrictions and habitat choice. *Evolution* 53: 1446–1457.

Via, S. 2001. Sympatric speciation in animals: The ugly duckling grows up. *Trends Ecol. Evol.* 16: 381–390.

Via, S. 2002. The ecological genetics of speciation. *Am. Nat.* 159: S1-S7.

Via, S. and D. J. Hawthorne. 1998. The genetics of speciation: Promises and prospects of quantitative trait locus mapping. Pp. 352–364 in *Endless Forms: Species and Speciation*, D. J. Howard and S. H. Berlocher (eds.). Oxford University Press, New York.

Victor, B. C. 1986. Duration of the planktonic larval stage of one hundred species of Pacific and Atlantic wrasses (family Labridae). *Mar. Biol.* 90: 317–326.

Vielle-Calzada, J. P., C. F. Crane and D. M. Stelly. 1996. Apomixis: The asexual revolution. *Science* 274: 1322–1323.

Vigilant, L., R. Pennington, H. Harpending, T. D. Kocher and A. C. Wilson. 1989. Mitochondrial DNA sequences in single hairs from a southern African population. *Proc. Natl. Acad. Sci. USA* 86: 9350–9354.

Vigilant, L., M. Stoneking, H. Harpending, K. Hawkes and A. C. Wilson. 1991. African populations and the evolution of human mitochondrial DNA. *Science* 253: 1503–1507.

Vigneault, G. and E. Zouros. 1986. The genetics of asymmetrical male sterility in *Drosophila mojavensis* and *Drosophila arizonensis* hybrids: Interactions between the Y-chromosome and autosomes. *Evolution* 40: 1160–1170.

Vila, C. and R. K. Wayne. 1999. Hybridization between wolves and dogs. *Cons. Biol.* 13: 195–198.

Vila, C. and 7 others. 2001. Widespread origins of domestic horse lineages. *Science* 291: 474–477.

Villanueva, E., K. R. Luehrsen, J. Gibson, N. Delihas and G. E. Fox. 1985. Phylogenetic origins of the plant mitochondrion based on a comparative analysis of 5S ribosomal RNA sequences. *J. Mol. Evol.* 22: 46–52.

Visscher, P. K. 1986. Kinship discrimination in queen rearing by honey bees (*Apis mellifera*). *Behav. Ecol. Sociobiol.* 18: 453–460.

Voelker, G. 1999. Molecular evolutionary relationships in the avian genus *Anthus* (pipits: Motacillidae). *Mol. Phylogen. Evol.* 11: 84–94.

Vogler, A. P. and R. DeSalle. 1993. Phylogeographic patterns in coastal North American tiger beetles (*Cicindela dorsalis* Say) inferred from mitochondrial DNA sequences. *Evolution* 47: 1192–1202.

Vogler, A. P. and R. DeSalle. 1994. Evolution and phylogenetic information content of the ITS-1 region in the tiger beetle *Cicindela dorsalis*. *Mol. Biol. Evol.* 11: 393–405.

Vogler, G. 1994. Extinction and the formation of phylogenetic lineages: Diagnosing units of conservation management in the tiger beetle *Cicindela dorsalis*. Pp. 261–274 in *Molecular Ecology and Evolution: Approaches and Applications*, B. Schierwater, B. Streit, G. P. Wagner and R. DeSalle (eds.). Birkhauser-Verlag, Basel, Switzerland.

Volckaert, G., R. Aert, M. Voet, S. Van Campenhout and P. Verhasselt. 1998. Pp. 119–130 in *Molecular Tools for Screening Biodiversity*, A. Karp, P. G. Isaac and D. S. Ingram (eds.). Chapman & Hall, London.

von Nickisch-Rosenegk, M., W. M. Brown and J. L. Boore. 2001. Complete sequence of the mitochondrial genome of the tapeworm *Hymenolopis diminuta*: Gene arrangements indicate that Platyhelminths are Eutrochozoans. *Mol. Biol. Evol.* 18: 721–730.

Vos, P. and 10 others. 1995. AFLP: A new technique for DNA fingerprinting. *Nucl. Acids Res.* 23: 4407–4414.

Vrijenhoek, R. C. 1985. Homozygosity and interstrain variation in the self-fertilizing hermaphroditic fish, *Rivulus marmoratus*. *J. Heredity* 76: 82–84.

Vrijenhoek, R. C. 1996. Conservation genetics of North American desert fishes. Pp. 367–397 in *Conservation Genetics: Case Histories from Nature*, J. C. Avise and J. L. Hamrick (eds.). Chapman & Hall, New York.

Vrijenhoek, R. C. and M. A. Graven. 1992. Population genetics of Egyptian *Biomphalaria alexandrina* (Gastropoda, Planorbidae). *J. Heredity* 83: 255–261.

Vrijenhoek, R. C. and P. L. Leberg. 1991. Let's not throw the baby out with the bathwater: A comment on management for MHC diversity in captive populations. *Cons. Biol.* 5: 252–254.

Vrijenhoek, R. C. and R. J. Schultz. 1974. Evolution of a trihybrid unisexual fish (*Poeciliopsis*, Poeciliidae). *Evolution* 28: 306–319.

Vrijenhoek, R. C., M. E. Douglas and G. K. Meffe. 1985. Conservation genetics of endangered fish populations in Arizona. *Science* 229: 400–402.

Vrijenhoek, R. C., R. M. Dawley, C. J. Cole and J. P. Bogart. 1989. A list of known unisexual vertebrates. Pp. 19–23 in *Evolution and Ecology of Unisexual Vertebrates*, R. Dawley and J. Bogart (eds.). New York State Museum, Albany, NY.

Vulliamy, T. J. and 6 others. 1991. Polymorphic sites in the African population detected by sequence analysis of the glucose-6-phosphate dehydrogenase gene outline the evolution of the variants A and A-. *Proc. Natl. Acad. Sci. USA* 88: 8568–8571.

Vuorinen, J. and O. K. Berg. 1989. Genetic divergence of anadromous and nonanadromous Atlantic salmon (*Salmo salar*) in the River Namsen, Norway. *Can. J. Fish. Aquat. Sci.* 46: 406–409.

Vyas, D. K., C. Moritz, D. M. Peccinini-Seale, J. W. Wright and W. M. Brown. 1990. The evolutionary history of parthenogenetic *Cnemidophorus lemniscatus* (Sauria: Teiidae). II. Maternal origin and age inferred from mitochondrial DNA analyses. *Evolution* 44: 922–932.

Wada, S., T. Kobayashi and K.-I. Numachi. 1991. Genetic variability and differentiation of mitochondrial DNA in Minke whales. Pp. 203–215 in *Genetic Ecology of Whales and Dolphins*, A. R. Hoelzel (ed.). Black Bear Press, Cambridge, England.

Wada, H., K. W. Makabe, M. Nakauchi and N. Satoh. 1992. Phylogenetic relationships between solitary and colonial ascidians, as inferred from the sequence of the central region of their respective 18S rDNAs. *Biol. Bull.* 183: 448–455.

Waddell, P. J., Y. Cao, M. Hasegawa and D. P. Mindell. 1999. Assessing the Cretaceous superordinal divergence times within birds and placental mammals by using whole mitochondrial protein sequences and an extended statitstical framework. *Syst. Biol.* 48: 119–137.

Wade, M. J. 1982. The effect of multiple inseminations on the evolution of social behaviors in diploid and haplodiploid organisms. *J. Theoret. Biol.* 95: 351–368.

Wade, M. J. and D. McCauley. 1984. Group selection: The interaction of local deme size and migration in the differentiation of small populations. *Evolution* 38: 1047–1058.

Wagner, D. B., G. R. Furnier, M. A. Saghai-Maroof, S. M. Williams, B. P. Dancik and R. W. Allard. 1987. Chloroplast DNA polymorphisms in lodgepole and jack pines and their hybrids. *Proc. Natl. Acad. Sci. USA* 84: 2097–2100.

Wain, R. P. 1983. Genetic differentiation during speciation in the *Helianthus debilis* complex. *Evolution* 37: 1119–1127.

Wainscoat, J. S. and 10 others. 1986. Evolutionary relationships of human populations from an analysis of nuclear DNA polymorphisms. *Nature* 319: 491–493.

Waits, L. P., S. L. Talbot, R. H. Ward and G. F. Shields. 1998. Mitochondrial DNA phylogeography of the North American brown bear and implications for conservation. *Cons. Biol.* 12: 408–417.

Wake, D. B. and K. P. Yanev. 1986. Geographic variation in allozymes in a "ring species," the plethodontid salamander *Ensatina eschscholtzii* of western North America. *Evolution* 40: 702–715.

Wake, D. B., K. P. Yanev and C. W. Brown. 1986. Intraspecific sympatry in a "ring species," the plethodontid salamander *Ensatina eschscholtzii*, in southern California. *Evolution* 40: 866–868.

Wake, D. B., K. P. Yanev and M. M. Frelow. 1989. Sympatry and hybridization in a "ring species": The plethodontid salamander *Ensatina eschscholtzii*. Pp. 134–157 in *Speciation and Its Consequences*, D. Otte and J. A. Endler (eds.). Sinauer Associates, Sunderland, MA.

Wakeley, J. and N. Aliacar. 2001. Gene genealogies in a metapopulation. *Genetics* 159: 893–905.

Wakeley, J. and J. Hey. 1997. Estimating ancestral population parameters. *Genetics* 145: 847–855.

Waldman, B. 1988. The ecology of kin recognition. *Annu. Rev. Ecol. Syst.* 19: 543–571.

Waldman, B. 1991. Kin recognition in amphibians. Pp. 162–219 in *Kin Recognition*, PG. Hepper (ed.). Cambridge University Press, Cambridge, England.

Waldman, B., J. E. Rice and R. L. Honeycutt. 1992. Kin recognition and incest avoidance in toads. *Am. Zool.* 32: 18–30.

Walker, D. and J. C. Avise. 1998. Principles of phylogeography as illustrated by freshwater and terrestrial turtles in the southeastern United States. *Annu. Rev. Ecol. Syst.* 29: 23–58.

Walker, D., V. J. Burke, I. Barák and J. C. Avise. 1995. A comparison of mtDNA restriction sites vs. control region sequences in phylogeographic assessment of the musk turtle (*Sternotherus minor*). *Mol. Ecol.* 4: 365–373.

Walker, D., W. S. Nelson, K. A. Buhlmann and J. C. Avise. 1997. Mitochondrial DNA phylogeography and subspecies issues in the monotypic freshwater turtle *Sternotherus odoratus*. *Copeia* 1997: 16–21.

Walker, D., P. E. Moler, K. A. Buhlmann and J. C. Avise. 1998a. Phylogeographic patterns in *Kinosternon subrubrum* and *K. baurii* based on mitochondrial DNA restriction analyses. *Herpetologica* 54: 174–184.

Walker, D., P. E. Moler, K. A. Buhlmann and J. C. Avise. 1998b. Phylogeographic uniformity in mitochondrial DNA of the snapping turtle (*Chelydra serpentina*). *Anim. Cons.* 1: 55–60.

Walker, D., G. Ortí and J. C. Avise. 1998c. Phylogenetic distinctiveness of a threatened aquatic turtle (*Sternotherus depressus*). *Cons. Biol.* 12: 639–645.

Walker, D., B. A. Porter and J. C. Avise. 2002. Genetic parentage assessment in the crayfish *Orconectes placidus*, a high-fecundity invertebrate with extended maternal brood care. *Mol. Ecol.* 11: 2115–2122.

Wallace, B. 1958. The role of heterozygosity in *Drosophila* populations. *Proc. 10th Int. Congr. Genet.* 1: 408–419.

Wallace, B. 1970. *Genetic Load*. Prentice-Hall, Engelwood Cliffs, NJ.

Wallace, B. 1991. *Fifty Years of Genetic Load: An Odyssey*. Cornell University Press, Ithaca, NY.

Wallace, D. C. 1982. Structure and evolution of organelle genomes. *Microbiol. Rev.* 46: 208–240.

Wallace, D. C. 1986. Mitochondrial genes and disease. *Hospital Practice* 21: 77–92.

Wallace, D. C. 1992. Mitochondrial genetics: A paradigm for aging and degenerative diseases? *Science* 256: 628–632.

Wallace, D. C., G. Stugard, D. Murdock, T. Schurr and M. D. Brown. 1997. Ancient mtDNA sequences in the human nuclear genome: A potential source of errors in identifying pathogenic mutations. *Proc. Natl. Acad. Sci. USA* 94: 14900–14905.

Wallace, D. C., M. D. Brown and M. T. Lott. 1999. Mitochondrial DNA variation in human evolution and disease. *Gene* 238: 211–230.

Waller, D. M., D. M. O'Malley and S. C. Gawler. 1987. Genetic variation in the extreme endemic *Pedicularis furbishiae* (Scrophulariaceae). *Cons. Biol.* 1: 335–340.

Wallis, G. P. and J. W. Arntzen. 1989. Mitochondrial-DNA variation in the crested newt superspecies: Limited cytoplasmic gene flow among species. *Evolution* 43: 88–104.

Walters, M. 1992. *A Shadow and a Song: Extinction of the Dusky Seaside Sparrow*. Chelsea Green Publ. Co., Post Mills, VT.

Walton, C. and 6 others. 2001. Genetic population structure and introgression in *Anopheles dirus* mosquitoes in South-east Asia. *Mol. Ecol.* 10: 569–580.

Wan, Q.-H. and 8 others. 2003. Use of oligonucleotide fingerprinting and faecal DNA in identifying the distribution of the Chinese tiger (*Panthera tigris amoyensis* Hilzheimer). *Biodiv. and Cons.* 12: 1641–1648.

Wang, D. G. and 26 others. 1998. A large-scale identification, mapping, and genotyping of single nucleotide polymorphisms in the human genome. *Science* 280: 1077–1082.

Wang, D. Y.-C., S. Kumar and S. B. Hedges. 1999. Divergence time estimates for the early history of animal phyla and the origin of plants, animals and fungi. *Proc. R. Soc. Lond. B* 266: 163–171.

Wang, H., E. D. McArthur, S. C. Sanderson, J. H. Graham and D. C. Freeman. 1997. Narrow hybrid zone between two subspecies of big sagebrush (*Artemisia tridentata*: Asteraceae). IV. Reciprocal transplant experiments. *Evolution* 51: 95–102.

Wang, J. 2002. An estimator of pairwise relatedness using molecular markers. *Genetics* 160: 1203–1215.

Wang, S., J. J. Hard and F. Utter. 2002. Genetic variation and fitness in salmonids. *Cons. Genet.* 3: 321–333.

Waples, R. S. 1987. A multispecies approach to the analysis of gene flow in marine shore fishes. *Evolution* 41: 385–400.

Waples, R. S. 1991a. Heterozygosity and life-history variation in bony fishes: An alternative view. *Evolution* 45: 1275–1280.

Waples, R. S. 1991b. Pacific salmon, *Oncorhynchus* spp., and the definition of a "species" under the Endangered Species Act. *Mar. Fish. Rev.* 53: 11–22.

Waples, R. S. 1998. Separating the wheat from the chaff: Patterns of genetic differentiation in high gene flow species. *J. Heredity* 89: 438–450.

Ward, B. L., R. S. Anderson and A. J. Bendich. 1981. The mitochondrial genome is large and variable in a family of plants (Cucurbitaceae). *Cell* 25: 793–803.

Ward, P. S. 1983. Genetic relatedness and colony organization in a species complex of ponerine ants, I: Phenotypic and genotypic composition of colonies. *Behav. Ecol. Sociobiol.* 12: 285–299.

Ward, P. S. and R. W. Taylor. 1981. Allozyme variation, colony structure and genetic relatedness in the primitive ant *Nothomyrmecia macrops* Clark (Hymenoptera: Formicidae). *J. Austral. Entomol. Soc.* 20: 177–183.

Ward, R. D. 1977. Relationship between enzyme heterozygosity and quaternary structure. *Biochem. Genet.* 15: 123–135.

Ward, R. D., D. O. F. Skibinski and M. Woodward. 1992. Protein heterozygosity, protein structure, and taxonomic differentiation. *Evol. Biol.* 26: 73–159.

Ward, R. D., M. Woodward and D. O. F. Skibinski. 1994. A comparison of genetic diversity levels in marine, freshwater, and anadromous fishes. *J. Fish. Biol.* 44: 213–227.

Ward, R. H., B. L. Frazier, K. Dew-Jager and S. Pääbo. 1991. Extensive mitochondrial diversity within a single Amerindian tribe. *Proc. Natl. Acad. Sci. USA* 88: 8720–8724.

Waring, G. L., W. G. Abrahamson and D. J. Howard. 1990. Genetic differentiation among host-associated populations of the gallmaker *Eurosta solidagnus* (Diptera: Tephritidae). *Evolution* 44: 1648–1655.

Warren, K. S. and 6 others. 2001. Speciation and intrasubspecific variation of Bornean orangutans, *Pongo pygmaeus pygmaeus*. *Mol. Biol. Evol.* 18: 472–480.

Waser, P. M. and C. Strobeck. 1998. Genetic signatures of interpopulation dispersal. *Trends Ecol. Evol.* 13: 43–44.

Wasser, S. K., C. S. Houston, G. M. Koehler, G. G. Cadd and S. R. Fain. 1997. Techniques for application of faecal DNA methods to field studies of Ursids. *Mol. Ecol.* 6: 1091–1097.

Waterman, M. S., T. F. Smith, M. Singh and W. A. Beyer. 1977. Additive evolutionary trees. *J. Theoret. Biol.* 64: 199–213.

Watson, J. D. and F. H. C. Crick. 1953. A structure for DNA. *Nature* 171: 736–738.

Watson, J. D., M. Gilman, J. Witkowski and M. Zoller. 1992. *Recombinant DNA* (2nd Edition). W. H. Freeman and Co., New York.

Watt, E. M. and M. B. Fenton. 1995. DNA fingerprinting provides evidence of discriminate suckling and non-random mating in little brown bats *Myotis lucifugus*. *Mol. Ecol.* 4: 261–264.

Watt, W. B. 1977. Adaptation at specific loci. I. Natural selection on phosphoglucose isomerase of *Colias* butterflies: Biochemical and population aspects. *Genetics* 87: 177–194.

Watt, W. B., R. C. Cassin and M. S. Swan. 1983. Adaptation at specific loci, III. Field behavior and survivorship differences among *Colias* PGI genotypes are predictable from in vitro biochemistry. *Genetics* 103: 725–739.

Watt, W. B., P. A. Carter and S. M. Blower. 1985. Adaptation at specific loci. IV. Differential mating success among glycolytic allozyme genotypes of *Colias* butterflies. *Genetics* 109: 157–194.

Watterson, G. A. 1975. On the number of segregating sites in genetical models without recombination. *Theoret. Pop. Biol.* 10: 256–276.

Watts, R. J., M. S. Johnson and R. Black. 1990. Effects of recruitment on genetic patchiness in the urchin *Echinometra mathaei* in Western Australia. *Mar. Biol.* 105: 145–151.

Wayne, R. K. 1992. On the use of morphologic and molecular genetic characters to investigate species status. *Cons. Biol.* 6: 590–592.

Wayne, R. K. 1996. Conservation genetics in the Canidae. Pp. 75–118 in *Conservation Genetics: Case Histories from Nature*, J. C. Avise and J. L. Hamrick (eds.). Chapman & Hall, New York.

Wayne, R. K. and S. M. Jenks. 1991. Mitochondrial DNA analysis implying extensive hybridization of the endangered red wolf *Canis rufus*. *Nature* 351: 565–568.

Wayne, R. K., B. Van Valkenburgh and S. J. O'Brien. 1991a. Molecular distance and divergence time in carnivores and primates. *Mol. Biol. Evol.* 8: 297–319.

Wayne, R. K. and 10 others. 1991b. Conservation genetics of the endangered Isle Royale gray wolf. *Cons. Biol.* 5: 41–51.

Wayne, R. K., N. Lehman, M. W. Allard and R. L. Honeycutt. 1992. Mitochondrial DNA variability of the gray wolf: Genetic consequences of population decline and habitat fragmentation. *Cons. Biol.* 6: 559–569.

Wayne, R. K., J. A. Leonard and A. Cooper. 1999. Full of sound and fury: The recent history of ancient DNA. *Annu. Rev. Ecol. Syst.* 30: 457–477.

Weatherhead, P. J. and R. D. Montgomerie. 1991. Good news and bad news about DNA fingerprinting. *Trends Ecol. Evol.* 6: 173–174.

Weber, J. L. and P. E. May. 1989. Abundant class of human DNA polymorphisms which can be typed using the polymerase chain reaction. *Am. J. Human Genet.* 44: 388–396.

Weber, J. L. and C. Wong. 1993. Mutation of human short tandem repeats. *Human Mol. Genet.* 2: 1123–1128.

Webster, A. J., R. J. H. Payne and M. Pagel. 2003. Molecular phylogenies link rates of evolution and speciation. *Science* 301: 478.

Weiblen, G. D. and G. L. Bush. 2002. Speciation in fig pollinators and parasites. *Mol. Ecol.* 11: 1573–1578.

Weigel, D. E., J. T. Peterson and P. Spruell. 2002. A model using phenotypic characteristics to detect introgressive hybridization in wild westslope cutthroat trout and rainbow trout. *Trans. Am. Fish. Soc.* 131: 389–403.

Weigel, D. E., J. T. Peterson and P. Spruell. 2003. Introgressive hybridization between native cutthroat trout and introduced rainbow trout. *Ecol. Appl.* 13(1): 38–50.

Weiner, A. S., M. Lederer and S. H. Polayes. 1930. Studies in isohemagglutination: IV. On the chances of proving nonpaternity with special reference to blood groups. *J. Immunol.* 19: 259–282.

Weir, B. S. 1990. *Genetic Data Analysis*. Sinauer Associates, Sunderland, MA.

Weir, B. S. 1992. Independence of VNTR alleles defined as fixed bins. *Genetics* 130: 873–887.

Weir, B. S. 1996. *Genetic Data Analysis II*. Sinauer Associates, Sunderland, MA.

Weir, B. S. and C. C. Cockerham. 1973. Mixed self and random mating at two loci. *Genet. Res.* 21: 247–262.

Weir, B. S. and C. C. Cockerham. 1984. Estimating *F*-statistics for the analysis of population structure. *Evolution* 38: 1358–1370.

Weiss, K. M. and A. G. Clark. 2002. Linkage disequilibrium and the mapping of complex human traits. *Trends Genet.* 18: 19–24.

Weiss, R. A. 2001. Polio vaccines exonerated. *Nature* 410: 1035–1036.

Weissenbach, J. and 7 others. 1992. A second-generation linkage map of the human genome. *Nature* 359: 794–801.

Weller, J., M. Sober and T. Brody. 1988. Linkage analysis of quantitative traits in an interspecific cross of tomato (*Lycopersicon esculentum* X *Lycopersicon pimpinellifolium*) by means of genetic markers. *Genetics* 118: 329–339.

Weller, R. and D. M. Ward. 1989. Selective recovery of 16S rRNA sequences from natural microbial communities in the form of cDNA. *Appl. Environ. Microbiol.* 55: 1818–1822.

Weller, R., J. W. Weller and D. M. Ward. 1991. 16S rRNA sequences of uncultivated hot spring cyanobacterial mat inhabitants retrieved as randomly primed cDNA. *Appl. Environmen. Microbiol.* 57: 1146–1151.

Wells, K. D. 1977. The social behaviour of anuran amphibians. *Anim. Behav.* 25: 666–693.

Wells, M. M. and C. S. Henry. 1998. Songs, reproductive isolation, and speciation in cryptic species of insects. Pp. 217–233 in *Endless Forms: Species and Speciation*. D. J. Howard and S. H. Berlocher (eds.). Oxford University Press, New York.

Wells, R. S. 1996. Nucleotide variation at the *Gpdh* locus in the genus *Drosophila*. *Genetics* 143: 375–384.

Welsh, J., C. Petersen and M. McClelland. 1991. Polymorphisms generated by arbitrarily primed PCR in the mouse: Application to strain identification and genetic mapping. *Nucleic Acids Res.* 19: 303–306.

Wendel, J. F. and V. A. Albert. 1992. Phylogenetics of the cotton genus (*Gossypium* L.): Character-state weighted parsimony analysis of chloroplast-DNA restriction site data and its systematic and biogeographic implications. *Syst. Bot.* 17: 115–143.

Wendel, J. F., J. M. Stewart and J. H. Rettig. 1991. Molecular evidence for homoploid reticulate evolution among Australian species of *Gossypium*. *Evolution* 45: 694–711.

Wenink, P. W. and A. J. Baker. 1996. Mitochondrial DNA lineages in composite flocks of migratory and wintering dunlins (*Calidris alpina*). *Auk* 113: 744–756.

Wenink, P. W., A. J. Baker and M. G. J. Tilanus. 1993. Hypervariable-control-region sequences reveal global population structuring in a long-distance migrant shorebird, the Dunlin (*Caladris alpina*). *Proc. Natl. Acad. Sci. USA* 90: 94–98.

Wenink, P. W., A. J. Baker, H.-U. Rösner and M. G. J. Tilanus. 1996. Global mitochondrial DNA phylogeography of holarctic breeding dunlins (*Calidris alpina*). *Evolution* 50: 318–330.

Wennerberg, L. 2001. Breeding origin and migration pattern of dunlin (*Calidris alpina*) revealed by mitochondrial DNA analysis. *Mol. Ecol.* 10: 1111–1120.

Werman, S. D., M. S. Springer and R. J. Britten. 1996. Nucleic acids I: DNA-DNA hybridization. Pp. 169–203 in *Molecular Systematics* (2nd Edition), D. M. Hillis, C. Moritz and B. K. Maple (eds.). Sinauer Associates, Sunderland, MA.

Werren, J. H. 1998. *Wolbachia* and speciation. Pp. 245–260 in *Endless Forms: Species and Speciation*, D. Howard and S. Berlocher (eds.). Oxford University Press, Oxford, England.

Werth, C. R., S. I. Guttman and W. H. Eshbaugh. 1985. Recurring origins of allopolyploid species in *Asplenium*. *Science* 228: 731–733.

West-Eberhard, M. J. 1983. Sexual selection, social competition, and speciation. *Q. Rev. Biol.* 58: 155–183.

West-Eberhard, M. J. 1990. The genetic and social structure of polygynous social wasp colonies (Vespidae: Polistinae). Pp. 254–255 in *Social Insects and the Environment*, G. K. Veeresh, B. Mallik and C. A. Viraktamath (eds.). Oxford and IBH, New Delhi, India.

West-Eberhard, M. J. 2003. *Developmental Plasticity and Evolution*. Oxford University Press, New York.

Westerdahl, H., S. Bensch, B. Hansson, D. Hasselquist and T. von Schantz. 1997. Sex ratio variation among broods of great reed warblers *Acrocephalus arundinaceus*. *Mol. Ecol.* 6: 543–548.

Westerman, M., M. S. Springer, J. Dixon and C. Krajewski. 1999. Molecular relationships of the extinct pig-footed bandicoot *Chaeropus ecaudatus* (Marsupalia: Perameloidea) using 12S rRNA sequences. *J. Mamm. Evol.* 6: 271–288.

Westneat, D. F. 1987. Extra-pair fertilizations in a predominantly monogamous bird: Genetic evidence. *Anim. Behav.* 35: 877–886.

Westneat, D. F. 1990. Genetic parentage in the indigo bunting: A study using DNA fingerprinting. *Behav. Ecol. Sociobiol.* 27: 67–76.

Westneat, D. F. and P. W. Sherman. 1997. Density and extra-pair fertilizations in birds: A comparative analysis. *Behav. Ecol. Sociobiol.* 41: 205–215.

Westneat, D. F. and I. R. K. Stewart. 2003. Extra-pair paternity in birds: Causes, correlates, and conflict. *Annu. Rev. Ecol. Evol. Syst.* 34: 365–396.

Westneat, D. F., P. C. Frederick and R. H. Wiley. 1987. The use of genetic markers to estimate the frequency of successful alternative reproductive tactics. *Behav. Ecol. Sociobiol.* 21: 35–45.

Westneat, D. F., P. W. Sherman and M. L. Morton. 1990. The ecology and evolution of extra-pair copulations in birds. Pp. 331–369 in *Current Ornithology*, Vol. 7, D. M. Power (ed.). Plenum Press, New York.

Wetton, J. H. and D. T. Parkin. 1991. An association between fertility and cuckoldry in the house sparrow, *Passer domesticus*. *Proc. R. Soc. Lond. B* 245: 227–233.

Wheat, T. E., G. S. Whitt and W. F. Childers. 1973. Linkage relationships of six enzyme loci in interspecific sunfish hybrids (genus *Lepomis*). *Genetics* 74: 343–350.

Wheeler, M. R. 1986. Additions to the catalog of the world's Drosophilidae. Pp. 395–409 in *The Genetics and Biology of Drosophila*, Vol. 3, M. Ashburner, H. L. Carson and J. N. Thompson Jr. (eds.). Academic Press, New York.

Wheeler, N. C. and R. P. Guries. 1982. Population structure, genetic diversity and morphological variation in *Pinus contorta* Dougl. *Can. J. Forest Res.* 12: 595–606.

Wheeler, Q. D. and R. Meier (eds.). 2000. *Species Concepts and Phylogenetic Theory*. Columbia University Press, New York.

Wheelis, M. L., O. Kandler and C. R. Woese. 1992. On the nature of global classification. *Proc. Natl. Acad. Sci. USA* 89: 2930–2934.

Whisenant, E. C., B. K. A. Rasheed, H. Ostrer and Y. M. Bhatnagar. 1991. Evolution and sequence analysis of a human Y-chromosomal DNA fragment. *J. Mol. Evol.* 33: 133–141.

Whitaker, R. J., D. W. Grogan and J. W. Taylor. 2003. Geographic barriers isolate endemic populations of hyperthermophilic Archaea. *Science* 301: 976–978.

White, G. M., D. H. Bossier and W. Powell. 2002. Increased pollen flow counteracts fragmentation in a tropical dry forest: An example from *Swietenia humilis* Zuccarini. *Proc. Natl. Acad. Sci. USA* 99: 2038–2042.

White, M. J. D. 1978a. *Modes of Speciation*. W. H. Freeman and Co., San Francisco.

White, M. J. D. 1978b. Cytogenetics of the parthenogenetic grasshopper *Warramaba* (formerly *Moraba*) *virgo* and its bisexual relatives III. Meiosis of male "synthetic *virgo*" individuals. *Chromosoma* 67: 55–61.

White, T. J., N. Arnheim and H. A. Erlich. 1989. The polymerase chain reaction. *Trends Genet.* 5: 185–188.

Whitehead, H. 1998. Cultural selection and genetic diversity in matrilineal whales. *Science* 282: 1708–1711.

Whitfield, L. S., J. E. Suiston and P. N. Goodfellow. 1995. Sequence variation of the human Y chromosome. *Nature* 378: 379–380.

Whiting, M. F., S. Bradler and T. Maxwell. 2003. Loss and recovery of wings in stick insects. *Nature* 421: 264–267.

Whitlock, M. C. and D. E. McCauley. 1999. Indirect measures of gene flow and migration: F_{ST} not equal $1/4Nm+1$). *Heredity* 82: 117–125.

Whitt, G. S. 1983. Isozymes as probes and participants in developmental and evolutionary genetics. *Isozymes* X: 1–40.

Whitt, G. S. 1987. Species differences in isozyme tissue patterns: Their utility for systematic and evolutionary analysis. *Isozymes* XV: 1–26.

Whitt, G. S., J. B. Shaklee and C. L. Markert. 1975. Evolution of the lactate dehydrogenase isozymes of fishes. *Isozymes* IV: 381–400.

Whitt, G. S., W. F. Childers, J. B. Shaklee and J. Matsumoto. 1976. Linkage analysis of the multilocus glucosephosphate isomerase isozyme system in sunfish (Centrarchidae, Teleostii). *Genetics* 82: 35–42.

Whittaker, R. H. 1959. On the broad classification of organisms. *Q. Rev. Biol.* 34: 210–226.

Whittam, T. and M. Nei. 1991. Neutral mutation hypothesis test. *Nature* 354: 115–116.

Whittam, T. S. and A. C. Bumbaugh. 2002. Inferences from whole-genome sequences of bacterial pathogens. *Curr. Opin. Genet. Develop.* 12: 719–725.

Whittam, T. S., H. Ochman and R. K. Selander. 1983a. Geographic components of linkage disequilibrium in natural populations of *Escherichia coli. Mol. Biol. Evol.* 1: 67–83.

Whittam, T. S., H. Ochman and R. K. Selander. 1983b. Multilocus genetic structure in natural populations of *Escherichia coli. Proc. Natl. Acad. Sci. USA* 80: 1751–1755.

Whittam, T. S., A. G. Clark, M. Stoneking, R. L. Cann and A. C. Wilson. 1986. Allelic variation in human mitochondrial genes based on patterns of restriction site polymorphism. *Proc. Natl. Acad. Sci. USA* 83: 9611–9615.

Whittemore, A. T. and B. A. Schaal 1991. Interspecific gene flow in oaks. *Proc. Natl. Acad. Sci. USA* 88: 2540–2544.

Whitty, P. W., W. Powell and J. I. Sprent. 1994. Molecular separation of genera in Cassiinae (Leguminosae), and analysis of variation in the nodulating species of *Chamaecrista. Mol. Ecol.* 3: 507–515.

Widmer, A., S. Cozzolino, G. Pellegrino, M. Soliva and A. Dafni. 2000. Molecular analysis of orchid pollinaria and pollinaria-remains found on insects. *Mol. Ecol.* 9: 1911–1914.

Wiebes, J. T. 1979. Co-evolution of figs and their insect pollinators. *Annu. Rev. Ecol. Syst.* 10: 1–12.

Wiens, J. J. 1999. Polymorphism in systematics and comparative biology. *Annu. Rev. Ecol. Syst.* 30: 327–362.

Wiens, J. J. and T. A. Penkrot. 2002. Delimiting species using DNA and morphological variation and discordant species limits in spiny lizards (*Sceloporus*). *Syst. Biol.* 51: 69–91.

Wikström, N., V. Savolainen and M. W. Chase. 2001. Evolution of the angiosperms: Calibrating the family tree. *Proc. R. Soc. Lond. B* 268: 2211–2220.

Wildt, D. E. and C. Wemmer. 1999. Sex and wildlife: The role of reproductive science in conservation. *Biodiversity and Cons.* 8: 965–976.

Wildt, D. E. and 7 others. 1987. Reproductive and genetic consequences of founding isolated lion populations. *Nature* 329: 328–331.

Wiley, E. O. 1981. *Phylogenetics*. John Wiley and Sons, New York.

Wiley, E. O. 1988. Vicariance biogeography. *Annu. Rev. Ecol. Syst.* 19: 513–542.

Willers, B. 1992. Toward a science of letting things be. *Cons. Biol.* 6: 605–607.

Williams, G. C. 1975. *Sex and Evolution*. Princeton University Press, Princeton, NJ.

Williams, G. C. and R. K. Koehn. 1984. Population genetics of North Atlantic catadromous eels (*Anguilla*). Pp. 529–560 in *Evolutionary Genetics of Fishes*, B. J. Turner (ed.). Plenum Press, New York.

Williams, G. C. and D. C. Williams. 1957. Natural selection of individually harmful social adaptations among sibs with special reference to social insects. *Evolution* 11: 32–39.

Williams, G. C., R. K. Koehn and J. B. Mitton. 1973. Genetic differentiation without isolation in the American eel, *Anguilla rostrata*. *Evolution* 27: 192–204.

Williams, J. G. K., A. R. Kubelik, J. Livak, J. A. Rafalski and S. V. Tingey. 1990. DNA polymorphisms amplified by arbitrary primers are useful as genetic markers. *Nucleic Acids Res.* 18: 6531–6535.

Williams, P. H., C. J. Humphries and R. I. Vane-Wright. 1991. Measuring biodiversity: Taxonomic relatedness for conservation priorities. *Aust. J. Syst. Bot.* 4: 665–679.

Williams, S. A. and M. Goodman. 1989. A statistical test that supports a human/chimpanzee clade based on noncoding DNA sequence data. *Mol. Biol. Evol.* 6: 325–330.

Williams, S. M., R. DeSalle and C. Strobeck. 1985. Homogenization of geographical variants at the non-transcribed spacer of rDNA in *Drosophila mercatorum*. *Mol. Biol. Evol.* 2: 338–346.

Williams, S. M., G. R. Furnier, E. Fuog and C. Strobeck. 1987. Evolution of the ribosomal DNA spacers of *Drosophila melanogaster*: Different patterns of variation on X and Y chromosomes. *Genetics* 116: 225–232.

Williams, S. T. and J. A. H. Benzie. 1998. Evidence of a biogeographic break between populations of a high dispersal starfish: Congruent regions within the Indo-West Pacific defined by color morphs, mtDNA, and allozyme data. *Evolution* 52: 87–99.

Williams, S. T. and N. Knowlton. 2001. Mitochondrial pseudogenes are pervasive and often insidious in the snapping shrimp genus *Alpheus*. *Mol. Biol. Evol.* 18: 1484–1493.

Wilmer, J. W., L. Hall, E. Barratt and C. Moritz. 1999. Genetic structure and male-mediated gene flow in the ghost bat (*Macroderma gigas*). *Evolution* 53: 1582–1591.

Wilson, A. B., K. Noack-Kunnmann and A. Meyer. 2000. Incipient speciation in sympatric Nicaraguan crater lake cichlid fishes: Sexual selection versus ecological diversification. *Proc. R. Soc. Lond. B* 267: 2133–2141.

Wilson, A. B., A. Vincent, I. Ahnesjö and A. Meyer. 2001. Male pregnancy in seahorses and pipefishes (Family Syngnathidae): Rapid diversification of paternal brood pouch morphology inferred from a molecular phylogeny. *J. Heredity* 92: 159–166.

Wilson, A. B., I. Ahnesjö, A. C. J. Vincent and A. Meyer. 2003. The dynamics of male brooding, mating patterns, and sex roles in pipefishes and seahorses (family Syngnathidae). *Evolution* 57: 1374–1386.

Wilson, A. C. 1976. Gene regulation in evolution. Pp. 225–234 in *Molecular Evolution*, F. J. Ayala (ed.). Sinauer Associates, Sunderland, MA.

Wilson, A. C. 1985. The molecular basis of evolution. *Sci. Am.* 253(4): 164–173.

Wilson, A. C., L. R. Maxson and V. M. Sarich. 1974a. Two types of molecular evolution. Evidence from studies of interspecific hybridization. *Proc. Natl. Acad. Sci. USA* 71: 2843–2847.

Wilson, A. C., V. M. Sarich and L. R. Maxson. 1974b. The importance of gene rearrangement in evolution: Evidence from studies on rates of chromosomal, protein, and anatomical evolution. *Proc. Natl. Acad. Sci. USA* 71: 3028–3030.

Wilson, A. C., S. S. Carlson and T. J. White. 1977. Biochemical evolution. *Annu. Rev. Biochem.* 46: 473–639.

Wilson, A. C. and 10 others. 1985. Mitochondrial DNA and two perspectives on evolutionary genetics. *Biol. J. Linn. Soc.* 26: 375–400.

Wilson, A. C., H. Ochman and E. M. Prager. 1987. Molecular time scale for evolution. *Trends Genet.* 3: 241–247.

Wilson, A. C., M. Stoneking and R. L. Cann. 1991. Ancestral geographic states and the peril of parsimony. *Syst. Zool.* 40: 363–365.

Wilson, C. and L. Bernatchez. 1998. The ghost of hybrids past: Fixation of arctic charr (*Salvelinus alpinus*) mitochondrial DNA in an introgressed population of lake trout (*S. namaycush*). *Mol. Ecol.* 7: 127–132.

Wilson, D. S. 1980. *The Natural Selection of Populations and Communities*. Addison-Wesley, Reading, MA.

Wilson, E. O. 1971. *The Insect Societies*. Harvard University Press, Cambridge, MA.

Wilson, E. O. 1975. *Sociobiology*. Belknap Press, Cambridge, MA.

Wilson, E. O. 1984. *Biophilia*. Harvard University Press, Cambridge, MA.

Wilson, E. O. 1987. Kin recognition: An introductory synopsis. Pp. 7–18 in *Kin Recognition in Animals*, D. J. C. Fletcher and C. D. Michener (eds.). John Wiley and Sons, New York.

Wilson, E. O. 1992. *The Diversity of Life*. W. H. Norton, New York.

Wilson, E. O. 2002. *The Future of Life*. Alfred Knopf, New York.

Wilson, E. O. 2003. The encyclopedia of life. *Trends Ecol. Evol.* 18: 77–80.

Wilson, E. O. and W. L. Brown. 1953. The subspecies concept and its taxonomic application. *Syst. Zool.* 2: 97–111.

Wilson, G. A. and C. Strobeck. 1999. Genetic variation within and relatedness among wood and plains bison populations. *Genome* 42: 483–496.

Wilson, G. M., W. K. Thomas and A. T. Beckenbach. 1987. Mitochondrial DNA analysis of Pacific Northwest populations of *Oncorhynchus tshawytscha*. *Can. J. Fish. Aquat. Sci.* 44: 1301–1305.

Wilson, P. J. and 14 others. 2000. DNA profiles of the eastern Canadian wolf and the red wolf provide evidence for a common evolutionary history independent of the gray wolf. *Can. J. Zool.* 78: 2156–2166.

Wimpee, C. F., T.-L. Nadeau and K. H. Nealson. 1991. Development of species-specific hybridization probes for marine luminous bacteria by using *in vitro* DNA amplification. *Appl. Environ. Microbiol.* 57: 1319–1324.

Winans, G. A. 1980. Geographic variation in the milkfish *Chanos chanos*. I. Biochemical evidence. *Evolution* 34: 558–574.

Winchell, C. J., J. Sullivan, C. B. Cameron, B. J. Swalla and J. Mallet. 2002. Evaluating hypotheses of deuterostome phylogeny and chordate evolution with new LSU and SSU ribosomal DNA data. *Mol. Biol. Evol.* 19: 762–776.

Wink, M. 1995. Phylogeny of Old and New World vultures (Aves: Accipitridae and Cathartidae) inferred from nucleotide sequences of the mitochondrial cytochrome *b* gene. *J. Biosci.* 50: 868–882.

Wink, M. and A. Dyrcz. 1999. Mating systems in birds: A review of molecular studies. *Acta Ornithologica* 34: 91–109.

Winkler, C., A. Schultz, S. Cevario and S. O'Brien. 1989. Genetic characterization of *FLA*, the cat major histocompatibility complex. *Proc. Natl. Acad. Sci. USA* 86: 943–947.

Winkler, D. W. and F. W. Sheldon. 1993. Evolution of nest construction in swallows (Hirundinidae): A molecular phylogenetic perspective. *Proc. Natl. Acad. Sci. USA* 90: 5705–5709.

Winnepenninckx, B. M. H., Y. Van de Peer and T. Backeljau. 1998. Metazoan relationships on the basis of 18S rRNA sequences: A few years later... *Am. Zool.* 38: 888–906.

Winzeler, E. A. and 10 others. 1998. A direct allelic variation scanning of the yeast genome. *Science* 281: 1194–1197.

Wirth, T. and L. Bernatchez. 2001. Genetic evidence against panmixia in the European eel. *Nature* 409: 1037–1040.

Wirth, T. and L. Bernatchez. 2003. Decline of North Atlantic eels: A fatal synergy? *Proc. R. Soc. Lond. B* 270: 681–688.

Wisely, S. M., S. W. Buskirk, M. A. Fleming, D. B. McDonald and E. A. Ostrander. 2002. Genetic diversity and fitness in black-footed ferrets before and during a bottleneck. *J. Heredity* 93: 231–237.

Witter, M. S. and G. D. Carr. 1988. Adaptive radiation and genetic differentiation in the Hawaiian silversword alliance (Compositae: Madiinae). *Evolution* 42: 1278–1287.

Woese, C. R. 1987. Bacterial evolution. *Microbiol. Rev.* 51: 221–271.

Woese, C. R. 1998a. Default taxonomy: Ernst Mayr's view of the microbial world. *Proc. Natl. Acad. Sci. USA* 95: 11043–11046.

Woese, C. R. 1998b. The universal ancestor. *Proc. Natl. Acad. Sci. USA* 95: 6854–6859.

Woese, C. R. 2002. On the evolution of cells. *Proc. Natl. Acad. Sci. USA* 99: 8742–8747.

Woese, C. R. and G. E. Fox. 1977. Phylogenetic structure of the prokaryotic domain: The primary kingdoms. *Proc. Natl. Acad. Sci. USA* 74: 5088–5090.

Woese, C. R., O. Kandler and M. L. Wheelis. 1990. Towards a natural system of organisms: Proposal for the domains Archaea, Bacteria, and Eucarya. *Proc. Natl. Acad. Sci. USA* 87: 4576–4579.

Woese, C. R., O. Kandler and M. L. Wheelis. 1991. A natural classification. *Nature* 351: 528–529.

Woischnik, M. and C. T. Moraes. 2002. Pattern of organization of human mitochondrial pseudogenes in the nuclear genome. *Genome Res.* 12: 885–893.

Wolf, Y. I., I. B. Rogozin, N. V. Grishin and E. V. Koonin. 2002. Genome trees and the tree of life. *Trends Genet.* 18: 472–479.

Wolfe, K. H., W.-H. Li and P. M. Sharp. 1987. Rates of nucleotide substitution vary greatly among plant mitochondrial, chloroplast, and nuclear DNAs. *Proc. Natl. Acad. Sci. USA* 84: 9054–9058.

Wolfe, K. H., M. Gouy, Y.-W. Yang, P. M. Sharp and W.-H. Li. 1989. Date of the monocot-dicot divergence estimated from chloroplast DNA sequence data. *Proc. Natl. Acad. Sci. USA* 86: 6201–6205.

Wollenberg, K., J. Arnold and J. C. Avise. 1996. Recognizing the forest for the trees: Testing temporal patterns of cladogenesis using a null model of stochastic diversification. *Mol. Biol. Evol.* 13: 833–849.

Wolpoff, M. H. 1989. Multiregional evolution: The fossil alternative to Eden. Pp. 62–108 in *The Human Revolution: Behavioural and Biological Perspectives on the Origins of Modern Humans*, P. Mellars and C. Stringer (eds.). Princeton University Press, Princeton, NJ.

Wolpoff, M. H., X. Z. Yu and A. G. Thorne. 1984. Modern *Homo sapiens* origins: A general theory of hominid evolution involving the fossil evidence from East Asia. Pp. 411–483 in *The Origins of Modern Humans: A World Survey of the Fossil Evidence*, F. H. Smith and F. Spencer (eds.). Alan R. Liss, New York.

Wolstenholme, D. R. 1992. Animal mitochondrial DNA: Structure and evolution. *Int. Rev. Cytol.* 141: 173–216.

Wolstenholme, D. R., D. O. Clary, J. L. MacFarlane, J. A. Wahleithner and L. Wilcox. 1985. Organization and evolution of invertebrate mitochondrial-genomes. Pp. 61–69 in *Achievements and Perspectives of Mitochondrial Research Vol. II: Biogenesis*, E. Quagliariello, E. C. Slater, F. Palmieri, C. Saccone and A. M. Kroon (eds.). Elsevier, Amsterdam.

Wolstenholme, D. R., J. L. MacFarlane, R. Okimoto, D. O. Clary and J. A. Wahleithner. 1987. Bizarre tRNAs inferred from DNA sequences of mitochondrial genomes of nematode worms. *Proc. Natl. Acad. Sci. USA* 84: 1324–1328.

Wong, Z., V. Wilson, A. J. Jeffreys and S. L. Thein. 1986. Cloning a selected fragment from a human DNA fingerprint: Isolation of an extremely polymorphic minisatellite. *Nucleic Acids Res.* 14: 4605–4616.

Wong, Z., V. Wilson, I. Patel, S. Povey and A. J. Jeffreys. 1987. Characterization of a panel of highly variable minisatellites cloned from human DNA. *Ann. Human Genet.* 51: 269–288.

Wood, T. K., K. J. Tilmon, A. B. Shantz, C. K. Harris and J. Pesek. 1999. The role of host-plant fidelity in intiating insect race formation. *Evol. Ecol. Research* 1: 317–332.

Woodin, S. A. 1986. Settlement of infauna: Larval choice? *Bull. Mar. Sci.* 39: 401–407.

Woodruff, D. S. and S. J. Gould. 1987. Fifty years of interspecific hybridization: Genetics and morphometrics of a controlled experiment on the land snail *Cerion* in the Florida Keys. *Evolution* 41: 1022–1045.

Woodruff, D. S., M. Mulvey and M. W. Yipp. 1985. Population genetics of *Biomphalaria straminea* in Hong Kong. *J. Heredity* 76: 355–360.

Woodruff, R. C., H. Huai and J. N. Thompson Jr. 1996. Clusters of identical new mutation in the evolutionary landscape. *Genetica* 98: 149–160.

Woods, J. G., D. Paetkau, D. Lewis, B. N. McLellan, M. Proctor and C. Strobeck. 1999. Genetic tagging of free-ranging black and brown bears. *Wildl. Soc. Bull.* 27: 616–627.

Woodward, S. R., N. J. Weyand and M. Bunnell. 1994. DNA sequence from Cretaceous bone fragments. *Science* 266: 1229–1232.

Woolfenden, G. E. and J. W. Fitzpatrick. 1984. *The Florida Scrub Jay*. Princeton University Press, Princeton, NJ.

Wooten, M. C. and C. Lydeard. 1990. Allozyme variation in a natural contact zone between *Gambusia affinis* and *Gambusia holbrooki*. *Biochem. Syst. Ecol.* 18: 169–173.

Wooten, M. C., K. T. Scribner and M. H. Smith. 1988. Genetic variability and systematics of *Gambusia* in the southeastern United States. *Copeia* 1988: 283–289.

Wörheide, G., J. N. A. Hooper and B. M. Degnan. 2002. Phylogeography of western Pacific *Leucetta 'chagosensis'* (Porifera: Calcarea) from ribosomal DNA sequences: Implications for population history and conservation of the Great Barrier Reef World Heritage Area (Australia). *Mol. Ecol.* 11: 1753–1768.

Wrege, P. H. and S. T. Emlen. 1987. Biochemical determination of parental uncertainty in white-fronted bee-eaters. *Behav. Ecol. Sociobiol.* 20: 153–160.

Wright, C. A. 1974. *Biochemical and Immunological Taxonomy of Animals*. Academic Press, New York.

Wright, J. W. 1983. The evolution and biogeography of the lizards of the Galapagos archipelago: Evolutionary genetics of *Phyllodactylus* and *Tropidurus* populations. Pp. 123–156 in *Patterns of Evolution in Galapagos Organisms*, R. I. Bowman, M. Berson and A. E. Leviton (eds.). American Association for the Advancement of Science, San Francisco, CA.

Wright, J. W., C. Spolsky and W. M. Brown. 1983. The origin of the parthenogenetic lizard *Cnemidophorus laredoensis* inferred from mitochondrial DNA analysis. *Herpetologica* 39: 410–416.

Wright, S. 1931. Evolution in Mendelian populations. *Genetics* 16: 97–159.

Wright, S. 1951. The genetical structure of populations. *Ann. Eugen.* 15: 323–354.

Wrischnik, L. A., R. G. Higuchi, M. Stoneking, H. A. Erlich, N. Arnheim and A. C. Wilson. 1987. Length mutations in human mitochondrial DNA: Direct sequencing of enzymatically amplified DNA. *Nucleic Acids Res.* 15: 529–542.

Wu, C.-I. 1991. Inferences of species phylogeny in relation to segregation of ancient polymorphisms. *Genetics* 127: 429–435.

Wu, C.-I. 2001. The genic view of the process of speciation. *J. Evol. Biol.* 14: 851–865.

Wu, C.-I. and A. W. Davis. 1993. Evolution of postmating reproductive isolation: The composite nature of Haldane's rule and its genetic bases. *Am. Nat.* 142: 187–212.

Wu, C.-I. and H. Hollocher. 1998. Subtle is nature: The genetics of species differentiation and speciation. Pp. 339–351 in *Endless Forms: Species and Speciation*, D. J. Howard and S. H. Berlocher (eds.). Oxford University Press, New York.

Wu, C.-I. and W.-H. Li. 1985. Evidence for higher rates of nucleotide substitution in rodents than in man. *Proc. Natl. Acad. Sci. USA* 82: 1741–1745.

Wyatt, R. 1988. Phylogenetic aspects of the evolution of self-pollination. Pp. 109–131 in *Plant Evolutionary Biology*, L. D. Gottlieb and S. K. Jain (eds.). Chapman & Hall, New York.

Wyatt, R. 1997. Reproductive ecology of granite outcrop plants from the southeastern United States. *J. R. Soc. Western Australia* 80: 123–129.

Wyatt, R., I. J. Odrzykoski, A. Stoneburner, H. W. Bass and G. A. Galau. 1988. Allopolyploidy in bryophytes: Multiple origins of *Plagiomnium medium*. *Proc. Natl. Acad. Sci. USA* 85: 5601–5604.

Wyatt, R., E. A. Evans and J. C. Sorenson. 1992. The evolution of self-pollination in granite outcrop species of *Arenaria* (Caryophyllaceae). VI. Electrophoretically detectable genetic variation. *Syst. Bot.* 17: 201–209.

Wyckoff, R. W. G. 1972. *The Biochemistry of Animal Fossils*. Scientechnica, Bristol, England.

Wyles, J. S. and V. M. Sarich. 1983. Are the Galapagos iguanas older than the Galapagos? Pp. 177–199 in *Patterns of Evolution in Galapagos Organisms*, R. I. Bowman, M. Berson and A. E. Leviton (eds.). American Association for the Advancement of Science, San Francisco, CA.

Wyles, J. S., J. G. Kunkel and A. C. Wilson. 1983. Birds, behavior, and anatomical evolution. *Proc. Natl. Acad. Sci. USA* 80: 4394–4397.

Xia, X. and J. S. Millar. 1991. Genetic evidence of promiscuity in *Peromyscus leucopus*. *Behav. Ecol. Sociobiol.* 28: 171–178.

Xiong, W. and 6 others. 1991. No severe bottleneck during human evolution: Evidence from two apolipoprotein C-II deficiency alleles. *Am. J. Human Genet.* 48: 383–389.

Xiong, Y. and T. H. Eickbush. 1990. Origin and evolution of retroelements based upon their reverse transcriptase sequences. *The EMBO J.* 9: 3353–3362.

Xu, J., T. G. Mitchell and R. Vilgalys. 1999. PCR-restriction fragment length polymorphism (RFLP) analyses reveal both extensive clonality and local genetic differences in *Candida albicans*. *Mol. Ecol.* 8: 59–73.

Xu, S., C. J. Kobak and P. E. Smouse. 1994. Constrained least squares estimation of mixed population stock composition from mtDNA haplotype frequency data. *Can. J. Fish. Aquat. Sci.* 51: 417–425.

Xu, X. and U. Arnason. 1996. The mitochondrial DNA molecule of Sumatran orangutan and a molecular proposal for two (Bornean and Sumatran) species of orangutan. *J. Mol. Evol.* 43: 431–437.

Yahnke, C. J. and 9 others. 1996. Darwin's fox: A distinct endangered species in a vanishing habitat. *Cons. Biol.* 10: 366–375.

Yang, D., Y. Oyaizu, H. Oyaizu, G. J. Olsen and C. R. Woese. 1985. Mitochondrial origins. *Proc. Natl. Acad. Sci. USA* 82: 4443–4447.

Yang, H., E. M. Golenberg and J. Shoshani. 1996. Phylogenetic resolution within the Elephantidae using fossil DNA sequences from the American mastodon (*Mammut americanum*) as an outgroup. *Proc. Natl. Acad. Sci. USA* 93: 1190–1194.

Yang, S. Y. and J. L. Patton. 1981. Genic variability and differentiation in the Galápagos finches. *Auk* 98: 230–242.

Yang, T. W., Y. A. Yang and Z. Xiong. 2000. Paternal inheritance of chloroplast DNA in interspecific hybrids in the genus *Larrea* (Zygophyllaceae). *Am. J. Bot.* 87: 1452–1458.

Yoder,, J. A., C. P. Walsh and T. H. Bestor. 1997. Cytosine methylation and the ecology of intragenomic parasites. *Trends Genet.* 13: 335–340.

Yokoyama, S. and T. Gojobori. 1987. Molecular evolution and phylogeny of the human AIDS viruses LAV, HTLV-III, and ARV. *J. Mol. Evol.* 24: 330–336.

Young, D. L., M. W. Allard, J. A. Moreno, M. M. Miyamoto, C. R. Ruiz and R. A. Pérez-Rivera. 1998. DNA fingerprint variation and reproductive fitness in the plain pigeon. *Cons. Biol.* 12: 225–227.

Yuhki, N. and S. J. O'Brien. 1990. DNA variation of the mammalian major histocompatibility complex reflects genomic diversity and population history. *Proc. Natl. Acad. Sci. USA* 87: 836–840.

Zambryski, P., J. Tempe and J. Schell. 1989. Transfer and function of T-DNA genes from *Agrobacterium* Ti and Ri plasmids in plants. *Cell* 56: 193–201.

Zamudio, K. R. and H. W. Greene. 1997. Phylogeography of the bushmaster (*Lachesis muta*: Viperidae): Implications for neotropical biogeography, systematics, and conservation. *Biol. J. Linn. Soc.* 62: 421–442.

Zamudio, K. R. and B. Sinervo. 2000. Polygyny, mate-guarding, and posthumous fertilization as alternative male strategies. *Proc. Natl. Acad. Sci. USA* 97: 14427–14432.

Zane, L., W. S. Nelson, A. G. Jones and J. C. Avise. 1999. Microsatellite assessments of multiple paternity in natural populations of a live-bearing fish, *Gambusia holbrooki. J. Evol. Biol.* 12: 61–69.

Zane, L., L. Bargelloni and T. Patarnello. 2002. Strategies for microsatellite isolation: A review. *Mol. Ecol.* 11: 1–16.

Zardoya, R. and A. Meyer. 1996a. Evolutionary relationships of the coelacanth, lungfishes, and tetrapods based on the 28S ribosomal RNA gene. *Proc. Natl. Acad. Sci. USA* 93: 5449–5454.

Zardoya, R. and A. Meyer. 1996b. The complete nucleotide sequence of the mitochondrial genome of the lungfish (*Protopterus dolloi*) supports its phylogenetic position as a close relative of land vertebrates. *Genetics* 142: 1249–1263.

Zardoya, R. and A. Meyer. 1997. The complete DNA sequence of the mitochondrial genome of a 'living fossil', the coelacanth *(Latimeria chalumnae)*. *Genetics* 146: 995–1010.

Zardoya, R., D. M. Vollmer, C. Craddock, J. T. Streelman, S. A. Karl and A. Meyer. 1996. Evolutionary conservation of microsatellite flanking regions and their use in resolving the phylogeny of cichlid fishes (Pisces: Perciformes). *Proc. R. Soc. Lond. B* 263: 1589–1598.

Zardoya, R., Y. Cao, M. Hasegawa and A. Meyer. 1998. Searching for the closest living relative(s) of tetrapods through evolutionary analyses of mitochondrial and nuclear data. *Mol. Biol. Evol.* 15: 506–517.

Zeh, D. W., J. A. Zeh, M. A. Coffroth and E. Bermingham. 1992. Population-specific DNA fingerprints in a neotropical pseudoscorpion (*Cordylochernes scorpioides*). *Heredity* 69: 201–208.

Zera, A. J. 1981. Genetic structure of two species of waterstriders (Gerridae: Hemiptera) with differing degrees of winglessness. *Evolution* 35: 218–225.

Zhang, D.-X. and G. M. Hewitt. 1996. Nuclear integrations: Challenges for mitochondrial DNA markers. *Trends Ecol. Evol.* 11: 247–251.

Zhang, D.-X. and G. M. Hewitt. 1997. Assessment of the universality and utility of a set of conserved mitochondrial primers in insects. *Insect Mol. Biol.* 6: 143–150.

Zhang, D.-X. and G. M. Hewitt. 2003. Nuclear DNA analyses in genetic studies of populations: Practice, problems and prospects. *Mol. Ecol.* 12: 563–584.

Zhang, J., H. F. Rosenberg and M. Nei. 1998. Positive Darwinian selection after gene duplication in primate ribonuclease genes. *Proc. Natl. Acad. Sci. USA* 95: 3708–3713.

Zhang, Q., M. Tibayrenc and F. J. Ayala. 1988. Linkage disequilibrium in natural populations of *Trypanosoma cruzi* (Flagellate), the agent of Chagas' disease. *J. Protozool.* 35: 81–85.

Zhi, L. and 8 others. 1996. Genomic differentiation among natural populations of orangutan (*Pongo pygmaeus*). *Curr. Biol.* 6: 1326–1336.

Zhu, D., S. Degnan and C. Moritz. 1998. Evolutionary distinctiveness and status of the endangered Lake Eucham rainbowfish (*Melanotaenia eachamensis*). *Cons. Biol.* 12: 80–93.

Zhu, T., B. T. Korber, A. J. Nahmias, E. Hooper, P. M. Sharp and D. D. Ho. 1998. An African HIV-1 sequence from 1959 and implications for the origin of the epidemic. *Nature* 391: 594–597.

Zhu, Y., D. C. Queller and J. E. Strassmann. 2000. A phylogenetic perspective on sequence evolution in microsatellite loci. *J. Mol. Evol.* 50: 324–338.

Zimmer, E. A., S. L. Martin, S. M. Beverley, Y. W. Kan and A. C. Wilson. 1980. Rapid duplication and loss of genes coding for the α chains of hemoglobin. *Proc. Natl. Acad. Sci. USA* 77: 2158–2162.

Zimmerman, E. G., C. W. Kilpatrick and B. J. Hart. 1978. The genetics of speciation in the rodent genus *Peromyscus*. *Evolution* 32: 565–579.

Zink, R. M. 1991. Geography of mitochondrial DNA variation in two sympatric sparrows. *Evolution* 45: 329–339.

Zink, R. M. 1996. Comparative phylogeography in North American birds. *Evolution* 50: 308–317.

Zink, R. M. 1997. Phylogeographic studies of North American birds. Pp. 301–324 in Avian *Molecular Evolution and Systematics*, D. P. Mindell (ed.). Academic Press, New York.

Zink, R. M. 2002. Towards a framework for understanding the evolution of avian migration. *J. Avian Biol.* 33: 433–436.

Zink, R. M. and N. K. Johnson. 1984. Evolutionary genetics of flycatchers. I. Sibling species in the genera *Empidonax* and *Contopus. Syst. Zool.* 33: 205–216.

Zink, R. M., M. F. Smith and J. L. Patton. 1985. Associations between heterozygosity and morphological variance. *J. Heredity* 76: 415–420.

Zink, R. M., S. Rohwer, A. V. Andreev and D. L. Dittmann. 1995. Trans-Beringia comparisons of mitochondrial DNA differentiation in birds. *Condor* 97: 639–649.

Zink, R. M., R. C. Blackwell and O. Rojassoto. 1997. Species limits in the LeConte's thrasher. *Condor* 99: 132–138.

Zink, R. M., G. F. Barrowclough, J. L. Atwood and R. C. Blackwell-Rago. 2000. Genetics, taxonomy, and conservation of the threatened California gnatcatcher. *Cons. Biol.* 14: 1394–1405.

Zinkham, W. H., H. Isensee and J. H. Renwick. 1969. Linkage of lactate dehydrogenase B and C loci in pigeons. *Science* 164: 185–187.

Zouros, E. 1976. Hybrid molecules and the superiority of the heterozygote. *Nature* 262: 227–229.

Zouros, E. 2000. The exceptional mitochondrial DNA system of the mussel family Mytilidae. *Genes Genet. Syst.* 75: 313–318.

Zouros, E., S. M. Singh and M. E. Miles. 1980. Growth rate in oysters: An overdominant phenotype and its possible explanations. *Evolution* 34: 856–867.

Zouros, E., K. R. Freeman, A. O. Ball and G. H. Pogson. 1992. Direct evidence for extensive paternal mitochondrial DNA inheritance in the marine mussel *Mytilus*. *Nature* 359: 412–414.

Zouros, E., A. O. Ball, C. Saavedra and K. R. Freeman. 1994. Mitochondrial DNA inheritance. *Nature* 368: 818.

Zuckerkandl, E. and L. Pauling. 1965. Evolutionary divergence and convergence in proteins. Pp. 97–166 in *Evolving Genes and Proteins*, V. Bryson and H. J. Vogel (eds.). Academic Press, New York.

Zurawski, G. and M. T. Clegg. 1987. Evolution of higher-plant chloroplast DNA-encoded genes: Implications for structure-function and phylogenetic studies. *Annu. Rev. Plant Physiol.* 38: 391–418.

Taxonomic Index

Subject Index

About the Book

Editor: Andrew D. Sinauer
Project Editor: Carol Wigg
Copy Editor: Norma Roche
Production Manager: Christopher Small
Book Design: Joan Gemme
Book Layout and Composition: Joanne Delphia
Art Studio: Michele Ruschhaupt/The Format Group, LLC
Book and Cover Manufacture: Courier Companies, Inc.